Important Formulas

Quadratic Formula

If $ax^2 + bx + c = 0$, for $a \neq 0$, then

$$x = \frac{-b \pm \sqrt{b^2 - 4ac}}{2a}$$

Distance Formula

The distance d between $P_1 = (x_1, y_1)$ and $P_2 = (x_2, y_2)$ is

$$d = \sqrt{(x_2 - x_1)^2 + (y_2 - y_1)^2}$$

Slope Formula

The slope m of the line segment $\overline{AB}$, where $A = (x_1, y_1)$ and $B = (x_2, y_2)$ is

$$m = \frac{y_2 - y_1}{x_2 - x_1}$$

Geometry

Assume A = area, C = circumference, V = volume, S = surface area, r = radius, h = altitude, l = length, w = width, b (or a) = length of a base, and s = length of a side.

1 Square $A = s^2$

2 Rectangle $A = lw$

3 Parallelogram $A = bh$

4 Triangle $A = \frac{1}{2}bh$

5 Circle $A = \pi r^2$; $C = 2\pi r$

6 Trapezoid $A = \frac{1}{2}(a + b)h$

7 Cube $S = 6s^2$; $V = s^3$

8 Rectangular Box $S = 2(lw + wh + lh)$; $V = lwh$

9 Cylinder $S = 2\pi rh$; $V = \pi r^2 h$

10 Sphere $S = 4\pi r^2$; $V = \frac{4}{3}\pi r^3$

11 Cone $S = \pi r\sqrt{r^2 + h^2}$; $V = \frac{1}{3}\pi r^2 h$

Algebra and Trigonometry

SECOND EDITION

Algebra and Trigonometry

with Applications

M. A. Munem

Macomb County Community College

D. J. Foulis

University of Massachusetts

Worth Publishers, Inc.

ALGEBRA AND TRIGONOMETRY WITH APPLICATIONS SECOND EDITION

LIBRARY OF CONGRESS CATALOG CARD NO. 85-51291
ISBN: 0-87901-281-1
SECOND PRINTING, JULY 1988

EDITOR: ANNE VINNICOMBE
PRODUCTION: MARGIE BRASSIL
DESIGN: MALCOLM GREAR DESIGNERS
ILLUSTRATOR: FELIX COOPER
COPYEDITOR: TRUMBULL ROGERS
TYPOGRAPHER: SYNTAX INTERNATIONAL
PRINTING AND BINDING: R.R. DONNELLEY & SONS
COVER: COMPUTER GRAPHICS BY THOMAS BANCHOFF AND DAVID MARGOLIS

Photo Credits: P. 46: Hale Observatories, P. 52: The Bettmann Archive, P. 63: Culver Pictures, P. 84: T.D. Lovering/Stock, Boston, P. 86: George Bellerose/Stock, Boston, P. 87: Donald Dietz/Stock, Boston, P. 110: Frank Siteman/Stock, Boston, P. 125: Barbara Alper/Stock, Boston P. 129: The Bettmann Archive, P. 157: Mark Antman/The Image Works, P. 181: Ellis Herwig/Stock, Boston, P. 197: Allen Rokach/Animals Animals, P. 198: Charles Kennard/Stock, Boston, P. 205: Peter Menzel/Stock, Boston, P. 204: Ralph Kubner/Black Star, P. 210: George Bellerose/Stock, Boston, P. 276: (left) AP/World Wide Photos; (right) Marty Stouffer/Animals Animals, P. 296: Reproduced with permission of AT&T Corporate Archives, P. 313: The Bettman Archive, P. 363: Culver Pictures, P. 366: NASA, P. 419: Leonard Lee Rue III/Animals Animals, P. 433: IRA Kirschenbaum/Stock, Boston, P. 471: The Granger Collection, P. 532: Charles Gatewood/ Stock Boston, P. 538: Joel Gordon, P. 545: Architect of the Capitol, P. 547: (left) Fundamental Photographs; (center) Jeff Albertson/Stock, Boston; (right) Owen Franden /Stock, Boston, P. 553: (left) The Marley Company, Mission, Kansas; (center) Lick Observatory, University of California at Santa Cruz; (right) Adapted from Loran-C User Handbook, United States Coast Guard, P. 579: UPI/Bettmann Newsphotos

WORTH PUBLISHERS, INC.
33 IRVING PLACE
NEW YORK, NEW YORK 10003

Preface

The second edition of *Algebra and Trigonometry* is a refinement of the first. Again we have attempted to provide students with a straightforward, readable text that presents important concepts and skills in an appealing manner. Although some new material has been added, the major effort has been devoted to less obvious improvements, such as adding more graphs, problems, and illustrative examples, and including brief historical notes and portraits of prominent mathematicians.

Prerequisites

The student we had in mind as we wrote has had the equivalent of two years of college-preparatory mathematics in high school, including algebra and some plane geometry, or has taken a college-level course in introductory algebra. Determined students with less preparation should be able to master this material, particularly if they use the accompanying *Study Guide*.

Objectives

The aim of the book remains the same—to provide students with the working knowledge of college algebra and trigonometry, including functions and graphs, that they will need for their later study and work. We have written and revised the book with the objective of complementing and enhancing effective classroom teaching. The student is provided with every opportunity to learn how to solve routine problems successfully, to develop computational skills, and to build confidence. We continue to attune the textbook to the two outstanding mathematical trends of our time—the computer revolution and the burgeoning use of mathematical models.

Presentation

Topics are presented in brief sections that progress logically from basic to more difficult mathematical concepts and skills. The discussions of new concepts include succinct explanations and numerous illustrative examples worked out in detail. Wherever appropriate, specific problem-solving procedures are given. To develop students' geometric intuition, graphing techniques and illustrations are emphasized.

Problem Sets

Problems at the end of each section begin with simple drill-type problems and progress gradually to more challenging ones. Most of the problems are in the middle range of difficulty.

Odd-numbered problems.   Most of the odd-numbered problems are similar in scope to the worked-out examples in the text. Answers to virtually all odd-numbered problems, with appropriate graphs, are given in the back of the book.

Even-numbered problems.   Although most of the even-numbered problems are also similar to the worked-out examples, some are considerably more challenging and probe for a deeper understanding of concepts.

Homework.   You will notice at the end of each section that certain problem numbers are printed in color. This indicates a group of problems that could serve as a homework assignment for the section.

Review Problem Sets.   The review problems at the end of each chapter can be used in a variety of ways: Instructors may wish to use them for supplementary or extra-credit assignments or for quizzes and exams; students may wish to scan them to pinpoint areas where further study is needed. In some places, these problems are not arranged by section so that students can gain experience in recognizing types of problems as well as in solving them.

Special Features

Use of Calculators.   In keeping with the recommendations of the National Council of Teachers of Mathematics (NCTM) and the Mathematical Association of America (MAA), we continue to de-emphasize the use of logarithmic and trigonometric tables in favor of the use of scientific calculators. As classroom teachers, we feel obliged to prepare our students to function in the real world, where virtually everyone who uses mathematics in a practical way—from the actuary to the zoologist—routinely employs a calculator. Today, a scientific calculator can be purchased inexpensively and used throughout the student's tenure in college and beyond.

Logarithmic and Trigonometric Tables.   For those teachers who feel that some instruction in the use of tables is desirable, we continue to include in the appendices tables of logarithmic, exponential, and trigonometric functions, along with examples illustrating their use and the technique of linear interpolation.

Mathematical Models and Applications.   Many applications and mathematical models in the life sciences, physics, chemistry, engineering, social sciences, economics, and business are to be found throughout the book.

Trigonometric Functions.   As in the first edition, trigonometric functions are first defined in terms of right triangles. This simple, geometrically appealing approach provides a firm foundation for the ensuing treatment of trigonometric functions of general angles and of circular functions and their graphs.

Major Changes

We have rewritten and reorganized the presentation of some topics and added new material to reflect the needs and interests of students and the goals of the course. The main changes are as follows:

Complex Numbers are introduced in Chapter 1; they are used in subsequent sections throughout the book.

Polynomials and Their Zeros.   Chapter 4 has been reorganized and rewritten. Emphasis is now given to polynomials and their zeros. Descartes' rule of signs and the upper and lower bound rule for real zeros of polynomial functions have been added to Section 4.4. New material on locating irrational zeros of polynomial functions by using the *bisection method* has been added to Section 4.5. Complex polynomials are now covered in Section 4.7.

Rational Functions.   Guidelines are listed in Section 4.6 to help students proceed in a systematic manner when sketching graphs of rational functions. The concept of *oblique asymptote* has also been added to this section.

Trigonometric Form for Complex Numbers and DeMoivre's Theorem are now covered in Section 8.10. This insertion makes it unnecessary to include a separate chapter on complex numbers.

Analytic Geometry and The Conics.   Chapter 10 has been extensively rewritten and covers the standard topics in Analytic Geometry. New material on translation and rotation of axes has been added to the chapter. This allows us to convert a general degree equation of a rotated conic to standard form.

Student Aids

SOFTWARE, OF COURSE!   A microcomputer diskette, written by Robert J. Weaver of Mount Holyoke College and designed to be used in conjunction with the textbook, is available for users with access to Apple and IBM microcomputers. The programs include (1) Synthetic Division, (2) Division of Polynomials, (3) Function Plotters (including polynomial, rational, trigonometric, exponential, and logarithmic functions), (4) Bisection Method, (5) General Triangles, and (6) Standard Conic Sections.

Study Guide.   The *Study Guide* is a supplementary learning resource, offering students tutorial aid on each topic in the textbook. It is a useful source of drill problems for students who need additional reinforcement and practice. The *Study Guide* contains study objectives, fill-in statements, problems (broken down into simple steps), and a self-test for each chapter. Answers to all problems and tests in the *Study Guide* are included.

Instructor Aids

Instructor's Resource Manual.   This manual provides a comprehensive testing program, coordinated with the textbook. It includes: (1) a diagnostic examination, (2) a syllabus with teaching suggestions, (3) two examinations for each chapter, one of which is multiple-choice, and (4) a multiple-choice final examination.

Solutions Manual.   A systematic solution to each problem in the textbook is available in this manual for instructors. A glance at the worked-out solutions will assist instructors in selecting the appropriate problems to assign.

Acknowledgments

Again we have drawn on our own experience in teaching from *Algebra and Trigonometry* and on the feedback provided by students. The suggestions obtained from instructors using the first edition and the appraisals offered by reviewers were essential. We wish to thank all of these people and, in particular, to express our gratitude to the following: Richelle Blair, *Lakeland Community College*; Murray Blose, *Oklahoma State University*; Carol A. Edwards, *Saint Louis Community College at Florissant Valley*; David Ellenbogen, *Cape Cod Community College*; Li-ren Fong, *Johnson County Community College*; Mel Hamburger, *Laramie County Community College*; Daniel A. Hogan, *Hinds Junior College*; E. John Hornsby, Jr., *University of New Orleans*; Diane Johnson, *University of San Diego*; Stanley Lukawecki, *Clemson University*; DeWayne S. Nymann, *University of Tennessee at Chattanooga*; Vivian Savoy, *University of Southwestern Louisiana*; Kathleen Shay, *Middlesex County College*; Dorothy Sulock, *University of North Carolina at Asheville*; Ken Stewart Wagman, *University of California, Santa Cruz*; John Wagner, *Michigan State University*; Chris Watkiss, *Capilano College*; Judith L. Willoughby, *Minneapolis Community College*.

We are deeply indebted to Professor Steve Fasbinder of Oakland University for assisting in the proofreading and to Hyla Gold Foulis for reviewing each successive stage of the manuscript, proofreading pages, and preparing the *Solutions Manual* for the book. We also thank the people at Worth Publishers for their assistance.

December, 1985

M. A. Munem
D. J. Foulis

A Note to the Instructor on the Use of Calculators

Problems and examples for which the use of a calculator is recommended are marked with the symbol $\boxed{c}$. Answers to these problems and solutions for these examples were obtained using an HP-67 calculator—other calculators may give slightly different results because they use different internal routines.

The rule for rounding off numbers presented in Section 1.9 of Chapter 1 is consistent with the operation of most calculators with round-off capability. Some instructors may wish to mention the popular alternative round-off rule: If the first dropped digit is 5 and there are no nonzero digits to its right, round off so that the digit retained is even.

Because there are so many different calculators on the market, we have made no attempt to give detailed instructions for calculator operation in this textbook. Students should be urged to consult the instruction manuals furnished with their calculators.

Conscientious instructors will wish to encourage their students to learn to use calculators *efficiently*; for instance, to do chain calculations using the memory features of the calculator. In some of our examples we have shown the intermediate results of chain calculations so the students can check their calculator work; however, it should be emphasized that it is not necessary to write down these intermediate results when using the calculator.

As important as it is to encourage students to use their calculators for the examples and problems marked $\boxed{c}$, it is perhaps more important to *restrain* them from attempting to use their calculators when the symbol $\boxed{c}$ is not present. For instance, a student who routinely uses a calculator to determine the algebraic sign of a trigonometric function of an angle may fail to learn the connection between the algebraic sign and the quadrant containing the angle.

Finally, we have made no attempt to provide a systematic discussion of the inaccuracies inherent in computations with a calculator. However, in order to make students aware that such inaccuracies exist, we have carried out some computations to the full ten places available on our calculator. (See, for instance, Example 4 on page 266.) An excellent account of calculator inaccuracy can be found in "Calculator Calculus and Roundoff Errors," by George Miel in *The American Mathematical Monthly*, 1980, Vol. 87, No. 4, pp. 243–52.

Contents

Chapter 1

Concepts of Algebra

This chapter is designed as a review of the basic concepts and methods of algebra. Its purpose is to help you attain the algebraic skills that are required throughout the textbook. Topics covered include the language and symbols of algebra, polynomials, fractions, exponents, radicals, complex numbers, and the use of a calculator.

1.1 THE LANGUAGE AND NOTATION OF ALGEBRA

Algebra begins with a systematic study of the operations and rules of arithmetic. The operations of addition, subtraction, multiplication, and division serve as a basis for all arithmetic calculations. In order to achieve generality, letters of the alphabet are used in algebra to represent numbers. A letter such as x, y, a, or b can stand for a particular number (known or unknown), or it can stand for any number at all. The sum, difference, product, and quotient of two numbers, x and y, can then be written as

$$x + y, \qquad x - y, \qquad x \times y, \qquad \text{and} \qquad x \div y.$$

In algebra, the notation $x \times y$ for the product of x and y is not often used because of the possible confusion of the letter x with the multiplication sign $\times$. The preferred notation is $x \cdot y$ or simply xy. Similarly, the notation $x \div y$ is usually avoided in favor of the fraction $\dfrac{x}{y}$ or x/y.

Algebraic notation—the "shorthand" of mathematics—is designed to clarify ideas and simplify calculations by permitting us to write expressions compactly and efficiently. For instance, $x + x + x + x + x$ can be written simply as $5x$. The use of exponents provides an economy of notation for products; for instance, $x \cdot x$ can be written simply as x^2 and $x \cdot x \cdot x$ as x^3. In general, if n is a positive integer,

$$x^n \quad \text{means} \quad \overbrace{x \cdot x \cdot x \cdots x}^{n \text{ times}}.$$

In using the **exponential notation x^n,** we refer to x as the **base** and n as the **exponent,** or the **power** to which the base is raised.

Example 1   Rewrite each expression using exponential notation.

(a) $3 \cdot 3 \cdot 3 \cdot 3 \cdot 3 \cdot 3 \cdot 3$ (b) $(-y)(-y)(-y)(-y)(-y)$

(c) $d \cdot d \cdot d \cdot e \cdot e \cdot e \cdot e$ (d) $(3a - 2b)(3a - 2b)(3a - 2b)$

Solution   (a) 3^7 (b) $(-y)^5$

(c) $d^3 e^4$ (d) $(3a - 2b)^3$

By writing an **equals sign** ($=$) between two algebraic expressions, we obtain an **equation,** or **formula,** stating that the two expressions represent the same number. Using equations and formulas, we can express mathematical facts in compact, easily remembered forms.

The important **principle of substitution** states that, *if $a = b$, we may substitute b for a in any formula or expression that involves a.* A second important rule, called the **reflexive principle,** states that *any expression may be set equal to itself.* Many standard algebraic procedures, such as *adding or subtracting the same quantity to both sides of an equation,* can be justified on the basis of the reflexive and substitution principles.

Example 2   Given that $a = b$, show that $a + c = b + c$.

Solution   By the reflexive principle,

$$a + c = a + c.$$

Since $a = b$, we can substitute b for a on the right side of this equation to obtain

$$a + c = b + c.$$

In such fields as geometry, physics, engineering, statistics, geology, business, medicine, economics, and the life sciences, formulas are used to express relationships among various quantities.

substitution $s = x$

Example 3   Write a formula for the volume V of a cube that has sides of length x units.

Solution $V = x \cdot x \cdot x = x^3$ cubic units.

$V = x^3$

■

Example 4   A certain type of living cell divides every hour. Starting with one such cell in a culture, the number N of cells present at the end of t hours is given by the formula $N = 2^t$. Find the number of cells in the culture after 6 hours.

Solution   Substituting $t = 6$ in the formula $N = 2^t$, we find that

$$N = 2^6 = 2 \cdot 2 \cdot 2 \cdot 2 \cdot 2 \cdot 2 = 64 \text{ cells.}$$

■

Basic Algebraic Properties of Real Numbers

The numbers used to measure real-world quantities such as length, area, volume, speed, electrical charge, efficiency, probability of rain, intensity of earthquakes, profit, body temperature, gross national product, growth rate, and so forth are called **real numbers.** They include such numbers as

$$5, \quad -17, \quad \frac{17}{13}, \quad -\frac{2}{3}, \quad 0, \quad 2.71828, \quad \sqrt{2}, \quad -\frac{\sqrt{3}}{2}, \quad 3 \times 10^8, \quad \text{and} \quad \pi.$$

The basic algebraic properties of the real numbers can be expressed in terms of the two fundamental operations of addition and multiplication.

Basic Algebraic Properties of Real Numbers

Let a, b, and c denote real numbers.

1. *The Commutative Properties*

> (i) $a + b = b + a$ (ii) $a \cdot b = b \cdot a$

The commutative properties say that the *order* in which we either add or multiply real numbers doesn't matter.

2. *The Associative Properties*

> (i) $a + (b + c) = (a + b) + c$ (ii) $a \cdot (b \cdot c) = (a \cdot b) \cdot c$

The associative properties tell us that the way real numbers are *grouped* when they are either added or multiplied doesn't matter. Because of the associative properties, expressions such as $a + b + c$ or $a \cdot b \cdot c$ make sense without parentheses.

3. *The Distributive Properties*

(i) $a \cdot (b + c) = a \cdot b + a \cdot c$ **(ii)** $(b + c) \cdot a = b \cdot a + c \cdot a$

The distributive properties can be used to expand a product into a sum, such as

$$a(b + c + d) = ab + ac + ad,$$

or the other way around, to rewrite a sum as a product:

$$ax + bx + cx + dx + ex = (a + b + c + d + e)x.$$

4. *The Identity Properties*

(i) $a + 0 = 0 + a = a$ **(ii)** $a \cdot 1 = 1 \cdot a = a$

We call 0 the **additive identity** and 1 the **multiplicative identity** for the real numbers.

5. *The Inverse Properties*

(i) For each real number a, there is a real number $-a$, called the **additive inverse** of a, such that

$$a + (-a) = (-a) + a = 0.$$

(ii) For each real number $a \neq 0$, there is a real number $1/a$, called the **multiplicative inverse** of a, such that

$$a \cdot \frac{1}{a} = \frac{1}{a} \cdot a = 1.$$

Although the additive inverse of a, namely $-a$, is usually called the **negative** of a, you must be careful because $-a$ *isn't necessarily a negative number*. For instance, if $a = -2$, then $-a = -(-2) = 2$. Notice that *the multiplicative inverse $1/a$ is assumed to exist only if $a \neq 0$*. The real number $1/a$ is also called the **reciprocal** of a and is often written as a^{-1}.

Example 5   State one basic algebraic property of the real numbers to justify each statement.

(a) $7 + (-2) = (-2) + 7$ **(b)** $x + (3 + y) = (x + 3) + y$

(c) $a + (b + c)d = a + d(b + c)$ **(d)** $x[y + (z + w)] = xy + x(z + w)$

(e) $(x + y) + [-(x + y)] = 0$ **(f)** $(x + y) \cdot 1 = x + y$

(g) If $x + y \neq 0$, then $(x + y)[1/(x + y)] = 1$

Solution   **(a)** Commutative property for addition   **(b)** Associative property for addition
(c) Commutative property for multiplication
(d) Distributive property   **(e)** Additive inverse property
(f) Multiplicative identity property   **(g)** Multiplicative inverse property   ∎

Many of the important properties of the real numbers can be *derived* as results of the basic properties, although we shall not do so here. Among the more important **derived properties** are the following.

6. *The Cancellation Properties*

> **(i)** If $a + x = a + y$, then $x = y$. $a = 2$
>
> **(ii)** If $a \neq 0$ and $ax = ay$, then $x = y$. $a = 1$

7. *The Zero-Factor Properties*

> **(i)** $a \cdot 0 = 0 \cdot a = 0$
>
> **(ii)** If $a \cdot b = 0$, then $a = 0$ or $b = 0$ (or both).

8. *Properties of Negation*

> **(i)** $-(-a) = a$ **(ii)** $(-a)b = a(-b) = -(ab)$
>
> **(iii)** $(-a)(-b) = ab$ **(iv)** $-(a + b) = (-a) + (-b)$

The operations of subtraction and division are defined as follows.

Definition 1   **Subtraction and Division**

> Let a and b be real numbers.
>
> **(i)** The **difference** $a - b$ is defined by $a - b = a + (-b)$.
>
> **(ii)** The **quotient** $a \div b$ or $\dfrac{a}{b}$ is defined only if $b \neq 0$. If $b \neq 0$, then by definition
>
> $$\frac{a}{b} = a \cdot \frac{1}{b}.$$

For instance, $7 - 4$ means $7 + (-4)$ and $\frac{7}{4}$ means $7 \cdot \frac{1}{4}$. Because 0 has no multiplicative inverse, a/b is undefined when $b = 0$. Thus:

> *Division by zero is not allowed.*

When $a \div b$ is written in the form a/b, it is called a **fraction** with **numerator a** and **denominator b**. Although the denominator can't be zero, there's nothing wrong with having a zero in the numerator. In fact, if $b \neq 0$,

$$\frac{0}{b} = 0 \cdot \frac{1}{b} = 0$$

by the zero-factor property 7(i).

We can now state the following additional derived property for fractions.

9. *The Negative of a Fraction*

If $b \neq 0$, then $\dfrac{-a}{b} = \dfrac{a}{-b} = -\dfrac{a}{b}$. $-1 \cdot \dfrac{a}{b} = \dfrac{-a}{b} = \dfrac{a}{-b} = -\dfrac{a}{b}$

Because Properties 1 through 9 provide a foundation for the algebra of real numbers, it is essential for you to become familiar with them.

Problem Set 1.1

In each problem set, problems with colored numbers constitute a good representation of the main ideas of the section.

In Problems 1 to 10, rewrite each expression using exponential notation.

1. $5 \cdot 5 \cdot 5 \cdot 5 \cdot 5 \cdot 5 \cdot 5 \cdot 5$ 5^8

2. $(-7)(-7)(-7)(-7)$ 3. $3 \cdot 3 \cdot 3 \cdot 3 \cdot 4 \cdot 4 \cdot 4$

4. $8 \cdot 8 + (-6)(-6)(-6)$

5. $x \cdot x \cdot x \cdot x \cdot y \cdot y \cdot y$ $x^4 y^3$ 6. $4 \cdot 4 \cdot 4 \cdot 4 \cdot y \cdot y \cdot y$

7. $(-x)(-x)(-y)(-y)(-y)$ $(-x)^2(-y)^3$

8. $y^2 y^2 y^2 z^3 z^3$ $x^2 y^2$

9. $(2a + 1)(2a + 1)(2a + 1)$ $(2a+1)^3$

10. $(x + \frac{1}{2})(x + \frac{1}{2})(x - \frac{1}{2})(x - \frac{1}{2})(x - \frac{1}{2})$

In Problems 11 to 22, write a formula for the given quantity.

✳ 11. The number z that is twice the sum of x and y.

12. The area A of a rectangle with length a units and width b units.

✳ 13. The area A of a circle of radius r units.

14. The perimeter P of a rectangle with length a units and width b units.

✳ 15. The number x that is 5% of a number n.

16. The volume V of a rectangular box with length L units, width W units, and height H units.

✳ 17. The surface area A of a cube with sides of length x units. [*Hint:* A cube has six faces, each of which is a square.]

18. The volume V of a sphere of radius r units. [Consult a geometry book if you don't know the formula.]

✳ 19. The area A of a triangle with base b units and height h units.

20. The number N of living cells in a culture after t hours if there are N_0 cells when $t = 0$, if each cell divides into two cells at the end of each hour, and if no cells die.

21. The amount A dollars you owe after t years if you borrow p dollars at a simple interest rate r per year.

22. The number L of board feet of lumber in a tree d feet in diameter and h feet high. Assume for simplicity that the tree is a right circular cyclinder. [*Note:* One board foot is the volume of a board with dimensions 1 foot by 1 foot by 1 inch.]

c 23. If p dollars is invested at a nominal annual interest rate r compounded n times per year, the investment will be worth $p\left(1 + \dfrac{r}{n}\right)^{nt}$ dollars at the end of t years. Suppose you invest \$5000 at 10% interest ($r = 0.10$) compounded semiannually ($n = 2$). What is your investment worth at the end of 2 years? [The c indicates that you may use a calculator if you wish.]

24. The *half-life* of a radioactive substance is the period of time T during which exactly half of the substance will undergo radioactive disintegration. If q represents the quantity of such a substance at time t, then $q = \dfrac{q_0}{2^{t/T}}$, where q_0 is the original amount of the substance when $t = 0$. Potassium 42, which is used as a biological tracer, has a half-life of $T = 12.5$ hours. If a certain quantity of potassium 42 is injected into an organism, and if none is lost by excretion, what percentage will remain after $t = 50$ hours?

In Problems 25 to 40, state one basic algebraic property of the real numbers to justify each statement.

25. $5 + (-3) = (-3) + 5$ Comm. for add.

26. $5 \cdot (3 + 7) = (3 + 7) \cdot 5$

27. $(-13)(-12) = (-12)(-13)$ Comm. for mult.

28. $5 \cdot (3 + 7) = 5 \cdot 3 + 5 \cdot 7$

29. $1 \cdot (3 + 7) = 3 + 7$ Identity for x

30. $[(3)(4)](5) = (3)[(4)(5)]$

31. $4 + (x + z) = (4 + x) + z$ Assoc. for add.

32. $5 \cdot \frac{1}{5} = 1$ 33. $4 \cdot (x + 0) = 4x$

34. $(x - y) + [-(x - y)] = 0$

35. $x[(-y) + y] = x(-y) + xy$ Distributive

36. $(-4) \cdot \dfrac{1}{(-4)} = 1$

37. $(4 + x) + [-(4 + x)] = 0$

38. $x + [y + (z + w)] = (x + y) + (z + w)$

39. $(ab)(cd) = [(ab)c]d$ Assoc for mult.

40. $(a + b)(c + d) = a(c + d) + b(c + d)$

In Problems 41 to 50, state one of the derived algebraic properties to justify each statement.

41. $-(-5) = 5$ negation

42. $(-3)(-x) = 3x$

43. $5(-6) = -(5)(6)$ negation

44. $(x + y) \cdot 0 = 0$

45. If $7 + x = 7 + x^2$, then $x = x^2$.

46. If $2y = 2y^{-1}$, then $y = y^{-1}$.

47. If $(2x + 3)(x + 1) = 0$, then $2x + 3 = 0$ or $x + 1 = 0$.   zero factor

48. If $x^2 = 0$, then $x = 0$.

49. $\dfrac{x + 1}{-2} = -\dfrac{x + 1}{2}$

50. $\dfrac{-3}{2 - x} = \dfrac{3}{-(2 - x)}$

In Problems 51 and 52 a *mistake* has been made. Find the mistake and make the correct calculation.

51. $(3 + 5)^2 = 3^2 + 5^2 = 9 + 25 = 34$? $8^2 = 64$

52. $\dfrac{1}{3 + 5} = \dfrac{1}{3} + \dfrac{1}{5} = \dfrac{5}{15} + \dfrac{3}{15} = \dfrac{8}{15}$?

53. Does the operation of subtraction have the associative property; that is, is it always true that $a - (b - c) = (a - b) - c$?

54. Does the operation of division have the associative property?

55. Prove that both sides of an equation can be multiplied by the same quantity; that is, prove that if $a = b$, then $c \cdot a = c \cdot b$. [*Hint:* Use the reflexive and substitution principles.]

56. Prove that both sides of an equation can be interchanged; that is, prove that if $a = b$, then $b = a$. This is called the **symmetric property of equality.** [*Hint:* Use the reflexive and substitution principles.]

1.2 SETS OF REAL NUMBERS

Grouping or **classifying** is a familiar technique in the natural sciences for dealing with the immense diversity of things in the real world. For instance, in biology, plants and animals are divided into various phyla, and then into classes, orders, families, genera, and species. In much the same way, real numbers can be grouped or classified by singling out important features possessed by some numbers but not by others. By using the idea of a *set*, classification of real numbers can be accomplished with clarity and precision.

A **set** may be thought of as a collection of objects. Most sets considered in this textbook are sets of real numbers. Any one of the objects in a set is called an **element,** or **member,** of the set. Sets are denoted either by capital letters such as A, B, C or else by braces $\{\cdots\}$ enclosing symbols for the elements in the set. Thus, if we write $\{1, 2, 3, 4, 5\}$, we mean the set whose elements are the numbers 1, 2, 3, 4, and 5. Two sets are said to be **equal** if they contain precisely the same elements.

Example 1   Write the set A consisting of all the odd numbers between 2 and 8.

Solution $A = \{3, 5, 7\}$

Figure 1

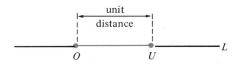

Figure 2

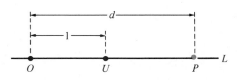

Figure 3

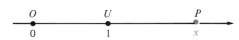

Sets of numbers and relations among such sets can often be visualized by the use of a **number line** or **coordinate axis.** A number line is constructed by fixing a point O called the *origin* and another point U called the **unit point** on a straight line L (Figure 1). The distance between O and U is called the **unit distance,** and may be 1 inch, 1 centimeter, or 1 unit of whatever measure you choose. If the line L is horizontal, it is customary to place U to the right of O.

Each point P on the line L is now assigned a "numerical address" or **coordinate** x representing its signed distance from the origin, measured in terms of the given unit. Thus, $x = \pm d$, where d is the distance between O and P; the plus sign or minus sign is used to indicate whether P is to the right or left of O (Figure 2). Of course, the origin O is assigned the coordinate 0 (zero), and the unit point U is assigned the coordinate 1. On the resulting number scale (Figure 3), each point P has a corresponding numerical coordinate x and each real number x is the coordinate of a uniquely determined point P. It is convenient to use an arrowhead on the number line to indicate the direction in which the numerical coordinates are increasing (to the right in Figure 3).

A set of numbers can be illustrated on a number line by shading or coloring the points whose coordinates are members of the set. For instance:

Figure 4

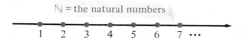

$\mathbb{N}$ = the natural numbers

Figure 5

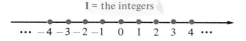

$\mathbf{I}$ = the integers

Figure 6

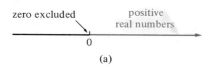

zero excluded

positive real numbers

0

(a)

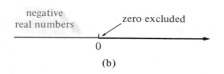

negative real numbers

zero excluded

0

(b)

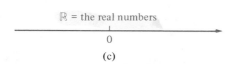

$\mathbb{R}$ = the real numbers

0

(c)

1. The **natural numbers,** also called the **counting numbers,** or **positive integers,** are the numbers 1, 2, 3, 4, 5, and so on, obtained by adding 1 over and over again. The set $\{1, 2, 3, 4, 5, \cdots\}$ of all natural numbers, denoted by the symbol $\mathbb{N}$, is illustrated in Figure 4.

2. The **integers** consist of all the natural numbers, the negatives of the natural numbers, and zero. The set $\{\cdots -4, -3, -2, -1, 0, 1, 2, 3, 4, \cdots\}$ of all integers, denoted by the symbol $\mathbf{I}$, is illustrated in Figure 5.

3. The **positive real numbers** correspond to points to the right of the origin (Figure 6a), and the **negative real numbers** correspond to points to the left of the origin (Figure 6b). The set of all real numbers is denoted by the symbol $\mathbb{R}$ (Figure 6c).

4. The **rational numbers** are those real numbers that can be written in the form a/b, where a and b are integers and $b \neq 0$. Since b may equal 1, every integer is a rational number. Other examples of rational numbers are $\frac{13}{2}$, $\frac{3}{4}$, and $-\frac{22}{7}$. The set of all rational numbers is denoted by the symbol $\mathbb{Q}$ (which reminds us that rational numbers are *quotients* of integers). Rational numbers in decimal form either **terminate** or begin to **repeat** the same pattern indefinitely.

5. The **irrational numbers** are the real numbers that are not rational. A real number is irrational if and only if its decimal representation is **nonterminating** and **nonrepeating.** Examples are $\sqrt{2} = 1.4142135 \cdots$, $\sqrt{3} = 1.7320508 \cdots$, and $\pi = 3.1415926 \cdots$.

Inequalities and Intervals

If the point with coordinate x lies to the left of the point with coordinate y (Figure 7), we say that y is **greater than** x (or equivalently, that x is **less than** y) and we write $y > x$ (or $x < y$). In other words, $y > x$ (or $x < y$) means that $y - x$ is positive. A statement of the form $y > x$ (or $x < y$) is called an **inequality.**

Sometimes we know only that a certain inequality does *not* hold. If $x < y$ does not hold, then either $x > y$ or $x = y$. In this case, we say that x is **greater than or equal to** y, and we write $x \geq y$ (or $y \leq x$). If $y \leq x$, we say that y is **less than or equal to** x. Statements of the form $y < x$ (or $x > y$) are called **strict** inequalities, whereas those of the form $y \leq x$ (or $x \geq y$) are called **nonstrict** inequalities.

Figure 7

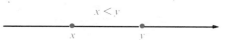

$x < y$

x y

If we write $x < y < z$, we mean that $x < y$ *and* $y < z$. Likewise, $x \geq y > z$ means that $x \geq y$ *and* $y > z$. Notice that this notation for combined inequalities is only used when the inequalities run in the *same direction*. If you are ever tempted to write something like $x < y > z$, resist the urge—such notation is improper and confusing.

In Section 5 of Chapter 2 we shall study inequalities in more detail. Here we use inequalities to define sets called **intervals.**

Definition 1   **Bounded Intervals**

Let a and b be fixed real numbers with $a < b$.

(i)   The **open interval** (a, b) with **endpoints** a and b is the set of all real numbers x such that $a < x < b$.

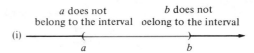

(ii)   The **closed interval** $[a, b]$ with **endpoints** a and b is the set of all real numbers x such that $a \leq x \leq b$.

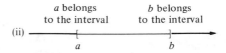

(iii) The **half-open interval** $[a, b)$ with **endpoints** a and b is the set of all real numbers x such that $a \leq x < b$.

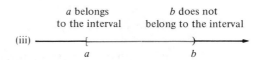

(iv) The **half-open interval** $(a, b]$ with **endpoints** a and b is the set of all real numbers x such that $a < x \leq b$.

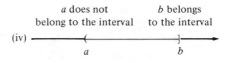

Notice that a closed interval contains its endpoints, but an open interval does not. A half-open interval (also called a **half-closed** interval) contains one of its endpoints, but not the other.

Unbounded intervals, which extend indefinitely to the right or left, are written with the aid of the special symbols $+\infty$ and $-\infty$, called **positive infinity** and **negative infinity.**

Definition 2   **Unbounded Intervals**

Let a be a fixed real number.

(i) $(a, +\infty)$ is the set of all real numbers x such that $a < x$.

(ii) $(-\infty, a)$ is the set of all real numbers x such that $x < a$.

(iii) $[a, +\infty)$ is the set of all real numbers x such that $a \leq x$.

(iv) $(-\infty, a]$ is the set of all real numbers x such that $x \leq a$.

It must be emphasized that $+\infty$ and $-\infty$ are just convenient symbols—*they are not real numbers* and should not be treated as if they were. In the notation for unbounded intervals, we usually write ∞ rather than $+\infty$. For instance, $(5, \infty)$ denotes the set of all real numbers that are greater than 5.

Example 2   Illustrate each set on a number line.

(a) $(2, 5]$

(b) $(3, \infty)$

(c) The set A of all real numbers that belong to both intervals $(2, 5]$ and $(3, \infty)$.

(d) The set B of all real numbers that belong to at least one of the intervals $(2, 5]$ and $(3, \infty)$.

Solution   **(a)** The interval $(2, 5]$ consists of all real numbers between 2 and 5, including 5, but excluding 2.

(b) The interval $(3, \infty)$ consists of all real numbers that are greater than 3.

(c) The set A of all real numbers that belong to both intervals $(2, 5]$ and $(3, \infty)$ is the interval $(3, 5]$.

$A = (3, 5]$

(d) The set B consists of all numbers in the interval $(2, 5]$ together with all numbers in the interval $(3, \infty)$, so $B = (2, \infty)$.

$B = (2, \infty)$

Rational Numbers and Decimals

By using long division, you can express a rational number as a decimal. For instance, if you divide 2 by 5, you will obtain $\frac{2}{5} = 0.4$, a terminating decimal. Similarly, if you divide 2 by 3, you will obtain $\frac{2}{3} = 0.66666 \cdots$, a nonterminating, repeating decimal. A repeating decimal, such as $0.66666 \cdots$, is often written as $0.\overline{6}$, where the overbar indicates the digit or digits that repeat; hence $\frac{2}{3} = 0.\overline{6}$.

Example 3   Express each rational number as a decimal.

(a) $-\frac{3}{5}$ **(b)** $\frac{3}{8}$ **(c)** $\frac{17}{6}$ **(d)** $\frac{3}{7}$

Solution   **(a)** $-\frac{3}{5} = -0.6$

(b) $\frac{3}{8} = 0.375$

(c) $\frac{17}{6} = 2.83333 \cdots = 2.8\overline{3}$

(d) $\frac{3}{7} = 0.428571428571428571 \cdots = 0.\overline{428571}$

Every terminating or repeating decimal represents a rational number. The following example illustrates how you can rewrite a terminating decimal as a quotient of integers.*

Example 4   Express each terminating decimal as a quotient of integers.

(a) 0.7   (b) −0.63   (c) 1.075

Solution   (a) $0.7 = \frac{7}{10}$ (b) $-0.63 = -\frac{63}{100}$ (c) $1.075 = \frac{1075}{1000} = \frac{43}{40}$ ■

Fractions or decimals are often expressed as percents; for instance, 3% means $\frac{3}{100}$ or 0.03.

Example 5   Rewrite each percent as a decimal.

(a) 5.7%   (b) 0.003%

Solution   (a) $5.7\% = \frac{5.7}{100} = 0.057$ (b) $0.003\% = \frac{0.003}{100} = 0.00003$ ■

Example 6   Rewrite each rational number as a percent.

(a) $\frac{4}{5}$ (b) $\frac{1}{3}$

Solution   (a) $\frac{4}{5} = 0.8 = 0.8 \times 100\% = 80\%$ (b) $\frac{1}{3} = 0.\overline{3} = 0.\overline{3} \times 100\% = 33.\overline{3}\%$ ■

Example 7   What percent is 40 of 2000?

Solution $\frac{40}{2000} = 0.02 = 0.02 \times 100\% = 2\%$ ■

If a number increases, then the **percent of increase** is given by

$$\frac{\text{amount of increase}}{\text{original value}} \times 100\%.$$

If a number decreases, then the **percent of decrease** is given by

$$\frac{\text{amount of decrease}}{\text{original value}} \times 100\%.$$

Example 8   Juanita's weekly salary increases from \$205 to \$213.20. What is the percent of increase?    Find difference & divide by original value. ×100%

* For the technique for rewriting a repeating decimal as a quotient of integers, see Example 6 on page 74.

Solution   The amount of increase is $213.20 − $205 = $8.20. Since the original salary before the increase was $205, the percent of increase is given by

$$\frac{8.20}{205} \times 100\% = 0.04 \times 100\% = 4\%.$$

Example 9   A small town decreases its annual budget from $800,000 to $600,000. What is the percent of decrease?

Solution $$\frac{800,000 - 600,000}{800,000} \times 100\% = \frac{200,000}{800,000} \times 100\% = 0.25 \times 100\% = 25\%$$

Problem Set 1.2

In Problems 1 to 4, list all of the elements that belong to each set. $\{4, 6, 8, 10\}$

1. A is the set of all even integers between 3 and 11.

2. B is the set of all natural numbers x such that $-\frac{1}{2} \le x \le \frac{7}{2}$. $\{2, 4\}$

✳ 3. C is the set of all real numbers that are in the closed interval $[2, 4]$ but not in the open interval $(2, 4)$. $\{2, 4\}$

4. D is the set of all real numbers x such that $x(x - 1)(x + 1) = 0$.

In Problems 5 to 16, indicate whether the statement is true or false.

5. $\frac{1}{2}$ is an element of $\mathbb{N}$. F   6. $\frac{1}{2}$ is an element of $\mathbb{Q}$.

7. $\frac{1}{2}$ is an element of $(\frac{1}{2}, \frac{2}{3})$. F

8. $\frac{1}{2}$ is an element of $(-\frac{2}{3}, \frac{2}{3})$.

9. $\sqrt{2}$ is an element of $\mathbb{R}$. T

10. $\sqrt{2}$ is an element of $\mathbb{Q}$.

11. $\frac{1}{2} + \frac{2}{3}$ is an element of $\mathbb{Q}$.

12. $\frac{1}{2} + \sqrt{2}$ is an element of $\mathbb{Q}$.

13. Every natural number is an integer. T

✳ 14. Every natural number is a rational number. F

15. Every positive real number is a rational number.

16. $\sqrt{\pi}$ is a rational number.

In Problems 17 to 24, illustrate each set on a number line.

17. The set of odd positive integers.

18. The set of even integers that are not positive.

✳ 19. (a) $(2, 5)$ (b) $[-1, 3)$ (c) $[-4, 0]$
      (d) $(-\infty, -3)$ (e) $[1, \infty)$ (f) $(-\frac{1}{2}, \infty)$
      (g) $[-\frac{3}{2}, \frac{5}{2}]$ (h) $(-4, 0]$

20. The set of real numbers x such that $-3 < x < 3$ and $2x$ is in $\mathbf{I}$.

✳ 21. The set A of all real numbers that belong to both of the intervals $(0, 3]$ and $(1, \infty)$.

22. The set B of all real numbers that belong to both of the intervals $(-1, \frac{1}{2}]$ and $[\frac{1}{2}, \frac{3}{4}]$.

23. The set C of all real numbers that belong to at least one of the intervals $[-2, -1]$ and $[1, 2]$.

24. The set D of all real numbers that belong to at least one of the intervals $(-1, 0]$ and $(0, 1]$.

In Problems 25 to 38, express each rational number as a decimal.

25. $\frac{4}{5}$ 0.8   26. $-\frac{7}{25}$ 27. $-\frac{3}{8}$ −0.38   28. $\frac{3}{50}$

29. $-\frac{13}{50}$ 30. $\frac{7}{100}$ 31. $\frac{-17}{200}$ 32. $\frac{200}{500}$

33. $-\frac{5}{8}$ **34.** $\frac{75}{30}$ **35.** $-\frac{5}{3}$ **36.** $\frac{11}{12}$

37. $\frac{15}{7}$ *2¼*   **38.** $\frac{-23}{7}$

In Problems 39 to 48, express each terminating decimal as a quotient of integers.

39. 0.41   *$\frac{41}{100}$*   **40.** -0.54

41. -0.032 *$\frac{0.032}{1000}$*   **42.** 22.61

43. -0.581 **44.** 2.691

45. -0.913 **46.** 0.00012

47. 1.0451   *$\frac{10451}{10000}$*   **48.** -2.00002

In Problems 49 to 56, rewrite each percent as a decimal.

49. 11%   *$\frac{11}{100} = 0.11$*   **50.** 99.44%

51. 1.03%   *$\frac{1.03}{100} = 0.01$*   **52.** 315%

53. 432%   *4.32%*   **54.** 0.001%

55. 0.0006%   **56.** 1.00001%

In Problems 57 to 64, rewrite each rational number as a percent.

.02 < 2.40%

57. $\frac{1}{2}$ **58.** $\frac{7}{50}$ **59.** $\frac{3}{125}$ **60.** 0.042

61. $\frac{2}{3}$ *66.67%*   **62.** $\frac{5}{8}$ **63.** $\frac{2}{7}$ **64.** 1.245

65. What percent is 3 of 20?   *15%*

66. In the United States, 5 people out of every 1000 are in the army and 8 people out of every 10,000 are army officers. What percent of army personnel are officers?

67. The cost of a book increases from \$20 to \$24. What is the percent of increase?   *$\frac{4}{20} = .20 \times 100 = 20\%$*

68. In April, Pedro's salary is \$180 per week. In May, his April salary is increased by 5%. In June, his May salary is decreased by 5%. What is Pedro's salary in June?

69. The cost of a calculator decreases from \$24 to \$20. What is the percent of decrease?   *$\frac{4}{24} = .17 \times 100 = 17\%$*

1.3 POLYNOMIALS

An **algebraic expression** is an expression formed from any combination of numbers and variables by using the operations of addition, subtraction, multiplication, division, exponentiation (raising to powers), or extraction of roots. For instance,

$$7, \quad x, \quad 2x - 3y + 1, \quad \frac{5x^3 - 1}{4xy + 1}, \quad \pi r^2, \quad \text{and} \quad \pi r \sqrt{r^2 + h^2}$$

are algebraic expressions. By an algebraic expression *in* certain variables, we mean an expression that contains only those variables, and by a **constant,** we mean an algebraic expression that contains no variables at all. If numbers are substituted for the variables in an algebraic expression, the resulting number is called the **value** of the expression for these values of the variables.

Example 1   Find the value of $\dfrac{2x - 3y + 1}{xy^2}$ when $x = 2$ and $y = -1$.

Solution   Substituting $x = 2$ and $y = -1$, we obtain

$$\frac{2(2) - 3(-1) + 1}{2(-1)^2} = \frac{4 + 3 + 1}{2(1)} = \frac{8}{2} = 4.$$ ∎

If an algebraic expression consists of parts connected by plus or minus signs, it is called an **algebraic sum,** and each of the parts, together with the sign preceding it, is called a **term.** For instance, in the algebraic sum

$$3x^2y - \frac{4xz^2}{y} + \pi x^{-1}y,$$

the terms are $3x^2y$, $-4xz^2/y$, and $\pi x^{-1}y$.

Any part of a term that is multiplied by the remaining part is called a **coefficient** of the remaining part. For instance, in the term $-4xz^2/y$, the coefficient of z^2/y is $-4x$, whereas the coefficient of xz^2/y is -4. A coefficient such as -4, which involves no variables, is called a **numerical coefficient.** Terms such as $5x^2y$ and $-12x^2y$, which differ only in their numerical coefficients, are called **like terms** or **similar terms.**

An algebraic expression such as $4\pi r^2$ can be considered an algebraic sum consisting of just one term. Such a one-termed expression is called a **monomial.** An algebraic sum with two terms is called a **binomial,** and an algebraic sum with three terms is called a **trinomial.** For instance, the expression $3x^2 + 2xy$ is a binomial, whereas $-2xy^{-1} + 3\sqrt{x} - 4$ is a trinomial. An algebraic sum with two or more terms is called a **multinomial.**

A **polynomial** is an algebraic sum in which no variables appear in denominators or under radical signs, and all variables that do appear are raised only to positive-integer powers. For instance, the trinomial

$$-2xy^{-1} + 3\sqrt{x} - 4$$

is *not* a polynomial; however, the trinomial

$$3x^2y^4 + \sqrt{2}xy - \tfrac{1}{2}$$

is a polynomial in the variables x and y. A term such as $-\tfrac{1}{2}$, which contains no variables, is called a **constant term** of the polynomial. The numerical coefficients of the terms in a polynomial are called the **coefficients** of the polynomial. The coefficients of the polynomial above are

$$3, \quad \sqrt{2}, \quad \text{and} \quad -\tfrac{1}{2}.$$

The **degree of a term** in a polynomial is the sum of all the exponents of the variables in the term. In adding exponents, you should regard a variable with no exponent as being a first power. For instance, in the polynomial

$$9xy^7 - 12x^3yz^2 + 3x - 2,$$

the term $9xy^7$ has degree $1 + 7 = 8$, the term $-12x^3yz^2$ has degree $3 + 1 + 2 = 6$, and the term $3x$ has degree 1. The constant term, if it is nonzero, is always regarded as having degree 0.

The highest degree of all terms that appear with nonzero coefficients in a polynomial is called the **degree of the polynomial.** For instance, the polynomial considered above has degree 8. Although the constant monomial 0 is regarded as a polynomial, this particular polynomial is not assigned a degree.

Example 2 In each case, identify the algebraic expression as a monomial, binomial, trinomial, multinomial, and/or polynomial, and specify the variables involved. For any polynomials, give the degree and the coefficients.

[handwritten annotations: mon. x, y   Bi x+y multi   TRI, Multi, Poly in x+y of degree 3   Coeff. -⅝, 8, 11   3, 8, 11]

(a) $\dfrac{4x}{y}$ **(b)** $4x + 3y^{-1}$ **(c)** $-\frac{5}{3}x^2y + 8xy - 11$

Solution **(a)** Monomial in x and y (*not* a polynomial because of the variable y in the denominator)

(b) Binomial in x and y, multinomial (*not* a polynomial because of the negative exponent on y)

(c) Trinomial, multinomial, polynomial in x and y of degree 3 with coefficients $-\frac{5}{3}$, 8, and -11 ∎

A polynomial of degree n in a single variable x can be written in the **general form**

$$a_n x^n + a_{n-1} x^{n-1} + \cdots + a_2 x^2 + a_1 x + a_0,$$

in which $a_n, a_{n-1}, \ldots, a_2, a_1, a_0$ are the numerical coefficients, $a_n \neq 0$ (although any of the other coefficients can be zero), and a_0 is the constant term.

Addition and Subtraction of Polynomials

To find the sum of two or more polynomials, we use the associative and commutative properties of addition to group like terms together, and then we combine the like terms by using the distributive property (see Problem 74).

Example 3 Find the following sum: $(2x^2 + 7x - 5) + (3x^2 - 11x + 8)$.

Solution $(2x^2 + 7x - 5) + (3x^2 - 11x + 8) = (2x^2 + 3x^2) + (7x - 11x) + (-5 + 8)$
$$= 5x^2 - 4x + 3 \qquad ∎$$

To find the difference of two polynomials, we change the signs of all the terms in the polynomial being subtracted, and then add.

Example 4 Find the following difference: $(3x^3 - 5x^2 + 8x - 3) - (5x^3 - 7x + 11)$.

Solution $(3x^3 - 5x^2 + 8x - 3) - (5x^3 - 7x + 11)$
$$= (3x^3 - 5x^2 + 8x - 3) + (-5x^3 + 7x - 11)$$
$$= (3x^3 - 5x^3) - 5x^2 + (8x + 7x) + (-3 - 11)$$
$$= -2x^3 - 5x^2 + 15x - 14 \qquad ∎$$

In adding or subtracting polynomials, you may prefer to use a vertical arrangement with like terms in the same columns.

Example 5   Perform the indicated operations:

$$(4x^3 + 7x^2y + 2xy^2 - 2y^3) + (2x^3 + xy^2 + 4y^3) + (4x^2y - 8xy^2 - 9y^3) - (y^3 - 7x^2y).$$

Solution   First we use a vertical arrangement to perform the addition.

$$
\begin{array}{r}
4x^3 + 7x^2y + 2xy^2 - 2y^3 \\
(+) \quad 2x^3 \qquad\qquad + xy^2 + 4y^3 \\
(+) \qquad\qquad 4x^2y - 8xy^2 - 9y^3 \\
\hline
6x^3 + 11x^2y - 5xy^2 - 7y^3
\end{array}
$$

Then we perform the subtraction vertically.

$$
\begin{array}{r}
6x^3 + 11x^2y - 5xy^2 - 7y^3 \\
(-) \qquad - 7x^2y \qquad\qquad + y^3 \\
\hline
6x^3 + 18x^2y - 5xy^2 - 8y^3
\end{array}
$$

Multiplication of Polynomials

To multiply two or more monomials, we use the commutative and associative properties of multiplication along with the following properties of exponents.

Properties of Exponents

Let a and b denote real numbers. Then, if m and n are positive integers,

(i) $a^m a^n = a^{m+n}$	**(ii)** $(a^m)^n = a^{mn}$	**(iii)** $(ab)^n = a^n b^n$

We verify (i) as follows and leave (ii) and (iii) as exercises (Problems 75 and 76):

$$a^m a^n = \overbrace{(a \cdot a \cdots a)}^{m\text{ factors}}\overbrace{(a \cdot a \cdots a)}^{n\text{ factors}} = \overbrace{a \cdot a \cdots a \cdot a \cdot a \cdots a}^{m + n\text{ factors}} = a^{m+n}.$$

Properties (i), (ii), and (iii) are useful for simplifying algebraic expressions containing exponents. In general, when the properties of real numbers are used to rewrite an algebraic expression as compactly as possible, or in a form so that further calculations are made easier, we say that the expression has been **simplified.** Thus, although the word "simplify" has no precise mathematical definition, its meaning is usually clear from the context in which it is used.

Example 6 Use the properties of exponents to simplify each expression.

(a) x^5x^4 X^9

(b) $2y^4y^6y^2z^2$ $2y^{12}z^2$

(c) $(5x^3y)(-3x^2y^4)$ -15^5y^5

(d) $(3x^2y^4)^2$ $9x^4y^8$

✱ (e) $(-2x^4y^2)(3x^2y^3)(5xy^4)$ $-30x^7y^9$

(f) $(x + 3)^2(x + 3)^4$ $x+3^6$

Solution   (a) $x^5x^4 = x^{5+4} = x^9$ [by Exponent Property (i)]

(b) $2y^4y^6y^2z^2 = 2y^{4+6+2}z^2 = 2y^{12}z^2$ [by Exponent Property (i)]

(c) $(5x^3y)(-3x^2y^4) = (5)(-3)(x^3x^2)(y^1y^4)$

            $= -15x^5y^5$ [by Exponent Property (i)]

(d) $(3x^2y^4)^2 = 3^2(x^2)^2(y^4)^2$ [by Exponent Property (iii)]

         $= 9x^{(2)(2)}y^{(4)(2)}$ [by Exponent Property (ii)]

         $= 9x^4y^8$

(e) $(-2x^4y^2)(3x^2y^3)(5xy^4) = (-2)(3)(5)(x^4x^2x)(y^2y^3y^4) = -30x^7y^9$

(f) $(x + 3)^2(x + 3)^4 = (x + 3)^{2+4} = (x + 3)^6$ ■

To multiply a polynomial by a monomial we use the distributive property. Thus, we multiply each term of the polynomial by the monomial, and then simplify the resulting products by using the properties of exponents.

Example 7 Find the product $(-3x^2y^3 + 5xy + 7)(4x^3y)$.

Solution $(-3x^2y^3 + 5xy + 7)(4x^3y) = (-3x^2y^3)(4x^3y) + (5xy)(4x^3y) + 7(4x^3y)$

                      $= -12x^5y^4 + 20x^4y^2 + 28x^3y$ ■

To multiply two polynomials, we again employ the distributive property. Thus, we multiply each term of the first polynomial by each term of the second and combine like terms.

Example 8 Find the product $(x^4 - 5x^2 + 7)(3x^2 + 2)$.

Solution $(x^4 - 5x^2 + 7)(3x^2 + 2) = (x^4 - 5x^2 + 7)(3x^2) + (x^4 - 5x^2 + 7)(2)$

                      $= (3x^6 - 15x^4 + 21x^2) + (2x^4 - 10x^2 + 14)$

                      $= 3x^6 - 13x^4 + 11x^2 + 14$

The same work is arranged vertically as follows:

$$x^4 - 5x^2 + 7$$
$$(\times) \quad\quad\quad 3x^2 + 2$$

$$3x^6 - 15x^4 + 21x^2 \longleftarrow (x^4 - 5x^2 + 7)(3x^2)$$
$$(+) \quad\quad\quad 2x^4 - 10x^2 + 14 \longleftarrow (x^4 - 5x^2 + 7)(2)$$

$$3x^6 - 13x^4 + 11x^2 + 14 \longleftarrow \text{the sum of the above}$$

   ■

Certain products of polynomials occur so often that it is useful to know the expanded forms by heart. The following list contains some of these **special products.**

Special Products

If a, b, c, d, x, and y are real numbers, then:

1. $(a - b)(a + b) = a^2 - b^2$

2. $(a + b)^2 = a^2 + 2ab + b^2$

3. $(a - b)^2 = a^2 - 2ab + b^2$

4. $(a + b)(a^2 - ab + b^2) = a^3 + b^3$

5. $(a - b)(a^2 + ab + b^2) = a^3 - b^3$

6. $(a + b)^3 = a^3 + 3a^2b + 3ab^2 + b^3$

7. $(a - b)^3 = a^3 - 3a^2b + 3ab^2 - b^3$

8. $(ax + by)(cx + dy) = acx^2 + (ad + bc)xy + bdy^2.$

You should verify the expansions in the list above by actually doing the multiplication (Problem 78). They should become so familiar that you use them automatically in your calculations.

Example 9   Use Special Products 1 to 8 to perform each multiplication.

(a) $(3x - 2y)(3x + 2y)$ (b) $(7x^2 + 3y^3)^2$ (c) $(3x + 4)(9x^2 - 12x + 16)$

(d) $[4(p + q) - 5]^2$ (e) $(5r + 6s)^3$ (f) $(2p - 3q + 4r)^2$

Solution   (a) $(3x - 2y)(3x + 2y) = (3x)^2 - (2y)^2$ (by Special Product 1)
$$= 9x^2 - 4y^2$$

(b) $(7x^2 + 3y^3)^2 = (7x^2)^2 + 2(7x^2)(3y^3) + (3y^3)^2$ (by Special Product 2)
$$= 49x^4 + 42x^2y^3 + 9y^6$$

(c) $(3x + 4)(9x^2 - 12x + 16)$
$$= (3x + 4)[(3x)^2 - (3x)(4) + 4^2]$$
$$= (3x)^3 + 4^3 \qquad \text{(by Special Product 4)}$$
$$= 27x^3 + 64$$

(d) $[4(p + q) - 5]^2$
$$= 16(p + q)^2 - 2[4(p + q)(5)] + 5^2 \qquad \text{(by Special Product 3)}$$
$$= 16(p^2 + 2pq + q^2) - 40(p + q) + 25 \qquad \text{(by Special Product 2)}$$
$$= 16p^2 + 32pq + 16q^2 - 40p - 40q + 25$$

(e) $(5r + 6s)^3$
$$= (5r)^3 + 3(5r)^2(6s) + 3(5r)(6s)^2 + (6s)^3 \qquad \text{(by Special Product 6)}$$
$$= 125r^3 + 450r^2s + 540rs^2 + 216s^3$$

(f) $(2p - 3q + 4r)^2$
$$= [(2p - 3q) + 4r]^2$$
$$= (2p - 3q)^2 + 2(2p - 3q)(4r) + (4r)^2 \qquad \text{(by Special Product 2)}$$
$$= 4p^2 - 12pq + 9q^2 + 16pr - 24qr + 16r^2 \qquad \text{(Why?)}$$ ∎

Problem Set 1.3

In Problems 1 to 4, find the value of the algebraic expression for the given values of the variables.

1. $-3xy + 5x + 2$ when $x = 2$ and $y = -1$

2. prt when $p = 1000$, $r = 0.08$, and $t = 3$

3. $\frac{1}{2}gt^2$ when $g = 32$ and $t = 5$

4. $\frac{1}{2}(-b + \sqrt{b^2 - 4c})$ when $b = 1$ and $c = -2$

In Problems 5 to 14, identify the algebraic expression as a monomial, binomial, trinomial, multinomial and/or polynomial, and specify the variables involved. For the polynomials, give the degree and the coefficients.

5. $-4x^2$

6. $2\dfrac{x^4}{y} - 3\dfrac{xy^2}{z} + 2xyz - 17$

7. $5x^3y^{-1} + 8x^2 + 1$

8. $\sqrt{x} + \sqrt{y} + \sqrt{x^2 + w^2}$

9. $3xz^2 - \frac{6}{11}x^2z - x - z + 2$

10. $-\dfrac{b}{2a} + \dfrac{\sqrt{b^2 - 4ac}}{2a}$

11. $\sqrt{2}x^5 + \pi x^4 - \sqrt{\pi}x^3 + \frac{12}{13}x^2 - 5$

12. $pq - \pi$

13. $x^3 + y^3 + z^3 + w^3$

14. $x^{-1}y^{-1}z^{-1}$

In Problems 15 to 28, perform the indicated operations.

15. $(-5x + 1) + (7x + 11)$

16. $(4z^2 + 2) + (-3z^2 + z)$

17. $(7x^2 - x + 9) + (11x^2 + 2x - 4)$

18. $5s^2 + (\pi r^2 + 2s^2) - 2s^2$

19. $(3z^2 + 7z + 5) - (-z^2 - 3z + 2)$

20. $3\pi r^2 + (2\pi^2 h - \pi r^2)$

21. $(n^3 - 5n^2 + 3n + 4) - (-2n^3 + 4n - 3)$

22. $(4t^3 - 3t^2 + 2t + 1) + (-2t^3 + 5t^2 + 3t + 4)$

23. $(5x^2y - 3xy^2 + 7xy - 11) +$
$$(-3x^2y + 7xy^2 + 4xy + 8)$$

24. $(-4pq^2 + 5p^2q + 11pq + 7) - (-7pq^2 + 4pq - 5p + 4)$

25. $(5t^2 + 4t - 3) + (-9t^2 + 2t + 1) - (3t^2 - 8t + 7)$

26. $(5uv - 3u + 5v) - (2uv^2 + 7u + 2v) +$
$$(6uv - 3uv^2 - 8v + 7u)$$

27. $(3x^2 - xy + y^2 - 2x + y - 5) +$
$$(7x^2 + 2y^2 - x + 2) - (8x^2 + 4xy - 3y^2 - x)$$

28. $x - \{[xy + (1 + x)] - [-xy + x^2]\}$

In Problems 29 to 44, use the properties of exponents to simplify each expression.

29. $(3x^4)(2x^3)$

30. $(-y)^4(-y)^5$

31. $2^{3n}2^{7n}$

32. $(4t^{2n})(3t)^n$

33. $(3x + y)^2(3x + y)^4$

34. $(5r - 3t)^5(3t - 5r)^2(3t - 5r)$

35. $(t^4)^{12}$

36. $(-x^4)^{11}$

37. $(u^n)^{5n}$

38. $[-(3x + y)^4]^3$

39. $(3v)^4(2v)^2$

40. $(-5r^4)^3$

41. $(5a^2b)^2(3ab^2)^3$

42. $(rs)^n(r^2s)^{2n}(rs^2t)^{3n}$

43. $[2(x + 3y)^2]^5[3(x + 3y)^4]^3$

44. $(n^n)^n$

In Problems 45 to 56, expand each product.

45. $5x^3y(3x - 4xy + 4z)$

46. $-a^2b(-3a^2 + 4b + 2b^2)$

47. $(3x + 2y)(-4x + 5y)$ **48.** $(2p^3 + 5q)(3p^2 - q)$

49. $(x + 2y)(x^2 - 2xy + 4y^2)$

50. $(r^2 + 3r + s^2)(r^2 + 3r - s^2)$

51. $(5c + d)(5c - d)(25c^2 + d^2)$

52. $(6x^3 + 2x^2 - 3x + 1)(2x^2 - 4x + 3)$

53. $(x^2 + 7xy - y^2)(3x^2 - xy + y^2 - 2x - y + 2)$

54. $(2x^{2n} - x^n - 1)(3x^{2n} + 2x^n + 2)$

55. $(4p^2q - 3pq^2 + 5pq)(p^2q - 2pq^2 - 3pq)$

56. $(3x^3 - 2x^2 + x - 1)(x^4 - x^3 + 5x^2 - 2x + 1)$

In Problems 57 to 72, use the appropriate special products to perform each multiplication.

57. $(4 + 3x)^2$ **58.** $(5p^2 - 2q^2)^2$

59. $[(4t^2 + 1) - s]^2$ **60.** $(5u + 2v + w)^2$

61. $(3r - 2s)(3r + 2s)$ **62.** $(4a^n - b^n)(4a^n + b^n)$

63. $[(2a - 3b) - 4c][(2a - 3b) + 4c]$

64. $[(x + 3y) - 2z][(x + 3y) + 2z]$

65. $(2 + t)(4 - 2t + t^2)$ **66.** $(3 - u)(9 + 3u + u^2)$

67. $(2x^2 + 3y)^3$ **68.** $(p^n - 2q^n)^3$

69. $(2x - y + z)^2$

70. $[1 + y + z][(1 + y)^2 - (1 + y)z + z^2]$

71. $(t^3 - 2t^2 + 5t + 2)^2$ **72.** $[(x + y)^2]^3$

73. Explain why the degree of the product of two polynomials is the sum of their degrees.

74. Explain how the procedure of combining like terms (for instance, $3x^2y + 4x^2y = 7x^2y$) is justified by the distributive property.

75. If a is a real number and n and m are positive integers, show that $(a^m)^n = a^{mn}$.

76. If a and b are real numbers and n is a positive integer, show that $(ab)^n = a^nb^n$.

77. Obtain a formula for the expansion of $(a + b)^4$.

78. Verify Special Products 1 to 8 by actually carrying out the multiplication.

79. If a and b are real numbers, $x = a^2 - b^2$, $y = 2ab$, and $z = a^2 + b^2$, show that $x^2 + y^2 = z^2$.

80. Using the result of Problem 79, find several triples of positive integers x, y, z such that $x^2 + y^2 = z^2$. [*Hint:* Substitute positive integers for a and b.]

1.4 FACTORING POLYNOMIALS

When two or more algebraic expressions are multiplied, each expression is called a **factor** of the product. For instance, in the product

$$(x - y)(x + y)(2x^2 - y)x,$$

the factors are $x - y$, $x + y$, $2x^2 - y$, and x. Often we are given a product in its expanded form and we need to find the original factors. The process of finding these factors is called **factoring.**

In this section, we confine our study of factoring to polynomials with integer coefficients. Thus, we shall not yet consider such possibilities as $5x^2 - y^2 = (\sqrt{5}x - y)(\sqrt{5}x + y)$, because $\sqrt{5}$ isn't an integer.

Of course, we can factor any polynomial "trivially" by writing it as 1 times itself or as -1 times its negative. A polynomial with integer coefficients that cannot be factored (except trivially) into two or more polynomials with integer coefficients is said to be **prime.** When a polynomial is written as a product of prime factors, we say that it is **factored completely.**

Removing a Common Factor

The distributive property can be used to factor a polynomial in which all the terms contain a common factor. The following example illustrates how to "*remove the common factor.*"

Example 1   Factor each polynomial by removing the common factor.

(a) $20x^2y + 8xy$ **(b)** $u(v + w) + 7v(v + w)$

Solution   **(a)** Here $4xy$ is a common factor of the two terms, since

$$20x^2y = (4xy)(5x) \qquad \text{and} \qquad 8xy = (4xy)(2).$$

Therefore,

$$20x^2y + 8xy = (4xy)(5x) + (4xy)(2) = 4xy(5x + 2).$$

(b) Here the common factor is $v + w$, and we have

$$u(v + w) + 7v(v + w) = (u + 7v)(v + w).$$ ■

Factoring by Recognizing Special Products

Success in factoring depends on your ability to recognize patterns in the polynomials to be factored—an ability that grows with practice. Special Products 1 to 8 in Section 1.3, read from right to left, suggest useful patterns; for instance, Special Product 5, read as $a^3 - b^3 = (a - b)(a^2 + ab + b^2)$, shows that a **difference of two cubes** can always be factored.

Example 2   Use the special products on page 20 to factor each expression.

(a) $25t^2 - 16s^2$ **(b)** $x^3 + 64y^3$ **(c)** $27r^3 - 8c^3$
(d) $16x^4 - (3y + 2z)^2$ **(e)** $81b^4 - 1$ **(f)** $(x + y)^3 - (z - w)^3$

Solution   **(a)** Notice that $25t^2 = (5t)^2$ and $16s^2 = (4s)^2$, so the given expression is the **difference of two squares;** that is,

$$25t^2 - 16s^2 = (5t)^2 - (4s)^2.$$

Using Special Product 1 in the form $a^2 - b^2 = (a - b)(a + b)$, with $a = 5t$ and $b = 4s$, we have

$$25t^2 - 16s^2 = (5t)^2 - (4s)^2 = (5t - 4s)(5t + 4s).$$

(b) Since $64 = 4^3$, the expression $x^3 + 64y^3$ is a **sum of two cubes.** Using Special Product 4 in the form $a^3 + b^3 = (a + b)(a^2 - ab + b^2)$, with $a = x$ and $b = 4y$, we have

$$x^3 + 64y^3 = x^3 + (4y)^3 = (x + 4y)(x^2 - 4xy + 16y^2).$$

(c) Here we have a difference of two cubes, so we use Special Product 5:

$$27r^3 - 8c^3 = (3r)^3 - (2c)^3 = (3r - 2c)[(3r)^2 + (3r)(2c) + (2c)^2]$$
$$= (3r - 2c)(9r^2 + 6rc + 4c^2).$$

(d) The expression is a difference of two squares, so

$$16x^4 - (3y + 2z)^2 = (4x^2)^2 - (3y + 2z)^2$$
$$= [4x^2 + (3y + 2z)][4x^2 - (3y + 2z)]$$
$$= (4x^2 + 3y + 2z)(4x^2 - 3y - 2z).$$

(e) Using Special Product 1 *twice*, we have

$$81b^4 - 1 = (9b^2)^2 - 1$$
$$= (9b^2 - 1)(9b^2 + 1)$$
$$= (3b - 1)(3b + 1)(9b^2 + 1).$$

(f) $(x + y)^3 - (z - w)^3$
$$= [(x + y) - (z - w)][(x + y)^2 + (x + y)(z - w) + (z - w)^2]$$
$$= (x + y - z + w)(x^2 + 2xy + y^2 + xz - xw + yz - yw + z^2 - 2zw + w^2). \quad\blacksquare$$

Factoring Trinomials of the Type $ax^2 + bx + c$

If a, b, and c are integers and $a \neq 0$, it may be possible to factor a trinomial of the type $ax^2 + bx + c$ into a product of two binomials. For instance,

$$8x^2 + 22x + 15 = (2x + 3)(4x + 5).$$

Notice the relationship between the coefficients of the trinomial and the coefficients of the factors. The coefficients 8 and 15 are obtained as follows:

$$8x^2 + 22x + 15 = (2x + 3)(4x + 5).$$

The coefficient of the middle term in the trinomial, namely 22, is obtained as follows:

$$8x^2 + (10 + 12)x + 15 = (2x + 3)(4x + 5).$$

Thus, to factor a trinomial $ax^2 + bx + c$, begin by writing

$$ax^2 + bx + c = (\ \ x + \ \ )(\ \ x + \ \ ),$$

where the blanks are to be filled in with integers. The product of the unknown coefficients of x must be the integer a, and the product of the unknown constant terms must be the integer c. Just try all possible choices of such integers until you find a combination that gives the desired middle coefficient b. If all of the coefficients of the trinomial are positive, you need only try combinations of positive integers.

Example 3   Factor the trinomial $2x^2 + 9x + 4$.

Solution   Since the only two positive integers whose product is 2 are 2 and 1, we can begin by writing

$$2x^2 + 9x + 4 = (2x + \underset{\underset{?}{\uparrow}}{\,}\,)(x + \underset{\underset{?}{\uparrow}}{\,}\,).$$

Because 4 can be factored as $4 = 4 \cdot 1$ or as $4 = 2 \cdot 2$ or as $4 = 1 \cdot 4$, there are just three possible ways to fill the remaining blanks:

$$(2x + 4)(x + 1) \quad \text{or} \quad (2x + 2)(x + 2) \quad \text{or} \quad (2x + 1)(x + 4).$$

Now, $(2x + 4)(x + 1) = 2x^2 + 6x + 4$, and that isn't what we want. Also, $(2x + 2)(x + 2) = 2x^2 + 6x + 4$, and that isn't what we want either. But, $(2x + 1)(x + 4) = 2x^2 + 9x + 4$, and that *is* what we want! Therefore,

$$2x^2 + 9x + 4 = (2x + 1)(x + 4).$$ ∎

If some of the coefficients of the trinomial $ax^2 + bx + c$ are negative, you will have to try combinations with negative integers.

Example 4   Factor the trinomial $6x^2 + 13x - 5$.

Solution   Because -5 can be factored as $-5 = 5 \cdot (-1)$ or as $-5 = (-5) \cdot 1$, we can begin by writing

$$6x^2 + 13x - 5 = (\underset{\underset{?}{\uparrow}}{\,}x \pm 5)(\underset{\underset{?}{\uparrow}}{\,}x \mp 1),$$

where the two blanks must still be filled in, and we must choose the correct algebraic signs. The possibilities for the blanks are given by $6 = 6 \cdot 1 = 3 \cdot 2 = 2 \cdot 3 = 1 \cdot 6$. Hence, the possible combinations are

	$(6x + 5)(x - 1)$	or	$(6x - 5)(x + 1)$
or	$(3x + 5)(2x - 1)$	or	$(3x - 5)(2x + 1)$
or	$(2x + 5)(3x - 1)$	or	$(2x - 5)(3x + 1)$
or	$(x + 5)(6x - 1)$	or	$(x - 5)(6x + 1)$.

We try each of these, one at a time, until we find that $(2x + 5)(3x - 1)$ works. Therefore,

$$6x^2 + 13x - 5 = (2x + 5)(3x - 1).$$ ∎

The method illustrated above can also be used to factor trinomials of the form $ax^2 + bxy + cy^2$.

Example 5   Factor the trinomial $6x^2 - 19xy + 3y^2$.

Solution   We begin by writing

$$6x^2 - 19xy + 3y^2 = (\ \ x - 3y)(\ \ x - y).$$

The two minus signs provide for the positive coefficient $+3$ of y^2 and for the negative coefficient -19 of the middle term. Since $6 = 6 \cdot 1 = 3 \cdot 2 = 2 \cdot 3 = 1 \cdot 6$, there are four possible ways to fill in the two blanks. We try them one at a time until we find that $(x - 3y)(6x - y)$ works. Therefore,

$$6x^2 - 19xy + 3y^2 = (x - 3y)(6x - y).$$ ∎

If, in attempting to factor a trinomial $ax^2 + bx + c$ or $ax^2 + bxy + cy^2$, you find that none of the possible combinations works, you can conclude that the trinomial is prime. However, you can test to see if it is prime without bothering to try all these combinations just by evaluating the expression

$$b^2 - 4ac.$$

If $b^2 - 4ac$ is the square of an integer, then the trinomial can be factored; otherwise, it is prime. (You will see why this test works in Section 2.3.)

Example 6   Test each trinomial to see whether it is factorable or prime. If it can be factored, do so.

(a) $4x^2 - 8x + 3$ **(b)** $3x^2 - 5xy + y^2$

Solution   **(a)** Here $a = 4$, $b = -8$, $c = 3$, and

$$b^2 - 4ac = (-8)^2 - 4(4)(3) = 64 - 48 = 16 = 4^2.$$

Therefore, the trinomial can be factored. Indeed,

$$4x^2 - 8x + 3 = (2x - 1)(2x - 3).$$

(b) Here $a = 3$, $b = -5$, $c = 1$, and

$$b^2 - 4ac = (-5)^2 - 4(3)(1) = 25 - 12 = 13.$$

Because 13 is not the square of an integer, $3x^2 - 5xy + y^2$ is prime. ∎

Factoring by Grouping

Sometimes we have to group the terms of a polynomial in a certain way in order to see how it can be factored.

Example 7   Factor each expression by grouping the terms in a suitable way.

(a) $ac + d - c - ad$ **(b)** $x^2 + 4xy - 9c^2 + 4y^2$

(c) $4x^3 - 8x^2 - x + 2$ **(d)** $x^4 + 6x^2y^2 + 25y^4$

Solution   **(a)** $ac + d - c - ad = (ac - c) + (d - ad) = c(a - 1) + d(1 - a)$
$$= c(a - 1) - d(a - 1) = (c - d)(a - 1)$$

(b) $x^2 + 4xy - 9c^2 + 4y^2 = (x^2 + 4xy + 4y^2) - 9c^2 = (x + 2y)^2 - (3c)^2$
$$= [(x + 2y) - 3c][(x + 2y) + 3c]$$
$$= (x + 2y - 3c)(x + 2y + 3c)$$

(c) $4x^3 - 8x^2 - x + 2 = 4x^2(x - 2) - x + 2 = 4x^2(x - 2) - (x - 2)$
$$= (4x^2 - 1)(x - 2) = (2x - 1)(2x + 1)(x - 2)$$

(d) If the middle term were $10x^2y^2$ rather than $6x^2y^2$, we could factor the expression as $x^4 + 10x^2y^2 + 25y^4 = (x^2 + 5y^2)^2$. But the middle term can be changed to $10x^2y^2$ by adding $4x^2y^2$ to the $6x^2y^2$ already there and then subtracting $4x^2y^2$ at the end of the expression. Thus,

$$x^4 + 6x^2y^2 + 25y^4 = (x^4 + 10x^2y^2 + 25y^4) - 4x^2y^2$$
$$= (x^2 + 5y^2)^2 - (2xy)^2$$
$$= [(x^2 + 5y^2) - 2xy][(x^2 + 5y^2) + 2xy]$$
$$= (x^2 - 2xy + 5y^2)(x^2 + 2xy + 5y^2). \qquad \blacksquare$$

Problem Set 1.4

In Problems 1 to 8, factor the polynomial by removing the common factor.

1. $10x^3y - 5x^2y^2$

2. $x^3y^2z - x^2y^3z + xy^4z^3$

3. $a^3b + 2a^2b + a^2b^2$

4. $14a^2b - 35ab - 63ab^2$

5. $2(x - y)r^2 + 2(x - y)rh$

6. $a^2(s + 2t)^2 + a(-s - 2t)$

7. $t(a - b) - r(b - a)$ **8.** $9a^{n+1}b^2 - 3a^nb$

In Problems 9 to 24, use the special products (page 20) to factor each expression.

9. $x^2 - 36$ **10.** $4u^2 - 25v^2$

11. $p^2q^2 - 64$ **12.** $49s^2 - 25t^2$

13. $n^2 - 36m^2$ **14.** $121s^{2n} - 81t^2$

15. $(x + y)^2 - 49z^2$ **16.** $100a^4 - (3a + 2b)^2$

17. $81y^4 - z^4$

18. $(x - 2y)^4 - (2x - y)^4$

19. $8x^3 + 1000$ **20.** $8a^3 + (b + c)^3$

21. $a^3 - (3b + c)^3$

22. $m^9 + n^9$

23. $64x^6 + (p - q)^3$

24. $8r^{12} - 27(s + 5t)^3$

In Problems 25 to 42, factor the trinomial.

25. $x^2 - 8x + 15$

26. $r^2 - 12r + 35$

27. $y^2 - 3y - 10$

28. $u^2 + 2uv - 35v^2$

29. $t^2 - 7t - 18$

30. $8 - 2t - t^2$

31. $x^2 - xy - 20y^2$

32. $x^{2n} - 9x^n - 22$

33. $8x^2 + 10x - 7$

34. $30 - 49w + 6w^2$

35. $2v^2 + 3v - 20$

36. $8t^4 - 14t^2 - 15$

37. $4r^2 - 12rs + 9s^2$

38. $6(u + v)^2 - 5(u + v) - 6$

39. $6x^4 + 13x^2 + 6$

40. $8x^2 + 22x(y + 2z) + 5(y + 2z)^2$

41. $3(3a - 2b)^2 + (3a - 2b) - 14$

42. $3s^2t^{2n} - st^n - 10$

In Problems 43 to 48, test the trinomial to see whether it is factorable or prime. If it can be factored, do so. [The ⓒ indicates you may use a calculator if you wish.]

43. $x^2 + x + 1$

44. $16x^2 - 12xy + y^2$

45. $12r^2 - 11r + 2$

ⓒ **46.** $52a^2 - 37ab - 35b^2$

47. $2a^2 - 6a + 5$

ⓒ **48.** $33(x + y)^2 - 22(x + y) + 3$

In Problems 49 to 58, factor each expression by grouping the terms in a suitable way.

49. $3x - 2y - 6 + xy$

50. $3rs - 2t - 3rt + 2s$

51. $a^2x + a^2d - x - d$

52. $x^2 + 3x - y^2 + 3y$

53. $4x^2 + 12x - 9y^2 + 9$

54. $9a^4 + 8a^2b^2 + 4b^4$

55. $12u^2v + 9 - 4u^2 - 27v$

56. $p^4 + q^4 - pq^3 - p^3q$

57. $x^2 + x - 9y^2 - 3y$

58. $a^3b^3 + a^3 - b^3 - 1$

In Problems 59 to 68, factor each expression completely.

59. $x^4 + x^2y^2 + y^4$

60. $9r^4 + 2r^2s^2 + s^4$

61. $t^4 - 8t^2 + 16$

62. $u^4 + 5u^2v^2 + 9v^4$

63. $(2s + 1)a^3 - (2s + 1)b^3$

64. $s^8 - 82s^4 + 81$

65. $x^8 - 256y^8$

66. $4(x + y)^2z - 25w - 25z + 4(x + y)^2w$

67. $10x^2y^3z - 13xy^4z - 3y^5z$

68. $x^6 - 4 - x^4 + 4x^2$

69. The following expressions were obtained as solutions to problems in a popular calculus textbook. Factor each expression and then simplify the factors.

(a) $2(3x^2 + 7)(6x)(5 - 3x)^3 + 3(3x^2 + 7)^2(-3)(5 - 3x)^2$

(b) $2(5t^2 + 1)(10t)(3t^4 + 2)^4 + 4(5t^2 + 1)(12t^3)(3t^4 + 2)^3$

1.5 FRACTIONS

If p and q are algebraic expressions, the quotient p/q is called a **fractional expression** (or simply a *fraction*) with **numerator p** and **denominator q.** Always remember that *the denominator of a fraction cannot be zero.* If $q = 0$, the expression p/q simply has no meaning. Therefore, whenever we use a fractional expression, we shall automatically assume that the variables involved are restricted to numerical values that will give a nonzero denominator.

A fractional expression in which both the numerator and the denominator are polynomials is called a **rational expression.** Examples of rational expressions are

$$\frac{2x}{1}, \quad \frac{3}{y}, \quad \frac{5x + 3}{17}, \quad \frac{2st - t^3}{3s^4 - t^4}, \quad \text{and} \quad \frac{x^2 - 1}{y^2 - 1}.$$

Notice that a fraction such as $3x/\sqrt{y}$ isn't a rational expression because its denominator isn't a polynomial. We say that a rational expression is **reduced to lowest terms** or **simplified** if its numerator and denominator have no common factors (other than 1 and -1). Thus, to simplify a rational expression, we factor both the numerator and denominator into prime factors and then **cancel** common factors by using the following property.

The Cancellation Property for Fractions

If $q \neq 0$ and $k \neq 0$, then $\dfrac{pk}{qk} = \dfrac{p}{q}$.

Cancellation is usually indicated by slanted lines drawn through the canceled factors; for instance,

$$\frac{x^2 - 1}{x^2 - 3x + 2} = \frac{(x - 1)(x + 1)}{(x - 1)(x - 2)} = \frac{x + 1}{x - 2}.$$

If one fraction can be obtained from another by canceling common factors or by multiplying numerator and denominator by the same nonzero expression, then the two fractions are said to be **equivalent.** Thus, the calculation above shows that

$$\frac{x^2 - 1}{x^2 - 3x + 2} \quad \text{and} \quad \frac{x + 1}{x - 2}$$

are equivalent fractions.

Example 1   Reduce each fraction to lowest terms.

(a) $\dfrac{14x^7 y}{56x^3 y^2}$ (b) $\dfrac{cd - c^2}{c^2 - d^2}$

(c) $\dfrac{5x^2 - 14x - 3}{2x^2 + x - 21}$ (d) $\dfrac{16x - 32}{8y(x - 2) - 4(x - 2)}$

Solution   (a) $\dfrac{14x^7 y}{56x^3 y^2} = \dfrac{14x^3 y \cdot x^4}{14x^3 y \cdot 4y} = \dfrac{x^4}{4y}$

(b) First, we factor the numerator and denominator, and then we use the fact that $d - c = -(c - d)$:

$$\frac{cd - c^2}{c^2 - d^2} = \frac{c(d - c)}{(c - d)(c + d)} = \frac{-c(c - d)}{(c - d)(c + d)} = \frac{-c}{c + d}.$$

(c) $\dfrac{5x^2 - 14x - 3}{2x^2 + x - 21} = \dfrac{(5x + 1)(x - 3)}{(2x + 7)(x - 3)} = \dfrac{5x + 1}{2x + 7}$

(d) $\dfrac{16x - 32}{8y(x - 2) - 4(x - 2)} = \dfrac{16(x - 2)}{(x - 2)(8y - 4)} = \dfrac{\overset{4}{16(x - 2)}}{4(x - 2)(2y - 1)} = \dfrac{4}{2y - 1}$ ∎

Multiplication and Division of Fractions

The following rules for multiplication and division of fractions can be derived from the basic algebraic properties of the real numbers and the definition of a quotient.

1. *Rule for Multiplication of Fractions*

> If p, q, r, and s are real numbers, $q \neq 0$, and $s \neq 0$, then
>
> $$\frac{p}{q} \cdot \frac{r}{s} = \frac{pr}{qs}.$$

For instance,

$$\frac{2}{3} \cdot \frac{5}{7} = \frac{2 \cdot 5}{3 \cdot 7} = \frac{10}{21}.$$

2. *Rule for Division of Fractions*

> If p, q, r, and s are real numbers, $q \neq 0$, $r \neq 0$, and $s \neq 0$, then
>
> $$\frac{p}{q} \div \frac{r}{s} = \frac{p/q}{r/s} = \frac{p}{q} \cdot \frac{s}{r} = \frac{ps}{qr}.$$

For instance,

$$\frac{2}{3} \div \frac{5}{7} = \frac{2}{3} \cdot \frac{7}{5} = \frac{14}{15} \quad \text{and} \quad \frac{3/4}{4/5} = \frac{3}{4} \cdot \frac{5}{4} = \frac{15}{16}.$$

Of course, the same rules apply to multiplying or dividing fractional expressions in general. Before multiplying fractions, it's a good idea to factor the numerators and denominators completely to reveal any factors that can be canceled.

Example 2   Perform the indicated operation and simplify the result.

(a) $\dfrac{x + 4}{y} \cdot \dfrac{y^3}{5x + 20}$

(b) $\dfrac{x^3 - 1}{5x^2 - 26x + 5} \cdot \dfrac{5x^2 + 9x - 2}{x^2 + x - 2}$

(c) $\dfrac{t^2 - 49}{t^2 - 5t - 14} \div \dfrac{2t^2 + 15t + 7}{2t^2 - 13t - 7}$

Solution   **(a)** $\dfrac{x+4}{y} \cdot \dfrac{y^3}{5x+20} = \dfrac{x+4}{y} \cdot \dfrac{y \cdot y^2}{5(x+4)} = \dfrac{\cancel{(x+4)}y \cdot y^2}{\cancel{y}(5)\cancel{(x+4)}} = \dfrac{y^2}{5}$

(b) $\dfrac{x^3-1}{5x^2-26x+5} \cdot \dfrac{5x^2+9x-2}{x^2+x-2} = \dfrac{(x-1)(x^2+x+1)}{(5x-1)(x-5)} \cdot \dfrac{(5x-1)(x+2)}{(x+2)(x-1)}$

$$= \dfrac{\cancel{(x-1)}(x^2+x+1)\cancel{(5x-1)}\cancel{(x+2)}}{\cancel{(5x-1)}(x-5)\cancel{(x+2)}\cancel{(x-1)}}$$

$$= \dfrac{x^2+x+1}{x-5}$$

(c) $\dfrac{t^2-49}{t^2-5t-14} \div \dfrac{2t^2+15t+7}{2t^2-13t-7} = \dfrac{t^2-49}{t^2-5t-14} \cdot \dfrac{2t^2-13t-7}{2t^2+15t+7}$

$$= \dfrac{(t-7)(t+7)}{(t+2)(t-7)} \cdot \dfrac{(2t+1)(t-7)}{(2t+1)(t+7)}$$

$$= \dfrac{\cancel{(t-7)}\cancel{(t+7)}\cancel{(2t+1)}(t-7)}{(t+2)\cancel{(t-7)}\cancel{(2t+1)}\cancel{(t+7)}} = \dfrac{t-7}{t+2}$$ ∎

Addition and Subtraction of Fractions

Two or more fractions with the same denominator are said to have a **common denominator.** The following rules for adding and subtracting fractions with a common denominator can be derived from the basic algebraic properties of the real numbers and the definition of a quotient.

Rules for Addition and Subtraction of Fractions with a Common Denominator

If p, q, and r are real numbers and $q \neq 0$, then

$$\frac{p}{q} + \frac{r}{q} = \frac{p+r}{q} \qquad \text{and} \qquad \frac{p}{q} - \frac{r}{q} = \frac{p-r}{q}.$$

For instance,

$$\frac{3}{7} + \frac{2}{7} = \frac{3+2}{7} = \frac{5}{7} \qquad \text{and} \qquad \frac{3}{7} - \frac{2}{7} = \frac{3-2}{7} = \frac{1}{7}.$$

Again, the same rules apply to adding or subtracting fractional expressions.

Example 3   Perform each operation.

(a) $\dfrac{5x}{2x-1} + \dfrac{3x}{2x-1}$ **(b)** $\dfrac{5x}{(3x-2)^2} - \dfrac{3x}{(3x-2)^2}$

Solution   **(a)** $\dfrac{5x}{2x-1} + \dfrac{3x}{2x-1} = \dfrac{5x+3x}{2x-1} = \dfrac{8x}{2x-1}$

(b) $\dfrac{5x}{(3x-2)^2} - \dfrac{3x}{(3x-2)^2} = \dfrac{5x-3x}{(3x-2)^2} = \dfrac{2x}{(3x-2)^2}$ ∎

To add or subtract fractions that do not have a common denominator, you must rewrite the fractions so they do have the same denominator. To do this, multiply the numerator and denominator of each fraction by an appropriate quantity. For instance,

$$\frac{2}{3} + \frac{4}{5} = \frac{2\cdot5}{3\cdot5} + \frac{3\cdot4}{3\cdot5} = \frac{10}{15} + \frac{12}{15} = \frac{10+12}{15} = \frac{22}{15}.$$

More generally, if $q \neq s$, you can always add p/q and r/s as follows:

$$\frac{p}{q} + \frac{r}{s} = \frac{p\cdot s}{q\cdot s} + \frac{q\cdot r}{q\cdot s} = \frac{ps+qr}{qs}.$$

Example 4   Add the expressions $\dfrac{3x}{4x-1}$ and $\dfrac{2x}{3x-5}$.

Solution $\dfrac{3x}{4x-1} + \dfrac{2x}{3x-5} = \dfrac{3x(3x-5)}{(4x-1)(3x-5)} + \dfrac{(4x-1)(2x)}{(4x-1)(3x-5)}$

$= \dfrac{3x(3x-5) + (4x-1)(2x)}{(4x-1)(3x-5)} = \dfrac{9x^2 - 15x + 8x^2 - 2x}{(4x-1)(3x-5)}$

$= \dfrac{17x^2 - 17x}{(4x-1)(3x-5)} = \dfrac{17x(x-1)}{(4x-1)(3x-5)}$ ∎

When adding or subtracting fractions, it's usually best to use the **least common denominator,** abbreviated LCD. The LCD of two or more fractions is found as follows:

Step 1. Factor each denominator completely.

Step 2. Form the product of all the different prime factors in the denominators of the fractions, each taken the greatest number of times it occurs in any of the denominators.

Example 5   Find the LCD of the fractions, perform the indicated operations, and simplify the result.

(a) $\dfrac{8x}{x^2-9} + \dfrac{4}{5x-15}$ **(b)** $\dfrac{x}{x^3+x^2+x+1} - \dfrac{1}{x^3+2x^2+x} - \dfrac{1}{x^2+2x+1}$

Solution   **(a)** We factor the denominators to obtain

$$\frac{8x}{x^2 - 9} + \frac{4}{5x - 15} = \frac{8x}{(x - 3)(x + 3)} + \frac{4}{5(x - 3)}.$$

Here, the LCD is $5(x - 3)(x + 3)$. Now we rewrite each fraction as an equivalent fraction with denominator $5(x - 3)(x + 3)$. We do this by multiplying the numerator and denominator of the first fraction by 5 and the numerator and denominator of the second fraction by $x + 3$. Thus,

$$\frac{8x}{x^2 - 9} + \frac{4}{5x - 15} = \frac{8x}{(x - 3)(x + 3)} + \frac{4}{5(x - 3)}$$

$$= \frac{5(8x)}{5(x - 3)(x + 3)} + \frac{4(x + 3)}{5(x - 3)(x + 3)}$$

$$= \frac{5(8x) + 4(x + 3)}{5(x - 3)(x + 3)} = \frac{44x + 12}{5(x - 3)(x + 3)} = \frac{4(11x + 3)}{5(x - 3)(x + 3)}.$$

(b) Factoring the denominators, we have

$$\frac{x}{x^3 + x^2 + x + 1} - \frac{1}{x^3 + 2x^2 + x} - \frac{1}{x^2 + 2x + 1}$$

$$= \frac{x}{x^2(x + 1) + (x + 1)} - \frac{1}{x(x^2 + 2x + 1)} - \frac{1}{x^2 + 2x + 1}$$

$$= \frac{x}{(x^2 + 1)(x + 1)} - \frac{1}{x(x + 1)^2} - \frac{1}{(x + 1)^2},$$

so the LCD is $x(x^2 + 1)(x + 1)^2$. Therefore,

$$\frac{x}{(x^2 + 1)(x + 1)} - \frac{1}{x(x + 1)^2} - \frac{1}{(x + 1)^2}$$

$$= \frac{x \cdot x(x + 1)}{x(x^2 + 1)(x + 1)^2} - \frac{x^2 + 1}{x(x^2 + 1)(x + 1)^2} - \frac{x(x^2 + 1)}{x(x^2 + 1)(x + 1)^2}$$

$$= \frac{x^2(x + 1) - (x^2 + 1) - x(x^2 + 1)}{x(x^2 + 1)(x + 1)^2} = \frac{x^3 + x^2 - x^2 - 1 - x^3 - x}{x(x^2 + 1)(x + 1)^2}$$

$$= \frac{-1 - x}{x(x^2 + 1)(x + 1)^2} = -\frac{(x + 1)}{x(x^2 + 1)(x + 1)^2}$$

$$= -\frac{1}{x(x^2 + 1)(x + 1)}.$$

Complex Fractions

A fraction that contains one or more fractions in its numerator or denominator is called a **complex fraction.*** Examples are

$$\frac{\dfrac{3}{xy^2}}{\dfrac{2}{x^2y}}, \qquad \frac{\dfrac{x}{y} - \dfrac{y}{x}}{\dfrac{1}{x} + \dfrac{1}{y}}, \qquad \text{and} \qquad \frac{x^2 - \dfrac{1}{x}}{x + \dfrac{1}{x} + 1}.$$

A complex fraction may be simplified by reducing its numerator and denominator (separately) to simple fractions and then dividing.

Example 6   Simplify the complex fraction $\dfrac{1 + \dfrac{1}{x}}{x - \dfrac{1}{x}}$.

Solution

$$\frac{1 + \dfrac{1}{x}}{x - \dfrac{1}{x}} = \frac{\dfrac{x+1}{x}}{\dfrac{x^2-1}{x}} = \frac{x+1}{x} \div \frac{x^2-1}{x} = \frac{x+1}{x} \cdot \frac{x}{x^2-1}$$

$$= \frac{(x+1)x}{(x^2-1)x} = \frac{x+1}{(x-1)(x+1)} = \frac{1}{x-1}$$

An alternative method for simplifying a complex fraction is to multiply its numerator and denominator by the LCD of all fractions occurring in its numerator *and* denominator. The resulting fraction may then be simplified by cancellation.

Example 7   Simplify the complex fraction $\dfrac{\dfrac{1}{x^3} + \dfrac{2}{x^2y} + \dfrac{1}{xy^2}}{\dfrac{y}{x^2} - \dfrac{1}{y}}$.

---

* There is no particular relationship between complex fractions and the complex numbers introduced later (in Section 1.8).

Solution   The LCD of x^3, x^2y, xy^2, x^2, and y is x^3y^2. Therefore,

$$\frac{\dfrac{1}{x^3} + \dfrac{2}{x^2y} + \dfrac{1}{xy^2}}{\dfrac{y}{x^2} - \dfrac{1}{y}} = \frac{x^3y^2\left(\dfrac{1}{x^3} + \dfrac{2}{x^2y} + \dfrac{1}{xy^2}\right)}{x^3y^2\left(\dfrac{y}{x^2} - \dfrac{1}{y}\right)} = \frac{y^2 + 2xy + x^2}{xy^3 - x^3y} = \frac{(y + x)^2}{xy(y^2 - x^2)}$$

$$= \frac{(y + x)^2}{xy(y - x)(\cancel{y + x})} = \frac{y + x}{xy(y - x)}. \qquad \blacksquare$$

Problem Set 1.5

In Problems 1 to 18, reduce each fraction to lowest terms.

1. $\dfrac{2a^2bxy^2}{6a^2xy}$

2. $\dfrac{18x^4(-y)(-z)^5}{30x(-y)^2(-z)}$

3. $\dfrac{cy + cz}{2y + 2z}$

4. $\dfrac{(2t + 6)^2}{4t^2 - 36}$

5. $\dfrac{t - 5}{25 - t^2}$

6. $\dfrac{c^2 - cd}{bc - bd}$

7. $\dfrac{r - 3}{r^2 + 3r - 18}$

8. $\dfrac{a^2 - 9a + 18}{3a^2 - 5a - 12}$

9. $\dfrac{6y^2 - y - 1}{2y^2 + 9y - 5}$

10. $\dfrac{r^2 - s^2}{s^4 - r^4}$

11. $\dfrac{(a + b)^2 - 4ab}{(a - b)^2}$

12. $\dfrac{(x - y)^2 - z^2}{(x + z)^2 - y^2}$

13. $\dfrac{6c^2 - 7c - 3}{4c^2 - 8c + 3}$

14. $\dfrac{6m(m - 1) - 12}{m^3 - 8 - (m - 2)^2}$

15. $\dfrac{3(r + t)^2 + 17(r + t) + 10}{2(r + t)^2 + 7(r + t) - 15}$

16. $\dfrac{c^2 - d^2 + c - d}{c^2 + 2cd + d^2 - 1}$

17. $\dfrac{(c + d)^2 - 4(c + d)x + 3x^2}{c^2 + 2cd + d^2 - x^2}$

18. $\dfrac{x^{2n+2} + x^{2n+1}y + x^{2n}y^2}{x^{n+3} - x^ny^3}$

In Problems 19 to 36, perform the indicated operations and simplify the result.

19. $\dfrac{c^2 - 8c}{c - 8} \cdot \dfrac{c + 2}{c}$

20. $\dfrac{81 - p^2}{p + q} \cdot \dfrac{p}{9 - p}$

21. $\dfrac{2u - v}{4u} \cdot \dfrac{2u - v}{4u^2 - 4uv + v^2}$

22. $\dfrac{r^4 - 16}{(r - 2)^2} \cdot \dfrac{r - 2}{4 - r^2}$

23. $\dfrac{3x^2 + 15}{x^2 + 6x + 15} \cdot \dfrac{x^2 + 2x + 1}{x^2 - 1}$

24. $\dfrac{x(2x - 9) - 5}{20 + x(1 - x)} \cdot \dfrac{x(3x + 14) + 8}{x(2x - 11) - 6}$

25. $\dfrac{c^2}{a^2 - 1} \div \dfrac{c^2}{a - 1}$

26. $\dfrac{z^2 - 49}{z^2 - 5z - 14} \div \dfrac{z + 7}{2z^2 - 13z - 7}$

27. $\dfrac{2t^2 - t}{4t^2 - 4t + 1} \div \dfrac{t^2}{8t - 4}$

28. $\dfrac{y - 1}{y^2 + 1} \div \dfrac{(y - 1)^2}{y^4 - 1}$

29. $\dfrac{x^3 - y^3}{x - y} \div \dfrac{x^2 + xy + y^2}{x^2 - 2xy + y^2}$

30. $\dfrac{xy^3 - 4x^2y^2}{y - x} \div \dfrac{16x^2y^2 - y^4}{4x^2 - 3xy - y^2}$

31. $\dfrac{14u^2 + 23u + 3}{2u^2 + u - 3} \div \dfrac{7u^2 + 15u + 2}{2u^2 - 3u + 1}$

32. $\dfrac{(t+1)^2 - 9}{(t+\frac{1}{2})^2 - \frac{49}{4}} \div \dfrac{(t-\frac{1}{2})^2 - \frac{9}{4}}{t^2 - 3t}$

33. $\dfrac{x(3x+22)-16}{x(x+11)+24} \div \dfrac{x(3x+13)-10}{x(2x+13)+15}$

34. $\dfrac{w(3w-17)+10}{3w(3w-4)+4} \div \dfrac{20+w(1-w)}{w(6w-7)+2}$

35. $\left(\dfrac{2t^2+9t-5}{t^2+10t+21} \div \dfrac{2t^2-13t+6}{3t^2+11t+6}\right) \cdot \dfrac{t^2+3t-28}{3t^2-10t-8}$

36. $\left(\dfrac{a(3a+17)+10}{a(a+10)+25} \cdot \dfrac{a^2-25}{a(3a+11)+6}\right) \div \dfrac{a(5a-24)-5}{a(5a+16)+3}$

In Problems 37 to 44, perform the indicated operations and simplify the result.

37. $\dfrac{a}{b} + \dfrac{2a-1}{b}$

38. $\dfrac{m}{m+n} - \dfrac{m+n}{m+n}$

39. $\dfrac{x^2}{x+1} - \dfrac{1}{x+1}$

40. $\dfrac{x^2}{x-2} + \dfrac{3x}{x-2} - \dfrac{10}{x-2}$

41. $\dfrac{2}{x+2} + \dfrac{1}{x-5}$

42. $1 - \dfrac{1}{x+1}$

43. $\dfrac{4u}{2u+1} - \dfrac{3}{2u-1} + \dfrac{1}{4}$

44. $\dfrac{3+p}{2-p} + \dfrac{1+2p}{1+3p+2p^2}$

In Problems 45 to 54, find the LCD of the fractions, perform the indicated operations, and simplify the result.

45. $\dfrac{y^2-2}{y^2-y-2} + \dfrac{y+1}{y-2}$

46. $\dfrac{4z}{xy} - \dfrac{9x}{yz}$

47. $\dfrac{3}{x^2-9} - \dfrac{2}{x^2+6x+9}$

48. $\dfrac{t+3}{t^2-t-2} + \dfrac{2t-1}{t^2+2t-8}$

49. $\dfrac{2}{t-2} + \dfrac{3}{t+1} - \dfrac{t-8}{t^2-t-2}$

50. $\dfrac{y}{y-z} - \dfrac{2y}{y+z} + \dfrac{3yz}{z^2-y^2}$

51. $\dfrac{u}{(u^2+3)(u-1)} - \dfrac{3u^2}{(u-1)^2(u+2)}$

52. $\dfrac{1}{(a-b)(a-c)} + \dfrac{1}{(b-a)(c-b)} - \dfrac{1}{(a-c)(b-c)}$

53. $\dfrac{2x}{4x^3-4x^2+x} - \dfrac{x}{2x^3-x^2} + \dfrac{1}{x^3}$

54. $\dfrac{3}{t-3} - \dfrac{2}{t^2+3t} + \dfrac{10}{t^3-9t}$

In Problems 55 to 64, simplify each complex fraction.

55. $\dfrac{\frac{1}{d} - \frac{1}{c}}{c^2 - d^2}$

56. $\dfrac{\frac{x}{y} + 1}{(x+y)^2}$

57. $\dfrac{(a+b)^2}{\frac{1}{a} + \frac{1}{b}}$

58. $\dfrac{\frac{1}{a^2} + \frac{2}{ab} + \frac{1}{b^2}}{\frac{1}{a^2} - \frac{1}{b^2}}$

59. $\dfrac{\frac{1}{p^2} - \frac{1}{q^2}}{\frac{1}{p^3} - \frac{1}{q^3}}$

60. $\dfrac{\frac{a^3}{b^2} + b}{1 - \frac{a}{b} + \frac{a^2}{b^2}}$

61. $\dfrac{\frac{1}{x+y} - \frac{1}{x-y}}{\frac{2y}{x^2-y^2}}$

62. $\dfrac{\frac{6}{p^2+3p-10} - \frac{1}{p-2}}{\frac{1}{p-2} + 1}$

63. $\dfrac{\frac{x-y}{x+y} - \frac{y}{y-x}}{1 - y\left(\frac{3}{x-y} - \frac{2}{x+y}\right)}$

64. $\dfrac{\frac{x^2-y^2}{x^2+y^2} + \frac{x^2+y^2}{x^2-y^2}}{\frac{x-y}{x+y} - \frac{x+y}{x-y}}$

The expressions in Problems 65 and 66 were obtained as solutions to problems in a popular calculus textbook. Simplify each expression.

65. $\dfrac{\frac{1}{(x+h)^2} - \frac{1}{x^2}}{h}$

66. $3\left(\dfrac{x+1}{x^2+1}\right)^2 \dfrac{x^2+1-(x+1)(2x)}{(x^2+1)^2}$

67. If $q \neq 0$ and n is a positive integer, explain why $(p/q)^n = p^n/q^n$.

68. If $q \neq 0$ and m and n are positive integers, explain why

$$\frac{q^n}{q^m} = \begin{cases} q^{n-m} & \text{if } n > m \\ 1 & \text{if } n = m \\ \dfrac{1}{q^{m-n}} & \text{if } n < m \end{cases}$$

69. In optics, the distance x of an object from the focus of a thin converging lens of focal length f and the distance y of the corresponding image from the focus of the lens satisfy the relation $xy = f^2$ (Figure 1). If $p = x + f$ and $q = y + f$ are the distances from the object to the lens and from the image to the lens, show that

$$\frac{1}{p} + \frac{1}{q} = \frac{1}{f}.$$

Figure 1

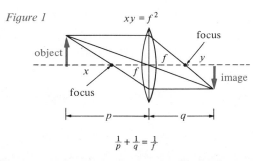

$$\frac{1}{p} + \frac{1}{q} = \frac{1}{f}$$

70. In electronics, if two resistors of resistance R_1 and R_2 ohms are connected in parallel (Figure 2), then the resistance R of the combination is given by

$$R = \frac{1}{(1/R_1) + (1/R_2)}.$$

Simplify this formula for R.

Figure 2

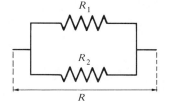

In Problems 71 to 75, *an error in calculation has been made*. Find the error.

71. $\dfrac{4 - x}{4} = 1 - x$?

72. $\dfrac{3 + 4x}{5 - 3x} = \dfrac{7}{2}$?

73. $\dfrac{(a + b)c}{d + c} = \dfrac{a + b}{d}$?

74. $\dfrac{a}{x + y} = \dfrac{a}{x} + \dfrac{a}{y}$?

75. $(a + b)\left(\dfrac{1}{a} + \dfrac{1}{b}\right) = a\left(\dfrac{1}{a}\right) + b\left(\dfrac{1}{b}\right) = 2$?

1.6 RADICAL EXPRESSIONS

If n is a positive integer and a is a real number, then any real number x such that $x^n = a$ is called an **nth root of a**. For $n = 2$ an nth root is called a **square root** and for $n = 3$ it is called a **cube root**. For instance, 4 is a cube root of 64, since $4^3 = 64$, and -3 is a cube root of -27, since $(-3)^3 = -27$. Because $5^2 = 25$ and $(-5)^2 = 25$, *both 5 and -5 are square roots of 25*.

If n is an *odd* positive integer, each real number a has exactly one (real) nth root x, and x has the same algebraic sign as a. If n is an *even* positive integer, each positive number has *two* (real) nth roots—one positive and one negative. However, if n is even, negative numbers have no real nth roots (why?). Thus, we make the following definition.

Definition 1   **The Principal nth Root of a Number**

Let n be a positive integer and suppose that a is a real number.

(i)   If a is a positive number, the **principal nth root of a,** $\sqrt[n]{a}$, is the positive nth root of a.

(ii)   The **principal nth root of zero** is zero, that is, $\sqrt[n]{0} = 0$.

(iii) If a is a negative number, the **principal nth root of a,** $\sqrt[n]{a}$, is defined as a real number only when n is odd, in which case it is the negative number whose nth power is a.

Notice that $\sqrt[n]{a}$ is defined except when a is negative and n is even. Furthermore if $\sqrt[n]{a}$ is defined, it represents just *one* real number. For instance, $\sqrt{4} = 2$,* $\sqrt[3]{27} = 3$, $\sqrt[5]{32} = 2$, and $\sqrt[5]{-32} = -2$, whereas $\sqrt{-4}$ is undefined as a real number. Although it is correct to say that 2 and -2 are square roots of 4, it is *incorrect* to write $\sqrt{4} = \pm 2$ because the symbol $\sqrt{}$ denotes the **principal square root.**

In the expression $\sqrt[n]{a}$, the symbol $\sqrt[n]{}$ is called the **radical sign** (or simply the **radical**); the positive integer n is called the **index**; and the real number a under the radical is called the **radicand:**

$$\underset{\text{radical}}{\overset{\text{index}}{\sqrt[4]{625}}} = 5 \quad \text{principal fourth root of 625}$$
$$\text{radicand}$$

Algebraic expressions containing radicals are called **radical expressions.** Examples are

$$\sqrt[4]{625}, \qquad 5 + \sqrt{x}, \qquad \text{and} \qquad \sqrt{t} - \sqrt[7]{t^4 - 1}.$$

A positive integer is called a **perfect nth power** if it is the nth power of an integer. Obviously, the nth root of a perfect nth power is an integer. For instance,

$$25 = 5^2 \text{ is a perfect square and } \sqrt{25} = 5$$
$$27 = 3^3 \text{ is a perfect cube and } \sqrt[3]{27} = 3$$
$$625 = 5^4 \text{ is a perfect fourth power and } \sqrt[4]{625} = 5.$$

Example 1   Find each principal root (if it is defined).

(a) $\sqrt{36}$ **(b)** $\sqrt[3]{64}$

(c) $\sqrt[3]{-64}$ **(d)** $\sqrt[6]{-64}$

Solution   **(a)** $\sqrt{36} = 6$ **(b)** $\sqrt[3]{64} = 4$

(c) $\sqrt[3]{-64} = -4$ **(d)** $\sqrt[6]{-64}$ is undefined.   ∎

---

* We usually write $\sqrt{}$ rather than $\sqrt[2]{}$.

If both of the positive integers h and k are perfect nth powers, then $\sqrt[n]{h/k} = \sqrt[n]{h}/\sqrt[n]{k}$, and it follows that $\sqrt[n]{h/k}$ is a rational number. However, if the fraction h/k is reduced to lowest terms and if either h or k (or both) fails to be a perfect nth power, it can be shown that $\sqrt[n]{h/k}$ is an irrational number.

Example 2   Determine which of the indicated principal roots are rational numbers and evaluate those that are.

(a) $\sqrt[3]{\frac{8}{125}}$ **(b)** $\sqrt{\frac{2}{50}}$ **(c)** $\sqrt[4]{75}$ ‡

Solution   **(a)** Both the numerator $8 = 2^3$ and the denominator $125 = 5^3$ are perfect cubes, so $\sqrt[3]{\frac{8}{125}} = \frac{2}{5}$.

(b) The fraction $\frac{2}{50}$ isn't reduced to lowest terms. But $\frac{2}{50} = \frac{1}{25}$ and both $1 = 1^2$ and $25 = 5^2$ are perfect squares, so $\sqrt{\frac{2}{50}} = \sqrt{\frac{1}{25}} = \frac{1}{5}$.

(c) The number 75 isn't a perfect 4th power, so $\sqrt[4]{75}$ isn't a rational number. ∎

Radical expressions can often be simplified by using the following properties.

Properties of Radicals

Let a and b be real numbers and suppose that m and n are positive integers. Then, provided that all expressions are defined:

(i) $\sqrt[n]{ab} = \sqrt[n]{a}\,\sqrt[n]{b}$ **(ii)** $\sqrt[n]{\dfrac{a}{b}} = \dfrac{\sqrt[n]{a}}{\sqrt[n]{b}}$ **(iii)** $\sqrt[n]{\sqrt[m]{a}} = \sqrt[nm]{a}$

(iv) $\sqrt[n]{a^m} = (\sqrt[n]{a})^m$ **(v)** $\sqrt[n]{a^n} = (\sqrt[n]{a})^n = a$

Furthermore, *if n is odd*,

$$\sqrt[n]{-a} = -\sqrt[n]{a}.$$

In using these properties to simplify radical expressions, it's a good idea to factor the radicand in order to reveal any exponents that are multiples of the index.

Example 3   Use the properties of radicals to simplify each expression. Assume that variables are restricted to values for which all expressions are defined.

(a) $\sqrt{9x^3} + \sqrt{x^2}$ **(b)** $\sqrt[3]{-16x^7y^4}$ **(c)** $\sqrt{\dfrac{8x^7}{9y^4}}$

(d) $\sqrt[3]{(-x^5)^2}$ **(e)** $\sqrt[7]{\sqrt[5]{x^{35}}}$ **(f)** $\sqrt[5]{a^4}\,\sqrt[5]{a^3}$

Solution   **(a)** If $\sqrt{9x^3}$ is defined, then x cannot be negative and it follows that $\sqrt{x^2} = x$. Therefore,

$$\sqrt{9x^3} + \sqrt{x^2} = \sqrt{9x^2 \cdot x} + x = \sqrt{9x^2}\sqrt{x} + x = 3x\sqrt{x} + x.$$

(b) $\sqrt[3]{-16x^7y^4} = -\sqrt[3]{16x^7y^4} = -\sqrt[3]{8x^6y^3 \cdot 2xy} = -\sqrt[3]{8x^6y^3}\sqrt[3]{2xy} = -2x^2y\sqrt[3]{2xy}$

[Here we used $\sqrt[3]{x^6} = \sqrt[3]{(x^2)^3} = x^2$.]

(c) $\sqrt{\dfrac{8x^7}{9y^4}} = \sqrt{\dfrac{4x^6}{9y^4}(2x)} = \sqrt{\dfrac{4x^6}{9y^4}}\sqrt{2x} = \dfrac{\sqrt{4x^6}}{\sqrt{9y^4}}\sqrt{2x} = \dfrac{2x^3}{3y^2}\sqrt{2x}$

(d) $\sqrt[3]{(-x^5)^2} = \sqrt[3]{x^{10}} = \sqrt[3]{x^9x} = \sqrt[3]{x^9}\sqrt[3]{x} = \sqrt[3]{(x^3)^3}\sqrt[3]{x} = x^3\sqrt[3]{x}$

(e) $\sqrt[7]{\sqrt[5]{x^{35}}} = \sqrt[35]{x^{35}} = x$

(f) $\sqrt[5]{a^4}\sqrt[5]{a^3} = \sqrt[5]{a^4a^3} = \sqrt[5]{a^7} = \sqrt[5]{a^5a^2} = \sqrt[5]{a^5}\sqrt[5]{a^2} = a\sqrt[5]{a^2}$ ∎

Two or more radical expressions are said to be **like** or **similar** if, after being simplified, they contain the same index and radicand. For instance, $3\sqrt{5}$, $x\sqrt{5}$, and $\sqrt{5}/2$ are similar, but $3\sqrt{5}$ and $5\sqrt[3]{5}$ are not similar. To simplify a sum whose terms are radical expressions, we simplify each term and then combine similar terms.

Example 4   Simplify each sum. Assume that variables are restricted to values for which all expressions are defined.

(a) $7\sqrt{12} + \sqrt{75} - 5\sqrt{27}$

(b) $\sqrt{2x^2y} + x\sqrt{18y} - 3\sqrt{x^3y}$

(c) $\sqrt[3]{4a^5} + a\sqrt[3]{32a^2}$

Solution   **(a)** $7\sqrt{12} + \sqrt{75} - 5\sqrt{27} = 7\sqrt{4 \cdot 3} + \sqrt{25 \cdot 3} - 5\sqrt{9 \cdot 3}$

$$= 7\sqrt{4}\sqrt{3} + \sqrt{25}\sqrt{3} - 5\sqrt{9}\sqrt{3}$$
$$= 7 \cdot 2\sqrt{3} + 5\sqrt{3} - 5 \cdot 3\sqrt{3}$$
$$= 14\sqrt{3} + 5\sqrt{3} - 15\sqrt{3}$$
$$= (14 + 5 - 15)\sqrt{3} = 4\sqrt{3}$$

(b) $\sqrt{2x^2y} + x\sqrt{18y} - 3\sqrt{x^3y} = \sqrt{x^2}\sqrt{2}\sqrt{y} + x\sqrt{9}\sqrt{2}\sqrt{y} - 3\sqrt{x^2}\sqrt{x}\sqrt{y}$

$$= x\sqrt{2}\sqrt{y} + 3x\sqrt{2}\sqrt{y} - 3x\sqrt{x}\sqrt{y}$$
$$= 4x\sqrt{2}\sqrt{y} - 3x\sqrt{x}\sqrt{y}$$
$$= x\sqrt{y}(4\sqrt{2} - 3\sqrt{x})$$

(c) $\sqrt[3]{4a^5} + a\sqrt[3]{32a^2} = \sqrt[3]{a^3 \cdot 4a^2} + a\sqrt[3]{8 \cdot 4a^2} = \sqrt[3]{a^3}\sqrt[3]{4a^2} + a\sqrt[3]{8}\sqrt[3]{4a^2}$

$$= a\sqrt[3]{4a^2} + 2a\sqrt[3]{4a^2} = 3a\sqrt[3]{4a^2}$$ ∎

Sums involving radicals are multiplied the same way polynomials are.

Example 5   Expand each product and simplify the result. Assume that $x \geq 0$ and $y \geq 0$.

(a) $\sqrt{3}(\sqrt{27} + \sqrt{15})$

(b) $(3\sqrt{x} - 7\sqrt{y})(5\sqrt{x} + 2\sqrt{y})$

(c) $(\sqrt[3]{s} - \sqrt[3]{2})(\sqrt[3]{s^2} + \sqrt[3]{2s} + \sqrt[3]{4})$

Solution   **(a)** $\sqrt{3}(\sqrt{27} + \sqrt{15}) = \sqrt{3}\sqrt{27} + \sqrt{3}\sqrt{15}$

$$= \sqrt{3}\sqrt{3}\sqrt{9} + \sqrt{3}\sqrt{3}\sqrt{5} = 3 \cdot 3 + 3\sqrt{5} = 9 + 3\sqrt{5}$$

(b) $(3\sqrt{x} - 7\sqrt{y})(5\sqrt{x} + 2\sqrt{y}) = 15(\sqrt{x})^2 + (3 \cdot 2 - 7 \cdot 5)\sqrt{x}\sqrt{y} - 14(\sqrt{y})^2$

$$= 15x - 29\sqrt{xy} - 14y$$

(c) $(\sqrt[3]{s} - \sqrt[3]{2})(\sqrt[3]{s^2} + \sqrt[3]{2s} + \sqrt[3]{4}) = (\sqrt[3]{s} - \sqrt[3]{2})[(\sqrt[3]{s})^2 + \sqrt[3]{s}\sqrt[3]{2} + (\sqrt[3]{2})^2]$

$$= (\sqrt[3]{s})^3 - (\sqrt[3]{2})^3 = s - 2$$

[Here we used Special Product 5 on page 20, with $a = \sqrt[3]{s}$ and $b = \sqrt[3]{2}$.]   ∎

In part (c) of Example 5, the product of two radical expressions contains no radicals. Whenever the product of two radical expressions is free of radicals, we say that the two expressions are **rationalizing factors** for each other. For instance,

$$(2\sqrt{3} - 3\sqrt{x})(2\sqrt{3} + 3\sqrt{x}) = (2\sqrt{3})^2 - (3\sqrt{x})^2 = 12 - 9x,$$

so $2\sqrt{3} - 3\sqrt{x}$ is a rationalizing factor for $2\sqrt{3} + 3\sqrt{x}$.

Fractions containing radicals are sometimes easier to deal with if their denominators are free of radicals. To rewrite a fraction so that there are no radicals in the denominator, we multiply the numerator and denominator by a rationalizing factor for the denominator. This is called **rationalizing the denominator.**

Example 6   Rationalize the denominator of $\dfrac{3}{\sqrt{5}}$.

Solution $\dfrac{3}{\sqrt{5}} = \dfrac{3\sqrt{5}}{\sqrt{5}\sqrt{5}} = \dfrac{3\sqrt{5}}{5}$.   ∎

Rationalizing factors can often be found by using the special products that yield the difference of two squares or the sum or difference of two cubes.

Example 7   Rationalize the denominator of each fraction and simplify the result. Assume that x is restricted to values for which all expressions are defined.

(a) $\dfrac{\sqrt{3} - 1}{\sqrt{2} + 1}$ **(b)** $\dfrac{3}{7\sqrt[3]{x}}$ **(c)** $\dfrac{\sqrt{x} - 3}{\sqrt{x} + 3}$ **(d)** $\dfrac{2}{\sqrt[3]{x} + 2}$

Solution   **(a)** Since $(\sqrt{2} + 1)(\sqrt{2} - 1) = (\sqrt{2})^2 - 1^2 = 2 - 1 = 1$, it follows that $\sqrt{2} - 1$ is a rationalizing factor for the denominator. Thus,

$$\frac{\sqrt{3} - 1}{\sqrt{2} + 1} = \frac{(\sqrt{3} - 1)(\sqrt{2} - 1)}{(\sqrt{2} + 1)(\sqrt{2} - 1)} = \frac{\sqrt{6} - \sqrt{3} - \sqrt{2} + 1}{1}$$

$$= \sqrt{6} - \sqrt{3} - \sqrt{2} + 1.$$

(b) Since $\sqrt[3]{x}(\sqrt[3]{x^2}) = \sqrt[3]{x^3} = x$, we have

$$\frac{3}{7\sqrt[3]{x}} = \frac{3\sqrt[3]{x^2}}{7\sqrt[3]{x}\,\sqrt[3]{x^2}} = \frac{3\sqrt[3]{x^2}}{7x}.$$

(c)
$$\begin{aligned}
\frac{\sqrt{x} - 3}{\sqrt{x} + 3} &= \frac{(\sqrt{x} - 3)(\sqrt{x} - 3)}{(\sqrt{x} + 3)(\sqrt{x} - 3)} \\
&= \frac{(\sqrt{x})^2 - 6\sqrt{x} + 9}{(\sqrt{x})^2 - 9} \\
&= \frac{x - 6\sqrt{x} + 9}{x - 9}
\end{aligned}$$

(d) Using Special Product 4 on page 20, with $a = \sqrt[3]{x}$ and $b = 2$, we see that a rationalizing factor for the denominator is provided by

$$a^2 - ab + b^2 = (\sqrt[3]{x})^2 - (\sqrt[3]{x})(2) + 2^2 = \sqrt[3]{x^2} - 2\sqrt[3]{x} + 4.$$

Thus,

$$\begin{aligned}
(\sqrt[3]{x} + 2)(\sqrt[3]{x^2} - 2\sqrt[3]{x} + 4) &= (a + b)(a^2 - ab + b^2) = a^3 + b^3 \\
&= (\sqrt[3]{x})^3 + 2^3 = x + 8,
\end{aligned}$$

and we have

$$\frac{2}{\sqrt[3]{x} + 2} = \frac{2(\sqrt[3]{x^2} - 2\sqrt[3]{x} + 4)}{(\sqrt[3]{x} + 2)(\sqrt[3]{x^2} - 2\sqrt[3]{x} + 4)} = \frac{2\sqrt[3]{x^2} - 4\sqrt[3]{x} + 8}{x + 8}.$$

In calculus, it is sometimes necessary to rationalize the *numerator* of a fraction.

Example 8   Rationalize the numerator of $\dfrac{\sqrt{x + h + 1} - \sqrt{x + 1}}{h}$.

Solution

$$\begin{aligned}
\frac{\sqrt{x + h + 1} - \sqrt{x + 1}}{h} &= \frac{(\sqrt{x + h + 1} - \sqrt{x + 1})(\sqrt{x + h + 1} + \sqrt{x + 1})}{h(\sqrt{x + h + 1} + \sqrt{x + 1})} \\
&= \frac{(\sqrt{x + h + 1})^2 - (\sqrt{x + 1})^2}{h(\sqrt{x + h + 1} + \sqrt{x + 1})} \\
&= \frac{(x + h + 1) - (x + 1)}{h(\sqrt{x + h + 1} + \sqrt{x + 1})} \\
&= \frac{\cancel{h}}{\cancel{h}(\sqrt{x + h + 1} + \sqrt{x + 1})} \\
&= \frac{1}{\sqrt{x + h + 1} + \sqrt{x + 1}}
\end{aligned}$$

Problem Set 1.6

Throughout this problem set, assume that variables are restricted to values for which all radical expressions are defined.

In Problems 1 to 12, determine which of the indicated principal roots are rational numbers and evaluate those that are.

1. $\sqrt{121}$

2. $\sqrt[4]{256}$

3. $\sqrt[3]{\dfrac{27}{64}}$

4. $\sqrt[5]{-32}$

5. $\sqrt{24}$

6. $\sqrt{0.04}$

7. $\sqrt[3]{\dfrac{16}{250}}$

8. $\sqrt[3]{-\dfrac{18}{54}}$

9. $\sqrt[7]{\dfrac{1}{128}}$

10. $\sqrt[4]{\dfrac{-1}{81}}$

11. $\sqrt[4]{\dfrac{81}{256}}$

12. $\sqrt[4]{(81)(256)}$

In Problems 13 to 41, simplify each radical expression.

13. $\sqrt{27x^3}$

14. $\sqrt[3]{4a^5}$

15. $\sqrt{75x^4y^9}$

16. $\sqrt[4]{81x^5y^{16}}$

17. $\sqrt[3]{24a^4b^5}$

18. $\sqrt[5]{-u^6v^7}$

19. $(\sqrt{x-1})^4$

20. $(\sqrt[3]{2a+b})^6$

21. $\dfrac{\sqrt{16x^4y^3}}{\sqrt{64x^{12}y}}$

22. $\sqrt[5]{\dfrac{1}{243b^5}}$

23. $\sqrt{\dfrac{6y}{7}}\sqrt{\dfrac{35}{72y^2}}$

24. $\sqrt[3]{\dfrac{u^3v^6}{(3uv^2)^3}}$

25. $\sqrt[3]{\dfrac{-27x^4}{y^{21}}}$

26. $\dfrac{\sqrt[6]{u^{15}v^{21}}}{\sqrt[6]{u^3v^6}}$

27. $(2\sqrt[3]{11})^3(-\tfrac{1}{2}\sqrt[5]{24})^5$

28. $\dfrac{(\tfrac{1}{2}\sqrt[3]{9}\sqrt{3})^4}{(3\sqrt[5]{3}\sqrt{3})^5}$

29. $\sqrt[4]{125t^2}\sqrt{25t^4}$

30. $\sqrt[5]{\sqrt[3]{x^{30}}}$

31. $\sqrt[4]{x^2y^{10}}\sqrt[4]{x^6y^9}$

32. $\sqrt[2]{s^3t^5}\sqrt[2]{st^5}\sqrt[2]{st}$

33. $\sqrt[3]{ab^2}\sqrt[3]{a^2b^5}$

34. $\sqrt{\dfrac{5a+b}{3a^2b^2}}\sqrt{\dfrac{54a^5b^5}{25a^2-b^2}}$

35. $\sqrt{\dfrac{u-v}{u+v}}\sqrt{\dfrac{u^2+2uv+v^2}{u^2-v^2}}$

36. $\dfrac{\sqrt{3xy^3z}\sqrt{2x^2yz^4}}{\sqrt{6x^3y^4z^3}}$

37. $\dfrac{\sqrt[3]{54x^2yz^4}}{\sqrt[3]{16xy^4z^3}}$

38. $\sqrt[5]{\dfrac{(a+b)^4(c+d)^3}{8}}\sqrt[5]{\dfrac{(a+b)^6}{4(c+d)^8}}$

39. $\dfrac{\sqrt[3]{x^2y^3}\sqrt[3]{125x^3y^2}}{\sqrt[3]{8x^3y^4}}$

40. $\sqrt[4]{u^3}\sqrt{u}\sqrt[3]{u^3}$

41. $\dfrac{\sqrt[3]{a^2b^4}\sqrt[3]{a^4b}\sqrt[3]{a^3b^4}}{\sqrt[3]{ab^2}\sqrt[3]{a^2b^7}}$

42. If n is an odd positive integer, show that $\sqrt[n]{-a} = -\sqrt[n]{a}$.

In Problems 43 to 52, simplify each algebraic sum.

43. $6\sqrt{2} - 3\sqrt{8} + 98$

44. $2\sqrt{54} - \sqrt{216} - 7\sqrt{24}$

45. $5\sqrt[3]{81} - 3\sqrt[3]{24} + \sqrt[3]{192}$

46. $8xy\sqrt{x^2y} - 3\sqrt{x^4y^3} + 5x^2\sqrt{y^3}$

47. $4t^3\sqrt{180t} - 2t^2\sqrt{20t^3} - \sqrt{5t^7}$

48. $\sqrt[3]{375} - \sqrt[6]{576} - 4\sqrt[9]{27}$

49. $\sqrt[4]{\dfrac{625}{216}} - \sqrt[4]{\dfrac{32}{27}}$

50. $\sqrt{\dfrac{a^3}{3b^3}} + ab\sqrt{\dfrac{a}{3b^5}} - \dfrac{b^2}{3}\sqrt{\dfrac{3a^3}{b^7}}$

51. $\sqrt[3]{27x^4y^5} + xy\sqrt[3]{-8xy^2}$

52. $4w^3\sqrt{180w} - 2w^2\sqrt{20w^3} - \sqrt{5w^7}$

In Problems 53 to 66, expand each product and simplify the result.

53. $\sqrt{6}(2\sqrt{3} - 3\sqrt{2})$

54. $\sqrt{2x}(x\sqrt{2} - 2\sqrt{x})$

55. $(2 + \sqrt{3})(1 - \sqrt{3})$

56. $(5\sqrt{x} - 1)(3\sqrt{x} + 2)$

57. $(\sqrt{x} + \sqrt{3})^2$

58. $(\sqrt{5} - \sqrt{3})(\sqrt{7} - 2)\sqrt{3}$

59. $(\sqrt{5} - 3\sqrt{2})(\sqrt{5} + 3\sqrt{2})$

60. $(3\sqrt{x} - 2\sqrt{y})(3\sqrt{x} + 2\sqrt{y})$

61. $\sqrt{\sqrt{21} + \sqrt{5}} \sqrt{\sqrt{21} - \sqrt{5}}$

62. $\sqrt[3]{\sqrt{33} - \sqrt{6}} \sqrt[3]{\sqrt{33} + \sqrt{6}}$

63. $(\sqrt{3} + \sqrt{2} + 1)(\sqrt{3} + \sqrt{2} - 1)$

64. $(\sqrt{a} + \sqrt{b} - \sqrt{c})(\sqrt{a} + \sqrt{b} + \sqrt{c})$

65. PQ, where
$P = \sqrt[3]{a + b} + \sqrt[3]{a - b}$ and
$Q = (\sqrt[3]{a + b})^2 - \sqrt[3]{a^2 - b^2} + (\sqrt[3]{a - b})^2$

66. RS, where
$R = 2\sqrt{p^2 - 1} - \sqrt{p^2 + 1}$ and
$S = (2\sqrt{p^2 - 1} + \sqrt{p^2 + 1})\left(\dfrac{1}{\sqrt{3}p - \sqrt{5}}\right)$

In Problems 67 to 88, rationalize the denominator and simplify the result.

67. $\dfrac{8}{\sqrt{11}}$

68. $\dfrac{1}{\sqrt[4]{2}}$

69. $\dfrac{4}{\sqrt{8}}$

70. $\dfrac{\sqrt{t}}{\sqrt{7tu}}$

71. $\dfrac{5\sqrt{2}}{3\sqrt{5}}$

72. $\dfrac{3x}{\sqrt{21x}}$

73. $\dfrac{\sqrt{2}}{3\sqrt{x + 2y}}$

74. $\dfrac{10t}{3\sqrt{5(3t + 7)}}$

75. $\dfrac{20}{\sqrt[3]{5}}$

76. $\dfrac{a^2b^3}{3\sqrt[4]{ab}}$

77. $\dfrac{\sqrt{3}}{\sqrt{3} + 5}$

78. $\dfrac{5}{1 - \sqrt{p}}$

79. $\dfrac{\sqrt{5} + \sqrt{2}}{\sqrt{5} - \sqrt{2}}$

80. $\dfrac{2\sqrt{x} - 3\sqrt{y}}{3\sqrt{x} - 4\sqrt{y}}$

81. $\dfrac{3\sqrt{p} + \sqrt{q}}{4\sqrt{p} - 3\sqrt{q}}$

82. $\dfrac{5\sqrt{3} - 4\sqrt{5}}{2\sqrt{3} + 3\sqrt{5}}$

83. $\dfrac{\sqrt{a + 5} + \sqrt{a}}{\sqrt{a + 5} - \sqrt{a}}$

84. $\dfrac{2\sqrt{x^2 - 1} + \sqrt{x^2 + 1}}{3\sqrt{x^2 - 1} + \sqrt{x^2 + 1}}$

85. $\dfrac{5}{2 - \sqrt[3]{x}}$

86. $\dfrac{2}{1 + \sqrt{x} - \sqrt{y}}$

87. $\dfrac{\sqrt[3]{a} - \sqrt[3]{b}}{\sqrt[3]{a} + \sqrt[3]{b}}$

88. $\dfrac{1}{\sqrt{x} + \sqrt{y} + \sqrt{z}}$

In Problems 89 to 92 rationalize the *numerator* of each fraction and simplify the result.

89. $\dfrac{\sqrt{x + h} - \sqrt{x}}{h}$

90. $\dfrac{\sqrt[3]{x} - \sqrt[3]{a}}{x - a}$

91. $\dfrac{\sqrt{(x + h)^2 + 1} - \sqrt{x^2 + 1}}{h}$

92. $\dfrac{\dfrac{1}{\sqrt{x + h}} - \dfrac{1}{\sqrt{x}}}{h}$

93. Show that the expression

$$\frac{\sqrt{x}}{\sqrt{x + a}} - \frac{\sqrt{x + a}}{\sqrt{x}}$$

can be rewritten in the form

$$\frac{-a}{\sqrt{x(x + a)}}.$$

94. In astronomy, the relativistic Doppler shift formula

$$v = v_0 \frac{\sqrt{1 - (v^2/c^2)}}{1 + (v/c)}$$

is used to determine the frequency v of light from a distant galaxy as measured at an observatory on earth. (v is the Greek letter nu.) In this formula, c is the speed of light, v is the speed with which the galaxy is receding from the earth, and v_0 is the frequency of the light emitted by the galaxy. The fact that the observed frequency v is smaller than the original frequency is popularly known as the "red shift." If $b = v/c$, show that

$$\frac{v}{v_0} = \sqrt{\frac{1 - b}{1 + b}}.$$

Spiral galaxy in Ursa Major

1.7 RATIONAL EXPONENTS

In Section 1.3 we established the following properties of exponents:

(i) $a^m a^n = a^{m+n}$ **(ii)** $(a^m)^n = a^{mn}$ **(iii)** $(ab)^n = a^n b^n$

These properties follow from the definition

$$a^n = \overbrace{aaa \cdots a}^{n \text{ factors}},$$

which makes sense only if n is a positive integer. In the present section we extend this definition so that zero, negative integers, and rational numbers can be used as exponents. The key idea is to extend the definition in such a way that Properties (i), (ii), and (iii) continue to hold.

We begin with the question of how to define a^0. If zero as an exponent is to obey Property (i), we must have

$$a^0 a^n = a^{0+n} = a^n.$$

If $a \neq 0$, this equation can hold only if $a^0 = 1$, which leads us to the following definition.

Definition 1   **Zero as an Exponent**

> If a is any nonzero real number, we define $a^0 = 1$.

Notice that 0^0 *is not defined.*

Now we consider how to define a^{-n} when n is a positive integer. Again, if Property (i) is to hold, we must have

$$a^{-n}a^n = a^{-n+n} = a^0 = 1 \qquad (a \neq 0).$$

If $a \neq 0$, this equation can hold only if $a^{-n} = 1/a^n$, which leads us to the following definition.

Definition 2 **Negative Integer Exponents**

> If a is any nonzero real number and n is a positive integer,
>
> $$a^{-n} = \frac{1}{a^n}.$$

In other words, a^{-n} is the reciprocal of a^n. In particular,

$$a^{-1} = \frac{1}{a^1} = \frac{1}{a}.$$

Definitions 1 and 2 enable us to use *any integers*—positive, negative, or zero—as exponents, with the exception that 0^n is defined only when n is positive. You can verify that Properties (i), (ii), and (iii) hold for all integer exponents if $a \neq 0$ and $b \neq 0$. The following properties also hold for all integers m and n and all nonzero real numbers a and b (Problems 79 to 85).

> (iv) $\left(\dfrac{a}{b}\right)^n = \dfrac{a^n}{b^n}$ (v) $\dfrac{a^m}{a^n} = a^{m-n}$ (vi) $\dfrac{a^{-m}}{b^{-n}} = \dfrac{b^n}{a^m}$ (vii) $\left(\dfrac{a}{b}\right)^{-1} = \dfrac{b}{a}$

Property (vi) permits us to move a factor in a numerator to the denominator of a fraction, or vice versa, simply by changing the sign of the exponent of the factor. By Property (vii), the reciprocal of a nonzero fraction is obtained by inverting the fraction.

Example 1   Rewrite each expression so it contains only positive exponents and simplify the result.

(a) 2^{-3} (b) $(7x^0)^{-2}$ (c) $(x^4)^{-2}$

(d) $\left(\dfrac{3}{x}\right)^{-4}$ (e) $\left(\dfrac{3}{2^{-1}}\right)^{-1}$ (f) $(x - y)^{-4}(x - y)^{13}$

(g) $(x^{-2}y^{-3})^{-4}$ (h) $\dfrac{5x^{-3}(a + b)^2}{15x^4(a + b)^{-5}}$ (i) $\dfrac{x^{-1} - y^{-1}}{x - y}$

Solution   (a) $2^{-3} = \dfrac{1}{2^3} = \dfrac{1}{8}$ (by Definition 2)

(b) $(7x^0)^{-2} = (7 \cdot 1)^{-2}$ (by Definition 1)

$\qquad = 7^{-2} = \dfrac{1}{7^2} = \dfrac{1}{49}$ (by Definition 2)

(c) $(x^4)^{-2} = x^{-8} = \dfrac{1}{x^8}$ $\qquad\qquad$ [by Property (ii) and Definition 2]

(d) $\left(\dfrac{3}{x}\right)^{-4} = \dfrac{3^{-4}}{x^{-4}} = \dfrac{x^4}{3^4} = \dfrac{x^4}{81}$ $\qquad$ [by Properties (iv) and (vi)]

(e) $\left(\dfrac{3}{2^{-1}}\right)^{-1} = \dfrac{2^{-1}}{3} = \dfrac{1}{2\cdot 3} = \dfrac{1}{6}$ $\qquad$ [by Properties (vii) and (vi)]

(f) $(x-y)^{-4}(x-y)^{13} = (x-y)^{-4+13} = (x-y)^9$ $\qquad$ [by Property (i)]

(g) $(x^{-2}y^{-3})^{-4} = (x^{-2})^{-4}(y^{-3})^{-4}$ $\qquad\qquad$ [by Property (iii)]
$\qquad\qquad = x^{(-2)(-4)}y^{(-3)(-4)} = x^8 y^{12}$ $\qquad$ [by Property (ii)]

(h) $\dfrac{5x^{-3}(a+b)^2}{15x^4(a+b)^{-5}} = \dfrac{(a+b)^2(a+b)^5}{3x^4x^3} = \dfrac{(a+b)^7}{3x^7}$ $\qquad$ [by Properties (vi) and (i)]

(i) $\dfrac{x^{-1}-y^{-1}}{x-y} = \dfrac{\dfrac{1}{x}-\dfrac{1}{y}}{x-y} = \dfrac{\dfrac{y-x}{xy}}{x-y} = \dfrac{y-x}{xy}\cdot\dfrac{1}{x-y} = \dfrac{y-x}{xy(x-y)}$
$\qquad\qquad = \dfrac{-(x-y)}{xy(x-y)} = \dfrac{-1}{xy}$ $\qquad\qquad\qquad$ ■

If m and n are integers with $n \neq 0$, how shall we define $a^{m/n}$? If Property (ii) is to hold, we must have

$$a^{m/n} = a^{(1/n)m} = (a^{1/n})^m,$$

so the basic question is how to define $a^{1/n}$. But, again, if Property (ii) is to hold, we must have

$$(a^{1/n})^n = a^{(1/n)n} = a^1 = a;$$

in other words, $a^{1/n}$ must be an nth root of a. Thus, we have the following definition.

Definition 3   **Rational Exponents**

Let a be a nonzero real number. Suppose that m and n are integers, that n is positive, and that the fraction m/n is reduced to lowest terms. Then, if $\sqrt[n]{a}$ exists,

$$a^{1/n} = \sqrt[n]{a} \qquad \text{and} \qquad a^{m/n} = (\sqrt[n]{a})^m = (a^{1/n})^m.$$

Also, if m/n is a positive rational number,

$$0^{m/n} = 0.$$

Notice that $a^{1/n}$ is just an alternative notation for $\sqrt[n]{a}$, the principal nth root of a. For instance,

$$25^{1/2} = \sqrt{25} = 5, \qquad 0^{1/7} = \sqrt[7]{0} = 0, \qquad (-27)^{1/3} = \sqrt[3]{-27} = -3,$$

and so forth. It is important to keep in mind that Definition 3 is to be used *only* when $n > 0$ *and m/n is reduced to lowest terms.* For instance, it is *incorrect* to write $(-8)^{2/6} = (\sqrt[6]{-8})^2$ because 2/6 isn't reduced to lowest terms. Instead, we must first reduce 2/6 to lowest terms and write

$$(-8)^{2/6} = (-8)^{1/3} = \sqrt[3]{-8} = -2.$$

Example 2   Find the value of each expression (if it is defined).

(a) $8^{4/3}$ (b) $81^{-3/4}$ (c) $(-7)^{3/2}$ (d) $(-64)^{8/12}$

Solution   Using Definition 3, we have

(a) $8^{4/3} = (\sqrt[3]{8})^4 = 2^4 = 16$

(b) $81^{-3/4} = 81^{(-3)/4} = (\sqrt[4]{81})^{-3} = 3^{-3} = \dfrac{1}{3^3} = \dfrac{1}{27}$

(c) $(-7)^{3/2}$ is undefined since $\sqrt{-7}$ does not exist (as a real number).

(d) $(-64)^{8/12} = (-64)^{2/3} = (\sqrt[3]{-64})^2 = (-4)^2 = 16$

For convenience, we now summarize the properties of rational exponents.

Properties of Rational Exponents

Let a and b be real numbers, suppose that p and q are rational numbers, and let n be a positive integer. Then, provided that all expressions are defined (as real numbers):

(i) $a^p a^q = a^{p+q}$ (ii) $(a^p)^q = a^{pq}$, $a > 0$ (iii) $(ab)^p = a^p b^p$

(iv) $\left(\dfrac{a}{b}\right)^p = \dfrac{a^p}{b^p}$ (v) $\dfrac{a^p}{a^q} = a^{p-q}$ (vi) $\dfrac{a^{-p}}{b^{-q}} = \dfrac{b^q}{a^p}$

(vii) $\left(\dfrac{a}{b}\right)^{-1} = \dfrac{b}{a}$ (viii) $\sqrt[n]{a^p} = (\sqrt[n]{a})^p = a^{p/n}$ (ix) $a^{-p} = \dfrac{1}{a^p}$

As illustrated by the following example, these properties are especially useful for simplifying algebraic expressions containing rational exponents.

Example 3   Simplify each expression and write the answer so that it contains only positive exponents. You may assume that variables are restricted to values for which the properties of rational exponents hold.

(a) $\left(\dfrac{3}{2}\right)^{-2}$ (b) $\dfrac{x^3 y^{-2}}{x^{-1} y}$

(c) $(64x^{-3})^{-2/3} + 25^{-0.5}$

(d) $\left[\dfrac{(27x^2y^3)^{1/3}}{9x^{-2}y^4}\right]^{-1}$

(e) $\dfrac{(3x+2)^{1/2}(3x+2)^{-1/4}}{(3x+2)^{-3/4}}$

(f) $(a^{-1/2} - b^{-1/2})(a^{-1/2} + b^{-1/2})$

Solution **(a)** $\left(\dfrac{3}{2}\right)^{-2} = \left(\dfrac{2}{3}\right)^2 = \dfrac{4}{9}$

(b) $\dfrac{x^3y^{-2}}{x^{-1}y} = \dfrac{x^3x}{y^2y} = \dfrac{x^4}{y^3}$

(c) $(64x^{-3})^{-2/3} + 25^{-0.5} = 64^{-2/3}(x^{-3})^{-2/3} + \dfrac{1}{25^{1/2}}$

$$= (\sqrt[3]{64})^{-2}x^{(-3)(-2/3)} + \dfrac{1}{5}$$

$$= 4^{-2}x^2 + \dfrac{1}{5} = \dfrac{x^2}{4^2} + \dfrac{1}{5} = \dfrac{x^2}{16} + \dfrac{1}{5} = \dfrac{5x^2 + 16}{80}$$

(d) $\left[\dfrac{(27x^2y^3)^{1/3}}{9x^{-2}y^4}\right]^{-1} = \dfrac{9x^{-2}y^4}{(27x^2y^3)^{1/3}} = \dfrac{9x^{-2}y^4}{27^{1/3}x^{2/3}y^1}$

$$= \dfrac{9y^4y^{-1}}{3x^{2/3}x^2} = \dfrac{3y^{4-1}}{x^{2/3+2}} = \dfrac{3y^3}{x^{8/3}}$$

(e) $\dfrac{(3x+2)^{1/2}(3x+2)^{-1/4}}{(3x+2)^{-3/4}} = (3x+2)^{1/2}(3x+2)^{-1/4}(3x+2)^{3/4}$

$$= (3x+2)^{1/2-1/4+3/4} = (3x+2)^1 = 3x+2$$

(f) $(a^{-1/2} - b^{-1/2})(a^{-1/2} + b^{-1/2}) = (a^{-1/2})^2 - (b^{-1/2})^2$

$$= a^{-1} - b^{-1}$$

$$= \dfrac{1}{a} - \dfrac{1}{b} = \dfrac{b-a}{ab}$$

To factor an algebraic sum in which each term contains a rational power of the same expression, begin by factoring out the expression with the *smallest* rational power.

Example 4 Factor $(1 - 3x)(3x + 4)^{-1/2} + (3x + 4)^{1/2}$ and simplify the result.

Solution We factor out $(3x + 4)^{-1/2}$ because $-1/2$ is the smaller exponent of $3x + 4$. Thus,

$$(1 - 3x)(3x + 4)^{-1/2} + (3x + 4)^{1/2} = (3x + 4)^{-1/2}[(1 - 3x) + (3x + 4)^{1/2 - (-1/2)}]$$
$$= (3x + 4)^{-1/2}[1 - 3x + (3x + 4)^1]$$
$$= (3x + 4)^{-1/2}(1 - 3x + 3x + 4)$$
$$= 5(3x + 4)^{-1/2}$$

Problem Set 1.7

In Problems 1 to 34, rewrite each expression so it contains only positive exponents, and simplify the result. Assume that n is a positive integer.

1. $\left(\dfrac{1}{3}\right)^{-3}$

2. $\dfrac{1}{7^{-2}}$

3. $(8x^0)^{-2}$

4. $(2^0y^{-2})^{-5}$

5. $x^{-2}y^4z^{-1}$

6. $[(3^0)/(4^{-2})]^{-1}$

7. $(-1)^{-1}$

8. $(x^{-3})^6(x^0)^{-2}$

9. $(4c^{-4})(-7c^6)$

10. $(a+b)^{-4}(a+b)^9$

11. $(m^4)^{-2}m^{11}$

12. $[(c+3d)^{-1}]^{-5}$

13. $\dfrac{3x^{-5}}{6y^{-2}}$

14. $\dfrac{c^{-1}+d^{-1}}{cd}$

15. $(t^{-1}+3^{-2})^{-1}$

16. $\dfrac{a^{-1}-b^{-1}}{(a+b)^{-2}}$

17. $\dfrac{a^{-1}}{b^{-1}}+\left(\dfrac{b}{a}\right)^{-1}$

18. $(1+x^{-1})^{-1}+(1+x)^{-1}$

19. $x^{2n-3}x^{3-3n}$

20. $t^nt^{1-n}t^{2n-4}$

21. $[(x^{-1})^{-1}]^{-1}$

22. $[(-y^{-2})^{-1}]^{-1}$

23. $1-(p-1)^{-1}+(p+1)^{-1}$

24. $\dfrac{x^{-2}-y^{-2}}{x^{-1}+y^{-1}}$

25. $(t+2)^{-1}(t+2)^{-2}(t^2-4)^2$

26. $(x^{-2}y^{-2}z^{-3})^{-2n}$

27. $\left(\dfrac{a^{-3}}{b^{-3}}\right)^{-n}$

28. $\left(\dfrac{5x^{-1}}{y}\right)^{-1}\dfrac{y}{5x^{-1}}$

29. $\dfrac{(a+8b)^3}{(a+8b)^{-n}}$

30. $\left[\dfrac{(cd)^{-2n}}{c^{-2n}d^{-2n}}\right]^{5n}$

31. $\dfrac{(c+d)^{-2}}{(r+s)^{-2}}\left(\dfrac{r+s}{c+d}\right)^2$

32. $\left[\dfrac{(5x)^{-1}(3x)^2y^{-3}}{15x^{-2}(25y^{-4})}\right]^{-4}$

33. $\dfrac{(p+q)^{-1}(p-q)^{-1}}{(p+q)^{-1}-(p-q)^{-1}}$

34. $\dfrac{2(x+5y)^{-1}+3(x+5y)^{-1}z^{-2}}{4(x+5y)^{-2}-9(x+5y)^{-2}z^{-4}}$

In Problems 35 to 46, find the value of each expression (if it is defined). Do not use a calculator.

35. $9^{3/2}$

36. $16^{-5/4}$

37. $(-8)^{5/3}$

38. $(-4)^{7/8}$

39. $32^{0.6}$

40. $(-1)^{-10/6}$

41. $(-8)^{0.3}$

42. $\left(\dfrac{-8}{27}\right)^{4/6}$

43. $\left(-\dfrac{1}{8}\right)^{-6/9}$

44. $\left(\dfrac{-1}{32}\right)^{1.8}$

45. $6^{1/2}15^{1/2}10^{1/2}$

46. $(-0.125)^{-2/6}$

In Problems 47 to 70, simplify each expression and write the answer so it contains only positive exponents. (You may assume that variables are restricted to values for which the properties of rational exponents hold.)

47. $a^{1/2}a^{3/2}$

48. $m^{-1.4}m^{2.4}m^{-2}$

49. $y^{1/3}y^{2/3}y^{4/3}$

50. $(x+3)^{-1/2}(x+3)^{5/2}$

51. $(8p^9)^{4/3}$

52. $(81m^{12})^{-3/4}$

53. $(32u^{-5})^{-3/5}$

54. $[(3t+5)^{-7/5}]^{-10/7}$

55. $(x^{-7/9})^{27/14}$

56. $[(a^4b^{-3})^{1/5}]^{-10}$

57. $(16x^{-4})^{-3/4}$

58. $\left(\dfrac{2a^{3/2}b^{7/2}}{4a^2b^{-1}}\right)^{-4}$

59. $\left(\dfrac{x^{-1/3}}{x^{3/2}}\right)^6$

60. $\left(\dfrac{x^{1/2}}{y^2}\right)^4\left(\dfrac{y^{-1/3}}{x^{2/3}}\right)^3$

61. $\left(\dfrac{81r^{-12}}{16s^8}\right)^{-1/4}$

62. $\left(\dfrac{25x^{-16}y^{-8}}{4x^4y^{-2}}\right)^{-3/2}$

63. $\left(\dfrac{x^{m/3}y^{-3m/2}}{x^{-2m/3}y^{m/2}}\right)^{-2/m}$, $\quad m>0$

64. $(x^{3/2}-y^{3/2})^2$

65. $(2p+q)^{-1/4}(2p+q)^{1/2}(2p+q)^{3/4}$

66. $[(s+2t)^{1/n}(s+2t)^{1/m}]^{nm/(n+m)}$, $\quad n>0$, $\quad m>0$

67. $\left[\dfrac{(3x+2y)^{1/2}(4r+3t)^{1/3}}{(4r+3t)^{1/2}(3x+2y)^{1/3}}\right]^6$

68. $[(2x+7)^{-3/4}]^{4/3}-[(2x+7)^{4/3}]^{-3/4}$

69. $(m+n)^{1/3}(m-n)^{1/3}(m^2-n^2)^{-2/3}$

70. $[(x^2 + 1)^{1/3} - 1][(x^2 + 1)^{2/3} + (x^2 + 1)^{1/3} + 1]$

In Problems 71 to 78, factor each expression and simplify the result.

71. $x^{3/2} + 2x^{1/2}y + x^{-1/2}y^2$

72. $(2x - 1)^{-1/2}(6x - 3) + (2x - 1)^{1/2}$

73. $(p - 1)^{-2} - 2(p - 1)^{-3}(p + 1)$
[*Hint:* -3 is smaller than -2.]

74. $-2(1 - 5a)^{-3} + 3(3a - 4)^{-1}(1 - 5a)^{-4}$

75. $2(4t - 1)^{-1}(2t + 1)^{-2} + 4(4t - 1)^{-2}(2t + 1)^{-1}$
[*Hint:* $-2 < -1$.]

76. $2(2t + 3)(4t - 3)^{-1/4} + 4(4t - 3)^{3/4}$

77. $x^{-2/3}(x - 1)^{2/3} + 2x^{1/3}(x - 1)^{-1/3}$

78. $3(1 - x)^{-1/5}(1 + x^2)^{-2/3} + 5x(1 - x)^{4/5}(1 + x^2)^{-5/3}$

In Problems 79 to 84, assume that a and b are nonzero real numbers and that m and n are integers. By considering all possible cases in which m or n are positive, negative, or zero, verify each property.

79. $a^n a^m = a^{n+m}$

80. $(a^n)^m = a^{nm}$

81. $(ab)^n = a^n b^n$

82. $\left(\dfrac{a}{b}\right)^n = \dfrac{a^n}{b^n}$

83. $\dfrac{a^m}{a^n} = a^{m-n}$

84. $\dfrac{a^{-m}}{b^{-n}} = \dfrac{b^n}{a^m}$

85. If $a \neq 0$ and $b \neq 0$, show that $\left(\dfrac{a}{b}\right)^{-1} = \dfrac{b}{a}$.

C **86.** Using a calculator, find approximate numerical values for

(a) $2^{1/2}$ (b) $3^{5/2}$ (c) $5^{-1.5}$ (d) π^{-2}

The expressions in Problems 87 to 90 were obtained as answers to problems in a popular calculus textbook. Factor each expression and simplify the result.

87. $2(3x + x^{-1})(3 - x^{-2})(6x - 1)^5 + 30(3x + x^{-1})^2(6x - 1)^4$

88. $-3(3t - 1)^{-2}(2t + 5)^{-3} - 6(3t - 1)^{-1}(2t + 5)^{-4}$

89. $-14(7y + 3)^{-3}(2y - 1)^4 + 8(7y + 3)^{-2}(2y - 1)^3$

90. $-5(6u + u^{-1})^{-6}(6 - u^{-2})(2u - 2)^7 + 14(6u + u^{-1})(2u - 2)^6$

In Problems 91 to 95, an *error in calculation* has been made. Find the error.

91. $(-1)^{2/4} = [(-1)^2]^{1/4} = 1^{1/4} = 1$?

92. $\sqrt[4]{(-4)^2} = (\sqrt[4]{-4})^2$ is undefined?

93. $[(-2)(-8)]^{3/2} = (-2)^{3/2}(-8)^{3/2}$ is undefined?

94. $(-32)^{0.2} = (-32)^{2/10} = (\sqrt[10]{-32})^2$ is undefined?

95. $[(-1)^2]^{1/2} = (-1)^{2(1/2)} = (-1)^1 = -1$?

1.8 COMPLEX NUMBERS

Because the square of a real number is nonnegative, there is no real number whose square is -1. Pondering this fact, the Italian physician and mathematician Geronimo Cardano (1501–1576) declared that numbers such as $\sqrt{-1}$ are "of hidden nature." Nevertheless, he calculated formally with these "hidden numbers" in a remarkable and influential book, *Ars Magna* (*The Great Art*) published in 1545. Gradually, other mathematicians began to accept the idea of calculating with square roots of negative numbers, although they regarded such numbers as being fictitious or imaginary. During the nineteenth century these so-called imaginary numbers were linked with "real" objects in various ways, and it was shown that they are as legitimate as numbers of any other kind.

   To launch our study of square roots of negative numbers we introduce the **imaginary unit**

Geronimo Cardano

$$i = \sqrt{-1}.$$

Leaving aside for now the question of just what i is,* let's work with it a bit and see what happens. For the time being, we simply regard i or $\sqrt{-1}$ as an "invented number" with the property that

$$i^2 = -1.$$

Using i, we can define the **principal square root** of a negative number as follows:

If $c > 0$, then $\sqrt{-c} = \sqrt{c(-1)} = \sqrt{c}\sqrt{-1} = \sqrt{c}\,i = i\sqrt{c}.$

Example 1   Find each principal square root.

(a) $\sqrt{-3}$ (b) $\sqrt{-4}$

Solution   (a) $\sqrt{-3} = \sqrt{3(-1)} = \sqrt{3}\sqrt{-1} = \sqrt{3}\,i = i\sqrt{3}$
(b) $\sqrt{-4} = \sqrt{4(-1)} = \sqrt{4}\sqrt{-1} = 2i$

Notice that, if we multiply i by a real number such as $\sqrt{3}$ or 2, we simply have to write the result as $\sqrt{3}\,i$ or $2i$. More generally, we write the product of i and a real number b as bi or as ib. By further such multiplication, we get nothing new. For instance, if c is a real number and we multiply bi by c, we just get $(cb)i$, which again has the same form—a real number times i. On the other hand, if we multiply bi by ci, we get

$$(bi)(ci) = (bc)i^2 = (bc)(-1) = -bc$$

which is a real number!

Adding two numbers of the form bi and ci, we obtain

$$bi + ci = (b + c)i,$$

which is another number of the same form—a real number times i. However, if we add a real number a to bi, we simply have to write the result as $a + bi$, which *is* something new! A number of the form

$$a + bi$$

where a and b are real numbers, is called a **complex number.** Examples of complex numbers are

$$5 + 3i, \qquad -4 + 7i, \qquad \text{and} \qquad 2 + \sqrt{5}\,i.$$

Let's agree to write $a + (-b)i$ in the simpler form $a - bi$; for instance,

$$3 + (-6)i = 3 - 6i.$$

* For an interpretation of i, see Section 8.7.

If we suppose that the basic algebraic properties of the real numbers continue to operate for complex numbers, then, for any real numbers a, b, c, and d, we have:

> **1.** $(a + bi) + (c + di) = a + c + bi + di = (a + c) + (b + d)i$
>
> **2.** $(a + bi) - (c + di) = a - c + bi - di = (a - c) + (b - d)i$
>
> **3.** $(a + bi)(c + di) = ac + adi + bic + bidi = ac + bdi^2 + (ad + bc)i$
> $= ac + bd(-1) + (ad + bc)i = (ac - bd) + (ad + bc)i$

We take 1, 2, and 3 above as definitions of **addition, subtraction,** and **multiplication** of complex numbers. Notice that the sum, difference, and product of complex numbers is again a complex number. In dealing with complex numbers, we understand that the real number 0 has its usual additive and multiplicative properties, so that

$$0 + bi = bi \qquad \text{and} \qquad 0i = 0.$$

Finally, let's agree that

$$a + bi = c + di \qquad \text{means that} \qquad a = c \qquad \text{and} \qquad b = d.$$

The real numbers a and b are called the **real part** and the **imaginary part** of the complex number $a + bi$. Thus, to say that two complex numbers are equal is to say that their real parts are equal and their imaginary parts are equal. In other words, a single equation involving complex numbers represents *two* equations involving real numbers!

Example 2   Express each complex number in the form $a + bi$, where a and b are real numbers.

(a) $(3 + 5i) + (6 + 5i)$ **(b)** $(2 - 3i) - (6 + 4i)$

(c) $(4 + 3i)(2 + 4i)$

Solution   **(a)** $(3 + 5i) + (6 + 5i) = 3 + 6 + 5i + 5i = 9 + 10i$

(b) $(2 - 3i) - (6 + 4i) = 2 - 6 - 3i - 4i = -4 - 7i$

(c) $(4 + 3i)(2 + 4i) = 8 + 16i + 6i + 12i^2 = 8 + 22i + 12(-1) = -4 + 22i$ ∎

The set of all complex numbers, equipped with the algebraic operations of addition, subtraction, and multiplication, is called the **complex number system** and is denoted by the symbol $\mathbb{C}$. Note that a real number a can be regarded as a complex number whose imaginary part is zero: $a = a + 0i$; therefore, the real number system $\mathbb{R}$ forms part of the complex number system $\mathbb{C}$. It isn't difficult to show that the complex numbers have the same basic algebraic properties—commutative, associative, distributive, identity, and inverse—as the real numbers (Section 1.1, pages 3 and 4). Of these, the most intriguing is certainly the multiplicative inverse property—the fact that every nonzero complex number has a multiplicative inverse or reciprocal.

If $a + bi \neq 0$, you can obtain the **reciprocal** $1/(a + bi)$ by a technique similar to that for rationalizing the denominator of a fraction: Just *multiply numerator and denominator by a − bi*. Thus:

$$\frac{1}{a + bi} = \frac{a - bi}{(a + bi)(a - bi)} = \frac{a - bi}{a^2 - (bi)^2} = \frac{a - bi}{a^2 - b^2 i^2} = \frac{a - bi}{a^2 - b^2(-1)}$$

$$= \frac{a - bi}{a^2 + b^2} = \left(\frac{a}{a^2 + b^2}\right) + \left(\frac{-b}{a^2 + b^2}\right) i.$$

The complex number $a - bi$ used in the calculation above is called the **complex conjugate** of $a + bi$. Complex conjugates are also useful in dealing with quotients of complex numbers. If $a + bi \neq 0$, the **quotient** $(c + di)/(a + bi)$ is defined to be the product of $c + di$ and $1/(a + bi)$.

To find the real and imaginary parts of a quotient of complex numbers, multiply the numerator and denominator by the complex conjugate of the denominator.

Example 3   Express each complex number in the form $a + bi$, where a and b are real numbers.

(a) $\dfrac{1}{4 + 3i}$ (b) $\dfrac{2 - 3i}{1 - 4i}$

Solution   **(a)** Multiplying numerator and denominator by $4 - 3i$, the complex conjugate of $4 + 3i$, we obtain

$$\frac{1}{4 + 3i} = \frac{4 - 3i}{(4 + 3i)(4 - 3i)} = \frac{4 - 3i}{16 - 9i^2} = \frac{4 - 3i}{25} = \frac{4}{25} - \frac{3}{25} i.$$

(b) Multiplying numerator and denominator by $1 + 4i$, the complex conjugate of $1 - 4i$, we obtain

$$\frac{2 - 3i}{1 - 4i} = \frac{(2 - 3i)(1 + 4i)}{(1 - 4i)(1 + 4i)} = \frac{2 + 8i - 3i - 12i^2}{1 - 16i^2} = \frac{2 + 5i - 12(-1)}{1 - 16(-1)}$$

$$= \frac{14 + 5i}{17} = \frac{14}{17} + \frac{5}{17} i.$$

Complex numbers, like real numbers, can be denoted by letters of the alphabet and treated as variables or unknowns. The letters z and w are special favorites for this purpose. If $z = a + bi$, where a and b are real numbers, the complex conjugate of z is often written as $\bar{z} = a - bi$. Notice that

$$z\bar{z} = (a + bi)(a - bi) = a^2 - (bi)^2 = a^2 - b^2 i^2 = a^2 - b^2(-1) = a^2 + b^2.$$

Thus, $z\bar{z}$ is the sum of the squares of the real and imaginary parts of z, and so it is *always a nonnegative real number.*

Example 4   If $z = 3 + 4i$, find

(a) $\bar{z}$ **(b)** $z + \bar{z}$ **(c)** $z - \bar{z}$ **(d)** $z\bar{z}$

Solution   **(a)** $\bar{z} = 3 - 4i$

(b) $z + \bar{z} = (3 + 4i) + (3 - 4i) = 6$

(c) $z - \bar{z} = (3 + 4i) - (3 - 4i) = 8i$

(d) $z\bar{z} = (3 + 4i)(3 - 4i) = 3^2 + 4^2 = 25$ ∎

As we have mentioned, a real number a can be regarded as a complex number $a + 0i$ with imaginary part zero. Thus, $\bar{a} = a - 0i = a$, so *each real number is its own complex conjugate*. Complex conjugation "preserves" all of the algebraic operations; that is, if z and w are complex numbers, then:

(i) $\overline{z + w} = \bar{z} + \bar{w}$ **(ii)** $\overline{z - w} = \bar{z} - \bar{w}$

(iii) $\overline{zw} = \bar{z}\bar{w}$ **(iv)** $\overline{\left(\dfrac{z}{w}\right)} = \dfrac{\bar{z}}{\bar{w}}, \qquad w \neq 0.$

We leave the verification of (i)–(iv) as an exercise (Problem 73). Properties (i) and (iii) can be extended to more than two complex numbers. For instance, if u, v, and w are three complex numbers, we can apply (i) twice to obtain

$$\overline{u + v + w} = \overline{(u + v) + w} = \overline{(u + v)} + \bar{w} = (\bar{u} + \bar{v}) + \bar{w} = \bar{u} + \bar{v} + \bar{w}.$$

Thus, for complex numbers, *the conjugate of a sum is the sum of the conjugates*, and a similar result holds for products.

Of course, integer powers of complex numbers are defined just as they are for real numbers: $z^2 = zz$, $z^3 = zzz$, and so on; and, if $z \neq 0$, $z^0 = 1$, $z^{-1} = 1/z$, $z^{-2} = 1/z^2$, and so on. Notice that

$$\begin{array}{lll} i^1 = i & i^5 = i & i^9 = i \\ i^2 = -1 & i^6 = -1 & i^{10} = -1 \\ i^3 = -i & i^7 = -i & i^{11} = -i \\ i^4 = 1 & i^8 = 1 & i^{12} = 1 \end{array}$$

and so on. Thus, the positive integer powers of i endlessly repeat the pattern of the first four. Hence, if n is a positive integer and r is the remainder when n is divided by 4, then $i^n = i^r$.

Example 5   Find i^{59}.

Solution $i^{59} = i^{56+3} = i^{56}i^3 = (i^4)^{14}i^3 = 1^{14}(-i) = -i$ ∎

*U.S. postage
stamp honoring
Charles P. Steinmetz*

Although complex numbers were originally introduced by Cardano and others to provide solutions for certain algebraic equations (such as $x^2 + 1 = 0$), they now have a wide variety of important applications in engineering and physics. For example, in 1893, Charles P. Steinmetz (1865–1923), an American electrical engineer born in Germany, developed a theory of alternating currents based on the complex numbers.

In direct-current theory, **Ohm's law**

$$E = IR$$

relates the electromotive force (voltage) E, the current I, and the resistance R. Because of inductive and capacitative effects, voltage and current may be out of phase in alternating current circuits, and the equation $E = IR$ may no longer hold. Steinmetz saw that, by representing voltage and current with *complex numbers E and I*, he could deal algebraically with phase differences. Furthermore, he combined the resistance R, the inductive effect X_L (called **inductive reactance**), and the capacitative effect X_C (called **capacitative reactance**) in a single complex number

$$Z = R + (X_L - X_C)i,$$

called **complex impedance.** Using the complex numbers E, I, and Z, Steinmetz showed that Ohm's law for alternating currents takes the form

$$E = IZ.$$

Today, these ideas of Steinmetz are used routinely by electrical engineers all over the world. It has been said that Steinmetz "generated electricity with the square root of minus one."

Problem Set 1.8

In Problems 1 to 4, find each principal square root.

1. $\sqrt{-2}$

2. $\sqrt{-9}$

3. $\sqrt{-27}$

4. $\sqrt{\sqrt[3]{-64}}$

In Problems 5 to 60, express each complex number in the form $a + bi$, where a and b are real numbers.

5. $3 + \sqrt{-16}$

6. $-2\sqrt{-8}$

7. $5\sqrt{-72}$

8. $-7 - \sqrt{-64}$

9. $(2 + 3i) + (7 - 2i)$

10. $(-1 + 2i) + (3 + 4i)$

11. $(4 + i) + 2(3 - i)$

12. $(4 + 2i) + 3(2 - 5i)$

13. $(5 - 4i) + 14$

14. $(\frac{1}{2} - \frac{2}{3}i) + (\frac{3}{4} + \frac{1}{6}i)$

15. $(3 + 2i) - (5 + 4i)$

16. $(2 - 3i) - i$

17. $3(1 + 2i) - 4(2 + i)$

18. $(\frac{1}{2} - \frac{4}{3}i) - \frac{1}{6}(5 + 7i)$

19. $2(5 + 4i) - 3(7 + 4i)$

20. $i - \frac{1}{2}(1 + 5i)$

21. $-(-3 + 5i) - (4 + 9i)$

22. $(\sqrt{2} + \sqrt{3}i) - \left(\dfrac{\sqrt{2}}{2} - \dfrac{\sqrt{3}}{3}i\right)$

23. $(2 + i)(1 + 5i)$

24. $(4 + 3i)(-1 + 2i)$

25. $(7 + 4i)(3 + 6i)$

26. $(7 - 6i)(-5 - i)$

27. $(3 - 2i)(-3 + i)$

28. $i(3 + 7i)$

29. $(-7 + 3i)(-3 + 2i)$

30. $(\frac{2}{3} + \frac{3}{5}i)(\frac{1}{2} - \frac{1}{3}i)$

31. $-8i(5 + 8i)$

32. $(\sqrt{2} + \sqrt{3}i)(\sqrt{2} - \sqrt{3}i)$

33. $(-4i)(-5i)$ **34.** $(-2i)(3i)(-4i)$

35. $(\frac{1}{2} + \frac{1}{3}i)(\frac{1}{2} - \frac{1}{3}i)$ **36.** $(\sqrt{2} + \sqrt{3}i)^2$

37. i^{21} **38.** i^{41}

39. i^{201} **40.** $(1 + i)^3$

41. $(4 + 2i)^2$ **42.** $(1 + i)^4$

43. $\left(\dfrac{1}{2} + \dfrac{\sqrt{3}}{2}i\right)^2$ **44.** $(1 - i)^4$

45. $\dfrac{1}{2 + 3i}$ **46.** $\dfrac{1}{3 - 4i}$

47. $\dfrac{3}{7 + 2i}$ **48.** $\dfrac{-4i}{6 - i}$

49. $\dfrac{3 - 4i}{4 + 2i}$ **50.** $\dfrac{\pi + 4i}{2 - i}$

51. $\dfrac{7 + 2i}{3 - 5i}$ **52.** $\dfrac{1}{i}$

53. $\dfrac{4 + i}{(3 - 2i) + (4 - 3i)}$ **54.** $\dfrac{2 + 3i}{4 - 3i} + \dfrac{3 + 5i}{1 - 2i}$

55. $\dfrac{2 - 6i}{3 + i} - \dfrac{4 + i}{3 + i}$ **56.** $\dfrac{(1 - i)(2 + i)}{(2 - 3i)(3 - 4i)}$

57. $\dfrac{3 - 2i}{3 + i} + \dfrac{4i}{3 - 7i}$ **58.** $\dfrac{3i^3 - 5}{1 + i^5}$

59. $\dfrac{3 - 2i}{(2 + i)(5 + 2i)}$ **60.** $\left(\dfrac{3 - i^7}{i^9 - 3}\right)^2$

In Problems 61 to 72, calculate (a) $\bar{z}$, (b) $z + \bar{z}$, (c) $z - \bar{z}$, and (d) $z\bar{z}$.

61. $z = 2 + i$ **62.** $z = i$

63. $z = -i$ **64.** $z = (1 + i)^2$

65. $z = \dfrac{1 + i}{1 - i}$ **66.** $z = -3i^5$

67. $z = -12 + 5i$ **68.** $z = 7i^{101}$

69. $z = \dfrac{4 - 3i}{2 + 4i}$ **70.** $z = (2 + i)^{-1}$

71. $z = 5$ **72.** $z = 1 + i + i^2 + i^3$

73. Show that (a) $\overline{z + w} = \bar{z} + \bar{w}$, (b) $\overline{z - w} = \bar{z} - \bar{w}$, (c) $\overline{zw} = \bar{z}\bar{w}$, and (d) if $w \neq 0$, $\overline{z/w} = \bar{z}/\bar{w}$.

74. Assume that $a + bi \neq 0$. By multiplying, show that

$$(a + bi)\left[\left(\dfrac{a}{a^2 + b^2}\right) + \left(\dfrac{-b}{a^2 + b^2}\right)i\right] = 1.$$

75. If z is a nonzero complex number, show that $z^{-1} = (\bar{z})^{-1}$.

76. If z is a complex number, show that (a) $\frac{1}{2}(z + \bar{z})$ is the real part of z and that (b) $\dfrac{1}{2i}(z - \bar{z})$ is the imaginary part of z.

1.9 CALCULATORS, SCIENTIFIC NOTATION, AND APPROXIMATIONS

Today many students own or have access to an electronic calculator. A scientific calculator with keys for exponential, logarithmic, and trigonometric functions costs less than many college textbooks and will expedite some of the calculations required in this book. Problems or groups of problems for which the use of a calculator is recommended are marked with the symbol ⌷. If you don't have access to a calculator, you can still work most of these problems by using the tables provided in the appendixes at the back of the book.

There are two types of calculators available, those using **algebraic notation (AN)** and those using **reverse Polish notation (RPN)**. Advocates of AN claim that it is more "natural," while supporters of RPN say that RPN is just as "natural" and avoids the parentheses required when sequential calculations are made in AN. Before purchasing a scientific calculator, you should familiarize yourself with both AN and RPN so that you can make an intelligent decision based on your own preferences.

After acquiring any calculator, learn to use it properly by studying the instruction booklet furnished with it. In particular, practice performing chain calculations so you can do them as efficiently as possible, using whatever "memory" features your calculator may possess to store intermediate results. After you learn *how* to use a calculator, it is imperative that you learn *when* to use it, and especially when *not* to use it.

> Attempts to use a calculator for problems that are *not* marked with the symbol Ⓒ can lead to bad habits, which not only waste time, but actually hinder understanding.

Scientific Notation

In applied mathematics, very large and very small numbers are written in compact form by using integer powers of 10. For instance, the speed of light in vacuum,

$$c = 300{,}000{,}000 \text{ meters per second (approximately)}$$

can be written more compactly as

$$c = 3 \times 10^8 \text{ meters per second.}$$

More generally, a real number x is said to be expressed in **scientific notation** if it is written in the form

$$x = \pm p \times 10^n,$$

where n is an integer and $1 \le p < 10$.

Many calculators automatically switch to scientific notation whenever the number is too large or too small to be displayed in ordinary decimal form. When a number such as 2.579×10^{-13} is displayed, the multiplication sign and the base 10 usually do not appear and the display shows simply

$$2.579 \qquad -13.$$

To change a number from ordinary decimal form to scientific notation, move the decimal point to obtain a number between 1 and 10 and multiply by 10^n or by 10^{-n}, where n is the number of places the decimal point was moved to the left or to the right, respectively. Final zeros after the decimal point can be dropped unless it is necessary to retain them to indicate the accuracy of an approximation.

Example 1   Rewrite each statement so that all numbers are expressed in scientific notation.

(a) The volume of the earth is approximately

$$1{,}087{,}000{,}000{,}000{,}000{,}000{,}000 \text{ cubic meters.}$$

(b) The earth rotates about its axis with an angular speed of approximately 0.00417 degree per second.

Solution **(a)** We move the decimal point 21 places to the left

$$1.087\,0\,0\,0\,0\,0\,0\,0\,0\,0\,0\,0\,0\,0\,0\,0\,0\,0\,0\,0$$

to obtain a number between 1 and 10 and multiply by 10^{21}, so that

$$1{,}087{,}000{,}000{,}000{,}000{,}000{,}000 = 1.087 \times 10^{21}.$$

Thus, the volume of the earth is approximately 1.087×10^{21} cubic meters.
(b) We move the decimal point three places to the right

$$0\,0\,0\,4.1\,7$$

to obtain a number between 1 and 10 and multiply by 10^{-3}, so that

$$0.00417 = 4.17 \times 10^{-3}.$$

Thus, the earth rotates about its axis with an angular speed of approximately 4.17×10^{-3} degree per second.

The procedure above can be reversed whenever a number is given in scientific notation and we wish to rewrite it in ordinary decimal form.

Example 2 Rewrite the following numbers in ordinary decimal form:
(a) 7.71×10^5 **(b)** 6.32×10^{-8}

Solution **(a)** $7.71 \times 10^5 = 7\,7\,1\,0\,0\,0. = 771{,}000$
(b) $6.32 \times 10^{-8} = 0.0\,0\,0\,0\,0\,0\,0\,6\,3\,2 = 0.000{,}000{,}063{,}2$

Approximations

Numbers produced by a calculator are often inexact, because the calculator can work only with a finite number of decimal places. For instance, a 10-digit calculator gives

$$2 \div 3 = 6.666666667 \times 10^{-1} \quad \text{and} \quad \sqrt{2} = 1.414213562,$$

both of which are **approximations** of the true values. Therefore:

> Unless we explicitly ask for numerical approximations or indicate that a calculator is recommended, it's usually best to leave answers in fractional form or as radical expressions.

Don't be too quick to pick up your calculator—answers such as $2/3, \sqrt{2}, (\sqrt{2} + \sqrt{3})/7,$ *and* $\pi/4$ *are often preferred to much more lengthy decimal expressions that are only approximations.*

Most numbers obtained from measurements of real-world quantities are subject to error and also have to be regarded as approximations. If the result of a measurement (or any calculation involving approximations) is expressed in scientific notation, $p \times 10^n$, it is usually understood that p should contain only **significant digits,** that is, digits that, except possibly for the last, are known to be correct or reliable. (The last digit may be off by one unit because the number was rounded off.) For instance, if we read in a physics textbook that

$$\text{one electron volt} = 1.60 \times 10^{-19} \text{ joule,}$$

we understand that the digits 1, 6, and 0 are significant and we say that, *to an accuracy of three significant digits,* one electron volt is 1.60×10^{-19} joule.

To emphasize that a numerical value is only an approximation, we often use a wave-shaped equal sign, $\approx$. For instance,

$$\sqrt{2} \approx 1.414.$$

However, we sometimes use ordinary equals signs when dealing with inexact quantities, simply because it becomes tiresome to indicate repeatedly that approximations are involved.

Rounding Off

Some scientific calculators can be set to round off all displayed numbers to a particular number of decimal places or significant digits. However, it's easy enough to round off numbers without a calculator: Simply drop all unwanted digits to the right of the digits that are to be retained, and increase the last retained digit by 1 if the first dropped digit is 5 or greater. It may be necessary to replace dropped digits by zeros in order to hold the decimal point; for instance, we round off 5157.3 to the nearest hundred as 5200.

Rounding off should be done in one step, rather than digit by digit. Digit-by-digit rounding off may produce an incorrect result. For instance, if 8.2347 is rounded off to four significant digits as 8.235, which in turn is rounded off to three significant digits, the result would be 8.24. However, 8.2347 is *correctly* rounded off in one step to three significant digits as 8.23.

Example 3   Round off the given number as indicated.

(a) 1.327 to the nearest tenth

(b) $-19.873,5$ to the nearest thousandth

(c) 4671 to the nearest hundred

(d) $9.223,45 \times 10^7$ to four significant digits

Solution   **(a)** To the nearest tenth,

hundredths place

$$1.327 \approx 1.3$$

tenths place   thousandths place

(b) To the nearest thousandth,

$$-19.8735 \approx -19.874$$

thousandths place

(c) To the nearest hundred,

$$4671 \approx 4700$$

hundredths place

(d) To four significant digits, $9.22345 \times 10^7 \approx 9.223 \times 10^7$ ∎

If approximate numbers expressed in ordinary decimal form are added or subtracted, the result should be considered accurate only to as many decimal places as the least accurate of the numbers, and it should be rounded off accordingly. To add or subtract approximate numbers expressed in scientific notation, first convert the numbers to ordinary decimal form, then round off the result as above, and finally rewrite the answer in scientific notation to obtain the appropriate number of significant digits.

Ⓒ **Example 4**   Suppose that each of the quantities $a = 1.7 \times 10^{-2}$, $b = 2.711 \times 10^{-2}$, and $c = 6.213455 \times 10^2$ is accurate only to the number of displayed digits. Find $a - b + c$ and express the result in scientific notation rounded off to an appropriate number of significant digits.

Solution   Since we are adding and subtracting, we begin by rewriting

$$a = 0.017, \qquad b = 0.02711, \qquad \text{and} \qquad c = 621.3455$$

in ordinary decimal form. The least accurate of these numbers is a, which is accurate only to the nearest thousandth. Using a calculator, we find that

$$a - b + c = 621.33539,$$

but we must round off this result to the nearest thousandth and write

$$a - b + c = 621.335.$$

Finally, we rewrite the answer in scientific notation, so that

$$a - b + c = 621.335 = 6.21335 \times 10^2.$$ ∎

If approximate numbers are multiplied or divided, the result should be considered accurate only to as many significant digits as the least accurate of the numbers, and it should be rounded off accordingly.

© **Example 5**   Suppose that each of the quantities $a = 2.15 \times 10^{-3}$ and $b = 2.874 \times 10^2$ is accurate only to the number of displayed digits. Calculate the indicated quantity and express it in scientific notation rounded off to an appropriate number of significant digits.

(a) ab **(b)** b/a **(c)** b^2

Solution   **(a)** Using a calculator, we find that $ab = 6.1791 \times 10^{-1}$. Since a, the least accurate of the two factors, is accurate only to three significant digits, we must round off our answer to three significant digits and write $ab = 6.18 \times 10^{-1}$.

(b) Here we have $b/a = 1.336744186 \times 10^5$, but again we must round off our answer to three significant digits and write $b/a = 1.34 \times 10^5$.

(c) Here we have $b^2 = 8.259876 \times 10^4$, but, since b is accurate only to four significant digits, we must round off our answer to four significant digits and write $b^2 = 8.260 \times 10^4$.

Problem Set 1.9

In Problems 1 to 8, rewrite each number in scientific notation.

1. 15,500

2. 0.0043

3. 58,761,000

4. 77 million

5. 186,000,000,000

6. 420 trillion

7. 0.000,000,901

8. $(0.025)^{-5}$

In Problems 9 to 14, rewrite each number in ordinary decimal form.

9. 3.33×10^4

10. 1.732×10^{10}

11. 4.102×10^{-5}

12. -8.255×10^{-11}

13. 1.001×10^7

14. -2.00×10^9

In Problems 15 to 20, rewrite each statement so that all numbers are expressed in scientific notation.

15. The image of one frame in a motion-picture film stays on the screen approximately 0.062 second.

16. One liter is defined to be 0.001,000,028 cubic meter.

17. An *astronomical unit* is defined to be the average distance between the earth and the sun, 92,900,000 miles, and a *parsec* is the distance at which one astronomical unit would subtend one second of arc, about 19,200,000,000,000 miles.

18. A *light year* is the distance that light, traveling at approximately 186,200 miles per second, traverses in one year. Thus, a light year is approximately 5,872,000,000,000 miles.

19. In physics, the average lifetime of a lambda particle is estimated to be 0.000,000,000,251 second.

20. In thermodynamics, the Boltzmann constant is 0.000,000,000,000,000,000,000,0138 joule per degree Kelvin.

In Problems 21 to 24, convert the given numbers to scientific notation and calculate the indicated quantity. (Use the properties of exponents.) Do not round off your answers.

21. $(8,000)(2,000,000,000)(0.000,03)$

22. $(0.000,006)^3(500,000,000)^{-4}$

23. $\dfrac{(7,000,000,000)^3}{0.0049}$

C **24.** $\dfrac{(0.000,000,039)^2(591,000)^3}{(197,000)^2}$

C **25.** In electronics, $P = I^2R$ is the formula for the power P in watts dissipated by a resistance of R ohms through which a current of I amperes flows. Calculate P if $I = 1.43 \times 10^{-4}$ ampere and $R = 3.21 \times 10^4$ ohms.

C **26.** The mass of the sun is approximately 1.97×10^{29} kilograms and our galaxy (the Milky Way) is estimated to have a total mass of 1.5×10^{11} suns. The mass of the known universe is at least 10^{11} times the mass of our galaxy. Calculate the approximate mass of the known universe.

In Problems 27 to 30, specify the accuracy of the indicated value in terms of significant digits.

27. A drop of water contains 1.7×10^{21} molecules.

28. The binding energy of the earth to the sun is 2.5×10^{33} joules.

29. One mile $= 6.3360 \times 10^4$ inches.

30. One atmosphere $= 1.01 \times 10^5$ newtons per square meter.

In Problems 31 to 38, round off the given number as indicated.

31. 5280 to the nearest hundred

32. 9.29×10^7 to the nearest million

33. 0.0145 to the nearest thousandth

34. 999 to the nearest ten

35. 111111.11 to the nearest ten thousand

36. 5.872×10^{12} to three significant digits

37. 2.1448×10^{-13} to three significant digits

38. π to four significant digits

C In Problems 39 to 48, find the numerical value of the indicated quantity in scientific notation rounded off to an appropriate number of significant digits. Assume that the given values are accurate only to the number of displayed digits.

39. $a + b$ if $a = 2.0371 \times 10^2$ and $b = 2.7312 \times 10^1$

40. $a + b - c$ if $a = 1.450 \times 10^5$, $b = 7.63 \times 10^2$, and $c = 2.251 \times 10^3$

41. $a - b + c$ if $a = 4.900 \times 10^{-4}$, $b = 3.512 \times 10^{-6}$, and $c = 2.27 \times 10^{-7}$

42. $a + b - c + d$ if $a = 8.1370$, $b = 2.2 \times 10^1$, $c = 1 \times 10^{-3}$, and $d = 5.23 \times 10^{-4}$

43. ab if $a = 3.19 \times 10^2$ and $b = 4.732 \times 10^{-3}$

44. ab^2 if $a = 2.11 \times 10^4$ and $b = 1.009 \times 10^{-2}$

45. a^3 if $a = 1.02 \times 10^9$

46. $\dfrac{ab}{c}$ if $a = 7.71 \times 10^3$, $b = 3.250 \times 10^{-4}$, and $c = 1.09 \times 10^5$

47. $\dfrac{a^2}{b}$ if $a = 3.32 \times 10^2$ and $b = 3.18 \times 10^{-1}$

48. $\dfrac{a + b}{c}$ if $a = 4.163 \times 10^2$, $b = 2.142 \times 10^1$, and $c = 1.555 \times 10^3$

C **49.** According to the U.S. Bureau of Economic Analysis, the gross national product (GNP) of the United States in 1982 was 3.073×10^{12}. According to the U.S. Office of Management and Budget, the national debt at the end of fiscal 1982 was \$$1.147 \times 10^{12}$. Round off your answers to the following questions in an appropriate manner: (a) How much more was the GNP than the national debt in 1982? (b) If we estimate the population of the United States in 1982 as 2.3×10^8, find the per capita GNP (that is, GNP $\div$ population) in 1982.

50. In the BASIC language, often used for microcomputers, the product of X and Y is denoted by X*Y, 10 raised to the power N is denoted by 10^N, and INT(X) is notation for the largest integer that is less than or equal to X. If N is a positive integer, show that the formula INT(X*10^N + .5)/10^N gives the value of X rounded off to N decimal places.

51. Let $x = p \times 10^n$ in scientific notation, and suppose that p is rounded off to a certain number of decimal places. Explain why x is thereby rounded off to one more significant digit than the number of decimal places to which p was rounded off.

REVIEW PROBLEM SET, CHAPTER 1

In Problems 1 to 4, rewrite each expression using exponential notation.

1. $5 \cdot 5 \cdot 5 \cdot 5 \cdot 5 \cdot 5 \cdot 5 \cdot 5 \cdot x \cdot x \cdot x$

2. $w^2 \cdot w^2 \cdot w^2 \cdot w^2 \cdot z^3 \cdot z^3 \cdot z^3 \cdot z^3$

3. $(-4)(-4)(-4)(-4)(-4)yyyyyy$

4. $(-a^4)(-a^4)(-a^4)(-a^4)(-a^4)$

In Problems 5 to 10, write a formula for the given quantity.

5. w is three times the product of x and y, divided by z.

6. x is 7% less than the number n.

7. s is one-half of the perimeter of a triangle with sides a, b, and c.

8. The surface area A of a rectangular box, open at the top and closed at the bottom, with length l, width w, and height h.

9. The population P of a town n years from now, if the current population is 1000 and the population triples every year. [*Hint:* After $n = 1$ year, the population is 3000, after $n = 2$ years, it is 9000, and so forth.]

10. The number N of board feet of lumber in n "two-by-fours," each of which is l feet long. [*Note:* The dimensions of a cross section of a "two-by-four" are actually 1.5 inches by 3.5 inches; one board foot is the volume of a board with dimensions 1 foot by 1 foot by 1 inch.]

11. The formula $K = \frac{1}{2}mv^2$ gives the kinetic energy K, in joules, of an object of mass m kilograms moving with a speed of v meters per second. A jogger with a mass of 70 kilograms is running at a speed of 3 meters per second. Find the kinetic energy of the jogger.

Ⓒ 12. If P dollars is borrowed and paid back in n equal periodic installments of R dollars each, including interest at the rate r per period on the unpaid

balance, then

$$R = \frac{Pr}{1 - (1 + r)^{-n}}.$$

Find R if $20,000 is borrowed and paid back in 5 equal yearly installments including an interest of 10% per year ($r = 0.1$) on the unpaid balance.

In Problems 13 to 22, state one basic algebraic property of the real numbers to justify each statement.

13. $3 \cdot (-7) = (-7) \cdot 3$

14. $3(x + 2) = 3x + 3 \cdot 2$

15. $(-3) + (5 + \pi) = [(-3) + 5] + \pi$

16. $0 + y^2 = y^2$

17. $15 \cdot (x + y) = (x + y) \cdot 15$

18. $3(\pi + 0) = 3\pi$ 19. $1 \cdot (a - b) = a - b$

20. $6 \cdot (4 \cdot 3) = (6 \cdot 4) \cdot 3$

21. $(-3) \cdot \dfrac{1}{(-3)} = 1$ 22. $\pi + (-\pi) = 0$

In Problems 23 to 26, state one of the derived algebraic properties (6 to 9, pages 5 and 6) to justify each statement.

23. $-[-(x + y)] = x + y$

24. $(-4)(-\pi) = 4\pi$ 25. $(x^2 - y^2) \cdot 0 = 0$

26. If $(3x^2 - 5)(2x - 1) = 0$, then $3x^2 - 5 = 0$ or $2x - 1 = 0$.

In Problems 27 and 28, find the mistake and correct the calculation.

27. $\frac{1}{2} + \frac{1}{3} = \frac{2}{5}$?

28. $\sqrt{9 + 16} = 3 + 4 = 7$?

In Problems 29 and 30, list all the elements that belong to the set.

29. A is the set of all natural numbers x such that $-\frac{3}{2} \leq x \leq \frac{5}{2}$.

30. B is the set of all real numbers in $[0, 1]$, but not in $(0, 1)$.

In Problems 31 to 34, illustrate each set on a number line.

31. (a) $(-1, 3)$ (b) $[-2, 5]$ (c) $[-7, \infty)$
    (d) $(-\infty, 4]$ (e) $[-\frac{1}{3}, 5]$ (f) $(-\frac{5}{2}, \frac{3}{2})$
    (g) $[-\frac{2}{3}, \frac{1}{3})$ (h) $(-\infty, \sqrt{2})$

32. The set of all real numbers x such that $-3 \le x \le 0$ and $3x$ is an integer.

33. The set A of all real numbers that belong to both of the intervals $(-\infty, 5]$ and $(-\frac{2}{3}, 10]$.

34. The set B of all real numbers that belong to at least one of the intervals $(-3, -1]$ and $[1, 3)$.

In Problems 35 to 40, express each rational number (a) as a decimal and (b) as a percent.

35. $\frac{11}{50}$ **36.** $\frac{-3}{1000}$

37. $-\frac{17}{200}$ **38.** $-\frac{7}{8}$

39. $\frac{130}{40}$ **40.** $-\frac{7}{3}$

In Problems 41 to 46, express each percent (a) as a decimal and (b) as a quotient of integers.

41. 49.5%   **42.** 0.007%

43. 0.43%   **44.** 13.4%

45. 140%   **46.** 215%

47. The price of an automobile increases from $8000 to $8500. What is the percent of increase?

48. Employment at a factory decreases from 400 workers to 375 workers. What is the percent of decrease?

In Problems 49 to 60, specify the type of each algebraic expression (monomial, binomial, trinomial, multinomial, polynomial, constant, fraction, rational expression, or radical expression). For the polynomials, give the degree and the coefficients.

49. $-2x^2$ **50.** $\sqrt{7}$

51. $\dfrac{x + y}{x - y}$ **52.** $3x^3 - 2x^2 + 6x - 4$

53. $3x^2 - 5x - 1$ **54.** xy

55. $xy + \sqrt{x}$ **56.** $xy + x^{-1} - 1$

57. $\sqrt{2}x^3 - \sqrt[5]{7}$ **58.** $\dfrac{x^2 - y^2}{x^2 + 1}$

59. $\sqrt{\pi} + \dfrac{x}{y}$ **60.** $x^2 + x + \sqrt{x^2 + 1}$

In Problems 61 to 64, perform the indicated operations.

61. $(x^3 - 2x^2 + 7x - 5) + (2x^3 - x^2 - 5x + 11)$

62. $(3x^3 - 3x^2 - 8x - 17) - (4x^3 + 5x - 7)$

63. $(a^3 + 3a^2b + 2ab^2 + b^3) + (2a^3 + ab^2 - 3b^3) - (4a^2b - 3ab^2 - 5b^3)$

64. $(u^3 + 3u^2v + v^2) - (2u^3 - u^2v - 3uv - v^3) + (u^3 - 2uv - 4v^2)$

In Problems 65 to 74, use the properties of exponents to simplify each expression.

65. $5y^2 \cdot 4y^3$ **66.** $(-6t^4)(-5t^6)t^{10}$

67. $(-x^2y)(x^4y^3)$ **68.** $t^{3n}t^{2n}t^n$

69. $(-p^2)^4$ **70.** $(-q^3)^5$

71. $(-x^2y)^7$ **72.** $[-(x + y)^2]^3$

73. $(ab^n)(ab)^n$ **74.** $(2x^n)^4$

In Problems 75 to 84, expand each product.

75. $3x^2(x^3 + 2x - 3)$ **76.** $x^2y(2x + y + 7)$

77. $(2t + 3)(t - 4)$ **78.** $(2u^2 - v)(u^2 + 3v)$

79. $(xy^2 + 3)(2xy^2 + 1)$ **80.** $(s^3 + t^2)(s^3 - 2t^2)$

81. $(2p + 3)(p^2 - 4p + 1)$

82. $(x^2 + x - 2)(x^2 + 3x + 1)$

83. $(2x + 3)(x - 1)(x - 2)$

84. $(2x - y)(x + 3y)(3x + y)(3x - y)$

In Problems 85 to 94, use the Special Products 1 to 8, page 20, to perform each multiplication.

85. $(3x + 5y)^2$ **86.** $(2q + 7r)^2$

87. $(2x^2 - 5yz)^2$ **88.** $(3a^n - b^n)^2$

89. $(2x - y + 3z)^2$ **90.** $(p - q)^2(p + q)^2$

91. $(3t^n - 11)(3t^n + 11)$ **92.** $(2x^2 + y^2)^3$

93. $(3x^3 - 2xy)^3$

94. $(p^n - 3)(p^{2n} + 3p^n + 9)$

In Problems 95 to 124, factor each polynomial completely.

95. $9x^2y^2 - 12xy^4$ **96.** $18r^2s + 12r^3s^4$

97. $(a + b)^2c^2 - (a + b)c^4$

98. $(3p + q)^3 - (3p + q)^2u$

99. $36c^2 - d^4$ **100.** $x^6 - 25y^4$

101. $(x - y)^2 - z^2$ **102.** $(t + 2s)^2 - 9u^2$

103. $x^{2n} - y^2$ **104.** $x^8y^8 - 1$

105. $25x^2 - 49x^2y^2$

106. $(c - 2d)^4 - (3c - d)^4$

107. $8p^3 + 27q^3$ **108.** $125x^3 - 64y^3$

109. $(a + 2b)^3 - (a - 2b)^3$

110. $t^3 + 125(u + v)^3$ **111.** $x^2 + 2x - 24$

112. $x^2 - 13x + 40$ **113.** $a^6 + 8a^3 + 16$

114. $t^4 - 5t^2 + 4$ **115.** $6x^2 + 5xy - 6y^2$

116. $x^{2n} + x^n - 6$ **117.** $4u^2v^2 - 7uv - 2$

118. $4t^2 + 19tu - 30u^2$ **119.** $20 + 7x - 6x^2$

120. $2a^2 + 4ab + 2b^2 - a - b - 10$

121. $p^2 + 9q^2 - 4 + 6pq$

122. $x^2 + 2xy - 4x - 4y + y^2 + 4$

123. $(a + b)^4 - 7(a + b)^2 + 1$

124. $(x - y)^{2n} - 2(x - y)^n(x^2 + y^2) + (x^2 + y^2)^2$

In Problems 125 to 130, reduce each fraction to lowest terms.

125. $\dfrac{(x + 3)^2(x - 1)}{5(x + 3)(x - 1)}$ **126.** $\dfrac{3(x - 2)^3(x + 1)^2}{12(x - 2)^2(x + 1)}$

127. $\dfrac{t^2 + 5t + 6}{t^2 + 4t + 4}$ **128.** $\dfrac{2a^2 - 3a - 2}{2a^2 + 3a + 1}$

129. $\dfrac{c(c - 2) - 3}{(c - 2)(c + 1)}$ **130.** $\dfrac{(b - 2)(b + 1) - 4}{(b + 2)(b - 2)}$

In Problems 131 to 148, perform the indicated operations and simplify the result.

131. $\dfrac{x^2 - 9}{6x - 3} \cdot \dfrac{10x - 5}{x^2 + 3x}$ **132.** $\dfrac{2a + b}{a^2 - 2ab} \cdot \dfrac{a^3 - 2a^2b}{4a^2 - b^2}$

133. $\dfrac{x^2 + 6x + 5}{2x^2 - 2x - 12} \cdot \dfrac{4x^2 - 36}{x^2 + 8x + 15}$

134. $\dfrac{2y^2 - 7y - 15}{5y^2 - 24y - 5} \cdot \dfrac{20y^2 + 14y + 2}{2y^2 + 11y + 12} \cdot \dfrac{3y^2 + y - 2}{10y^2 + 35y + 15}$

135. $\dfrac{x^2y - xy^2}{3x^2 - 9xy + 6y^2} \div \dfrac{x^3 + x^2y}{6x^3 - 6x^2y - 12xy^2}$

136. $\dfrac{-p^3 + p}{p^2 - p - 2} \div \dfrac{p^3 - p^2}{p^2 - 5p + 6}$

137. $\dfrac{3t^2 + 9t - 54}{2t^2 - 2t - 12} \div \dfrac{3t^2 + 21t + 18}{4t^2 - 12t - 40}$

138. $\dfrac{14a^2 + 23a + 3}{2a^2 + a - 3} \div \dfrac{7a^2 + 15a + 2}{2a^2 - 3a + 1}$

139. $\dfrac{2x - y}{3x^2} + \dfrac{4x + y}{3x^2}$ **140.** $\dfrac{x}{x - 2} - \dfrac{x + 2}{x + 1}$

141. $\dfrac{t^2 - 2t + 1}{t^2 + t} - \dfrac{t - 3}{t + 1}$

142. $\dfrac{a - 2b}{2ab - 6b^2} - \dfrac{b}{a^2 - 4ab + 3b^2}$

143. $\dfrac{2}{c - 2} - \dfrac{1}{c + 3} - \dfrac{10}{c^2 + c - 6}$

144. $\dfrac{3}{p + 1} - \dfrac{3}{p^2 + p} + \dfrac{6}{p^2 - 1}$

145. $\dfrac{1 + \dfrac{6}{a - 3}}{a + 3}$ **146.** $\dfrac{2 - \dfrac{3}{y + 2}}{\dfrac{x}{y - 1} + \dfrac{x}{y + 2}}$

147. $\dfrac{\dfrac{6}{a^2 + 3a - 10} - \dfrac{1}{a - 2}}{\dfrac{1}{a - 2} + 1}$

148. $\dfrac{\dfrac{1}{x + y} - \dfrac{1}{x - y}}{\dfrac{2y}{x^2 - y^2}}$

In Problems 149 to 154, determine which of the indicated principal roots are rational numbers and evaluate those that are. Do not use a calculator.

149. $\sqrt{169}$

150. $\sqrt[3]{\dfrac{375}{24}}$

151. $\sqrt[3]{0.6}$

152. $\sqrt[3]{-\dfrac{108}{32}}$

153. $\sqrt[4]{-81}$

154. $\sqrt[3]{0.216}$

In Problems 155 to 162, use the properties of radicals to simplify each expression. Assume that variables are restricted to values for which all expressions are defined.

155. $\sqrt[3]{4x^2}\,\sqrt[3]{2x^4}$

156. $\sqrt[5]{(c+2d)^4}\,\sqrt[5]{c+2d}$

157. $\sqrt[3]{\dfrac{(a+b)^9}{27a^3}}$

158. $\dfrac{\sqrt[4]{64a^3b^2c}}{\sqrt[8]{16a^2b^{12}c^{10}}}$

159. $\sqrt{\sqrt[4]{p^{24}}}\,\sqrt[n]{p^{4n}}$

160. $\sqrt[n]{(a+b)^{4n}c^{2n}}\,\sqrt[m]{(a+b)^{2m}c^m}$

161. $\sqrt[3]{\dfrac{a^{14}\sqrt{a^6}}{a^7}}\,\sqrt[3]{\dfrac{5a^{12}}{a^{15}}}$

162. $\dfrac{\sqrt[4]{x^6y^3z^2}\,\sqrt[4]{x^3yz^6}}{\sqrt[4]{xy^2}}$

In Problems 163 to 180, perform the indicated operations and simplify the result. Rationalize all denominators (whenever possible). Assume that variables are restricted to values for which all expressions are defined.

163. $\sqrt{50a} + 2\sqrt{32a} - \sqrt{2a}$

164. $\sqrt{8a^3} - 2\sqrt{18a^3} + 3\sqrt{50a^3}$

165. $5\sqrt[3]{2p} + 4\sqrt[3]{16p}$

166. $\sqrt[3]{250x^2} - 6\sqrt[3]{16x^2}$

167. $\sqrt{6}(5 - \sqrt{6}) + \sqrt[3]{216}$

168. $(\sqrt{y} + 1)(\sqrt{y} - 2)$

169. $(\sqrt{2a} - \sqrt{3})(\sqrt{2a} + \sqrt{3})$

170. $(\sqrt{y+z} - 3\sqrt{x})(\sqrt{y+z} + 3\sqrt{x})$

171. $(\sqrt{a+b} - \sqrt{a})^2$

172. $(\sqrt[3]{x+1} - \sqrt[3]{x-1})(\sqrt[3]{(x+1)^2} + \sqrt[3]{x^2-1} + \sqrt[3]{(x-1)^2})$

173. $\dfrac{6}{\sqrt{2x}}$

174. $\dfrac{5}{\sqrt[3]{3p}}$

175. $\dfrac{\sqrt{a}}{\sqrt{a} - \sqrt{b}}$

176. $\dfrac{\sqrt{c}}{\sqrt{c} + \sqrt{d}}$

177. $\dfrac{\sqrt{a-1}}{1 + \sqrt{a-1}}$

178. $\dfrac{(x+1)^2}{\dfrac{x\sqrt{x+1}}{2x\sqrt{x}} - \dfrac{\sqrt{x-1}}{2\sqrt{x}}}$

179. $\dfrac{5}{\sqrt[3]{2} - 1}$

180. $\dfrac{6}{\sqrt[3]{x+y} - \sqrt[3]{x}}$

In Problems 181 and 182, rationalize the *numerator*.

181. $\dfrac{3\sqrt{x} + \sqrt{y}}{5}$

182. $\dfrac{\sqrt{x+h+2} - \sqrt{x+2}}{h}$

In Problems 183 to 194, simplify each expression and write it in a form containing only positive exponents.

183. $[(ab^{-1})^{-2} + (c^{-1}d)^{-3}]^0$

184. $[(-5)^{-2} + 3^{-1}]^{-1}$

185. $\left(\dfrac{x}{y^{-2}}\right)^{-1} + \left(\dfrac{y}{x^{-2}}\right)^{-1}$

186. $\dfrac{(a+2)^{-1} - (a-2)^{-1}}{(a+2)^{-1} + (a-2)^{-1}}$

187. $x^{-3}(x - x^{-1})$

188. $\dfrac{c^{-1} + c^{-2}}{c^{-3}}$

189. $(-3a^{-3})(-a^{-1})^3$

190. $(a^2b^{-4})^{-1}(a^{-3}b^2)^{-2}$

191. $\left(\dfrac{x^{-2}}{y^3}\right)^{-2}\left(\dfrac{x^{-3}}{y^{-4}}\right)^{-3}$

192. $\dfrac{(xy^{-1})^{-2}}{x} \cdot \left(\dfrac{x}{y^{-1}}\right)^{-3}$

193. $\dfrac{(5p^2)^{-2}(5p^5)^{-2}}{(5^{-1}p^{-2})^2}$

194. $\dfrac{(a^3b^2c^4)^{-2}(a^4b^2c)^{-1}}{(abc)^{-1}(a^2bc^3)^2}$

In Problems 195 to 198, find the value of each expression without using a calculator.

195. $\left(\dfrac{8}{27}\right)^{2/3}$

196. $243^{0.6}$

197. $32^{-1.8}$

198. $(64^{1/6} + 4096^{1/12})^{-2}$

In Problems 199 to 206, use the properties of rational exponents to simplify each expression and write it in a form containing only positive exponents. Assume that variables are restricted to values for which the properties of rational exponents hold.

199. $y^{-3/4}y^{2/3}y^{4/3}y^{-1/4}$

200. $x^{-1/2}(x^{3/2} + x^{1/2})$

201. $(a^{-1/4})^8(a^{-1/15})^{-45}a^2$

202. $a^{1/3}b^{1/3}\left[\left(\dfrac{a+b}{2}\right)^2 - \left(\dfrac{a-b}{2}\right)^2\right]^{-2/3}$

203. $(x^2y^{-1})^{-1/2}(x^{-3})^{-1/3}(y^{-2})^{-1/2}$

204. $(a^{1/m}b^{-m})^{-m}(a^{-m}b^{1/m})^m, \qquad m > 0$

205. $\left(\dfrac{-64a^3}{b^6c^4}\right)^{-2/3}\left(\dfrac{8a^{1/3}b^{3/2}}{c^{1/3}}\right)^6$

206. $\left(\dfrac{a^{-3/5}b^{-1/3}c^{2/5}}{a^{-1/5}b^{-2/3}c^{1/5}}\right)^{15}$

In Problems 207 to 210, factor each expression and simplify the result.

207. $y^{-12}(x-y)(x+y)^{-3} + y^{-10}(x+y)^{-4}$

208. $a^{7/5}b^{-2/3} - a^{2/5}b^{1/3}$

209. $(y+2)^{-2/3}(y+1)^{2/3} + 2(y+2)^{1/3}(y+1)^{-1/3}$

210. $2(x+1)^{5/3}(x-2)^{-1/3} + (x^2 - x - 2)^{2/3}$

In Problems 211 to 230, express each complex number in the form $a + bi$, where a and b are real numbers.

211. $(3 + 2i) + (7 + 3i)$

212. $(2 - 3i) + (1 + 2i)$

213. $(5 - 7i) - (4 + 2i)$

214. $(3 + 5i) - (5 - 3i)$

215. $(7 - 4i) - (-6 + 4i)$

216. $(-\frac{5}{2} + 6i) + (-\frac{7}{2} - 3i)$

217. $(5 - 11i)(5 + 2i)$

218. $(5 + 2i)(7 + 3i)$

219. $(2 + 5i)(-2 + 4i)$

220. $(7 - 2i)(2 + 3i)$

221. $\dfrac{2 + 5i}{3 + 2i}$

222. $\dfrac{1 + 4i}{\sqrt{3} + 2i}$

223. $\dfrac{4 + 2i}{(2 + 3i)(4 + i)}$

224. $\dfrac{5 + 15i}{(3 - i)(1 + i)}$

225. i^{403}

226. i^{-21}

227. $\left(\dfrac{1}{3i}\right)^3$

228. $\dfrac{1}{(3 + 2i)^2}$

229. $(2 - 3i)\overline{(3 - 2i)}$

230. $\overline{(3 + 5i)}(3 + 5i)$

In Problems 231 to 234, rewrite each number in scientific notation.

231. 57,120,000,000

232. 731 billion

233. 0.000,000,714

234. 33 millionths

In Problems 235 to 238, rewrite each number in ordinary decimal form.

235. 1.732×10^7

236. -1.066×10^4

237. 3.12×10^{-8}

238. -3.05×10^{-11}

In Problems 239 to 242, rewrite each statement so that all numbers are expressed in scientific notation.

239. An amoeba weighs about 5 millionths of a gram.

240. A tobacco mosaic virus weighs about 0.000,000,000,000,000,066 gram.

241. The diameter of the star Betelgeuse is approximately 358,400,000 kilometers.

242. In a game of bridge there is one chance in approximately 158,800,000,000 that a player will be dealt a hand containing all cards of the same suit, and there is one chance in approximately 2,235,000,000,000,000,000,000,000 that all four players will be dealt such a hand.

C In Problems 243 and 244, convert the given numbers to scientific notation and calculate the indicated quantity. Do not round off your answer.

243. $\dfrac{(40,320,000,000)(0.000,007,703)}{21,000}$

244. $\dfrac{(97,400,000)(705,000)(1,410,000)^2}{0.000,000,220,9}$

In Problems 245 to 248, specify the accuracy of the indicated value in terms of significant digits.

245. The chances of being dealt a "full house" in five-card poker are about one in 6.94×10^2.

246. Five thousand miles is about 3×10^8 inches.

247. One British thermal unit is about 6.6×10^{21} electron volts.

248. The standard value of the acceleration of gravity is $g = 9.80665$ meters/second^2.

In Problems 249 to 252, round off the given number as indicated.

249. 17,450 to the nearest thousand

250. 0.00251 to the nearest thousandth

251. 7.2283×10^5 to three significant digits

252. 2.71828 to four significant digits

[C] In Problems 253 to 258, find the numerical value in scientific notation of the indicated quantity rounded off to an appropriate number of significant digits. Assume that the given values are accurate only to the number of displayed digits.

253. $R_1 + R_2 + R_3$ if $R_1 = 2.7 \times 10^4$, $R_2 = 1.5 \times 10^3$, and $R_3 = 7 \times 10^3$

254. $(R_1^{-1} + R_2^{-1} + R_3^{-1})^{-1}$ if $R_1 = 1.7 \times 10^3$, $R_2 = 3.1 \times 10^4$, and $R_3 = 5 \times 10^3$

255. $\frac{1}{2}mv^2$ if $m = 5.98 \times 10^{24}$ and $v = 2.9770 \times 10^4$

256. mc^2 if $c = 3.00 \times 10^8$ and $m = 9.11 \times 10^{-31}$

257. $\dfrac{IB}{nex}$ if $I = 2.05 \times 10^2$, $B = 1.5$, $n = 8.4 \times 10^{28}$, $e = 1.6 \times 10^{-19}$, and $x = 1.3 \times 10^{-3}$

258. $\frac{4}{3}\pi r^3$ if $r = 6.4 \times 10^6$

259. In adding approximate numbers, does it ever matter whether you round the numbers off (to the number of decimal places in the least accurate of them) before adding them, instead of adding them and then rounding off the sum?

Chapter 2

Equations and Inequalities

The basic algebraic skills developed in Chapter 1 are especially useful in solving the equations and inequalities that arise in practical applications of mathematics. In this chapter we discuss methods for solving equations and inequalities that contain just one variable. Equations and inequalities containing more than one variable are considered later in the book.

2.1 EQUATIONS

An equation containing a variable is neither true nor false until a particular number is substituted for the variable. If a true statement results from such a substitution, we say that the substitution **satisfies** the equation. For instance, the substitution $x = 3$ satisfies the equation $x^2 = 9$, but the substitution $x = 4$ does not.

An equation that is satisfied by every substitution for which both sides are defined is called an **identity.** For instance, $(x + 1)^2 = x^2 + 2x + 1$ is an identity, as is $(\sqrt{x})^2 = x$. An equation that is not an identity is called a **conditional equation.** For instance, $2x = 6$ is a conditional equation because there is at least one substitution (say, $x = 4$) that produces a false statement.

If the substitution $x = a$ satisfies an equation, we say that the number a is a **solution** or a **root** of the equation. Thus, 3 is a root of the equation $2x = 6$, but 4 is not. Two equations are said to be **equivalent** if they have exactly the same roots. Thus, the equation $2x - 6 = 0$ is equivalent to the equation $2x = 6$ because both equations have one and the same root, namely, $x = 3$.

You can change an equation into an equivalent one by performing any of the following operations:

1. *Add or subtract the same quantity on both sides of the equation.*

2. *Multiply or divide both sides of the equation by the same nonzero quantity.*

3. *Simplify one or both sides of the equation by using the methods described in Chapter 1.*

4. *Interchange the two sides of the equation.*

To **solve** an equation means to find all of its roots. The usual method for solving an equation is to write a sequence of equations, starting with the given one, in which each equation is equivalent to the previous one, but "simpler" in some sense. The last equation should either express the solution directly, or be so simple that its solution is obvious. For example, to solve the equation

$$2x - 6 = 0,$$

we begin by adding 6 to both sides to get the equivalent equation

$$2x = 6,$$

then we divide both sides by 2 to produce the equivalent equation

$$x = 3.$$

The last equation shows that the root is 3.

Variables representing quantities whose value or values we wish to find by solving equations are called **unknowns.** A common practice is to use letters toward the end of the alphabet for unknowns, and letters toward the front of the alphabet for **constants** whose value we can assign at will. In particular, the letter x is often used for an "unknown quantity," and the letters a, b, and c are used for constants. A **literal,** or **general equation** is an equation containing, in addition to one or more unknowns, at least one letter that stands for a constant. For instance,

$$ax + b = 0$$

is a literal equation in which x is the unknown and the constant coefficients a and b can be assigned whatever values we please. If we let $a = 2$ and $b = -6$, we obtain

$$2x - 6 = 0,$$

whose solution is $x = 3$.

In applied mathematics, we cannot always follow the convention that unknowns are represented by letters toward the end of the alphabet, because certain symbols are reserved for special quantities. For instance, in physics, c is used for the speed of light, m is used for mass, v is used for velocity, and so on. We shall specify which letters represent unknowns to be solved for, whenever it isn't clear from the context.

An equation such as $7x^3 + 3x^2 + x + 1 = 2x - 5$, in which both sides are polynomials in the unknown, is called a **polynomial equation.** By subtracting the polynomial on the right from both sides of the polynomial equation, we obtain an equivalent polynomial equation in **standard form** with zero on the right side:

$$7x^3 + 3x^2 + x + 1 = 2x - 5$$

$$7x^3 + 3x^2 + x + 1 - (2x - 5) = 0 \qquad \text{(We subtracted } 2x - 5 \text{ from both sides.)}$$

$$7x^3 + 3x^2 - x + 6 = 0 \qquad \text{(We combined like terms.)}$$

The last equation is in standard form. The **degree** of a polynomial equation is defined as the degree of the polynomial on the left side when the equation is in standard form. For instance, $7x^3 + 3x^2 + x + 1 = 2x - 5$ is a third-degree polynomial equation because, after it is rewritten in standard form, $7x^3 + 3x^2 - x + 6 = 0$, the polynomial on the left side has degree 3.

First-Degree or Linear Equations

A **first-degree** or **linear equation** in x is written in standard form as

$$ax + b = 0 \qquad \text{with } a \neq 0.$$

This equation is solved as follows:

$$ax + b = 0$$

$$ax = -b \qquad \text{(We subtracted } b \text{ from both sides.)}$$

$$x = \frac{-b}{a} \qquad \text{(We divided both sides by } a.)$$

In many cases, simple first-degree equations can be solved mentally. For example,

$$\text{the solution of} \qquad 5x = 10 \qquad \text{is} \qquad x = 2,$$

and $\qquad$ the solution of $\qquad 2x + 3 = 0 \qquad$ is $\qquad x = -\frac{3}{2}.$

Sometimes it is convenient to use a calculator.

ⓒ **Example 1**   Solve the equation $2.35x - 3.337 = 0$.

Solution   The solution is $x = 3.337/2.35$. Using a calculator, we find that $x = 1.42$. ∎

In Examples 2 to 4, solve each equation.

Example 2 $29 - 2x = 15x - 5$

Solution

$$29 - 2x = 15x - 5$$

$$29 - 2x - 15x = -5 \qquad \text{(We subtracted } 15x \text{ from both sides.)}$$

$$29 - 17x = -5 \qquad \text{(We combined like terms.)}$$

$$-17x = -34 \qquad \text{(We subtracted 29 from both sides.)}$$

$$17x = 34 \qquad \text{(We multiplied both sides by } -1.)$$

$$x = \tfrac{34}{17} \qquad \text{(We divided both sides by 17.)}$$

$$x = 2$$

∎

Example 3 $(2n + 3)(6n - 1) - 9 = 15n^2 - (3n - 2)(n - 2)$

Solution   We begin by expanding the products on both sides of the equation:

$$(12n^2 + 16n - 3) - 9 = 15n^2 - (3n^2 - 8n + 4)$$

$12n^2 + 16n - 12 = 12n^2 + 8n - 4$	(We collected like terms.)
$16n - 12 = 8n - 4$	(We subtracted $12n^2$ from both sides.)
$8n - 12 = -4$	(We subtracted $8n$ from both sides.)
$8n = 8$	(We added 12 to both sides.)
$n = 1$	(We divided both sides by 8.) ■

Example 4 $\frac{1}{7}(3x - 1) - \frac{1}{5}(2x - 4) = 1$

Solution   In order to clear the equation of fractions, we begin by multiplying both sides by 35, the LCD of the two fractions:

$5(3x - 1) - 7(2x - 4) = 35$	
$15x - 5 - 14x + 28 = 35$	(We expanded the products.)
$x + 23 = 35$	(We collected like terms.)
$x = 12$	(We subtracted 23 from both sides.) ■

If, in solving an equation, you multiply both sides by an expression *containing the unknown*, you must always check the solution. The following example shows why.

Example 5   Solve the equation $\dfrac{1}{y(y - 1)} - \dfrac{1}{y} = \dfrac{1}{y - 1}$.

Solution   Multiplying both sides of the equation by the LCD $y(y - 1)$ and simplifying, we have

$$\cancel{y(y-1)}\frac{1}{\cancel{y(y-1)}} - \cancel{y}(y - 1)\frac{1}{\cancel{y}} = y\cancel{(y-1)}\frac{1}{\cancel{y-1}},$$

that is,

$$1 - (y - 1) = y \qquad \text{or} \qquad 2 - y = y.$$

Adding y to both sides of the last equation, we obtain

$$2 = 2y, \qquad \text{that is,} \qquad 2y = 2,$$

from which it follows that $y = 1$. We now check by substituting $y = 1$ in the original equation to obtain

$$\frac{1}{1(1 - 1)} - \frac{1}{1} = \frac{1}{1 - 1},$$

an equation in which neither side is defined because of the zeros in the denominators. In other words, the substitution $y = 1$ doesn't make the equation true—it makes the equation meaningless! We conclude that the equation *has no root*. ∎

In Example 5, a solution $y = 1$ was found for the *final* equation, but this was not a solution of the *original* equation. What happened? Well, we multiplied both sides of the original equation by $y(y - 1)$, a quantity that equals zero when $y = 1$. But, multiplication of both sides of an equation by zero does not produce an equivalent equation!

A fake "root" that doesn't satisfy the original equation (as in Example 5) is called an **extraneous root.** Extraneous roots can be introduced when both sides of an equation are multiplied by a quantity containing the unknown or when both sides are raised to an even power (for instance, when both sides are squared). It's always a good idea to check your solution by substituting into the original equation, but when extraneous roots could be involved, you *must* make such a check.

The following example illustrates an interesting application of linear equations.

Example 6 Express the repeating decimal $0.32\overline{57}$ as a quotient of integers.

Solution Let $x = 0.32\overline{57}$. Then $100x = 32.57\overline{57}$. If we subtract $0.32\overline{57}$ from $32.57\overline{57}$, the repeating portion of the decimals cancels out:

$$
\begin{array}{r}
32.57\overline{57} \\
(-) \quad 0.32\overline{57} \\
\hline
32.25
\end{array}
$$

Therefore,

$$100x - x = 32.57\overline{57} - 0.32\overline{57} = 32.25,$$

that is,

$$99x = 32.25 \quad \text{or} \quad 9900x = 3225.$$

It follows that

$$x = \tfrac{3225}{9900} = \tfrac{43}{132}.$$

∎

Literal Equations That Can Be Reduced to First-Degree Form

Literal equations containing one unknown can often be solved by using the methods illustrated above. It's usually a good idea to begin by trying to bring all terms containing the unknown to one side of the equation, and all terms not containing the unknown to the opposite side. As always, you must be careful not to divide by zero.

Example 7   Solve the equation $ax + 4c = b - 2x$ for x.

Solution

$$ax + 4c = b - 2x$$

$ax + 2x + 4c = b$ (We added $2x$ to both sides so that all terms containing x are on the left side.)

$ax + 2x = b - 4c$ (We subtracted $4c$ from both sides so that all terms not containing x are on the right side.)

$(a + 2)x = b - 4c$ (We used the distributive property.)

Now, provided that $a + 2 \neq 0$, we can divide both sides of the last equation by $a + 2$ to obtain the solution

$$x = \frac{b - 4c}{a + 2} \qquad \text{for } a \neq -2.$$

You can check this solution by substituting it into the original equation (Problem 71).   ∎

Example 8   The formula $S = 2\pi r^2 + 2\pi rh$ gives the total surface area S of a closed right circular cylinder of radius r and height h (Figure 1). Solve for h in terms of S and r.

Figure 1

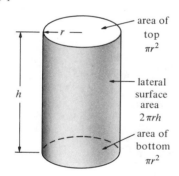

- area of top πr^2
- lateral surface area $2\pi rh$
- area of bottom πr^2

Solution

$$S = 2\pi r^2 + 2\pi rh$$

$2\pi r^2 + 2\pi rh = S$ (We interchanged the two sides.)

$2\pi rh = S - 2\pi r^2$ (We subtracted $2\pi r^2$ from both sides to isolate the term containing h on the left side.)

$h = \dfrac{S - 2\pi r^2}{2\pi r}$ (We divided both sides by $2\pi r$.)

Because the radius of a cylinder must be positive the denominator $2\pi r$ is nonzero.   ∎

Problem Set 2.1

In each problem set, problems with colored numbers constitute a good representation of the main ideas of the section.

In Problems 1 to 6, solve each equation mentally for x.

1. $3x + 6 = 0$

2. $5x = -4$

3. $6x - 8 = 0$

4. $\frac{1}{2}x + 2 = 0$

5. $\frac{2}{3}x - 3 = 0$

6. $cx = d, c \neq 0$

© In Problems 7 to 10, solve each equation with the aid of a calculator. Round off all answers to the correct number of significant digits.

7. $31.02x + 47.71 = 0$

8. $2713x + (7.412 \times 10^4) = 0$

9. $0.1559x - 6.637 = 0$

10. $(3.442 \times 10^{-14})x + (2.193 \times 10^9) = 0$

In Problems 11 to 30, solve each equation.

11. $3x - 2 = 7 + 2x$

12. $3x + 8 = 9 - 2x$

13. $2t + 3 = t + 6$

14. $-2c + 18 = 3c + 3$

15. $9 - 2y = 12 - 3y$

16. $5p + 6 = 3p + 5$

17. $3(y + 6) = y - 1$

18. $10x - 1 - 7x + 3 = 7x - 10$

19. $14 - (3x - 30) = 15x - 10$

20. $7(2n + 5) - 6(n + 8) = 7$

21. $\dfrac{3}{4}x - \dfrac{5}{2}x = -7$

22. $\dfrac{a}{3} + \dfrac{13}{6} = 3 - \dfrac{a}{2}$

23. $\dfrac{1}{2}y - \dfrac{2}{3}y = 7 - \dfrac{3}{4}y$

24. $\dfrac{n - 1}{3} + 3 = \dfrac{n + 14}{9}$

25. $\dfrac{5 + x}{6} - \dfrac{10 - x}{3} = 1$

26. $\dfrac{2(4x - 5)}{3} + 9 = \dfrac{3(x + 2)}{4} - \dfrac{13}{6}$

27. $(2x + 3)^2 = (2x - 1)(2x + 1)$

28. $(3x - 1)^2 - 2x(x + 1) = 7x^2 - 5x + 2$

29. $(u - 1)^2 - (u + 1)^2 = 1 - 5u$

30. $(t - 1)(2t + 3) + (t + 1)(t - 4) = 3t^2$

In Problems 31 and 32, rewrite each repeating decimal as a quotient of integers.

31. (a) $0.\overline{21}$ (b) $3.41\overline{21}$ (c) $0.0\overline{39}$
(d) $-1.00\overline{17}$ (e) $0.00\overline{7}$

32. (a) $0.\overline{121}$ (b) $-3.\overline{321}$
(c) $0.1\overline{523}$ (d) $0.\overline{285714}$

In Problems 33 to 40, solve each equation. Be sure to check for extraneous roots.

33. $\dfrac{10}{x} - 2 = \dfrac{5 - x}{4x}$

34. $\dfrac{3 - y}{3y} + \dfrac{1}{4} = \dfrac{1}{2y}$

35. $\dfrac{t}{t + 4} = \dfrac{1}{2}$

36. $\dfrac{u - 5}{u + 5} + \dfrac{u + 15}{u - 5} = \dfrac{25}{25 - u^2} + 2$

37. $\dfrac{2}{x - 2} + \dfrac{1}{x + 1} = \dfrac{1}{(x - 2)(x + 1)}$

38. $\dfrac{2n}{n + 7} - 1 = \dfrac{n}{n + 3} + \dfrac{1}{(n + 7)(n + 3)}$

39. $\dfrac{1}{y - 3} - \dfrac{1}{3 - y} = \dfrac{1}{y^2 - 9}$

40. $\dfrac{5}{y - 1} + \dfrac{1}{4 - 3y} = \dfrac{3}{6y - 8}$

In Problems 41 to 50, solve each literal equation for the indicated unknown. Be careful not to divide by zero.

41. $5(2x + a) = bx - c$ for x

42. $7(2t + 5a) - 6(t + b) = 3a$ for t

43. $\dfrac{ax + b}{c} = d + \dfrac{x}{4c}$ for x, if $c \neq 0$

44. $\dfrac{y - 3a}{b} = \dfrac{2a}{b} + y$ for y, if $b \neq 0$

45. $\dfrac{x}{m} - \dfrac{a - x}{m} = d$ for x, if $m \neq 0$

46. $\dfrac{3ap - 2b}{3b} - \dfrac{ap - a}{2b} = \dfrac{ap}{b} - \dfrac{2}{3}$ for p, if $b \neq 0$

47. $\dfrac{mn}{x} - bc = d + \dfrac{1}{x}$ for x

48. $\dfrac{1}{a} + \dfrac{a}{a + x} = \dfrac{a + x}{ax}$ for x, if $a \neq 0$

49. $\dfrac{2x}{x - b} = 3 - \dfrac{x - b}{x}$ for x

50. $\dfrac{x - 2r}{25 + x} + \dfrac{x + 2r}{25 - x} = \dfrac{4rs}{625 - x^2}$ for x

In Problems 51 to 58, a formula used in the specified field of applied mathematics is given. In each case, solve for the indicated unknown.

51. $V = \pi r^2 h$ for h (geometry)

52. $F = \dfrac{mv^2}{r}$ for m (mechanics)

53. $A = P(1 + rt)$ for t (finance)

54. $PV = nRT$ for T (physics)

55. $\dfrac{1}{p} + \dfrac{1}{q} = \dfrac{1}{f}$ for f (optics)

56. $S = \dfrac{rl - a}{r - 1}$ for r (economics)

57. $\dfrac{P_1 V_1}{T_1} = \dfrac{P_2 V_2}{T_2}$ for T_2 (thermodynamics)

58. $I = \dfrac{nE}{nr + R}$ for n (electrical engineering)

In Problems 59 to 64, determine whether each equation is a conditional equation or an identity.

59. $(4x + 3)^2 = 16x^2 + 24x + 9$

60. $\dfrac{1}{(x + 1)^2} = \dfrac{x}{x^3 + 2x^2 + x}$

61. $\sqrt{x^2} = x$ **62.** $\sqrt{1 + x^2} = 1 + x$

63. $\dfrac{1}{x} + \dfrac{1}{2} = \dfrac{2}{x + 2}$

64. $\dfrac{1 - x}{x^2 - 1} + \dfrac{x}{x + 1} = \dfrac{x - 1}{x + 1}$

In Problems 65 to 69, determine whether the given equations are equivalent. Give reasons for your answers.

65. $x = 6$ and $x^2 = 36$

66. $(x - 1)(x + 2) = x^2$ and $(x - 1)(x + 2)x = x^3$

67. $x = 3$ and $x^3 = 27$

68. $\dfrac{t^2 - 1}{t + 1} = t$ and $t^2 - 1 = t(t + 1)$

69. $x^2 = 1$ and $x^3 = 1$

70. For what value of the constant a is $x = -1$ a solution of $2(ax + 2) - ax = 1$?

71. Check the solution of Example 7.

72. The formula $H = (A + B\sqrt{V} - CV)(S - T)$ gives the heat loss (wind chill) H in Btu's per square foot of skin per hour if the air temperature is T degrees Fahrenheit and the wind speed is V miles per hour. Here $S = 91.4°$F represents neutral skin temperature, and A, B, and C are constants determined experimentally to be $A = 2.14$, $B = 1.37$, and $C = 0.0916$. The *equivalent temperature* (**wind chill index**) is defined to be the air temperature T_E degrees Fahrenheit at which the same heat loss would occur if the wind speed were 4 miles per hour (the speed of a brisk walk). Thus,

$$(A + B\sqrt{V} - CV)(S - T)$$
$$= [A + B\sqrt{4} - C(4)](S - T_E).$$

(a) By solving the last equation, find a formula for T_E in terms of T, V, S, A, B, and C.

© (b) Find the equivalent temperature T_E if $T = 25°$F and $V = 20$ miles per hour.

© In Problems 73 and 74, solve each equation with the aid of a calculator. Round off your answers to the correct number of significant digits.

73. $2.72x + 2.24 = 2.45x - 2.65$

74. $(6.86 \times 10^{-5})w - (7.14 \times 10^9) =$
$(7.28 \times 10^{-5})w + (1.05 \times 10^9)$

2.2 APPLICATIONS INVOLVING FIRST-DEGREE EQUATIONS

Questions that arise in the real world are usually expressed in words, rather than in mathematical symbols. For example: "What will be the monthly payment on my mortgage?" "How much insulation must I use in my house?" "What course should I fly to Boston?" "How safe is this new product?" In order to answer such questions, it is necessary to have certain pertinent information. For instance, to determine the monthly payment on a mortgage, you need to know the amount of the mortgage, the interest rate, and the time period involved.

Problems in which a question is asked and pertinent information is supplied in the form of words are called "word problems" or "story problems" by students and teachers alike. In this section, we study word problems that can be worked by setting up an equation containing the unknown and solving it by the methods illustrated in Section 2.1. For working these problems, we recommend the following systematic procedure:

Step 1. Begin by reading the problem carefully, several times if necessary, until you understand it well. Draw a diagram whenever possible. Look for the question or questions you are to answer.

Step 2. List all of the unknown numerical quantities involved in the problem. It may be useful to arrange these quantities in a table or chart along with related known quantities. Select one of the unknown quantities in your list, one that seems to play a prominent role in the problem, and call it x. (Of course, any other letter will do as well.)

Step 3. Using information given or implied in the wording of the problem, write algebraic relationships among the numerical quantities listed in step 2. Relationships that express some of these quantities in terms of x are especially useful. Reread the problem, sentence by sentence, to make sure you have rewritten all the given information in algebraic form.

Step 4. Combine the algebraic relationships written in step 3 into a single equation containing only x and known numerical constants.

Step 5. Solve the equation for x. Use this value of x to answer the question or questions in step 1.

Step 6. Check your answer to make certain that it agrees with the facts in the problem.

Of course, a calculator is often useful to expedite arithmetic.

Example 1   One number is 15 less than a second number. Three times the first number added to twice the second number is 80. Find the two numbers.

Solution   We follow the procedure just outlined.

Step 1. Question: What are the two numbers?

Step 2. Unknown quantities: *The first number* and *the second number.* Let $x = $ *the first number.* (See the alternative solution below, where we choose x to represent the second number.)

Step 3. Information given:

(i)   *The first number* = *the second number* $- 15$, that is,
$$x = \text{the second number} - 15.$$

(ii)   3 (*the first number*) + 2(*the second number*) = 80, that is,
$$3x + 2(\text{the second number}) = 80.$$

Step 4. From relationship (i) in step 3, we have

$$the\ second\ number = x + 15.$$

Therefore, relationship (ii) can be written as

$$3x + 2(x + 15) = 80.$$

Step 5. Solving the equation $3x + 2(x + 15) = 80$, we obtain

$$3x + 2x + 30 = 80$$

$$5x = 50$$

$$x = 10.$$

Therefore $$the\ first\ number = x = 10$$

and $$the\ second\ number = x + 15 = 10 + 15 = 25.$$

Step 6. *Check:* Indeed, 10 is 15 less than 25 and $3(10) + 2(25) = 80$.

Alternative Solution   In step 2 above we could have let $x = the\ second\ number$. With this assignment of the variable, relationship (i) in step 3 becomes *the first number* $= x - 15$, and relationship (ii) becomes $3(x - 15) + 2x = 80$, or $5x = 125$. Hence, *the second number* $= x = 25$ and *the first number* $= x - 15 = 10$. ■

Example 2   A suit is on sale for $195. What was the original price of the suit if it has been discounted 25%?

Solution   **Step 1.** Question: What was the original price of the suit?

Step 2. Unknown quantities: The *original price of the suit* and the *amount of the discount in dollars*. Let $x = the\ original\ price$.

Step 3. (*original price*) $- discount =$ sale price $= 195$ dollars, that is,

$$x - discount = 195.$$

$$Discount = 25\%\ of\ original\ price = 0.25x.$$

Step 4. $x - 0.25x = 195$

Step 5. $0.75x = 195$

$$x = \frac{195}{0.75}$$

$$= 260.$$

The original price of the suit was $260.

Step 6. *Check:* If a $260 suit is discounted by 25%, the discount is $(0.25)(\$260) = \$65$ and the sale price is $\$260 - \$65 = \$195$. ■

In solving the following problem, we use the **simple interest formula**

$$I = Prt$$

where I is the **simple interest,** P is the **principal** (the amount invested), r is the **rate** of interest per interest period, and t is the number of interest periods.

Example 3   A businesswoman has invested a total of $30,000 in two certificates. The first certificate pays 10.5% annual simple interest, and the second pays 9% annual simple interest. At the end of one year, her combined interest on the two certificates is $2970. How much did she originally invest in each certificate?

Solution   **Step 1.** Question: What was the principal for each of the two certificates?

Step 2. Let $x =$ *the principal for the first certificate* in dollars. Quantities involved in the problem appear in the following table. Here the interest period is $t = 1$ year, so that $I = Pr$.

Certificate	Principal	Rate	Time	Simple Interest
First	x dollars	0.105	1	0.105x dollars
Second	30,000 − x dollars	0.09	1	0.09(30,000 − x) dollars

Step 3. Most of the information given in the problem appears in the table. The only remaining fact is that

the combined simple interest = 2970 dollars.

Step 4. Because the sum of the simple interest on the two certificates is the combined simple interest,

$$0.105x + 0.09(30{,}000 - x) = 2970.$$

Step 5.
$$0.105x + 0.09(30{,}000 - x) = 2970$$
$$0.105x + 2700 - 0.09x = 2970$$
$$0.105x - 0.09x = 2970 - 2700$$
$$0.015x = 270$$
$$15x = 270{,}000$$
$$x = \frac{270{,}000}{15} = 18{,}000.$$

Therefore, $18,000 was invested in the first certificate and

$$\$30,000 - \$18,000 = \$12,000$$

was invested in the second.

Step 6. *Check:* The simple interest on $18,000 for one year at 10.5% is $0.105(\$18,000) = \$1890$. The simple interest on $12,000 for one year at 9% is $0.09(\$12,000) = \$1080$. The total amount invested is

$$\$18,000 + \$12,000 = \$30,000$$

and the combined interest is $\$1890 + \$1080 = \$2970$. ∎

Many word problems involving mixtures of substances or items can be worked by solving first-degree equations. Examples 4 and 5 illustrate how to solve typical **mixture problems.**

Example 4   A chemist has one solution containing a 10% concentration of acid and a second solution containing a 15% concentration of acid. How many milliliters of each should be mixed in order to obtain 10 milliliters of a solution containing a 12% concentration of acid?

Solution   Let $x =$ *the number of milliliters of the first solution,* so that $10 - x =$ *the number of milliliters of the second solution.* The following table summarizes the given information:

	Milliliters of Solution	Acid Concentration	Milliliters of Acid in Solution
First Solution	x	0.10	$0.10x$
Second Solution	$10 - x$	0.15	$0.15(10 - x)$
Mixture	10	0.12	$0.12(10) = 1.2$

Since the amount of acid in the mixture is the sum of the amounts in the two solutions,

$$0.10x + 0.15(10 - x) = 1.2$$
$$0.10x + 1.5 - 0.15x = 1.2$$
$$0.10x - 0.15x = 1.2 - 1.5$$
$$-0.05x = -0.3$$
$$5x = 30$$
$$x = 6.$$

Hence, 6 milliliters of the first solution and $10 - 6 = 4$ milliliters of the second solution should be mixed.

Check: In 6 milliliters of the first solution there is $(0.10)6 = 0.6$ milliliter of acid. In 4 milliliters of the second solution there is $(0.15)4 = 0.6$ milliliter of acid. Thus, there are $0.6 + 0.6 = 1.2$ milliliters of acid in the $6 + 4 = 10$ milliliters of the mixture. Therefore, the acid concentration of the mixture is $\frac{1.2}{10} = 0.12 = 12\%$.   ■

Example 5   A vending machine for chewing gum accepts nickels, dimes, and quarters. When the coin box is emptied, the total value of the coins is found to be $24.15. Find the number of coins of each kind in the box if there are twice as many nickels as quarters and five more dimes than nickels.

Solution   Let $n = $ *the number of nickels.* The following table summarizes the given information:

	Number of Coins	Individual Value	Total Value in Dollars
Nickels	n	$0.05	$0.05n$
Dimes	$n + 5$	$0.10	$0.10(n + 5)$
Quarters	$\frac{1}{2}n$	$0.25	$0.25(\frac{1}{2}n)$

Since the coins have a total value of $24.15,

$$0.05n + 0.10(n + 5) + 0.25(\tfrac{1}{2}n) = 24.15.$$

To solve this equation, we begin by multiplying both sides by 100 to remove the decimals:

$$5n + 10(n + 5) + 25(\tfrac{1}{2}n) = 2415$$

$$15n + \frac{25}{2}n = 2415 - 50 = 2365$$

$$30n + 25n = 4730$$

$$55n = 4730$$

$$n = \frac{4730}{55} = 86.$$

Therefore, there are 86 nickels, $86 + 5 = 91$ dimes, and $\frac{1}{2}(86) = 43$ quarters. *Check:* $\$0.05(86) + \$0.10(91) + \$0.25(43) = \$24.15$.   ■

Another type of applied problem involves objects that move a distance d at a constant rate r (also called speed) in t units of time. To solve these problems, use the formula

$$\boxed{d = rt \qquad (distance = rate \times time).}$$

This formula can be rewritten as $r = d/t$ or as $t = d/r$.

Example 6   A jogger and a bicycle rider leave a field house at the same time and set out for a nearby town. The jogger runs at a constant speed of 16 kilometers per hour. At the end of 2 hours, the bicycle rider is 19.2 kilometers ahead of the jogger (Figure 1). How fast is the bicycle rider traveling, assuming that his speed is constant?

Figure 1

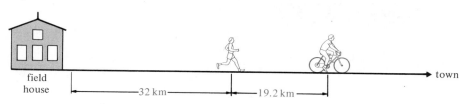

Solution   Let $x =$ *the speed of the bicycle rider*. The following table summarizes the given information:

	Rate	Time	Rate × Time = Distance
Jogger	16 km/hr	2 hr	2(16) = 32 km
Bicyclist	x km/hr	2 hr	2x km

Since the distance the bicyclist has traveled is 19.2 kilometers more than the jogger has run during the 2 hours,

$$2x = 32 + 19.2$$
$$2x = 51.2$$
$$x = 25.6.$$

Therefore, the speed of the bicyclist is 25.6 kilometers per hour.

Check: 2(25.6) = 51.2 = 32 + 19.2.   ∎

Problems concerning a job that is done at a constant rate can often be solved by using the following principle:

> If a job can be done in time t, then $1/t$ of the job can be done in one unit of time.

Example 7   At a factory, smokestack A pollutes the air 1.25 times as fast as smokestack B. How long would it take smokestack B, operating alone, to pollute the air by as much as both smokestacks do in 20 hours?

Solution   Here the job in question is polluting the air by as much as both smokestacks do in 20 hours. Let t be the time in hours required for smokestack B operating alone to do this job. Then $1/t$ of the job is done by smokestack B in one hour, and $1.25(1/t)$ of the job is done by smokestack A in one hour. The two smokestacks together accomplish

$$\frac{1}{t} + 1.25\left(\frac{1}{t}\right) = \frac{2.25}{t}$$

of the job in one hour. We also know that both smokestacks accomplish $\frac{1}{20}$ of the job in one hour. Therefore,

$$\frac{2.25}{t} = \frac{1}{20}.$$

Solving this equation, we find that

$$20\!\!\!\not{t}\,\frac{2.25}{\not{t}} = 20\!\!\!\not{t}\,\frac{1}{\not{20}}$$

$$20(2.25) = t$$

$$45 = t.$$

Therefore, it requires 45 hours for smokestack B to do the job alone.

Check: Is it true that $\frac{1}{45} + 1.25(\frac{1}{45}) = \frac{1}{20}$? Yes.

Problem Set 2.2

Ⓒ In many of the following problems, a calculator may be useful to expedite the arithmetic.

1. The difference between two numbers is 12. If 2 is added to seven times the smaller number, the result is the same as if 2 is subtracted from three times the larger. Find the numbers.

2. Psychologists define the intelligence quotient (IQ) of a person to be 100 times the person's mental age divided by the chronological age. What is the chronological age of a person with an IQ of 150 and a mental age of 18?

3. At the end of the model year, a car dealer advertises that the list prices on all of last year's models have been discounted by 20%. What was the original list price of a car that has a discounted price of $6800?

4. A retail outlet sells wood-burning stoves for $675. At this price, the profit on the stove is one-third of its cost to the retailer. What is the amount of the retailer's profit on the stove? [*Hint:* cost + profit = selling price.]

5. A person invests part of $75,000 in a certificate that yields 8.5% simple annual interest, and the rest in a certificate that yields 9.2% simple annual interest. At the end of the year, the combined interest on the two certificates is $6606. How much was invested in each certificate?

6. To reduce their income tax, Jay and Joan invest a total of $140,000 in two municipal bonds, one that pays 6% tax-free simple annual interest and one that pays 6.5%. The total nontaxable income from both investments at the end of one year is $8775. How much was invested in each bond?

7. A family takes advantage of a state income tax credit of 15% of the cost of installing solar-heating equipment and 8% of the cost of upgrading insulation in their home. After spending a total of $6510 on insulation and solar heating, the family receives a state income tax credit of $854. How much was spent on solar heating and how much on insulation?

8. The manager of a trust fund invests $210,000 in three enterprises. She invests three times as much at 8% as she does at 9%, and commits the rest at 10%. Her total annual income from the three investments will be $17,850. How much does she invest at each rate?

9. A factory pays time and a half for all hours worked above 40 hours per week. An employee who makes $7.30 per hour grossed $478.15 in one week. How many hours did the employee work that week?

10. Because of inflation, the price of units in a condominium increases by 3% in April. The price increases again, by 2%, in July. What was the price of a unit before the April increase if its price after the July increase is $78,795?

11. A petroleum distributor has two gasohol storage tanks, the first containing 9% alcohol and the second containing 12% alcohol. An order is received for 300,000 gallons of gasohol containing 10% alcohol. How can this order be filled by mixing gasohol from the two storage tanks?

12. The cooling system of an automobile engine holds 16 liters of fluid. The system is filled with a mixture of 80% water and 20% antifreeze. It is necessary to increase the amount of antifreeze to 40%. This is to be done by draining some of the mixture and adding pure antifreeze to bring the total amount of fluid back up to 16 liters. How much of the mixture should be drained?

13. To generate hydrogen in a chemistry laboratory, a 40% solution of sulfuric acid is needed. How many milliters of water must be mixed with 25 milliliters of an 88% solution of sulfuric acid to dilute it to the required 40% of acid?

14. At a certain factory, twice as many men as women apply for work. If 5% of the people who apply are hired and 3% of the men who apply are hired, what percent of the women who apply are hired?

15. A bill of $7.45 was paid with 32 coins: half dollars, quarters, dimes, and nickels. If there were seven more dimes than nickels and twice as many dimes as quarters, how many coins of each type were used? [*Hint:* Let x be the number of quarters.]

16. Three brothers, Joe, Jamal, and Gus, decided to contribute toward a present for their mother. Their father agreed to match their combined contribution and purchase the gift. Joe's contribution was entirely in nickels, Jamal's was in dimes, and Gus's was in quarters. Jamal contributed three times as many coins as Gus, and Joe contributed 10 more coins than Jamal. The gift cost $10.80. How much did each person contribute?

17. A cross-country skier starts from a certain point and travels at a constant speed of 4 kilometers per hour. A snowmobile starts from the same point 45 minutes later, follows the skier's tracks at a constant speed, and catches up to the skier in 10 minutes. What is the speed of the snowmobile in kilometers per hour?

18. Two airplanes leave an airport at the same time and travel in opposite directions. One plane is traveling 64 kilometers per hour faster than the other. After 2 hours, they are 3200 kilometers apart. How fast is each plane traveling?

19. A jogger takes 3.5 minutes to run the same distance that a second jogger can run in 3 minutes. What is this distance, if the second jogger runs 2 feet per second faster than the first jogger?

20. A driver plans to average 50 miles per hour on a trip from A to B. Her average speed for the first half of the distance from A to B is 45 miles per hour. How fast must she drive for the rest of the way?

21. Factory A pollutes a lake twice as fast as factory B. The two factories operating together emit a certain amount of pollutant in 18 hours. How long would it take for factory A, operating alone, to produce the same amount of pollutant?

22. Student activists at Curmudgeon College have planned to distribute leaflets on the campus. The leaflets were to be run off on two machines, one electrically driven and the other operated by a hand crank. The electric machine produces copies four times as fast as the hand-cranked machine. The students figured that with both machines operating together they could run off the number of leaflets they needed in a total of 1.5 hours. However, because of a power blackout, they can use only the hand-cranked machine. How long will it take to run off the leaflets?

23. A computer can do a biweekly payroll in 10 hours. If a second computer is added, the two computers can do the job in only 3 hours. How long would it take the second computer alone to do the payroll?

24. A solar collector can generate 50 Btu's in 8 minutes. (One Btu is the amount of energy necessary to raise the temperature of one pound of water by one degree Fahrenheit.) A second solar collector can generate 50 Btu's in 5 minutes. The first collector is operated by itself for 1 minute, then the second collector is activated and both operate together. How long, after the second collector is activated, will it take to generate a total of 50 Btu's?

25. The length of a rectangular playground is twice its width and the perimeter is 900 meters. Find the length and width of the playground.

26. The circumference of the earth at the equator is 1.315×10^8 feet. Suppose that a steel belt is fitted tightly around the equator. If an additional 10 feet is added onto this belt and the slack is uniformly distributed around the earth, would you be able to crawl under the belt?

27. The Fahrenheit temperature corresponding to a particular Celsius temperature can be found by adding 32 to $\frac{9}{5}$ of the Celsius temperature. Find the temperature at which the reading is the same on both the Fahrenheit and the Celsius scales.

28. A commuter is picked up by her husband at the train station every afternoon. The husband leaves the house at the same time every day, always drives at the same speed, and regularly arrives at the station just as his wife's train pulls in. One day she takes a different train and arrives at the station one hour earlier than usual. She starts immediately to walk home at a constant speed. Her husband sees her along the road, picks her up, and drives her the rest of the way home. They arrive there 10 minutes earlier than usual. How many minutes did she spend walking?

29. In Problem 27, when is the Celsius reading three times the Fahrenheit reading?

30. In Problem 28, if the wife walks 4 miles an hour, how fast does the husband drive?

31. The primary (P) wave of an earthquake travels 1.7 times faster than the secondary (S) wave. Assuming that the S wave travels at 275 kilometers per minute and that the P wave is recorded at a seismic station 5.07 minutes before the S wave, how far from the station was the earthquake?

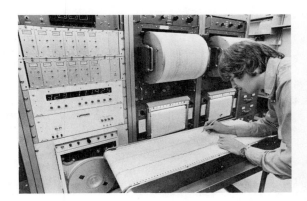

Seismograph

2.3 SECOND-DEGREE OR QUADRATIC EQUATIONS

A **second-degree** or **quadratic equation** in x is written in standard form as

$$ax^2 + bx + c = 0, \qquad \text{with } a \neq 0.$$

We discuss three methods for solving quadratic equations: *factoring*, *completing the square*, and using the *quadratic formula*.

Solution by Factoring

When a quadratic equation is in standard form, it may be possible to factor its left side as a product of two first-degree polynomials. The equation can then be solved by setting each factor equal to zero and solving the resulting first-degree equations. This procedure is justified by the fact that a product of real numbers is zero if and only if at least one of the factors is zero. (See the zero-factor properties on page 5.)

In Examples 1 to 3, solve each equation by factoring.

Example 1 $15x^2 + 14x = 8$

Solution   We begin by subtracting 8 from both sides of the equation to change it into standard form

$$15x^2 + 14x - 8 = 0.$$

Factoring the polynomial on the left, we obtain

$$(3x + 4)(5x - 2) = 0.$$

Now we set each factor equal to zero and solve the resulting first-degree equations:

$$
\begin{array}{c|c}
3x + 4 = 0 & 5x - 2 = 0 \\
x = -\frac{4}{3} & x = \frac{2}{5}
\end{array}
$$

Therefore, the roots are $-\frac{4}{3}$ and $\frac{2}{5}$.   ■

Example 2 $\dfrac{x - 6}{3x + 4} - \dfrac{2x - 3}{x + 2} = 0$

Solution   We begin by multiplying both sides of the equation by $(3x + 4)(x + 2)$, the LCD of the two fractions:

$$(3x+4)(x + 2)\frac{x - 6}{3x+4} - (3x + 4)(x+2)\frac{2x - 3}{x+2} = 0$$

$$(x + 2)(x - 6) - (3x + 4)(2x - 3) = 0$$

$$x^2 - 4x - 12 - (6x^2 - x - 12) = 0$$

$$x^2 - 4x - 12 - 6x^2 + x + 12 = 0$$

$$-5x^2 - 3x = 0$$

$$5x^2 + 3x = 0.$$

Factoring the left side of the last equation and setting the factors equal to zero, we obtain:

$$x(5x + 3) = 0$$

$$x = 0 \qquad \qquad 5x + 3 = 0$$

$$x = -\tfrac{3}{5}$$

Because we multiplied both sides of the original equation by the expression $(3x + 4)(x + 2)$, which contains the unknown, we must check to be sure that $x = 0$ and $x = -\tfrac{3}{5}$ aren't extraneous roots. Substituting $x = 0$ in the original equation, we obtain

$$\frac{0 - 6}{3(0) + 4} - \frac{2(0) - 3}{0 + 2} = \frac{-6}{4} - \frac{-3}{2} = -\frac{3}{2} + \frac{3}{2} = 0,$$

so $x = 0$ is a solution. Substituting $x = -\tfrac{3}{5}$, we have

$$\frac{(-\tfrac{3}{5}) - 6}{3(-\tfrac{3}{5}) + 4} - \frac{2(-\tfrac{3}{5}) - 3}{(-\tfrac{3}{5}) + 2} = \frac{-33}{11} - \frac{-21}{7} = -3 + 3 = 0,$$

so $x = -\tfrac{3}{5}$ is also a solution. Hence, the roots are $-\tfrac{3}{5}$ and 0.   ■

Example 3 $x^2 - 4x + 4 = 0$

Solution   Factoring, and setting the factors equal to zero, we have

$$x^2 - 4x + 4 = 0$$

$$(x - 2)(x - 2) = 0$$

$$x - 2 = 0 \qquad \qquad x - 2 = 0$$

$$x = 2 \qquad \qquad x = 2$$

Therefore, there is only one solution, namely, $x = 2$.   ■

When the two first-degree polynomials obtained by factoring have the same root, as in Example 3 above, we call the result a **double root**.

Solution by Completing the Square

A second method for solving quadratic equations is based on the idea of a **perfect square**—a polynomial that is the square of another polynomial. For example, $x^2 + 6x + 9$ is a perfect square because it is the square of $x + 3$. The polynomial $x^2 + 6x$ isn't a perfect square, but if we add 9 to it, we get the perfect square $x^2 + 6x + 9$.

More generally, by adding $(k/2)^2$ to $x^2 + kx$ we obtain a perfect square:

$$x^2 + kx + \left(\frac{k}{2}\right)^2 = \left(x + \frac{k}{2}\right)^2.$$

Thus, to create a perfect square from an expression of the form $x^2 + kx$:

Add the square of half the coefficient of x.

This is called **completing the square.***

Example 4   Complete the square by adding a constant to each expression.

(a) $x^2 + 8x$ **(b)** $x^2 - 4x$ **(c)** $x^2 - 3x$

Solution

(a) $x^2 + 8x + \left(\dfrac{8}{2}\right)^2 = x^2 + 8x + 16 = (x + 4)^2$

(b) $x^2 - 4x + \left(\dfrac{-4}{2}\right)^2 = x^2 - 4x + 4 = (x - 2)^2$

(c) $x^2 - 3x + \left(\dfrac{-3}{2}\right)^2 = x^2 - 3x + \dfrac{9}{4} = \left(x - \dfrac{3}{2}\right)^2$

The following example illustrates how to solve a quadratic equation by completing the square.

Example 5   Solve the equation $2x^2 - 6x - 5 = 0$.

Solution   We begin by adding 5 to both sides:

$$2x^2 - 6x = 5.$$

Next, we divide both sides of the equation by 2 so that the coefficient of x^2 will be 1:

$$x^2 - 3x = \tfrac{5}{2}.$$

* Note that this method of completing the square works only *when the coefficient of x^2 is 1.*

Now we can complete the square for the expression on the left by adding $\left(\dfrac{-3}{2}\right)^2 = \dfrac{9}{4}$ to both sides of the equation:

$$x^2 - 3x + \frac{9}{4} = \frac{5}{2} + \frac{9}{4}$$

$$\left(x - \frac{3}{2}\right)^2 = \frac{19}{4}.$$

It follows that $x - \dfrac{3}{2}$ is a square root of $\dfrac{19}{4}$. But there are two square roots of $\dfrac{19}{4}$: the principal square root $\sqrt{\dfrac{19}{4}} = \dfrac{\sqrt{19}}{2}$, and its negative $-\sqrt{\dfrac{19}{4}} = -\dfrac{\sqrt{19}}{2}$. Therefore,

$$x - \frac{3}{2} = \frac{\sqrt{19}}{2} \qquad \text{or else} \qquad x - \frac{3}{2} = -\frac{\sqrt{19}}{2},$$

that is,

$$x = \frac{3}{2} + \frac{\sqrt{19}}{2} = \frac{3 + \sqrt{19}}{2} \qquad \text{or else} \qquad x = \frac{3}{2} - \frac{\sqrt{19}}{2} = \frac{3 - \sqrt{19}}{2}.$$

These two solutions can be written in the compact form

$$x = \frac{3 \pm \sqrt{19}}{2}.$$

$\blacksquare$

The Quadratic Formula

Let's apply the method of completing the square to solve for the unknown x in the literal equation

$$ax^2 + bx + c = 0, \qquad \text{with } a \neq 0.$$

First, we subtract c from both sides:

$$ax^2 + bx = -c.$$

Then, to prepare for completing the square, we divide both sides by a:

$$x^2 + \frac{b}{a}x = -\frac{c}{a}.$$

To complete the square, we add $\left[\dfrac{1}{2}\left(\dfrac{b}{a}\right)\right]^2 = \dfrac{b^2}{4a^2}$ to both sides:

$$x^2 + \frac{b}{a}x + \frac{b^2}{4a^2} = \frac{b^2}{4a^2} - \frac{c}{a}$$

$$\left(x + \frac{b}{2a}\right)^2 = \frac{b^2 - 4ac}{4a^2}.$$

It follows that

$$x + \frac{b}{2a} = \pm\sqrt{\frac{b^2 - 4ac}{4a^2}}$$

$$x = -\frac{b}{2a} \pm \frac{\sqrt{b^2 - 4ac}}{2a}.$$

Therefore, the roots of the quadratic equation $ax^2 + bx + c = 0$ can be found by using the **quadratic formula:**

$$x = \frac{-b \pm \sqrt{b^2 - 4ac}}{2a}.$$

Although the method of factoring is the quickest way to solve a quadratic equation when the factors are easily recognized, *we recommend using the quadratic formula in all other cases.* Quadratic equations are rarely solved by completing the square—this method was introduced primarily to show how the quadratic formula is derived.

Example 6   Use the quadratic formula to solve the equation $2x^2 - 5x + 1 = 0$.

Solution   The equation is in the standard form $ax^2 + bx + c = 0$, with $a = 2$, $b = -5$, and $c = 1$. Substituting these values into the quadratic formula, we obtain

$$x = \frac{-b \pm \sqrt{b^2 - 4ac}}{2a} = \frac{-(-5) \pm \sqrt{(-5)^2 - 4(2)(1)}}{2(2)}$$

$$= \frac{5 \pm \sqrt{25 - 8}}{4} = \frac{5 \pm \sqrt{17}}{4}.$$

In other words, the two roots are $\dfrac{5 - \sqrt{17}}{4}$ and $\dfrac{5 + \sqrt{17}}{4}$.   ∎

Ⓒ **Example 7**   Using the quadratic formula and a calculator, find the roots of the quadratic equation $-1.32x^2 + 2.78x + 9.37 = 0$. Round off the answers to two decimal places.

Solution   According to the quadratic formula, the roots are

$$x = \frac{-2.78 - \sqrt{(2.78)^2 - 4(-1.32)(9.37)}}{2(-1.32)} \approx 3.92$$

and $$x = \frac{-2.78 + \sqrt{(2.78)^2 - 4(-1.32)(9.37)}}{2(-1.32)} \approx -1.81.$$ ∎

Quadratic equations have many applications in the sciences, business, economics, medicine, and engineering.

Example 8   A telephone company is placing telephone poles along a road. If the distance between successive poles were increased by 10 meters, 5 fewer poles per kilometer would be required. How many telephone poles is the company now placing along each kilometer of the road?

Solution   Let x be the number of poles now being placed along each kilometer. Then the distance between successive poles is

$$\frac{1}{x} \text{ kilometer} = \frac{1000}{x} \text{ meters.}$$

If the distance between poles were increased by 10 meters, then the (new) number of poles per kilometer would be $x - 5$. Therefore, the (new) distance between poles would be

$$\frac{1}{x - 5} \text{ kilometer} = \frac{1000}{x - 5} \text{ meters.}$$

We know that this new distance between poles would be 10 meters more than the current distance, so the new distance would also be given by $(1000/x) + 10$ meters. Therefore,

$$\frac{1000}{x - 5} = \frac{1000}{x} + 10.$$

We multiply both sides of this equation by the LCD $x(x - 5)$, and simplify:

$$x(x-5)\frac{1000}{x-5} = x(x - 5)\frac{1000}{x} + 10x(x - 5)$$

$$1000x = 1000x - 5000 + 10x^2 - 50x$$

$$10x^2 - 50x - 5000 = 0$$

$$x^2 - 5x - 500 = 0.$$

Using the quadratic formula, with $a = 1$, $b = -5$, and $c = -500$, we find that

$$x = \frac{-b \pm \sqrt{b^2 - 4ac}}{2a} = \frac{-(-5) \pm \sqrt{(-5)^2 - 4(1)(-500)}}{2(1)}$$

$$= \frac{5 \pm \sqrt{25 + 2000}}{2} = \frac{5 \pm \sqrt{2025}}{2} = \frac{5 \pm 45}{2}.$$

The two solutions of the quadratic equation are

$$x = \frac{5 - 45}{2} = -20 \quad \text{and} \quad x = \frac{5 + 45}{2} = 25.$$

Since we cannot have a negative number of poles per kilometer, only the second solution has meaning, so we conclude that 25 poles are currently being placed per kilometer. ∎

The Discriminant and Complex Roots

The expression $b^2 - 4ac$, which appears under the radical sign in the quadratic formula

$$x = \frac{-b \pm \sqrt{b^2 - 4ac}}{2a},$$

is called the **discriminant** of the quadratic equation

$$ax^2 + bx + c = 0.$$

If a, b, and c are real numbers, you can use the algebraic sign of the discriminant to determine the number and the nature of the roots of the quadratic equation.

If $b^2 - 4ac > 0$, the equation has two real and unequal roots.

(See Examples 6 and 7.)

If $b^2 - 4ac = 0$, the equation has only one root—a double root.

(See Example 3.)

If $b^2 - 4ac < 0$, the equation has no real root—its roots are two complex numbers that are complex conjugates of each other.

(See Example 10 below.)

Example 9   Use the discriminant to determine the nature of the roots of each quadratic equation without actually solving it.

(a) $5x^2 - x - 3 = 0$ **(b)** $9x^2 + 42x + 49 = 0$ **(c)** $x^2 - x + 1 = 0$

Solution   **(a)** Here $a = 5$, $b = -1$, $c = -3$, and $b^2 - 4ac = (-1)^2 - 4(5)(-3) = 61 > 0$; hence, there are two unequal real roots.

(b) Here $a = 9$, $b = 42$, $c = 49$, and $b^2 - 4ac = 42^2 - 4(9)(49) = 0$; hence, the equation has just one root—a double root—and this root is a real number.

(c) Here $a = 1$, $b = -1$, $c = 1$, and $b^2 - 4ac = (-1)^2 - 4(1)(1) = -3 < 0$; hence, the equation has no real root.   ■

Example 10   Use the quadratic formula to find the roots of the quadratic equation $x^2 - x + 1 = 0$.

Solution   Using the quadratic formula, with $a = 1$, $b = -1$, and $c = 1$, we have

$$x = \frac{-(-1) \pm \sqrt{(-1)^2 - 4(1)(1)}}{2(1)}$$

$$= \frac{1 \pm \sqrt{-3}}{2}$$

$$= \frac{1 \pm \sqrt{3}i}{2} = \frac{1}{2} \pm \frac{\sqrt{3}}{2} i.$$

Thus, the roots are the complex conjugates

$$\frac{1}{2} - \frac{\sqrt{3}}{2} i \quad \text{and} \quad \frac{1}{2} + \frac{\sqrt{3}}{2} i.$$

A more detailed and more general discussion of complex roots of polynomial equations can be found in Section 4.7.

Problem Set 2.3

In Problems 1 to 24, solve each equation by factoring.

1. $x^2 - 7x = 0$

2. $2x^2 - 5x = 0$

3. $3t^2 - 48 = 0$

4. $9 - y^2 = 2y^2$

5. $x^2 + 2x = 3$

6. $z^2 - 2z = 35$

7. $x^2 - 4x = 21$

8. $2t^2 + 5t = -3$

9. $2z^2 - 7z - 15 = 0$

10. $10r^2 + 19r + 6 = 0$

11. $6y^2 - 13y = 5$

12. $54z^2 - 9z = 30$

13. $15x^2 + 4 = 23x$

14. $24u^2 + 94u = 25$

15. $(8x + 19)x = 27$

16. $30(y^2 + 1) = 61y$

17. $25x - 40 + \dfrac{16}{x} = 0$

18. $36 + \dfrac{60}{s} = \dfrac{-25}{s^2}$

19. $x - 16 = \dfrac{105}{x}$

20. $\dfrac{10y + 19}{y} = \dfrac{15}{y^2}$

21. $\dfrac{5}{x + 4} - \dfrac{3}{x - 2} = 4$

22. $\dfrac{2y + 11}{2y + 8} = \dfrac{3y - 1}{y - 1}$

23. $\dfrac{2y - 5}{2y + 1} + \dfrac{6}{2y - 3} = \dfrac{7}{4}$

24. $\dfrac{t - 4}{t + 1} - \dfrac{15}{4} = \dfrac{t + 1}{t - 4}$

In Problems 25 to 30, complete the square by adding a constant to each expression.

25. $x^2 + 6x$

26. $x^2 - 6x$

27. $x^2 - 5x$

28. $x^2 + \dfrac{b}{a} x$

29. $x^2 + \dfrac{3}{4} x$

30. $x^2 + \sqrt{2} x$

In Problems 31 to 36, complete the square to solve each equation.

31. $x^2 + 4x - 15 = 0$

32. $x^2 - 5x - 5 = 0$

33. $x^2 + 4 = 6x$

34. $2y^2 + y = 3$

35. $3r^2 + 6r = 4$

36. $5u^2 + 9u = 3$

In Problems 37 to 48, use the quadratic formula to solve each equation.

37. $5x^2 - 7x - 6 = 0$

38. $6x^2 - x - 2 = 0$

39. $3x^2 + 5x + 1 = 0$

40. $12y^2 + y - 1 = 0$

41. $2x^2 - x - 2 = 0$

42. $4u^2 - 11u + 3 = 0$

43. $4y^2 - 3y - 3 = 0$

44. $5x^2 + 17x - 3 = 0$

45. $9x^2 - 6x + 1 = 0$

46. $\dfrac{2}{3}x^2 - \dfrac{8}{9}x = 1$

47. $\dfrac{x+2}{x} + \dfrac{x}{x-2} = 5$

48. $\dfrac{7u+4}{u^2 - 6u + 8} - \dfrac{5}{2-u} = \dfrac{u+5}{u-4}$

Ⓒ In Problems 49 to 52, use the quadratic formula and a calculator to find the roots of each quadratic equation. Round off all answers to two decimal places.

49. $44.04x^2 + 64.72x - 31.23 = 0$

50. $8.85x^2 - 71.23x + 94.73 = 0$

51. $4.59x^2 - 90.29x + 118.85 = 0$

52. $-1.47x^2 + 9.06x + 6.57\pi = 0$

In Problems 53 to 64, solve each quadratic equation by any method you wish.

53. $x^2 = 5$

54. $25y^2 - 20y + 4 = 0$

55. $4x(2x - 1) = 3$

56. $2x^2 + 7x + 6 = 0$

57. $x^2 + 6x = -9$

58. $3z^2 - 2z = 5$

59. $x^2 + 4x + 1 = 0$

60. $15 - 7u - 4u^2 = 0$

61. $9t^2 - 6t + 1 = 0$

62. $0.2x^2 - 1.2x + 1.7 = 0$

63. $z^2 + 2z = 8$

64. $8t^2 - 10t + 3 = 0$

65. The staff members of an athletic department at a college agreed to contribute equal amounts to make up a scholarship fund of $200. Since then, two new members have been added to the staff; as a result, each member's share has been reduced by $5. How many members are now on the staff?

Ⓒ **66.** A cable television company plans to begin operations in a small town. The company foresees that about 600 people will subscribe to the service if the price per subscriber is $5 per month, but that for each 5-cent increase in the monthly subscription price, 4 of the original 600 people will decide not to subscribe. The company begins operations, and its total revenue for the first month is $1500. How many people have subscribed to the cable television service?

67. A gardener sets 180 plants in rows. Each row contains the same number of plants. If there were 40 more plants in each row, the gardener would need 6 fewer rows. How many rows are there?

68. In a medical laboratory, the quantity x milligrams of antigen present during an antigen–antibody reaction is related to the time t in minutes required for the precipitation of a fixed amount of antigen–antibody by the equation

$$3t = 140 - 50x + 5x^2.$$

Find x if $t = 20$ minutes and $x > 2$.

69. In order to support a solar collector at the correct angle, the roof trusses for a house are designed as right triangles. Rafters form the right angle, and the base of the truss is the hypotenuse (Figure 1). If the rafter on the same side as the solar collector is 10 feet shorter than the other rafter and if the base of each truss is 50 feet long, how long are each of the rafters? [*Hint:* Use the Pythagorean theorem.*]

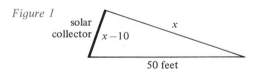

Figure 1

solar collector

$x - 10$

x

50 feet

70. Carlos, who is training to run in the Boston Marathon, runs 18 miles every Saturday afternoon. His goal is to cut his running time by one-half hour; he figures that to accomplish this he will have to increase his average speed by 1.2 miles per hour. What is his current average speed for the 18 mile run?

Ⓒ **71.** Neglecting air resistance, a projectile shot straight upward with an initial velocity of v meters per second will be at a height of $vt - 4.9t^2$ meters t seconds later. If $v = 30$ meters per second, how long will it take the projectile to reach a height of 28 meters? (Use a calculator and round off your answer to the nearest tenth of a second.)

* See the Appendix on Analytical Geometry for a review of basic geometry, including the Pythagorean Theorem.

In Problems 72 to 74, a formula from applied mathematics is given. Solve each equation for the indicated unknown variable. State any necessary restrictions on the values of the remaining variables.

72. $s = vt + \frac{1}{2}gt^2$ for t (mechanics)

73. $LI^2 + RI + \dfrac{1}{C} = 0$ for I (electronics)

74. $P = EI - RI^2$ for I (electrical engineering)

In Problems 75 to 80, use the discriminant to determine the nature of the roots of each quadratic equation without solving it.

75. $x^2 - 5x - 7 = 0$

76. $6t^2 - 2 = 7t$

77. $9x^2 - 6x + 1 = 0$

78. $4x^2 + 20x + 25 = 0$

79. $6y^2 + y + 3 = 0$

80. $10r^2 + 2r + 8 = 0$

In Problems 81 to 88, use the quadratic formula to find the complex roots of each quadratic equation.

81. $6x^2 + x + 3 = 0$

82. $10x^2 + 2x + 8 = 0$

83. $5x^2 - 4x + 2 = 0$

84. $4x^2 + 4x + 5 = 0$

85. $x^2 + x + 1 = 0$

86. $x^2 - 2x + \frac{3}{2} = 0$

87. $4x^2 - 2x + 1 = 0$

88. $x^2 - 6x + 13 = 0$

2.4 MISCELLANEOUS EQUATIONS

In this section, we discuss special types of equations that can be solved using slight variations of the methods presented in previous sections. Here we consider only *real* roots of equations.

Equations of the Form $x^p = a$

If the exponent p is an integer, you can solve the equation $x^p = a$ as shown in the following example.

Example 1   Solve each equation.

(a) $x^3 = 5$ (b) $x^{-3} = 5$ (c) $x^4 = 5$ (d) $x^4 = -5$

Solution   (a) Since the exponent is odd, there is just one solution, $x = \sqrt[3]{5}$.

(b) We begin by rewriting the equation in the equivalent form

$$\frac{1}{x^3} = 5 \quad \text{or} \quad x^3 = \frac{1}{5}.$$

The solution is

$$x = \sqrt[3]{\frac{1}{5}} = \frac{1}{\sqrt[3]{5}} = \frac{\sqrt[3]{25}}{5}.$$

(c) Since the exponent is even, there are two solutions, $x = \sqrt[4]{5}$ and $x = -\sqrt[4]{5}$.

(d) A real number raised to an even power cannot be negative, so the equation $x^4 = -5$ has no real root.

The following example shows how to solve the equation $x^p = a$ when the exponent p is the reciprocal of an integer.

Example 2   Solve each equation.

    (a) $x^{1/6} = 5$ **(b)** $x^{1/3} = -5$ **(c)** $x^{-1/3} = 5$ **(d)** $x^{1/6} = -5$

Solution   **(a)** The equation can be rewritten as $\sqrt[6]{x} = 5$, so that $x = 5^6 = 15{,}625$.
(b) Here we have $\sqrt[3]{x} = -5$, so that $x = (-5)^3 = -125$.
(c) We begin by rewriting the equation in the equivalent form

$$\frac{1}{x^{1/3}} = 5 \quad \text{or} \quad x^{1/3} = \frac{1}{5},$$

that is, $\sqrt[3]{x} = \frac{1}{5}$. Thus, $x = (\frac{1}{5})^3 = \frac{1}{125}$.
(d) The equation can be rewritten as $\sqrt[6]{x} = -5$. Since a principal sixth root cannot be negative, the equation has no real solution.   ■

If n and m are positive integers and the fraction n/m is reduced to lowest terms, an equation of the form

$$x^{n/m} = a$$

can be rewritten as

$$\sqrt[m]{x^n} = a$$

and solved using the methods illustrated above.

Example 3   Solve each equation.

    (a) $x^{3/2} = -8$ **(b)** $(t-3)^{2/5} = 4$ **(c)** $x^{-5/2} = -\frac{1}{4}$

Solution   **(a)** The equation is equivalent to $\sqrt{x^3} = -8$. Because a principal square root of a real number cannot be negative, the equation has no real solution.
(b) The equation can be rewritten as $\sqrt[5]{(t-3)^2} = 4$ or $(t-3)^2 = 4^5$. Hence,

$$t - 3 = \pm\sqrt{4^5} = \pm(\sqrt{4})^5 = \pm 2^5 = \pm 32.$$

Therefore, $t = 3 \pm 32$, and the two roots are $t = -29$ and $t = 35$.
(c) We begin by rewriting the equation as

$$\frac{1}{x^{5/2}} = -\frac{1}{4} \qquad \text{or} \qquad x^{5/2} = -4.$$

The last equation is equivalent to $\sqrt{x^5} = -4$, so it has no real solution.   ■

Radical Equations

An equation in which the unknown appears in a radicand is called a **radical equation.** For instance,

$$\sqrt[4]{x^2 - 5} = x + 1 \qquad \text{and} \qquad \sqrt{3t + 7} + \sqrt{t + 2} = 1$$

are radical equations.

To solve a radical equation, begin by isolating the most complicated radical expression on one side of the equation, and then eliminate the radical by raising both sides of the equation to a power equal to the index of the radical. You may have to repeat this technique in order to eliminate all radicals. When the equation is radical-free, simplify and solve it. Since extraneous roots may be introduced when both sides of an equation are raised to an even power, all roots must be checked in the original equation whenever a radical with an *even index* is involved.

In Examples 4 and 5, solve each equation.

Example 4 $\sqrt[3]{x - 1} - 2 = 0$

Solution   To isolate the radical, we add 2 to both sides of the equation:

$$\sqrt[3]{x - 1} = 2.$$

Now we raise both sides to the power 3 and obtain

$$(\sqrt[3]{x - 1})^3 = 2^3 \qquad \text{or} \qquad x - 1 = 8.$$

It follows that $x = 9$. Since we did not raise both sides of the equation to an even power, it isn't necessary to check our solution, but it's good practice to do so anyway. Substituting $x = 9$ in the original equation, we obtain

$$\sqrt[3]{9 - 1} - 2 = \sqrt[3]{8} - 2 = 2 - 2 = 0;$$

hence, the solution $x = 9$ is correct.   ∎

Example 5 $\sqrt{3t + 7} + \sqrt{t + 2} = 1$

Solution   We add $-\sqrt{t + 2}$ to both sides of the equation to isolate $\sqrt{3t + 7}$ on the left side. Thus,

$$\sqrt{3t + 7} = 1 - \sqrt{t + 2}.$$

Now we square both sides of the equation to obtain

$$(\sqrt{3t + 7})^2 = (1 - \sqrt{t + 2})^2$$
$$3t + 7 = 1 - 2\sqrt{t + 2} + (t + 2).$$

The equation still contains a radical, so we simplify and isolate this radical:

$$2t + 4 = -2\sqrt{t + 2}$$
$$-(t + 2) = \sqrt{t + 2}.$$

Again, we square both sides, so

$$[-(t + 2)]^2 = (\sqrt{t + 2})^2$$
$$t^2 + 4t + 4 = t + 2$$
$$t^2 + 3t + 2 = 0$$
$$(t + 2)(t + 1) = 0$$

$$t + 2 = 0 \qquad\qquad t + 1 = 0$$
$$t = -2 \qquad\qquad\quad t = -1$$

Check: For $t = -2$,

$$\sqrt{3t + 7} + \sqrt{t + 2} = \sqrt{3(-2) + 7} + \sqrt{-2 + 2} = \sqrt{1} + \sqrt{0} = 1;$$

hence, $t = -2$ is a solution. For $t = -1$,

$$\sqrt{3t + 7} + \sqrt{t + 2} = \sqrt{3(-1) + 7} + \sqrt{-1 + 2} = \sqrt{4} + \sqrt{1} \neq 1.$$

Therefore, $t = -1$ is an extraneous root that was introduced by squaring both sides of the equation. The only solution is $t = -2$. ∎

Equations of Quadratic Type

An equation with x as the unknown is said to be of **quadratic type** if it can be rewritten in the form

$$au^2 + bu + c = 0,$$

where u is an expression containing x.

Example 6   Show that the equation $x^4 - 5x^2 + 4 = 0$ is of quadratic type.

Solution   The given equation can be rewritten as

$$(x^2)^2 - 5(x^2) + 4 = 0,$$

so it is of quadratic type with $u = x^2$. ∎

Further examples of equations of quadratic type are

$$x^{2/3} - x^{1/3} - 12 = 0 \qquad (u = x^{1/3}),$$
$$(x^2 - x)^2 - 8(x^2 - x) + 12 = 0 \qquad (u = x^2 - x),$$
$$x^2 + 3x + \sqrt{x^2 + 3x} = 6 \qquad (u = \sqrt{x^2 + 3x}).$$

If such an equation is first solved for u, any resulting values can be used to solve for x; thus the solution of the original equation can be found.

In Examples 7 and 8, solve each equation.

Example 7 $x^{2/3} - x^{1/3} - 12 = 0$

Solution   The given equation can be rewritten as

$$(x^{1/3})^2 - x^{1/3} - 12 = 0.$$

Let $u = x^{1/3}$. Then the equation becomes

$$u^2 - u - 12 = 0$$

$$(u - 4)(u + 3) = 0.$$

Therefore, $u = 4$ or $u = -3$. Since $u = x^{1/3}$, we have

$$x^{1/3} = 4 \qquad \qquad x^{1/3} = -3$$

$$x = 4^3 = 64 \qquad \qquad x = (-3)^3 = -27$$

Hence, the two solutions are $x = 64$ and $x = -27$.   ■

Example 8 $t^2 + 3t - 2 + \sqrt{t^2 + 3t - 2} - 20 = 0$

Solution   Let $u = \sqrt{t^2 + 3t - 2}$. Then the equation becomes

$$u^2 + u - 20 = 0.$$

$$(u - 4)(u + 5) = 0.$$

Therefore, $u = 4$ or $u = -5$. Since $u = \sqrt{t^2 + 3t - 2}$, we have

$$\sqrt{t^2 + 3t - 2} = 4 \qquad \text{or} \qquad \sqrt{t^2 + 3t - 2} = -5.$$

The second equation has no roots because a principal square root cannot be negative. Squaring both sides of the first equation, we have

$$t^2 + 3t - 2 = 16$$

$$t^2 + 3t - 18 = 0$$

$$(t - 3)(t + 6) = 0.$$

Hence, $t = 3$ or $t = -6$. Because we squared both sides of an equation, we must check for extraneous roots. Substituting $t = 3$ in the original equation, we obtain

$$t^2 + 3t - 2 + \sqrt{t^2 + 3t - 2} - 20 = 3^2 + 3(3) - 2 + \sqrt{3^2 + 3(3) - 2} - 20$$

$$= 9 + 9 - 2 + \sqrt{9 + 9 - 2} - 20$$

$$= 16 + \sqrt{16} - 20 = 0,$$

so $t = 3$ is a solution. Substituting $t = -6$ in the original equation, we get

$$t^2 + 3t - 2 + \sqrt{t^2 + 3t - 2} - 20 = (-6)^2 + 3(-6) - 2$$
$$+ \sqrt{(-6)^2 + 3(-6) - 2} - 20$$
$$= 36 - 18 - 2 + \sqrt{36 - 18 - 2} - 20$$
$$= 16 + \sqrt{16} - 20 = 0,$$

so $t = -6$ is also a solution. Hence, the two solutions are $t = 3$ and $t = -6$. ∎

Nonquadratic Equations That Can Be Solved by Factoring

The method of solving an equation by rewriting it as an equivalent equation with zero on the right side, then factoring the left side and equating each factor to zero, is not limited to quadratic equations. Indeed, this method works whenever the required factorization can be accomplished.

In Examples 9 and 10, solve each equation by factoring.

Example 9 $4x^3 - 8x^2 - x + 2 = 0$

Solution   We begin by grouping the terms so the common factor $x - 2$ emerges:

$$4x^2(x - 2) - (x - 2) = 0$$
$$(x - 2)(4x^2 - 1) = 0$$
$$(x - 2)(2x - 1)(2x + 1) = 0.$$

Setting each factor equal to zero, we have

$$
\begin{array}{c|c|c}
x - 2 = 0 & 2x - 1 = 0 & 2x + 1 = 0 \\
x = 2 & x = \frac{1}{2} & x = -\frac{1}{2}
\end{array}
$$

Therefore, the three solutions are $x = 2$, $x = \frac{1}{2}$, and $x = -\frac{1}{2}$. ∎

Example 10 $x^{5/6} - 3x^{1/3} - 2x^{1/2} + 6 = 0$

Solution   Again, we group and factor:

$$x^{1/3}[x^{(5/6)-(1/3)} - 3] - 2[x^{1/2} - 3] = 0$$
$$x^{1/3}(x^{1/2} - 3) - 2(x^{1/2} - 3) = 0$$
$$(x^{1/2} - 3)(x^{1/3} - 2) = 0.$$

Setting each factor equal to zero, we have

$$
\begin{array}{c|c}
x^{1/2} - 3 = 0 & x^{1/3} - 2 = 0 \\
x^{1/2} = 3 & x^{1/3} = 2 \\
x = 3^2 = 9 & x = 2^3 = 8
\end{array}
$$

Therefore, the two solutions are $x = 9$ and $x = 8$. ∎

Problem Set 2.4

In this problem set, we consider only real roots of equations.

In Problems 1 to 24, solve each equation.

1. $x^2 = 9$

2. $x^4 = -16$

3. $x^6 = 2$

4. $(x - 1)^4 = 81$

5. $x^{-2} = 4$

6. $u^{-4} = \frac{1}{9}$

7. $x^3 = 27$

8. $x^5 = -3$

9. $x^7 = -4$

10. $t^{-3} = -\frac{1}{8}$

11. $x^{-5} = \frac{-1}{32}$

12. $x^{-7} = 3$

13. $y^{1/3} = 2$

14. $x^{-1/2} = -4$

15. $x^{-1/3} = -\frac{1}{3}$

16. $t^{-2/3} = 9$

17. $z^{5/2} = 3$

18. $(x - 3)^{3/2} = 8$

19. $(2t - 1)^{-2/5} = 4$

20. $(3z - 7)^{4/3} = -1$

21. $(3x)^{-5/3} = \frac{1}{32}$

22. $(x^{1/2})^{3/5} = \frac{1}{27}$

23. $(x^2 - 7x)^{4/3} = 16$

24. $(x^2 + 12x)^{2/3} = 9$

In Problems 25 to 46, solve each equation. Check for extraneous roots.

25. $\sqrt{2t + 9} = t - 13$

26. $\sqrt[3]{2y - 3} = \sqrt[3]{y + 1}$

27. $\sqrt{10 - x} - 8 = 2x$

28. $t = -\sqrt{19 - 2t} - 8$

29. $\sqrt{16 - m} + 4 = m$

30. $\sqrt{u + \sqrt{u - 4}} = 2$

31. $\sqrt[3]{2x + 3} = -1$

32. $\sqrt[5]{3x - 1} = 2$

33. $\sqrt[4]{y^4 + 2y - 6} = y$

34. $\sqrt[3]{x^3 + 2x^2 - 9x - 26} = x + 1$

35. $\sqrt{3t + 12} - 2 = \sqrt{2t}$

36. $\sqrt{5y + 1} = 1 + \sqrt{3y}$

37. $6\sqrt{x} - 3 = \sqrt{10x - 1}$

38. $2\sqrt{x - 1} - (x/2) = \frac{3}{2}$

39. $\sqrt{2y - 5} - \sqrt{y - 2} = 2$

40. $\sqrt{3t + 8} = 3\sqrt{t} - 2\sqrt{2}$

41. $\sqrt{5x + 6} = \sqrt{x + 1} + 1$

42. $(2\sqrt{n})^{-1} + (\sqrt{4n + 9})^{-1} = 9(\sqrt{16n^2 + 36n})^{-1}$

43. $\sqrt{2x + 5} - \sqrt{5x - 1} = \sqrt{2x - 4}$

44. $x(\sqrt{x^2 + 1})^{-1} = (4 - x)(\sqrt{x^2 - 8x + 17})^{-1}$

45. $\sqrt{5t - 10} = \sqrt{8t + 2} - \sqrt{3t + 12}$

46. $\dfrac{\sqrt{x + 1} + \sqrt{x - 1}}{\sqrt{x + 1} - \sqrt{x - 1}} = \dfrac{4x - 1}{2}$

In Problems 47 to 64, solve each equation.

47. $x^4 - 7x^2 + 12 = 0$

48. $6x^4 + 3 = 19x^2$

49. $(x^2 + x)^2 + (x^2 + x) - 6 = 0$

50. $(y^2 + y - 2)^2 + (y^2 + y - 2) = 20$

51. $t^{1/3} - 4t^{1/6} + 3 = 0$

52. $y^{-2/3} + 2y^{-1/3} = 8$

53. $x^{1/4} + 2x^{1/2} = 3$

54. $10t^{-2} - 9t^{-1} - 1 = 0$

55. $9x^{-4} - 145x^{-2} + 16 = 0$

56. $z^2 + z + \dfrac{12}{z^2 + z} = 8$

57. $2x^2 - 5x + 7 - 8\sqrt{2x^2 - 5x + 7} + 7 = 0$

58. $26 - 8x = 6x^2 - \sqrt{3x^2 + 4x + 1}$

59. $y^2 + 2y - 3\sqrt{y^2 + 2y + 4} = 0$

60. $t^2 + 6t - 6\sqrt{t^2 + 6t - 2} + 3 = 0$

61. $\sqrt{t + 20} - 4\sqrt[4]{t + 20} + 3 = 0$

62. $3(1 - x)^{1/6} + 2(1 - x)^{1/3} = 2$

63. $2r^2 - 3r - 2\sqrt{2r^2 - 3r} - 3 = 0$

64. $2s^2 + s - 4\sqrt{2s^2 + s + 4} = 1$

In Problems 65 to 72, solve each equation by factoring.

65. $x^3 + x^2 - x - 1 = 0$

66. $x^5 - 4x^3 - x^2 + 4 = 0$

67. $x^6 - x^4 + 4x^2 - 4 = 0$

68. $z^7 - 27z^4 - z^3 + 27 = 0$

69. $x^{5/6} + 2x^{1/2} - x^{1/3} - 2 = 0$

70. $x^{7/12} - 2x^{1/3} + 5x^{1/4} - 10 = 0$

71. $x + x^{2/3} - x^{1/3} - 1 = 0$

72. $4x^{3/2} - 8x - x^{1/2} + 2 = 0$

73. The formula for the lateral surface area A of a right circular cone with height h and base radius r is $A = \pi r \sqrt{r^2 + h^2}$ (Figure 1). Solve for r in terms of A and h.

Figure 1

74. Prove that the equation $x^3 = a$ has exactly one real solution, $x = \sqrt[3]{a}$, by rewriting the equation in the form $x^3 - (\sqrt[3]{a})^3 = 0$ and factoring.

75. Find the dimensions of a rectangle that has an area of 60 square centimeters and a diagonal that is 13 centimeters long.

76. The frequency ω of a certain electronic circuit is given by

$$\omega = \frac{1}{2\pi \sqrt{\dfrac{LC_1C_2}{C_1 + C_2}}}.$$

Solve for C_2 in terms of ω, L, and C_1.

77. The deflection d of a certain beam is given by $d = l\sqrt{a/(2l + a)}$. Solve for l in terms of a and d.

78. In meteorology the heat loss H (wind chill) is given by

$$H = (A + B\sqrt{V} - CV)(S - T).$$

Solve for the wind speed V in terms of H, the neutral skin temperature S, the air temperature T, and the constants A, B, and C.

2.5 INEQUALITIES

We introduced the inequality symbols $<$, $\leq$, $>$, and $\geq$, in Section 1.2. By writing one of these symbols between two expressions, we obtain an **inequality,** and the two expressions are called the **members** or **sides** of the inequality. By connecting more than two expressions with these symbols, we can form **compound inequalities** such as $x < y \leq z$ and $5 \geq 2x - 3 \geq 1$. A compound inequality can always be broken down into simple inequalities: For instance, $x < y \leq z$ means $x < y$ and $y \leq z$.

In Section 1.1 we presented the basic algebraic properties of the real numbers. All of our work up to now has been founded upon these properties. To deal with inequalities, we must introduce the following properties:

Basic Order Properties of the Real Numbers

Let a, b, and c denote real numbers.

    1. *The Trichotomy Property*

> One and only one of the following is true:
>
> $a < b$ or $b < a$ or $a = b$.

The trichotomy property has a simple geometric interpretation on the number line; namely, if a and b are distinct, then one must lie to the left of the other.

2. *The Transitivity Property*

If $a < b$ and $b < c$, then $a < c$.

The transitivity property means that if a lies to the left of b and b lies to the left of c, then a lies to the left of c (Figure 1).

Figure 1

3. *The Addition Property*

If $a < b$, then $a + c < b + c$.

The addition property permits us to add the same number c to both sides of an inequality $a < b$.

4. *The Multiplication Property*

If $a < b$ and $c > 0$, then $ac < bc$.

The multiplication property permits us to multiply both sides of an inequality $a < b$ by the same *positive* number c.

Example 1   State one basic order property of the real numbers to justify each statement.

(a) The discriminant of a quadratic equation is either positive, negative, or zero.
(b) $-1 < 0$ and $0 < 1$, so it follows that $-1 < 1$.
(c) $5 < 6$, so it follows that $5 + (-11) < 6 + (-11)$.
(d) $4 < 7$ and $\frac{1}{2} > 0$, so it follows that $2 < \frac{7}{2}$.

Solution   (a) The trichotomy property.   (b) The transitivity property.
(c) The addition property.   (d) The multiplication property.

Many of the important properties of inequalities can be *derived* from the basic order properties, although we shall not derive them here. Among the more important derived properties are the following.

5. *The Subtraction Property*

If $a < b$, then $a - c < b - c$

6. *The Division Property*

> If $a < b$ and $c > 0$, then $\dfrac{a}{c} < \dfrac{b}{c}$.

According to the multiplication and division properties, you can multiply or divide both sides of an inequality by a *positive* number. According to the following property, if you multiply or divide both sides of an inequality by a *negative* number, you must **reverse the inequality sign.**

7. *The Order-Reversing Properties*

> **(i)** If $a < b$ and $c < 0$, then $ac > bc$.
>
> **(ii)** If $a < b$ and $c < 0$, then $\dfrac{a}{c} > \dfrac{b}{c}$.

Example 2   What equivalent inequality is obtained if both sides of the inequality $2 < 5$ are multiplied by -3?

Solution   If we multiply both sides of $2 < 5$ by the negative number -3, we must reverse the inequality sign and write $-6 > -15$, or $-15 < -6$. (Indeed, on a number line, -15 lies to the left of -6.) ∎

Statements analogous to Properties 2–7 can be made for nonstrict inequalities (those involving $\leq$ or $\geq$) and for compound inequalities. For instance, if you know that

$$-1 \leq 3 - \frac{x}{2} < 5,$$

you can multiply all members by -2, provided that you reverse the inequality signs:

$$2 \geq x - 6 > -10$$

or $$-10 < x - 6 \leq 2.$$

An inequality containing a variable is neither true nor false until a particular number is substituted for the variable. If a true statement results from such a substitution, we say that the substitution **satisfies** the inequality. If the substitution $x = a$ satisfies an inequality, we say that the real number a is a **solution** of the inequality. The set of all solutions of an inequality is called its **solution set.** Two inequalities are said to be **equivalent** if they have exactly the same solution set.

To **solve** an inequality—that is, to find its solution set—you proceed in much the same way as in solving equations, except of course that when you multiply or divide all members by a negative quantity, you must reverse all signs of inequality. The usual approach is to try to isolate the unknown on one side of the inequality.

In Examples 3 to 6, solve each inequality and sketch its solution set on a number line.

Example 3 $5x + 3 < 18$

Solution

$$5x + 3 < 18$$
$$5x < 15 \qquad \text{(We subtracted 3 from both sides.)}$$
$$x < 3 \qquad \text{(We divided both sides by 5.)}$$

Figure 2

Since the last inequality is equivalent to the original one, we conclude that the solution of $5x + 3 < 18$ consists of all numbers x such that $x < 3$. In other words, the solution set is the interval $(-\infty, 3)$ (Figure 2). ∎

Example 4 $3 - 2x \geq 4x - 9$

Solution

$$3 - 2x \geq 4x - 9$$
$$3 - 6x \geq -9 \qquad \text{(We subtracted } 4x \text{ from both sides.)}$$
$$-6x \geq -12 \qquad \text{(We subtracted 3 from both sides.)}$$
$$x \leq 2 \qquad \text{(We divided both sides by } -6 \text{ and reversed the inequality.)}$$

Figure 3

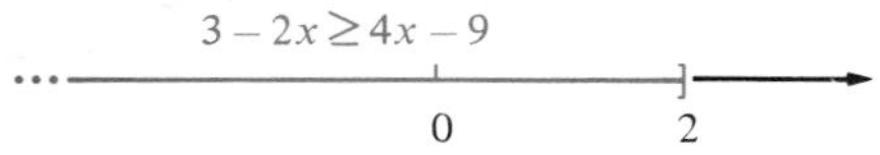

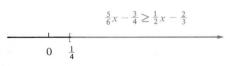

Therefore, the solution set is $(-\infty, 2]$ (Figure 3). ∎

Example 5 $\frac{5}{6}x - \frac{3}{4} \geq \frac{1}{2}x - \frac{2}{3}$

Solution   We begin by multiplying both sides by 12, the LCD of the fractional coefficients:

$$10x - 9 \geq 6x - 8$$
$$4x - 9 \geq -8 \qquad \text{(We subtracted } 6x \text{ from both sides.)}$$
$$4x \geq 1 \qquad \text{(We added 9 to both sides.)}$$
$$x \geq \tfrac{1}{4} \qquad \text{(We divided both sides by 4.)}$$

Figure 4

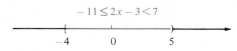

Therefore, the solution set is $[\frac{1}{4}, \infty)$ (Figure 4). ∎

Example 6 $-11 \leq 2x - 3 < 7$

Solution   We begin by adding 3 to all members to help isolate x in the middle:

$$-8 \leq 2x < 10$$
$$-4 \leq x < 5 \qquad \text{(We divided all members by 2.)}$$

Figure 5

Therefore, the solution set is $[-4, 5)$ (Figure 5). ∎

To solve a compound inequality in which the unknown appears only in the middle member, you can always proceed as in Example 6 above. Otherwise, you can break the compound inequality up into simple inequalities and proceed as in the following example.

Example 7   Solve the inequality $3x - 10 \leq 5 < x + 3$ and sketch its solution set on a number line.

Solution   The given inequality holds if and only if *both* the inequalities $3x - 10 \leq 5$ and $5 < x + 3$ hold.

Solution of $3x - 10 \leq 5$	Solution of $5 < x + 3$
$3x - 10 \leq 5$	$5 < x + 3$
$3x \leq 5 + 10$	$5 - 3 < x$
$3x \leq 15$	$2 < x$
$x \leq 5$	
Here, the solution set is $(-\infty, 5]$ (Figure 6a).	Here, the solution set is $(2, \infty)$ (Figure 6b).

Figure 6

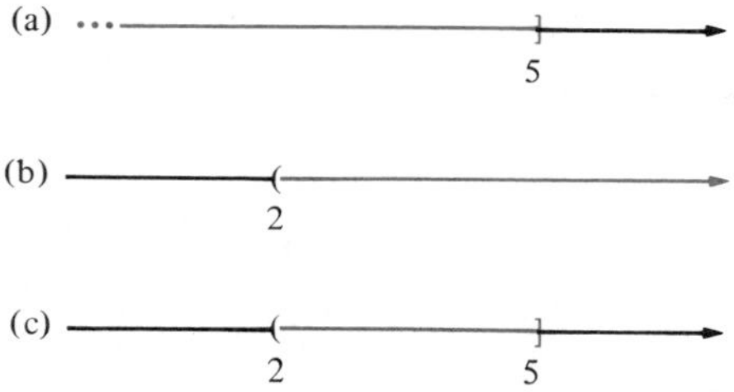

(a)
5

(b)
2

(c)
2   5

Therefore, the solution set of $3x - 10 \leq 5 < x + 3$ consists of all real numbers x that belong to *both* the intervals $(-\infty, 5]$ and $(2, \infty)$; that is, the solution set is the interval $(2, 5]$ (Figure 6c). ■

The following example illustrates one of the many practical applications of inequalities.

Example 8   A technician determines that an electronic circuit fails to operate because the resistance between points A and B, 1200 ohms, exceeds the specifications, which call for a resistance of no less than 400 ohms and no greater than 900 ohms. The circuit can be made to satisfy the specifications by adding a shunt resistor R ohms, $R > 0$ (Figure 7). After adding the shunt resistor, the resistance between points A and B will be $\dfrac{1200R}{1200 + R}$ ohms. What are the possible values of R?

Solution   To satisfy the specifications, we must have

$$400 \leq \frac{1200R}{1200 + R} \leq 900 \qquad \text{where } R > 0.$$

Because R is positive, it follows that $1200 + R$ is positive, so we can multiply all members of the inequality by $1200 + R$:

$$400(1200 + R) \leq 1200R \leq 900(1200 + R)$$

$$480{,}000 + 400R \leq 1200R \leq 1{,}080{,}000 + 900R.$$

Figure 7

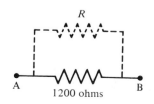

R

A   1200 ohms   B

Now we break the compound inequality into two simple inequalities.

$480{,}000 + 400R \leq 1200R$	$1200R \leq 1{,}080{,}000 + 900R$
$480{,}000 \leq 1200R - 400R$	$1200R - 900R \leq 1{,}080{,}000$
$480{,}000 \leq 800R$	$300R \leq 1{,}080{,}000$
$\dfrac{480{,}000}{800} \leq R$	$R \leq \dfrac{1{,}080{,}000}{300}$
$600 \leq R$	$R \leq 3600$

Therefore, we must have both $600 \leq R$ and $R \leq 3600$; that is,

$$600 \leq R \leq 3600.$$

The shunt resistance R must be no less than 600 ohms and no greater than 3600 ohms. ∎

Problem Set 2.5

1. State one basic order property of the real numbers to justify each statement.

(a) $3 < 5$, so it follows that $3 + 2 < 5 + 2$.

(b) $-3 < -2$ and $-2 < 2$, so it follows that $-3 < 2$.

(c) $-3 < -2$ and $\frac{1}{3} > 0$, so it follows that $-1 < -\frac{2}{3}$.

(d) We know that $\sqrt{2} \neq 1.414$, so it follows that either $\sqrt{2} < 1.414$ or else $\sqrt{2} > 1.414$.

2. State one derived order property of the real numbers to justify each statement.

(a) $-1 < 0$, so it follows that $-2 < -1$.

(b) $-5 < -3$ and $5 > 0$, so it follows that $-1 < -\frac{3}{5}$.

(c) $3 < 5$ and $-2 < 0$, so it follows that $-6 > -10$.

(d) $3 < 5$ and $-2 < 0$, so it follows that $-\frac{3}{2} > -\frac{5}{2}$.

3. Given that $x > -3$, what equivalent inequality is obtained if (a) 4 is added to both sides? (b) 4 is subtracted from both sides? (c) Both sides are multiplied by 5? (d) Both sides are multiplied by -5?

4. Given that $-2 \leq 3 - 4x < 1$, what equivalent inequality is obtained if (a) 3 is subtracted from all members? (b) 2 is added to all members? (c) All members are divided by 2? (d) All members are divided by -4?

In Problems 5 to 32, solve each inequality and sketch its solution set on a number line.

5. $2x - 5 < 5$

6. $7 - 3x < 22$

7. $2x - 10 \leq 15 - 3x$

8. $5x + 3 \leq 8x - 6$

9. $15 + 6x > 9x + 9$

10. $x - 11 > 2x - 13$

11. $-5(1 - x) \geq 15$

12. $4(x - 4) - 3 \geq 17$

13. $7(x + 2) \geq 2(2x - 4)$

14. $21 - 3(x - 7) \leq x + 20$

15. $\dfrac{x}{3} < 2 - \dfrac{x}{4}$

16. $\dfrac{2x}{3} > \dfrac{23}{24} - \dfrac{5x}{4}$

17. $\frac{1}{2}(x + 1) \geq \frac{1}{3}(x - 5)$

18. $\frac{1}{5}(x + 1) \geq \frac{1}{3}(x - 5)$

19. $\frac{1}{3}(x + 1) \geq -\frac{1}{3}(2 - x)$

20. $\frac{2}{3}x - \frac{3}{5}(5 - x) \geq 0$

21. $\frac{1}{6}(x + 3) - \frac{1}{2}(x - 1) \leq \frac{1}{3}$

22. $\dfrac{x + 5}{6} - \dfrac{10 - x}{-3} < \pi$

23. $5 \leq 2x + 3 \leq 13$

24. $14 \le 5 - 3x < 20$

25. $4 \le 5x - 1 < 9$

26. $0 < 3x - 5 \le 16$

27. $-3 < 5 - 4x \le 17$

28. $3 + x < 3(x - 1) \le x - 7$

29. $4x - 1 \le 3 \le 7 + 2x$

30. $3x + 1 \le 2 - x < 18 + 7x$

31. $1 - 2x < 5 - x \le 25 - 6x$

32. $2x - 1 \le 5 \le x + 2$

In Problems 33 to 36, use the method illustrated in the solution of Example 8 to solve the inequalities. Illustrate each solution set on a number line.

33. $1 \le \dfrac{10x}{10 + x} \le 5$ where $x > 0$

34. $1 \le \dfrac{4x}{4 + x} \le 3$ where $x > -4$

35. $\dfrac{4}{x + 1} \le 3 \le \dfrac{6}{x + 1}$ where $x > 0$

36. $\dfrac{8}{x + 3} \le 2 \le \dfrac{13}{x + 3}$ where $x > -3$

37. A student's scores on the first three tests in a sociology course are 65, 78, and 84. What range of scores on a fourth test will give the student an average that is less than 80 but no less than 70 for the four tests?

38. An object thrown straight upward with an initial velocity of 100 feet per second will have a velocity v, given by $v = 100 - 32t$, exactly t seconds later. Over what interval of time will its velocity be greater than 50 feet per second but no greater than 60 feet per second?

39. A shunt resistor R ohms is to be added to a resistance of 600 ohms in order to reduce the resistance to a value strictly between 540 ohms and 550 ohms. After the shunt is added, the resistance is given by $600R/(600 + R)$. Find the possible values of R. (Assume that $R > 0$.)

C **40.** A realtor charges the seller of a house a 7% commission on the price for which the house is sold. Suppose that the seller wants to clear at least $60,000 after paying the commission, and that similar houses are selling for $70,000 or less.

(a) What is the range of selling prices for the house?

(b) If the house is sold, what is the range of amounts for the realtor's commission?

41. A rectangular solar collector is to have a height of 1.5 meters, but its length is still to be determined. What is the range of values for this length if the collector provides 400 watts per square meter, and if it must provide a total of between 2000 and 3500 watts?

42. Suppose that families with an income, after deductions, of over $19,200 but not over $23,200 must pay an income tax of $3260 plus 28% of the amount over $19,200. What is the range of possible values for the income, after deductions, of families whose income tax liability is between $3484 and $4044?

43. An operator-assisted station-to-station phone call from Boston, MA, to Tucson, AZ, costs $2.25 plus $0.38 for each additional minute after the first 3 minutes. A group of such calls were made, each costing between $6.05 and $8.71, inclusive. Give the range of minutes for the lengths of these calls.

44. Sketch diagrams to illustrate the addition and multiplication order properties of the real numbers.

45. If $a \le b$, prove that $a + c \le b + c$. [*Hint:* Consider separately the two possibilities $a < b$ and $a = b$.]

46. If $c \ne 0$, prove that $c^2 > 0$. [*Hint:* Consider separately the two cases $c > 0$ and $c < 0$.]

47. Give an example to show that, in general, you *cannot* square both sides of an inequality and get an equivalent inequality.

48. If $c > 0$, show that $1/c > 0$. [*Hint:* If $1/c$ is not positive it must be negative, since it cannot be zero.]

2.6 EQUATIONS AND INEQUALITIES INVOLVING ABSOLUTE VALUES

Many calculators have an *absolute-value* key marked $|x|$ or ABS. If you enter a *nonnegative* number and press this key, the same number appears in the display. However, if you enter a *negative* number and press the absolute-value key, the display shows the number without the negative sign. For instance,

Enter	Press ABS Key	Display Shows
3.2	$\longrightarrow$	3.2
-3.2	$\longrightarrow$	3.2

This feature is useful whenever you are interested only in the "magnitude" of a number, without regard to its algebraic sign.

On a calculator without an absolute-value key, you can obtain the absolute value $|x|$ of a number x by entering x, pressing the x^2 key, and then pressing the square-root key. In other words, if you square a number and take the principal square root of the result, you will remove any negative sign that may have been present, so that

$$\sqrt{x^2} = |x|.$$

For instance, $\sqrt{(-5)^2} = \sqrt{25} = 5 = |-5|$.

The absolute value is defined formally as follows.

Definition 1   **Absolute Value**

> If x is a real number, then $|x|$, the **absolute value** of x, is defined by
> $$|x| = \begin{cases} x, & \text{if } x \geq 0 \\ -x, & \text{if } x < 0. \end{cases}$$

For instance,

$$|5| = 5 \quad \text{because } 5 \geq 0,$$

$$|0| = 0 \quad \text{because } 0 \geq 0,$$

$$|-5| = -(-5) = 5 \quad \text{because } -5 < 0.$$

Notice that $|x| \geq 0$ is always true. (Why?)

On a number line, $|x|$ *is the number of units of distance between the point with coordinate x and the origin*, regardless of whether the point is to the right or left of the origin (Figure 1a). For instance, both the point with coordinate -3 and the point with coordinate 3 are 3 units from the origin (Figure 1b), so $|-3| = 3 = |3|$.

Figure 1

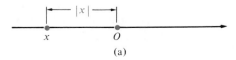

(a)

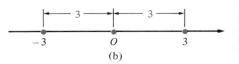

(b)

More generally,

> $|x - y|$ *is the number of units of distance between the point with coordinate x and the point with coordinate y.*

This statement holds no matter which point is to the left of the other (Figure 2a and b), and of course it also holds when $x = y$. For simplicity, we refer to the distance $|x - y|$ as the **distance between the numbers x and y.**

Figure 2

(a)   (b)

Example 1   Find the distance between the numbers.

(a) 4 and 7   (b) 7 and 4   (c) -4 and 7
(d) -4 and -7 (e) -4 and 0   (f) -4 and 4

Solution   (a) $|4 - 7| = |-3| = 3$ (b) $|7 - 4| = |3| = 3$
(c) $|-4 - 7| = |-11| = 11$ (d) $|-4 - (-7)| = |-4 + 7| = |3| = 3$
(e) $|-4 - 0| = |-4| = 4$ (f) $|-4 - 4| = |-8| = 8$ ■

The absolute value of a product of two numbers is equal to the product of their absolute values; that is,

$$|xy| = |x| |y|.$$

This rule can be derived as follows:

$$|xy| = \sqrt{(xy)^2} = \sqrt{x^2 y^2} = \sqrt{x^2} \sqrt{y^2} = |x| |y|.$$

A similar rule applies to quotients; if $y \neq 0$,

$$\left| \frac{x}{y} \right| = \frac{|x|}{|y|}$$

(Problem 75). However, there is no such rule for sums or differences:

$$|x + y| \quad \textit{need not be the same as} \quad |x| + |y|;$$

for instance,

$$|5 + (-2)| = |3| = 3, \quad \text{but} \quad |5| + |-2| = 5 + 2 = 7.$$

Similarly,

$$|x - y| \quad \textit{need not be the same as} \quad |x| - |y|.$$

(See Problems 74 and 78.)

Example 2   Let $x = -3$ and $y = 5$. Find the value of each expression.

    (a) $|6x + y|$ **(b)** $|6x| + |y|$ **(c)** $|xy|$

    (d) $|x||y|$ **(e)** $\dfrac{|x|}{x}$ **(f)** $\left|\dfrac{x}{y}\right|$

Solution   **(a)** $|6x + y| = |6(-3) + 5| = |-18 + 5| = |-13| = 13$

    (b) $|6x| + |y| = |6(-3)| + |5| = |-18| + |5| = 18 + 5 = 23$

    (c) $|xy| = |(-3)5| = |-15| = 15$

    (d) $|x||y| = |-3||5| = 3 \cdot 5 = 15$

    (e) $\dfrac{|x|}{x} = \dfrac{|-3|}{-3} = \dfrac{3}{-3} = -1$

    (f) $\left|\dfrac{x}{y}\right| = \left|\dfrac{-3}{5}\right| = \dfrac{3}{5}$

Equations Involving Absolute Values

Geometrically, the equation $|x| = 3$ means that the point with coordinate x is 3 units from 0 on the number line. Obviously, the number line contains *two* points that are 3 units from the origin—one to the right of the origin and the other to the left (Figure 3). Thus, the equation $|x| = 3$ has two solutions, $x = 3$ and $x = -3$. More generally, we have the following theorem, whose proof is left as an exercise (Problem 77).

Figure 3

Theorem 1   **Solution of the Equation $|u| = a$**

Let a be a real number.

    (i)   If $a > 0$, then $|u| = a$ if and only if $u = -a$ or $u = a$.

    (ii) If $a < 0$, then the equation $|u| = a$ has no solution.

    (iii) $|u| = 0$ if and only if $u = 0$.

In Examples 3 to 7, solve each equation.

Example 3 $|4x| = 12$

Solution   By part (i) of Theorem 1, with $u = 4x$, we have $4x = -12$ or $4x = 12$; that is, $x = -3$ or $x = 3$.

Example 4 $|3t| - 1 = 17$

Solution   The equation is equivalent to $|3t| = 18$, so that $3t = -18$ or $3t = 18$; that is, $t = -6$ or $t = 6$. ∎

Example 5 $|7x + 1| = -3$

Solution   By part (ii) of Theorem 1, the equation has no solution. ∎

Example 6 $|1 - 2x| = 7$

Solution   The equation holds if and only if

$$1 - 2x = 7 \qquad \text{or} \qquad 1 - 2x = -7$$
$$-2x = 6 \qquad \text{or} \qquad -2x = -8$$
$$x = -3 \qquad \text{or} \qquad x = 4.$$

∎

Example 7 $|x^2 + x - 6| = 0$

Solution   By part (iii) of Theorem 1, the equation is equivalent to

$$x^2 + x - 6 = 0, \qquad \text{that is,} \qquad (x + 3)(x - 2) = 0.$$

The solutions are $x = -3$ and $x = 2$. ∎

An equation of the form $|u| = |v|$ can be solved by using the following theorem, whose proof is left as an exercise (Problem 79).

Theorem 2   **Solution of the Equation $|u| = |v|$**

$|u| = |v|$ if and only if $u = -v$ or $u = v.$

Example 8   Solve the equation $|p - 5| = |3p + 7|$.

Solution   By Theorem 2 with $u = p - 5$ and $v = 3p + 7$, the given equation holds if and only if

$$p - 5 = 3p + 7 \qquad \text{or} \qquad p - 5 = -(3p + 7)$$
$$-2p = 12 \qquad \text{or} \qquad 4p = -2$$
$$p = -6 \qquad \text{or} \qquad p = -\tfrac{1}{2}.$$

∎

Inequalities Involving Absolute Values

Referring once again to the number line (Figure 4), we see that the inequality $|x| < 3$ means that the point with coordinate x is less than 3 units from 0, that is, $-3 < x < 3$. More generally, we have the following theorem (Problem 80).

Figure 4

|x|<3 if and only if
x is in this interval
−5 −4 −3 −2 −1 0 1 2 3 4 5

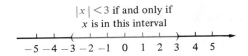

Theorem 3 **Solution of the Inequalities $|u| < a$ and $|u| \leq a$**

If $a > 0$, then:

(i) $|u| < a$ if and only if $-a < u < a$.

(ii) $|u| \leq a$ if and only if $-a \leq u \leq a$.

In Examples 9 and 10, solve each inequality and sketch its solution set on a number line.

Example 9 $|2x + 3| < 5$

Solution

$|2x + 3| < 5$

$-5 < 2x + 3 < 5$ [By part (i) of Theorem 3.]

$-8 < 2x < 2$ (We subtracted 3 from all members.)

$-4 < x < 1$ (We divided all members by 2.)

Hence, the solution set is the interval $(-4, 1)$ (Figure 5). ■

Figure 5

|2x+3|<5
−6 −5 −4 −3 −2 −1 0 1 2 3 4

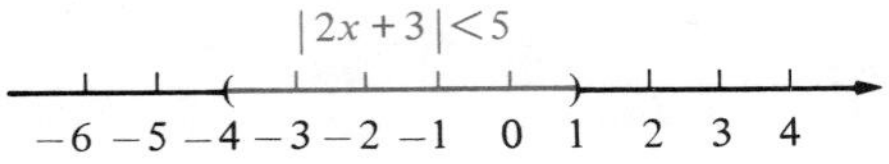

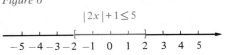

Example 10 $|2x| + 1 \leq 5$

Solution

$|2x| \leq 4$ (We subtracted 1 from both sides.)

$-4 \leq 2x \leq 4$ [By part (ii) of Theorem 3.]

$-2 \leq x \leq 2$ (We divided all members by 2.)

Hence, the solution set is the interval $[-2, 2]$ (Figure 6). ■

Figure 6

|2x|+1≤5
−5 −4 −3 −2 −1 0 1 2 3 4 5

Figure 7 $|x| > 3$ if and only if x
belongs to one of these intervals

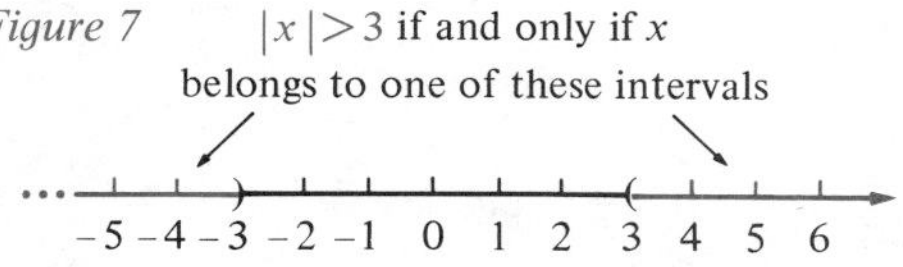

On a number line, the inequality $|x| > 3$ means that the point with coordinate x is more than 3 units from 0 (Figure 7); that is, either $x > 3$ or $x < -3$. More generally, we have the following theorem (Problem 82).

Theorem 4   **Solution of the Inequalities $|u| > a$ and $|u| \geq a$**

> If $a > 0$, then:
>
> **(i)** $|u| > a$ if and only if $u < -a$ or $u > a$.
>
> **(ii)** $|u| \geq a$ if and only if $u \leq -a$ or $u \geq a$.

Example 11   Solve the inequality $\left|\frac{1}{2}x - 1\right| \geq 2$ and sketch its solution set on a number line.

Solution   By part (ii) of Theorem 4 with $u = \frac{1}{2}x - 1$, we have

$$\frac{1}{2}x - 1 \leq -2 \quad \text{or} \quad \frac{1}{2}x - 1 \geq 2$$

$$\frac{1}{2}x \leq -1 \quad \text{or} \quad \frac{1}{2}x \geq 3$$

$$x \leq -2 \quad \text{or} \quad x \geq 6.$$

Thus, the solution set consists of all numbers that belong to either of the intervals $(-\infty, -2]$ or $[6, \infty)$ (Figure 8).

Figure 8

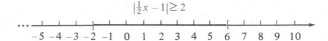

$\left|\frac{1}{2}x - 1\right| \geq 2$

If A and B are sets, then the set consisting of all elements that belong to A or to B (or to both) is called the **union** of A and B and is written as $A \cup B$. Thus, in Example 11, the solution set (Figure 8) is the union of the two intervals $(-\infty, -2]$ and $[6, \infty)$, and it can be written concisely as the set

$$(-\infty, -2] \cup [6, \infty).$$

Problem Set 2.6

In Problems 1 to 10, find the distance between the numbers and illustrate on a number line.

1. 5 and 3

2. -3 and 0

3. -3 and -5

4. -5 and 3

5. $\frac{5}{2}$ and $\frac{3}{2}$

Ⓒ **6.** 412.7 and 152.3

Ⓒ **7.** -3.311 and 2.732

Ⓒ **8.** $-\sqrt{3}$ and $\sqrt{2}$

Ⓒ **9.** π and 1

Ⓒ **10.** π and $\frac{22}{7}$

In Problems 11 to 20, find the value of each expression if $x = -4$ and $y = 6$.

11. $|5x + 2y|$

12. $|5x| + |2y|$

13. $|5x - 2y|$

14. $|5x| - |2y|$

15. $||5x| - |2y||$

16. $||2y| - |5x||$

17. $|3xy| - 3|x||y|$

18. $\left|\dfrac{5x}{2y}\right|$

19. $\left|\dfrac{5x}{2y}\right| - \dfrac{5|x|}{2|y|}$

20. $\dfrac{|x|}{x} - \dfrac{x}{|x|}$

In Problems 21 to 44, solve each equation.

21. $|4x| = 16$

22. $|-5y| = 15$

23. $|5y| + 2 = 17$

24. $|2t| - 3 = 5$

25. $|2x| + |-4| = 4$

26. $|7x| = |-14|$

27. $|x^2 - x - 20| = 0$

28. $|2p - 1| = 3$

29. $|4 - y| = 2$

30. $|11 - 5x| = 16$

31. $|3p| = 3p$

32. $|7y - 3| = 3 - 7y$

33. $\left|\dfrac{3|x|}{5} + 1\right| = 0$

34. $\left|\dfrac{5c}{2} - 3\right| = -1$

35. $\left|\dfrac{x}{2} - \dfrac{3}{4}\right| = \dfrac{1}{12}$

36. $\left|\dfrac{13}{12} + \dfrac{2x}{3}\right| = \dfrac{11}{12}$

37. $|t + 1| = |4t|$

38. $|3x - 1| = x + 5$

39. $|1 - x| = |2x + 3|$

40. $|2t - 7| = t^2$

41. $|4 - 3y| = |10 - 5y|$

42. $|2x - 1| = x^2$

43. $|4x - 7| = |2 + x|$

44. $|x + 1| + |x - 1| = 4$

In Problems 45 to 68, solve each inequality and sketch its solution set on a number line.

45. $|2x| < 4$

46. $|x| - 3 \le 2$

47. $|5x| - 1 \le 9$

48. $|3t| > 15$

49. $|x - 3| < 2$

50. $|x| + 2 \ge 3$

51. $|x| - 3 \le 2$

52. $|-3y| - 2 \ge 4$

53. $|x| \ge 3$

54. $|2x - 5| < 1$

55. $|4x - 3| > 9$

56. $|3y + 5| \le |-2|$

57. $|5x - 8| \le 2$

58. $|5 - 3p| \ge |-11|$

59. $|2x + 3| \ge \frac{1}{4}$

60. $|x + 5| \ge x + 1$

61. $|1 - \frac{7}{6}x| \le \frac{1}{3}$

62. $|3x + 6| > 0$

63. $|7 - 4x| \ge |-3|$

64. $|6y - 3| > -3$

65. $|\frac{11}{2} - 2x| > \frac{3}{2}$

66. $|3x + 5| < |2x + 1|$

67. $|2x - 5| \ge 0$

68. $|5x + 2| < |3x - 4|$

In Problems 69 to 72, describe each interval as the set of all real numbers x that satisfy an inequality containing an absolute value.

69. $(-1, 1)$

70. $(-3, 7)$

71. $[-3, 3]$

72. $[-2, 3]$

73. Show that $-|x| \le x \le |x|$ holds for all real numbers x.

74. Although the absolute value of a sum need not be the same as the sum of the absolute values, the inequality $|x + y| \le |x| + |y|$, called the **triangle inequality,** always holds. Prove the triangle inequality by considering the case in which $x + y \ge 0$ separately from the case in which $x + y < 0$. Use the facts that $-|x| \le x \le |x|$ and $-|y| \le y \le |y|$. (See Problem 73.)

75. If $y \ne 0$, show that $\left|\dfrac{x}{y}\right| = \dfrac{|x|}{|y|}$.

76. Use the triangle inequality (Problem 74) to establish the following:

(a) If $|x - 2| < \frac{1}{2}$ and $|y + 2| < \frac{1}{3}$, then $|x + y| < \frac{5}{6}$.

(b) If $|x - 2| < \frac{1}{2}$ and $|z - 2| < \frac{1}{3}$, then $|x - z| < \frac{5}{6}$.

77. Prove Theorem 1.

78. Using the triangle inequality (Problem 74), prove that $||x| - |y|| \le |x - y|$.

79. Prove Theorem 2.

80. Prove Theorem 3.

81. Show that if $a < b$, then $a < \dfrac{a + b}{2} < b$ and the distance between a and $\dfrac{a + b}{2}$ is the same as the distance between $\dfrac{a + b}{2}$ and b.

82. Prove Theorem 4.

83. Show that the expression $\frac{1}{2}(x + y + |x - y|)$ always equals the larger of the two numbers x and y (or their common value if they are equal).

84. Find an expression that is similar to the one given in Problem 83, but that gives the smaller of the two numbers x and y (or their common value if they are equal).

85. Show that if $a < b$, then the solution set of $|a + b - 2x| < b - a$ is the open interval (a, b).

86. In calculus, it is necessary to show that a condition of the form $|x + y| < \varepsilon$ follows from the conditions $|x| < \varepsilon/2$ and $|y| < \varepsilon/2$. (ε is the Greek letter *epsilon*.) Use the triangle inequality (Problem 74) to make this argument.

87. Statisticians often report their findings in terms of so-called *confidence intervals*—intervals that they believe (with a certain level of confidence) contain the unknown quantity with which they are concerned. Suppose that a statistician determines that (with a certain level of confidence) the average weight x pounds of newborn babies in a large city satisfies the inequality $|x - 6.8| < 0.85$. Find the confidence interval for x by solving this inequality.

88. Let x be a positive number and let r be the result of rounding off x to the nearest 10^{-n}, where n is an integer. Prove that $|x - r|$, the absolute value of the error made in estimating x by r, cannot exceed $5 \times 10^{-n-1}$.

2.7 POLYNOMIAL AND RATIONAL INEQUALITIES

Up to this point, we have dealt mainly with inequalities containing only first-degree polynomials in the unknown. In this section, we study inequalities containing higher degree polynomials or rational expressions in the unknown.

Polynomial Inequalities

An inequality such as $4x^3 - 3x^2 + x - 1 > x^2 + 3x - x$, in which both sides are polynomials in the unknown, is called a **polynomial inequality.** Before solving such inequalities, we shall rewrite them in **standard form** with zero on the right side and a polynomial on the left side.

Suppose we have to solve the polynomial inequality

$$4x^2 - x - 8 < 3x^2 - 4x + 2.$$

We begin by subtracting $3x^2 - 4x + 2$ from both sides to bring it into the standard form

$$x^2 + 3x - 10 < 0.$$

Now, let's consider how the algebraic sign of the polynomial on the left changes as we vary the values of the unknown. As we move x along the number line, the quantity $x^2 + 3x - 10$ is sometimes positive, sometimes negative, and sometimes zero. To solve the inequality, we must find the values of x for which $x^2 + 3x - 10$ is negative.

It seems plausible (and is, in fact, true—see Problem 52) that *intervals where $x^2 + 3x - 10$ is positive are separated from intervals where it is negative by values of x for which it is zero.* To locate these intervals, we begin by solving the equation

$$x^2 + 3x - 10 = 0.$$

Factoring, we obtain $(x + 5)(x - 2) = 0$, so that $x = -5$ or $x = 2$. The two roots -5 and 2 divide the number line into three open intervals, $(-\infty, -5)$, $(-5, 2)$, and $(2, \infty)$ (Figure 1). Therefore, *the quantity $x^2 + 3x - 10$ will have a constant algebraic sign over each of these intervals.*

Figure 1

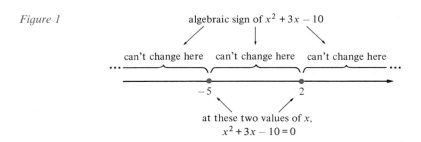

To find out whether $x^2 + 3x - 10$ is positive or negative over the first interval, $(-\infty, -5)$, we select any convenient test number in this interval, say, $x = -6$, and substitute it into $x^2 + 3x - 10$:

$$(-6)^2 + 3(-6) - 10 = 36 - 18 - 10 = 8 > 0.$$

Because $x^2 + 3x - 10$ is positive at one number in $(-\infty, -5)$ and cannot change its algebraic sign over this interval, we can conclude that $x^2 + 3x - 10 > 0$ for all values of x in $(-\infty, -5)$.

Similarly, to find the algebraic sign of $x^2 + 3x - 10$ over the interval $(-5, 2)$, we substitute a test number, say $x = 0$, to obtain

$$0^2 + 3(0) - 10 = -10 < 0,$$

and conclude that $x^2 + 3x - 10 < 0$ for all values of x in the interval $(-5, 2)$. Finally, using another test number, say, $x = 3$, from the interval $(2, \infty)$, we find that

$$3^2 + 3(3) - 10 = 9 + 9 - 10 = 8 > 0,$$

so that $x^2 + 3x - 10 > 0$ for all values of x in the interval $(2, \infty)$.

The information we now have about the algebraic sign of $x^2 + 3x - 10$ in each interval is summarized in Figure 2. Thus, $x^2 + 3x - 10 < 0$ if and only if $-5 < x < 2$; in other words, the solution set is the open interval $(-5, 2)$.

Figure 2

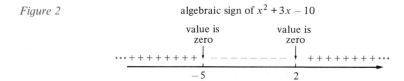

The method illustrated above can be used to solve any polynomial inequality in one unknown. This method is summarized in the following step-by-step procedure.

Procedure for Solving a Polynomial Inequality

Step 1. Bring the inequality into standard form, with zero on the right side and a polynomial on the left side.

Step 2. Set the polynomial equal to zero and find all real roots of the resulting equation.

Step 3. Arrange the roots obtained in step 2 in increasing order on a number line. These roots will divide the number line into open intervals.

Step 4. The algebraic sign of the polynomial cannot change over any of the intervals obtained in step 3. Determine this sign for each interval by selecting a test number in the interval and substituting it for the unknown in the polynomial. The algebraic sign of the resulting value is the sign of the polynomial over the entire interval.

Step 5. Draw a figure showing the information obtained in step 4 about the algebraic sign of the polynomial over the various open intervals. The solution set of the inequality can be read from this figure.

If it turns out in step 2 that the polynomial has no real roots, there is just one open interval involved, namely the entire number line $\mathbb{R}$.

In Examples 1 and 2, solve each polynomial inequality.

Example 1 $4x^2 + 8x \geq 5$

Solution   We carry out the steps in the procedure above.

Step 1. The given inequality is equivalent to $4x^2 + 8x - 5 \geq 0$.

Step 2. We write $4x^2 + 8x - 5 = 0$ and factor the left side to obtain

$$(2x + 5)(2x - 1) = 0.$$

The roots are $x = -\frac{5}{2}$ and $x = \frac{1}{2}$.

Step 3. The open intervals over which the algebraic sign of $4x^2 + 8x - 5$ is constant are $(-\infty, -\frac{5}{2})$, $(-\frac{5}{2}, \frac{1}{2})$, and $(\frac{1}{2}, \infty)$ (Figure 3).

Figure 3

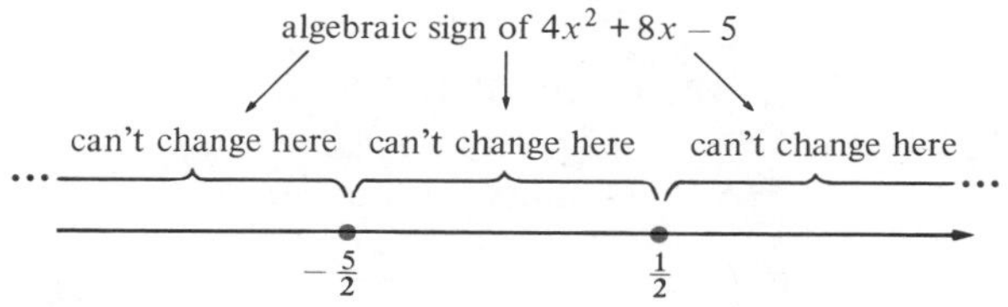

algebraic sign of $4x^2 + 8x - 5$

can't change here   can't change here   can't change here

$-\frac{5}{2}$ $\frac{1}{2}$

Step 4. For the interval $(-\infty, -\frac{5}{2})$, we substitute a test number, say, $x = -3$, in the polynomial $4x^2 + 8x - 5$ to obtain

$$4(-3)^2 + 8(-3) - 5 = 7 > 0.$$

We conclude that $4x^2 + 8x - 5 > 0$ for all values of x in the interval $(-\infty, -\frac{5}{2})$. For the interval $(-\frac{5}{2}, \frac{1}{2})$, we substitute a test number, say, $x = 0$, in the polynomial to get

$$4(0)^2 + 8(0) - 5 = -5 < 0.$$

Therefore, $4x^2 + 8x - 5 < 0$ for all values of x in the interval $(-\frac{5}{2}, \frac{1}{2})$. For the interval $(\frac{1}{2}, \infty)$, we substitute a test number, say, $x = 1$, and find that

$$4(1)^2 + 8(1) - 5 = 7 > 0.$$

Hence, $4x^2 + 8x - 5 > 0$ for all values of x in the interval $(\frac{1}{2}, \infty)$.

Step 5. The information obtained in step 4 is summarized in Figure 4. From this figure, we see that $4x^2 + 8x - 5 \geq 0$ for $x \leq -\frac{5}{2}$ and also for $x \geq \frac{1}{2}$. In other words, the solution set consists of all real numbers that belong to either one of the intervals $(-\infty, -\frac{5}{2}]$ or $[\frac{1}{2}, \infty)$. (Note that the *nonstrict* inequality $4x^2 + 8x - 5 \geq 0$ requires that we include the values $-\frac{5}{2}$ and $\frac{1}{2}$ in the solution set.) Thus, the solution set is $(-\infty, -\frac{5}{2}] \cup [\frac{1}{2}, \infty)$. ∎

Figure 4

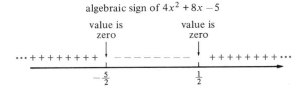

Example 2 $5x^2 + x + 16 < -4x^2 - 23x$

Solution   The given inequality is equivalent to

$$9x^2 + 24x + 16 < 0; \qquad \text{that is,} \qquad (3x + 4)^2 < 0,$$

which is impossible. Therefore, the inequality has no solution. ∎

Rational Inequalities

An inequality such as

$$\frac{2x + 5}{x + 1} < \frac{x + 1}{x - 1},$$

in which all members are rational expressions in the unknown, is called a **rational inequality.** To rewrite such an inequality in **standard form,** subtract the expression on the right side from both sides and then simplify the left side. Be careful—*don't multiply both sides of an inequality by an expression containing the unknown unless you are certain that the multiplier can have only positive values.* (Why not?)

Example 3   Rewrite the rational inequality $\dfrac{2x + 5}{x + 1} < \dfrac{x + 1}{x - 1}$ in standard form.

Solution   We begin by subtracting $\dfrac{x + 1}{x - 1}$ from both sides:

$$\frac{2x + 5}{x + 1} - \frac{x + 1}{x - 1} < 0.$$

Using the LCD $(x + 1)(x - 1)$, we combine the fractions on the left side to obtain

$$\frac{(2x + 5)(x - 1) - (x + 1)(x + 1)}{(x + 1)(x - 1)} < 0 \quad \text{or} \quad \frac{x^2 + x - 6}{(x + 1)(x - 1)} < 0.$$

Finally, factoring the numerator, we can write the inequality as

$$\frac{(x + 3)(x - 2)}{(x + 1)(x - 1)} < 0.$$ ∎

When you have rewritten a rational inequality in standard form, you can solve it by considering how the algebraic sign of the expression on the left side changes as the unknown is varied. Notice that a fraction can change its algebraic sign only if its numerator or denominator changes its algebraic sign. Hence:

> Intervals where a rational expression is positive are separated from intervals where it is negative by values of the variable for which the numerator or denominator* is zero.

It is thus possible to solve rational inequalities by a simple modification of the step-by-step procedure for solving polynomial inequalities. The method is illustrated in the following example.

Example 4   Solve the rational inequality $\dfrac{(x + 3)(x - 2)}{(x + 1)(x - 1)} < 0$.

---

* Of course, the fraction itself is *undefined* when its denominator is zero.

Solution   The numerator of $\dfrac{(x + 3)(x - 2)}{(x + 1)(x - 1)}$ is zero when $x = -3$ and when $x = 2$; its denominator is zero when $x = -1$ and when $x = 1$. We arrange these four numbers in order along a number line to determine five open intervals over which the algebraic sign of $\dfrac{(x + 3)(x - 2)}{(x + 1)(x - 1)}$ does not change, namely, $(-\infty, -3), (-3, -1), (-1, 1), (1, 2),$ and $(2, \infty)$ (Figure 5). We select test numbers, one from each of these intervals, say, $-4, -2, 0, \frac{3}{2},$ and 3. Substituting these test numbers into $\dfrac{(x + 3)(x - 2)}{(x + 1)(x - 1)}$, we obtain

$$\frac{(-4 + 3)(-4 - 2)}{(-4 + 1)(-4 - 1)} = \frac{2}{5} > 0, \qquad \frac{(-2 + 3)(-2 - 2)}{(-2 + 1)(-2 - 1)} = -\frac{4}{3} < 0,$$

$$\frac{(0 + 3)(0 - 2)}{(0 + 1)(0 - 1)} = 6 > 0, \qquad \frac{(\frac{3}{2} + 3)(\frac{3}{2} - 2)}{(\frac{3}{2} + 1)(\frac{3}{2} - 1)} = -\frac{9}{5} < 0,$$

$$\frac{(3 + 3)(3 - 2)}{(3 + 1)(3 - 1)} = \frac{3}{4} > 0.$$

Figure 5

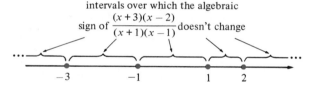

The algebraic signs obtained for the test numbers determine the algebraic signs of $\dfrac{(x + 3)(x - 2)}{(x + 1)(x - 1)}$ over each of the five open intervals, as shown in Figure 6. From this figure, we see that the given inequality $\dfrac{(x + 3)(x - 2)}{(x + 1)(x - 1)} < 0$ holds for values of x in the intervals $(-3, -1)$ and $(1, 2)$, and that it fails to hold for all other values of x. Therefore, the solution set is $(-3, -1) \cup (1, 2)$. ∎

Figure 6

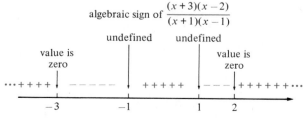

Problem Set 2.7

In Problems 1 to 4, rewrite the polynomial inequality in standard form.

1. $6x^2 - x + 3 < 5x^2 + 5x - 5$

2. $17x^2 - 12x + 33 > 14x^2 - 15x + 28$

3. $\dfrac{x^2}{2} + \dfrac{25}{2} \geq 5x$

4. $4x^3 + x \leq 3x^3 + 3x^2 - x$

In Problems 5 to 30, solve each polynomial inequality.

5. $(x - 2)(x - 4) < 0$

6. $x^2 - 2x < 15$

7. $(x + 4)(x + 6) > 0$

8. $x^2 > x + 6$

9. $x^2 + 49 \leq 14x$

10. $x^2 + 32 \geq 12x + 6$

11. $3x^2 + 22x + 35 \geq 0$

12. $x^2 + 25 \geq 10x$

13. $6x^2 + 1 \leq 5x$

14. $2t^2 + 3t \leq 5$

15. $25y^2 + 4 \geq 20y$

16. $x(6x - 13) > -6$

17. $x(10 - 3x) < 8$

18. $y(10y + 19) \leq 15$

19. $1 - y \geq 2y^2$

20. $2x^2 < x + 2$

21. $(x + 2)^2 > (3x + 1)^2$

22. $(x - 1)(x + 1)(x + 2) \geq 0$

23. $x^2 - x + 1 > 0$

24. $4x^3 - 8x^2 + 2 < x$

25. $5x^3 \geq x^2 + 3x$

26. $x^3 + x^2 \leq x + 1$

27. $x^3 > x$

28. $x^4 + 12 > 7x^2$

29. $(x + 4)(x - 2) \leq (2x - 3)(x + 1)$

30. $(2x + 1)(3x - 1)(x - 2) \leq 0$

In Problems 31 to 34, rewrite the rational inequality in standard form.

31. $\dfrac{2x}{x + 3} > \dfrac{1}{x + 3}$

32. $\dfrac{1 - 5x}{1 + 2x} \leq \dfrac{9 - 3x}{x - 3}$

33. $\dfrac{x^2 - 3}{x^2 + 5x + 6} \leq \dfrac{1}{x + 2}$

34. $\dfrac{3x + 1}{x + 2} > \dfrac{2x - 1}{x + 3}$

In Problems 35 to 48, solve the rational inequality.

35. $\dfrac{3x + 6}{x + 2} > 0$

36. $\dfrac{2x - 1}{3x - 2} < 0$

37. $\dfrac{3y - 5}{y + 3} \leq 0$

38. $\dfrac{(t - 4)^2}{3t + 1} \geq 0$

39. $\dfrac{8x}{2x - 3} > 0$

40. $\dfrac{x^2 - 2x - 15}{2x^2 + 3x - 5} \leq 0$

41. $\dfrac{x + 1}{x - 2} \leq 1$

42. $\dfrac{x + 2}{x - 3} \geq 1$

43. $\dfrac{6x + 8}{x + 2} > \dfrac{3x - 1}{x - 1}$

44. $\dfrac{x + 1}{x + 2} < \dfrac{x + 3}{x + 4}$

45. $\dfrac{(x - 3)(x + 3)}{(x + 1)(x - 2)} > 0$

46. $\dfrac{(x + 1)(x - 2)(x + 2)}{(x - 3)(x - 4)(x + 5)} \leq 0$

47. $\dfrac{(x - 1)(x + 2)(3 - x)}{(x + 3)(x^2 - x - 2)} \leq 0$

48. $\dfrac{(x - 2)^2(x + 1)(2x - 1)}{(x + 3)(x - 1)^2(x + 4)} \geq 0$

49. A projectile is fired straight upward from ground level with an initial velocity of 480 feet per second. Its distance above ground level t seconds later is given by the expression $480t - 16t^2$. For what time interval is the projectile more than 3200 feet above ground level?

50. In economics, the *supply equation* for a commodity relates the selling price, p dollars per unit, to the total number of units q that the manufacturers would be willing to put on the market at that selling price. If the manufacturers' cost of producing one unit is c dollars, then, assuming that all units are sold, the total profit P dollars to the manufacturers is given by $P = pq - cq$. The supply equation for a small manufacturer of souvenirs in a resort area is $q = 8p + 90$, and each souvenir costs the manufacturer $c = \$4$ to produce. Assuming that all souvenirs are sold, what price per souvenir should be charged to bring in a total profit of at least \$315?

53. In an experimental solar power plant, at least 60 mirrors that will focus sunlight on a boiler are to be arranged in rows, each row containing the same number of mirrors. For optimal operation, it is determined that the number of mirrors per row must be 2 more than twice the number of rows. At least how many rows must there be?

54. By examining the algebraic signs of $x + 3$, $x - 2$, $x + 1$, and $x - 1$, prove that the intervals where $\dfrac{(x + 3)(x - 2)}{(x + 1)(x - 1)}$ is positive are separated from intervals where it is negative by values of x for which it is zero or undefined.

51. David, Dean, and Scott have grown 100 pumpkins. They believe they can sell all their pumpkins at a price of 75 cents each, but that each 5-cent increase in the price per pumpkin will result in the sale of 4 fewer pumpkins. What is the maximum price they can charge per pumpkin and still each make $25 on the sale?

52. By examining the algebraic signs of the factors $x + 5$ and $x - 2$, prove that the intervals where $x^2 + 3x - 10$ is positive are separated from intervals where it is negative by values of x for which it is zero.

C In Problems 55 to 58, solve each inequality with the aid of a calculator. Round off your answer to three decimal places.

55. $23.8x^2 - 21.1x + 3.3 \geq 0$

56. $2.12x^2 - 9.87x + 6.79 < 0$

57. $\dfrac{3.4x - 2.7}{2.1x^2 + 8.9x - 1.3} < 0$

58. $\dfrac{22.71x^2 + 11.11x - 8.23}{7.23x^2 - 41.02x + 6.66} \leq 0$

REVIEW PROBLEM SET, CHAPTER 2

In Problems 1 to 16, solve each equation.

1. $6x - 30 = 50 - 10x$

2. $10(t + 1) = 6t - 3(2t + 5)$

3. $3(2y - 3) = 4(y + 1) - 3$

4. $7(w + 2) - 4(w + 1) = -21$

5. $5\left(\dfrac{t}{3} + \dfrac{2}{5}\right) = 2\left(\dfrac{4t}{5} + \dfrac{3}{2}\right)$

6. $\frac{1}{5}(6a - 7) = \frac{1}{2}(3a - 1)$

7. $\dfrac{5x - 2}{6} - \dfrac{2x - 5}{3} = 1$

8. $\dfrac{5 + u}{6} - \dfrac{10 - u}{3} = 1$

9. $\dfrac{3t}{4} - 12 = \dfrac{3(t - 12)}{5}$

10. $\dfrac{10 - r}{2} - \dfrac{5 - r}{3} = \dfrac{5}{2}$

11. $\dfrac{3 - y}{3y} + \dfrac{1}{4} = \dfrac{1}{2y}$

12. $\dfrac{5}{2x} = \dfrac{9 - 2x}{8x} + 3$

13. $\dfrac{n - 5}{n + 5} + \dfrac{n + 15}{n - 5} = \dfrac{25}{25 - n^2} + 2$

14. $\dfrac{2}{x - 2} + \dfrac{1}{x + 1} = \dfrac{1}{x^2 - x - 2}$

15. $\dfrac{2x}{x + 7} - 1 = \dfrac{x}{x + 3} + \dfrac{1}{x^2 + 10x + 21}$

16. $\dfrac{9}{y - 3} - \dfrac{4}{y - 6} = \dfrac{18}{y^2 - 9y + 18}$

In Problems 17 to 20, solve each equation for the indicated unknown.

17. $\dfrac{x}{a+b} - b = a$ for x

18. $\dfrac{ay}{b} + \dfrac{by}{a} = 2$ for y

19. $\dfrac{m}{x} + \dfrac{x-m}{x} - m = 1$ for x

20. $\dfrac{a-a^2}{x} = 1 - \dfrac{x-2a}{2x}$ for x

C In Problems 21 to 24, solve each equation with the aid of a calculator. Round off all answers to the correct number of significant digits.

21. $71.42x - 33.21 = 0$

22. $\dfrac{35x + 17.05}{-3x - 21.14} = 37.01$

23. $(3.004 \times 10^{17})x + (2.173 \times 10^{15}) = 0$

24. $13.031 = 2\pi(11.076 + x)$

25. A pharmacist has 8 liters of a 15% solution of acid. How much distilled water must she add to it to reduce the concentration of acid to 10%?

26. A family has spent a total of $3500 on a solar water heater and insulation for their home. Federal income tax credits of 30% of the cost of the solar water heat and 15% of the cost of the insulation will amount to $825. How much was spent on the water heater and how much on insulation?

27. A total of $10,000 is invested, part at 7% simple annual interest, and the other part at 8%. If a return of $735 is expected at the end of one year, how much was invested at each rate?

28. In a certain district, 25% of the voters eligible for grand-jury service are members of minorities. Records show that during a 5-year period, 3% of all eligible people in that district were selected for grand-jury service, whereas 4% of the eligible minority members were selected. What percent of the eligible nonminority members in that district were selected for grand-jury duty over the 5-year period?

29. By installing a $120 thermostat that automatically reduces the temperature setting at night, a family hopes to cut its annual bill for heating oil by 8%, and thereby recover the cost of the thermostat in fuel savings after 2 years. What was the family's annual fuel bill before installing the thermostat?

C **30.** An archaeologist discovers a crown weighing 800 grams in the tomb of an ancient king. Evidence indicates that the crown is made of a mixture of gold and silver. Gold weighs 19.3 grams per cubic centimeter; silver weighs 10.5 grams per cubic centimeter; and it is found that the crown weighs 16.2 grams per cubic centimeter. How many grams of gold does the crown contain?

31. Juanita and Pedro run a race from point A to point B and Juanita wins by 5 meters. She calculates that Pedro will have a fair chance to win a second race to point B if he starts at point A and she starts 6 meters behind him. What is the distance between point A and point B?

32. Plant A produces methane from biomass three times as fast as plant B. The two plants operating together yield a certain amount of methane in 12 hours. How long would it take plant A operating alone to yield the same amount of methane?

In Problems 33 to 36, complete the square by adding a constant to each quantity.

33. $x^2 + x$

34. $x^2 - 24x$

35. $x^2 - 9x$

36. $x^2 + \sqrt{3}\,x$

In Problems 37 to 50, find all roots (real or complex) of each equation. (In Problems 43 and 44, round off your answers to three decimal places.)

37. $2x^2 - 5x - 7 = 0$

38. $2t^2 - 7t + 5 = 0$

39. $10y^2 = 3 + y$

40. $r(3r + 1) = 4$

41. $5x^2 - 7x - 2 = 0$

42. $x^2 + 2x = 5$

C **43.** $2.35t^2 + 6.42t - 0.91 = 0$

C **44.** $1.7y^2 + 0.33y = \dfrac{\pi}{3}$

45. $4x^2 + 3 = 3x$

46. $4s^2 + \tfrac{11}{2}s + 4 = 0$

47. $2r = 1 - \dfrac{2}{r}$

48. $5x = \dfrac{3}{x-2}$

49. $\dfrac{1-6x}{2x} = 2x$

50. $x^2 + \sqrt{2}x - 1 = 0$

In Problems 51 to 54, use the discriminant to determine the nature of the roots of each quadratic equation without solving it.

51. $2x^2 - 3x + 1 = 0$

c **52.** $169x^2 + 286x + 121 = 0$

53. $4x^2 + 9x + 6 = 0$

54. $\sqrt{3}x^2 + 2x - \sqrt{3} = 0$

55. A commercial jet could decrease the time needed to cover a distance of 2475 nautical miles by one hour if it were to increase its present speed by 100 knots (1 knot = 1 nautical mile per hour). What is its present speed?

c **56.** A biologist finds that the number N of water mites in a sample of lake water is related to the water temperature T in degrees Celsius by the equation

$$N = 5.5T^2 - 19T.$$

At what temperature is $N = 2860$?

57. The diagonal of a rectangular television screen measures 20 inches and the height of the screen is 4 inches less than its width. Find the dimensions of the screen.

58. All of the electricity in an alternative-energy home is supplied by batteries that are charged by a combination of solar panels and a wind-powered generator. Working together on a sunny, windy day, the solar panels and the generator can charge the batteries in 2.4 hours. Working alone on a windless, sunny day, it takes the solar panels 2 hours longer to charge the batteries than it takes the generator to do the job alone on a windy night. How long does it take the solar panels to charge the batteries on a windless, sunny day?

In Problems 59 to 92, find all real roots of each equation. Check for extraneous roots when necessary.

59. $\sqrt{x-3} - 6 = 0$

60. $8 - \sqrt[3]{y-1} = 6$

61. $t + \sqrt{t+1} = 7$

62. $\sqrt{2y-5} = 10 - y$

63. $\sqrt[4]{x^2 + 5x + 6} = \sqrt{x+4}$

64. $\sqrt[3]{t+1} = \sqrt{t+1}$

65. $\sqrt{t+1} + \sqrt{2t+3} = 5$

66. $\sqrt{5x-1} - \sqrt{2x} = \sqrt{x-1}$

67. $\sqrt{t+4} + \sqrt{2t+10} = 3$

68. $\sqrt{2r+5} - \sqrt{r+2} = \sqrt{3r-5}$

69. $t^4 - 5t^2 + 4 = 0$

70. $24v^4 - v^2 - 10 = 0$

71. $9c^4 - 18c^2 + 8 = 0$

72. $6a^4 + a^2 - 15 = 0$

73. $y - 2y^{1/2} - 15 = 0$

74. $t^{2/3} + 2t^{1/3} - 48 = 0$

75. $\left(x - \dfrac{8}{x}\right)^2 + \left(x - \dfrac{8}{x}\right) = 42$

76. $(t^2 + 2t)^2 - 2(t^2 + 2t) = 3$

77. $z^{-2} + z^{-1} = 6$

78. $x^{-2} = 2x^{-1} + 8$

79. $y + 2 + \sqrt{y+2} = 2$

80. $x + 7 = 2 + \sqrt{x+7}$

81. $\sqrt{x+20} + 3 = 4\sqrt{x+20}$

82. $2x^2 + x - 4\sqrt{2x^2 + x + 4} = 1$

83. $x^5 - 2x^3 - x^2 + 2 = 0$

84. $(u-1)^3 - (u-1)^2 = u - 2$

85. $t^4 + t^3 - t - 1 = 0$

86. $x^{7/12} + x^{1/4} - x^{1/3} = 1$

87. $x^{-3/5} + x^{-2/5} - 2x^{-1/5} - 2 = 0$

88. $u - 4u^{5/7} + 32u^{2/7} = 128$

89. $\sqrt{5t + 2\sqrt{t^2 - t + 7}} - 4 = 0$

90. $\dfrac{\sqrt{v+16}}{\sqrt{4-v}} + \dfrac{\sqrt{4-v}}{\sqrt{v+16}} = \dfrac{5}{2}$

91. $\dfrac{\sqrt{1+x} + \sqrt{1-x}}{\sqrt{1+x} - \sqrt{1-x}} = 2$

92. $\sqrt{13 + 3\sqrt{y+5} + 4\sqrt{y+1}} = 5$

93. In engineering, the equation

$$L\sqrt[3]{F} = d\sqrt[3]{4bwY}$$

can be used to determine the bending of a beam with a rectangular cross section that is freely supported at the ends and loaded at the middle with a force F. Here L is the length of the beam, d is its depth, w is its width, b is the amount of bending in the middle, and Y is Young's modulus of elasticity for the material of the beam. Solve the equation for b in terms of L, F, d, w, and Y.

© **94.** The *equally tempered* musical scale, used by Bach in his "*Well-Tempered Clavier*" of 1722, consists of thirteen notes, C, C$^{\#}$, D, D$^{\#}$, E, F, F$^{\#}$, G, G$^{\#}$, A, A$^{\#}$, B, and high C. In this scale, the frequency of high C is twice the frequency of C, and the ratios of the frequencies of successive notes are equal. If the frequency of A is taken to be 440 Hz ("concert A"), find the frequencies of the remaining notes.

95. Square roots can be found on a four-function calculator as follows: To find $\sqrt{x}$, begin by making a reasonable estimate. Add half of the estimated number to x divided by twice the estimated number. Call the result the second estimate. Repeat the procedure with the second estimate to obtain a third estimate. Continue in this way until the estimate no longer changes when you carry out the procedure. The resulting number is $\sqrt{x}$, to within the calculator's limits of accuracy. Prove this algebraically.

96. Government economists monitoring the rate of inflation calculate the percent of increase in the cost of living at the end of each month. If m is the percent increase in a given month, then the equivalent annual percent increase a is given by

$$a = 100\left(1 + \frac{m}{100}\right)^{12} - 100.$$

(a) Solve this equation for m in terms of a.

© (b) Find the equivalent annual percent of increase corresponding to $m = 1\%$ increase per month.

97. Given that $-3 \leq 2 - 5x < 4$, what equivalent inequality is obtained if

(a) 3 is added to all members?

(b) All members are divided by -5?

98. If $0 \leq a < b$, prove that $\sqrt{a} < \sqrt{b}$.

In Problems 99 to 104, solve each inequality and sketch its solution set on a number line.

99. $3x + 1 \geq 16$

100. $-4x + 3 < 15$

101. $-9t - 3 < -30$

102. $-8y - 4 \geq -32$

103. $\dfrac{4x}{3} - 2 > \dfrac{x}{4}$

104. $\dfrac{3x}{2} + 13 \leq \dfrac{x}{8}$

105. A student on academic probation must earn a C in his algebra course in order to avoid being expelled from college. Since he is lazy and has no interest in earning a grade any higher than necessary, he wants his final numerical grade for algebra to lie on the interval $[70, 80)$. This grade will be determined by taking three-fifths of his classroom average, which is 68, plus two-fifths of his score on the final exam. Since he has always had trouble with inequalities, he can't figure out the range in which his score on the final exam should fall to allow him to stay in school. Can you help him?

© **106.** A runner leaves a starting point and runs along a straight road at a steady speed of 8.8 miles per hour. After a while, she turns around and begins to run back to the starting point. How fast must she run on the way back so that her average speed for the entire run will be greater than 8 miles per hour, but no greater than 8.5 miles per hour? (*Caution:* The average speed for her entire run *cannot* be found by adding 8.8 miles per hour to her return speed and dividing by 2.)

In Problems 107 to 112, find the distance between the numbers and illustrate on a number line.

107. -3 and 4

108. 0 and -5

© **109.** -2.735 and $-\pi$

110. -3.2 and 4.1

111. $-\frac{2}{3}$ and $\frac{5}{2}$

© **112.** 1.42 and $\sqrt{2}$

In Problems 113 to 118, determine which equations or inequalities are true for all values of the variables.

113. $|x^2 - 4| = |x - 2||x + 2|$

114. $|x^2 - 4| = x^2 + 4$

115. $|-x|^2 = x^2$

116. $|x - y| = |y - x|$

117. $|x^2 + 3x| \leq x^2 + |3x|$

118. $|x - y| \leq |x| + |y|$

In Problems 119 to 146, solve each equation or inequality. Sketch the solution set of each inequality on a number line.

119. $|2x - 3| = 6$ **120.** $|2y + 1| = 5$

121. $|t - 1| = 3t$ **122.** $|2t + 3| = |t - 2|$

123. $|3x - 6| > 3$ **124.** $|2x - 3| \leq 5$

125. $|4x - 3| \geq 5$ **126.** $|2x + 1| < 3$

127. $|2 - 3y| \leq 1$ **128.** $|x - 1| < |x + 1|$

129. $|5 - 3z| < \frac{1}{2}$ **130.** $|2 - s| > |5 - 3s|$

131. $6x^2 - 11x - 10 < 0$ **132.** $6x^2 - x \geq 2$

133. $6 - 5x - 6x^2 \geq 0$ **134.** $6x^2 \leq 15x$

135. $5 > x(3x + 14)$ **136.** $x^5 + 2 < 2x^4 + x$

137. $x(x + 3)(x + 5) < 0$ **138.** $x^4 + 1 \geq x^3 + x$

139. $\dfrac{3x - 1}{2x - 5} > 0$ **140.** $\dfrac{t + 1}{t - 2} \leq 0$

141. $\dfrac{3x - 1}{5x - 7} \geq 2$ **142.** $\dfrac{5x - 3}{7x + 1} \leq 1$

143. $\dfrac{x(x + 1)(x - 2)}{(x + 2)(x + 3)} < 0$

144. $\dfrac{(x - 1)(x - 2)(x + 3)}{(x - 4)(x + 5)(x + 1)} > 0$

145. $\dfrac{x^2(x^2 - 1)}{x^2 - x - 2} \geq 0$ **146.** $\dfrac{1 - 5x}{9 - 3x} \leq \dfrac{2x + 1}{x - 3}$

147. It has been predicted that over the next decade, the number Q of quadrillions of Btu's of energy used per year in the United States will satisfy the inequality $|Q - 95| \leq 10$. Find an interval containing Q according to this estimate.

148. It is predicted that the number N of cars to be produced by the automobile industry for the coming year will satisfy the inequality $|N - 4,000,000| < 500,000$. Describe the anticipated production as an interval of real numbers.

149. Engineers estimate the percentage p of the heat stored in wood that will be delivered by a free-standing stove with good air control satisfies the inequality $|p - 65| \leq 10$. Find an interval containing p according to this estimate.

150. The diameter d, in millimeters, of a human capillary satisfies $|d - 10^{-2}| \leq 5 \times 10^{-3}$. According to this inequality, what are the maximum and minimum diameters of a human capillary?

⊙ 151. A company specializes in buying complimentary copies of textbooks from professors and selling the books to students. The company has purchased 500 copies of a popular calculus book from professors at $5 per copy. Suppose that all of these copies will be sold to students if the selling price is $10 per copy, but that each $1 increase in the selling price will result in 100 fewer copies being sold. If the profit to the company from the sale of these calculus books is to be between $1000 and $1100, inclusive, what will the selling price be? [*Hint:* Profit = Revenue − Cost.]

152. Automakers in Detroit calculate that for each car they produce, they will have to replace a weight W of not less than 400 pounds and not more than 500 pounds of steel by light-weight materials, in order to meet mileage goals set by the government. Find numbers A and B such that W satisfies the inequality $|W - A| \leq B$.

153. A machine that fills one-liter milk cartons can make an error of at most 2%. If L is the number of liters of milk in a carton filled by this machine, find numbers A and B such that $|L - A| \leq B$.

154. Let x be a positive number, let n be a positive integer, and let r be the result of rounding off x to n significant digits. Prove that the *relative error* $|x - r|/x$ made in estimating x by r cannot exceed 5×10^{-n}. [*Hint:* See Problem 51 in Problem Set 1.9 and Problem 88 in Problem Set 2.6.]

Chapter 3

Functions and Their Graphs

Ever since the Italian scientist Galileo Galilei (1564–1642) first used quantitative methods in the study of dynamics, advances in our understanding of the natural world have become more and more dependent on the use of mathematics. Important quantities have been identified, methods of measurement have been developed, and relationships among quantities have been formulated as scientific laws and principles. The idea of a *function* enables us to express relationships among observable quantities with efficiency and precision. In this chapter we discuss the Cartesian coordinate system, equations of straight lines and circles, graphs of functions, the algebra of functions, and inverse functions.

3.1 THE CARTESIAN COORDINATE SYSTEM

In Section 1.2, we saw that a point P on a number line can be specified by a real number x called its *coordinate*. Similarly, by using a *Cartesian coordinate system*, named in honor of the French philosopher and mathematician René Descartes (1596–1650), we can specify a point P in the plane with *two* real numbers, also called *coordinates*.

René Descartes

Figure 1

y axis

x axis

O

A **Cartesian coordinate system** consists of two perpendicular number lines, called **coordinate axes,** which meet at a common origin O (Figure 1). Ordinarily, one of the number lines, called the *x* **axis,** is horizontal, and the other, called the **y axis,** is vertical. Numerical coordinates increase to the right along the *x* axis and upward along the *y* axis. We usually use the same scale (that is, the same unit distance) on the two axes, although in some of our figures, space considerations make it convenient to use different scales.

Figure 2

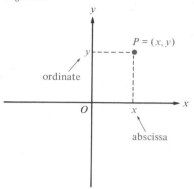

If *P* is a point in the plane, the **coordinates** of *P* are the coordinates *x* and *y* of the points where perpendiculars from *P* meet the two axes (Figure 2). The *x* coordinate is called the **abscissa** of *P* and the *y* coordinate is called the **ordinate** of *P*. The coordinates of *P* are traditionally written as an ordered pair (x, y) enclosed in parentheses, with the abscissa first and the ordinate second.*

To **plot** the point *P* with coordinates (x, y) means to draw Cartesian coordinate axes and to place a dot representing *P* at the point with abscissa *x* and ordinate *y*. You can think of the ordered pair (x, y) as the numerical "address" of *P*. The correspondence between *P* and (x, y) seems so natural that in practice we identify the point *P* with its "address" (x, y) by writing $P = (x, y)$. With this identification in mind, we call an ordered pair of real numbers (x, y) a **point,** and we refer to the set of all such ordered pairs as the **Cartesian plane** or the ***xy* plane.**

The *x* and *y* axes divide the plane into four regions called **quadrants I, II, III,** and **IV** (Figure 3). Quadrant I consists of all points (x, y) for which both *x* and *y* are positive, quadrant II consists of all points (x, y) for which *x* is negative and *y* is positive, and so forth, as shown in Figure 3. Notice that a point on a coordinate axis belongs to no quadrant.

Figure 3

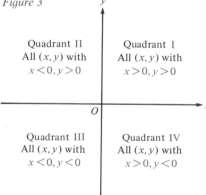

Example 1   Plot each point and indicate which quadrant or coordinate axis contains the point.

(a) $(4, 3)$ **(b)** $(-3, 2)$ **(c)** $(-5, -1)$ **(d)** $(2, -4)$
(e) $(-3, 0)$ **(f)** $(0, 4)$ **(g)** $(0, -\frac{3}{2})$ **(h)** $(0, 0)$

Solution   The points are plotted in Figure 4.

(a) $(4, 3)$ lies in quadrant I
(b) $(-3, 2)$ lies in quadrant II
(c) $(-5, -1)$ lies in quadrant III
(d) $(2, -4)$ lies in quadrant IV
(e) $(-3, 0)$ lies on the *x* axis
(f) $(0, 4)$ lies on the *y* axis
(g) $(0, -\frac{3}{2})$ lies on the *y* axis
(h) $(0, 0)$, the origin, lies on both axes

Figure 4

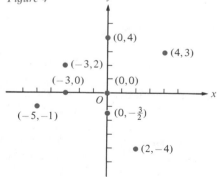

* Unfortunately, this is the same symbolism used for an open interval; however, it is usually clear from the context what is intended.

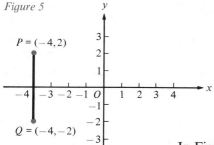

Figure 5

Example 2   Plot the point $P = (-4, 2)$ and determine the coordinates of the point Q if the line segment $\overline{PQ}$ is perpendicular to the x axis and is bisected by it.

Solution   We begin by plotting the point $P = (-4, 2)$ (Figure 5). We see that P is 2 units directly above the point with coordinate -4 on the x axis; hence, Q is 2 units directly below the same point. Therefore, $Q = (-4, -2)$.   ■

In Figure 5, the point $(-4, 0)$ is the midpoint of the line segment $\overline{PQ}$. More generally, the midpoint of a line segment can be obtained by using the following formula:

The Midpoint Formula

The point $\left(\dfrac{a + c}{2}, \dfrac{b + d}{2} \right)$ is the midpoint of the line segment joining the point (a, b) and the point (c, d).

We leave the proof of the midpoint formula as an exercise (Problem 44).

Example 3   Find the midpoint M of the line segment joining $R = (5, -4)$ and $S = (3, 6)$. Plot the points R, M, and S.

Solution   By the midpoint formula, $M = \left(\dfrac{5 + 3}{2}, \dfrac{-4 + 6}{2} \right) = (4, 1)$. In Figure 6, we have plotted R, M, and S.   ■

Figure 6

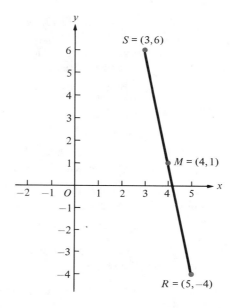

The Distance Formula

One of the attractive features of the Cartesian coordinate system is a simple formula that gives the distance between two points in terms of their coordinates. If P_1 and P_2 are two points in the Cartesian plane, we denote the distance between P_1 and P_2 by $|\overline{P_1P_2}|$.

The Distance Formula

> If $P_1 = (x_1, y_1)$ and $P_2 = (x_2, y_2)$ are two points in the Cartesian plane, then the distance between P_1 and P_2 is given by
> $$|\overline{P_1P_2}| = \sqrt{(x_2 - x_1)^2 + (y_2 - y_1)^2}.$$

To derive the distance formula, we reason as follows: The *horizontal* distance between P_1 and P_2 is the same as the distance between the points with coordinates x_1 and x_2 on the x axis (Figure 7a). From the discussion of absolute value in Section 2.6, it follows that the horizontal distance between P_1 and P_2 is $|x_2 - x_1|$ units.

Figure 7

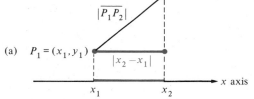

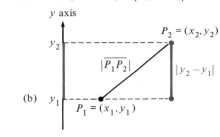

Figure 8

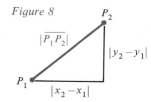

Similarly, the *vertical* distance between P_1 and P_2 is $|y_2 - y_1|$ units (Figure 7b). If the line segment $\overline{P_1P_2}$ is neither horizontal nor vertical, it forms the hypotenuse of a right triangle with legs of lengths $|x_2 - x_1|$ and $|y_2 - y_1|$ (Figure 8). Therefore, by the Pythagorean theorem,*

$$|\overline{P_1P_2}|^2 = |x_2 - x_1|^2 + |y_2 - y_1|^2 = (x_2 - x_1)^2 + (y_2 - y_1)^2,$$

and it follows that

$$|\overline{P_1P_2}| = \sqrt{(x_2 - x_1)^2 + (y_2 - y_1)^2}.$$

We leave it to you to check that the formula works even if the line segment $\overline{P_1P_2}$ is horizontal or vertical (Problem 36).

Because $(x_2 - x_1)^2 = (x_1 - x_2)^2$ and $(y_2 - y_1)^2 = (y_1 - y_2)^2$, the distance formula can also be written as

$$|\overline{P_1P_2}| = \sqrt{(x_1 - x_2)^2 + (y_1 - y_2)^2}.$$

In other words, the order in which you subtract the abscissas or the ordinates does not affect the result.

* See the Appendix on Analytical Geometry for a review of the Pythagorean theorem and its converse.

Example 4 Let $A = (-2, -1)$, $B = (1, 3)$, $C = (-1, 2)$, and $D = (3, -2)$. Find

(a) the distance $|\overline{AB}|$.   (b) the distance $|\overline{CD}|$.

Solution By the distance formula, we have

(a) $|\overline{AB}| = \sqrt{[1 - (-2)]^2 + [3 - (-1)]^2} = \sqrt{3^2 + 4^2} = \sqrt{25} = 5$ units.

(b) $|\overline{CD}| = \sqrt{[3 - (-1)]^2 + (-2 - 2)^2} = \sqrt{32} = 4\sqrt{2}$ units. ∎

© **Example 5** Let $P = (31.42, -17.04)$ and $Q = (13.75, 11.36)$. Using a calculator, find $|\overline{PQ}|$. Round off your answer to four significant digits.

Solution $$|\overline{PQ}| = \sqrt{(13.75 - 31.42)^2 + [11.36 - (-17.04)]^2} = 33.45 \text{ units.}$$ ∎

Example 6 Let $A = (-5, 3)$, $B = (6, 0)$, and $C = (5, 5)$.

(a) Plot the points A, B, and C, and draw the triangle ABC.

(b) Find the distances $|\overline{AB}|$, $|\overline{AC}|$, and $|\overline{BC}|$.

(c) Show that ACB is a right triangle.

(d) Find the area of triangle ACB.

Figure 9

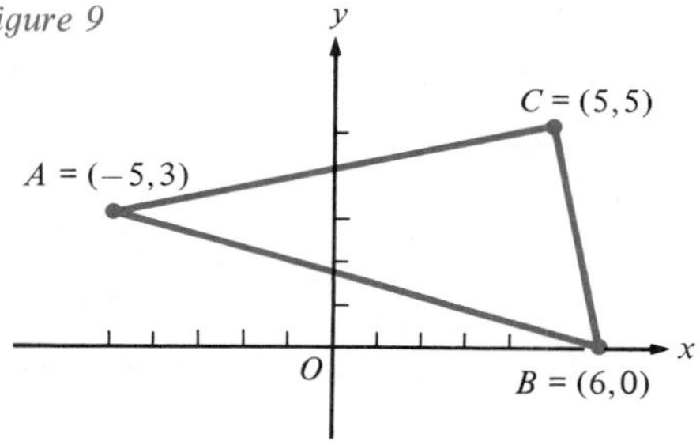

Solution   (a) The points A, B and C are plotted and the triangle ABC is drawn in Figure 9.

(b) $|\overline{AB}| = \sqrt{[6 - (-5)]^2 + (0 - 3)^2} = \sqrt{11^2 + (-3)^2} = \sqrt{130}$

$|\overline{AC}| = \sqrt{[5 - (-5)]^2 + (5 - 3)^2} = \sqrt{10^2 + 2^2} = \sqrt{104} = 2\sqrt{26}$

$|\overline{BC}| = \sqrt{(5 - 6)^2 + (5 - 0)^2} = \sqrt{1^2 + 5^2} = \sqrt{26}$

(c) Figure 9 leads us to suspect that the angle at vertex C is a right angle. To confirm this, we use the *converse* of the Pythagorean theorem; that is, we check to see if $|\overline{AC}|^2 + |\overline{BC}|^2 = |\overline{AB}|^2$. From (b), we have

$$|\overline{AC}|^2 + |\overline{BC}|^2 = 104 + 26 = 130 = |\overline{AB}|^2.$$

Therefore, ACB is, indeed, a right triangle.

(d) Taking $|\overline{AC}| = 2\sqrt{26}$ as the base of the triangle and $|\overline{BC}| = \sqrt{26}$ as its altitude, we find that

$$\text{area} = \tfrac{1}{2} \text{ base} \times \text{altitude} = \tfrac{1}{2}(2\sqrt{26})\sqrt{26} = 26 \text{ square units.}$$ ∎

Graphs in the Cartesian Plane

The **graph** of an equation or inequality in two unknowns x and y is defined to be the set of all points $P = (x, y)$ in the Cartesian plane whose coordinates x and y satisfy the equation or inequality. Many (but not all) equations in x and y have graphs that are smooth curves in the plane. For instance, consider the equation $x^2 + y^2 = 9$. We can rewrite this equation as $\sqrt{x^2 + y^2} = \sqrt{9}$ or as

$$\sqrt{(x - 0)^2 + (y - 0)^2} = 3.$$

Figure 10

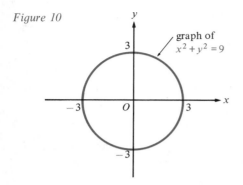

By the distance formula, the last equation holds if and only if the point $P = (x, y)$ is 3 units from the origin $O = (0,0)$. Therefore, the graph of $x^2 + y^2 = 9$ is a circle of radius 3 units with its center at the origin O (Figure 10).

If we are given a curve in the Cartesian plane, we can ask whether there is an equation for which it is the graph. Such an equation is called an **equation for the curve** or an **equation of the curve.** For instance, $x^2 + y^2 = 9$ is an equation for the circle in Figure 10. Two equations or inequalities in x and y are said to be **equivalent** if they have the same graph. For example, the equation $x^2 + y^2 = 9$ is equivalent to the equation $\sqrt{x^2 + y^2} = 3$. We often use an equation for a curve to designate the curve; for instance, if we speak of "the circle $x^2 + y^2 = 9$," we mean "the circle for which $x^2 + y^2 = 9$ is an equation."

If $r > 0$, then the circle of radius r with center (h, k) consists of all points (x, y) such that the distance between (x, y) and (h, k) is r units (Figure 11). Using the distance formula, we can write an equation for this circle as $\sqrt{(x - h)^2 + (y - k)^2} = r$, or, equivalently,

Figure 11

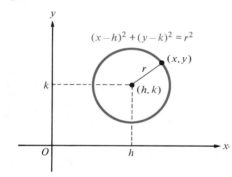

$$(x - h)^2 + (y - k)^2 = r^2.$$

This last equation is called the **standard form** for the equation of a circle in the xy plane.

Example 7   Find an equation for the circle of radius 5 with center at the point $(3, -2)$ (Figure 12).

Solution   Here $r = 5$ and $(h, k) = (3, -2)$, so, in standard form, the equation of the circle is

$$(x - 3)^2 + [y - (-2)]^2 = 5^2 \qquad \text{or} \qquad (x - 3)^2 + (y + 2)^2 = 25.$$

If desired, we can expand the squares, combine like terms, and rewrite the equation in the equivalent form

$$x^2 + y^2 - 6x + 4y - 12 = 0.$$

Figure 12

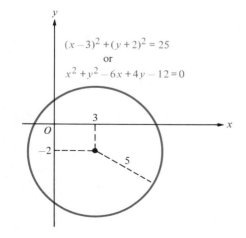

The last equation can be *restored* to standard form by completing the squares.* The work is arranged as follows:

$$x^2 + y^2 - 6x + 4y - 12 = 0$$
$$x^2 - 6x \qquad + y^2 + 4y \qquad = 12$$
$$x^2 - 6x + 9 + y^2 + 4y + 4 = 12 + 9 + 4$$
$$(x - 3)^2 + (y + 2)^2 = 25$$

*See page 89.

Figure 13

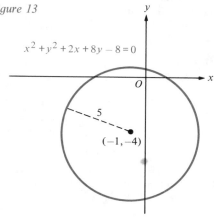

$$x^2 + y^2 + 2x + 8y - 8 = 0$$

Here, we added 9 to both sides of the equation to change $x^2 - 6x$ to the perfect square $x^2 - 6x + 9$, and we added 4 to both sides to change $y^2 + 4y$ to the perfect square $y^2 + 4y + 4$.

Example 8   Sketch the graph of $x^2 + y^2 + 2x + 8y - 8 = 0$.

Solution   Completing the square, we have

$$x^2 + 2x \qquad + y^2 + 8y \qquad = 8$$
$$x^2 + 2x \qquad + y^2 + 8y \qquad = 8$$
$$(x + 1)^2 + (y + 4)^2 = 25$$

Therefore, the graph is a circle of radius $\sqrt{25} = 5$ units with center $(-1, -4)$ (Figure 13). ∎

Problem Set 3.1

In each problem set, problems with colored numbers constitute a good representation of the main ideas of the section.

1. Plot each point and indicate which quadrant or coordinate axis contains it.

   (a) $(1, 6)$ (b) $(-2, 3)$ (c) $(4, -1)$
   (d) $(4, -2)$ (e) $(-1, -4)$ (f) $(0, 2)$
   (g) $(-3, 0)$ (h) $(0, -4)$

2. Plot the points $A = (\pi, \sqrt{2})$, $B = (-\sqrt{3}, \sqrt{2})$, $C = (\sqrt{5}, -\sqrt{2})$, and $D = (\frac{3}{4}, -\frac{27}{5})$.

In Problems 3 to 6, plot the point P and determine the coordinates of points Q, R, and S such that: (a) the line segment $\overline{PQ}$ is perpendicular to the x axis and is bisected by it; (b) the line segment $\overline{PR}$ is perpendicular to the y axis and is bisected by it; and (c) the line segment $\overline{PS}$ passes through the origin and is bisected by it.

3. $P = (3, 2)$ 4. $P = (-4, -3)$

5. $P = (-1, 3)$ 6. $P = \left(\dfrac{\sqrt{3}}{2}, -\dfrac{1}{2} \right)$

In Problems 7 to 12, use the midpoint formula to find the midpoint M of the line segment joining R and S. In 7 to 10, plot the points R, M, and S.

7. $R = (1, 2)$, $S = (3, 4)$ 8. $R = (2, 1)$, $S = (1, 4)$

9. $R = (-2, 3)$, $S = (8, 5)$

10. $R = (0, -9)$, $S = (-3, 7)$

11. $R = (1, \sqrt{2})$, $S = (-3, 2\sqrt{2})$

12. $R = (27.31, -42.62)$, $S = (13.27, 11.56)$

In Problems 13 to 26, find the distance between the two points.

13. $(7, 10)$ and $(1, 2)$ 14. $(-1, 7)$ and $(2, 11)$

15. $(7, -1)$ and $(7, 3)$ 16. $(-4, 7)$ and $(0, -8)$

17. $(-6, 3)$ and $(3, -5)$ 18. $(0, 4)$ and $(-4, 0)$

19. $(0, 0)$ and $(-8, -6)$ 20. $(t, 4)$ and $(t, 8)$

21. $(-3, -5)$ and $(-7, -8)$

22. $(-\frac{1}{2}, -\frac{3}{2})$ and $(-3, -\frac{5}{2})$

23. $(2, -t)$ and $(5, t)$

24. $(a, b + 1)$ and $(a + 1, b)$

25. $(2\sqrt{2}, -3)$ and $(4\sqrt{2}, 2)$

26. $(s + t, s - t)$ and $(t - s, t + s)$

Ⓒ In Problems 27 and 28, use a calculator to find the distance between the two points. Round off your answer to four significant digits.

27. $(-2.714, 7.111)$ and $(3.135, 4.982)$

28. $(\pi, \frac{53}{4})$ and $(-\sqrt{17}, \frac{211}{5})$

In Problems 29 to 32, (a) use the distance formula and the converse of the Pythagorean theorem to show that triangle ABC is a right triangle, and (b) find the area of triangle ABC (see Appendix III).

29. $A = (1, 1)$, $B = (5, 1)$, $C = (5, 7)$

30. $A = (-1, -2)$, $B = (3, -2)$, $C = (-1, -7)$

31. $A = (0, 0)$, $B = (-3, 3)$, $C = (2, 2)$

32. $A = (-2, -5)$, $B = (9, \frac{1}{2})$, $C = (4, \frac{21}{2})$

33. Show that the points $A = (-2, -3)$, $B = (3, -1)$, $C = (1, 4)$, and $D = (-4, 2)$ are the vertices of a square.

34. Show that the distance between the points (x_1, y_1) and (x_2, y_2) is the same as the distance between the point $(x_1 - x_2, y_1 - y_2)$ and the origin.

35. If $A = (-5, 1)$, $B = (-6, 5)$, and $C = (-2, 4)$, determine whether or not triangle ABC is isosceles.

36. Verify that the distance formula holds even if the line segment is horizontal or vertical.

37. Find all values of x for which the distance between the points $(x, 8)$ and $(-5, 3)$ is 13 units.

38. Find x if the distance between the points $(x, 0)$ and $(0, 3)$ is the same as the distance between the points $(x, 0)$ and $(7, 4)$.

39. Find all values of t for which the distance between the points $(-2, 3)$ and (t, t) is 5 units.

40. If P_1, P_2, and P_3 are points in the plane, then P_2 lies on the line segment $\overline{P_1P_3}$ if and only if $|\overline{P_1P_3}| = |\overline{P_1P_2}| + |\overline{P_2P_3}|$. Illustrate this geometric fact with diagrams.

In Problems 41 to 43, determine whether or not P_2 lies on the line segment $\overline{P_1P_3}$ by checking to see if $|\overline{P_1P_3}| = |\overline{P_1P_2}| + |\overline{P_2P_3}|$ (see Problem 40).

41. $P_1 = (1, 2)$, $P_2 = (0, \frac{5}{2})$, $P_3 = (-1, 3)$

42. $P_1 = (-\frac{7}{2}, 0)$, $P_2 = (-1, 5)$, $P_3 = (2, 11)$

43. $P_1 = (2, 3)$, $P_2 = (3, -3)$, $P_3 = (-1, -1)$

44. Show that the point $P_2 = \left(\dfrac{a + c}{2}, \dfrac{b + d}{2}\right)$ is the midpoint of the line segment joining $P_1 = (a, b)$ and $P_3 = (c, d)$. [*Hint:* Use the condition in Problem 40 to show that P_2 actually belongs to the line segment $\overline{P_1P_3}$. Then show that $|\overline{P_1P_2}| = |\overline{P_2P_3}|$.]

In Problems 45 to 50, find the equation in standard form of the circle satisfying the given conditions.

45. Radius 4, center $(0, 0)$.

46. Radius $\sqrt{3}$, center $(4, -\frac{3}{2})$.

47. Radius 2, center $(-1, 3)$.

48. Radius $\frac{2}{3}$, center $(\frac{3}{2}, -\sqrt{5})$.

49. Tangent to the x axis, center $(1, 3)$.

50. Radius $\sqrt{17}$, center on the x axis, and containing the point $(0, 1)$. [There are two such circles.]

In Problems 51 to 60, sketch the graph of each equation.

51. $x^2 + y^2 = 36$ **52.** $x^2 + y^2 = 3$

53. $(x - 3)^2 + (y - 5)^2 = 49$ **54.** $(x + 1)^2 + (y - 4)^2 = 4$

55. $(x + 2)^2 + (y - 1)^2 = 64$

56. $x^2 + 4x + y^2 - 4y = 0$

57. $x^2 - 4x + y^2 - 10y + 4 = 0$

58. $x^2 + y^2 - x - y - 1 = 0$

59. $x^2 + y^2 + 2x + 4y + 4 = 0$

60. $4x^2 + 4y^2 + 8x - 4y + 1 = 0$

61. On a Cartesian coordinate grid, an aircraft carrier is detected by radar at point $A = (52, 71)$ and a submarine is detected by sonar at point $S = (47, 83)$. If distances are measured in nautical miles, how far is the carrier from a point on the surface of the water directly over the submarine?

62. Sketch the graph of the inequality $x^2 + y^2 \leq 9$.

3.2 THE SLOPE OF A LINE

Figure 1

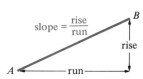

In ordinary language, the word "slope" refers to a steepness, an incline, or a deviation from the horizontal. For instance, we speak of a ski slope or the slope of a roof. In mathematics, the word "slope" has a precise meaning. Consider the line segment $\overline{AB}$ in Figure 1. The horizontal distance between A and B is called the **run** and the vertical distance between A and B is called the **rise**. The ratio of rise to run is called the **slope** of the line segment $\overline{AB}$ and is traditionally denoted by the symbol m:

$$m = \text{the slope of } \overline{AB} = \frac{\text{rise}}{\text{run}}.$$

Figure 2

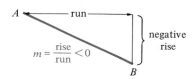

If the line segment $\overline{AB}$ is turned so that it becomes more nearly vertical, then the rise increases, the run decreases, and the slope $m = $ rise/run becomes larger. Therefore, the slope m really does give a numerical measure of the inclination or steepness of the line segment $\overline{AB}$—the greater the inclination, the greater the slope.

If the line segment $\overline{AB}$ is horizontal, its rise is zero, so its slope $m = $ rise/run is zero. Thus, *horizontal line segments have slope zero.* If $\overline{AB}$ slants downward to the right, as in Figure 2, its rise is considered to be negative; hence, its slope $m = $ rise/run is negative. (The run is always considered to be nonnegative.) Thus, *a line segment that slants downward from left to right has a negative slope.* Notice that *the slope $m = $ rise/run of a vertical line segment is undefined because the denominator is zero.*

Figure 3

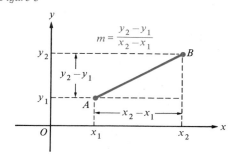

Now let $A = (x_1, y_1)$ and $B = (x_2, y_2)$, and consider the line segment $\overline{AB}$ (Figure 3). If B is above and to the right of A, the line segment $\overline{AB}$ has rise $= y_2 - y_1$ and run $= x_2 - x_1$, so its slope is

$$m = \frac{y_2 - y_1}{x_2 - x_1}$$

Even if B is not above and to the right of A, the slope m of $\overline{AB}$ is given by the same formula (Problem 25), and we have the following:

The Slope Formula

Let $A = (x_1, y_1)$ and $B = (x_2, y_2)$ be two points in the Cartesian plane. Then, if $x_1 \neq x_2$, the slope m of the line segment $\overline{AB}$ is

$$m = \frac{y_2 - y_1}{x_2 - x_1}.$$

Notice that $\dfrac{y_2 - y_1}{x_2 - x_1} = \dfrac{y_1 - y_2}{x_1 - x_2}$ (why?), so the slope of a line segment is the same regardless of which endpoint is called (x_1, y_1) and which is called (x_2, y_2).

Example 1   In each case, sketch the line segment $\overline{AB}$ and find its slope m by using the slope formula.

(a) $A = (-3, -2), B = (4, 1)$ **(b)** $A = (-2, 3), B = (5, 1)$

(c) $A = (-2, 4), B = (5, 4)$ **(d)** $A = (3, -1), B = (3, 6)$

Solution   The line segments are sketched in Figure 4.

(a) $m = \dfrac{y_2 - y_1}{x_2 - x_1} = \dfrac{1 - (-2)}{4 - (-3)} = \dfrac{1 + 2}{4 + 3} = \dfrac{3}{7}$

(b) $m = \dfrac{y_2 - y_1}{x_2 - x_1} = \dfrac{1 - 3}{5 - (-2)} = \dfrac{-2}{5 + 2} = -\dfrac{2}{7}$

(c) $m = \dfrac{y_2 - y_1}{x_2 - x_1} = \dfrac{4 - 4}{5 - (-2)} = \dfrac{0}{5 + 2} = 0$

(d) m is undefined, since $x_2 - x_1 = 0$. ∎

Figure 4

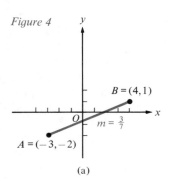

(a)

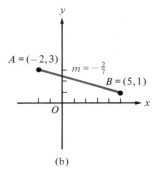

(b)

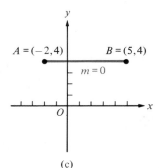

(c)

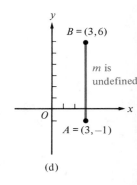

(d)

From the similar triangles APB and CQD in Figure 5, you can see that *two parallel line segments $\overline{AB}$ and $\overline{CD}$ have the same slope*. Likewise, *if two line segments $\overline{AB}$ and $\overline{CD}$ lie on the same straight line L, then they have the same slope* (Figure 6). The common slope of all the segments lying on a line L is called the **slope** of L.

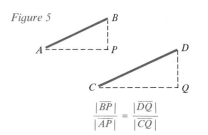

Figure 5

$$\dfrac{|\overline{BP}|}{|\overline{AP}|} = \dfrac{|\overline{DQ}|}{|\overline{CQ}|}$$

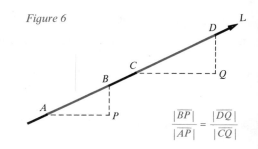

Figure 6

$$\dfrac{|\overline{BP}|}{|\overline{AP}|} = \dfrac{|\overline{DQ}|}{|\overline{CQ}|}$$

Example 2   Sketch the line L that contains the point $P = (1, 2)$ and has slope

(a) $m = \frac{2}{3}$ **(b)** $m = -\frac{2}{3}$.

Solution   **(a)** The condition $m = 2/3$ means that, for every 3 units we move to the right from a point on L, we must move up 2 units to get back to L. If we start at the point $P = (1, 2)$ on L, move 3 units to the right and 2 units up, we arrive at the point $Q = (1 + 3, 2 + 2) = (4, 4)$ on L. Because any two points on a line determine the line, we simply plot $P = (1, 2)$ and $Q = (4, 4)$, and use a straightedge to draw L (Figure 7a).

Figure 7

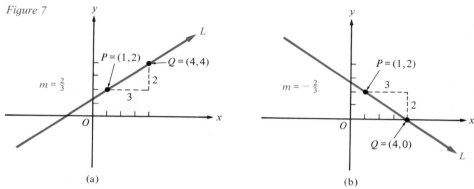

(a)   (b)

(b) The condition $m = -2/3$ means that, for every 3 units we move to the right from a point on L, we must move down 2 units to get back to L. If we start at the point $P = (1, 2)$ on L, move 3 units to the right and 2 units down, we arrive at the point $Q = (1 + 3, 2 - 2) = (4, 0)$ on L. Thus, we plot $P = (1, 2)$ and $Q = (4, 0)$, and use a straightedge to draw L (Figure 7b). ∎

From the fact that two parallel line segments have the same slope, it follows that *two parallel lines have the same slope.* Conversely, you can prove by elementary geometry that *two distinct lines with the same slope are parallel* (Problem 26). Thus, we have the following:

Parallelism Condition

> Two distinct nonvertical straight lines in the Cartesian plane are parallel if and only if they have the same slope.

Example 3   Sketch the line L that contains the point $P = (3, 4)$ and is parallel to the line segment $\overline{AB}$, where $A = (-1, 2)$ and $B = (4, -5)$.

Solution   By the slope formula, the line segment $\overline{AB}$ has slope

$$m = \frac{y_2 - y_1}{x_2 - x_1} = \frac{-5 - 2}{4 - (-1)} = -\frac{7}{5}.$$

Hence, by the parallelism condition, L also has slope $m = -\frac{7}{5}$. Starting at the point $P = (3, 4)$ on L, we move 5 units to the right and 7 units down to the point

$Q = (3 + 5, 4 - 7) = (8, -3)$. Using a straightedge, we draw the line L through P and Q (Figure 8). ■

Figure 8

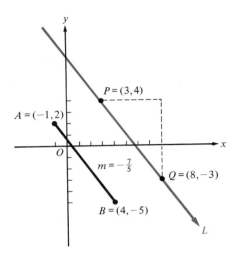

The following condition for two lines to be perpendicular is quite useful.

Perpendicularity Condition

> Two nonvertical straight lines in the Cartesian plane are perpendicular if and only if the slope of one of the lines is the negative of the reciprocal of the slope of the other.

To establish this result, let the two lines be L_1 and L_2, with slopes m_1 and m_2, respectively. The condition that the slope of either one of the lines is the negative of the reciprocal of the slope of the other can be written as $m_1 m_2 = -1$. Neither the angle between the lines nor their slopes are affected if we place the origin O at the point where L_1 and L_2 intersect (Figure 9). Starting at O on L_1, we move 1 unit to the right and $|m_1|$ units vertically to arrive at the point $A = (1, m_1)$ on L_1. Likewise, the point $B = (1, m_2)$ is on line L_2. By the Pythagorean theorem and its converse, triangle AOB is a right triangle if and only if

Figure 9

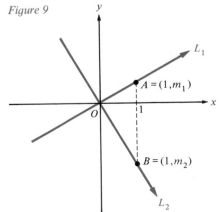

$$|\overline{AB}|^2 = |\overline{OA}|^2 + |\overline{OB}|^2.$$

Using the distance formula we find that

$$|\overline{AB}|^2 = (1 - 1)^2 + (m_1 - m_2)^2 = m_1^2 - 2m_1 m_2 + m_2^2,$$
$$|\overline{OA}|^2 = (1 - 0)^2 + (m_1 - 0)^2 = 1 + m_1^2, \quad \text{and}$$
$$|\overline{OB}|^2 = (1 - 0)^2 + (m_2 - 0)^2 = 1 + m_2^2.$$

Therefore, the condition $|\overline{AB}|^2 = |\overline{OA}|^2 + |\overline{OB}|^2$ is equivalent to

$$m_1^2 - 2m_1m_2 + m_2^2 = 1 + m_1^2 + 1 + m_2^2.$$

The last equation simplifies to $m_1m_2 = -1$, and the result is established.

Example 4   Find the slope m_1 of a line that is perpendicular to the line segment $\overline{AB}$, where $A = (-1, 2)$ and $B = (4, -5)$.

Solution   The slope m_2 of the line containing the points A and B is given by

$$m_2 = \frac{y_2 - y_1}{x_2 - x_1} = \frac{-5 - 2}{4 - (-1)} = -\frac{7}{5}.$$

Therefore, by the perpendicularity condition,

$$m_1 = -\frac{1}{m_2} = -\frac{1}{(-7/5)} = \frac{5}{7}.$$

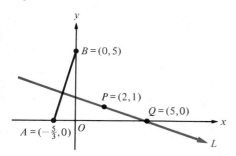

Figure 10

Example 5   Sketch the line L that contains the point $P = (2, 1)$ and is perpendicular to the line segment $\overline{AB}$, where $A = (-\frac{5}{3}, 0)$ and $B = (0, 5)$.

Solution   Slope of $\overline{AB} = \dfrac{y_2 - y_1}{x_2 - x_1} = \dfrac{5 - 0}{0 - (-5/3)} = \dfrac{5}{5/3} = 3.$

Therefore, the slope m of L is $m = -\frac{1}{3}$. Starting at the point $P = (2, 1)$ on L, we move 3 units to the right and 1 unit down, to the point $Q = (2 + 3, 1 - 1) = (5, 0)$ on L. Using a straightedge, we draw the line L through P and Q (Figure 10).

Problem Set 3.2

In Problems 1 to 10, sketch the line segment $\overline{AB}$ and find its slope using the slope formula.

1. $A = (-1, 8)$, $B = (4, 3)$

2. $A = (6, -1)$, $B = (-3, 4)$

3. $A = (-2, 2)$, $B = (2, -2)$

4. $A = (-3, 7)$, $B = (-1, 0)$

5. $A = (4, 0)$, $B = (0, -1)$

6. $A = (-1, 3)$, $B = (0, 0)$

7. $A = (-3, -\frac{1}{2})$, $B = (1, \frac{3}{2})$

8. $A = (-\frac{1}{3}, \frac{1}{4})$, $B = (\frac{1}{4}, -\frac{1}{3})$

Ⓒ 9. $A = (2.6, -5.3)$, $B = (1.7, -1.1)$

Ⓒ 10. $A = (73.24, 31.53)$, $B = (1.71, 2.4)$

In Problems 11 to 16, sketch the line L that contains the point P and has slope m.

11. $P = (1, 1)$, $m = 2$

12. $P = (-4, 1)$, $m = 0$

13. $P = (-3, 2)$, $m = -\frac{2}{5}$

14. $P = (\frac{1}{2}, 0)$, $m = \frac{1}{2}$

15. $P = (-\frac{2}{3}, -3)$, $m = -2$

16. $P = (-\frac{4}{3}, \frac{1}{2})$, $m = -\frac{3}{4}$

In Problems 17 and 18, sketch the line L that contains the point P and is parallel to the line segment $\overline{AB}$.

17. $P = (4, -3)$, $A = (-2, 3)$, $B = (3, -7)$

18. $P = (\frac{2}{3}, \frac{5}{3})$, $A = (\frac{1}{5}, \frac{3}{5})$, $B = (-\frac{2}{5}, \frac{4}{5})$

In Problems 19 and 20, find the slope m_1 of a line L that is perpendicular to the line segment $\overline{AB}$.

19. $A = (-1, 4)$, $B = (-2, -1)$

20. $A = (2, \frac{7}{5})$, $B = (1, \frac{12}{5})$

In Problems 21 to 24, sketch the line L that contains the point P and is perpendicular to the line segment $\overline{AB}$.

21. $P = (1, 2)$, $A = (-7, -3)$, $B = (-5, 0)$

22. $P = (5, 12)$, $A = (3, -2)$, $B = (\frac{4}{3}, -\frac{4}{3})$

23. $P = (\frac{3}{2}, \frac{5}{2})$, $A = (4, -\frac{1}{3})$, $B = (\frac{1}{2}, 6)$

24. $P = (0, 4)$, $A = (\frac{22}{7}, \sqrt{2})$, $B = (\frac{22}{7}, \sqrt{3})$

25. Show that the slope formula holds in all cases— even if the point B is not above and to the right of the point A.

26. Show that two distinct straight lines with the same slope are necessarily parallel. [*Hint:* If they weren't parallel, they would meet at some point P.]

27. Use slopes to show that the triangle with vertices $(-4, -2)$, $(2, -8)$, and $(4, 6)$ is a right triangle.

28. Use slopes to show that the quadrilateral with vertices $(-5, -2)$, $(1, -1)$, $(4, 4)$, and $(-2, 3)$ is a parallelogram. [*Hint:* Show that opposite sides have the same slope.]

29. (a) Determine d so that the line containing the points $(d, 3)$ and $(-2, 1)$ is perpendicular to the line containing the points $(5, -2)$ and $(1, 4)$.

   (b) Determine k so that the line containing the points $(k, 3)$ and $(-2, 1)$ is parallel to the line containing the points $(5, -2)$ and $(1, 4)$.

30. Consider the quadrilateral $ABCD$ with $A = (3, 1)$, $B = (2, 4)$, $C = (7, 6)$, and $D = (8, 3)$.

   (a) Use the concept of slope to determine whether or not the diagonals $\overline{AC}$ and $\overline{BD}$ are perpendicular.

   (b) Is the quadrilateral $ABCD$ a parallelogram? A rectangle? A square? A rhombus? Justify your answer.

3.3 EQUATIONS OF STRAIGHT LINES IN THE CARTESIAN PLANE

Consider a nonvertical line L having slope m and containing the point $P_0 = (x_0, y_0)$* (Figure 1). If $P = (x, y)$ is any other point on L, then, by the slope formula,

$$m = \frac{y - y_0}{x - x_0},$$

that is,

$$y - y_0 = m(x - x_0).$$

Notice that this last equation holds even if $P = P_0$, when it simply reduces to $0 = 0$. Using the parallelism condition and elementary geometry, you can see that, conversely, any point $P = (x, y)$ whose coordinates satisfy the equation $y - y_0 = m(x - x_0)$ lies on the line L (Problem 32). Therefore, we have the following:

Figure 1

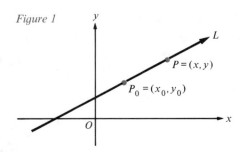

* P_0 is read "P sub zero," or, more often, "P naught." Similarly, we refer to x_0 and y_0 as "x naught" and "y naught."

Point-Slope Equation of a Straight Line

In the Cartesian plane, the straight line L that contains the point $P_0 = (x_0, y_0)$ and has slope m is the graph of the equation

$$y - y_0 = m(x - x_0).$$

You can write the equation $y - y_0 = m(x - x_0)$ as soon as you know the coordinates (x_0, y_0) of one point P_0 on the line L and the slope m of L. For this reason, the equation is called a **point-slope form.**

In Examples 1 and 2, find an equation in point-slope form of the given line L.

Example 1 L contains the point $(3, 4)$ and has slope $m = 5$.

Solution Substituting $x_0 = 3$, $y_0 = 4$, and $m = 5$ in the equation $y - y_0 = m(x - x_0)$, we have $y - 4 = 5(x - 3)$. ∎

Example 2 L contains the two points $(-5, 3)$ and $(4, -6)$.

Solution By the slope formula, L has slope

$$m = \frac{-6 - 3}{4 - (-5)} = \frac{-9}{9} = -1.$$

Since the point $(-5, 3)$ belongs to L, we can take $(x_0, y_0) = (-5, 3)$. The resulting point-slope equation of L is

$$y - 3 = -1[x - (-5)] \qquad \text{or} \qquad y - 3 = -(x + 5).$$

Of course, we could have chosen $(4, -6)$ as (x_0, y_0), in which case we would have obtained the equivalent equation

$$y + 6 = -(x - 4).$$ ∎

Figure 2

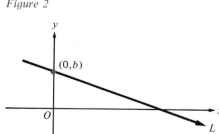

Now suppose that L is any nonvertical line with slope m. Since L is not parallel to the y axis, it must intersect this axis at some point $(0, b)$ (Figure 2). The ordinate b of the intersection point is called the **y intercept** of L. Since $(0, b)$ belongs to L, a point-slope equation of L is $y - b = m(x - 0)$. This equation simplifies to

$$y = mx + b$$

which is called the **slope-intercept form** of the equation for L. In this form, the coefficient of x is the slope and the constant term on the right is the y intercept.

Example 3   Show that the graph of the equation $3x - 5y - 15 = 0$ is a straight line by rewriting the equation in slope-intercept form. Find the slope m and the y intercept b, and sketch the graph.

Solution   We begin by solving the equation for y in terms of x:

$$3x - 5y - 15 = 0$$
$$-5y = -3x + 15$$
$$y = \tfrac{3}{5}x - 3.$$

Thus, we have the equation of a straight line in slope-intercept form with slope $m = \tfrac{3}{5}$ and y intercept $b = -3$. We can obtain the graph by drawing the line with slope $m = \tfrac{3}{5}$ through the point $(0, b) = (0, -3)$. An alternative method is to find the **x intercept** of the line; that is, the abscissa x of the point $(x, 0)$ where the line intersects the x axis. This is accomplished by putting $y = 0$ in the original equation $3x - 5y - 15 = 0$ and then solving for x:

$$3x - 15 = 0 \qquad \text{or} \qquad x = 5.$$

Thus, we obtain the graph by drawing the line through the point $(0, -3)$ on the y axis and the point $(5, 0)$ on the x axis (Figure 3). ∎

Figure 3

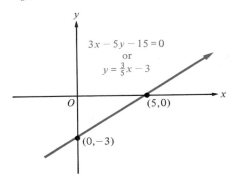

Figure 4

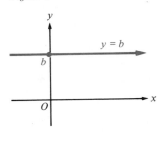

A horizontal line has slope $m = 0$; hence, in slope-intercept form, its equation is $y = 0(x) + b$, or simply $y = b$ (Figure 4). The equation $y = b$ places no restriction at all on the abscissa x of a point (x, y) on the horizontal line, but it requires that all of the ordinates y have the same value b.

A vertical line has an undefined slope, so you can't write its equation in slope-intercept form. However, since all points on a vertical line have the same abscissa, say, a, an equation of such a line is $x = a$ (Figure 5).

If A, B, and C are constants and if A and B are not both zero, an equation of the form

$$Ax + By + C = 0$$

Figure 5

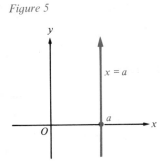

represents a straight line. If $B \neq 0$, the equation can be rewritten in the slope-intercept form as

$$y = \left(\frac{-A}{B}\right)x + \left(\frac{-C}{B}\right),$$

with slope $m = \dfrac{-A}{B}$ and y intercept $b = \dfrac{-C}{B}$. If $B = 0$, then $A \neq 0$ and the equation can be rewritten as

$$x = \frac{-C}{A},$$

the equation of a vertical line. The equation $Ax + By + C = 0$ is called the **general form** of the equation of a line.

Example 4   Let L be the line that contains the point $(-1, 2)$ and is perpendicular to the line L_1 whose equation is $3x - 2y + 5 = 0$ (Figure 6). Find an equation of L in **(a)** point-slope form, **(b)** slope-intercept form, and **(c)** general form.

Solution   We obtain the slope m_1 of L_1 by solving the equation $3x - 2y + 5 = 0$ for y in terms of x. The result is $y = \frac{3}{2}x + \frac{5}{2}$, so the slope of L_1 is $m_1 = \frac{3}{2}$. By the perpendicularity condition, the slope m of L is

$$m = -\frac{1}{m_1} = -\frac{1}{3/2} = -\frac{2}{3}.$$

Figure 6

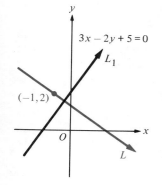

(a) Since L has slope $m = -\frac{2}{3}$ and contains the point $(-1, 2)$, its equation in point-slope form is

$$y - 2 = -\tfrac{2}{3}[x - (-1)]$$

or

$$y - 2 = -\tfrac{2}{3}(x + 1).$$

(b) Solving the equation in (a) for y in terms of x, we obtain the equation of L in slope-intercept form:

$$y = -\tfrac{2}{3}x + \tfrac{4}{3}.$$

(c) Multiplying both sides of the equation in (b) by 3 and rearranging terms, we obtain an equation of L in general form:

$$2x + 3y - 4 = 0. \qquad \blacksquare$$

Problem Set 3.3

In Problems 1 to 6, find an equation in point-slope form of the given line L.

1. L contains the point $(3, 2)$ and has slope $m = \frac{3}{4}$.

2. L contains the point $(0, 2)$ and has slope $m = -\frac{2}{3}$.

3. L contains the points $(-3, 2)$ and $(4, 1)$.

4. L contains the points $(1, 3)$ and $(-1, 1)$.

5. L contains the point $(-3, 5)$ and is parallel to the line segment $\overline{AB}$, where $A = (3, 7)$ and $B = (-2, 2)$.

6. L contains the point $(7, 2)$ and is parallel to the line segment $\overline{AB}$, where $A = (\frac{1}{3}, 1)$ and $B = (-\frac{2}{3}, \frac{3}{5})$.

In Problems 7 to 14, rewrite each equation in slope-intercept form, find the slope m and the y intercept b, and sketch the graph.

7. $3x - 2y = 6$

8. $5x - 2y - 10 = 0$

9. $y - 3x - 1 = 0$

10. $y + 1 = 0$

11. $x = -3y + 9$

12. $2x + y + 3 = 0$

13. $y - 2x - 3 = 0$

14. $x = -\frac{3}{5}y + \frac{7}{5}$

In Problems 15 to 25, find an equation of the line L in (a) point-slope form, (b) slope-intercept form, and (c) general form.

15. L contains the point $(-5, 2)$ and has slope $m = 4$.

16. L contains the point $(3, -1)$ and has slope $m = 0$.

17. L has slope $m = -3$ and y intercept $b = 5$.

18. L has slope $m = \frac{4}{5}$ and intersects the x axis at $(-3, 0)$.

19. L intersects the x and y axes at $(3, 0)$ and $(0, 5)$.

20. L contains the points $(\frac{7}{2}, \frac{5}{3})$ and $(\frac{2}{5}, -6)$.

21. L contains the point $(4, -4)$ and is parallel to the line that has the equation $2x - 5y + 3 = 0$.

22. L contains the point $(-3, \frac{2}{3})$ and is perpendicular to the y axis.

23. L contains the point $(-3, \frac{2}{3})$ and is perpendicular to the line that has the equation $5x + 3y - 1 = 0$.

24. L contains the point $(-6, -8)$ and has y intercept $b = 0$.

25. L contains the point $(\frac{2}{3}, \frac{5}{7})$ and is parallel to the line that has the equation $7x + 3y - 12 = 0$.

26. Find a real number B so that the graph of the equation $3x + By - 5 = 0$ has y intercept $b = -4$.

27. Find an equation of the perpendicular bisector of the line segment $\overline{AB}$, where $A = (3, -2)$ and $B = (7, 6)$. [*Hint:* Use the midpoint formula to find the midpoint of $\overline{AB}$.]

28. Suppose that the line L intersects the axes at $(a, 0)$ and $(0, b)$. Show that the equation of L can be

written in **intercept form**

$$\frac{x}{a} + \frac{y}{b} = 1.$$

29. A car-rental company leases automobiles for a charge of $22 per day plus $0.20 per mile. Write an equation for the cost y dollars in terms of the distance x miles driven if the car is leased for N days. If $N = 3$, sketch a graph of the equation.

30. If a piece of property is *depreciated linearly* over a period of n years, then its value y dollars at the end of x years is given by $y = c[1 - (x/n)]$, where c dollars is the original value of the property. An apartment building built in 1975 and originally worth $400,000 is being depreciated linearly over a period of 40 years. Sketch a graph showing the value y dollars of the apartment building x years after it was built and determine its value in the year 1995.

31. In 1984 the Solar Electric Company showed a profit of $3.45 per share, and it expects this figure to increase annually by $0.25 per share. If the year 1984 corresponds to $x = 0$, and successive years correspond to $x = 1, 2, 3$, and so on, find the equation $y = mx + b$ of the line that allows the company to predict its profit y dollars per share in future years. Sketch the graph of this equation and find the predicted profit per share in 1992.

32. Let L be the line that contains the point (x_0, y_0) and has slope m. Suppose that (x, y) is a point such that $y - y_0 = m(x - x_0)$. Show that (x, y) lies on the line L.

33. In 1984, tests showed that water in a lake was polluted with 7 milligrams of mercury compounds per 1000 liters of water. Cleaning up the lake became an immediate priority, and environmentalists determined that the pollution level would drop at the rate 0.75 milligram of mercury compounds per 1000 liters of water per year if all of their recommendations were followed. If 1984 corresponds to $x = 0$ and successive years correspond to $x = 1, 2, 3$, and so on, find the equation $y = mx + b$ of the line that allows the environmentalists to predict the pollution level y in future years if their recommendations are followed. Sketch the graph of the equation and determine when the lake will be free of mercury pollution according to this graph.

3.4 FUNCTIONS

Advances in our understanding of the world often result from the discovery that things depend on one another in definite ways. For instance, the pitch of a guitar string depends on its tension, the area of a circle depends on its radius, and the gravitational attraction between two material bodies depends on the distance between them. The idea that a quantity y depends on another quantity x is nicely symbolized by the **mapping notation**

$$x \longmapsto y.$$

This notation indicates that to each value of x there corresponds a uniquely determined value of y, or that each value of x is "mapped onto" a corresponding value of y. Another name for a mapping is a *function*.

Definition 1   **Function***

> A **function** is a rule, correspondence, or mapping
>
> $$x \longmapsto y$$
>
> that assigns to each real number x in a certain set D one and only one real number y.

In Definition 1, the set D is called the **domain** of the function, and y is called the **dependent variable,** since its value depends on the value of x. Because x is permitted to have any value in the domain D, we refer to x as the **independent variable.** The set of values assumed by y as x runs through all values in D is called the **range** of the function. If $x \longmapsto y$, we say that y is the **image** of x under the function.

Scientific calculators have special keys for some of the more important functions. By entering a number x and touching, for instance the $\sqrt{x}$ key, you obtain a vivid demonstration of the mapping $x \longmapsto \sqrt{x}$ as the display changes from x to its image $\sqrt{x}$ under the square-root function. For instance,

$$4 \longmapsto 2,$$

$$25 \longmapsto 5,$$

$$2 \longmapsto \sqrt{2} \approx 1.414.$$

Programmable calculators and microcomputers have "user-definable" keys that can be programmed for whatever function $x \longmapsto y$ may be required. The program for the required function is the actual rule whereby y is to be calculated from x. Each user-definable key is marked with a letter of the alphabet or other symbol, so that, after the key has been programmed for a particular function, the letter or symbol can be used as the "name" of the function.

---

* It is also possible to define a complex-valued function of a complex variable, or, more generally, a function from any set D into any set R.

The use of letters of the alphabet to designate functions is not restricted exclusively to calculating machines. Although any letters of the alphabet can be used to designate functions, the letters f, g, and h as well as F, G, and H are most common. (Letters of the Greek alphabet are also used.) For instance, if we wish to designate the square-root function $x \longmapsto \sqrt{x}$ by the letter f, we write

$$x \overset{f}{\longmapsto} \sqrt{x} \qquad \text{or} \qquad f : x \longmapsto \sqrt{x}.$$

If $f : x \longmapsto y$ is a function, it is customary to write the value of y that corresponds to x as $f(x)$, read "f of x." In other words, $f(x)$ is the image of x under the function f. For instance, if $f : x \longmapsto \sqrt{x}$ is the square-root function, then

$$f(4) = \sqrt{4} = 2,$$
$$f(25) = \sqrt{25} = 5,$$
$$f(2) = \sqrt{2} \approx 1.414,$$

and, in general, for any nonnegative value of x,

$$f(x) = \sqrt{x}.$$

Note carefully that $f(x)$ does *not* mean f times x.

If $f : x \longmapsto y$, then, for every value of x in the domain of f, we have

$$y = f(x),$$

an equation relating the dependent variable y to the independent variable x. Conversely, if an equation of the form

$$y = \text{an expression involving } x$$

determines a function $f : x \longmapsto y$, we say that the function f is **defined by** or **given by** the equation. For instance, the equation

$$y = 3x^2 - 1$$

defines a function $f : x \longmapsto y$, so that

$$y = f(x) = 3x^2 - 1$$

or simply

$$f(x) = 3x^2 - 1.$$

When a function f is defined by an equation, you can determine, by substitution, the image $f(a)$ corresponding to a particular value $x = a$. For instance, if f is defined by

$$f(x) = 3x^2 - 1,$$

then

$$f(2) = 3(2)^2 - 1 = 11,$$
$$f(0) = 3(0)^2 - 1 = -1,$$
$$f(-1) = 3(-1)^2 - 1 = 2,$$

and

$$f(t + 1) = 3(t + 1)^2 - 1 = 3(t^2 + 2t + 1) - 1$$
$$= 3t^2 + 6t + 2.$$

Example 1   Let g be the function defined by $g(x) = 5x^2 + 3x$. Find the indicated values.

(a) $g(1)$ (b) $g(-2)$ (c) $g(t^3)$

(d) $[g(-1)]^2$ (e) $g(t + h)$ (f) $g(-x)$

Solution   (a) $g(1) = 5(1)^2 + 3(1) = 5 + 3 = 8$

(b) $g(-2) = 5(-2)^2 + 3(-2) = 20 - 6 = 14$

(c) $g(t^3) = 5(t^3)^2 + 3(t^3) = 5t^6 + 3t^3$

(d) $[g(-1)]^2 = [5(-1)^2 + 3(-1)]^2 = (5 - 3)^2 = 2^2 = 4$

(e) $g(t + h) = 5(t + h)^2 + 3(t + h) = 5t^2 + 10th + 5h^2 + 3t + 3h$

(f) $g(-x) = 5(-x)^2 + 3(-x) = 5x^2 - 3x$ ■

Example 2   A small company finds that its profit depends on the amount of money it spends on advertising. If x dollars are spent on advertising, the corresponding profit $p(x)$ dollars is given by

$$p(x) = \frac{x^2}{100} - \frac{x}{50} + 100.$$

Determine the profit if the amount spent on advertising is

(a) \$50   (b) \$100   (c) \$300

Solution   (a) $p(50) = \dfrac{50^2}{100} - \dfrac{50}{50} + 100 = 25 - 1 + 100 = 124$ dollars

(b) $p(100) = \dfrac{100^2}{100} - \dfrac{100}{50} + 100 = 100 - 2 + 100 = 198$ dollars

(c) $p(300) = \dfrac{300^2}{100} - \dfrac{300}{50} + 100 = 900 - 6 + 100 = 994$ dollars   ■

Whenever a function $f : x \longmapsto y$ is defined by an equation, you may assume (unless we say otherwise) that *its domain consists of all values of x for which the equation makes sense and determines a corresponding real number y*. The range of the function is then automatically determined, since it consists of the set of all values of y that correspond, by the equation that defines the function, to values of x in the domain.

In Examples 3 to 5, find the domain of the function defined by each equation.

Example 3 $h(x) = \dfrac{1}{x - 1}$

Solution   The domain of h is the set of all real numbers except 1; that is, it is the set $(-\infty, 1) \cup (1, \infty)$.   ■

Example 4 $G(x) = \sqrt{4 - x}$

Solution   The expression $\sqrt{4 - x}$ represents a real number if and only if $4 - x \geq 0$; that is, if and only if $x \leq 4$. Therefore, the domain of G is the interval $(-\infty, 4]$.   ∎

Example 5 $F(x) = 3x - 5$

Solution   Since the expression $3x - 5$ is defined for all real values of x, the domain of F is the set $\mathbb{R}$ of all real numbers.   ∎

Functions that arise in applied mathematics may have restrictions imposed on their domains by physical or geometrical circumstances. For instance, the function $x \longmapsto \pi x^2$ that expresses the correspondence between the radius x and the area πx^2 of a circle would have its domain restricted to the interval $(0, \infty)$, since a circle must have a positive radius.

For a function f, the expression

$$\frac{f(x + h) - f(x)}{h}, \qquad h \neq 0,$$

called the **difference quotient,** plays an important role in calculus.

Example 6   Find the difference quotient for the function f defined by $f(x) = \sqrt{x}$ and simplify the result.

Solution   Assuming that $h \neq 0$, $x \geq 0$, and $x + h \geq 0$, we have

$$\frac{f(x + h) - f(x)}{h} = \frac{\sqrt{x + h} - \sqrt{x}}{h}.$$

In calculus, this is "simplified" by rationalizing the numerator, so that

$$\frac{f(x + h) - f(x)}{h} = \frac{(\sqrt{x + h} - \sqrt{x})(\sqrt{x + h} + \sqrt{x})}{h(\sqrt{x + h} + \sqrt{x})}$$

$$= \frac{(x + h) - x}{h(\sqrt{x + h} + \sqrt{x})} = \frac{h}{h(\sqrt{x + h} + \sqrt{x})}$$

$$= \frac{1}{\sqrt{x + h} + \sqrt{x}}.$$ ∎

In dealing with a function f, it is important to distinguish among the *function itself*

$$f : x \longmapsto y,$$

which is a rule or correspondence; the *image*

$$y \quad \text{or} \quad f(x),$$

which is a number depending on x; and the *equation*

$$y = f(x),$$

which relates the dependent variable y to the independent variable x. Nevertheless, people tend to take shortcuts and speak, incorrectly, of "the function $f(x)$" or "the function $y = f(x)$." Similarly, in applied mathematics, people often say that "y is a function of x," for instance, "current is a function of voltage." Although we avoid these practices when absolute precision is required, we indulge in them whenever it seems convenient and harmless.

The particular letters used to denote the dependent and independent variables are of no importance in themselves—the important thing is the rule by which a definite value of the dependent variable is assigned to each value of the independent variable. In applied work, variables other than x and y are often used because physical and geometrical quantities are designated by conventional symbols.

Example 7 The radius of a circle is often denoted by r and its area by A. Write an equation for the function $f : r \longmapsto A$ that assigns to each positive value of r the corresponding value of A.

Solution
$$A = f(r) = \pi r^2.$$ ∎

The Graph of a Function

The **graph** of a function f is defined to be the graph of the corresponding equation $y = f(x)$. In other words, the graph of f is the set of all points (x, y) in the Cartesian plane such that x is in the domain of f and $y = f(x)$.

For instance, if m and b are constants, then the graph of the function

$$f(x) = mx + b$$

is the same as the graph of the equation

$$y = mx + b,$$

a straight line with slope m and y intercept b. For this reason, a function of the form $f(x) = mx + b$ is called a **linear function.**

Example 8 Sketch the graph of the function $f(x) = \frac{3}{4}x + 2$.

Solution The graph of f is the same as the graph of the equation

$$y = \frac{3}{4}x + 2,$$

a line with slope $m = \frac{3}{4}$ and y intercept $b = 2$ (Figure 1). ∎

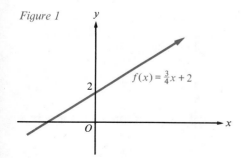

Figure 1

$f(x) = \frac{3}{4}x + 2$

Figure 2

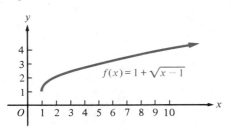

Graphs of functions that are not linear are often (but not always) smooth curves in the Cartesian plane. For instance, the graph of $f(x) = 1 + \sqrt{x-1}$ is shown in Figure 2. Sketching such a graph can be considerably more challenging than sketching a straight line, although the use of a calculator to determine points on the graph will often give a good indication of its general shape. Of course, most microcomputers also have graph-plotting capabilities. In Sections 3.5 and 3.6, we consider graph sketching in more detail.

In scientific work, a graph showing the relationship between two variable quantities is often obtained by means of actual measurement. For instance, Figure 3 shows the blood pressure p (in millimeters of mercury) in an artery of a healthy person plotted against time t in seconds. Such a curve may be regarded as the graph of a function $p = f(t)$, even though it may not be clear how to write a "mathematical formula" giving p in terms of t.

Figure 3

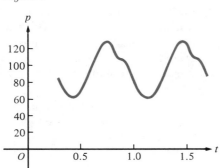

It is important to realize that *not every curve in the Cartesian plane is the graph of a function*. Indeed, the definition of a function (Definition 1) requires that there be one and *only one* value of y corresponding to each value of x in the domain. Thus, on the graph of a function, *we cannot have two points (x, y_1) and (x, y_2) with the same abscissa x and different ordinates y_1 and y_2*. Hence, we have the following test.

Vertical-Line Test

> A set of points in the Cartesian plane is the graph of a function if and only if no vertical straight line intersects the set more than once.

Example 9 Which of the curves in Figure 4 is the graph of a function?

Solution By the vertical-line test, the curve in Figure 4a is the graph of a function, but the curve in Figure 4b is not. ∎

Figure 4

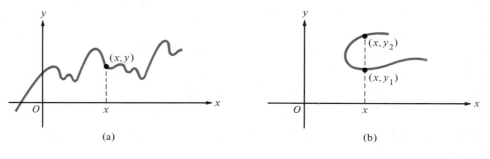

(a)   (b)

Problem Set 3.4

In Problems 1 to 40, let

$$f(x) = 2x + 1 \qquad g(x) = x^2 - 3x - 4$$

$$h(x) = \sqrt{3x + 5} \qquad F(x) = \frac{x - 2}{3x + 7}$$

$$G(x) = \sqrt[3]{x^3 - 4} \qquad H(x) = |2 - 5x|$$

Find the indicated values.

1. $f(-3)$

2. $g(-2)$

3. $h\left(-\dfrac{1}{3}\right)$

4. $F\left(\dfrac{7}{3}\right)$

5. $G(\sqrt[3]{31})$

6. $H(-4)$

7. $g(0)$

8. $G(-5)$

9. $F\left(-\dfrac{1}{2}\right)$

10. $f(-7)$

© **11.** $g(4.718)$

© **12.** $h(2.003)$

13. $F\left(-\dfrac{1}{3}\right)$

14. $\left[H\left(-\dfrac{1}{2}\right)\right]^2$

15. $[h(-1)]^2$

16. $[G(2)]^3$

17. $\sqrt{f(4)}$

18. $\sqrt{h(1)}$

19. $f(a + 1)$

20. $f\left(\dfrac{1}{a}\right)$

21. $g(-b)$

22. $F\left(\dfrac{a}{3}\right)$

23. $H(c + 2)$

24. $g(-\tfrac{1}{2}ab)$

25. $H\left(\dfrac{1}{a}\right)$

26. $h(2x - 1)$

27. $f(-x)$

28. $G(\sqrt{b})$

29. $h(x^4)$

30. $g(-x)$

31. $f\left(\dfrac{x}{2}\right)$

32. $g(x + t)$

33. $f\left(\dfrac{x - 1}{2}\right)$

34. $f(1) - f(0)$

35. $g(a) - g(b)$

36. $g(a) - g(-a)$

37. $h(x) - h(0)$

38. $F(x + t) - F(x)$

39. $\dfrac{f(x) - f(0)}{x}$

© **40.** $f(g(2.341))$

In Problems 41 to 54, find the domain of each function.

41. $f(x) = \dfrac{1}{x}$

42. $g(x) = |x|$

43. $f(x) = 1 - 4x^2$

44. $g(x) = (x + 2)^{-1}$

45. $h(x) = \sqrt{x}$

46. $F(x) = \sqrt{5 - 3x}$

47. $f(x) = -3x^{-3}$

48. $G(x) = 7/(5 - 6x)$

49. $p(x) = \sqrt{9 - x^2}$

50. $K(x) = (4 - 5x)^{-1/2}$

51. $g(x) = (9 + x^4)^{3/4}$

52. $h(x) = \dfrac{1}{x + |x|}$

53. $F(x) = \dfrac{x^3 - 8}{x^2 - 4}$

54. $h(x) = \sqrt{\dfrac{x - 2}{x - 4}}$

In Problems 55 to 60, find the difference quotient $\dfrac{f(x + h) - f(x)}{h}$ and simplify the result.

55. $f(x) = 4x - 1$

56. $f(x) = 5$

57. $f(x) = x^2 + 3$

58. $f(x) = mx + b, \qquad m \neq 0$

59. $f(x) = 1/\sqrt{x}$

60. $f(x) = 1/x$

In Problems 61 to 66, sketch the graph of the function.

61. $f(x) = \tfrac{2}{3}x - 5$

62. $f(x) = 2x \qquad$ for $x \geq 0$

63. $f(x) = x$

64. $f(x) = 4$

65. $f(x) = -2x + 1$

66. $f(x) = \sqrt{9 - x^2}$

67. Use the vertical-line test to determine which of the curves in Figure 5 are graphs of functions.

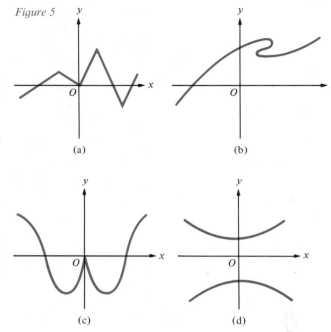

Figure 5

(a)   (b)

(c)   (d)

68. Is the graph of the equation $x^2 + y^2 = 9$ the graph of a function? Why or why not?

69. The function $f : C \longmapsto F$ given by the equation $F = \frac{9}{5}C + 32$ converts the temperature C in degrees Celsius to the corresponding temperature F in degrees Fahrenheit. Find $f(0)$, $f(15)$, $f(-10)$, and $f(55)$, and write the results using both function and mapping notation.

70. An ecologist investigating the effect of air pollution on plant life finds that the percentage, $p(x)$ percent, of diseased trees and shrubs at a distance of x kilometers from an industrial city is given by $p(x) = 32 - (3x/50)$ for $50 \le x \le 500$. Sketch a graph of the function p and find $p(50)$, $p(100)$, $p(200)$, $p(400)$, and $p(500)$.

71. An airline chart shows that the temperature T in degrees Fahrenheit at an altitude of $h = 15,000$ feet is $T = 5°$. At an altitude of $h = 20,000$ feet, $T = -15°$. Supposing that T is a linear function of h, obtain an equation that defines this function, sketch its graph, and find the temperature at an altitude $h = 30,000$ feet.

ⓒ **72.** In physics, the (absolute) pressure P in newtons per square meter at a point h meters below the surface of a body of water is shown to be a linear function of h. When $h = 0$, $P = 1.013 \times 10^5$ newtons per square meter. When $h = 1$ meter, $P = 2.003 \times 10^5$ newtons per square meter. Obtain the equation that defines P as a function of h, and use it to find the pressure at a depth of $h = 100$ meters.

73. A rectangle of length x units and width y units has a perimeter of 24 units. Express the area A of the rectangle as a function of x.

ⓒ **74.** The period $T(l)$ seconds of a simple pendulum of length l meters swinging along a small arc is given by $T(l) = 2\pi\sqrt{l/9.807}$. Using a calculator and rounding off your answer to four significant digits, find $T(0.1)$, $T(1)$, $T(1.5)$, and $T(0.2484)$.

3.5 GRAPH SKETCHING AND PROPERTIES OF GRAPHS

In this section, we consider methods for sketching graphs of functions that may not be linear, properties of graphs, and graphs of some important functions. The basic graph-sketching procedure is as follows.

The Point-Plotting Method

> To sketch the graph of $y = f(x)$, select several values of x in the domain of f, calculate the corresponding values of $f(x)$, plot the resulting points, and connect the points with a smooth curve. The more points you plot, the more accurate your sketch will be.

Example 1   Use the point-plotting method to sketch the graph of $f(x) = x^2$, where the domain of f is restricted by the condition that $x > 0$.

Solution   The domain of f is the interval, $(0, \infty)$, so we begin by selecting several values of x in this interval and calculating the corresponding values of $f(x) = x^2$, as in the table in Figure 1. We then plot the points $(x, f(x))$ from the table and connect them with a smooth curve (Figure 1). Because the domain of f consists only of positive numbers, the point $(0, 0)$ is excluded from the graph. This excluded point is indicated by a small open circle. ∎

Figure 1

x	$f(x) = x^2$
1	1
2	4
3	9
4	16

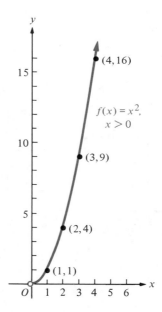

The point-plotting method requires us to *guess* about the shape of the graph between or beyond known points, and *therefore must be used with caution* (see Problems 43 and 44). If the function is fairly simple, the point-plotting method usually works pretty well; however, more complicated functions may require more advanced methods that are studied in calculus.

Geometric Properties of Graphs

The graph in Figure 2a is always *rising* as we move to the right, a geometric indication that the function f is **increasing;** that is, as x increases, so does the value of $f(x)$. On the other hand, the graph in Figure 2b is always *falling* as we move to the right, indicating that the function g is **decreasing;** that is, as x increases, the value of $g(x)$ decreases. In Figure 2c, the graph doesn't rise or fall, indicating that h is a **constant function** whose values $h(x)$ do not change as we increase x. The following definition makes these ideas more precise.

Figure 2

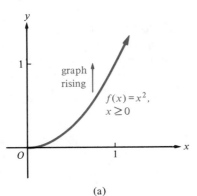

(a)

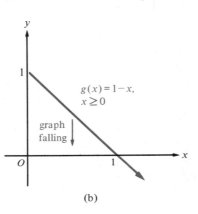

(b)

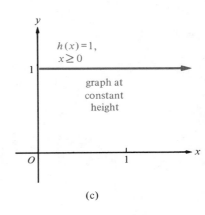

(c)

Figure 3

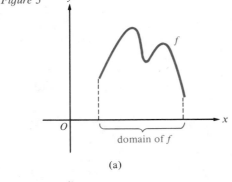

(a)

Definition 1   Increasing, Decreasing, and Constant Functions

Let the interval I be contained in the domain of the function f.

(i) f is **increasing** on I if for every two numbers a and b in I with $a < b$ we have $f(a) < f(b)$.

(ii) f is **decreasing** on I if for every two numbers a and b in I with $a < b$ we have $f(a) > f(b)$.

(iii) f is **constant** on I if for every two numbers a and b in I we have $f(a) = f(b)$.

As Figure 3 illustrates, the domain and range of a function are easily found from its graph.

The domain of a function is the set of all abscissas of points on its graph, and the range is the set of all ordinates of points on its graph.

(b)

Example 2   For the function f whose graph is shown in Figure 4, indicate the intervals over which f is increasing, decreasing, or constant. Also, find the domain and range of f.

Figure 4

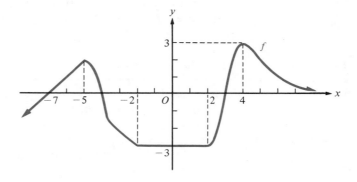

Solution   As we move from left to right, the function f is increasing over intervals where the graph is rising and decreasing over intervals where it is falling. Thus, assuming that the graph continues indefinitely to the left and right in the directions indicated by the arrowheads, we conclude that f is increasing on the intervals $(-\infty, -5]$ and $[2, 4]$ and that f is decreasing on the intervals $[-5, -2]$ and $[4, \infty)$. On the interval $[-2, 2]$, f is constant. The graph "covers" the entire x axis, so the domain of f is the set $\mathbb{R}$ of all real numbers. We assume that the graph keeps dropping as we move to the left of -5 on the x axis, but that it never climbs any higher than $y = 3$; hence, the range of f is the interval $(-\infty, 3]$.   ■

Consider the graphs in Figure 5. The graph of f (Figure 5a) is **symmetric about the y axis;** that is, the portion of the graph to the right of the y axis is the mirror image of the portion to the left of it. Specifically, if the point (x, y) belongs to the

Figure 5

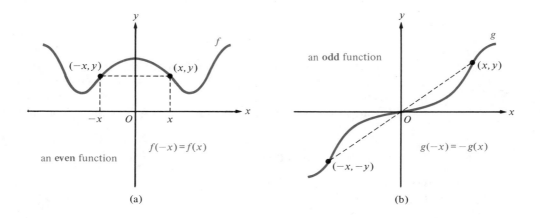

(a)   (b)

graph of f, then so does the point $(-x, y)$, so that $f(-x) = f(x)$. Similarly, the graph of g (Figure 5b) is **symmetric about the origin** because, if the point (x, y) belongs to the graph, then so does the point $(-x, -y)$; that is, $g(-x) = -g(x)$.

A function whose graph is symmetric about the y axis is called an **even function;** a function whose graph is symmetric about the origin is called an **odd function.** This is stated more formally in the following definition.

Definition 2   **Even and Odd Functions**

> **(i)** A function f is said to be **even** if, for every number x in the domain of f, $-x$ is also in the domain of f and
>
> $$f(-x) = f(x).$$
>
> **(ii)** A function f is said to be **odd** if, for every number x in the domain of f, $-x$ is also in the domain of f and
>
> $$f(-x) = -f(x).$$

Of course, there are many functions that are neither even nor odd.

Example 3   Determine which of the functions whose graphs are shown in Figure 6 are even, odd, or neither.

Figure 6

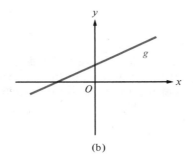

(a)   (b)

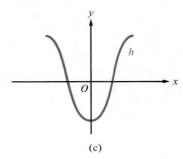

(c)

Solution   In Figure 6a, the graph of f is symmetric about the origin; thus, $f(-x) = -f(x)$ and f is an odd function. In Figure 6b, the graph of g is symmetric neither about the y axis nor about the origin, so g is neither even nor odd. In Figure 6c, the graph of h is symmetric about the y axis; thus, $h(-x) = h(x)$ and h is an even function. ■

Example 4   Determine whether each function is even, odd, or neither.

(a) $f(x) = x^4$ **(b)** $g(x) = x - 1$
(c) $h(x) = 2x^2 - 3|x|$ **(d)** $F(x) = x^5$

Solution   **(a)** $f(-x) = (-x)^4 = x^4 = f(x)$, so f is an even function.

(b) $g(-x) = -x - 1$, while $g(x) = x - 1$ and $-g(x) = -x + 1$. Since we have neither $g(-x) = g(x)$ nor $g(-x) = -g(x)$, g is neither even nor odd.

(c) $h(-x) = 2(-x)^2 - 3|-x| = 2x^2 - 3|x| = h(x)$, so h is an even function.

(d) $F(-x) = (-x)^5 = -x^5 = -F(x)$, so F is an odd function.   ■

Graphs of Some Particular Functions

The following examples illustrate the ideas discussed above and exhibit the graphs of some important functions.

In Examples 5 to 9, find the domain of f, sketch its graph, discuss the symmetry of the graph, indicate the intervals where f is increasing or decreasing, and find the range of f. (Do these in whatever order seems most convenient.)

Example 5 $f(x) = x$ (the **identity** function)

Solution   This is a special case of a linear function $f(x) = mx + b$ with slope $m = 1$ and y intercept $b = 0$ (Figure 7). Notice that the graph consists of all points for which the abscissa equals the ordinate. The set $\mathbb{R}$ of all real numbers is both the domain and range of f, and f is increasing over $\mathbb{R}$. Since $f(-x) = -x = -f(x)$, it follows that f is an odd function and that its graph is symmetric about the origin.   ■

Figure 7

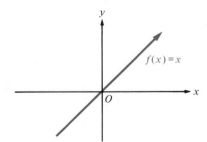

Example 6 $f(x) = x^2$ (the **squaring** function)

Solution   The domain of f is the set $\mathbb{R}$. Because $f(-x) = (-x)^2 = x^2 = f(x)$, the function f is even and its graph is symmetric about the y axis. We have already sketched the portion of this graph for $x > 0$ in Figure 1. The full graph includes the mirror image of Figure 1 on the other side of the y axis, and the point $(0,0)$ (Figure 8). From the graph we see that the range of f is the interval $[0, \infty)$, and that f is decreasing on the interval $(-\infty, 0]$ and increasing on the interval $[0, \infty)$.   ■

Figure 8

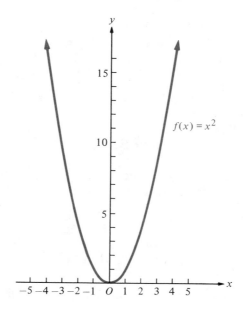

Example 7 $f(x) = \sqrt{x}$ (the **square-root** function*)

Solution   Here the domain of f is the interval $[0, \infty)$. We tabulate some points $(x, f(x))$, plot them, and connect them by a smooth curve to obtain a sketch of the graph (Figure 9). The function f is increasing on the interval $[0, \infty)$, and its graph rises higher and higher without bound, so the range of f is the interval $[0, \infty)$. The graph is neither symmetric about the y axis nor the origin, so f is neither even nor odd. ∎

Figure 9

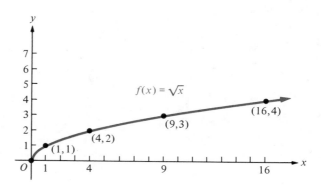

x	$f(x) = \sqrt{x}$
0	0
1	1
4	2
9	3
16	4

* In BASIC programming language, the square-root function is written SQR(X).

Example 8 $f(x) = b$ (a **constant** function)

Solution   This is a special case of the linear function $f(x) = mx + b$, with slope $m = 0$ and y intercept b. Its graph is a line parallel to the x axis and containing the point $(0, b)$ on the y axis. All points on this line have the same ordinate b (Figure 10). The domain of f is the set $\mathbb{R}$ of all real numbers and the range is the set $\{b\}$. The constant function f is neither increasing nor decreasing. Since $f(-x) = b = f(x)$, the function f is even and its graph is symmetric about the y axis.   ■

Figure 10

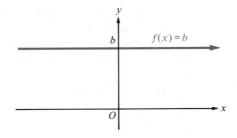

Example 9 $f(x) = |x|$ (the **absolute-value** function*)

Solution   The domain of f consists of all real numbers $\mathbb{R}$. For $x \geq 0$, $f(x) = x$, so the portion of the graph to the right of the y axis is the same as the graph of the identity function (Example 5). Since $f(-x) = |-x| = |x| = f(x)$, the function f is even and its graph is symmetric about the y axis. Reflecting the portion of the graph of the identity function for $x \geq 0$ across the y axis, we obtain the V-shaped graph shown in Figure 11. Evidently, the range of f is the interval $[0, \infty)$; f is decreasing on the interval $(-\infty, 0]$ and increasing on the interval $[0, \infty)$.   ■

Figure 11

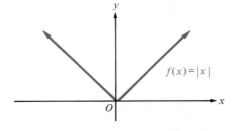

Sometimes a function is defined by using different equations on different intervals. Such a **piecewise-defined** function is illustrated in the following example.

* In BASIC programming language, the absolute-value function is written **ABS(X)**.

Example 10   Sketch the graph of the function f defined by

$$f(x) = \begin{cases} -2 & \text{if } x < 0 \\ x^2 & \text{if } 0 \le x < 2 \\ x & \text{if } x \ge 2. \end{cases}$$

Solution   For $x < 0$, we have $f(x) = -2$, so this portion of the graph of f will look like the graph of a constant function (Figure 10 with $b = -2$). In sketching this portion of the graph, we must be careful to leave out the point $(0, -2)$ on the y axis because of the strict inequality $x < 0$ (Figure 12). For $0 \le x < 2$, we have $f(x) = x^2$; this part of our graph coincides with the graph of the squaring function (Figure 8), starting at the origin and ending at the point $(2, 4)$, which does not belong to the graph of f (Figure 12). Finally, for $x \ge 2$, we have $f(x) = x$; this part of our graph will coincide with the graph of the identity function (Figure 7), starting at the point $(2, 2)$, which does belong to the graph, and continuing indefinitely to the right (Figure 12). ■

Figure 12

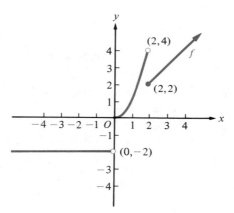

Problem Set 3.5

In Problems 1 to 8, use the point-plotting method to sketch the graph of each function. ⒸYou may wish to improve the accuracy of your sketch by using a calculator to determine the coordinates of some points on the graph.

1. $f(x) = x^3$ for $x \ge 0$

2. $g(x) = \sqrt[3]{x}$ for $-8 \le x \le 8$

3. $h(x) = \dfrac{1}{x}$ for $x \ge 1$

4. $F(x) = \dfrac{x-1}{x+1}$ for $x \ge 0$

5. $G(x) = x^2 + x + 1$

6. $H(x) = \dfrac{1}{x^2 + 1}$

7. $p(x) = \sqrt{25 - x^2}$

8. $q(x) = x + \sqrt{x}$

9. For each function in Figure 13, find the domain and range; indicate the intervals over which the function is increasing, decreasing, or constant; and determine whether the function is even, odd, or neither.

Figure 13

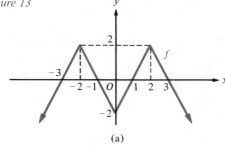

(a)

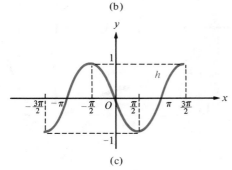

(b)

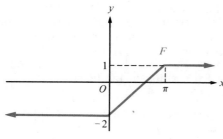

(c)

(d)

10. The graph of the function $f(x) = x^{1/3}(x-1)^{2/3}$ is shown in Figure 14. Indicate the intervals over which f is increasing or decreasing, and determine whether f is even, odd, or neither.

Figure 14

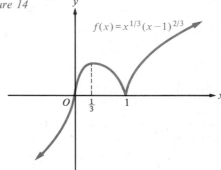

In Problems 11 to 19, determine without drawing a graph whether the function is even, odd, or neither, and discuss any symmetry of the graph.

11. $f(x) = x^4 + 3x^2$ **12.** $g(x) = -5x^2 + 4$

13. $h(x) = 4x^3 - 5x$ **14.** $F(x) = x^2 + |x|$

15. $G(x) = \sqrt{8x^4 + 1}$ **16.** $H(x) = 5x^2 - x^3$

17. $q(x) = \sqrt[3]{x^3 - 9}$ **18.** $p(x) = \dfrac{\sqrt{x^2 + 1}}{|x|}$

19. $r(x) = \dfrac{1}{x}$

20. Let $f(x) = 3x^4 - 2x^3 + x^2 - 1$. Define functions g and h by $g(x) = \frac{1}{2}[f(x) + f(-x)]$ and $h(x) = \frac{1}{2}[f(x) - f(-x)]$.

(a) Show that f is neither even nor odd.

(b) Show that g is even.

(c) Show that h is odd.

(d) Show that $f(x) = g(x) + h(x)$ holds for all values of x.

In Problems 21 to 35, find the domain of the function, sketch its graph, discuss the symmetry of the graph, indicate the intervals where the function is increasing or decreasing, and find the range of the function. Do these things in any convenient order. Ⓒ Use a calculator if you wish.

21. $f(x) = 3x + 1$ **22.** $g(x) = -4x + 3$

23. $h(x) = 5$

24. $F(x) = -2\sqrt{x}$

25. $H(x) = 2 - x^2$

26. $G(x) = -7$

27. $f(x) = x|x|$

28. $g(x) = \sqrt{x - 4} + x$

29. $F(x) = 1 + \sqrt{x - 1}$

30. $h(x) = x^2 - 4x$

31. $H(x) = \dfrac{x}{|x|}$

32. $g(x) = x - |\tfrac{1}{2}x|$

33. $f(x) = x^3 + 2x$

34. $f(x) = |x| - x$

35. $g(x) = \left|\dfrac{1}{x}\right|$

36. The symbol $[\![x]\!]$ is often used to denote the *greatest integer* less than or equal to x; that is, $[\![x]\!]$ is the one and only integer such that $[\![x]\!] \le x < [\![x]\!] + 1$. For instance, $[\![1.7]\!] = 1$, $[\![3.14]\!] = 3$, $[\![2]\!] = 2$, $[\![-1.3]\!] = -2$, $[\![-3]\!] = -3$, and so forth. The function defined by $f(x) = [\![x]\!]$ is called the **greatest integer function.*** Sketch a graph of the greatest integer function for $-4 \le x \le 4$.

In Problems 37 to 42, sketch the graph of each piecewise-defined function.

37. $f(x) = \begin{cases} 2 & \text{if } x < 0 \\ -1 & \text{if } x \ge 0 \end{cases}$

38. $g(x) = \begin{cases} x & \text{if } x \ge 0 \\ 2 & \text{if } x < 0 \end{cases}$

39. $F(x) = \begin{cases} 5 + x & \text{if } x \le 3 \\ 9 - \dfrac{x}{3} & \text{if } x > 3 \end{cases}$

40. $h(x) = \begin{cases} x^2 & \text{if } x < 0 \\ -x^2 & \text{if } x \ge 0 \end{cases}$

41. $g(x) = \begin{cases} -x & \text{if } x < 0 \\ x^2 & \text{if } 0 \le x < 2 \\ 8 - 2x & \text{if } x \ge 2 \end{cases}$

42. $G(x) = \begin{cases} 0 & \text{if } x < -2 \\ 3 + x^2 & \text{if } -2 \le x < 2 \\ 0 & \text{if } x \ge 2 \end{cases}$

C 43. A student sketches the graph of the function $f(x) = 4x^4 - 14x^3 + 20x^2 - 5x$ for $x \ge 0$ by plot-

* In BASIC the greatest integer function is written INT(X).

ting five points and connecting them with a smooth curve, as shown in Figure 15. However, the graph is not correct as shown. Find the error. [*Hint:* Plot more points for x between 0 and 0.5.]

Figure 15

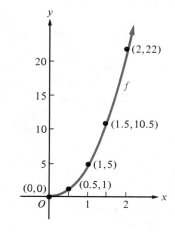

x	$f(x)$
0	0
0.5	1
1	5
1.5	10.5
2	22

C 44. A student sketches the graph of $f(x) = x^3 + 3x^2$ by plotting seven points and connecting them with smooth curve, as shown in Figure 16. Criticize the student's work.

Figure 16

x	$f(x)$
-2	4
-1.5	3.38
-1	2
-0.5	0.63
0	0
0.5	0.88
1	4

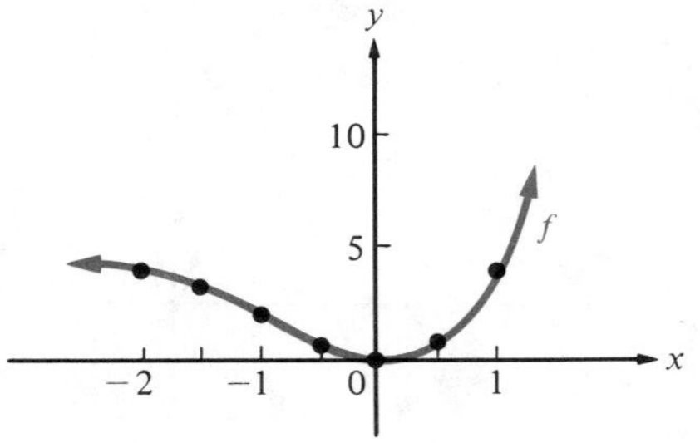

45. The management of a city parking lot charges its customers $0.50 per hour or part of an hour for up to 12 hours, with a minimum charge of $2.00 and a maximum charge of $5.00. If $C(x)$ denotes the charge in dollars for parking a car for x hours, sketch a graph of the function C for $0 \le x \le 12$ hours.

46. Explain how to use the graph of the squaring function (Figure 8) to see that, if $a < b < 0$, then $a^2 > b^2$.

3.6 SHIFTING, STRETCHING, AND REFLECTING GRAPHS

By altering a function f in certain ways we can obtain a new function F whose graph is closely related to the graph of the original function. In this section we study some of the basic ways in which functions can be altered and the resulting effects on their graphs.

Example 1   Sketch the graphs of $f(x) = x^2$ and $F(x) = x^2 + 1$ on the same coordinate system.

Solution   The graph of $f(x) = x^2$ was shown in Figure 8 of Section 3.5. The graph of $F(x) = x^2 + 1$ is obtained by adding 1 to the ordinate of each point on the graph of f. Thus, the graph of $F(x) = x^2 + 1$ is just the graph of $f(x) = x^2$ shifted upward by 1 unit (Figure 1). ∎

Figure 1

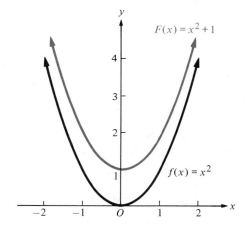

More generally, we have the following rule:

Vertical Shifting Rule

If f is a function and k is a constant, then the graph of the function F defined by

$$F(x) = f(x) + k$$

is obtained by **shifting** the graph of f **vertically** by $|k|$ units, **upward** if $k > 0$, and **downward** if $k < 0$ (Figure 2).

Figure 2

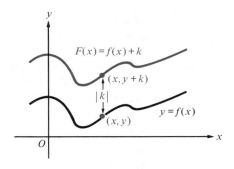

Example 2   Sketch the graphs of $F(x) = |x| + 2$ and $G(x) = |x| - 3$ on the same coordinate system.

Solution   The graph of

$$f(x) = |x|$$

was shown in Figure 11 of Section 3.5. We shift this graph upward by 2 units to obtain the graph of $F(x) = |x| + 2$; we shift it downward by 3 units to obtain the graph of $G(x) = |x| - 3$ (Figure 3). ∎

Figure 3

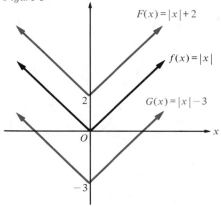

Figure 4

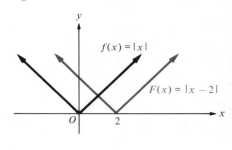

Example 3   Sketch the graphs of $f(x) = |x|$ and $F(x) = |x - 2|$ on the same coordinate system.

Solution   A point (x, y) belongs to the graph of F if and only if $y = |x - 2|$; that is, if and only if $(x - 2, y)$ belongs to the graph of f. For instance, the point $(2, 0)$ belongs to the graph of F because the point $(2 - 2, 0) = (0, 0)$ belongs to the graph of f. Because each point (x, y) on the graph of F is 2 units to the right of the point $(x - 2, y)$ on the graph of f, it follows that the graph of F is obtained by shifting the graph of f horizontally by 2 units to the right (Figure 4). ∎

More generally, we have the following rule:

Horizontal Shifting Rule

If f is a function and h is a constant, then the graph of the function F defined by

$$F(x) = f(x - h)$$

is obtained by **shifting** the graph of f **horizontally** by $|h|$ units, to the **right** if $h > 0$ and to the **left** if $h < 0$.

To understand the horizontal shifting rule, notice that a point (x, y) belongs to the graph of $F(x) = f(x - h)$ if and only if $y = f(x - h)$; that is, if and only if the point $(x - h, y)$ belongs to the graph of f (Figure 5).

Figure 5

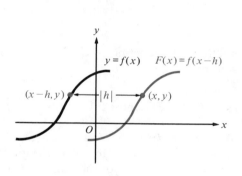

Figure 6

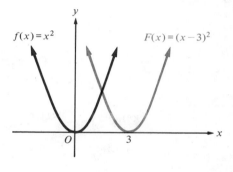

Example 4   Sketch the graph of $F(x) = (x - 3)^2$.

Solution   We begin by sketching the graph of $f(x) = x^2$. By shifting this graph 3 units to the right, we obtain the graph of $F(x) = (x - 3)^2$ (Figure 6).   ∎

Of course, we can combine vertical and horizontal shifts.

Combined Graph-Shifting Rule

If h and k are constants and f is a function, then the graph of

$$F(x) = f(x - h) + k$$

is obtained by shifting the graph of f horizontally by $|h|$ units and vertically by $|k|$ units. The horizontal shift is to the right if $h > 0$, and to the left if $h < 0$. The vertical shift is upward if $k > 0$, and downward if $k < 0$.

Example 5   Sketch the graph of $F(x) = |x + 2| - 3$.

Solution   We begin by sketching the graph of $f(x) = |x|$. In the combined graph-shifting rule we take $F(x) = f(x - h) + k$ with $h = -2$ and $k = -3$. Therefore, the graph of $F(x) = |x + 2| - 3$ is obtained by shifting the graph of $f(x) = |x|$ to the left by 2 units and downward by 3 units (Figure 7). ■

Figure 7

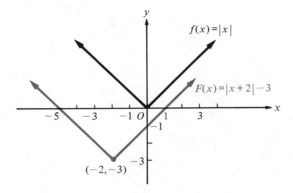

The effect of multiplying a function by a constant is to multiply the ordinate of each point on its graph by that constant.

Example 6   Sketch the graphs of $F(x) = 2x^2$ and $G(x) = \frac{1}{2}x^2$.

Solution   Figure 8a shows the graph of $f(x) = x^2$. In Figure 8b, we have doubled the ordinates of the points on the graph of $f(x) = x^2$ to obtain the graph of $F(x) = 2x^2$. In Figure 8c, we have multiplied the ordinates of the points on the graph of $f(x) = x^2$ by $\frac{1}{2}$ to obtain the graph of $G(x) = \frac{1}{2}x^2$. ■

Figure 8

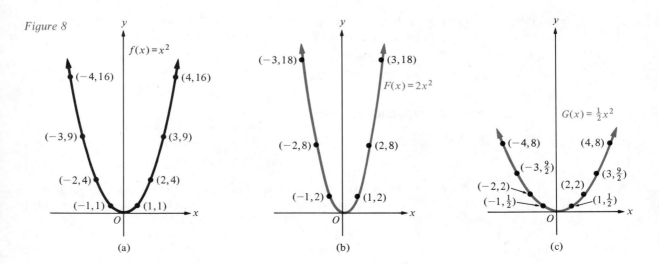

(a)   (b)   (c)

More generally, we have the following rule:

Stretching and Flattening Rule

> If f is a function and c is a positive constant, then the graph of the function F defined by
>
> $$F(x) = cf(x)$$
>
> is obtained from the graph of f by multiplying each ordinate by c.

If $c > 1$, the effect is to "stretch" the graph vertically by a factor of c. If $0 < c < 1$, the effect is to "flatten" the graph. The case in which $c < 0$ is addressed in Example 8.

Example 7   Sketch the graphs of $F(x) = 3|x|$ and $G(x) = \frac{1}{3}|x|$ on the same coordinate system.

Solution   In Figure 9 we have used the stretching and flattening rule to obtain the graphs of F and G from the graph of $f(x) = |x|$. ∎

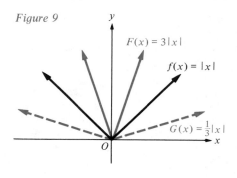

Figure 9

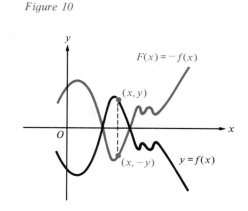

Figure 10

Figure 10 shows the effect of multiplying a function by -1 and illustrates the following rule:

x Axis Reflection Rule

> If f is a function, then the graph of the function
>
> $$F(x) = -f(x)$$
>
> is obtained by **reflecting** the graph of f across the **x axis.**

Example 8   Sketch the graphs of $F(x) = -|x|$ and $G(x) = 2 - |x|$.

Solution   We obtain the graph of $F(x) = -|x|$ by reflecting the graph of $f(x) = |x|$ across the x axis (Figure 11a). By shifting the graph of $F(x) = -|x|$ upward by 2 units, we obtain the graph of $G(x) = 2 - |x|$ (Figure 11b). ∎

Figure 11

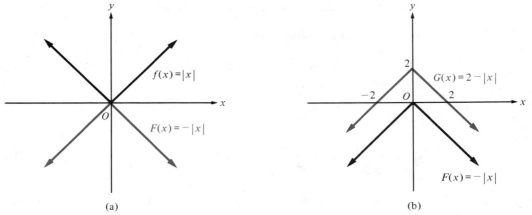

(a)   (b)

In Problem 31, we ask you to justify the following:

y Axis Reflection Rule

If f is a function, then the graph of the function

$$G(x) = f(-x)$$

is obtained by **reflecting** the graph of f across the **y axis.**

Example 9   Use the graph of the function f in Figure 12a to obtain the graph of $G(x) = f(-x)$.

Solution   We obtain the graph of $G(x) = f(-x)$ by reflecting the graph of f across the y axis (Figure 12b). ∎

Figure 12

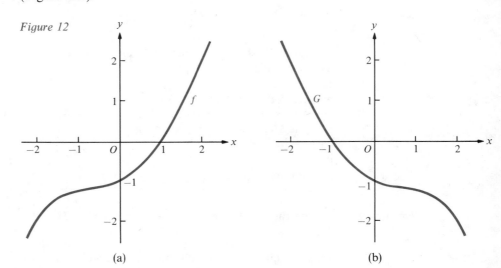

(a)   (b)

The techniques of shifting, stretching, and reflecting are often useful for determining the "general shape" of a graph before you begin to sketch it. Using this information and plotting a few points, you can rapidly sketch a reasonably accurate graph.

Problem Set 3.6

In Problems 1 to 18, use the techniques of shifting, stretching, and reflecting to sketch the graph of each function.

1. $F(x) = x^2 + 2$ **2.** $G(x) = x^2 - 1$

3. $H(x) = |x - 3|$ **4.** $f(x) = \sqrt{x - 1}$

5. $g(x) = (x - 2)^2$ **6.** $p(x) = \sqrt{x + 4} - 2$

7. $q(x) = |x - 1| + 1$ **8.** $F(x) = 3\sqrt{x - 1}$

9. $G(x) = 2|x|$ **10.** $H(x) = 3|x - 1| + 1$

11. $Q(x) = \frac{3}{4}x^2 + 1$ **12.** $R(x) = \frac{2}{3}(x - 1)^2 + 2$

13. $F(x) = -x^2$ **14.** $G(x) = 3 - \sqrt{x}$

15. $H(x) = 1 - 3|x|$ **16.** $f(x) = 1 - 2|x + 4|$

17. $g(x) = 1 - 2x^2$ **18.** $q(x) = 2 - \frac{3}{2}\sqrt{x + 4}$

In Problems 19 to 26, compare the graph of the first function with the graph of the second by sketching them both on the same coordinate system and explaining in words how they are related.

19. $f(x) = x$ and $F(x) = x + 2$

20. $g(x) = |x|$ and $G(x) = |x| - 5$

21. $p(x) = x^2$ and $P(x) = 1 - x^2$

22. $q(x) = |x|$ and $Q(x) = |x - 3| + 3$

23. $r(x) = |x|$ and $R(x) = 2|x| + 1$

24. $s(x) = x^2$ and $S(x) = -\frac{1}{2}(x - 1)^2$

25. $t(x) = \sqrt{x}$ and $T(x) = 2\sqrt{x - 2} + 4$

26. $f(x) = x^3$ and $F(x) = 1 - \frac{2}{3}(x - 1)^3$

In Problems 27 to 30, use the graph of the function f in the indicated figure to obtain the graph of (a) $F(x) = f(x) + 2$,   (b) $G(x) = f(x - 2)$,   (c) $H(x) = 2f(x)$, (d) $Q(x) = -f(x)$, and (e) $R(x) = f(-x)$.

27.

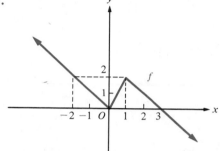

28.

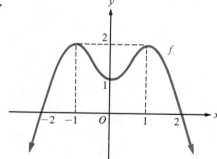

29.

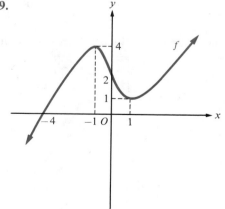

30.

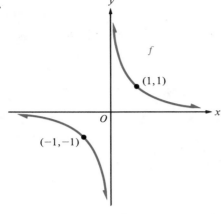

31. Justify the y axis reflection rule.

32. If c is a positive constant, how is the graph of $F(x) = f(cx)$ related to the graph of f?

33. If c is a constant and $c > 1$, the graph of $F(x) = cf(x/c)$ is said to be obtained from the graph of f by *magnification*, and c is called the *magnification factor*. Sketch the graph obtained if the graph of $f(x) = x^2$ is magnified by a factor of 2.

34. What happens if you magnify the graph of a linear function $f(x) = mx + b$ by a factor of c? [See Problem 33.]

3.7 ALGEBRA OF FUNCTIONS AND COMPOSITION OF FUNCTIONS

In the business world, the *profit P* from the sale of goods is related to the *revenue R* from the sale and to the *cost C* of producing the goods by the equation

$$P = R - C.$$

In other words, $R = P + C$. Often, the quantities P, R, and C are variables that depend on another variable, for instance, on the number x of units of goods produced; that is, P, R, and C are functions of x. Thus, business applications of mathematics may involve the *addition and subtraction of functions;* other applications require the *multiplication and division of functions.* For example, the function that describes an amplitude-modulated (AM) radio signal is the product of the function that describes the audio signal and the function that describes the carrier wave. Therefore, we make the following definition.

Definition 1   **Sum, Difference, Product, and Quotient of Functions**

Let f and g be functions whose domains overlap. We define functions $f + g$, $f - g$, $f \cdot g$, and f/g as follows:

$$(f + g)(x) = f(x) + g(x)$$

$$(f - g)(x) = f(x) - g(x)$$

$$(f \cdot g)(x) = f(x) \cdot g(x)$$

$$\left(\frac{f}{g}\right)(x) = \frac{f(x)}{g(x)}, \qquad g(x) \neq 0.$$

Example 1   Let $f(x) = 2x^3 - 1$ and $g(x) = x^2 + 5$. Find

(a) $(f + g)(x)$ (b) $(f - g)(x)$ (c) $(f \cdot g)(x)$ (d) $\left(\dfrac{f}{g}\right)(x)$

Solution   (a) $(f + g)(x) = f(x) + g(x) = (2x^3 - 1) + (x^2 + 5) = 2x^3 + x^2 + 4$
(b) $(f - g)(x) = f(x) - g(x) = (2x^3 - 1) - (x^2 + 5) = 2x^3 - x^2 - 6$
(c) $(f \cdot g)(x) = f(x) \cdot g(x) = (2x^3 - 1)(x^2 + 5) = 2x^5 + 10x^3 - x^2 - 5$
(d) $\left(\dfrac{f}{g}\right)(x) = \dfrac{f(x)}{g(x)} = \dfrac{2x^3 - 1}{x^2 + 5}$ ∎

Example 2   The Molar Brush Company finds that the total production cost for manufacturing x toothbrushes is given by the function

$$C(x) = 5000 + 30\sqrt{x} \text{ dollars.}$$

These toothbrushes sell for $1.50 each.

(a) Write a formula for the revenue $R(x)$ dollars to the company if x toothbrushes are sold.

(b) Write a formula for the profit $P(x)$ dollars if x toothbrushes are manufactured and sold.

(c) Find $P(40,000)$.

Solution   (a) If x toothbrushes are sold at $1.50 each, $R(x) = 1.50x = 1.5x$ dollars.
(b) $P(x) = R(x) - C(x) = 1.5x - (5000 + 30\sqrt{x})$ dollars.
(c) $P(40,000) = 1.5(40,000) - (5000 + 30\sqrt{40,000}) = 49,000$ dollars.   ∎

Geometrically, the graph of the sum, difference, product, or quotient of f and g has at each point an ordinate that is the sum, difference, product, or quotient, respectively, of the ordinates of f and g at the corresponding points. Thus, we can sketch graphs of $f + g$, $f - g$, $f \cdot g$, or f/g by the method of *adding*, *subtracting*, *multiplying*, or *dividing ordinates*, respectively. The following example illustrates the method of adding ordinates.

Example 3   Use the method of adding ordinates to sketch the graph of $h = f + g$ if $f(x) = x^2$ and $g(x) = x - 1$.

Solution   We begin by sketching the graph of f (Figure 1a) and the graph of g (Figure 1b). Then we add the ordinates $f(x)$ and $g(x)$ corresponding to selected values of x in order to obtain the corresponding ordinates $h(x) = f(x) + g(x)$. For instance, $f(2) = 2^2 = 4$ and $g(2) = 2 - 1 = 1$, so $h(2) = f(2) + g(2) = 4 + 1 = 5$. By calculating and plotting several such points and connecting them with a smooth curve, we obtain the graph of h (Figure 1c).   ∎

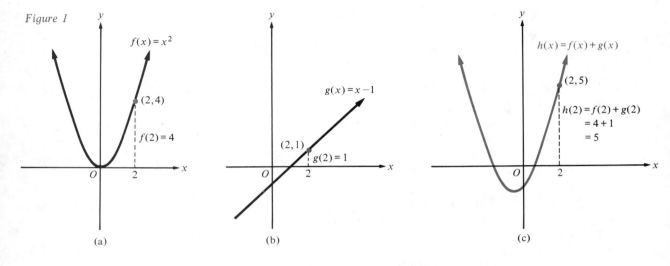

Figure 1

(a)   (b)   (c)

Composition of Functions

Suppose that we have three variable quantities y, t, and x. If y depends on t and if t, in turn, depends on x, then it is clear that y depends on x. In other words, if y is a function of t and t is a function of x, then y is a function of x. For instance, suppose that

$$y = t^2 \quad \text{and} \quad t = 3x - 1.$$

Then, substituting the value of t from the second equation into the first equation, we find that

$$y = (3x - 1)^2.$$

More generally, if

$$y = f(t) \quad \text{and} \quad t = g(x),$$

then, substituting t from the second equation into the first equation, we find that

$$y = f(g(x)).$$

To avoid a pileup of parentheses, we often replace the outside parentheses in the last equation by square brackets and write

$$y = f[g(x)].$$

If f and g are functions, the equation $y = f[g(x)]$ defines a new function h where

$$h(x) = f[g(x)].$$

The function h, obtained by "chaining" f and g together this way, is called the **composition** of f and g and is written as

$$h = f \circ g,$$

read, "h equals f composed with g." Thus, by definition,

$$(f \circ g)(x) = f[g(x)].$$

The domain of $f \circ g$ is the set of all numbers x in the domain of g such that $g(x)$ is in the domain of f.

Example 4   Let $f(x) = 3x - 1$ and $g(x) = x^3$. Find

(a) $(f \circ g)(2)$ (b) $(g \circ f)(2)$ (c) $(f \circ g)(x)$

(d) $(g \circ f)(x)$ (e) $(f \circ f)(x)$ C (f) $(f \circ g)(3.007)$

Solution   (a) $(f \circ g)(2) = f[g(2)] = f(2^3) = f(8) = 3(8) - 1 = 23$

(b) $(g \circ f)(2) = g[f(2)] = g[3(2) - 1] = g(5) = 5^3 = 125$

(c) $(f \circ g)(x) = f[g(x)] = f(x^3) = 3x^3 - 1$

(d) $(g \circ f)(x) = g[f(x)] = g(3x - 1) = (3x - 1)^3$

(e) $(f \circ f)(x) = f[f(x)] = f(3x - 1) = 3(3x - 1) - 1 = 9x - 4$

(f) $(f \circ g)(3.007) = f[g(3.007)] = 3(3.007)^3 - 1 \approx 80.568324$

Example 5   Let $f(x) = 3x - 1$, $g(x) = x^3$, and $h(x) = \sqrt{x}$. Find $[f \circ (g \circ h)](x)$.

Solution $[f \circ (g \circ h)](x) = f[(g \circ h)(x)] = f[g(h(x))] = f[g(\sqrt{x})] = f[(\sqrt{x})^3]$
$= f(x^{3/2}) = 3x^{3/2} - 1$

Example 6   If $f(x) = 3x - 1$ and $g(x) = \frac{1}{3}(x + 1)$, show that f and g "undo" each other in the sense that both $f \circ g$ and $g \circ f$ are the same as the identity function (page 159).

Solution $$(f \circ g)(x) = f[g(x)] = f[\tfrac{1}{3}(x + 1)] = 3[\tfrac{1}{3}(x + 1)] - 1 = x,$$

so $f \circ g$ is the same as the identity function. Also,

$$(g \circ f)(x) = g[f(x)] = g[3x - 1] = \tfrac{1}{3}[(3x - 1) + 1] = x.$$

Therefore, $g \circ f$ is also the identity function.

Example 7   The labor force $F(x)$ persons required by a certain industry to manufacture x units of a product is given by $F(x) = \frac{1}{2}\sqrt{x}$. At present, there is a demand for 40,000 units of the product, and this demand is increasing at a constant rate of 10,000 units per year.

(a) Write an equation for the function $u(t)$ that gives the number of units that will be demanded t years from now.

(b) Write a formula for $(F \circ u)(t)$ and interpret the resulting expression.

Solution   (a) $u(t) = 40,000 + 10,000t$

(b) $(F \circ u)(t) = F[u(t)] = F(40,000 + 10,000t)$
$$= \tfrac{1}{2}\sqrt{40,000 + 10,000t} = 50\sqrt{4 + t}$$

gives the labor force that will be required t years from now.

Using a programmable calculator, you can see a vivid demonstration of function composition. Suppose that the f and g keys are programmed with functions of your choice. If you enter a number x and touch the g key, you see the mapping

$$x \longmapsto g(x)$$

take place, and the number $g(x)$ appears in the display. Now, if you touch the f key, the mapping

$$g(x) \longmapsto f[g(x)]$$

will take place. Therefore, after you enter the number x, you can perform the composite mapping

$$x \longmapsto (f \circ g)(x)$$

by *touching first the g key, then the f key*. This fact is represented by the diagram in Figure 2.

In using a programmable calculator or a computer, it is important to be able to tell when a complicated function can be obtained as a composition of simpler functions. The same skill is essential in calculus. If

$$h(x) = (f \circ g)(x) = f[g(x)],$$

let's agree to call g the "inside function" and f the "outside function" because of the positions they occupy in the expression $f[g(x)]$. In order to see that h can be obtained as a composition $h = f \circ g$, you must be able to recognize the inside function g and the outside function f in the equation that defines h.

Figure 2

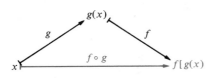

Example 8   Express the function $h(x) = (3x + 2)^2$ as a composition $h = f \circ g$ of two functions f and g.

Solution   Here we can take the inside function to be $g(x) = 3x + 2$ and the outside function to be $f(t) = t^2$, so that

$$(f \circ g)(x) = f[g(x)] = f(3x + 2) = (3x + 2)^2 = h(x). \qquad \blacksquare$$

In the solution of Example 8, we have chosen to write the squaring function as $f(t) = t^2$ rather than as $f(x) = x^2$, since x was used for the independent variable of the function $g(x) = 3x + 2$. Of course, we could just as well have written $f(x) = x^2$, since it is the *rule* f that is important, and no special significance is attached to the symbols used to denote dependent and independent variables. We would still obtain $(f \circ g)(x) = h(x)$.

Problem Set 3.7

In Problems 1 to 10, find (a) $(f + g)(x)$, (b) $(f - g)(x)$, (c) $(f \cdot g)(x)$, and (d) $\left(\dfrac{f}{g}\right)(x)$.

1. $f(x) = 5x + 2$, $g(x) = 2x - 5$

2. $f(x) = \frac{1}{2}x + 3$, $g(x) = 2x - 5$

3. $f(x) = x^2$, $g(x) = 4$

4. $f(x) = ax + b$, $g(x) = cx + d$

5. $f(x) = 2$, $g(x) = -3$

6. $f(x) = |x|$, $g(x) = x - |x|$

7. $f(x) = 2x - 5$, $g(x) = x^2 + 1$

8. $f(x) = 1 + \dfrac{1}{x}$, $g(x) = 1 - \dfrac{1}{x}$

9. $f(x) = \dfrac{5x}{2x - 1}$, $g(x) = \dfrac{3x + 1}{2x - 1}$

10. $f(x) = \sqrt{x - 3}$, $g(x) = \dfrac{1}{\sqrt{x - 3}}$

In Problems 11 to 14, use the method of adding or subtracting ordinates to sketch the graph of h.

11. $f(x) = x^2 - 2$, $g(x) = 1 - \dfrac{x}{2}$, $h = f + g$

12. $f(x) = 2x^2 + 1$, $g(x) = x$, $h = f + g$

13. $f(x) = 1 - x^2$, $g(x) = x - 1$, $h = f - g$

14. $f(x) = (x - 2)^2 + 2$, $g(x) = x^2 + 1$, $h = f - g$

In Problems 15 to 22, suppose that $f(x) = x - 3$ and $g(x) = x^2 + 4$. Find the indicated value.

15. $(f \circ g)(4)$

16. $(f \circ g)(\sqrt{2})$

© **17.** $(g \circ f)(4.73)$

© **18.** $(g \circ f)(-2.08)$

19. $(f \circ f)(3)$

20. $(g \circ g)(-3)$

21. $[f \circ (g \circ f)](2)$

22. $[(f \circ g) \circ f](2)$

In Problems 23 to 32, find (a) $(f \circ g)(x)$, (b) $(g \circ f)(x)$, and (c) $(f \circ f)(x)$.

23. $f(x) = 3x$, $g(x) = x + 1$

24. $f(x) = ax + b$, $g(x) = cx + d$

25. $f(x) = x^2$, $g(x) = \sqrt{x}$

26. $f(x) = x^3 + 1$, $g(x) = \sqrt[3]{x - 1}$

27. $f(x) = 1 - x$, $g(x) = x^2 + 2$

28. $f(x) = x + x^{-1}$, $g(x) = \sqrt{x - 1}$

29. $f(x) = 1 - 5x$, $g(x) = |2x + 3|$

30. $f(x) = 2$, $g(x) = 7$

31. $f(x) = \dfrac{1}{2x - 3}$, $g(x) = 2x - 3$

32. $f(x) = \dfrac{Ax + B}{Cx + D}$, $g(x) = \dfrac{ax + b}{cx + d}$

In Problems 33 to 40, let $f(x) = 4x$, $g(x) = x^2 - 3$, and $h(x) = \sqrt{x}$. Express each function as a composition of functions chosen from f, g, and h.

33. $F(x) = \sqrt{x^2 - 3}$

34. $G(x) = (\sqrt{x})^2 - 3$

35. $H(x) = 2\sqrt{x}$

36. $K(x) = 4x^2 - 12$

37. $Q(x) = 4\sqrt{x}$

38. $q(x) = 16x^2 - 3$

39. $r(x) = \sqrt[4]{x}$

40. $s(x) = x^4 - 6x^2 + 6$

In Problems 41 to 48, express each function h as a composition $h = f \circ g$ of two simpler functions f and g.

41. $h(x) = (2x^2 - 5x + 1)^{-7}$

42. $h(x) = |2x^2 - 3|$

43. $h(x) = \left(\dfrac{1 + x^2}{1 - x^2}\right)^5$

44. $h(x) = \dfrac{3(x^2 - 1)^2 + 4}{1 - (x^2 - 1)^2}$

45. $h(x) = \sqrt{\dfrac{x + 1}{x - 1}}$

46. $h(x) = \dfrac{1}{(4x + 5)^5}$

47. $h(x) = \dfrac{|x + 1|}{x + 1}$

48. $h(x) = \sqrt{1 - \sqrt{x - 1}}$

49. A company manufacturing integrated circuits finds that its production cost in dollars for manufacturing x of these circuits is given by the function $C(x) = 50{,}000 + 10{,}000 \sqrt[3]{x + 1}$. It sells the integrated circuits to distributors for \$10 each, and the demand is so high that all the manufactured circuits are sold.

(a) Write a formula for the revenue $R(x)$ in dollars to the company if x integrated circuits are sold.

(b) Write a formula for the profit function $P(x)$ in dollars.

© (c) Find the profit if $x = 46{,}655$.

50. An offshore oil well begins to leak and the oil slick begins to spread on the surface of the water in a circular pattern with a radius $r(t)$ kilometers. The radius depends on the time t in hours since the beginning of the leak, according to the equation $r(t) = 0.5 + 0.25t$.

(a) If $A(x) = \pi x^2$, obtain a formula for $(A \circ r)(t)$ and interpret the resulting expression.

(b) Find the area of the oil slick 4 hours after the beginning of the leak. Round off your answer to two decimal places.

51. The area A of an equilateral triangle depends on its side length x, and the side length x depends on the perimeter p of the triangle. In fact, $A = f(x)$ and $x = g(p)$, where $f(x) = (\sqrt{3}/4)x^2$ and $g(p) = (1/3)p$. Obtain a formula for $(f \circ g)(p)$ and interpret the resulting expression.

52. Suppose that user-definable keys f and g on a programmable calculator are programmed so that $f : x \longmapsto 3x^2 - 2$ and $g : x \longmapsto 5x - 3$. Draw a mapping diagram (see Figure 2, page 176) to show the effect of entering a number x and (a) touching first the g key, then the f key; (b) touching first the f key, then the g key; (c) touching the g key twice in succession.

53. Let $f(x) = 2x + 3$, $g(x) = x^2$, and $F(x) = \frac{1}{2}(x - 3)$. Find

(a) $(f \circ F)(x)$
(b) $(F \circ f)(x)$
(c) $[f \circ (g \circ F)](x)$
(d) $[(f \circ g) \circ F](x)$
(e) $[(f \circ g) + (f \circ F)](x)$

54. Give examples of (a) a function f for which $f \circ f$ is the same as $f \cdot f$, and (b) a function g for which $g \circ g$ is not the same as $g \cdot g$.

55. Give an example to show that $f \circ g$ need not be the same as $g \circ f$.

56. (a) Is $f \cdot (g + h)$ always the same as $(f \cdot g) + (f \cdot h)$?
(b) Is the function $f \circ (g + h)$ always the same as the function $(f \circ g) + (f \circ h)$?

57. If f, g, and h are three functions, show that $(f \circ g) \circ h$ is the same function as $f \circ (g \circ h)$.

58. Prove that the composition of two linear functions is again a linear function.

59. Let $f(x) = 5x + 3$ and $g(x) = 3x + k$, where k is a constant. Find a value of k for which $f \circ g$ and $g \circ f$ are the same function.

60. If $f(x) = mx + b$ with $m \neq 0$, find a function F such that $F \circ f$ is the same as the identity function.

61. Let $f(x) = cx + 1$, where c is a constant. Find a value of c so that $f \circ f$ is the same as the identity function.

3.8 INVERSE FUNCTIONS

Many calculators have both a squaring key (marked x^2) and a square–root key (marked $\sqrt{x}$). For *nonnegative numbers* x, the functions represented by these keys "undo each other" in the sense that

$$x \xrightarrow{\text{squaring function}} x^2 \xrightarrow{\text{square-root function}} \sqrt{x^2} = x$$

and

$$x \xrightarrow{\text{square-root function}} \sqrt{x} \xrightarrow{\text{squaring function}} (\sqrt{x})^2 = x.$$

Two functions related in such a way that each "undoes" what the other "does" are said to be **inverse** to one another. Most scientific calculators not only have keys corresponding to a number of important functions, but they also can calculate the values of the inverses of these functions.

In order that two functions f and g be inverses of one another, we must have the following: For every value of x in the domain of f, $f(x)$ is in the domain of g and

$$x \xmapsto{\;f\;} f(x) \xmapsto{\;g\;} g[f(x)] = x;$$

likewise, for every value of x in the domain of g, $g(x)$ is in the domain of f and

$$x \xmapsto{\;g\;} g(x) \xmapsto{\;f\;} f[g(x)] = x.$$

In other words:

> f and g are inverses of each other if and only if
>
> $$g[f(x)] = x$$
>
> for every value of x in the domain of f, and
>
> $$f[g(x)] = x$$
>
> for every value of x in the domain of g.

A function f for which such a function g exists is said to be **invertible.**

In Examples 1 and 2, show that the functions f and g are inverses of each other.

Example 1 $f(x) = 5x - 1$ and $g(x) = \dfrac{x + 1}{5}$.

Solution   The domain of f is the set $\mathbb{R}$ of all real numbers. For every real number x, we have

$$g[f(x)] = g(5x - 1) = \frac{(5x - 1) + 1}{5} = \frac{5x}{5} = x.$$

The domain of g is also the set $\mathbb{R}$ of all real numbers, and for every real number x, we have

$$f[g(x)] = f\left(\frac{x + 1}{5}\right) = 5\left(\frac{x + 1}{5}\right) - 1 = x + 1 - 1 = x. \qquad \blacksquare$$

Example 2 $f(x) = \dfrac{2x - 1}{x}$ and $g(x) = \dfrac{1}{2 - x}$.

Solution   The domain of f is the set of all nonzero real numbers. If $x \neq 0$, we have

$$g[f(x)] = g\left(\frac{2x - 1}{x}\right) = \frac{1}{2 - \left(\dfrac{2x - 1}{x}\right)} = \frac{x}{\left[2 - \left(\dfrac{2x - 1}{x}\right)\right]x}$$

$$= \frac{x}{2x - (2x - 1)} = \frac{x}{1} = x.$$

The domain of g is the set of all real numbers except 2. If $x \neq 2$, we have

$$f[g(x)] = f\left(\frac{1}{2-x}\right) = \frac{2\left(\dfrac{1}{2-x}\right) - 1}{\left(\dfrac{1}{2-x}\right)} = \left[2\left(\frac{1}{2-x}\right) - 1\right](2-x)$$

$$= 2 - (2 - x) = x. \qquad \blacksquare$$

Geometrically, f and g are inverses of each other if and only if the graph of g is the mirror image of the graph of f across the straight line $y = x$ (Figure 1).

This will be clear if you notice first that the mirror image of a point (a, b) across the line $y = x$ is the point (b, a). But, if f and g are inverses of each other and if (a, b) belongs to the graph of f, then $b = f(a)$, so $g(b) = g[f(a)] = a$; that is, (b, a) belongs to the graph of g. Similarly, you can show that, if (b, a) belongs to the graph of g, then (a, b) belongs to the graph of f (Problem 35).

Figure 1   *Figure 2*

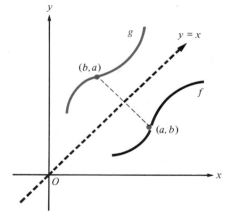
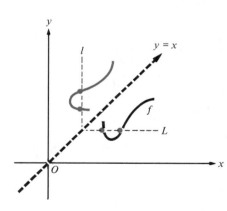

Not every function is invertible. Indeed, consider the function f whose graph appears in Figure 2. The mirror image of the graph of f across the line $y = x$ isn't the graph of a function, because there is a vertical line l that intersects it more than once. (Recall the vertical-line test, page 152.) Notice that the horizontal line L obtained by reflecting l across the line $y = x$ intersects the graph of f more than once. These considerations provide the basis of the following test (see Problem 36).

Horizontal-Line Test

A function f is invertible if and only if no horizontal straight line intersects its graph more than once.

Example 3   Use the horizontal-line test to determine whether or not the functions graphed in Figure 3 are invertible.

Figure 3

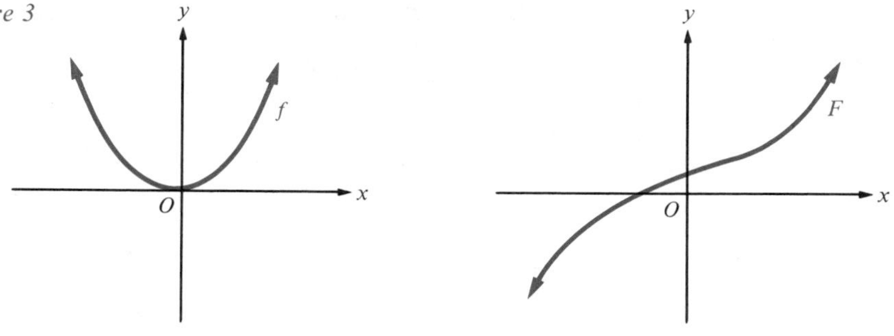

(a)   (b)

Solution   **(a)** Any horizontal line drawn above the origin will intersect the graph of f *twice*. Therefore, f is not invertible.

(b) No horizontal line intersects the graph of F more than once; hence, F is invertible.   ■

The horizontal line test can be restated in terms of the following condition:

One-to-One Function

A function f is said to be **one-to-one** if, whenever a and b are in the domain of f and $f(a) = f(b)$, it follows that $a = b$.

In Problem 37, we ask you to use the horizontal line test to show that *a function f is invertible if and only if it is one-to-one.*

Example 4   Show that $f(x) = 3x - 7$ is a one-to-one function.

Solution   Suppose that a and b are real numbers and $f(a) = f(b)$. Then

$$3a - 7 = 3b - 7$$

$$3a = 3b \qquad \text{(We added 7 to both sides.)}$$

$$a = b \qquad \text{(We divided both sides by 3.)}$$

Therefore, $f(x) = 3x - 7$ is one-to-one.   ■

If a function f is invertible, there is *exactly one* function g such that f and g are inverses of each other; indeed, g is the one and only function whose graph is the mirror image of the graph of f across the line $y = x$. We call g the **inverse** of the function f and we write $g = f^{-1}$. The notation f^{-1} is read "f inverse." If you imagine that the graph of f is drawn with wet ink, then the graph of f^{-1} would be the imprint obtained by folding the paper along the line $y = x$ (Figure 4). If f is invertible, then the domain of f^{-1} is the range of f and the range of f^{-1} is the domain of f. (Do you see why?)

Figure 4

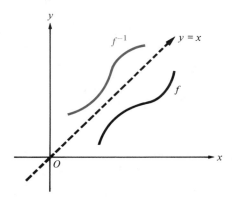

The fact that the graph of f^{-1} is the set of all points (x, y) such that (y, x) belongs to the graph of f provides the basis for the following procedure.

Algebraic Method for Finding f^{-1}

Step 1. Write the equation $y = f(x)$ that defines f.

Step 2. Interchange x and y in the equation obtained in step 1.

Step 3. Solve the equation in step 2 for y in terms of x to get $y = f^{-1}(x)$.

Figure 5

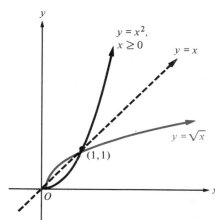

In step 2, you must change y to x and change all x's to y's so that the equation becomes $x = f(y)$. The equation $y = f^{-1}(x)$ obtained in step 3 then defines the inverse function f^{-1}.

Example 5   Use the algebraic method to find the inverse of the function $f(x) = x^2$, for $x \geq 0$, and sketch the graphs of f and f^{-1} on the same coordinate system.

Solution   We carry out the algebraic procedure:

Step 1. $y = x^2$ for $x \geq 0$

Step 2. $x = y^2$ for $y \geq 0$

Step 3. Solving the equation $x = y^2$ for y and using the condition that $y \geq 0$, we obtain $y = \sqrt{x}$; therefore, $f^{-1}(x) = \sqrt{x}$ (Figure 5). ∎

Example 6   Let $f(x) = 2x + 1$.

(a) Use the algebraic method to find f^{-1}.

(b) Check that $f^{-1}[f(x)] = x$ for all values of x in the domain of f.

(c) Check that $f[f^{-1}(x)] = x$ for all values of x in the domain of f^{-1}.

Solution   (a) We carry out the algebraic procedure:

Step 1. $y = 2x + 1$

Step 2. $x = 2y + 1$

Step 3. $2y + 1 = x$

$$2y = x - 1$$

$y = \frac{1}{2}(x - 1)$; therefore, $f^{-1}(x) = \frac{1}{2}(x - 1)$.

(b) $f^{-1}[f(x)] = f^{-1}(2x + 1) = \frac{1}{2}[(2x + 1) - 1] = x$

(c) $f[f^{-1}(x)] = f[\frac{1}{2}(x - 1)] = 2[\frac{1}{2}(x - 1)] + 1 = x$ ■

In working with inverses of functions, *you must be careful not to confuse $f^{-1}(x)$ and $[f(x)]^{-1}$*. Notice that $f^{-1}(x)$ is the value of the function f^{-1} at x, while $[f(x)]^{-1} = \dfrac{1}{f(x)}$ is the reciprocal of the value of the function f at x. For instance, in Example 6, above, $f^{-1}(x) = \frac{1}{2}(x - 1)$, whereas $[f(x)]^{-1} = \dfrac{1}{2x + 1}$.

Problem Set 3.8

In Problems 1 to 6, show that the functions f and g are inverses of each other.

1. $f(x) = 2x - 3$ and $g(x) = \dfrac{x + 3}{2}$

2. $f(x) = x^3$ and $g(x) = \sqrt[3]{x}$

3. $f(x) = \dfrac{1}{x}$ and $g(x) = \dfrac{1}{x}$

4. $f(x) = \dfrac{2x - 3}{3x - 2}$ and $g(x) = \dfrac{2x - 3}{3x - 2}$

5. $f(x) = \sqrt[3]{x + 8}$ and $g(x) = x^3 - 8$

6. $f(x) = x^2 + 1$ for $x \geq 1$ and $g(x) = \sqrt{x - 1}$

7. Use the horizontal-line test to determine whether

or not the functions graphed in Figure 6 are invertible.

Figure 6

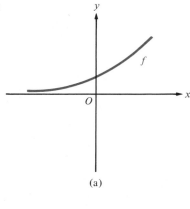

(a)

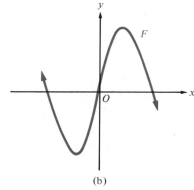

(b)

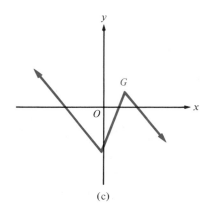

(c)

8. Under what conditions is a linear function $f(x) = mx + b$ invertible?

9. The three functions graphed in Figure 7 are invertible. In each case, obtain the graph of the inverse function by reflecting the given graph across the line $y = x$.

Figure 7

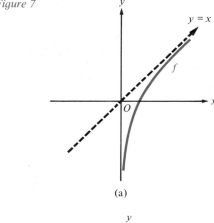

(a)

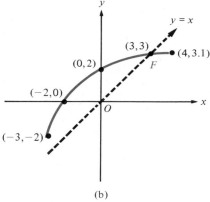

(b)

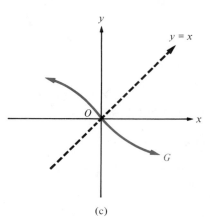

(c)

10. If a linear function f is invertible, show that f^{-1} is also a linear function.

In Problems 11 to 20, show that f is one-to-one, use the algebraic method to find $f^{-1}(x)$, and sketch the graphs of f and f^{-1} on the same coordinate system.

11. $f(x) = 7x - 13$

12. $f(x) = 5 + 7x$

13. $f(x) = -\frac{1}{2}x + 3$

14. $f(x) = x^3$

15. $f(x) = -x^2$ for $x \le 0$

16. $f(x) = 4 - x^2$ for $x \ge 0$

17. $f(x) = 1 + \sqrt{x}$

18. $f(x) = (x - 2)^2$ for $x \ge 2$

19. $f(x) = (x - 1)^2 + 1$ for $x \ge 1$

20. $f(x) = (x + 3)^2 + 2$ for $x \le -3$

In Problems 21 to 30, (a) use the algebraic method to find f^{-1}, (b) check that $f^{-1}[f(x)] = x$ for all values of x in the domain of f, and (c) check that $f[f^{-1}(x)] = x$ for all values of x in the domain of f^{-1}.

21. $f(x) = 2x - 5$

22. $f(x) = \dfrac{3x + 2}{5}$

23. $f(x) = (x + 2)^3$

24. $f(x) = (x + 1)^3 - 2$

25. $f(x) = 1 - 2x^3$

26. $f(x) = x^5 - 1$

27. $f(x) = -\dfrac{1}{x} - 1$

28. $f(x) = \dfrac{3}{x + 2}$

29. $f(x) = \dfrac{3x - 7}{x + 1}$

30. $f(x) = (x + 3)^5 - 2$

31. Sketch the graph of $f(x) = |x - 1|$ and determine whether or not f is invertible.

32. The linear function $f(x) = x$ is its own inverse. Find all linear functions that are their own inverses.

33. If a, b, c, and d are constants such that $ad \ne bc$, use the algebraic method to find the inverse of

$$f(x) = \frac{ax + b}{cx + d}.$$

34. If $f(x) = \sqrt{4 - x^2}$ for $0 \le x \le 2$, show that f is its own inverse.

35. If f and g are inverses of each other and (b, a) belongs to the graph of g, show that (a, b) belongs to the graph of f.

36. Complete the argument to justify the horizontal-line test by showing that, if no horizontal line intersects the graph of f more than once, then f is invertible.

37. By using the horizontal-line test, show that f is invertible if and only if it is one-to-one.

38. If f is invertible, show that the domain of f coincides with the range of f^{-1}.

39. Suppose that the concentration C of an anesthetic in body tissues satisfies an equation of the form $t = f(C)$, where t is the elapsed time since the anesthetic was administered. If f is invertible, write an equation for C in terms of t.

40. Show that the functions f and g are inverses of each other if and only if $(g \circ f)(x) = x$ for every x in the domain of f and $(f \circ g)(x) = x$ for every x in the domain of g.

41. Sketch the graph of any invertible function f; then turn the paper over, rotate it 90° clockwise, and hold it up to the light. Through the paper you will see the graph of f^{-1}. Why does this procedure work?

42. If bacteria are allowed to reproduce in a culture, the number N of bacteria at time t satisfies an equation of the form $N = f(kt + c)$, where k and c are constants, $k \ne 0$, and f is an invertible function. Write an equation involving f^{-1} that gives t in terms of N.

43. Let A, B, and C be constants with $A > 0$, and suppose that $f(x) = Ax^2 + Bx + C$ for $x \ge \dfrac{-B}{2A}$. Find $f^{-1}(x)$.

44. Is there any invertible function f such that $f^{-1}(x) = [f(x)]^{-1}$ for all values of x in the domain of f?

3.9 RATIO, PROPORTION, AND VARIATION

When we describe the ways in which real-world quantities are related, we do not ordinarily write or speak in equations. Thus, in order to apply mathematical methods to problems in engineering, medicine, business, economics, the earth sciences, the life sciences, and so forth, we must be able to translate verbal statements about the relationships among variable quantities into mathematical formulas, and vice versa.

For instance, in economics, we might learn that the price of a certain commodity varies linearly as the demand for it; in engineering, that the power consumed by an electrical transmission line varies directly as the square of the current; or in physics, that the gravitational attraction between two homogeneous spheres varies inversely as the square of the distance between their centers. In medicine, we might be told that the rate at which bacteria increase in a bladder infection is proportional to the number of bacteria already present; in respiratory physiology, that alveolar ventilation is inversely proportional to carbon dioxide pressure; or in chemistry, that the rate of an autocatalytic reaction is jointly proportional to the amount of unreacted substance and the amount of the product. In this section we study the exact mathematical meaning of statements such as these.

Ratio and Proportion

Two quantities a and b are often compared by considering their *ratio a/b*. For instance, we might compare the fuel efficiency of two automobiles by forming the ratio of the gas mileage of one to that of the other. If $b \neq 0$, the **ratio** of a to b is written as

$$\frac{a}{b} \quad \text{or} \quad a \div b \quad \text{or} \quad a{:}b.$$

Although the notation $a{:}b$ isn't as popular as it once was, it is still used occasionally.

Example 1   If the ratio of a to b is 2/3, find the ratio of $5a + 3b$ to $7a - 4b$.

Solution   Here, $\dfrac{a}{b} = \dfrac{2}{3}$. To find the value of $\dfrac{5a + 3b}{7a - 4b}$, we divide numerator and denominator by b:

$$\frac{5a + 3b}{7a - 4b} = \frac{\dfrac{5a}{b} + \dfrac{3b}{b}}{\dfrac{7a}{b} - \dfrac{4b}{b}} = \frac{5\left(\dfrac{a}{b}\right) + 3}{7\left(\dfrac{a}{b}\right) - 4} = \frac{5\left(\dfrac{2}{3}\right) + 3}{7\left(\dfrac{2}{3}\right) - 4} = \frac{\dfrac{10}{3} + 3}{\dfrac{14}{3} - 4} = \frac{10 + 9}{14 - 12} = \frac{19}{2}.$$

An equation such as

$$\frac{a}{b} = \frac{c}{d}, \quad \text{or} \quad a:b = c:d,$$

which expresses the equality of two ratios, is called a **proportion.** The proportion $a:b = c:d$ is often read, "a is to b as c is to d." If the proportion

$$a:b = c:d \quad \text{is written as} \quad \frac{a}{b} = \frac{c}{d}$$

and both sides of the equation are multiplied by bd, we have

$$\frac{a}{b} bd = \frac{c}{d} bd \quad \text{or} \quad ad = bc.$$

Conversely, if we divide both sides of the equation $ad = bc$ by bd, we obtain

$$\frac{ad}{bd} = \frac{bc}{bd},$$

which is just the original proportion $a:b = c:d$, so $a:b = c:d$ or $a/b = c/d$ is equivalent to the equation $ad = bc$. You may want to use the memory device

$$\frac{a}{b} \diagdown \frac{c}{d},$$

indicating that a is multiplied by d and that b is multiplied by c when

$$\boxed{\frac{a}{b} = \frac{c}{d} \quad \text{is rewritten as} \quad ad = bc.}$$

This procedure is called **cross-multiplication.**

Example 2   Solve the proportion $\dfrac{5}{x-1} = \dfrac{3}{x+1}$ for x.

Solution   Cross-multiplying, we have

$$5(x + 1) = 3(x - 1) \quad \text{or} \quad 5x + 5 = 3x - 3.$$

Thus, $2x = -8$, so $x = -4$.   ∎

Variation

Suppose that x and y are variable quantities and that y is a function of x, that is, $y = f(x)$. As the quantity x varies, the quantity y, which depends on x, is subject to a corresponding variation. One of the simplest and also most important types of variation is defined as follows.

Definition 1   **Direct Variation**

> If there is a constant k such that
>
> $$y = kx$$
>
> holds for all values of x, we say that **y is directly proportional to x** or that **y varies directly as x** (or *with x*). The constant k is called the **constant of proportionality.**

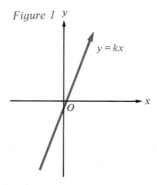

Figure 1

If $y = kx$, then $y = 0$ when $x = 0$. Also, if $x \neq 0$, we have

$$\frac{y}{x} = k.$$

Therefore, if y is directly proportional to x, it follows that the ratio y/x maintains the constant value k as x varies through nonzero values. A graph of $y = kx$ is, of course, a straight line that has slope k and contains the origin (Figure 1).

Example 3   Suppose that y varies directly as x and that $y = 6$ when $x = 2$. Express y as a function of x.

Solution   Let k be the constant of proportionality, so that $y = kx$. Substituting the given values $x = 2$ and $y = 6$, we find that

$$6 = k(2) \qquad \text{or} \qquad k = 3.$$

Therefore, $y = 3x$.   ∎

Example 4   Suppose that the rate r at which impulses are transmitted along a nerve fiber is directly proportional to the diameter d of the fiber. Given that $r = 20$ meters per second when $d = 6$ micrometers, express r as a function of d.

Solution   We have $r = kd$, and, in particular,

$$20 = k(6) \qquad \text{or} \qquad k = \tfrac{20}{6} = \tfrac{10}{3}.$$

Therefore, $r = \tfrac{10}{3}d$.   ∎

Figure 1 suggests the following natural extension of direct variation.

Definition 2   **Linear Variation**

> If there are constants k and b such that
>
> $$y = kx + b,$$
>
> then we say that **y varies linearly as x** (or *with x*).

Recall that the graph of the equation $y = kx + b$ is a straight line that has slope k and y intercept b (Figure 2). If you are given two distinct points on such a graph, you can use the methods in Section 3.3 to determine both the slope k and the y intercept b.

Example 5   In a certain income tax bracket, the amount y dollars of income tax that a family must pay to the federal government varies linearly with the amount x dollars of income after deductions. Suppose that $y = \$3260$ when $x = \$19,200$, and that $y = \$4380$ when $x = \$23,200$. Express y as a function of x and find y when $x = \$21,000$.

Solution   The graph of $y = f(x)$ is a straight line as in Figure 2. Its slope k is given by the slope formula (Section 3.2):

Figure 2

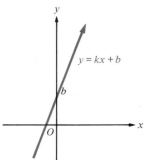

$$k = \frac{y_2 - y_1}{x_2 - x_1} = \frac{4380 - 3260}{23,200 - 19,200} = \frac{1120}{4000} = 0.28.$$

Thus, in point-slope form (page 143), an equation of the graph is

$$y - 3260 = 0.28(x - 19,200) = 0.28x - 0.28(19,200) = 0.28x - 5376,$$

or

$$y = 0.28x - 5376 + 3260 = 0.28x - 2116.$$

In particular, when $x = \$21,000$, we have

$$y = 0.28(21,000) - 2116 = 5880 - 2116 = 3764 \text{ dollars.} \qquad \blacksquare$$

If we say that y *varies directly as the square of* x, we mean that $y = kx^2$ for some constant k. More generally, if n is a fixed rational number and $y = kx^n$ for some constant k, we say that y **varies directly as the nth power of x** or that y **is proportional to the nth power of x.**

Ⓒ **Example 6**   Suppose that the weight w of men of a given body type varies directly as the cube of their height h. If such a man is 1.85 meters tall and weighs 80 kilograms, and if a basketball player who has this same body type is 2 meters tall, what is the basketball player's weight?

Solution   We have $w = kh^3$, where k is a constant. When $h = 1.85$, we have $w = 80$; that is,

$$80 = k(1.85)^3.$$

Therefore,

$$k = \frac{80}{(1.85)^3}.$$

Hence, when $h = 2$,

$$w = kh^3 = \frac{80}{(1.85)^3} (2)^3 \approx 101.08 \text{ kilograms.} \qquad \blacksquare$$

The situation in which y is proportional to the reciprocal of x arises so frequently that it is given a special name, **inverse variation.**

Definition 3   **Inverse* Variation**

If there is a constant k such that

$$y = \frac{k}{x}$$

for all nonzero values of x, then we say that y **is inversely proportional to** x or that y **varies inversely as** x (or *with* x).

Naturally, if $y = k/x^n$ for some constant k and some positive rational number n, we say that y **is inversely proportional to the nth power of x.**

Example 7   Express y as a function of x if y is inversely proportional to the cube of x, and $y = 2$ when $x = 3$. Find the value of y when $x = \frac{1}{2}$.

Solution   We have $y = k/x^3$, where k is a constant. Substituting the values $x = 3$ and $y = 2$, we obtain

$$2 = \frac{k}{3^3} = \frac{k}{27},$$

so

$$k = 2(27) = 54$$

and

$$y = \frac{54}{x^3}.$$

When $x = \frac{1}{2}$, we have

$$y = \frac{54}{(1/2)^3} = \frac{54}{(1/8)}$$

$$= 54(8) = 432. \qquad \blacksquare$$

Example 8   A company is informed by its bookkeeper that the number S of units of a certain item sold per month seems to be inversely proportional to the quantity $p + 20$, where p is the selling price per unit in dollars. Suppose that 800 units per month are sold when the price is \$10 per unit. How many units per month would be sold if the price were dropped to \$5 per unit?

Solution   Since S is inversely proportional to $p + 20$, there is a constant k such that

$$S = \frac{k}{p + 20}.$$

* This use of the word inverse should not be confused with the inverse of a function.

We know that $S = 800$ when $p = 10$, so we have

$$800 = \frac{k}{10 + 20} = \frac{k}{30}$$

or
$$k = 30(800) = 24{,}000.$$

Therefore,
$$S = \frac{24{,}000}{p + 20}$$

gives S as a function of p. For $p = 5$ dollars, we have

$$S = \frac{24{,}000}{5 + 20} = \frac{24{,}000}{25} = 960,$$

so 960 units per month would be sold at a price of $5 per unit.   ■

Joint and Combined Variation

Often the value of a variable quantity depends on the values of several other quantities; for instance, the amount of simple interest on an investment depends on the interest rate, the amount invested, and the period of time involved. For compound interest, the amount of interest depends on an additional variable: how often the compounding takes place. A situation in which one variable depends on several others is called **combined variation.** An important type of combined variation is defined as follows.

Definition 4   **Joint Variation**

> If a variable quantity y is proportional to the product of two or more variable quantities, we say that y **is jointly proportional** to these quantities, or that y **varies jointly as** (or *with*) these quantities.

Thus, if y is jointly proportional to u and v, then $y = kuv$ for some constant k. Sometimes the word "jointly" is omitted and we simply say that y is proportional to u and v.

Example 9   In chemistry, the absolute temperature T of a perfect gas varies jointly as its pressure P and its volume V. Given that $T = 500$ kelvins when $P = 50$ pounds per square inch and $V = 100$ cubic inches, find a formula for T in terms of P and V, and find T when $P = 100$ pounds per square inch and $V = 75$ cubic inches.

Solution   Since T varies jointly as P and V, there is a constant k such that $T = kPV$. Putting $T = 500$, $P = 50$, and $V = 100$, we find that

$$500 = k(50)(100) \qquad \text{or} \qquad k = \frac{500}{50(100)} = \frac{1}{10}.$$

Thus, the desired formula is

$$T = \frac{1}{10}PV.$$

When $P = 100$ and $V = 75$, we have

$$T = \frac{1}{10}(100)(75) = 750 \text{ kelvins.}$$ ∎

We sometimes have joint variation together with inverse variation. Perhaps the most important historical discovery of this type of combined variation is *Newton's law of universal gravitation*, which states that the gravitational force of attraction F between two particles is jointly proportional to their masses m and M, and inversely proportional to the square of the distance d between them. In other words,

$$F = G\frac{mM}{d^2},$$

where G is the constant of proportionality.

Ⓒ **Example 10**   Careful measurements show that two 1-kilogram masses 1 meter apart exert a mutual gravitational attraction of 6.67×10^{-11} newton. (One pound of force is approximately 4.45 newtons.) The earth has a mass of 5.98×10^{24} kilograms. Find the earth's gravitational force on a space capsule that has a mass of 1000 kilograms and that is 10^8 meters from the center of the earth.

Solution   Putting $F = 6.67 \times 10^{-11}$ newton, $m = 1$ kilogram, $M = 1$ kilogram, and $d = 1$ meter in the formula $F = G(mM/d^2)$, we find that

$$6.67 \times 10^{-11} = G\frac{1(1)}{1^2} = G,$$

so that, for arbitrary values of m, M, and d,

$$F = (6.67 \times 10^{-11})\frac{mM}{d^2}.$$

Now we substitute $M = 5.98 \times 10^{24}$, $m = 10^3$, and $d = 10^8$ to obtain

$$F = (6.67 \times 10^{-11})\frac{10^3(5.98 \times 10^{24})}{(10^8)^2}$$

$$= (6.67 \times 10^{-11})(5.98 \times 10^{24+3-16})$$
$$= (6.67)(5.98) \times 10^{24+3-16-11} \approx 39.9 \times 10^0 = 39.9 \text{ newtons}$$

(or less than 9 pounds of gravitational force). ∎

In solving Example 10 we used the fact—first proved by Newton himself using integral calculus—that in calculating the gravitational attraction of a homogeneous sphere, one can assume that all of the mass is concentrated at the center.

Problem Set 3.9

In Problems 1 to 6, find the ratio of the first quantity to the second. Write your answer in the form of a reduced fraction. If the quantities are measured in different units, you must first express both in the same units.

1. 51 pounds to 68 pounds

2. 700 grams to 21 kilograms

3. $\frac{3}{4}$ yard to 15 inches

4. 1 square foot to 1 square yard

5. 1500 cubic centimeters to $\frac{1}{10}$ cubic meter

6. 88 kilometers per hour to 11 miles per hour [Take 1 kilometer to be $\frac{5}{8}$ mile.]

In Problems 7 to 10, find the ratio of each pair of expressions if the ratio of a to b is 4/5.

7. $6a - 7b$ to $3a + 4b$ 8. $5a + 3b$ to $4a + 9b$

9. $3a + 11b$ to $2a - b$ 10. $-7a$ to $3a + 13b$

In Problems 11 to 16, solve each proportion for x.

11. $15:x = 10:7$ 12. $\dfrac{x-1}{10} = \dfrac{17}{4}$

13. 8 is to $3x$ as 7 is to 5   14. $5 - 3x: -8 = 3:2x$

15. $25/(x + 2) = 2/5$

16. $3x + 40:12 = 2x - 15:6$

In Problems 17 to 22, assume that $a/b = c/d$ and show by an algebraic calculation that the indicated proportion follows from it. Assume that the values of a, b, c, and d are restricted so that denominators are not zero.

17. $\dfrac{c}{a} = \dfrac{d}{b}$ 18. $b:a = d:c$

19. $c - a:a = d - b:b$ 20. $\dfrac{b+a}{a} = \dfrac{d+c}{c}$

21. $\dfrac{b}{a+b} = \dfrac{d}{c+d}$ 22. $\dfrac{a+b}{a-b} = \dfrac{c+d}{c-d}$

In Problems 23 to 34, relate the quantities by writing an equation involving a constant of proportionality k.

23. y is directly proportional to t.

24. A varies directly with r^2.

25. The volume V of a sphere varies directly as the cube of its radius r.

26. The distance d through which a body falls is directly proportional to the square of the amount of time t during which it falls.

27. The power P provided by a wind generator is directly proportional to the cube of the wind speed v.

28. The kinetic energy K of a moving body is jointly proportional to its mass m and the square of its speed v.

29. The heat energy E conducted per hour through the wall of a house is jointly proportional to the area A of the wall and the difference T degrees between the inside and outside temperatures, and inversely proportional to the thickness d of the wall.

30. The maximum safe load W on a horizontal beam supported at both ends varies directly as the width w of the beam and the square of its depth d, and inversely as the length l of the beam.

31. The force F of repulsion between two electrically charged particles is directly proportional to the charges Q_1 and Q_2 on the particles, and inversely proportional to the square of the distance d between them.

32. The intensity I of a sound wave is jointly proportional to the square of its amplitude A, the square of its frequency n, the speed of sound v, and the density d of the air.

33. The frequency n of a guitar string is directly proportional to the square root of the string tension F, and inversely proportional to the length l of the string and the square root of its linear density (mass per unit length) d.

34. The rate r at which an infectious disease spreads is jointly proportional to the fraction p of the population that has the disease and the fraction $1 - p$ of the population that does not have the disease.

In Problems 35 to 46, (a) find an equation relating the given quantities and (b) find the value of the indicated quantity under the given conditions.

35. y is directly proportional to x, and $y = 3$ when $x = 6$. Find y when $x = 12$.

36. y varies directly as x^3, and $y = 4$ when $x = 2$. Find y when $x = 4$.

37. w varies inversely as x, and $w = 3$ when $x = 4$. Find w when $x = 8$.

38. q varies inversely as $\sqrt{p}$, and $q = 2$ when $p = \frac{1}{4}$. Find q when $p = 9$.

39. u varies inversely as v^2, and $u = 2$ when $v = 10$. Find u when $v = 3$.

40. r is inversely proportional to $\sqrt[3]{t}$, and $r = 2$ when $t = \frac{1}{8}$. Find r when $t = 27$.

41. z is directly proportional to x and inversely proportional to y, and $z = 8$ when $x = 4$ and $y = 2$. Find z when $x = 8$ and $y = 4$.

42. z is directly proportional to x^3 and inversely proportional to y^2, and $z = 2$ when $x = 2$ and $y = 4$. Find z when $x = -2$ and $y = 2$.

43. H is directly proportional to t and inversely proportional to $\sqrt{s}$, and $H = 7$ when $t = 2$ and $s = 4$. Find H when $t = 4$ and $s = 9$.

44. s varies jointly as the product of x and $y^{2/3}$, and $s = 24$ when $x = 6$ and $y = 8$. Find s when $x = -3$ and $y = 1$.

45. V varies directly as T and inversely as P, and $V = 40$ when $T = 300$ and $P = 30$. Find V when $T = 324$ and $P = 24$.

46. y varies jointly as x^2 and z and inversely as w, and $y = 12$ when $x = 4$, $z = 3$, and $w = 2$. Find y when $x = -4$, $z = 5$, and $w = -1$.

47. In wildlife management, the population P of animals of a certain type that inhabit a particular area is often estimated by capturing M of these animals, banding or otherwise marking them, and releasing the marked animals. Later, after the marked animals have mixed with the rest of the animals in the area, a random sample consisting of S animals is captured and the number m of marked animals in the sample is counted. It is assumed that the ratio of marked animals in the population is the same as the ratio of marked animals in the sample, so that $M/P = m/S$. Solve for P in terms of M, S, and m.

Ecologists releasing banded duck

48. In order to estimate the number of fish in a lake, ecologists catch 100 fish, band them, and release the banded fish to mix with the rest of the fish. Later, 50 fish are caught from the lake, and 2 of them are found to be banded. Estimate the number of fish in the lake. [See Problem 47.]

49. At a constant speed v on level ground, the number N of gallons of gasoline used by an automobile is directly proportional to the amount of time T during which it travels. If the automobile uses 6 gallons in 1.5 hours, how many gallons will it use in 10 hours?

© 50. The distance d in which an automobile can stop when its brakes are applied varies directly as the square of its speed v. Skid marks from a car involved in an accident measured 173 feet. In a test, the same car going 40 miles per hour was braked to a panic stop in 88 feet. Was the driver exceeding the speed limit (55 miles per hour) when the accident occurred?

51. Fahrenheit temperature F varies linearly as Celsius temperature C. Find a formula for F in terms of C if $F = 212°$ when $C = 100°$, and $F = 32°$ when $C = 0°$.

© 52. A manufacturer finds that the cost C dollars of manufacturing N units of a certain product varies linearly with N. The manufacturer's *overhead* for the product is the cost if $N = 0$. Find the overhead, given that $C = \$4200$ when $N = 100$, and $C = \$13,000$ when $N = 500$.

53. A company estimates that its profit P dollars per month varies linearly with its sales revenue R. If $P = \$6000$ when $R = \$75,000$, and $P = \$16,000$ when $R = \$150,000$, find a formula for P in terms of R.

54. The shoe size of a normal man varies approximately as the 3/2 power of his height. If the average 6-foot man wears a size 11 shoe, what would you predict as the shoe size of a 7-foot basketball player?

55. The amount A of pollution entering the atmosphere in a certain region is found to be directly proportional to the 2/3 power of the number N of people in that region. If a population of $N = 8000$ people produces $A = 900$ tons of pollution per year, find a formula for A in terms of N.

56. In astronomy, Kepler's third law states that the time T required for a planet to make one revolution about the sun is directly proportional to the 3/2 power of the maximum radius of its orbit. If the maximum radius of the earth's orbit is 93 million miles and the maximum radius of Mars' orbit is 142 million miles, how many days are required for Mars to make one revolution about the sun?

57. The time t days required to pick up litter along a stretch of highway is inversely proportional to the number n of people on the work team. If 5 people can do the job in 3 days, how many days would be required for 8 people to do it?

58. The amount H of heat conducted per second through a cube is jointly proportional to the temperature difference T degrees between the opposite faces and the area A of one face, and inversely proportional to the length l of an edge of the cube. The constant of proportionality for this combined variation is called the *coefficient of conductivity* for the material of the cube. Find the coefficient of conductivity for a cube with an edge that is 45 centimeters long, if 1.26 joules of heat per second are conducted through the cube when the temperature difference between opposite faces is 31.5° Celsius.

59. The distance from the earth to the moon is approximately 3.8×10^8 meters, and the mass of the moon is approximately 7.3×10^{22} kilograms. Find the gravitational attraction of the moon on a person on earth whose mass is 70 kilograms. [See Example 10.]

60. Using the data in problem 59, find the ratio of a person's weight on the moon to the same person's weight on earth. Take the radius of the moon to be 1.7×10^6 meters and the radius of the earth to be 6.4×10^6 meters.

61. The frequency n hertz of a musical tone is directly proportional to the speed v of sound in air, and inversely proportional to the wavelength l of the sound wave. The speed of sound in air varies linearly as the temperature T in degrees Celsius. Write a formula for the frequency n in terms of l, T, and two constants.

62. Determine the numerical value of the two constants in problem 61, given the following data: $n = 440$ hertz when $T = 0°$ Celsius and $l = 0.75$ meter, and $n = 880$ hertz when $T = 20°$ Celsius and $l = 0.39$ meter.

63. Since ancient times, artists and artisans have recognized that figures whose proportions are in a certain ratio a/b, called the *golden ratio*, are especially pleasing. For instance, a rectangle whose width a and length b form the golden ratio is called a *golden rectangle*. The condition under which a/b is the golden ratio is that a is to b as b is to $a + b$. Calculate the golden ratio to three decimal places, and draw a golden rectangle.

64. In some European countries, the dimensions of a standard sheet of typing paper are chosen so that, if the paper is cut in half, the ratio of length to width of the half sheet is the same as the ratio of length to width of the original sheet. Find this ratio.

REVIEW PROBLEM SET, CHAPTER 3

In Problems 1 to 6, plot each pair of points and find the distance between them by using the distance formula.

1. $(1, 1)$ and $(4, 5)$

2. $(-1, 2)$ and $(5, -7)$

3. $(-3, 2)$ and $(2, 14)$

Ⓒ**4.** $(4.71, -3.22)$ and $(0, \pi)$

5. $(-2, -5)$ and $(-2, 3)$

Ⓒ**6.** $(\sqrt{\pi}, \pi)$ and $\left(\dfrac{1 + \sqrt{2}}{2}, \dfrac{31}{7}\right)$

Ⓒ**7.** What is the perimeter of the triangle with vertices $(-1, 5), (8, -7)$, and $(4, 1)$? Round off your answer to two decimal places.

8. Are the points $(-5, 4), (7, -11), (12, -11)$, and $(0, 4)$ the vertices of a parallelogram?

In Problems 9 and 10, sketch the graph of the equation.

9. $(x - 2)^2 + (y + 3)^2 = 25$

10. $x^2 + 2x + y^2 + 2y + 1 = 0$

In Problems 11 to 14, find the slope of the line segment $\overline{AB}$ and find an equation in point-slope form of the line containing A and B.

11. $A = (3, -5)$ and $B = (2, 2)$

12. $A = (0, 7)$ and $B = (5, 0)$

13. $A = (1, 2)$ and $B = (-3, -4)$

14. $A = (\frac{3}{2}, \frac{2}{3})$ and $B = (\frac{1}{6}, -\frac{5}{6})$

In Problems 15 and 16, sketch the line L that contains the point P and has slope m, and find an equation in point-slope form for L.

15. $P = (5, 2)$ and $m = -\frac{3}{5}$

16. $P = (-\frac{2}{3}, \frac{1}{2})$ and $m = \frac{3}{2}$

In Problems 17 and 18, (a) find an equation of the line that contains the point P and is parallel to the line segment $\overline{AB}$, and (b) find an equation of the line that contains the point P and is perpendicular to line segment $\overline{AB}$.

17. $P = (7, -5)$, $A = (1, 8)$, $B = (-3, 2)$

18. $P = (\frac{2}{5}, \frac{1}{3})$, $A = (\frac{7}{3}, -\frac{3}{5})$, $B = (1, \frac{2}{5})$

In Problems 19 and 20, rewrite each equation in slope-intercept form, find the slope m and the y intercept b of the graph, and sketch it.

19. $4x - 3y + 2 = 0$

20. $\frac{2}{3}x - \frac{1}{5}y + 3 = 0$

In Problems 21 to 24, find an equation of the line L in (a) point-slope form, (b) slope-intercept form, and (c) general form.

21. L contains the point $(-7, 1)$ and has slope $m = 3$.

22. L contains the points $(2, 5)$ and $(1, -3)$.

23. L contains the point $(1, -2)$ and is parallel to the line whose equation is $7x - 3y + 2 = 0$.

24. L contains the point $(3, -4)$ and is perpendicular to the line $2x - 5y + 4 = 0$.

25. If (a, b) is a point on the circle $x^2 + y^2 = 9$, find an equation of the line that is tangent to the circle at (a, b). [*Hint:* The line tangent to a circle at a point is perpendicular to the radius drawn from the center to the point.]

26. Show that the line containing the two points $P = (a, b)$ and $Q = (c, d)$ has an equation of the form $(b - d)x - (a - c)y + ad - bc = 0$.

27. In calculus, it is shown that the slope m of the line tangent to the graph of $y = x^2 + 4$ at the point (a, b) is given by $m = 2a$. (a) Write an equation of this tangent line. (b) Find an equation of the tangent line at $(2, 8)$. (c) Sketch the graph of $y = x^2 + 4$ and show the tangent line at $(2, 8)$.

28. An advertising agency claims that a furniture store's revenue will increase by $20 per month for each additional dollar spent on advertising. The current average monthly sales revenue is $140,000, with an expenditure of $100 per month for advertising. Find the equation that relates the store's expected average monthly sales revenue y to the total expenditure x for advertising. Find the value of y if $400 per month is spent for advertising.

In Problems 29 to 40, let $f(x) = 3x^2 - 4$, $g(x) = 6 - 5x$, and $h(x) = 1/x$. Find the indicated values.

29. $f(-3)$

30. $h\left(\dfrac{1}{2}\right)$

31. $g\left(\dfrac{6}{5}\right)$

32. $h[h(x)]$

33. $f(x) - f(2)$

34. $f(x + k) - f(x)$

35. $f[g(x)]$

36. $g\left(\dfrac{1}{4 + k}\right)$

37. $g(x) + g(-x)$

38. $\sqrt{f(-|x|)}$

39. $\dfrac{h(x + k) - h(x)}{k}$

40. $\dfrac{1}{h(4 + k)}$

In Problems 41 to 44, find the domain of each function.

41. $f(x) = \dfrac{1}{x - 1}$

42. $g(x) = \dfrac{1}{\sqrt{4 - x^2}}$

43. $h(x) = \sqrt{1 + x}$

44. $F(x) = \dfrac{3}{|x| - x}$

In Problems 45 and 46, find the difference quotient $\dfrac{f(x + h) - f(x)}{h}$ and simplify the resulting expression.

45. $f(x) = 3x^2 - 2x + 1$

46. $f(x) = x^{-1/3}$

47. Which of the graphs in Figure 1 are graphs of functions?

Figure 1

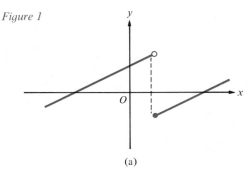

(a)

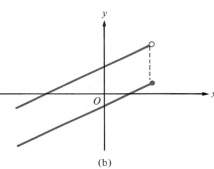

(b)

48. Which of the curves in Figure 2 are graphs of functions?

Figure 2

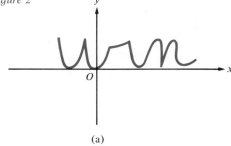

(a)

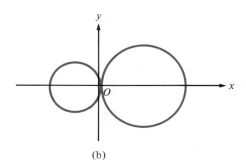

(b)

ⓒ 49. The function $f:x \mapsto y$ given by the equation $f(x) = 71.88 + 0.37x$ relates the atmospheric pressure x in centimeters of mercury to the boiling temperature $f(x)$ of water in degrees Celsius at that pressure. Find $f(74)$, $f(75)$, and $f(76)$. Round off your answers to two decimal places.

50. If $g(x) = \dfrac{2 + x}{2 - x}$, find and simplify $\dfrac{g(t) - g(-t)}{1 + g(t)g(-t)}$.

In Problems 51 and 52, use the point-plotting method to sketch the graph of each function. ⓒ You may wish to improve the accuracy of your sketch by using a calculator to determine the coordinates of some points on the graph.

51. $f(x) = \dfrac{x^4}{16}$ for $-2 \le x \le 2$

52. $f(x) = \dfrac{1 + x}{1 + x^2}$ for $-2 \le x \le 2$

53. For each function in Figure 3, find the domain and range; indicate the intervals over which the

function is increasing, decreasing, or constant; and determine whether the function is even, odd, or neither.

54. Let $g(x) = |x - 5| - |x + 5|$.

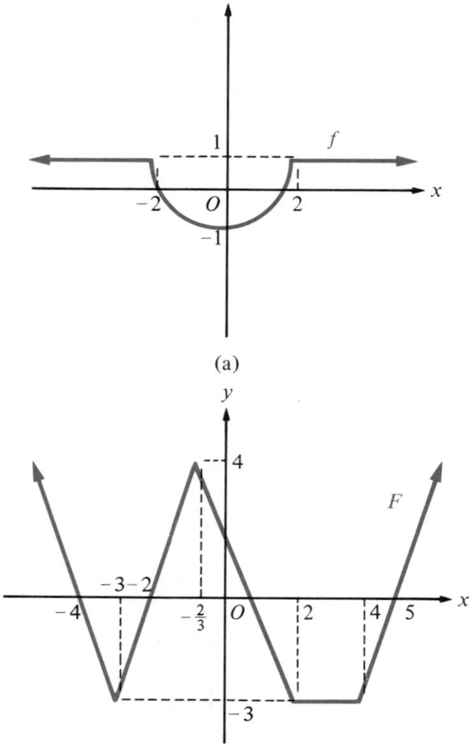

   (a) Sketch the graph of g.

   (b) Determine the intervals over which g is increasing or decreasing [*Hint:* Express $g(x)$ without absolute value symbols for the three cases $x \le -5$, $-5 < x < 5$, and $5 \le x$.]

In Problems 55 to 60, determine without drawing a graph whether the function is even, odd, or neither, and discuss any symmetry of the graph.

55. $f(x) = 5x^5 + 3x^3 + x$ **56.** $g(x) = (x^4 + x^2 + 1)^{-1}$

57. $h(x) = (x + 1)x^{-1}$ **58.** $F(x) = -x^3|x|$

59. $G(x) = x^{80} - 5x^6 + 9$ **60.** $H(x) = \dfrac{\sqrt{x}}{1 + x}$

In Problems 61 to 66, find the domain of the function, sketch its graph, discuss any symmetry of the graph, indicate the intervals where the function is increasing or decreasing, and find the range of the function. Do these things in any convenient order. ⓒ Use a calculator if you wish.

61. $f(x) = 5x - 3$ **62.** $g(x) = 3 - (x/5)$

63. $F(x) = 2\sqrt{x - 2}$ **64.** $G(x) = x^{1/3}$

65. $h(x) = \begin{cases} x^2 & \text{if } x > 0 \\ -x^2 & \text{if } x \le 0 \end{cases}$

66. $H(x) = \begin{cases} x & \text{if } x < 0 \\ 2x & \text{if } 0 \le x \le 1 \\ 3x^3 - 1 & \text{if } x > 1 \end{cases}$

67. The graph of a function f is shown in Figure 4.

Figure 3

(a)

(b)

Figure 4

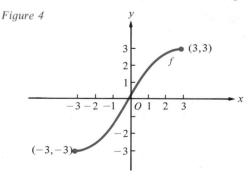

Sketch the graph of the function F defined by

(a) $F(x) = f(x) + 1$ (b) $F(x) = f(x) - 2$
(c) $F(x) = f(x - 1)$ (d) $F(x) = f(x + 2)$
(e) $F(x) = -f(x)$.

[C] **68.** The power delivered by a wind-powered generator is given by

$$P(x) = kx^3 \text{ horsepower,}$$

where x is the speed of the wind in miles per hour and k is a constant depending on the size of the generator and its efficiency. For a certain wind-powered generator, the constant $k = 3.38 \times 10^{-4}$. Sketch the graph of the function P for this generator and determine how many horsepower are generated when the wind speed is 35 miles per hour.

69. Compare the graph of the first function with the graph of the second by sketching them both on the same coordinate system and explaining in words how they are related.

(a) $F(x) = 3x + 2$, $f(x) = 3x - 1$
(b) $G(x) = x^2 - 5$, $g(x) = x^2$
(c) $H(x) = 1 - |x|$, $h(x) = |x|$
(d) $K(x) = \sqrt{x - 1} + 2$, $k(x) = \sqrt{x}$
(e) $Q(x) = \frac{1}{2}(x + 2)^2$, $q(x) = x^2$
(f) $R(x) = 1 - \frac{1}{2}|x + 1|$, $r(x) = |x|$

70. In economics, it is often assumed that the demand for a commodity is a function of selling price; that is, $q = f(p)$, where p dollars is the selling price per unit of the commodity and q is the number of units that will sell at that price.

(a) Would you expect f to be an increasing or a decreasing function?
(b) Write an equation for the total amount of money in dollars $F(p)$ spent by consumers for the commodity if the selling price per unit is p dollars.
(c) What would it mean to say that there is a value p_0 dollars for which $f(p_0) = 0$?

71. In economics, it is often assumed that the number of units s of a commodity that producers will sup-

ply to the market place is a function of the selling price p dollars per unit; that is, $s = g(p)$.

(a) Would you expect g to be an increasing or a decreasing function?
(b) Write an equation for the total amount of money in dollars $G(p)$ spent by consumers for the commodity if the selling price per unit is p dollars and all supplied units are purchased.
(c) What would it mean to say that there is a value p_1 dollars for which $g(p_1) = 0$?

72. A manufacturer finds that 100,000 programmable calculators are sold per month at a price of $50 each, but that only 60,000 are sold per month if the price is $75 each. Suppose that the demand function f for these calculators is linear (see Problem 70).

(a) Find a formula for $f(p)$, where p dollars is the selling price per calculator and $f(p)$ is the number that will sell at that price.
(b) Find the selling price per calculator if the monthly demand for calculators is 80,000.
(c) What price would be so high that no calculators would be sold?

In Problems 73 to 82, find (a) $(f + g)(x)$, (b) $(f - g)(x)$, (c) $(f \cdot g)(x)$, (d) $\left(\dfrac{f}{g}\right)(x)$, and (e) $(f \circ g)(x)$.

73. $f(x) = x + 2$, $g(x) = 3x - 4$

74. $f(x) = x^2 + 2x$, $g(x) = x^2 - 2x$

75. $f(x) = \dfrac{1}{x - 1}$, $g(x) = \dfrac{1}{x + 1}$

76. $f(x) = \dfrac{x + 3}{x - 2}$, $g(x) = \dfrac{x}{x - 2}$

77. $f(x) = x^4$, $g(x) = \sqrt{x + 1}$

78. $f(x) = x$, $g(x) = |x - 2| - x$

79. $f(x) = |x|$, $g(x) = -x$

80. $f(x) = \sqrt{1 + x^2}$, $g(x) = \pi|x|$

81. $f(x) = x^{2/3} + 1$, $g(x) = \sqrt{x}$

82. $f(x) = \dfrac{|x|}{x}$, $g(x) = \dfrac{-x}{|x|}$

© In Problems 83 to 86, let $f(x) = 1 + x^5$ and let $g(x) = 1 - \sqrt{x}$. Use a calculator to find each value. Round off to four decimal places.

83. $(f \circ g)(2.7746)$ **84.** $[(f \circ g) \circ f](\pi)$

85. $(g \circ f)(2.7746)$ **86.** $(g \circ g)(0.0007)$

In Problems 87 to 92, let $f(x) = x^2$, $g(x) = \sqrt{x}$, and $h(x) = x + 1$. Express each function as a composition of functions chosen from f, g, and h.

87. $F(x) = \sqrt{x + 1}$ **88.** $G(x) = |x|$

89. $H(x) = \sqrt{x} + 1$ **90.** $K(x) = x^4$

91. $P(x) = x^2 + 2x + 1$ **92.** $p(x) = x + 2$

In Problems 93 to 96, express each function h as a composition $h = f \circ g$ of two simpler functions f and g.

93. $h(x) = (4x^3 - 2x + 5)^{-3}$

94. $h(x) = \sqrt[3]{\dfrac{4 + x^3}{4 - x^3}}$

95. $h(x) = \dfrac{2(x^2 + 1)^2 + x^2 + 1}{\sqrt{x^2 + 1}}$

96. $h(x) = \dfrac{(2x^2 - 5x + 1)^{-2/3}}{(2x^2 - 5x + 1)^2 + 2}$

97. If $f(x) = x^2 + 1$, $g(x) = x^2 - 1$, and $h(x) = \sqrt{x}$, find

(a) $[(f + g) \circ h](x)$

(b) $[(f - g) \circ h](x)$

(c) $[(f \circ h) + (g \circ h)](x)$

(d) $[(f \circ h) - (g \circ h)](x)$.

98. Find $(f \circ g)(x)$ if

$$f(x) = \begin{cases} x & \text{if } x \geq 0 \\ 2x & \text{if } x < 0 \end{cases}$$

and $$g(x) = \begin{cases} -2x & \text{if } x \geq 0 \\ 4x & \text{if } x < 0. \end{cases}$$

99. A baseball diamond is a square that is 90 feet long on each side. A ball is hit from home plate directly toward third base at the rate of 50 feet per second. Let y denote the distance of the ball from first base in feet; let x denote its distance in feet from home plate; and let t denote the elapsed time in

seconds since the ball was hit. Here y is a function of x, say, $y = f(x)$, and x is a function of t, say, $x = g(t)$.

(a) Find formulas for $f(x)$ and $g(t)$.

(b) Explain why $y = (f \circ g)(t)$.

(c) Find a formula for $(f \circ g)(t)$.

100. Solar cells and a wind-powered generator charge batteries that supply electric power to an alternative-energy home. The batteries have a constant internal resistance of r ohms and provide a fixed voltage, E volts. The current I amperes drawn from the batteries depends on the net resistance R ohms of the appliances being used in the home according to the equation $I = \dfrac{E}{r + R}$. The electric power P watts being consumed by these appliances is given by the equation $P = I^2 R$. Find a function f such that $P = f(R)$.

101. Two college students earn extra money on weekends by delivering firewood in their pickup truck. They have found that they can sell x cords per weekend at a price of p dollars per cord, where $x = 75 - \frac{3}{5}p$. The students buy the firewood from a supplier who charges them C dollars for x cords according to the equation $C = 500 + 15x + \frac{1}{5}x^2$.

(a) Find a function f such that $P = f(p)$, where P dollars is the profit per weekend for the students if they charge p dollars per cord.

(b) Find the profit P dollars if $p = \$95$.

102. Which of the following functions is invertible?

(a) $f(x) = x$ (b) $f(x) = 1/x$

(c) $f(x) = 3x + 5$ (d) $f(x) = 3x^2 + 5$

In Problems 103 to 108, find $f^{-1}(x)$ for each function f, verify that $f^{-1}[f(x)] = x$ and $f[f^{-1}(x)] = x$, and sketch the graphs of f and f^{-1} on the same coordinate system.

103. $f(x) = 3x - 1$ **104.** $f(x) = (x - 1)^{-1}$

105. $f(x) = \frac{1}{5}x + 5$ **106.** $f(x) = \frac{1}{4}x^3$

107. $f(x) = 2\sqrt{x} - 1$

108. $f(x) = \dfrac{(x - 2)^2}{4} + 3$, $x \geq 2$.

109. Sketch the graph of the inverse of f, g, and h in Figure 5.

Figure 5

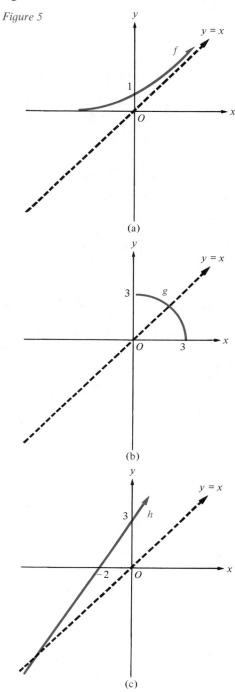

(a)

(b)

(c)

110. If $f(x) = x^2 - 3x + 2$ for $x \leq \frac{3}{2}$, find $f^{-1}(x)$.

111. If $f(x) = \dfrac{x-1}{x}$, find $f^{-1}(x)$.

112. If f is invertible, show that f^{-1} is also invertible and that $(f^{-1})^{-1}$ is the same as f.

113. Suppose that f is invertible, that the domain of f is $\mathbb{R}$, and that the range of f is $\mathbb{R}$. What function is $f^{-1} \circ f$?

114. Are there values of the constants a, b, c, and d for which the function $f(x) = \dfrac{ax+b}{cx+d}$ is its own inverse?

In Problems 115 to 120, find the ratio of the first quantity to the second. Write your answer as a reduced fraction. If the quantities are measured in different units, you must first express both in the same units.

115. Twelve dollars to sixty-five cents.

116. 100 milliwatts to 25 microwatts.

117. 250 turns in the primary winding of a transformer to 10,000 turns in the secondary winding (Figure 6).

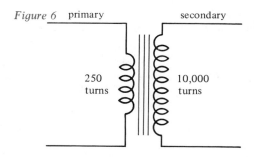

Figure 6   primary   secondary

250 turns   10,000 turns

118. 10 miles to 1000 feet.

119. 50 ounces to 2 pounds.

120. 1 kilometer to 1 mile.

In Problems 121 and 122, find the ratio of each pair of expressions if the ratio of a to b is $\frac{2}{3}$.

121. $3a + 2b$ to $4a - 7b$ **122.** $5a - 3b$ to $3b + 7a$

In Problems 123 and 124, solve each proportion for x.

123. $2 : x + 1 = 4 : 5$

124. $2x - 3$ is to 3 as $1 - x$ is to 5

125. If $a:x = x:b$, x is called the *mean proportional* between a and b. Find the mean proportional between 2 and 8.

126. The ratio of the surface areas of two spheres is $\frac{9}{4}$. Find the ratio of their volumes. [*Hint:* The volume V and surface area A of a sphere of radius r are given by $V = \frac{4}{3}\pi r^3$ and $A = 4\pi r^2$.]

127. The *voltage gain* of an electronic amplifier is defined to be the ratio of the output voltage to the input voltage. Find the voltage gain of an amplifier if the input voltage is 0.005 volt and the output voltage is 25 volts.

128. One of the indicators of the quality of instruction at an educational institution is the student-to-teacher ratio. Is it desirable that this ratio be large or small?

In Problems 129 to 138, relate the quantities by writing an equation involving at least one constant.

129. P is inversely proportional to V.

130. The surface area A of a sphere varies directly as the square of its diameter d.

131. For tax purposes, it is often assumed that the value V dollars of an article varies linearly with the time t years since it was purchased.

132. In a *voltage-controlled* electronic device, the change in output current, written ΔI, is directly proportional to the change in input voltage, written ΔV. [Δ is the capital Greek letter *delta*. It is often used to stand for "a change in."]

133. The force F of the wind on a blade of a wind-powered generator varies jointly with the area A of the blade and the square of the wind speed v.

134. The inductance L of a coil of wire is jointly proportional to the cross-sectional area A of the coil and the square of the number N of turns, and inversely proportional to the length l of the coil.

135. The rate r at which a rumor is spreading in a population of size P is jointly proportional to the number N of people who have heard the rumor and the number $P - N$ of people who have not.

136. The collector current I_C of a transistor is jointly proportional to its current ratio β and its base current I_B. [β is the small Greek letter *beta*.]

137. The power P provided by a jet of water is jointly proportional to the cross-sectional area A of the jet and the cube of the speed v of the water in the jet.

138. In geology, it is found that the erosive force E of a swiftly flowing stream is directly proportional to the sixth power of the speed v of flow of the water

In Problems 139 to 146, (a) find an equation relating the given quantities and, (b) find the value of the indicated quantity under the specified conditions.

139. y is inversely proportional to x, and $y = 1$ when $x = 5$. Find y when $x = 25$.

140. y is jointly proportional to u and $\sqrt{v}$, and $y = 6$ when $u = 1$ and $v = 4$. Find y when $u = \frac{1}{3}$ and $v = 9$.

141. w is directly proportional to x and inversely proportional to y, and $w = 7$ when $x/y = 3$. Find w when $x = 24$ and $y = 6$.

142. y varies linearly with x, $y = -1$ when $x = 1$, and $y = 5$ when $x = -1$. Find y when $x = 0$.

143. y is directly proportional to x^2 and inversely proportional to $z + 3$, and $y = 4$ when $x = 2$ and $z = 1$. Find y when $x = 3$ and $z = 6$.

144. y is jointly proportional to $\sqrt[4]{x}$ and z^3, and $y = 32$ when $x = 16$ and $z = 2$. Find y when $x = 81$ and $z = \frac{1}{3}$.

145. y varies linearly as $\sqrt{x}$, $y = 3$ when $x = 1$, and $y = 5$ when $x = 4$. Find y when $x = 9$.

©**146.** y is directly proportional to x and inversely proportional to z^2, and $y = 1.422$ when $x = 0.4181$ and $z = 0.7135$. Find y when $x = 2.133$ and $z = 5.357$.

©**147.** The volume V of a sphere is directly proportional to the cube of its diameter. If a sphere of diameter 2 meters has a volume of approximately 4.19 cubic meters, find the approximate volume of a sphere of diameter 10 meters.

148. Kelvin temperature K varies linearly with Fahrenheit temperature F. Find a formula for K in terms of F if $K = 0$ when $F = -459$, and $K = 273$ when $F = 32$.

©**149.** The volume V of a block of iron varies linearly with its temperature T in degrees Celsius. Find a formula for the volume V of a block of iron if $V = 0.25$ cubic meter when $T = 0°$, and $V = 0.25018$ cubic meter when $T = 20°$.

©**150.** A company estimates that its profit P dollars per month varies linearly with the number of items n manufactured per month. If $P = \$180{,}000$ when $n = 5000$, and $P = \$300{,}000$ when $n = 8000$, find P as a function of n.

©**151.** A company's sales volume S per month varies directly as the number A of dollars per month spent for advertising, and inversely as the product of the selling price x dollars per unit and the inflation index I. If S is 20,000 units when $A = \$10{,}000$, $x = \$400$, and $I = 10\%$, find S when $A = \$15{,}000$, $x = \$500$, and $I = 15\%$.

Chapter 4

Polynomial and Rational Functions

In this chapter, we continue the study of functions and their graphs. Here we explore polynomial and rational functions, their "zeros," and some of the properties of their graphs. The chapter also includes the remainder theorem, synthetic division, the rational zeros theorem, and Descartes' rule of signs; and it ends with a section on complex polynomials.

4.1 QUADRATIC FUNCTIONS

A function f of the form

$$f(x) = ax^2 + bx + c,$$

where a, b, and c are constants and $a \neq 0$, is called a **quadratic function.** Such functions often arise in applied mathematics. For instance, the height of a projectile is a quadratic function of time; the velocity of blood flow is a quadratic function of the distance from the center of the blood vessel; and the force exerted by the wind on the blades of a wind-powered generator is a quadratic function of the wind speed.

The simplest quadratic function is the squaring function $f(x) = x^2$, whose graph was sketched in Section 3.5 (page 160). As we saw in Section 3.6, the graph of $f(x) = ax^2$ is obtained from the graph of $y = x^2$ by vertical "stretching" if $a > 1$ or "flattening" if $0 < a < 1$. Furthermore, the graph of $f(x) = ax^2$ for negative values of a is obtained by reflecting the graph $y = |a|x^2$ across the x axis. Figure 1 shows the graph of $f(x) = ax^2$ for various values of a.

The graphs of equations of the form $y = ax^2$ (Figure 1) are examples of curves called **parabolas.** These parabolas* are symmetric about the y axis; they **open upward** and have a lowest point at $(0, 0)$ if $a > 0$ (Figure 2a), and they **open downward** and have a highest point at $(0, 0)$ if $a < 0$ (Figure 2b). The highest or lowest point of the

Illuminated fountain showing parabolic jets of water.

---

* We continue our study of parabolas in a later chapter on analytical geometry.

Figure 1

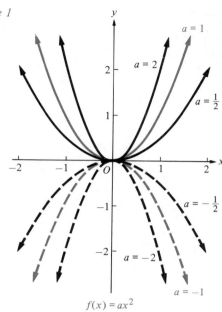

$f(x) = ax^2$

Figure 2

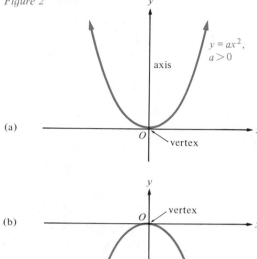

(a)

(b)

graph of $y = ax^2$ is called the **vertex** of the parabola, and its line of symmetry (in this case the y axis) is called the **axis of symmetry** or simply the **axis** of the parabola.

By the combined graph-shifting rule (Section 3.6, page 167), the graph of

$$f(x) = a(x - h)^2 + k$$

is obtained by shifting the parabola $y = ax^2$ horizontally by $|h|$ units and vertically by $|k|$ units. Hence its graph is a parabola with vertex at (h, k) (Figure 3). The parabola opens upward if $a > 0$ and downward if $a < 0$, and its axis is the vertical line $x = h$.

Figure 3

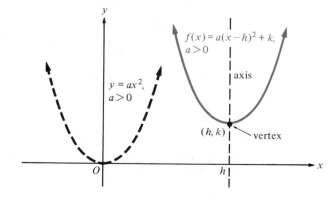

Figure 4

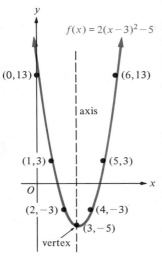

Example 1   Sketch the graph of $f(x) = 2(x - 3)^2 - 5$.

Solution   The function has the form $f(x) = a(x - h)^2 + k$, with $a = 2$, $h = 3$, and $k = -5$; hence, its graph is a parabola, opening upward, with vertical axis $x = 3$ and vertex $(h, k) = (3, -5)$. This information indicates the general appearance of the graph. By plotting a few points, we can obtain a reasonably accurate sketch of the parabola (Figure 4). In Figure 4 we have used different scales on the x and y axes to obtain a graph with reasonable proportions.   ∎

By completing the square as shown in the following example, you can always rewrite a quadratic function $f(x) = ax^2 + bx + c$ in the form $f(x) = a(x - h)^2 + k$. Then, using the technique illustrated in Example 1, you can sketch the graph of f.

Example 2   Rewrite $f(x) = 2x^2 + 12x + 17$ in the form $f(x) = a(x - h)^2 + k$ and sketch the graph of f.

Figure 5

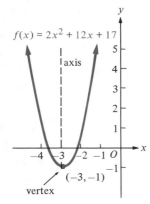

Solution   To prepare for completing the square (Section 2.3, page 89), we factor the coefficient 2 of x^2 out of the first two terms:

$$f(x) = 2x^2 + 12x + 17 = 2(x^2 + 6x \quad) + 17.$$

By adding $(\frac{6}{2})^2 = 9$ to $x^2 + 6x$, we obtain the perfect square

$$x^2 + 6x + 9 = (x + 3)^2.$$

Therefore, we have

$$f(x) = 2(x^2 + 6x + 9) + 17 - 2(9),$$

where we have subtracted $2(9)$ to compensate for the addition of 9 to $x^2 + 6x$. [Notice that the 9 added will be multiplied by 2, which is why we subtracted $2(9)$.] It follows that

$$f(x) = 2(x + 3)^2 - 1,$$

which has the form

$$f(x) = a(x - h)^2 + k$$

with $a = 2$, $h = -3$, and $k = -1$. Thus, the graph is a parabola, opening upward, with vertex $(h, k) = (-3, -1)$ and vertical axis $x = -3$ (Figure 5).   ∎

More generally, we have the following result:

Theorem 1   **Quadratic Function Theorem**

> The quadratic function
>
> $$f(x) = ax^2 + bx + c, \qquad a \neq 0$$
>
> can be rewritten in the form
>
> $$f(x) = a(x - h)^2 + k,$$
>
> where
>
> $$h = -\frac{b}{2a} \qquad \text{and} \qquad k = f(h).$$

Proof   We begin by writing

$$f(x) = ax^2 + bx + c = a\left(x^2 + \frac{b}{a}x \quad\right) + c.$$

By adding $\left[\frac{1}{2}\left(\frac{b}{a}\right)\right]^2 = \frac{b^2}{4a^2}$ to $x^2 + \frac{b}{a}x$, we obtain the perfect square

$$x^2 + \frac{b}{a}x + \frac{b^2}{4a^2} = \left(x + \frac{b}{2a}\right)^2.$$

Therefore,

$$f(x) = a\left(x^2 + \frac{b}{a}x + \frac{b^2}{4a^2}\right) + c - a\left(\frac{b^2}{4a^2}\right)$$

$$= a\left(x + \frac{b}{2a}\right)^2 + c - \frac{b^2}{4a}$$

$$= a(x - h)^2 + k,$$

where $h = -\dfrac{b}{2a}$ and $k = c - \dfrac{b^2}{4a}$. But, from the equation

$$f(x) = a(x - h)^2 + k,$$

we have

$$f(h) = a(h - h)^2 + k = a(0)^2 + k = k,$$

that is, $k = f(h)$.   ■

As a consequence of Theorem 1, we have the following formula for the coordinates of the vertex of a parabola:

Vertex Formula

If $a \neq 0$, then the graph of

$$f(x) = ax^2 + bx + c$$

is a parabola with vertex (h, k) where

$$h = -\frac{b}{2a} \qquad \text{and} \qquad k = f(h) = f\left(-\frac{b}{2a}\right).$$

If $a > 0$, the parabola opens upward and the vertex (h, k) is its lowest point; if $a < 0$, the parabola opens downward and the vertex (h, k) is its highest point.

It follows from the vertex formula that *the vertical line $x = h$ is the axis of symmetry of the graph of $f(x) = ax^2 + bx + c$.*

Example 3   Find the vertex and axis of symmetry of the graph of $f(x) = -3x^2 - 12x - 1$ and sketch the graph.

Solution   By the vertex formula, with $a = -3$ and $b = -12$, we have

$$h = -\frac{b}{2a} = -\frac{(-12)}{2(-3)} = -2$$

and

$$k = f(h) = f(-2)$$
$$= -3(-2)^2 - 12(-2) - 1 = 11.$$

Therefore, the graph is a parabola, opening downward, with vertex $(-2, 11)$ and vertical axis $x = -2$ (Figure 6). ∎

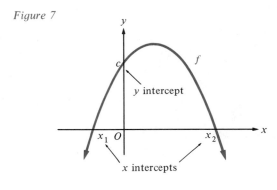

Figure 6

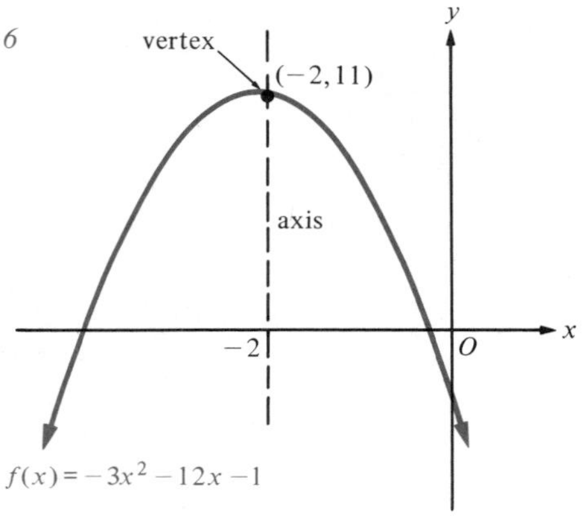

vertex

$(-2, 11)$

axis

-2

$f(x) = -3x^2 - 12x - 1$

Figure 7

y intercept

x_1 O x_2

x intercepts

f

Notice that the graph of $f(x) = ax^2 + bx + c$ intersects the y axis at the point $(0, c)$ (Figure 7).

> We call c the **y intercept** of the graph. If the graph intersects the x axis at the points $(x_1, 0)$ and $(x_2, 0)$, we call x_1 and x_2 its **x intercepts.**

Note that x_1 and x_2, if they exist, are the real roots of the quadratic equation $ax^2 + bx + c = 0$. (Do you see why?)

Example 4   Find the vertex and the y and x intercepts of the graph of $f(x) = -2x^2 - 5x + 3$, determine whether the graph opens upward or downward, sketch the graph, and find the domain and range of f.

Solution   Here we have $a = -2$, $b = -5$, and $c = 3$; hence, by the vertex formula,

$$h = -\frac{b}{2a} = -\frac{(-5)}{2(-2)} = -\frac{5}{4}$$

and

$$k = f(h) = f(-\tfrac{5}{4})$$
$$= -2(-\tfrac{5}{4})^2 - 5(-\tfrac{5}{4}) + 3 = \tfrac{49}{8}.$$

Figure 8

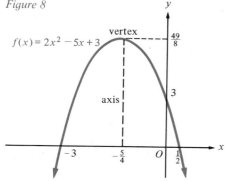

$f(x) = 2x^2 - 5x + 3$

vertex

axis

-3 $-\frac{5}{4}$ O $\frac{1}{2}$

$\frac{49}{8}$

3

Therefore, the vertex is

$$(h, k) = (-\tfrac{5}{4}, \tfrac{49}{8}).$$

The y intercept is $c = 3$, and the x intercepts are the solutions of the quadratic equation

$$-2x^2 - 5x + 3 = 0 \qquad \text{or} \qquad (x + 3)(-2x + 1) = 0.$$

It follows that the x intercepts are -3 and $\frac{1}{2}$. Because $a = -2 < 0$, the graph opens downward. Using this information, we can sketch the parabola (Figure 8). (If a more accurate sketch is desired, a calculator can be used to determine additional points on the graph.) The domain of f is $\mathbb{R}$ and, as Figure 8 shows, its range is the interval $(-\infty, \frac{49}{8}]$. ∎

Example 5   Use the graph of $f(x) = -2x^2 - 5x + 3$ to solve the quadratic inequalities:

(a) $-2x^2 - 5x + 3 < 0$ **(b)** $-2x^2 - 5x + 3 \geq 0$

Solution   **(a)** From the graph of $f(x) = -2x^2 - 5x + 3$, already sketched in Figure 8, we see that $f(x) < 0$ when $x < -3$ and when $x > \frac{1}{2}$. Hence, the solution set of the inequality $-2x^2 - 5x + 3 < 0$ is the union

$$(-\infty, -3) \cup (\tfrac{1}{2}, \infty)$$

of the two open intervals $(-\infty, -3)$ and $(\frac{1}{2}, \infty)$.

(b) The graph of $f(x)$ in Figure 8 shows that the solution set of the inequality $-2x^2 - 5x + 3 \geq 0$ is the closed interval $[-3, \frac{1}{2}]$. ∎

Maximum and Minimum Values of Quadratic Functions

Using the vertex formula, we can easily locate the maximum and minimum values of a quadratic function $f(x) = ax^2 + bx + c$. Indeed, if $h = -b/2a$, then the vertex $(h, f(h))$ is the *lowest point* on the graph if $a > 0$ (Figure 9a) and the *highest point* on the graph if $a < 0$ (Figure 9b). Therefore, we have the following useful fact:

Figure 9

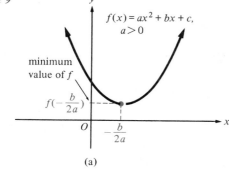

$f(x) = ax^2 + bx + c,$ $a > 0$

minimum value of f

$f(-\dfrac{b}{2a})$

O $-\dfrac{b}{2a}$

(a)

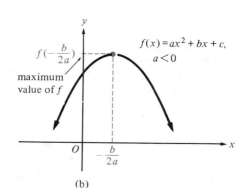

$f(x) = ax^2 + bx + c,$ $a < 0$

$f(-\dfrac{b}{2a})$

maximum value of f

O $-\dfrac{b}{2a}$

(b)

Maximum and Minimum Values of Quadratic Functions

Let $f(x) = ax^2 + bx + c$. If $a > 0$, the number

$$f\left(-\frac{b}{2a}\right)$$

is the **minimum** (smallest) value of the function f, and if $a < 0$, it is the **maximum** (largest) value of f.

Example 6   A manufacturer of synfuel (synthetic fuel) from coal estimates that the cost $f(x)$ in dollars per barrel for a production run of x thousand barrels is given by $f(x) = 9x^2 - 180x + 940$. How many thousands of barrels should be produced during each run to minimize the cost per barrel, and what is the minimum cost per barrel of the synfuel?

Solution   The cost per barrel is given by

$$f(x) = 9x^2 - 180x + 940$$
$$= ax^2 + bx + c,$$

where $a = 9$, $b = -180$, and $c = 940$. Therefore, $f(x)$ attains its minimum value when

$$x = -\frac{b}{2a} = -\frac{(-180)}{2(9)} = 10 \text{ thousand barrels,}$$

and the minimum cost per barrel is given by

$$f(10) = 9(10)^2 - 180(10) + 940 = 40 \text{ dollars.} \qquad \blacksquare$$

Example 7   An orchard contains 30 apple trees, each of which yields approximately 400 apples over the growing season. The owner plans to add more trees to the orchard, but the State Agricultural Service advises that because of crowding, each new tree will reduce the average yield per tree by about 10 apples over the growing season. How many trees should be added to maximize the total yield of apples, and what is the maximum yield?

Solution   Let x denote the number of trees added. After x trees are added, the orchard will contain $30 + x$ trees, but the average yield of each tree per season will be reduced from 400 apples to $400 - 10x$ apples. Therefore, the total yield of apples per season is given by

$$f(x) = (30 + x)(400 - 10x) = 12{,}000 + 100x - 10x^2$$
$$= ax^2 + bx + c,$$

where $a = -10$, $b = 100$, and $c = 12{,}000$. Thus, $f(x)$ attains its maximum value when

$$x = -\frac{b}{2a} = -\frac{100}{2(-10)} = 5 \text{ trees.}$$

If 5 trees are added to the orchard, the total yield per growing season will be

$$f(5) = 12{,}000 + 100(5) - 10(5)^2$$
$$= 12{,}250 \text{ apples.} \qquad \blacksquare$$

Problem Set 4.1

In each problem set, problems with colored numbers constitute a good representation of the main ideas of the section.

1. Sketch the graph of $f(x) = ax^2$ for (a) $a = 3$, (b) $a = -\frac{1}{3}$, and (c) $a = -3$.
2. Sketch the graph of $f(x) = 2x^2 + k$ for (a) $k = 0$, (b) $k = 1$, and (c) $k = -1$.

In Problems 3 to 6, sketch the graph of each function.

3. $f(x) = 2(x - 1)^2 + 3$
4. $g(x) = -2(x + 2)^2 - 3$
5. $h(x) = -\frac{1}{2}(x + 1)^2 - 4$
6. $p(x) = \frac{1}{2}(x + 1)^2 - \frac{1}{2}$

In Problems 7 to 18, rewrite each function f in the form $f(x) = a(x - h)^2 + k$ and sketch the graph of f.

7. $f(x) = x^2 - 4x - 1$
8. $f(x) = x^2 - 10x + 20$
9. $f(x) = x^2 + 2x - 4$
10. $f(x) = x^2 - 12x + 5$
11. $f(x) = 2x^2 - 4x + 1$
12. $f(x) = 3x^2 + 6x - 2$
13. $f(x) = 3x^2 - 10x - 2$
14. $f(x) = 2x^2 + 3x - 1$
15. $f(x) = -4x^2 + 8x - 5$
16. $f(x) = -6x^2 + 12x + 5$
17. $f(x) = -2x^2 + 6x + 3$
18. $f(x) = \frac{3}{2}x^2 - 6x - 7$

In Problems 19 to 32, find the vertex and the intercepts of the graph of each function, determine whether the graph opens upward or downward, sketch the graph, and find the domain and range of the function. ([c] You may use a calculator to determine additional points on the graph if you want to obtain a more accurate sketch.)

19. $f(x) = x^2 - 2x - 3$
20. $g(x) = \frac{1}{2}x^2 - \frac{3}{4}$
21. $h(x) = x^2 + 4x$
22. $K(x) = -x^2 - 8x - 15$
23. $Q(x) = -2x^2 + x - 15$
24. $f(t) = -t^2 - 2t + 8$
25. $p(x) = 6x^2 + x - 2$
26. $g(x) = 4x - 12x^2 + 21$

27. $f(x) = -3x^2 + 12x + 15$
28. $H(x) = 30 - x(1 + 14x)$
29. $F(x) = \frac{1}{2}x^2 + x + 2$
30. $G(t) = t - \frac{2}{3}t^2 - \frac{1}{3}$
31. $f(x) = 2x^2 - 20x + 57$, $\quad 0 \le x \le 10$
32. $g(x) = -3x^2 + 24x - 50$, $\quad 2 < x < 6$

In Problems 33 to 40, sketch the graph of each quadratic function and then use the graph to solve the accompanying quadratic inequality.

33. $f(x) = x^2 - 2x - 3$; $x^2 - 2x - 3 \ge 0$
34. $g(x) = -2x^2 + 4x + 3$; $-2x^2 + 4x + 3 < 0$
35. $h(x) = -x^2 - 2x + 8$; $-x^2 - 2x + 8 > 0$
36. $p(x) = x^2 - x + 1$; $x^2 > x - 1$
37. $f(x) = 8x - x^2$; $8x \le x^2$
38. $f(x) = x^2 + x + 1$; $x^2 + x + 1 < 0$
39. $g(x) = x^2 - 2x - 2$; $x^2 \le 2x + 2$
40. $h(x) = 3x^2 + 2x - 7$; $3x^2 + 2x \ge 7$

41. Find two positive numbers whose sum is 100 and whose product is as large as possible.

42. Find the minimum value of $g(x) = (ax^2 + bx + c)^2$ in terms of a, b, and c.

43. Find the length and width of a rectangular plot of land whose perimeter is 600 meters and whose area is the maximum for that perimeter.

44. If a manufacturer produces x thousand tons of a new lightweight alloy for engine blocks, each block will cost $3x^2 - 600x + 30{,}090$ dollars. How many thousand tons of the alloy should be produced to minimize the cost of the engine blocks, and what is the resulting minimum cost of one block?

45. If an object is projected straight upward from ground level with an initial speed of 96 feet per second, then (neglecting air resistance) its height $h(t)$ in feet after t seconds is given by $h(t) = 96t - 16t^2$. Find the maximum height reached by the object and determine the time t at which it returns to ground level.

46. In physics it is shown that, if air resistance is neglected, the stream of water projected from a fire hose satisfies the equation

$$y = mx - \frac{g}{2}(1 + m^2)\left(\frac{x}{v}\right)^2,$$

where m is the slope of the nozzle, v is the velocity of the stream at the nozzle, y is the height of the stream x units from the nozzle, and g is the acceleration of gravity (Figure 10). Assume that v and g are positive constants.

Figure 10

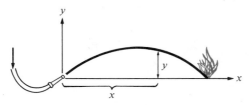

(a) For a fixed value of m, find the value of x for which the height y of the stream is maximum.

(b) For a fixed value of m, find the distance d from the nozzle at which the stream hits the ground.

(c) Find the value of m for which the water reaches the greatest height on a vertical wall x units from the nozzle.

47. In medicine, it is often assumed that a patient's *reaction* $R(x)$ to a drug dose of size x is given by an equation of the form $R(x) = Ax^2(B - x)$, where A and B are positive constants. It can then be shown that the body's *sensitivity* $S(x)$ to a dose of size x is given by $S(x) = 2ABx - 3Ax^2$. Find the reaction to the dose for which the sensitivity is maximum.

48. A real estate company manages an apartment building containing 80 units. When the rent for each unit is $250 per month, all apartments are occupied. However, for each $10 increase in monthly rent per unit, one of the units becomes vacant. Each vacant unit costs the management $15 per month for taxes and upkeep, and each occupied unit costs the management $65 per month for taxes, service, upkeep, and water. What rent should be charged for a maximum profit?

49. Suppose that the distance d in kilometers that a certain car can travel on one tank of gasoline at a speed of v kilometers per hour is given by $d = 12v - (v/4)^2$. What speed maximizes the distance d, and hence minimizes fuel consumption?

50. Ship A is 65 nautical miles due east of ship B and is sailing south at 15 knots (nautical miles per hour), while ship B is sailing east at 10 knots. Figure 11

Figure 11

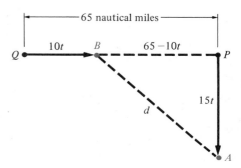

shows the original positions P and Q of ships A and B, their positions t hours later, and the distance d between the ships at time t. Find the minimum distance between the ships and the time when it occurs. [*Hint:* Use the Pythagorean theorem and the fact that d is minimum when d^2 is minimum.]

51. The management of a racquetball club foresees that 60 members will join the club if each membership is $5 per month, but that for each 50-cent increase in the membership price per month, 4 of the 60 potential members will decide not to join. The cost to the club per member is estimated to be $3.50 per month. What membership price per month will bring in the maximum profit to the club?

4.2 POLYNOMIAL FUNCTIONS

We have now discussed constant functions

$$f(x) = b,$$

Figure 1

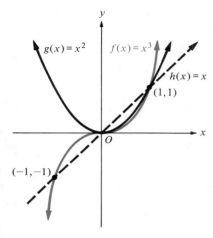

linear functions

$$f(x) = ax + b,$$

and quadratic functions

$$f(x) = ax^2 + bx + c.$$

These are all special cases of **polynomial functions;** that is, functions whose values are given by polynomials in the independent variable.

A polynomial function f of the form

$$f(x) = ax^n,$$

where $a \neq 0$ and n is a positive integer, is called a **power function** of **degree** n. For $n = 1$, the graph of f is a straight line that has slope a and contains the origin; for $n = 2$, the graph of f is a parabola that has its vertex at the origin and opens upward if $a > 0$ and downward if $a < 0$.

Figure 1 shows the graph of $f(x) = x^3$ and, for comparison, the graphs of $g(x) = x^2$ and $h(x) = x$. Notice that if x is less than 1 and positive, x^2 is smaller than x, and x^3 is even smaller than x^2. (For instance, if $x = \frac{1}{4}$, then $x^2 = \frac{1}{16} < \frac{1}{4}$ and $x^3 = \frac{1}{64} < \frac{1}{16}$.) It follows that on the open interval $(0, 1)$, the graph of $g(x) = x^2$ lies below the graph of $h(x) = x$, while the graph of $f(x) = x^3$ lies below the graph of $g(x) = x^2$. However, on the interval $(1, \infty)$, $x^3 > x^2 > x$, so the graph of $f(x) = x^3$ is above the graph of $g(x) = x^2$, which in turn is above the graph of $h(x) = x$. Notice that all three graphs contain the origin and the point $(1, 1)$. Since $f(x) = x^3$ is an odd function (Definition 2, page 158), its graph is symmetric about the origin.

Figure 2

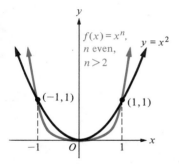

If n is an *even* positive integer, then the power function $f(x) = x^n$ is even (why?) and its graph is symmetric about the y axis (Figure 2). The graph contains the origin and the points $(-1, 1)$ and $(1, 1)$. It never falls below the x axis. If $n = 2$, the graph is a parabola; but, for $n > 2$, the graph of $f(x) = x^n$ falls below the parabola $y = x^2$ over the open intervals $(-1, 0)$ and $(0, 1)$ and rises above the parabola over the intervals $(-\infty, -1)$ and $(1, \infty)$.

Figure 3

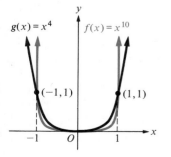

As the even integer n becomes larger and larger, the graph of $f(x) = x^n$ becomes flatter and flatter on both sides of the origin and rises more and more sharply through the two points $(-1, 1)$ and $(1, 1)$. Figure 3 shows the graph of $f(x) = x^{10}$ and contrasts it with the graph of $g(x) = x^4$. For large values of n, the graph of $f(x) = x^n$ may come so close to the x axis on both sides of the origin that it appears to coincide with a segment of the x axis. In reality, of course, the graph of $f(x) = x^n$ touches the x axis only at the origin.

If n is an *odd* positive integer, then the power function $f(x) = x^n$ is odd (why?) and its graph is symmetric about the origin (Figure 4). The graph contains the origin and the points $(-1, -1)$ and $(1, 1)$. For odd $n > 3$, the graph is similar to the graph of $y = x^3$, but is flatter near the origin, rises more sharply to the right of $x = 1$, and drops more rapidly to the left of $x = -1$ (Figure 4).

Figure 4

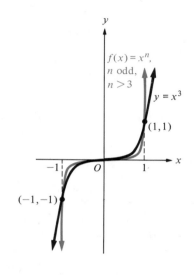

Figure 5

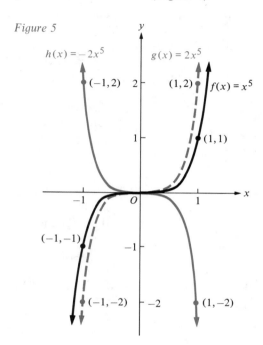

Example 1   Sketch each of the graphs of $f(x) = x^5$, $g(x) = 2x^5$, and $h(x) = -2x^5$ on the same coordinate system.

Solution   We begin by sketching the graph of $f(x) = x^5$. Then, by doubling the ordinates, we obtain the graph of $g(x) = 2x^5$. Finally, we reflect the graph of $g(x) = 2x^5$ across the x axis to obtain the graph of $h(x) = -2x^5$ (Figure 5). ∎

In Section 4.1 we studied the intercepts of the graph of a quadratic function. More generally, we have the following definition.

Definition 1   **Intercepts of a Graph**

> The **y intercept** of the graph of a function f is the ordinate $f(0)$ of the point $(0, f(0))$ where the graph intersects the y axis. Similarly, the abscissa of a point where the graph of f intersects the x axis is called an **x intercept** of the graph (Figure 6).

Figure 6

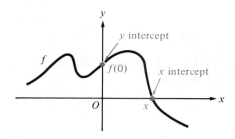

Thus, *the x intercepts of the graph of f are the real roots (if any) of the equation* $f(x) = 0$.

In Examples 2 and 3, find the y and x intercepts and sketch the graph of each function.

Example 2 $f(x) = -x^4 + 1$

Solution   The y intercept is $f(0) = 1$, and the x intercepts are the real roots of the equation $f(x) = 0$; that is, $-x^4 + 1 = 0$ or $x^4 = 1$. Hence, the x intercepts are $x = \sqrt[4]{1} = 1$ and $x = -\sqrt[4]{1} = -1$. By reflecting the graph of $y = x^4$ (Figure 3) across the x axis and then shifting it 1 unit upward, we obtain the graph of $f(x) = -x^4 + 1$ (Figure 7). ∎

Figure 7

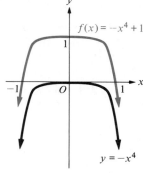

Figure 8

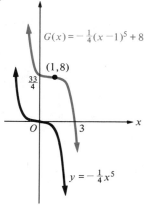

Example 3 $G(x) = -\frac{1}{4}(x - 1)^5 + 8$

Solution   The y intercept is $G(0) = -\frac{1}{4}(0 - 1)^5 + 8 = \frac{1}{4} + 8 = \frac{33}{4}$, and the x intercept is the real root of the equation $G(x) = 0$; that is,

$$-\tfrac{1}{4}(x - 1)^5 + 8 = 0 \qquad \text{or} \qquad (x - 1)^5 = 32.$$

Thus, $x - 1 = \sqrt[5]{32} = 2$, so the x intercept is given by $x = 3$. By the combined graph-shifting rule (page 167), the graph of $G(x) = -\frac{1}{4}(x - 1)^5 + 8$ is obtained by shifting the graph of $y = -\frac{1}{4}x^5$ upward by 8 units and to the right by 1 unit (Figure 8). ∎

Figure 9

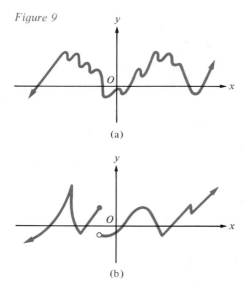

(a)

(b)

In calculus, it is shown that the graph of every polynomial function is a smooth curve with no jumps or breaks. Although the graph of a polynomial function f can wiggle up and down as in Figure 9a, it cannot have the sharp corners or jumps illustrated in Figure 9b. The x intercepts of the graph of f divide the x axis into open intervals. You can determine the algebraic signs of $f(x)$ over these open intervals by using convenient test numbers, just as you did in Section 2.7. The following example illustrates how you can use this information to help sketch the graph of f.

Example 4   Let $f(x) = (2x - 1)(x^2 - x - 2)$.

(a) Find the x intercepts of the graph of f.
(b) Sketch the graph of f.
(c) Use the graph to solve the inequality $f(x) < 0$.

Solution   **(a)** The x intercepts are the real roots of $f(x) = 0$; that is, $(2x - 1)(x^2 - x - 2) = 0$ or $(2x - 1)(x + 1)(x - 2) = 0$. Setting each factor equal to zero, we obtain

| $2x - 1 = 0$ | $x + 1 = 0$ | $x - 2 = 0$ |
| $x = \frac{1}{2}$ | $x = -1$ | $x = 2.$ |

Thus, in increasing order, the x intercepts are -1, $\frac{1}{2}$, and 2.

(b) The x intercepts divide the x axis into four open intervals, as shown in Figure 10. The graph of f touches the x axis only at the endpoints of these intervals; over each interval, the graph is either entirely above the x axis $[f(x) > 0]$ or entirely below it $[f(x) < 0]$. To see which is the case, we select convenient test numbers on each open interval (Figure 10) and evaluate $f(x) = (2x - 1)(x^2 - x - 2)$ at each test number. We obtain

$$f(-2) = -20, \qquad f(0) = 2, \qquad f(1) = -2, \qquad \text{and} \qquad f(3) = 20.$$

Figure 10

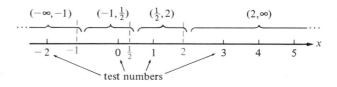

Since $f(-2) = -20$, the point $(-2, -20)$ belongs to the graph of f. This point is below the x axis, so the graph of f stays below the x axis over the first interval $(-\infty, -1)$. We apply similar reasoning to the remaining intervals and obtain the results in Figure 11a. Plotting the points $(-1, 0)$, $(\frac{1}{2}, 0)$, and $(2, 0)$ corresponding to the x intercepts and the points $(-2, -20)$, $(0, 2)$, $(1, -2)$, and $(3, 20)$ corresponding to the test numbers, and using the information in Figure 11a, we can sketch the graph of f (Figure 11b).

Figure 11

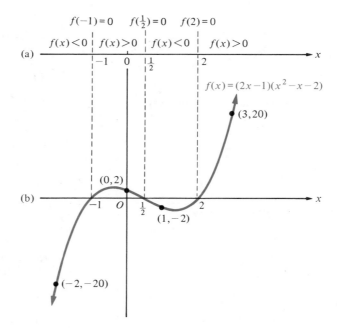

(c) From the graph of f (Figure 11b), it is clear that $f(x) < 0$ only for values of x in the two intervals $(-\infty, -1)$ and $(\frac{1}{2}, 2)$, so the solution set of the inequality $f(x) < 0$ is the union $(-\infty, -1) \cup (\frac{1}{2}, 2)$. ∎

When you use the method illustrated in Example 4, you should always ask whether the graph contains hidden peaks and valleys that are not indicated by the plotted points. Although methods studied in calculus are required to answer this question, the polynomial functions that we consider will be fairly simple, so that all peaks and valleys on their graphs can be detected by plotting a reasonable number of points.

Problem Set 4.2

In Problems 1 to 6, decide which functions are polynomial functions and which are not.

1. $f(x) = 2x^2 - 16x + 29$

2. $g(x) = \pi x - \sqrt{3}x^5 + \frac{1}{2}x^3$

3. $h(x) = \sqrt{x}$

4. $F(x) = (2x^2 - 1)(3x^3 + 2)$

5. $G(x) = 2x^{-3} + 7$ **6.** $H(x) = \dfrac{1}{x}$

In each of Problems 7 to 10, sketch graphs of the functions f, g, and h on the same coordinate system. ([C] You may use a calculator to determine additional points on the graph if you want to obtain a more accurate sketch.)

7. $f(x) = x^4$, $g(x) = 2x^4$, $h(x) = -2x^4$

8. $f(x) = x^6$, $g(x) = \dfrac{x^6}{3}$, $h(x) = -\dfrac{x^6}{3}$

9. $f(x) = x^5,$ $g(x) = \dfrac{1}{2}x^5,$ $h(x) = -\dfrac{1}{2}x^5$

10. $f(x) = x^7,$ $g(x) = \dfrac{x^7}{9},$ $h(x) = -\dfrac{x^7}{9}$

In Problems 11 to 22, find the y and x intercepts and sketch the graph of each function. (ℂ Use a calculator if you wish.)

11. $f(x) = x^4 + 1$

12. $g(x) = -\frac{2}{3}x^6 + 2$

13. $h(x) = 3x^5 - 1$

14. $F(x) = \frac{1}{16}(x + 2)^9$

15. $G(x) = 2(x + 1)^4$

16. $H(x) = -\frac{1}{2}(x - 1)^6$

17. $f(x) = -x^8 - 1$

18. $g(x) = -\frac{2}{3}x^7 + 1$

19. $h(x) = (x + \frac{1}{2})^3 - 1$

20. $F(x) = 7(x + \frac{1}{2})^3 + 2$

21. $G(x) = 2(x - 6)^4 + 1$

22. $H(x) = -20(x - 8)^8 + 8$

In Problems 23 to 32, find all x intercepts of the graph of each function.

23. $f(x) = 15x - 60$

24. $g(x) = 5x^2 + 17x - 12$

25. $h(x) = (x - 1)(4x^2 - 12x + 9)$

26. $k(x) = (3x + 1)(3x^2 - 4x)$

27. $p(x) = (4x^2 - 1)(6x^2 - 5x + 1)$

28. $P(x) = (x - 1)[x(x + 1) + 2x]$

29. $H(x) = (9x^2 - 25)(x^2 - 5x - 14)$

30. $q(x) = 3x^4 - 3x^2 - 6$

31. $F(x) = x^6 - x^4$

32. $f(x) = x^5 + 6x^3 + 9x$

In Problems 33 to 40, (a) find the x intercepts of the graph of the function, (b) sketch the graph, and (c)

use the graph to solve the given inequality. (ℂ Use a calculator if you wish.)

33. $f(x) = x(x - 1)(x + 1);\ f(x) > 0$

34. $g(x) = (x^2 - x - 2)(x + 3);\ g(x) \ge 0$

35. $p(x) = -x(x + 2)(x - 1);\ p(x) \ge 0$

36. $F(x) = -(2x + 1)(x - 1)(3x + 4);\ F(x) < 0$

37. $h(x) = \frac{1}{6}(x^2 - 1)(x - 3);\ h(x) < 0$

38. $G(x) = x^2(x - 1)(x + 3);\ G(x) \le 0$

39. $H(x) = (x + 4)(x^2 + x - 2);\ H(x) \le 0$

40. $g(x) = (x + 1)^2(x^2 - 9);\ g(x) \ge 0$

ℂ 41. Equal squares are cut off at each corner of a rectangular piece of cardboard 8 inches wide by 15 inches long, and an open-topped box is formed by turning up the sides (Figure 12). If x is the length of the sides of the cutoff squares, then the volume $V(x)$ of the resulting box is given by $V(x) = x(8 - 2x)(15 - 2x)$ cubic inches. Sketch the graph of the polynomial function V for $x > 0$ and use the graph to determine the approximate value of x for which $V(x)$ is maximum.

Figure 12

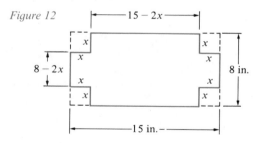

42. A child's sandbox is to be made by cutting equal squares from the corners of a square sheet of galvanized iron and turning up the sides. If each side of the sheet of galvanized iron is 2 meters long,

(a) find a formula for the volume $V(x)$ of the sandbox if x is the length in meters of the sides of the cutoff squares,

(b) sketch the graph of $V(x)$ for $x > 0$, and

(c) use the graph to determine the approximate value of x for which $V(x)$ is a maximum.

4.3 DIVISION OF ONE POLYNOMIAL BY ANOTHER

In elementary arithmetic, you learned to divide one integer by another to obtain a quotient and a remainder; for instance,

$$
\begin{array}{r}
71 \longleftarrow \text{quotient} \\
\text{divisor} \longrightarrow 32 \overline{\smash{)}\, 2277} \longleftarrow \text{dividend} \\
224 \\
\overline{37} \\
32 \\
\overline{5} \longleftarrow \text{remainder}
\end{array}
$$

The result of this calculation can be expressed as

$$2277 = (32)(71) + 5,$$

that is,

> dividend = (divisor)(quotient) + remainder.

In this section, we study a similar procedure, called **long division,** for dividing one polynomial by another, and we introduce a useful shortcut called **synthetic division.** These procedures will be used extensively in Section 4.4, where we continue our study of polynomial functions and their graphs.

When you divide one positive integer by another, you continue the procedure until you obtain a remainder that is less than the divisor. Likewise, when you divide one polynomial by another, you should continue the long-division procedure until the remainder is either the zero polynomial or a polynomial of lower degree than the divisor.

In Examples 1 and 2, perform the indicated long division to find a quotient polynomial and a remainder polynomial. Be sure that the remainder is either the zero polynomial or a polynomial of lower degree than the divisor. Check your work by verifying that

$$dividend = (divisor)(quotient) + remainder.$$

Example 1 $2x^2 + x - 1 \overline{\smash{)}\, 6x^4 + x^3 + 4x + 4}$

Solution   Note that both the divisor $2x^2 + x - 1$ and the dividend $6x^4 + x^3 + 4x + 4$ are arranged in descending powers of x. (If they weren't, we would begin by rewriting them so they were.) We divide $6x^4$, the leading term of the dividend, by $2x^2$, the leading term of the divisor, to obtain

$$\frac{6x^4}{2x^2} = 3x^2,$$

the first term of the quotient. Thus we write

$$2x^2 + x - 1 \overline{\smash{\big)}\ 6x^4 + x^3 + 4x + 4} \quad \overset{3x^2}{}$$

first term of the quotient

Now, we multiply $3x^2$ by the divisor $2x^2 + x - 1$, write this product under the dividend, and subtract to obtain a first trial remainder:

$$
\begin{array}{r}
3x^2 \\
2x^2 + x - 1 \overline{\smash{\big)}\ 6x^4 + x^3 + 4x + 4} \\
\text{subtract} \longrightarrow 6x^4 + 3x^3 - 3x^2 \\
\hline
-2x^3 + 3x^2 + 4x + 4
\end{array}
$$

first trial remainder

(To make the subtraction easier, we have spaced so that like terms are aligned vertically.) Because the degree of our first trial remainder isn't less than the degree of the divisor, we must repeat the procedure. Thus, we divide $-2x^3$, the leading term of the first trial remainder, by $2x^2$, the leading term of the divisor, to obtain $-x$, the second term of the quotient. We multiply $-x$ by divisor $2x^2 + x - 1$, write this product under the first trial remainder, and subtract to obtain a second trial remainder:

$$
\begin{array}{r}
3x^2 - x \\
2x^2 + x - 1 \overline{\smash{\big)}\ 6x^4 + x^3 + 4x + 4} \\
6x^4 + 3x^3 - 3x^2 \\
\hline
-2x^3 + 3x^2 + 4x + 4 \\
\text{subtract} \longrightarrow -2x^3 - x^2 + x \\
\hline
4x^2 + 3x + 4
\end{array}
$$

second term of the quotient

second trial remainder

The degree of the second trial remainder still isn't less than the degree of the divisor, so we divide its leading term $4x^2$ by $2x^2$ to obtain 2, the third term of the quotient. We multiply 2 by the divisor $2x^2 + x - 1$, write this product under the second trial remainder, and subtract to obtain a third trial remainder:

$$
\begin{array}{r}
3x^2 - x + 2 \\
2x^2 + x - 1 \overline{\smash{\big)}\ 6x^4 + x^3 + 4x + 4} \\
6x^4 + 3x^3 - 3x^2 \\
\hline
-2x^3 + 3x^2 + 4x + 4 \\
-2x^3 - x^2 + x \\
\hline
4x^2 + 3x + 4 \\
\text{subtract} \longrightarrow 4x^2 + 2x - 2 \\
\hline
x + 6
\end{array}
$$

third term of the quotient

third trial remainder (the remainder)

The third trial remainder is actually the remainder, since its degree is less than the degree of the divisor. We conclude that

$$3x^2 - x + 2 \quad \text{is the quotient polynomial}$$

and $\qquad x + 6 \quad$ is the remainder polynomial.

To check, we calculate

$$
\begin{aligned}
(\text{divisor})(\text{quotient}) + \text{remainder} &= (2x^2 + x - 1)(3x^2 - x + 2) + (x + 6) \\
&= 6x^4 + x^3 + 3x - 2 + x + 6 \\
&= 6x^4 + x^3 + 4x + 4 \\
&= \text{the dividend,}
\end{aligned}
$$

so our work is correct.

Example 2 $\quad x^3 - 2x^2 + 3x - 4\,\big)\,\overline{2x^5 - 7x^4 + 13x^3 - 19x^2 + 15x - 4}$

Solution

$$
\begin{array}{r}
2x^2 - 3x + 1 \qquad \leftarrow \text{quotient polynomial} \\
x^3 - 2x^2 + 3x - 4\,\big)\,\overline{2x^5 - 7x^4 + 13x^3 - 19x^2 + 15x - 4} \\
2x^5 - 4x^4 + 6x^3 - 8x^2 \qquad \leftarrow \text{first trial remainder} \\
\hline
-3x^4 + 7x^3 - 11x^2 + 15x - 4 \qquad \\
-3x^4 + 6x^3 - 9x^2 + 12x \qquad \leftarrow \text{second trial remainder} \\
\hline
x^3 - 2x^2 + 3x - 4 \qquad \\
x^3 - 2x^2 + 3x - 4 \qquad \leftarrow \text{third trial remainder (the remainder)} \\
\hline
0 \qquad
\end{array}
$$

(with "subtract" labels indicating subtraction at each stage)

The quotient polynomial is $2x^2 - 3x + 1$ and the remainder is the zero polynomial. To check, we calculate

$$
\begin{aligned}
(\text{divisor})(\text{quotient}) + \text{remainder} &= (x^3 - 2x^2 + 3x - 4)(2x^2 - 3x + 1) + 0 \\
&= 2x^5 - 7x^4 + 13x^3 - 19x^2 + 15x - 4 \\
&= \text{the dividend,}
\end{aligned}
$$

and our work is correct.

A systematic procedure that is guaranteed to work in a finite number of steps is called an **algorithm.** Long division is an algorithm because, as you carry it out, you accumulate the quotient polynomial term by term and you reduce the degree of the trial remainder at each stage. The process comes to an end when the trial remainder is either the zero polynomial or a polynomial of lower degree than the divisor. This is summarized by the following noteworthy theorem.

Theorem 1   **The Division Algorithm**

> Let f and g be polynomial functions and suppose that g is not the constant zero polynomial. Then there exist unique polynomial functions q and r such that
>
> $$f(x) = g(x)\,q(x) + r(x)$$
>
> holds for all values of x, and r is either the constant zero polynomial or a polynomial of degree lower than g.

In the division algorithm, $f(x)$ is the dividend polynomial, $g(x)$ the divisor polynomial, $q(x)$ the quotient polynomial, and $r(x)$ the remainder polynomial. The identity

$$f(x) = g(x)\, q(x) + r(x)$$

expresses the fact that

$$\text{dividend} = (\text{divisor})(\text{quotient}) + \text{remainder}.$$

We often refer to the function f/g or to a value $f(x)/g(x)$ of such a function as a *quotient*, but the word used in this way must not be confused with the *quotient polynomial* $q(x)$ in the division algorithm. The relationship between the two types of quotients is expressed in the following theorem.

Theorem 2   **The Quotient Theorem**

> Let $q(x)$ be the quotient polynomial and $r(x)$ be the remainder polynomial obtained by long division of the polynomial $f(x)$ by the nonzero polynomial $g(x)$. Then, for all values of x such that $g(x) \neq 0$,
>
> $$\frac{f(x)}{g(x)} = q(x) + \frac{r(x)}{g(x)}.$$

Proof   Divide both sides of the identity $f(x) = g(x)q(x) + r(x)$ by $g(x)$.   ∎

The quotient theorem is often used to write a rational expression $\dfrac{f(x)}{g(x)}$ as the sum of a polynomial $q(x)$ and a rational expression $\dfrac{r(x)}{g(x)}$ that is **proper** in the sense that its numerator is of lower degree than its denominator.

Example 3   Rewrite the rational expression $\dfrac{2x^3 - x^2 - 7}{x - 2}$ as the sum of a polynomial and a proper rational expression.

Solution   Let $f(x) = 2x^3 - x^2 - 7$ and $g(x) = x - 2$. By long division, we have

$$
\begin{array}{r}
2x^2 + 3x\ + 6 \qquad = q(x) \\
x - 2 \overline{\smash{\big)}\ 2x^3 - x^2 - 7} \\
\underline{2x^3 - 4x^2} \\
3x^2 - 7 \\
\underline{3x^2 - 6x} \\
6x - 7 \\
\underline{6x - 12} \\
5 = r(x).
\end{array}
$$

By Theorem 2, $\dfrac{f(x)}{g(x)} = q(x) + \dfrac{r(x)}{g(x)}$; that is,

$$\frac{2x^3 - x^2 - 7}{x - 2} = 2x^2 + 3x + 6 + \frac{5}{x - 2}. \qquad \blacksquare$$

In Example 3 above, the divisor is a *first-degree* polynomial of the form $g(x) = x - c$. In all such cases, the remainder will either be the zero polynomial or it will have degree zero—in other words, the remainder will be a *constant*. This particular type of division can be carried out by a shortcut called **synthetic division.**

Procedure for Synthetic Division

To find the quotient polynomial $q(x)$ and the remainder $r(x) = R$ when a dividend polynomial $f(x)$ of degree $n \geq 1$ is divided by a *first-degree* polynomial $g(x) = x - c$, do the following:

Step 1. *Arrange the polynomial $f(x)$ in descending powers of x:*

$$f(x) = a_n x^n + a_{n-1} x^{n-1} + a_{n-2} x^{n-2} + \cdots + a_1 x + a_0.$$

Represent all missing powers by using zero coefficients.

Step 2. *Write down the value of c, then draw a vertical line and after it list the coefficients of $f(x)$:*

$$c \quad \bigg| \quad a_n \quad a_{n-1} \quad a_{n-2} \quad \cdots \quad a_1 \quad a_0$$

Step 3. *Leave some space below the row of coefficients, draw a horizontal line, and copy the leading coefficient a_n below the line:*

$$
\begin{array}{c|cccccc}
c & a_n & a_{n-1} & a_{n-2} & \cdots & a_1 & a_0 \\
\hline
 & a_n
\end{array}
$$

Step 4. *Multiply a_n by c and write the product above the horizontal line under the second coefficient a_{n-1}; then add a_{n-1} to this product and write the result s_1 below the line:*

$$
\begin{array}{c|cccccc}
c & a_n & a_{n-1} & a_{n-2} & \cdots & a_1 & a_0 \\
 & & ca_n & & & & \\
\hline
 & a_n & s_1
\end{array}
$$

Now multiply s_1 by c and write the product above the line under the third coefficient a_{n-2}; then add a_{n-2} to this product and write the result s_2 below the line:

$$
\begin{array}{c|cccccc}
c & a_n & a_{n-1} & a_{n-2} & \cdots & a_1 & a_0 \\
 & & ca_n & cs_1 & & & \\
\hline
 & a_n & s_1 & s_2
\end{array}
$$

Continue in this way, multiplying each newly obtained number below the line by c, writing the product above the line under the next coefficient, and adding to produce the next number below the line. Do this until you have numbers below the line for every coefficient. Isolate the very last sum by drawing a short vertical line:

$$
\begin{array}{c|ccccccc}
c & a_n & a_{n-1} & a_{n-2} & \cdots & a_1 & a_0 \\
& & ca_n & cs_1 & \cdots & cs_{n-2} & cs_{n-1} \\
\hline
& a_n & s_1 & s_2 & \cdots & s_{n-1} & s_n
\end{array}
$$

Step 5. *Conclude that the numbers $a_n, s_1, s_2, \ldots, s_{n-1}$ are the coefficients of the quotient polynomial*

$$q(x) = a_n x^{n-1} + s_1 x^{n-2} + s_2 x^{n-3} + \cdots + s_{n-2} x + s_{n-1},$$

and that $s_n = R$, the remainder. [Note that $q(x)$ has degree 1 less than $f(x)$.]

In Examples 4 and 5, use synthetic division to obtain the quotient polynomial $q(x)$ and the remainder R upon division of $f(x)$ by the first-degree polynomial $g(x)$.

Example 4 $f(x) = 2x^3 - x^2 - 7; g(x) = x - 2$

Solution   The dividend $f(x) = 2x^3 - x^2 + 0x - 7$ has coefficients 2, -1, 0, and -7, and the divisor has the form $g(x) = x - c$ with $c = 2$. By synthetic division:

$$
\begin{array}{c|cccc}
2 & 2 & -1 & 0 & -7 \\
& & 4 & 6 & 12 \\
\hline
& 2 & 3 & 6 & 5
\end{array}
$$

Hence, the quotient polynomial is $q(x) = 2x^2 + 3x + 6$ and the remainder is $R = 5$. Note that this is the same problem that we did by long division in Example 3. Do you see how the shortcut works?   ∎

Example 5 $f(x) = 3x^4 - 2x^3 - 5x; g(x) = x + 2$

Solution   The dividend $f(x) = 3x^4 - 2x^3 + 0x^2 - 5x + 0$ has coefficients 3, -2, 0, -5, and 0, and the divisor has the form $g(x) = x - c$ with $c = -2$. By synthetic division:

$$
\begin{array}{c|ccccc}
-2 & 3 & -2 & 0 & -5 & 0 \\
& & -6 & 16 & -32 & 74 \\
\hline
& 3 & -8 & 16 & -37 & 74
\end{array}
$$

Hence, the quotient polynomial is $q(x) = 3x^3 - 8x^2 + 16x - 37$ and the remainder is $R = 74$.   ∎

A detailed proof that synthetic division always works is a bit tedious, so we shall not give it here. However, if you will work some simple examples by both long division and synthetic division, you will see clearly how every bit of the arithmetic required in the long division is accounted for in the synthetic division (see Problems 33 to 36).

Problem Set 4.3

In Problems 1 to 18, perform the indicated long division to find a quotient polynomial and a remainder polynomial. Be sure that the remainder is either the zero polynomial or a polynomial of lower degree than the divisor. Check your work by verifying the fact that dividend = (divisor)(quotient) + remainder.

1. $x - 5 \overline{\smash{\big)}\, x^2 + 3x - 10}$

2. $3x^2 - 1 \overline{\smash{\big)}\, 6x^4 + 10x^2 + 7}$ 3. $x - 1 \overline{\smash{\big)}\, x^3 - 1}$

4. $4x - 3 \overline{\smash{\big)}\, 4x^6 + 5x^3 - 6}$

5. $2x^2 - 4x + 1 \overline{\smash{\big)}\, 6x^4 - 31x^2 + 26x - 6}$

6. $x + 1 \overline{\smash{\big)}\, x^5 - 1}$

7. $2x - 3 \overline{\smash{\big)}\, 4x^4 - 12x^3 + 15x^2 - 17x}$

8. $2x^4 - 3x - 1 \overline{\smash{\big)}\, -6x^4 - 7x^3 + 14x^2 - 5x + 1}$

9. $x^2 + 3 \overline{\smash{\big)}\, x^3 + 3x^2 + 2x - 4}$

10. $1 - 4x - 2x^3 \overline{\smash{\big)}\, x^2 - 3x^4 - x^3 + 5}$

11. $x^3 + x^2 \overline{\smash{\big)}\, x^5 - 1}$ 12. $x - c \overline{\smash{\big)}\, x^2 - 2cx + 2}$

13. $2x - 1 \overline{\smash{\big)}\, x^2 + \frac{1}{2}x + 1}$

14. $t^2 - t + 1 \overline{\smash{\big)}\, 2t^4 + t^2 - 1}$

15. $3x^2 + 2x + 1 \overline{\smash{\big)}\, x^3 + x^2 + 1}$

16. $x - c \overline{\smash{\big)}\, x^3 - c^3}$

17. $5t^2 - t + 4 \overline{\smash{\big)}\, 10t^3 + 13t^2 + 5t + 2}$

18. $x^2 + x + 1 \overline{\smash{\big)}\, ax^2 + bx + c}$

In Problems 19 to 22, use the quotient theorem (Theorem 2) to rewrite each rational expression as the sum of a polynomial and a proper rational expression.

19. $\dfrac{5x^3 + 3x^2 - x + 2}{x - 4}$ 20. $\dfrac{x^2 - x + 1}{x^2 + x - 1}$

21. $\dfrac{5x^3 - 6x^2 - 68x - 16}{x^3 - 2x^2 - 8x}$ 22. $\dfrac{ax + b}{cx + d}$

In Problems 23 to 32, use synthetic division to obtain the quotient polynomial $q(x)$ and the remainder R upon division of the polynomial $f(x)$ by the first-degree polynomial $g(x)$.

23. $f(x) = 3x^3 - 2x^2 - x + 4; \; g(x) = x - 2$

24. $f(x) = x^6 - x^5 - x^2 - x - 1; \; g(x) = x - 2$

25. $f(x) = x^5 - 5x^3 + x - 16; \; g(x) = x + 2$

26. $f(x) = 5x^3 - 7x^2 + 3x - 2; \; g(x) = x - 1$

27. $f(x) = 3x^6 - 2x^4 + x^2; \; g(x) = x + 1$

28. $f(x) = 9x^2 - 3x + 1; \; g(x) = x - \frac{1}{3}$

29. $f(x) = -16x^3 - 12x^2 + 2x + 7; \; g(x) = x - \frac{1}{2}$

30. $f(x) = x^2 + 2x + 1; \; g(x) = x - c$

ⓒ 31. $f(x) = x^3 + x^2 + x + 1; \; g(x) = x - 1.1$

32. $f(x) = ax^2 + bx + c; \; g(x) = x - 1$

In Problems 33 to 36, find the quotient and remainder (a) by long division and (b) by synthetic division.

33. $x - 3 \overline{\smash{\big)}\, 5x^3 - 11x^2 - 14x - 10}$

34. $x + 3 \overline{\smash{\big)}\, 2x^3 + 3x^2 - 5x + 12}$

35. $x + 4 \overline{\smash{\big)}\, -2x^3 - 6x^2 + 18x + 20}$

36. $x + \frac{1}{2} \overline{\smash{\big)}\, 5x^3 + x - 9}$

37. Find a value of k such that $x^3 + 2x^2 - 3kx - 10$ divided by $x + 3$ has a remainder of 8.

38. Find a value of k such that $x + 5$ is a factor of $x^3 + kx + 125$.

39. Find a value of k such that $x - \frac{1}{3}$ is a factor of $3x^3 - x^2 + kx - 5$.

40. Suppose that the polynomial $f(x)$ is divided by the nonzero polynomial $g(x)$ to produce a quotient polynomial $q(x)$ and a remainder polynomial $r(x)$. Now, suppose that $q(x)$ is divided by the nonzero polynomial $G(x)$ to produce a quotient polynomial $Q(x)$ and a remainder polynomial $R(x)$. Show that, if $f(x)$ is divided by the product $g(x)G(x)$, the quotient polynomial is $Q(x)$ and the remainder polynomial is $r(x) + g(x)R(x)$.

41. In the division algorithm (Theorem 1), what are the polynomials q and r for the case in which the degree of f is less than the degree of g?

4.4 VALUES AND ZEROS OF POLYNOMIAL FUNCTIONS

In this section, we continue our study of polynomial functions and their graphs. We begin with the following important theorem, which shows that values of polynomial functions can be found by using the division algorithm.

Theorem 1   **The Remainder Theorem**

If f is a polynomial function and c is a constant, then
$$f(c) = R,$$
where R is the remainder upon division of $f(x)$ by $x - c$.

Proof   By the division algorithm (Theorem 1, Section 4.3) with $g(x) = x - c$, we have
$$f(x) = (x - c)q(x) + r(x),$$
where either $r(x)$ is the zero polynomial or its degree is less than 1 (the degree of $g(x)$). In either case, $r(x)$ is a constant polynomial, $r(x) = R$. Thus,
$$f(x) = (x - c)q(x) + R.$$
Substituting $x = c$ in the last equation, we find that
$$f(c) = (c - c)q(c) + R$$
$$= 0q(c) + R = R.$$ ∎

Example 1   Use the remainder theorem and synthetic division to find
$$f(-3) \quad \text{if} \quad f(x) = 4x^4 + 3x^3 - x^2 - 5x - 6.$$

Solution*   By the remainder theorem, $f(-3)$ is the remainder when $f(x)$ is divided by $x - (-3)$. Using synthetic division, we have

$$
\begin{array}{r|rrrrr}
-3 & 4 & 3 & -1 & -5 & -6 \\
   &   & -12 & 27 & -78 & 249 \\
\hline
   & 4 & -9 & 26 & -83 & 243 = R
\end{array}
$$

Therefore, $f(-3) = 243$. ∎

One of the noteworthy consequences of the remainder theorem is the following.

---

* This technique for evaluating a polynomial, which is sometimes called **Horner's method,** is readily carried out on a calculator or computer.

Theorem 2   **The Factor Theorem**

> Let f be a polynomial function and let c be a constant. Then $f(c) = 0$ if and only if $x - c$ is a factor of $f(x)$.

Proof   We have to prove that **(i)** if $f(c) = 0$, then $x - c$ is a factor of $f(x)$, and **(ii)** if $x - c$ is a factor of $f(x)$, then $f(c) = 0$.

(i) Suppose that $f(c) = 0$. Then, we know by the remainder theorem that 0 is the remainder when $f(x)$ is divided by $x - c$; hence, $x - c$ is a factor of $f(x)$.

(ii) Suppose that $x - c$ is a factor of $f(x)$. Then, we can write

$$f(x) = (x - c)q(x),$$

and it follows that

$$f(c) = (c - c)q(c) = 0q(c) = 0. \qquad \blacksquare$$

Example 2   Let $f(x) = x^3 + 2x^2 - 5x - 6$. Use the factor theorem to determine whether:

(a) $x + 1$ is a factor of $f(x)$

(b) $x - 3$ is a factor of $f(x)$

Solution   **(a)** We have $x + 1 = x - (-1)$, which has the form $x - c$ with $c = -1$. Since

$$f(-1) = (-1)^3 + 2(-1)^2 - 5(-1) - 6 = -1 + 2 + 5 - 6 = 0,$$

it follows from the factor theorem that $x + 1$ is a factor of $f(x)$.

(b) Here $x - 3$ has the form $x - c$ with $c = 3$. Since

$$f(3) = 3^3 + 2(3)^2 - 5(3) - 6 = 27 + 18 - 15 - 6 = 24 \neq 0,$$

it follows from the factor theorem that $x - 3$ is not a factor of $f(x)$. $\qquad \blacksquare$

If f is a function, then a root of the equation

$$f(x) = 0$$

is called a **zero** of f. By the factor theorem, each zero c of a polynomial function f corresponds to a first-degree factor $(x - c)$ of $f(x)$. When $f(x)$ is factored completely, the same factor $(x - c)$ may occur more than once, in which case c is called a **repeated** or **multiple** zero of f.

> If $(x - c)^m$ is a factor of $f(x)$, but $(x - c)^{m+1}$ is not, we say that c is a zero of **multiplicity** m.

For instance, if $(x - 5)^2$ is a factor of $f(x)$, but $(x - 5)^3$ is not, we say that 5 is a zero of multiplicity 2. Note that a zero of a function f is the same as an x intercept of the graph of f. At a zero that is not repeated, the graph of a polynomial function f simply cuts across the x axis (Figure 1a); however, the graph of f is very "flat" near a zero of multiplicity $m > 1$ (Figures 1b and 1c).*

Figure 1

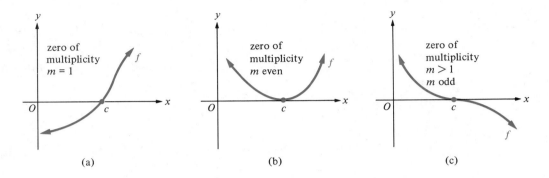

(a)   (b)   (c)

Example 3   Given that 2 is a zero of

$$f(x) = x^5 - 6x^4 + 11x^3 - 2x^2 - 12x + 8,$$

determine its multiplicity.

Solution   Using synthetic division, we divide $f(x)$ by $x - 2$ to obtain

$$f(x) = (x - 2)(x^4 - 4x^3 + 3x^2 + 4x - 4)$$

Again, we divide $x^4 - 4x^3 + 3x^2 + 4x - 4$ by $x - 2$ and find that

$$x^4 - 4x^3 + 3x^2 + 4x - 4 = (x - 2)(x^3 - 2x^2 - x + 2).$$

Combining the last two equations, we obtain

$$f(x) = (x - 2)(x - 2)(x^3 - 2x^2 - x + 2)$$
$$= (x - 2)^2(x^3 - 2x^2 - x + 2).$$

Now we divide $x^3 - 2x^2 - x + 2$ by $x - 2$ and find that

$$x^3 - 2x^2 - x + 2 = (x - 2)(x^2 - 1).$$

Therefore,

$$f(x) = (x - 2)^2(x - 2)(x^2 - 1)$$
$$= (x - 2)^3(x^2 - 1)$$

Because $x - 2$ is not a factor of $x^2 - 1$, our successive divisions terminate, and it follows that 2 is a zero of multiplicity 3 for the polynomial $f(x)$. ∎

Because a polynomial $f(x)$ can't have more first-degree factors than its degree, we have the following result.

* At a zero of multiplicity $m > 1$, the x axis is actually *tangent* to the graph of f.

Theorem 3   **The Maximum Number of Zeros of a Polynomial Function**

A polynomial function cannot have more zeros than its degree.

In using Theorem 3, you must count each zero as many times as its multiplicity. For instance, we found in Example 3 that 2 was a zero of multiplicity 3 for the fifth-degree polynomial function f; hence, we can conclude that f can have at most two more zeros. Indeed, f has exactly two more zeros: 1 and -1.

Descartes' Rule of Signs

The seventeenth-century French mathematician René Descartes discovered a simple and useful rule that helps to determine the number of positive and negative zeros of a polynomial with real coefficients. Before stating Descartes' rule, we must explain what is meant by a *variation of sign* for such a polynomial. If the terms of the polynomial are arranged in order of descending powers, we say that a **variation of sign** occurs whenever two successive terms have opposite signs. Missing terms (with zero coefficients) are ignored when counting the total number of variations of sign. In using Descartes' rule, it is necessary to count the variations of sign both for a polynomial $f(x)$ and for the related polynomial $f(-x)$.

Example 4   If $f(x) = 7x^5 - 3x^4 + 5x^2 + x - 2$, determine the total number of variations of sign for

(a) $f(x)$ **(b)** $f(-x)$

Solution   **(a)** $f(x) = 7x^5 - 3x^4 + 5x^2 + x - 2$ has three variations of sign.

(b) $f(-x) = 7(-x)^5 - 3(-x)^4 + 5(-x)^2 + (-x) - 2$

$$= -7x^5 - 3x^4 + 5x^2 - x - 2$$

has two variations of sign.

We can now state **Descartes' rule of signs** for a polynomial $f(x)$ with real coefficients. We assume that $f(x)$ is arranged in descending powers of x.

Descartes' Rule of Signs

(i) The number of *positive zeros* of $f(x)$ is either equal to the number of variations of sign for $f(x)$, or it is less than that number by an even integer.

(ii) The number of *negative zeros* of $f(x)$ is either equal to the number of variations of sign for $f(-x)$, or it is less than that number by an even integer.

Figure 2

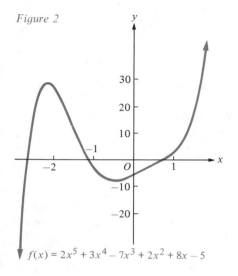

$f(x) = 2x^5 + 3x^4 - 7x^3 + 2x^2 + 8x - 5$

In using Descartes' rule, zeros of multiplicity m are to be counted m times. The proof of part (i) of the rule is somewhat technical and will be omitted here.* Part (ii) of the rule can be derived from part (i) by applying the results of Problems 55 and 57.

Example 5   Use Descartes' rule of signs to determine the possible number of positive and negative zeros of

$$f(x) = 2x^5 + 3x^4 - 7x^3 + 2x^2 + 8x - 5.$$

Solution   There are 3 variations of sign for $f(x)$, so $f(x)$ has either 3 or 1 positive zeros. There are 2 variations of sign for

$$f(-x) = -2x^5 + 3x^4 + 7x^3 + 2x^2 - 8x - 5,$$

so $f(x)$ has either 2 or 0 negative zeros. ∎

As a matter of fact, the function $f(x)$ in Example 5 has 2 negative zeros and 1 positive zero (Figure 2). It also has 2 *complex* zeros. In Section 4.5 we present some methods for finding the real zeros of a polynomial function, and we study complex zeros in Section 4.7.

The Upper-and-Lower-Bound Rule

A second rule, which can be used efficiently in conjunction with Descartes' rule of signs, determines a finite interval that contains all of the real zeros of a polynomial function $f(x)$. This rule involves the idea of *upper and lower bounds*. Any number that is greater than or equal to all of the real zeros of $f(x)$ is called an **upper bound** for these zeros. Likewise, any number that is less than or equal to all of the real zeros of $f(x)$ is called a **lower bound** for these zeros. Notice that such upper and lower bounds are not unique—any number greater than an upper bound is again an upper bound and any number less than a lower bound is again a lower bound. However, if L is a lower bound and U is an upper bound for the real zeros of $f(x)$, then we can be certain that all of these zeros belong to the finite interval $[L, U]$.

In the following upper-and-lower-bound rule for the real zeros of a polynomial function, it is assumed that $f(x)$ is a polynomial with real coefficients and that the *leading coefficient*[†] *of $f(x)$ is positive*.

---

* A proof of Descartes' rule of signs can be found in *A Modern Course on the Theory of Equations* by David E. Dobbs and Robert Hanks, Polygonal Publishing House, Passaic, N.J., 1980.

[†] The **leading coefficient** of a polynomial is the coefficient of its term of highest degree.

Upper-and-Lower-Bound Rule for Real Zeros

Let $f(x)$ be divided by $x - c$ using synthetic division.

(i) If $c > 0$ and all numbers in the last (quotient) row of the synthetic division are positive or zero, then c is an upper bound for the zeros of $f(x)$.

(ii) If $c < 0$ and the numbers in the last (quotient) row of the synthetic division alternate in sign, then c is a lower bound for the zeros of $f(x)$.

If 0 appears in one or more places in the quotient row, it can be regarded as being either positive or negative for purposes of applying part (ii) of this rule. Problem 56 outlines a proof of part (i) of the rule. Part (ii) can be derived from part (i) by applying the results of Problems 55 and 57.

Example 6 | Use the upper-and-lower-bound rule to find integers that are upper and lower bounds for the real zeros of

$$f(x) = 6x^4 - 19x^3 + 13x^2 + 4x - 4.$$

Solution | To locate an upper bound for the real zeros of $f(x)$ we use part (i) of the upper-and-lower-bound rule to test successive values of $c = 1, 2, 3, \ldots$ until the last row of the synthetic division becomes nonnegative:

$$
\begin{array}{r|rrrrr}
1 & 6 & -19 & 13 & 4 & -4 \\
  &   & 6 & -13 & 0 & 4 \\
\hline
  & 6 & -13 & 0 & 4 & 0
\end{array}
\qquad
\begin{array}{r|rrrrr}
2 & 6 & -19 & 13 & 4 & -4 \\
  &   & 12 & -14 & -2 & 4 \\
\hline
  & 6 & -7 & -1 & 2 & 0
\end{array}
$$

$$
\begin{array}{r|rrrrr}
3 & 6 & -19 & 13 & 4 & -4 \\
  &   & 18 & -3 & 30 & 102 \\
\hline
  & 6 & -1 & 10 & 34 & 98
\end{array}
\qquad
\begin{array}{r|rrrrr}
4 & 6 & -19 & 13 & 4 & -4 \\
  &   & 24 & 20 & 132 & 544 \\
\hline
  & 6 & 5 & 33 & 136 & 540
\end{array}
$$

The first value of c for which the last row becomes nonnegative is $c = 4$; hence, 4 is an upper bound for the real zeros of $f(x)$. Now we use part (ii) of the rule to test successive values of $c = -1, -2, -3, \ldots$ until the last row of the synthetic division alternates in sign:

$$
\begin{array}{r|rrrrr}
-1 & 6 & -19 & 13 & 4 & -4 \\
   &   & -6 & 25 & -38 & 34 \\
\hline
   & 6 & -25 & 38 & -34 & 30
\end{array}
$$

Here, on the very first trial with $c = -1$, we obtain alternating signs, and it follows that -1 is a lower bound for the real zeros of $f(x)$. ∎

By using the upper-and-lower-bound rule as in Example 6 above, we can conclude that all of the real zeros of the function

$$f(x) = 6x^4 - 19x^3 + 13x^2 + 4x - 4$$

lie on the interval $[-1, 4]$. This is quite correct, but, as the graph in Figure 3 shows, all of the real zeros of f actually lie on the smaller interval $[-1, 2]$. Thus, although the upper-and-lower-bound rule will give an interval that is guaranteed to contain all of the real zeros of a polynomial function, it does not necessarily produce the *smallest* such interval.

Figure 3

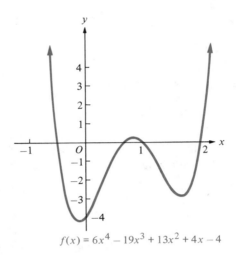

$$f(x) = 6x^4 - 19x^3 + 13x^2 + 4x - 4$$

Problem Set 4.4

In Problems 1 to 10, use the remainder theorem and synthetic division to find the value of $f(c)$ for the given polynomial function f and the indicated value of c. In each case, check your answer by a direct calculation of $f(c)$. ([C] Use a calculator if you wish.)

1. $f(x) = x^3 - 9x^2 + 23x - 15$, $c = 1$

2. $f(x) = 2x^3 - 2x^2 - x - 2$, $c = -1$

3. $f(x) = 8x^3 - 25x^2 + 4x - 3$, $c = -3$

4. $f(x) = 3x^4 - x^3 + 2x - 10$, $c = 2$

5. $f(x) = x^5 - 2x^4 + x^3 - 3x^2 + 8$, $c = 2$

6. $f(x) = x^5 + 2x^3 - 4x + 5$, $c = -2$

7. $f(x) = 3x^4 + 2x^2 - 5$, $c = -2$

8. $f(x) = 16x^4 - 8x^2 + 12x - 1$, $c = \frac{1}{2}$

9. $f(x) = x^3 - 4x + 8$, $c = \frac{1}{3}$

10. $f(x) = 5x^3 - 20x^2 + 2x - 1$, $c = 0.2$

[C] In Problems 11 to 18, use the factor theorem to determine (without actually dividing) whether or not the indicated binomial is a factor of the given polynomial. (You may want to use a calculator to find values of the polynomial functions.)

11. $f(x) = 4x^4 + 13x^3 - 13x^2 - 40x + 12$; $x + 2$

12. $g(x) = x^4 - 9x^3 + 18x^2 - 3$; $x + 1$

13. $H(x) = x^5 - 17x^3 + 75x + 9$; $x - 3$

14. $F(x) = 2x^4 - x^3 + x^2 + x - 3$; $x + 1$

15. $h(x) = 30x^3 - 20x^2 - 100x + 1000$; $x - 10$

16. $G(t) = t^4 + 2t^3 - 6t^2 - 14t - 7$; $t - 7$

17. $F(y) = 2y^3 + 4y - 2; \ y + \frac{1}{2}$

18. $f(z) = 81z^3 + 51z^2 - 153z + \frac{127}{3}; \ z - \frac{1}{3}$

In Problems 19 to 24, the indicated number is a zero of the polynomial function $f(x)$. Determine the multiplicity of this zero.

19. $1; f(x) = x^3 + x^2 - 5x + 3$

20. $1; f(x) = 3x^3 - 8x^2 + 7x - 2$

21. $2; f(x) = x^4 - 11x^3 + 42x^2 - 68x + 40$

22. $-2; f(x) = x^5 + 2x^4 - 9x^3 - 22x^2 + 4x + 24$

23. $-1; f(x) = 4x^4 + 9x^3 + 3x^2 - 5x - 3$

24. $-3; f(x) = x^5 + 10x^4 + 37x^3 + 63x^2 + 54x + 27$

In Problems 25 to 30, determine the total number of variations in sign for (a) $f(x)$ and (b) $f(-x)$.

25. $f(x) = 3x^3 - 2x^2 - 5x + 7$

26. $f(x) = 2x^4 + 5x^3 - 8x^2 - 5x + 4$

27. $f(x) = 12x^5 - 17x^4 + 7x^2 + 3x - 2$

28. $f(x) = -2x^5 - x^4 + 5x^3 + x^2$

29. $f(x) = 7x^7 - 8x^5 - 5x^3 + 11x + 6$

30. $f(x) = x^6 - 9x^4 - 3x^2 + 13$

In Problems 31 to 42, use Descartes' rule of signs to determine the possible number of positive and negative zeros of the polynomial function $f(x)$.

31. $f(x) = 2x^3 - 3x^2 - 4x - 5$

32. $f(x) = 2x^3 - 7x^2 + 12x - 4$

33. $f(x) = 2x^3 - 7x^2 + 17x - 4$

34. $f(x) = 3x^4 - 4x^3 - 5x - 1$

35. $f(x) = x^4 - 2x^3 - 2x^2 + 8x - 1$

36. $f(x) = 2x^3 - 7x^2 - x - 21$

37. $f(x) = 2x^5 + 4x^3 + 3x^2 - 1$

38. $f(x) = x^6 + x^2 + x + 1$

39. $f(x) = 2x^6 - x^4 + 3x + 13$

40. $f(x) = x^5 + x^2 - 3$

41. $f(x) = 2x^7 - x^5 + x^2 + 3$

42. $f(x) = x^6 - x^3 - x^2$

In Problems 43 to 54, use the upper-and-lower-bound rule to find integers that are upper and lower bounds for the real zeros of the polynomial function $f(x)$.

43. $f(x) = 2x^3 - 5x^2 - 4x + 3$

44. $f(x) = 2x^3 + 9x^2 - 5x - 41$

45. $f(x) = 2x^4 - 5x^3 - 8x^2 + 25x - 10$

46. $f(x) = 4x^4 - 8x^3 - 43x^2 + 29x + 60$

47. $f(x) = x^4 - x^3 - 10x^2 - 2x + 12$

48. $f(x) = 2x^5 - 2x^2 + x - 2$

49. $f(x) = x^3 - x^2 - x - 4$

50. $f(x) = x^3 + 6x^2 + 7x - 1$

51. $f(x) = x^4 - x^3 + x - 3$

52. $f(x) = 2x^4 - 5x^3 - 8x^2 + 25x - 10$

53. $f(x) = 4x^4 - 8x^3 - 43x^2 + 28x + 60$

54. $f(x) = 8x^5 - 44x^4 + 86x^3 - 73x^2 + 28x - 4$

55. How are the zeros of $f(x)$ related to the zeros of $f(-x)$?

56. Suppose that $c > 0$, that the polynomial $f(x)$ is divided by $x - c$ to produce a quotient polynomial $q(x)$ and a remainder R, and that R and all of the coefficients of $q(x)$ are nonnegative. Show that it is impossible for $f(x)$ to have a zero z with $z > c$. [*Hint:* Show that $f(z) > 0$ by using the fact that $f(x) = (x - c)q(x) + R$.]

57. How are the coefficients of $f(x)$ related to the coefficients of $f(-x)$?

58. Suppose $f(x) = a_n x^n + a_{n-1} x^{n-1} + \cdots + a_1 x + a_0$ is a polynomial function such that $f(x) = 0$ for all values of x. Prove that all of the coefficients $a_n, a_{n-1}, \ldots, a_1, a_0$ must be zero. [*Hint:* Use Theorem 3.]

59. If p and q are zeros of the polynomial function $f(x) = x^2 + bx + c$, show that $b = -(p + q)$ and $c = pq$.

60. Suppose that f and g are polynomial functions such that $f(x) = g(x)$ for all values of x. Prove that the coefficients of f are the same as the coefficients of g. [*Hint:* Use Problem 58.]

4.5 FINDING REAL ZEROS OF POLYNOMIALS

Although Descartes' rule of signs and the upper-and-lower-bound rule supply useful information about the zeros of polynomial functions, they do not provide the actual *numerical values* of these zeros. It is possible to obtain *rough approximations* to the real zeros of a polynomial function by sketching its graph; however, this method can be time-consuming.

In this section, we study some alternative methods for finding the zeros of a polynomial function with real coefficients. To begin with, suppose that all of the coefficients are rational numbers. Multiplying the polynomial by the least common denominator (LCD) of these coefficients, you obtain a polynomial function with *integer* coefficients and with the same zeros as the original function. Using the following theorem, you can then find all the *rational* zeros of the polynomial function.

Theorem 1   **The Rational-Zeros Theorem**

> Let $f(x) = a_n x^n + a_{n-1} x^{n-1} + \cdots + a_1 x + a_0$ be a polynomial function of degree $n > 0$ with integers as coefficients. Then, if p/q is any rational zero of f and if the fraction p/q is reduced to lowest terms, p must be a factor of a_0 and q must be a factor of a_n.

A rigorous proof* of the rational-zeros theorem can be found on page 280 of *A First Undergraduate Course in Abstract Algebra* by A.P. Hillman and G.L. Anderson (Wadsworth Publishing Co., Belmont, Calif., 1973).

In the following procedure, based on Theorem 1, we assume that $a_0 \neq 0$. (If $a_0 = 0$, then 0 is a rational root of $f(x)$.)

Procedure for Finding all Rational Zeros of a Polynomial Function

> **Step 1.** Check to see that all coefficients of $f(x)$ are integers. (If not, multiply by the LCD of the coefficients to clear fractions.) Find all factors p of a_0 (the constant term) and all factors q of a_n (the leading coefficient).
>
> **Step 2.** Form all possible ratios p/q. These rational numbers are all possible rational zeros of f.
>
> **Step 3.** Using synthetic division, check each[†] of the rational numbers p/q to see whether it is a zero of f. If none of them works, conclude that f has no rational zeros. If a rational zero p/q is found, proceed to step 4.
>
> **Step 4.** If $c = p/q$ is the rational zero produced in step 3, then, by the factor theorem, $f(x) = (x - c)Q(x)$, where the coefficients of $Q(x)$ were determined when you performed the synthetic division. The remaining zeros of f are the zeros of Q.

* The proof, although elementary, makes use of facts concerning relatively prime integers that are usually proved only in more advanced courses.

[†] For greater efficiency, you can use Descartes' rule of signs and the upper-and-lower-bound rule to reject all ratios p/q that cannot be zeros of f.

If the polynomial $Q(x)$ in step 4 is quadratic, you can use the quadratic formula (or factoring) to find its zeros. If $Q(x)$ has degree 3 or more, just repeat the whole procedure, starting with Q this time. To find all rational roots of the original polynomial may require several cycles through the procedure. If you encounter the same rational zero more than once, this just means that the zero has multiplicity greater than one.

Example 1   Find all rational zeros of $f(x) = 9x^3 + 6x^2 - 5x - 2$.

Solution   We follow the procedure above.

Step 1. All coefficients are integers, so there is no need to clear fractions. Here $a_0 = -2$ and $a_3 = 9$. The factors of a_0 are

$$p: \pm 1, \pm 2$$

and the factors of a_3 are

$$q: \pm 1, \pm 3, \pm 9.$$

Step 2. The possible zeros p/q are

$$\frac{p}{q}: \pm 1, \pm 2, \pm \tfrac{1}{3}, \pm \tfrac{2}{3}, \pm \tfrac{1}{9}, \pm \tfrac{2}{9}.$$

Step 3. Of the twelve possible rational zeros in step 2, three at most can be zeros of $f(x)$ (Theorem 3, Section 4.4), but it may be that none of them is. (The possibilities can be narrowed down by using Descartes' rule of signs and the upper-and-lower-bound rule, but we shall not do so here.) We check the possibilities, $1, -1, 2, -2$, and so forth, one at a time, using synthetic division. The first one that works is $\tfrac{2}{3}$:

$$
\begin{array}{r|rrrr}
\tfrac{2}{3} & 9 & 6 & -5 & -2 \\
  &   & 6 & 8 & 2 \\
\hline
  & 9 & 12 & 3 & 0 = R.
\end{array}
$$

Step 4. The synthetic division just performed yields the coefficients of the quotient polynomial $Q(x) = 9x^2 + 12x + 3$. We could now apply the same procedure to the polynomial function $Q(x)$, but it's easier to factor:

$$9x^2 + 12x + 3 = 3(x + 1)(3x + 1).$$

We conclude that the remaining zeros are -1 and $-\tfrac{1}{3}$. Therefore, the rational zeros of f are $\tfrac{2}{3}$, -1, and $-\tfrac{1}{3}$.   ■

By using the rational-zeros theorem and the factor theorem, you can often factor a polynomial function completely into prime factors. For instance, as a consequence of Example 1,

$$f(x) = 9x^3 + 6x^2 - 5x - 2 = (x - \tfrac{2}{3})(9x^2 + 12x + 3)$$
$$= (x - \tfrac{2}{3})(3)(x + 1)(3x + 1) = (3x - 2)(x + 1)(3x + 1).$$

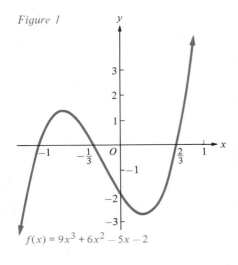

Figure 1

$f(x) = 9x^3 + 6x^2 - 5x - 2$

If you succeed in factoring a polynomial function completely, you can use the techniques discussed in Section 4.2 to sketch its graph. For instance, the graph of $f(x) = 9x^3 + 6x^2 - 5x - 2$ (Example 1) is shown in Figure 1.

Example 2   Factor $f(x) = 4x^4 - 4x^3 - 25x^2 + x + 6$ completely into prime factors and $\boxed{\text{C}}$ sketch its graph.

Solution   We begin by applying the procedure for finding rational zeros to the function f. Here $a_0 = 6$ and $a_4 = 4$. Thus

$$p: \pm 1, \pm 2, \pm 3, \pm 6 \qquad \text{(factors of 6)}$$

$$q: \pm 1, \pm 2, \pm 4 \qquad \text{(factors of 4)}$$

$$\frac{p}{q}: \pm 1, \pm 2, \pm 3, \pm 6, \pm \tfrac{1}{2}, \pm \tfrac{3}{2}, \pm \tfrac{1}{4}, \pm \tfrac{3}{4}.$$

We test each* of the possible rational zeros p/q, one by one, using synthetic division. The first one that works is -2:

$$
\begin{array}{r|rrrrr}
-2 & 4 & -4 & -25 & 1 & 6 \\
   &   & -8 & 24 & 2 & -6 \\
\hline
   & 4 & -12 & -1 & 3 & 0 = R.
\end{array}
$$

Therefore, -2 is a rational zero of f and the quotient polynomial is given by $Q(x) = 4x^3 - 12x^2 - x + 3$. Hence,

$$f(x) = (x + 2)(4x^3 - 12x^2 - x + 3).$$

Now, we repeat the procedure for $Q(x) = 4x^3 - 12x^2 - x + 3$. Here $a_0 = 3$ and $a_3 = 4$. Thus

$$p: \pm 1, \pm 3 \qquad \text{(factors of 3)}$$

$$q: \pm 1, \pm 2, \pm 4 \qquad \text{(factors of 4)}$$

$$\frac{p}{q}: \pm 1, \pm 3, \pm \tfrac{1}{2}, \pm \tfrac{1}{4}, \pm \tfrac{3}{2}, \pm \tfrac{3}{4}.$$

We test the possible rational zeros p/q of $Q(x)$, one by one, using synthetic division. The first one that works is 3:

$$
\begin{array}{r|rrrr}
3 & 4 & -12 & -1 & 3 \\
  &   & 12 & 0 & -3 \\
\hline
  & 4 & 0 & -1 & 0 = R.
\end{array}
$$

* By Descartes' rule of signs, f has either two or no positive zeros and either two or no negative zeros—although Descartes' rule alone gives no information about whether these zeros are rational numbers. By the upper-and-lower-bound rule, all zeros of f lie on the interval $[-3, 4]$. Thus, we can rule out ± 6 as possible zeros without bothering to test them by synthetic division.

Therefore, 3 is a rational zero of $Q(x)$ and

$$Q(x) = 4x^3 - 12x^2 - x + 3 = (x - 3)(4x^2 + 0x - 1)$$
$$= (x - 3)(4x^2 - 1) = (x - 3)(2x - 1)(2x + 1).$$

It follows that

$$f(x) = (x + 2)Q(x) = (x + 2)(x - 3)(2x - 1)(2x + 1).$$

Using the techniques discussed in Section 4.2, we can now sketch the graph of f (Figure 2). As usual, accuracy is enhanced if a calculator is used to help determine points on the graph.

Figure 2

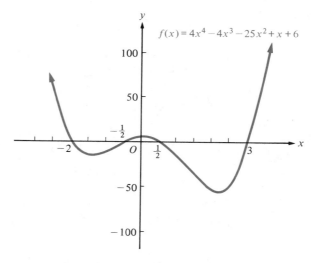

$$f(x) = 4x^4 - 4x^3 - 25x^2 + x + 6$$

Irrational Zeros of Polynomial Functions

A polynomial function with rational coefficients may have no rational zeros; for instance

$$f(x) = x^2 + x + 1$$

has no real zeros, rational or irrational. (Why?) There are polynomial functions, such as

$$g(x) = 2x^2 + 9x - 3,$$

all of whose zeros are irrational. Finally, it may happen that some of the zeros of a polynomial function are rational and some are irrational; for instance,

$$h(x) = (x - 1)(2x^2 + 9x - 3)$$

has the rational zero 1, but the other two zeros are irrational.

Most of the standard methods for finding the irrational zeros of polynomial functions involve a technique of **successive approximation;** that is, starting from a rough first approximation, they produce successively better and better approximations to a zero. Some of these methods are based on the idea of a *change of sign* of a function.

A function f is said to **change sign** on an interval I if there are numbers a and b in I such that $f(a)$ and $f(b)$ have opposite algebraic signs. In Section 2.7, we mentioned that *intervals where a polynomial with real coefficients is positive are separated from intervals where it is negative by values of the variable for which it is zero.* This fact can be restated as follows:

Figure 3

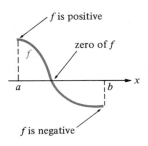

Change-of-Sign Property

> A polynomial function with real coefficients cannot change sign on an interval unless it has a zero in that interval.

The change-of-sign property is illustrated in Figure 3, which shows the graph of a polynomial function f that changes sign on an interval $[a, b]$. Because the graph is a continuous curve, with no jumps or breaks, it must cut across the x axis at least once between a and b.

Example 3   Show that the polynomial function $f(x) = x^3 + x - 1$ has a zero in the interval $[0, 1]$.

Figure 4.

Solution   Because $f(0) = -1$ and $f(1) = 1$, the polynomial function f changes its sign on the interval $[0, 1]$. Therefore, by the change-of-sign property, f has at least one zero in the interval $[0, 1]$.   ■

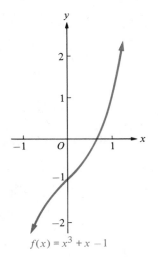

$f(x) = x^3 + x - 1$

The graph of $f(x) = x^3 + x - 1$ (Figure 4) shows that f has exactly one zero in the interval $[0, 1]$.

The change-of-sign property is used in the following *bisection method* for finding zeros of a polynomial function f with real coefficients:

The Bisection Method

> **Step 1.** By sketching a rough graph, or by trial and error using synthetic division (and a calculator if necessary), find a closed interval $[a, b]$ such that $f(a)$ and $f(b)$ have opposite algebraic signs. By the change-of-sign property, f must have a zero in this interval.
>
> **Step 2.** Using the midpoint $c = (a + b)/2$, divide the interval $[a, b]$ into two subintervals $[a, c]$ and $[c, b]$.
>
> **Step 3.** Calculate $f(c)$. If $f(a)$ and $f(c)$ have opposite algebraic signs, then f must have a zero in the interval $[a, c]$. If $f(c) = 0$, then c itself is a zero of f. The only other possibility is that $f(c)$ and $f(b)$ have opposite algebraic signs, in which case f must have a zero in the interval $[c, b]$.

Each time you apply the bisection method, either you find the exact value of a zero of f, or else you isolate a zero of f in an interval half the size of the preceding one. Using the bisection method over and over again, you can locate a zero of f as accurately as you wish by confining it in successively shorter and shorter intervals.

© **Example 4**   In Example 3, we used the change-of-sign property to show that the polynomial function

$$f(x) = x^3 + x - 1$$

has a zero between 0 and 1. Starting with the interval $[0, 1]$, use the bisection method twice in succession to locate a zero of f with more accuracy.

Solution   The midpoint of the interval $[0, 1]$ is $(0 + 1)/2 = 0.5$. A calculation reveals that

$$f(0) = -1 \quad \text{and} \quad f(0.5) = -0.375,$$

so $f(0)$ and $f(0.5)$ have the same algebraic sign. However, $f(0.5)$ and $f(1)$ have opposite algebraic signs:

$$f(0.5) = -0.375 \quad \text{and} \quad f(1) = 1,$$

so f has a zero between 0.5 and 1. Thus, for our second application of the bisection method, we start with the interval $[0.5, 1]$. The midpoint of this interval is $(0.5 + 1)/2 = 0.75$. Now we have

$$f(0.5) = -0.375 \quad \text{and} \quad f(0.75) = 0.171875,$$

so $f(0.5)$ and $f(0.75)$ have opposite algebraic signs. It follows that f has a zero between 0.5 and 0.75.   ■

After the bisection method is used k times in succession, starting from an interval of length one unit, the midpoint of the final interval provides a numerical approximation to a zero of the polynomial function. This approximation can be considered accurate to n decimal places, where n is the greatest integer less than or equal to $(3k - 1)/10$. For instance, $k = 7$ successive bisections will produce an interval whose midpoint approximates the zero to within $n = 2$ decimal places.

© **Example 5**   Find a zero of $f(x) = x^3 + x - 1$ on the interval $[0, 1]$ by using the bisection method seven times in succession. Round off your answer to two decimal places.

Solution   In Example 4 we applied the bisection method twice and located a zero of f on the interval $[0.5, 0.75]$. Five more bisections locate the zero on the interval $[0.6796875, 0.6875]$ with midpoint $(0.6796875 + 0.6875)/2 = 0.68359375$. Rounding off the last number to 2 decimal places, we obtain 0.68 as an approximation to a zero of f.   ■

As Example 5 illustrates, the computation of real zeros of a polynomial function by the bisection method, even with the aid of a calculator, can involve tedious repetition. However, the method is easily adapted to a programmable calculator or computer, and thus provides a reasonably efficient technique for finding real zeros. For instance, after 24 successive applications of the bisection method, a programmable calculator quickly produces 0.6823278 as an approximation, accurate to seven decimal places, for a zero of the polynomial function in Example 5. Appendix IV contains a **BASIC** computer program for carrying out the bisection method.

Problem Set 4.5

In Problems 1 to 14, find all rational zeros of each polynomial function.

1. $f(x) = x^3 - 9x^2 + 23x - 15$

2. $g(x) = x^3 - 3x^2 - 4x + 12$

3. $h(x) = 8x^3 - 25x^2 + 4x - 3$

4. $p(x) = 2x^3 - 15x^2 + 27x - 10$

5. $F(x) = 2x^3 - 3x^2 + 2x + 2$

6. $G(x) = x^3 - 2x^2 - 13x - 10$

7. $f(x) = x^4 - 4x^3 + 7x^2 - 12x + 12$

8. $H(x) = 4x^4 + 4x^3 + 9x^2 + 8x + 2$

9. $f(x) = x^4 - 5x^3 + \frac{1}{4}x^2 + \frac{9}{2}x + \frac{3}{2}$

10. $g(x) = x^5 - x^3 + 27x^2 - 27$

11. $Q(x) = x^5 - 17x^3 + 75x - 9$

12. $f(x) = \frac{1}{2}x^5 + 2x^4 + \frac{1}{2}x^2 - \frac{3}{2}x - 14$

13. $g(x) = 20x^5 - 9x^4 - 74x^3 + 30x^2 + 42x - 9$

14. $q(x) = x^5 + x^4 - \frac{5}{4}x^3 + \frac{25}{4}x^2 - 21x + 9$

In Problems 15 to 18, factor each polynomial function completely into prime factors and ⓒ sketch its graph.

15. $f(x) = x^3 - 6x^2 - x + 6$

16. $g(x) = x^3 - 8x^2 + 2x - 16$

17. $Q(x) = 2x^4 - 3x^3 - 7x^2 + 12x - 4$

18. $G(x) = 16x^4 - 40x^2 + 9$

In Problems 19 to 24, show that the polynomial function has a zero in the interval by using the change-of-sign property. (You need not try to find the value of the zero.)

19. $f(x) = x^5 - 2x^3 - 1$ in $[1, 2]$

ⓒ 20. $g(x) = x^4 + 6x^3 - 18x^2$ in $[2.1, 2.2]$

ⓒ 21. $f(x) = x^5 - 2x^3 - 1$ in $[1.5, 1.6]$

ⓒ 22. $g(x) = x^4 + 6x^3 - 18x^2$ in $[-8.2, -8.1]$

23. $h(x) = x^5 - 2x + 3$ in $[-2, -1]$

ⓒ 24. $p(x) = 6x^4 + 19x^3 - 7x^2 - 26x + 12$ in $[-1.54, -1.53]$

ⓒ 25. In Problem 19, use the bisection method twice in succession, starting with the interval $[1, 2]$, to locate a zero of $f(x) = x^5 - 2x^3 - 1$ with greater accuracy.

ⓒ 26. In Problem 20, use the bisection method twice in succession, starting with the interval $[2.1, 2.2]$, to locate a zero of $g(x) = x^4 + 6x^3 - 18x^2$ with greater accuracy.

ⓒ 27. In Problem 23, use the bisection method three times in succession, starting with the interval $[-2, -1]$, to locate a zero of $h(x) = x^5 - 2x + 3$ with greater accuracy.

ⓒ 28. In Problem 24, find a zero of the function $p(x) = 6x^4 + 19x^3 - 7x^2 - 26x + 12$, between -1.54 and -1.53, rounded off to the third decimal place, by using the bisection method four times in succession, starting with the interval $[-1.54, -1.53]$.

ⓒ 29. As the graph in Figure 5 shows, $f(x) = x^3 - 3x + 1$ has three real zeros. Locate each of these zeros on intervals of length 0.25 units by using the information shown in Figure 5 and the bisection method.

Figure 5

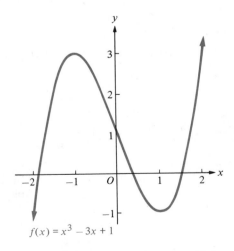

$f(x) = x^3 - 3x + 1$

ⓒ 30. As Figure 6 shows, $g(x) = x^4 + 3x - 13$ has two real zeros. Find these zeros, rounded off to the third decimal place, by using the information shown in Figure 6 and the bisection method eleven times in succession.

Figure 6

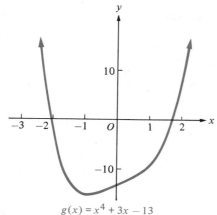

$g(x) = x^4 + 3x - 13$

change-of-sign property, then find this zero, rounded off to two decimal places, by applying the bisection method seven times in succession.

33. $f(x) = 2x^4 - 2x - 3$

34. $g(x) = x^4 - 3x^3 + 3x^2 - 2x - 2$

35. $h(x) = x^4 - 3x^2 - 6x - 2$

36. $p(x) = 2x^5 - 3x^4 + x^2 - 3x + 1$

37. A box is to be constructed so that the length is twice the width and the depth exceeds the width by 2 inches. Find the box dimensions if its volume is 350 cubic inches. [*Hint:* Let x denote the box width, express the length and height in terms of x, and solve the equation length × height × width = 350.]

38. Suppose that c is a real zero of a polynomial function $f(x)$, the coefficients of $f(x)$ are integers, and the leading coefficient of $f(x)$ is 1. Using the rational-zeros theorem, show that either c is an integer, or else c is an irrational number.

39. A plot of land has the shape of a right triangle with a hypotenuse 1 kilometer longer than one of the sides. Find the lengths of the sides of the plot of land if its area is 6 square kilometers.

40. Using the result of Problem 38, show that $\sqrt{2}$ is an irrational number.

c In Problems 31 and 32, the given polynomial has exactly one real zero. Find this zero, rounded off to two decimal places, by using the bisection method seven times in succession.

31. $f(x) = x^3 + x + 1$

32. $g(x) = x^3 + x^2 + x + 5$

c In Problems 33 to 36, locate the largest zero of each function between two successive integers by using the

4.6 RATIONAL FUNCTIONS

Although the sum $f + g$, the difference $f - g$, and the product $f \cdot g$ of polynomial functions f and g are again polynomial functions, the quotient f/g can only be a polynomial function if g is a factor of f. A function of the form $R = f/g$, where f and g are polynomial functions, is called a **rational function.*** In other words, a rational function is a function such as

$$R(x) = \frac{x^3 - 2x^2 + 5x - 4}{7x^2 - 3x + 2},$$

whose values are given by a rational expression in the independent variable. *The domain of a rational function is the set of all real numbers except the zeros of the denominator function.*

* In this section, we consider only rational functions for which the numerator and denominator polynomials have *real* coefficients.

Example 1   Determine the domain of each rational function.

(a) $R(x) = \dfrac{3x^2 + 2}{x + 1}$ (b) $F(x) = \dfrac{x + 2}{x}$

(c) $H(x) = \dfrac{x^3 - 3x^2 - 2x - 1}{4}$ (d) $h(x) = \dfrac{1}{(x + 2)(x - 3)}$

(e) $T(x) = \dfrac{3x - 2}{x^2 + 1}$

Solution   (a) All real numbers except -1.   (b) All real numbers except 0.
(c) All real numbers.   (d) All real numbers except -2 and 3.
(e) All real numbers.   ■

Figure 1

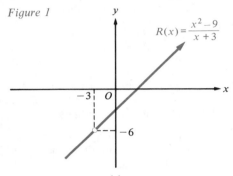

(a)

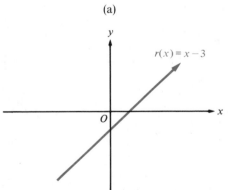

(b)

Recall that two rational expressions are said to be equivalent if one can be obtained from the other by canceling common factors or by multiplying numerator and denominator by the same non-zero polynomial. Strictly speaking, two rational functions defined by equivalent rational expressions may not be the same because their domains may be different. For instance, the rational function R defined by

$$R(x) = \frac{x^2 - 9}{x + 3} = \frac{(x - 3)(x + 3)}{x + 3}$$

isn't really the same as the rational function r defined by $r(x) = x - 3$ because

$$R(x) = \frac{(x - 3)(x + 3)}{x + 3} = x - 3 = r(x)$$

only if $x \neq -3$. Notice that -3 does not belong to the domain of R, but it does belong to the domain of r. The graph of R (Figure 1a) is the same as the graph of r (Figure 1b), except that the point $(-3, -6)$ does not belong to the graph of R.

Functions of the Form $R(x) = 1/x^n$

If n is a positive integer, the domain of the rational function $R(x) = 1/x^n$ consists of all real numbers except 0. Here we consider the graphs of such functions.

Example 2   Sketch the graph of each function.

(a) $F(x) = 1/x$

(b) $G(x) = 1/x^2$

Solution

(a) The function F is odd,

$$F(-x) = \frac{1}{(-x)} = -\frac{1}{x} = -F(x),$$

so its graph is symmetric about the origin. Therefore, we begin by sketching the portion of the graph for $x > 0$, and then we obtain the complete graph by using the symmetry. For small positive values of x, the reciprocal $1/x$ will be large. As x increases, $1/x$ decreases, coming closer and closer to 0 as x gets larger. This variation is shown in Table 1. By using this information and plotting a few points, we obtain the graph of F (Figure 2a). Notice that although the graph comes closer and closer to the x and y axes, it never *touches* them; in other words, it has no x or y intercepts.

(b) The graph of $G(x) = 1/x^2$ is obtained in much the same way as the graph of $F(x) = 1/x$ in part (a). The main differences are that $G(x)$ is always positive and that G is an even function,

$$G(-x) = \frac{1}{(-x)^2} = \frac{1}{x^2} = G(x),$$

so its graph is symmetric about the y axis. Also, the graph of G is steeper than the graph of F near the origin, and it approaches the x axis more rapidly as x becomes large. By using this information and plotting a few points, we obtain the graph of G (Figure 2b). Like the graph of F in part (a), the graph of G has no x or y intercepts. ∎

Table 1

x	$1/x$	$1/x^2$
$\frac{1}{100}$	100	10,000
$\frac{1}{10}$	10	100
$\frac{1}{3}$	3	9
$\frac{1}{2}$	2	4
1	1	1
2	$\frac{1}{2}$	$\frac{1}{4}$
3	$\frac{1}{3}$	$\frac{1}{9}$
10	$\frac{1}{10}$	$\frac{1}{100}$
100	$\frac{1}{100}$	$\frac{1}{10,000}$

Figure 2

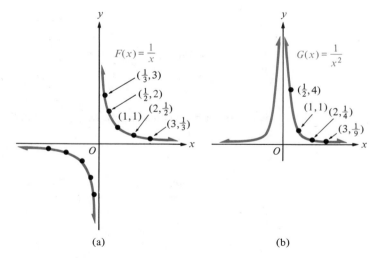

(a)   (b)

The graphs in Figure 2 are typical of the graphs of functions of the form $R(x) = 1/x^n$, where n is a positive integer. If n is odd, the graph resembles that of $F(x) = 1/x$ (Figure 2a), whereas if n is even, the graph resembles that of $G(x) = 1/x^2$ (Figure 2b). For larger values of n, the graphs become steeper near the origin and they approach the x axis more rapidly as one moves away from the origin.

Figure 3

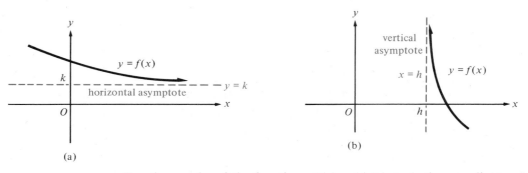

(a)

(b)

Figure 4

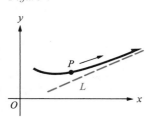

For the graphs of the functions $R(x) = 1/x^n$, $n \geq 1$, the coordinate axes play a very special role. Indeed, they are **asymptotes** of these graphs in the following sense: If the values $f(x)$ of a function f approach a fixed number k as x (or $-x$) gets larger and larger without bound, we say that the line $y = k$ is a **horizontal asymptote** of the graph of f (Figure 3a). Similarly, if $|f(x)|$ gets larger and larger without bound as x approaches a fixed number h, the line $x = h$ is called a **vertical asymptote** of the graph of f (Figure 3b). The general idea of an asymptote is illustrated in Figure 4, which shows a curve "approaching" a line L in the sense that, as the point P moves away from the origin along the curve, the distance between P and L becomes arbitrarily small. An asymptote that is neither horizontal nor vertical is called an **oblique** or **slant asymptote.** Although an asymptote isn't really part of the curve, it helps us to visualize how the curve behaves in distant regions of the xy plane.

Shifting and Stretching

From our discussion in Section 3.6, if a, h, and k are constants and n is a positive integer, then the graph of

$$f(x) = \frac{a}{(x - h)^n} + k$$

is obtained by vertically "stretching" or "flattening" the graph of

$$R(x) = \frac{1}{x^n}$$

by a factor of $|a|$, reflecting it across the x axis if $a < 0$, shifting it horizontally by $|h|$ units, and shifting it vertically by $|k|$ units. After the vertical stretching, flattening, or reflecting, the x and y axes will still be asymptotes. However, *when you shift a graph, you will shift its asymptotes as well; hence, $x = h$ is a vertical asymptote, and $y = k$ is a horizontal asymptote of the graph of f.*

Example 3   For the function

$$f(x) = \frac{3}{x - 2} + 4,$$

find the horizontal and vertical asymptotes of the graph, determine any y and x intercepts, and sketch the graph.

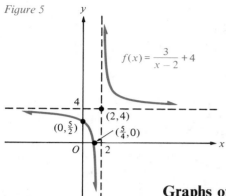

Figure 5

$$f(x) = \frac{3}{x-2} + 4$$

$(2,4)$

$(0,\frac{5}{2})$

$(\frac{5}{4},0)$

Solution The graph will have the same general shape as the graph of $F(x) = 1/x$ in Figure 2a, except that it is stretched vertically by a factor of 3 and then shifted 2 units to the right and 4 units upward (Figure 5). The asymptotes are $x = 2$ and $y = 4$. The y intercept is given by $f(0) = 3/(0 - 2) + 4 = \frac{5}{2}$ and the x intercept is obtained by solving the equation $f(x) = 0$, that is,

$$\frac{3}{x-2} + 4 = 0 \qquad \text{or} \qquad 3 + 4(x - 2) = 0.$$

Thus, $4x - 5 = 0$, so $x = \frac{5}{4}$. ∎

Graphs of General Rational Functions

Graphs of rational functions other than the simple types already considered may be quite difficult to sketch without using calculus. However, in some cases, you can sketch an accurate graph of a rational function

$$f(x) = \frac{p(x)}{q(x)}$$

by carrying out the following systematic procedure:

Step 1. Factor the numerator $p(x)$ and the denominator $q(x)$ and reduce the fraction $p(x)/q(x)$ by canceling all common factors. (Be careful though—as we saw in Figure 1, such common factors might produce "holes" in the graph of f.)

Step 2. Find the intercepts of the graph: If $q(0) \neq 0$, the y intercept is $f(0) = p(0)/q(0)$. Find the real zeros of the numerator p. These real zeros, which are the same as the real zeros of the function f, are the x intercepts of the graph.

Step 3. Find the real zeros of the denominator q. If $q(c) = 0$, then the line $x = c$ is a vertical asymptote of the graph of f. Note that the graph of f can have more than one vertical asymptote, but the graph cannot intersect any of these vertical asymptotes.

Step 4. Check whether the fraction is proper, that is, whether the degree of the numerator is smaller than the degree of the denominator. If the fraction is proper, then the x axis is the only horizontal asymptote of the graph. If the fraction is improper, perform a long division and use the quotient theorem (Section 4.3, Theorem 2) to write $f(x)$ as a sum of a polynomial and a proper fraction. If this polynomial has the form $ax + b$, then the line $y = ax + b$ is an asymptote* of the graph of f; otherwise f has no asymptotes other than the vertical ones found in Step 3.

---

* Note that a graph can intersect its own horizontal or oblique asymptote—see Problems 58 and 64.

Step 5. Check whether f has the form

$$f(x) = \frac{a}{(x-h)^n} + k.$$

If it does, proceed as in Example 3 and omit the following steps.

Step 6. Use the x intercepts and the vertical asymptotes to divide the x axis into intervals; then use convenient test numbers as in Section 2.7 to determine the algebraic sign of $f(x)$ on each of these intervals. This shows where the graph of f lies above or below the x axis.

Step 7. Check to see whether f is either an even or an odd function; if it is, then its graph is symmetric about the y axis or the origin, respectively.

Step 8. Sketch the graph of f by using the information obtained in the preceding steps together with the point-plotting method. Here a calculator may be useful.

© *In Examples 4 to 8, sketch the graph of each rational function.*

Example 4 $f(x) = \dfrac{-1}{x^2 + 4x + 4}$

Solution   We follow the steps in the graph-sketching procedure for rational functions:

Step 1. Here the denominator is a perfect square, and we have

$$f(x) = \frac{-1}{(x+2)^2}.$$

Step 2. The y intercept is given by $f(0) = -\frac{1}{4}$. Because the equation $f(x) = 0$ has no solution, there is no x intercept.

Step 3. The denominator is zero when $x = -2$; hence, the line $x = -2$ is a vertical asymptote.

Figure 6

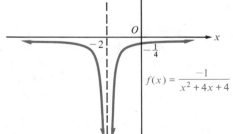

$$f(x) = \frac{-1}{x^2 + 4x + 4}$$

Step 4. The fraction is proper, so the x axis is a horizontal asymptote.

Step 5. The function f has the form

$$f(x) = \frac{a}{(x-h)^n} + k,$$

with $a = -1$, $h = -2$, $n = 2$, and $k = 0$: hence, the graph has the same shape as the graph of $G(x) = 1/x^2$ in Figure 2b, except that it is reflected across the x axis and then shifted 2 units to the left (Figure 6). ■

Example 5 $g(x) = \dfrac{2x - 1}{x + 1}$

Solution

Step 1. The fraction is in reduced form.

Step 2. The y intercept is given by $g(0) = -1$. The x intercept is $\frac{1}{2}$, since $2x - 1 = 0$ when $x = \frac{1}{2}$.

Step 3. The denominator is zero when $x = -1$; hence, the line $x = -1$ is a vertical asymptote.

Step 4. The fraction is improper because its numerator has the same degree as its denominator. We must divide and apply the quotient theorem:

$$x + 1 \overline{\smash{\big)}\ 2x - 1} \qquad \text{so} \qquad g(x) = 2 + \frac{-3}{x + 1}.$$
$$\underline{2x + 2}$$
$$-3$$

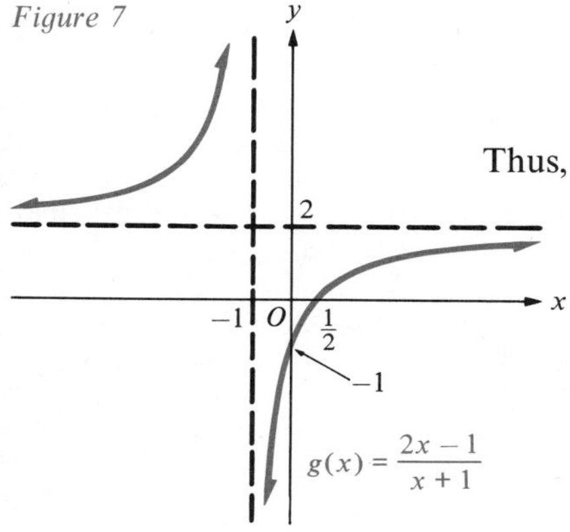

Thus, $y = 2$ is a horizontal asymptote.

Figure 7

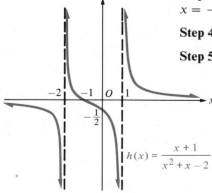

$$g(x) = \frac{2x - 1}{x + 1}$$

Step 5. The function g now has the form

$$g(x) = \frac{a}{(x - h)^n} + k,$$

with $a = -3$, $h = -1$, $n = 1$ and $k = 2$; hence, the graph has the same shape as the graph of $F(x) = 1/x$ in Figure 2a, except that it is stretched vertically by a factor of 3, reflected across the x axis, shifted 1 unit to the left, and shifted 2 units upward (Figure 7). ■

Example 6 $h(x) = \dfrac{x + 1}{x^2 + x - 2}$

Solution

Step 1. Factoring the denominator, we have $h(x) = \dfrac{x + 1}{(x + 2)(x - 1)}.$

Step 2. The y intercept is given by $h(0) = -\frac{1}{2}$. To find the x intercept, we set $x + 1 = 0$ to obtain $x = -1$.

Figure 8

Step 3. The denominator is zero when $x = -2$ and when $x = 1$; hence, the lines $x = -2$ and $x = 1$ are vertical asymptotes.

Step 4. The fraction is proper, so the x axis is a horizontal asymptote.

Step 5. The function does not have the form given in step 5 of the procedure.

Step 6. The x intercept and the vertical asymptotes divide the x axis into the intervals $(-\infty, -2), (-2, -1), (-1, 1)$, and $(1, \infty)$, on which the function values $h(x)$ cannot change algebraic sign. Selecting, say, the test numbers $-3, -1.5, 0$, and 2 on these intervals and evaluating h at these numbers, we find that $h(-3) = -0.5$, $h(-1.5) = 0.4$, $h(0) = -0.5$, and $h(2) = 0.75$. Therefore, the graph of h is above the x axis on the intervals $(-2, -1)$ and $(1, \infty)$, and it is below the x axis on the intervals $(-\infty, -2)$ and $(-1, 1)$.

$$h(x) = \frac{x + 1}{x^2 + x - 2}$$

Step 7. The function h is neither even nor odd.

Step 8. Using the information above and plotting a few points, we obtain the graph shown in Figure 8. ∎

Example 7 $F(x) = \dfrac{x^2 + 2x - 3}{x - 2}$

Solution

Step 1. Factoring the numerator, we have

$$F(x) = \frac{(x - 1)(x + 3)}{x - 2}.$$

Step 2. The y intercept is given by $F(0) = \frac{3}{2}$. To find the x intercepts, we set $(x - 1)(x + 3) = 0$ and conclude that 1 and -3 are the x intercepts.

Step 3. The denominator is zero when $x = 2$; hence, the line $x = 2$ is a vertical asymptote.

Step 4. The fraction is improper because its numerator has degree greater than the degree of the denominator. We must divide and apply the quotient theorem:

$$
\begin{array}{r}
x + 4 \\
x - 2 \overline{\smash{\big)}\ x^2 + 2x - 3} \\
\underline{x^2 - 2x} \\
4x - 3 \\
\underline{4x - 8} \\
5
\end{array}
\qquad \text{so} \qquad F(x) = (x + 4) + \frac{5}{x - 2}.
$$

Thus, $y = x + 4$ is an oblique asymptote.

Step 5. The function does not have the form given in step 5 of the procedure.

Step 6. The x intercepts and the vertical asymptote divide the x axis into the intervals $(-\infty, -3)$, $(-3, 1)$, $(1, 2)$, and $(2, \infty)$, on which the function values $F(x)$ cannot change algebraic sign. Using suitable test numbers, we find that the graph is above the x axis on $(-3, 1)$ and $(2, \infty)$, and below the x axis on $(-\infty, -3)$ and $(1, 2)$.

Step 7. The function F is neither even nor odd.

Step 8. Using the information above and plotting a few points, we obtain the graph shown in Figure 9. ∎

Figure 9

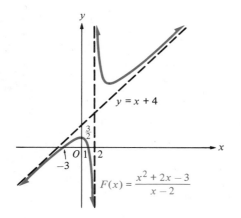

$y = x + 4$

$F(x) = \dfrac{x^2 + 2x - 3}{x - 2}$

Example 8 $G(x) = \dfrac{1}{x^2 + 1}$

Solution   **Step 1.** The denominator $x^2 + 1$ will not factor (over the real numbers) and the fraction is in reduced form.

Step 2. The y intercept is given by $G(0) = 1$. The equation $G(x) = 0$ has no real solution, so there are no x intercepts.

Step 3. The denominator has no real zeros, so there is no vertical asymptote.

Step 4. The fraction is proper, so the x axis is a horizontal asymptote.

Step 5. G does not have the form given in step 5 of the procedure.

Step 6. Since there are no x intercepts and no vertical asymptotes, there is only one interval to consider, namely $(-\infty, \infty)$. Evidently, $G(x)$ is always positive, so the graph is entirely above the x axis.

Step 7. G is an even function, so its graph is symmetric about the y axis.

Step 8. Using the information above and plotting a few points, we obtain the graph shown in Figure 10. ∎

Figure 10

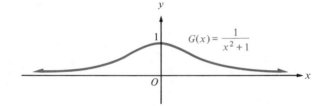

$$G(x) = \frac{1}{x^2 + 1}$$

Problem Set 4.6

In Problems 1 to 10, determine the domain of each rational function.

1. $f(x) = \dfrac{-5}{x^4}$

2. $r(x) = x - \dfrac{1}{x}$

3. $g(x) = \dfrac{-5}{x^2 + 5}$

4. $R(x) = \dfrac{2x^2 + 3x - 1}{(x - 1)(x + 2)}$

5. $G(x) = \dfrac{x^3 - 8}{40}$

6. $H(x) = \dfrac{40}{x^3 - 8}$

7. $P(x) = \dfrac{x^4 - 16}{2x^2 - x + 3}$

8. $T(x) = \dfrac{x^4 + 1}{2x^2 + 9x - 3}$

9. $F(x) = \dfrac{x^2 + x + 1}{x^2 + 8x + 15}$

10. $p(x) = 1 + \dfrac{x^2 + x}{x^3 - x}$

In Problems 11 to 28, the graph of each rational function can be obtained by stretching, flattening, reflecting, and/or shifting the graph of a function of the form $R(x) = 1/x^n$, for $n \geq 1$. In each case, find the horizontal and vertical asymptotes of the graph, determine any y and x intercepts, and sketch the graph (© use a calculator, if you wish, to determine more points on the graph and thus improve the accuracy of your sketch).

11. $H(x) = \dfrac{3}{x}$

12. $g(x) = \dfrac{7}{x^2}$

13. $F(x) = -\dfrac{3}{x}$

14. $G(x) = -7x^{-2}$

15. $p(x) = \dfrac{4}{x^4}$

16. $T(x) = -5x^{-5}$

17. $P(x) = \dfrac{-4}{x^4}$

18. $Q(x) = -7x^{-8}$

19. $f(x) = \dfrac{1}{x} + 1$

20. $F(x) = 2 - \dfrac{1}{x}$

21. $g(x) = \dfrac{1}{x^2} - 4$

22. $G(x) = -1 - \dfrac{1}{x^2}$

23. $p(x) = \dfrac{1}{x - 3}$

24. $P(x) = \dfrac{2}{x + 2}$

25. $Q(x) = \dfrac{-2}{x + 1}$

26. $K(x) = \dfrac{-3}{(x + 5)^2}$

27. $f(x) = \dfrac{3}{x - 2} + 6$

28. $k(x) = \dfrac{-3}{(x + 2)^3} - 4$

ⓒ In Problems 29 to 56, sketch the graph of each rational function by following the steps in the graph-sketching procedure.

29. $f(x) = \dfrac{-4x}{x - 1}$

30. $H(x) = \dfrac{2x + 1}{x - 4}$

31. $G(x) = \dfrac{4x^2 + 24x + 41}{x^2 + 6x + 9}$

32. $F(x) = \dfrac{x^2 - 2x + 1}{x - 1}$

33. $g(x) = \dfrac{1}{x^2 + 4}$

34. $H(x) = \dfrac{4}{x^2 + 9}$

35. $K(x) = \dfrac{-2}{x^2 + 9}$

36. $k(x) = \dfrac{3x^2 + 1}{x^2 - 1} + 1$

37. $Q(x) = \dfrac{1}{2x^2 + 1}$

38. $G(x) = \dfrac{x^2 + 5}{x^2 - 4}$

39. $p(x) = \dfrac{x + 2}{x^2 + 2x - 15}$

40. $F(x) = \dfrac{3}{x^2 - 2x + 1}$

41. $f(x) = \dfrac{2x^2 + 4x + 7}{x^2 + 2x + 1}$

42. $H(x) = \dfrac{-2}{x^2 - 4x + 4} + 5$

43. $g(x) = \dfrac{x^2 - x - 2}{x - 3}$

44. $P(x) = \dfrac{3x^2 - 18x + 28}{x^2 - 6x + 9}$

45. $H(x) = \dfrac{x^2 - 9}{x^2 - 1}$

46. $R(x) = \dfrac{x^2 - 4}{(x - 3)^2}$

47. $f(x) = \dfrac{2x^2 - 3x - 2}{x^2 + x - 2}$

48. $P(x) = \dfrac{x - 1}{x^2 - 2x - 3}$

49. $q(x) = \dfrac{2x}{x^2 + 1}$

50. $F(x) = \dfrac{2x^2 - 3x + 5}{x^2 - x - 2}$

51. $g(x) = \dfrac{x^2 - 1}{x^2 + 5x + 4}$

52. $h(x) = \dfrac{x^2 + x - 6}{x^2 - x - 12}$

53. $Q(x) = \dfrac{x^3 + 5x^2 + 6x}{x + 3}$

54. $G(x) = \dfrac{3x^3 + 5x^2 - 26x + 8}{x^2 + 2x - 8}$

55. $r(x) = \dfrac{x^3 + x^2 + 6x}{x^2 + x - 2}$

56. $K(x) = \dfrac{x^3 - x^2 - 6x}{x^2 - 3x + 2}$

ⓒ **57.** According to Boyle's law, the pressure P and the volume V of a sample of gas maintained at constant temperature are related by an equation of the form $P = K/V$, where K is a numerical constant. If $K = 3.5$, (a) sketch a graph of P as a function of V, (b) determine the domain of this function, and (c) indicate whether P increases or decreases when V increases.

ⓒ **58.** Sketch the graph of the rational function $f(x) = (2x^2 + 1)/(2x^2 - 3x)$. [*Hint:* The graph has a strange bend near $(-\frac{1}{3}, f(-\frac{1}{3}))$. Plot several points for values of x close to $-\frac{1}{3}$ to get an accurate sketch.]

ⓒ **59.** The electrical resistance R per unit length of a wire depends on the diameter d of the wire according to the formula $R = K/d^2$, where K is a constant. If $K = 0.1$, (a) sketch a graph of R as a function of d, (b) determine the domain of this function, and (c) indicate whether R increases or decreases when d decreases.

60. A function of the form $f(x) = (ax + b)/(cx + d)$ where a, b, c, and d are constants and c and d are not both zero, is called a *fractional linear function.* Discuss the graph of f with regard to (a) its x and y intercepts, (b) its horizontal and vertical asymptotes, and (c) its general shape.

© 61. A person learning to type has an achievement record given by $N(t) = 60[(t-2)/t]$ for $3 \leq t \leq 10$, where $N(t)$ is the number of words per minute at the end of t weeks. Sketch the graph of the rational function N.

© 62. The electrical power P in watts consumed by a certain electrical circuit is related to the load resistance R in ohms by the equation $P = 100R(0.5 + R)^{-2}$ for $R \geq 0$. Sketch a graph of P as a function of R and use it to estimate the value of R for which P is maximum.

63. Explain why the graph in Figure 11 cannot be the graph of a rational function.

Figure 11

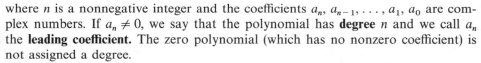

64. Find the point where the graph of $f(x) = \dfrac{2x^3 + x^2 + 3x}{x^2 + 1}$ intersects its own oblique asymptote.

4.7 COMPLEX POLYNOMIALS

If z represents a complex variable, then by a **complex polynomial** in z, we mean an expression of the form

$$f(z) = a_n z^n + a_{n-1} z^{n-1} + \cdots + a_1 z + a_0,$$

where n is a nonnegative integer and the coefficients $a_n, a_{n-1}, \ldots, a_1, a_0$ are complex numbers. If $a_n \neq 0$, we say that the polynomial has **degree** n and we call a_n the **leading coefficient.** The zero polynomial (which has no nonzero coefficient) is not assigned a degree.

Because the complex numbers have the same basic algebraic properties as the real numbers, much of our previous discussion of polynomials in Sections 1.3, 1.4, 2.1, 2.3, 4.3, and 4.4* extends almost verbatim to complex polynomials. In particular, the long-division procedure (page 219) works for complex polynomials, and it follows that the remainder theorem (page 226) and the factor theorem (page 227) hold as well for complex polynomials.

The following theorem, which was proved in 1799 by the great German mathematician Carl Friedrich Gauss (1777–1855), is so important that it is called the **fundamental theorem of algebra.**

Federal Republic of Germany, five deutschemark coin honoring Gauss (1977)

Theorem 1   **The Fundamental Theorem of Algebra**

> Every complex polynomial of degree $n \geq 1$ has a complex zero.

A proof of the fundamental theorem of algebra can be found in *A Modern Course on the Theory of Equations* by David Dobbs and Robert Hanks (Polygonal Publishing House, Passaic, N.J., 1980) on pages 111 to 112. By combining Theorem 1 with the factor theorem, we can obtain the following useful result.

---

* Of course, complex polynomials cannot be graphed in the Cartesian plane, and Descartes' rule of signs does not apply to such polynomials.

Theorem 2   **The Complete Linear Factorization Theorem**

> If $f(z)$ is a complex polynomial of degree $n \geq 1$, then there is a nonzero complex number a and there are complex numbers $c_1, c_2, \ldots, c_n$ such that
>
> $$f(z) = a(z - c_1)(z - c_2) \cdots (z - c_n).$$

Proof   By Theorem 1, there is a complex number c_1 such that

$$f(c_1) = 0.$$

Hence, by the factor theorem, $z - c_1$ is a factor of $f(z)$; that is,

$$f(z) = (z - c_1)Q_1(z),$$

where $Q_1(z)$ is a complex polynomial of degree $n - 1$. If $n - 1 \geq 1$, we can again apply Theorem 1 and the factor theorem to $Q_1(z)$ to obtain

$$Q_1(z) = (z - c_2)Q_2(z),$$

for some complex number c_2, where $Q_2(z)$ is a complex polynomial of degree $n - 2$; therefore,

$$f(z) = (z - c_1)(z - c_2)Q_2(z).$$

If we continue this process, then, after n steps, we arrive at

$$f(z) = (z - c_1)(z - c_2) \cdots (z - c_n)Q_n,$$

where Q_n has degree $n - n = 0$; that is, Q_n is a (complex) constant. Letting $a = Q_n$, we obtain

$$f(z) = a(z - c_1)(z - c_2) \cdots (z - c_n). \qquad \blacksquare$$

According to Theorem 2, every complex polynomial $f(z)$ of degree $n \geq 1$ can be factored completely into a nonzero complex constant a and exactly n linear (first-degree) factors

$$(z - c_1), (z - c_2), \ldots, \text{ and } (z - c_n).$$

Evidently, the complex number a is the leading coefficient of $f(z)$ (why?) and the complex numbers $c_1, c_2, \ldots, c_n$ are the zeros of $f(z)$. There can be no other zeros (Problem 48). The zeros $c_1, c_2, \ldots, c_n$ need not be distinct from one another, but *a complex polynomial $f(z)$ of degree n can have no more than n zeros.*

That the zeros $c_1, c_2, \ldots, c_n$ of $f(z)$ need not be distinct is illustrated by

$$f(z) = z^3 - 3z^2 + 4 = (z + 1)(z - 2)(z - 2),$$

which has the number 2 as a **double zero.** More generally, a complex zero c of a complex polynomial $f(z)$ is called a **zero of multiplicity** m if the linear factor $(z - c)$ appears exactly m times in the complete factorization of $f(z)$. Thus, if a zero of multiplicity m is counted m times, we can conclude from Theorem 2 that *a complex polynomial of degree $n \geq 1$ has exactly n zeros.* For instance,

$$f(z) = 7(z - 3)^4(z - i)^3(z + i)$$

is a complex polynomial of degree 8 with the following zeros:

$$\underbrace{4}_{} \quad \underbrace{3}_{} \quad \underbrace{1}_{} \quad \longleftarrow \text{ multiplicity}$$
$$3, 3, 3, 3, i, i, i, -i.$$

Example 1   Form a polynomial $f(z)$ that has zeros 2, 2, $1 + i$. (Zeros are repeated to show multiplicity.)

Solution
$$f(z) = (z - 2)(z - 2)[z - (1 + i)] = (z^2 - 4z + 4)[z - (1 + i)]$$
$$= z^3 - (5 + i)z^2 + (8 + 4i)z - (4 + 4i)$$

■

As we have seen in Section 2.3, the zeros of a quadratic polynomial with *real* coefficients and a negative discriminant are complex conjugates of each other. More generally, we have the following theorem.

Theorem 3   **The Conjugate Zeros Theorem**

> Let $f(z)$ be a polynomial with real coefficients. Then, if c is a zero of $f(z)$, so is $\bar{c}$, the complex conjugate of c.

Proof   Suppose that
$$f(z) = a_n z^n + a_{n-1} z^{n-1} + \cdots + a_1 z + a_0,$$

where the coefficients $a_n, a_{n-1}, \ldots, a_1$, and a_0 are real numbers. If c is a zero of $f(z)$, then $f(c) = 0$, so
$$a_n c^n + a_{n-1} c^{n-1} + \cdots + a_1 c + a_0 = 0.$$

Using the fact that complex conjugation "preserves" addition and multiplication (see page 55), we take the complex conjugate of both sides of the last equation to obtain
$$\bar{a}_n (\bar{c})^n + \bar{a}_{n-1} (\bar{c})^{n-1} + \cdots + \bar{a}_1 (\bar{c}) + \bar{a}_0 = \bar{0}.$$

Because all of the coefficients are real numbers, they are equal to their own conjugates, and likewise $\bar{0} = 0$, so we have
$$a_n (\bar{c})^n + a_{n-1} (\bar{c})^{n-1} + \cdots + a_1 (\bar{c}) + a_0 = 0.$$

In other words, $\bar{c}$ is also a complex zero of $f(z)$, and our proof is complete.   ■

As a consequence of Theorem 3, we have the following result:

> The nonreal zeros of a polynomial $f(z)$ with real coefficients must occur in conjugate pairs.

Therefore, *a polynomial with real coefficients must have an even number of nonreal zeros.* (*Note:* It may have 0 such nonreal zeros; but 0 is an even integer.) It follows that *a polynomial of odd degree with real coefficients must have at least one real zero* (Problem 49).

Example 2 Form a polynomial in z that has real coefficients, that has the smallest possible degree, and that has 2, -1, and $1 - i$ as zeros.

Solution Since we want a polynomial of the smallest possible degree, we assume that each zero has multiplicity 1. Because we want the coefficients to be real numbers, we must include $1 + i$, the complex conjugate of $1 - i$, as a zero. The polynomial is

$$
\begin{aligned}
f(z) &= (z - 2)[z - (-1)][z - (1 - i)][z - (1 + i)] \\
&= (z - 2)(z + 1)[z^2 - (1 + i)z - (1 - i)z + (1 - i)(1 + i)] \\
&= (z^2 - z - 2)(z^2 - 2z + 2) \\
&= z^4 - 3z^3 + 2z^2 + 2z - 4.
\end{aligned}
$$

■

If w is a complex number, then any complex number z such that $z^2 = w$ is called a *complex square root* of w. Since a complex square root of w is a zero of the polynomial $z^2 - w$, it follows that there are two such roots. These two complex square roots can be calculated by using the following theorem. The proof of this theorem, which is mainly computational, is left as an exercise (Problem 52).

Theorem 4 **Complex Square Roots**

Suppose that $w = p + qi$, where p and q are real numbers, and let $r = \sqrt{p^2 + q^2}$. Then $r + p \geq 0$ and $r - p \geq 0$. Furthermore, if

$$
x = \sqrt{(r + p)/2} \qquad \text{and} \qquad y = \pm\sqrt{(r - p)/2},
$$

where the $\pm$ sign is chosen so that y and q have the same algebraic sign, then

$$
x + yi \qquad \text{and} \qquad -x - yi
$$

are the two complex square roots of w.

Example 3 Find the two complex square roots of $w = 3 + 4i$.

Solution We use Theorem 4 with $p = 3$ and $q = 4$. Then $r = \sqrt{3^2 + 4^2} = 5$,

$$
x = \sqrt{(5 + 3)/2} = 2 \qquad \text{and} \qquad y = \pm\sqrt{(5 - 3)/2} = \pm 1.
$$

Since q is positive, we choose the plus sign, so that $y = 1$. Thus, the two complex square roots of w are

$$
x + yi = 2 + i \qquad \text{and} \qquad -x - yi = -2 - i.
$$

■

The quadratic formula (Section 2.3) is easily extended to complex quadratic polynomials, since the proof by completing the square still works. With the understanding that $\pm\sqrt{b^2 - 4ac}$ stands for two complex square roots of $b^2 - 4ac$, we can still write the quadratic formula

$$
z = \frac{-b \pm \sqrt{b^2 - 4ac}}{2a}
$$

for the roots of the complex quadratic equation

$$az^2 + bz + c = 0.$$

Example 4   Use the quadratic formula to find the complex roots of the quadratic equation

$$z^2 + (4 + 2i)z + (3 + 3i) = 0.$$

Solution   Here $a = 1$, $b = 4 + 2i$, and $c = 3 + 3i$. Thus,

$$b^2 - 4ac = (4 + 2i)^2 - 4(1)(3 + 3i) = 16 + 16i + 4i^2 - 12 - 12i$$
$$= 16 + 16i - 4 - 12 - 12i = 4i.$$

We are going to use Theorem 4 to find $\pm\sqrt{b^2 - 4ac} = \pm\sqrt{4i}$; that is, to find the two complex square roots of $w = 4i = 0 + 4i$. Putting $p = 0$ and $q = 4$ in Theorem 4, we find that $r = \sqrt{0^2 + 4^2} = 4$, $x = \sqrt{(4 + 0)/2} = \sqrt{2}$, and $y = \pm\sqrt{(4 - 0)/2} = \pm\sqrt{2}$. Since q is positive, we have $y = \sqrt{2}$. Therefore, the two complex square roots of $4i$ are

$$x + yi = \sqrt{2} + \sqrt{2}i \qquad \text{and} \qquad -x - yi = -\sqrt{2} - \sqrt{2}i.$$

Hence,

$$z = \frac{-b \pm \sqrt{b^2 - 4ac}}{2a} = \frac{-(4 + 2i) \pm \sqrt{4i}}{2(1)}$$
$$= \frac{-(4 + 2i) \pm (\sqrt{2} + \sqrt{2}i)}{2}.$$

In other words, the two roots are the complex numbers

$$z = \frac{-(4 + 2i) + (\sqrt{2} + \sqrt{2}i)}{2} = \frac{-4 + \sqrt{2}}{2} + \frac{-2 + \sqrt{2}}{2}i$$

and

$$z = \frac{-(4 + 2i) - (\sqrt{2} + \sqrt{2}i)}{2} = \frac{-4 - \sqrt{2}}{2} + \frac{-2 - \sqrt{2}}{2}i.$$ ∎

Example 5   Find all complex zeros of $f(z) = z^3 - 2z^2 - 3z + 10$ and factor $f(z)$ completely into linear factors.

Solution   Since $f(z)$ is a polynomial of degree 3 with real coefficients, we know that it has 3 zeros, at least one of which is a real number. By the rational zeros theorem (Section 4.5), the only possible rational zeros are ± 1, ± 2, ± 5, and ± 10. Testing these possibilities one at a time (using synthetic division for efficiency), we find that -2 is, in fact, a zero:

$$\begin{array}{r|rrrr} -2 & 1 & -2 & -3 & 10 \\ & & -2 & 8 & -10 \\ \hline & 1 & -4 & 5 & 0 \end{array}$$

From this, we also see that the quotient polynomial is

$$z^2 - 4z + 5;$$

that is,

$$z^3 - 2z^2 - 3z + 10 = (z + 2)(z^2 - 4z + 5).$$

By the quadratic formula, the zeros of $z^2 - 4z + 5$ are given by

$$z = \frac{4 \pm \sqrt{16 - 20}}{2} = \frac{4 \pm \sqrt{-4}}{2}$$

$$= \frac{4 \pm 2i}{2} = 2 \pm i.$$

Therefore, the zeros of $z^3 - 2z^2 - 3z + 10$ are

$$-2, \quad 2 + i, \quad \text{and} \quad 2 - i,$$

and the polynomial factors completely into linear factors as follows:

$$z^3 - 2z^2 - 3z + 10 = [z - (-2)][z - (2 + i)][z - (2 - i)]$$
$$= (z + 2)(z - 2 - i)(z - 2 + i).$$ ∎

Problem Set 4.7

In Problems 1 to 6, form a polynomial $f(z)$ that has the indicated zeros. (Zeros are repeated to show multiplicity.)

1. 3, 3, i

2. 3, $1 + i$, $1 + i$

3. 1, $-i$, $-i$

4. 1, 1, $1 + i$, $1 + i$

5. -2, -2, i, $-i$

6. -1, -1, $1 + i$, $1 - i$

In Problems 7 to 16, form a polynomial $f(z)$ that has real coefficients, that has the smallest possible degree, and that has the indicated zeros. (Zeros are repeated to show multiplicity.)

7. 1, 2, i

8. 2, 2, -1, $2 + i$

9. -1, -1, i

10. 0, 2, $-i$, $-i$, $3i$

11. 1, -1, i, $1 + 3i$

12. $\frac{1}{2}$, $2 + 3i$

13. 0, 0, 3, i

14. $\frac{1}{2}$, $\frac{1}{3}$, $\frac{1}{4}$, $\frac{1}{4}$, $-i$

15. $\frac{1}{3}$, i, $1 + i$

16. $\frac{2}{3}$, $\frac{3}{4}$, i, $3 - 2i$

In Problems 17 to 24, find the two complex square roots of each complex number.

17. i

18. $5 + 12i$

19. $4 - 3i$

20. $1 + i$

21. $-i$

22. $1 - i$

23. $\sqrt{3} + i$

24. $2 - \sqrt{5}i$

In Problems 25 to 32, use the quadratic formula to find the complex roots of each quadratic equation.

25. $z^2 - 2z + 2 = 0$

26. $z^2 - 2z + 10 = 0$

27. $z^2 + iz + 1 = 0$

28. $4w^2 - 4w + 3i = 0$

29. $z^2 + iz - 1 = 0$

30. $iz^2 + 2z + i = 0$

31. $z^2 - (3 + i)z + (2 + 2i) = 0$

32. $\frac{\sqrt{3}}{8} w^2 + \frac{\sqrt{2}}{2} w + i = 0$

In Problems 33 to 46, (a) find all of the complex zeros of each complex polynomial $f(z)$, (b) factor $f(z)$ completely into linear factors, and (c) determine the multiplicity of each zero of $f(z)$.

33. $f(z) = z^4 - 1$

34. $f(z) = z^4 - 81$

35. $f(z) = (z + 2)^2(z^2 + 4)$

36. $f(z) = (3z + 1)^5(z^2 - 5z - 6)$

37. $f(z) = (2z + 1)^3(z^2 - 2z - 3)$

38. $f(z) = (z^2 - 4)^2(z^2 + 4z + 13)$

39. $f(z) = z^3 - z^2 + z - 1$

40. $f(z) = z^4 + 4z^3 - 4z^2 - 40z - 33$

41. $f(z) = z^3 - 5z^2 + 8z - 6$

42. $f(z) = z^4 + 4z^3 + 5z^2 + 4z + 4$

43. $f(z) = z^4 - 3z^3 + z^2 + 4$

44. $f(z) = z^4 - 3z^3 + 3z^2 - 3z + 2$

45. $f(z) = 3z^4 - 5z^3 + 2z^2 + 6z - 4$

46. $f(z) = z^4 + 3z^3 + 5z^2 + 12z + 4$

47. If z_1 and z_2 are the complex roots of the quadratic equation $az^2 + bz + c = 0$, show that:
(a) $z_1 + z_2 = -b/a$ (b) $z_1 z_2 = c/a$

48. If $f(z) = a(z - c_1)(z - c_2) \cdots (z - c_n)$, where $a \neq 0$, and if c is a complex zero of $f(z)$, show that c must be one of the complex numbers $c_1, c_2, \ldots,$ or c_n.

49. Show that a polynomial with real coefficients and with odd degree must have at least one real zero.

50. Show that a nonconstant polynomial with real coefficients can be factored into linear and quadratic polynomials with real coefficients, where the quadratic factors have negative discriminants.

51. Given that $-\dfrac{1}{2} + \dfrac{\sqrt{3}}{2} i$ is a zero of the polynomial $f(z) = z^3 - 1$, find the remaining two zeros.

52. Prove Theorem 4.

REVIEW PROBLEM SET, CHAPTER 4

In Problems 1 to 12, find the vertex and the y and x intercepts of the graph of the quadratic function, determine whether the graph opens upward or downward, sketch the graph, and find the domain and range of the function.

1. $f(x) = 4x^2$

2. $g(x) = \frac{1}{4}x^2$

3. $h(x) = -\frac{1}{4}x^2$

4. $F(x) = 3x^2 + 2$

5. $G(x) = 3(x - 2)^2 + 1$

6. $H(x) = \frac{1}{3}[(x + 3)^2 + 2]$

7. $f(x) = x^2 - 3x + 2$

8. $g(x) = 6x^2 + 13x - 5$

9. $h(x) = -6x^2 - 7x + 20$

10. $G(x) = -2x^2 + x + 10$

11. $F(x) = 10x - 25 - x^2$

12. $H(x) = 7x + 2x^2 - 39$

13. Use the graph obtained in Problem 7 to find the solution set of $x^2 - 3x + 2 \geq 0$.

14. Suppose that x_1 and x_2 are the zeros of the quadratic function f. Show that the x coordinate of the vertex of the graph of f is $\frac{1}{2}(x_1 + x_2)$.

15. Use the graph obtained in Problem 9 to find the solution set of $-6x^2 - 7x + 20 \leq 0$.

16. Suppose that f is a quadratic function and that there is a number x_0 with $x_0 \neq 0$ and $f(x_0) = f(-x_0)$. Prove that f is an even function.

17. Find two numbers whose sum is 21 and whose product is maximum.

18. Find all values of K such that the graph of $g(x) = 3x^2 + Kx - 4K$ has no x intercepts.

19. The strength S of a new plastic is given by $S = 500 + 600T - 20T^2$, where T is the temperature in degrees Fahrenheit. At what temperature is the strength maximum?

20. The height of a projectile fired straight upward with an initial speed v_0 is given by the function $h = v_0 t - \frac{1}{2}gt^2$, where t is the elapsed time since the projectile was fired, and the constant g is the acceleration of gravity. At what time does the projectile reach its maximum height, and what is this height?

21. A manufacturer of sports trophies knows that the total cost C dollars of making x thousand trophies is given by $C = 600 + 60x$, and that the corresponding sales revenue R dollars is given by $R = 300x - 4x^2$. Find the number of trophies (in thousands) that will maximize the manufacturer's profit. [*Hint:* Profit = revenue − cost.]

22. A homeowner is planning a rectangular flower garden surrounded by an ornamental fence. The fencing for three sides of the garden costs $20 per meter, but the fencing for the fourth side, which faces the house, costs $30 per meter. If the homeowner has $1200 to spend on the fence, what dimensions of the garden will give it the maximum area?

In Problems 23 to 26, decide which functions are polynomial functions and which are not.

23. $f(x) = \sqrt{3}x - 4x^4 + \frac{1}{2}$

24. $g(x) = 2x^{-2} + 3x^{-1} + x + 1$

25. $h(x) = 2\sqrt{x} + x - 4$

26. $H(x) = (x - 1)(x - 2)(x - 3)$

In each of Problems 27 to 30, sketch the graphs of the functions f, g, F, and G on the same coordinate system.

27. $f(x) = \frac{1}{3}x^3$; $g(x) = -\frac{1}{3}x^3$; $F(x) = \frac{1}{3}x^3 + 1$; $G(x) = \frac{1}{3}(x + 1)^3 + 4$

28. $f(x) = \frac{1}{5}x^4$; $g(x) = -\frac{1}{5}x^4$; $F(x) = \frac{1}{5}x^4 - 1$; $G(x) = \frac{1}{5}(x - 1)^4 + 2$

29. $f(x) = \frac{1}{4}x^6$; $g(x) = -\frac{1}{4}x^6$; $F(x) = \frac{1}{4}x^6 + 4$; $G(x) = \frac{1}{4}(x + 1)^6 + 3$

30. $f(x) = \frac{1}{2}x^7$; $g(x) = -\frac{1}{2}x^7$; $F(x) = \frac{1}{2}x^7 + 2$; $G(x) = \frac{1}{2}(x + 2)^7 - 1$

In Problems 31 to 34, find the y and x intercepts and sketch the graph of each function.

31. $f(x) = (x - 2)^3 + 8$

32. $g(x) = \frac{1}{3}(x + 3)^4 + 2$

33. $F(x) = \frac{1}{4}(x + 1)^5 - 8$

34. $G(x) = -\frac{2}{3}(x - 5)^6 - \frac{1}{2}$

In Problems 35 to 42, (a) find the x intercepts of the graph of the function, (b) sketch the graph, and (c) use the graph to solve the given inequality. ([C] Use a calculator if you wish.)

35. $f(x) = x(x^2 - 9)$; $f(x) \le 0$

36. $g(x) = -(3x - 1)(x + 1)(x + 2)$; $g(x) \ge 0$

37. $F(x) = x(x^2 + 7x + 10)$; $F(x) > 0$

38. $H(x) = -x(3x^2 - 7x + 2)$; $H(x) < 0$

39. $G(x) = (x + 1)(10x^2 - 3x - 18)$; $G(x) \ge 0$

40. $f(x) = -(x^2 - 4)(x^2 - 9)$; $f(x) < 0$

41. $h(x) = -(x + 3)(2 - x)^2$; $h(x) > 0$

42. $g(x) = x^2(x - 1)^2(x + 2)^2$; $g(x) \le 0$

In Problems 43 to 48, use long division to find the quotient polynomial $q(x)$ and the remainder polynomial $r(x)$ when $f(x)$ is divided by $g(x)$. Check your work by verifying that $r(x)$ is either the zero polynomial or a polynomial of degree less than $g(x)$, and that $f(x) = g(x)q(x) + r(x)$.

43. $f(x) = x^2 + 5x + 2$; $g(x) = x + 3$

44. $f(x) = 6x^4 + 38x^3 + 44x^2 - 96x + 27$; $g(x) = 2x + 6$

45. $f(x) = x^5 - 32$; $g(x) = x^3 - 8$

46. $f(x) = 8x^3 - 5x^2 - 51x - 18$; $g(x) = 2x^3 + x^2 + 1$

47. $f(x) = 6x^6 + 9x^5 + x^4 - 3x^3 + 3x^2 + 6x + 1$; $g(x) = 2x^2 + 3x + 1$

48. $f(x) = 3x^4 - 4x^3 + 5x^2 + x + 7$; $g(x) = 2x^2 + x + 2$

In Problems 49 to 52, use the quotient theorem to rewrite each rational expression as the sum of a polynomial and a proper rational expression.

49. $\dfrac{4x^4 + x^3 - 2x + 3}{x + 1}$

50. $\dfrac{x^3 + 1}{x^2 + 1}$

51. $\dfrac{x^3 + 6x^2 + 10x}{x^2 + x + 2}$

52. $\dfrac{2x^4 + 3x^2 + 4x + 2}{x^4 + x^3 + x^2 + x + 1}$

In Problems 53 and 54, divide $f(x)$ by $g(x)$ to obtain a quotient polynomial $q(x)$ and a constant remainder R, (a) using long division and (b) using synthetic division.

53. $f(x) = 3x^4 + 2x^3 + x^2 + x + 2$; $g(x) = x + 2$

54. $f(x) = x^5 - 3x^3 + 5x^2 - 12$; $g(x) = x - \frac{1}{2}$

In Problems 55 to 60, (a) use synthetic division to obtain the quotient polynomial $q(x)$ and the constant remainder R upon division of the given polynomial by $x - c$ for the indicated value of c, and Ⓒ (b) verify by direct substitution that R is the value of the polynomial when $x = c$.

55. $x^3 - 2x^2 + 3x - 5$; $c = -2$

56. $x^3 - 2x^2 + 7x - 1$; $c = -3$

57. $2x^3 - x^2 + 7$; $c = 4$

58. $5x^4 - 10x^3 - 12x - 7$; $c = -4$

59. $x^5 + 5x - 13$; $c = -1$

60. $3x^5 + 5x^4 - 2x^3 + x^2 - x + 1$; $c = 2$

Ⓒ In Problems 61 and 62, use synthetic division, the remainder theorem, and a calculator to find the indicated number rounded off to two decimal places.

61. $f(2.72)$ if
$f(x) = 17.1x^4 + 33.3x^3 - 2.75x^2 + 11.1x + 21.8$

62. $g(3.14)$ if
$g(x) = 13.5x^8 - 31.7x^6 + 22.1x^4 - 35.7x^2 + 21.2$

In Problems 63 to 68, Ⓒ (a) use the remainder theorem to find the remainder R when each division is performed, and (b) verify your result using synthetic division.

63. $x^3 - 2x^2 + 3x - 5$ divided by $x + 2$

64. $x^3 - 2x^2 + 7x - 1$ divided by $x + 3$

65. $2x^3 - x^2 + 7$ divided by $x + 4$

66. $5x^4 - 10x^3 - 12x - 7$ divided by $x - 4$

67. $x^5 + 5x - 13$ divided by $x - 1$

68. $3x^5 + 5x^4 - 2x^3 + x^2 - x + 1$ divided by $x - 2$

In Problems 69 to 72, use the factor theorem to determine (without actually dividing) whether or not the indicated binomial is a factor of the given polynomial.

69. $x^4 - 4x - 69$; $x - 3$

70. $x^3 - 2x^2 + 1$; $x - 1$

71. $x^3 - 2x^2 + 3x + 4$; $x - 2$

72. $2x^4 + 3x - 26$; $x + 3$

In Problems 73 to 76, the indicated number is a zero of the polynomial function $f(x)$. Determine the multiplicity of this zero.

73. 1; $f(x) = x^4 - 4x^3 + 7x^2 - 6x + 2$

74. -2; $f(x) = x^5 + 7x^4 + 16x^3 + 8x^2 - 16x - 16$

75. -1; $f(x) = 3x^4 + 5x^3 - 3x^2 - 9x - 4$

76. $\frac{2}{3}$; $f(x) = 27x^4 - 27x^3 - 18x^2 + 28x - 8$

In Problems 77 to 80, determine the number of variations in sign for $f(x)$ and $f(-x)$; use Descartes' rule of signs to determine the possible number of positive and negative zeros of $f(x)$; and find integers that are upper and lower bounds for the real zeros of $f(x)$.

77. $f(x) = x^3 + 2x^2 - x - 2$

78. $f(x) = x^3 - x^2 - 5x - 3$

79. $f(x) = 5x^4 - 2x^3 - 5x + 2$

80. $f(x) = 6x^4 - 7x^3 - 7x^2 + 6x + 1$

In Problems 81 to 86, find all rational zeros of each polynomial function.

81. $f(x) = x^3 - 8x^2 + 5x + 14$

82. $g(x) = x^3 - 4x^2 - 5x + 14$

83. $F(x) = x^4 - 4x^3 - 5x^2 + 36x - 36$

84. $G(x) = 4x^3 - 19x^2 + 32x - 15$

85. $h(x) = 4x^4 - 4x^3 - 7x^2 + 4x + 3$

86. $H(x) = 4x^4 - 2x^3 + 2x^2 + 10x + 3$

In Problems 87 to 90, factor each polynomial function completely into prime factors, and Ⓒ sketch the graph of the function.

87. $f(x) = x^3 + 2x^2 - x - 2$

88. $F(x) = x^3 - x^2 - 14x + 24$

89. $g(x) = x^4 - 4x^3 - 14x^2 + 36x + 45$

90. $G(x) = 2x^4 - x^3 - 14x^2 + 19x - 6$

© In Problems 91 to 94, locate the largest zero of each function between two successive integers by using the change-of-sign property, then find this zero, rounded off to two decimal places, by applying the bisection method seven times in succession.

91. $f(x) = x^3 - x^2 + 3x - 2$

92. $g(x) = 2x^3 - 5x^2 + 4x + 1$

93. $h(x) = 2x^4 - 3x^3 - 2x^2 - 4x + 1$

94. $p(x) = x^5 + 2x^4 + 4x^3 - 2x^2 + 3x - 4$

© **95.** Find the original length of the edge of a cube if, after a slice 1 centimeter thick is cut from one side, the volume of the remaining solid is 448 cubic centimeters.

96. Corresponding to each polynomial function f is another polynomial function f' called its *derivative*, such that for each real number c, $f'(c) = q(c)$, where $q(x)$ is the quotient polynomial obtained by dividing $f(x)$ by $x - c$.

   (a) Find the derivative of $f(x) = 3x^2 + 2x + 7$.

   (b) Find the derivative of
   $f(x) = Ax^3 + Bx^2 + Cx + D$.

97. Let f be a polynomial function and suppose that a and b are constant real numbers with $a \neq b$. Show that the remainder upon division of $f(x)$ by $(x - a)(x - b)$ has the form

$$Ax + B,$$

where $A = \dfrac{f(b) - f(a)}{b - a}$ and $B = \dfrac{bf(a) - af(b)}{b - a}$.

98. Let f be a polynomial function and suppose that f' is the derivative of f, as in Problem 96. If c is a constant, show that the remainder upon division of $f(x)$ by $(x - c)^2$ has the form

$$Ax + B,$$

where $A = f'(c)$ and $B = f(c) - cf'(c)$.

In Problems 99 to 112, indicate the domain of each rational function, find the y and x intercepts of its graph, determine all the asymptotes of the graph, and © sketch it.

99. $f(x) = -\dfrac{2}{x}$

100. $g(x) = 1 - \dfrac{4}{x}$

101. $G(x) = 1 - \dfrac{3}{x^2}$

102. $P(x) = 2 + \dfrac{4}{x^2}$

103. $F(x) = \dfrac{2x}{x - 1}$

104. $H(x) = \dfrac{3x + 2}{2 - x}$

105. $p(x) = \dfrac{(x + 2)^2}{x^2 + 2x}$

106. $G(x) = \dfrac{4x^2}{x^2 - 4x}$

107. $T(x) = \dfrac{x^2 + 1}{x^2 - 3x}$

108. $k(x) = \dfrac{x^2 + 5x + 4}{x^2 + 5x}$

109. $f(x) = \dfrac{x - 1}{x^2 - 9}$

110. $g(x) = \dfrac{2x^2 - 5x - 3}{x^2 + 3x + 2}$

111. $h(x) = \dfrac{x^2 - 2x - 3}{x - 2}$

112. $F(x) = \dfrac{x^3 + x^2 - 2}{x^2 - 2x - 8}$

© **113.** If a resistor of resistance x ohms, $x \geq 0$, is connected in parallel with a resistor of resistance 1 ohm, the resulting net resistance y is given by $y = x/(1 + x)$ (Figure 1).

   (a) Sketch a graph of y as a function of x.

   (b) Determine the domain and the range of this function.

   (c) Indicate whether y increases or decreases when x increases.

Figure 1

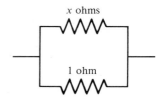

© **114.** According to the Doppler effect in physics, if a source of sound of frequency n is moving away from an observer with speed u, the frequency N of the sound heard by the observer is given by

$$N = \dfrac{n}{1 + \dfrac{u}{v}}, \text{ where } v \text{ is the speed of sound in air.}$$

If $v = 768$ miles per hour and $n = 440$ hertz, sketch the graph of N as a function of u for $0 \leq u \leq 100$ miles per hour.

In Problems 115 to 120, (a) find all complex zeros of each complex polynomial $f(z)$, (b) factor $f(z)$ completely into linear factors, and (c) determine the multiplicity of each zero of $f(z)$.

115. $z^3 - 9z - 10$ **116.** $z^4 - 8z^2 + 9z - 2$

117. $z^4 + 5z^3 + 8z^2 + z - 15$

118. $z^4 - 8z^2 - 4z + 3$

119. $z^4 - 6z^3 + 3z^2 + 24z - 28$

120. $z^4 - 4z^3 + 5z^2 - 4z + 4$

In Problems 121 to 126, form a polynomial in z that has real coefficients, that has the smallest possible degree, and that has the indicated zeros. (Zeros are repeated to show multiplicity.)

121. $1, -2, 1 + i$ **122.** $0, 0, i$

123. $2, 2, 2, i$ **124.** $0, 2i, 4, -3i$

125. $3 + \sqrt{2}i, 1 - \sqrt{2}i$ **126.** $-1 - i, \sqrt{2}, -\sqrt{2}$

Chapter 5

Exponential and Logarithmic Functions

Until now, we have considered only functions defined by equations involving algebraic expressions. Although such *algebraic functions* are useful and important, they are not sufficient for all the requirements of applied mathematics. In this chapter we begin our study of the *transcendental functions*—that is, functions that transcend (go beyond) purely algebraic methods. Here we consider the exponential and logarithmic functions and give some of their many applications to the life sciences, finance, earth sciences, engineering, electronics, and other fields.

5.1 EXPONENTIAL FUNCTIONS

In Section 4.2, we studied power functions

$$p(x) = x^n$$

in which the base is the variable x and the exponent n is constant. In this section we shall study functions of the form

$$f(x) = b^x$$

in which the exponent is the variable x and the base b is constant. Such a function is called an **exponential function.***

If b is a positive real number, then b^x is defined as in Section 1.7 for all **rational** values of x. It is possible to extend the definition so that **irrational** numbers can also be used as exponents. Although the technical details of the extended definition depend on methods studied in calculus, the basic idea is quite simple: *If $b > 0$ and x is an irrational number, then*

$$b^x \approx b^r,$$

* In BASIC, the exponential function with base B is written B^X.

where r is a rational number obtained by rounding off x to a finite number of decimal places. Better and better approximations to b^x are obtained by rounding off x to more and more decimal places. For instance,

$$b^\pi \approx b^{3.14},$$

and a better approximation is given by

$$b^\pi \approx b^{3.14159}.$$

If b is a positive constant, and if you plot several points (x, b^x) for rational values of x, you will notice that these points seem to lie along a smooth curve. This curve is the graph of the exponential function $f(x) = b^x$ with base b. For instance, taking $b = 2$ and plotting several points $(x, 2^x)$ for rational values of x, we obtain Figure 1a. In Figure 1b, we have connected these points with a smooth curve to obtain the graph of $f(x) = 2^x$.

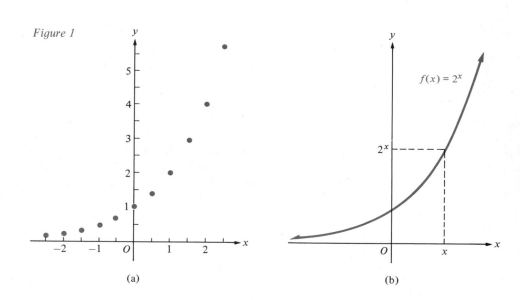

Figure 1

(a)   (b)

Example 1   Sketch the graphs of $g(x) = 3^x$ and $G(x) = (\frac{1}{3})^x$.

Solution   We begin by calculating values of $g(x) = 3^x$ and of $G(x) = (\frac{1}{3})^x$ for several integer values of x, as shown in the table in Figure 2. Then we plot the corresponding points and connect them by smooth curves to obtain the graphs of $g(x) = 3^x$ (Figure 2a) and $G(x) = (\frac{1}{3})^x$ (Figure 2b). Because

$$G(x) = \left(\frac{1}{3}\right)^x = \frac{1}{3^x} = 3^{-x} = g(-x),$$

these curves are reflections of each other across the y axis.

Figure 2

x	3^x	$(\frac{1}{3})^x$
-2	$\frac{1}{9}$	9
-1	$\frac{1}{3}$	3
0	1	1
1	3	$\frac{1}{3}$
2	9	$\frac{1}{9}$

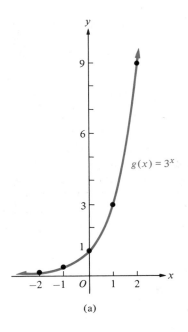

(a)

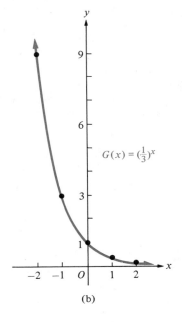

(b)

In general, graphs of exponential functions have shapes similar to the graphs in Figure 2. Thus, if $b > 1$, the graph of $f(x) = b^x$ is rising to the right (Figure 3a), while if $0 < b < 1$, the graph is falling to the right (Figure 3b). Of course, when $b = 1$, the graph is neither rising nor falling (Figure 3c). Notice that the graph of $f(x) = b^x$ always contains the point $(1, b)$ (because $b^1 = b$) and that its y intercept is always 1 (because $b^0 = 1$).

Figure 3

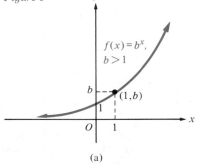

(a)

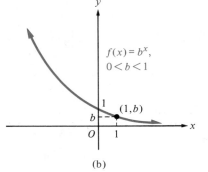

(b)

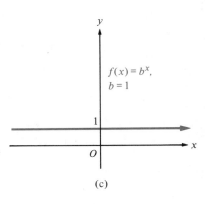

(c)

If $b > 0$, the domain of the exponential function $f(x) = b^x$ is $\mathbb{R}$. If $b > 1$ (Figure 3a), the graph of $f(x) = b^x$ comes as close to the x axis as we please if we move far enough to the left of the origin, but the curve never reaches the axis; in other words, the x axis is a horizontal asymptote. There is no vertical asymptote. As we move farther and farther to the right of the origin, the graph climbs higher and higher without bound; hence, the range of $f(x) = b^x$ is the interval $(0, \infty)$. Similar remarks apply to the graph of $f(x) = b^x$ for $0 < b < 1$ (Figure 3b).

Example 2   Sketch the graphs of $f(x) = 2^x$ and $F(x) = (\frac{1}{2})^x$ on the same coordinate system.

Solution   The graph of $f(x) = 2^x$ has already been sketched in Figure 1b. Notice that this graph is similar to the graph of $g(x) = 3^x$ in Figure 2a, except that it does not rise as rapidly. Because

$$F(x) = (\tfrac{1}{2})^x = 2^{-x} = f(-x),$$

it follows that the graph of $F(x) = (\frac{1}{2})^x$ is obtained by reflecting the graph of $f(x) = 2^x$ across the y axis (Figure 4). In both cases, the x axis is a horizontal asymptote and the y intercept is 1.   ■

Figure 4

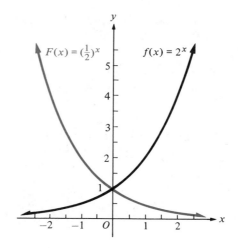

Of course, you can sketch graphs of exponential functions more accurately if you plot more points. For this purpose, a calculator with a y^x key is a most useful tool.

© **Example 3**   Using a calculator with y^x key, evaluate:

(a) $\sqrt{2}^{\sqrt{3}}$ (b) $\pi^{-\sqrt{2}}$

Solution   On a 10-digit calculator, we obtain

(a) $\sqrt{2}^{\sqrt{3}} = 1.414213562^{1.732050808} = 1.822634654$

(b) $\pi^{-\sqrt{2}} = 3.141592654^{-1.414213562} = 0.198117987$ ■

The properties of rational exponents (page 48) continue to hold for all real exponents, provided that all bases are positive. For instance, if a and b are positive, we have

$$a^x a^y = a^{x+y}, \qquad (a^x)^y = a^{xy}, \qquad \text{and} \qquad (ab)^x = a^x b^x,$$

for all real values of x and y.

© **Example 4** Using a calculator with a y^x key, verify that $\pi^{\sqrt{2}} \pi^{\sqrt{3}} = \pi^{\sqrt{2} + \sqrt{3}}$.

Solution On a 10-digit calculator, we obtain

$$\pi^{\sqrt{2}} \pi^{\sqrt{3}} = (5.047497266)(7.262545040) = 36.65767623$$

and

$$\pi^{\sqrt{2} + \sqrt{3}} = 3.141592654^{3.146264370} = 36.65767624.$$

The discrepancy in the last decimal place is due to accumulated error caused by the rounding off. ∎

The techniques of graph sketching presented in Section 3.6 and illustrated throughout Chapter 4 can be applied to exponential functions.

In Examples 5 and 6, sketch the graph of the given function; determine its domain, its range, and any horizontal or vertical asymptotes; and indicate where the function is increasing or decreasing. © *Use a calculator if you wish to increase the accuracy of your sketch.*

Example 5 $h(x) = 3^{x+2}$

Solution The graph of $g(x) = 3^x$ has already been sketched in Figure 2a. Since $g(x + 2) = 3^{x+2}$, we have

$$h(x) = g(x + 2).$$

Therefore, the graph of h is obtained by shifting the graph of g two units to the left. Because

$$h(x) = 3^{x+2} = 3^x \cdot 3^2 = 9 \cdot 3^x = 9g(x),$$

the graph of h can also be obtained from the graph of g by multiplying each ordinate by 9. Using this information, and plotting a few points with the aid of a calculator, we obtain the graph shown in Figure 5. Evidently, the domain of h is $\mathbb{R}$, the range is the interval $(0, \infty)$, the x axis is a horizontal asymptote, there is no vertical asymptote, and h is increasing throughout its domain. ∎

Figure 5

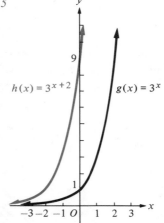

$h(x) = 3^{x+2}$

$g(x) = 3^x$

Example 6 $k(x) = -2^x + 1$

Solution   The graph of $y = -2^x$ is obtained by reflecting the graph of $y = 2^x$ (Figure 1b) across the x axis (Figure 6), and the graph of $k(x) = -2^x + 1$ is obtained by shifting the graph of $y = -2^x$ one unit upward (Figure 6). Evidently, the domain of k is $\mathbb{R}$, the range is the interval $(-\infty, 1)$, the line $y = 1$ is a horizontal asymptote, there is no vertical asymptote, and k is decreasing throughout its domain. ■

Figure 6

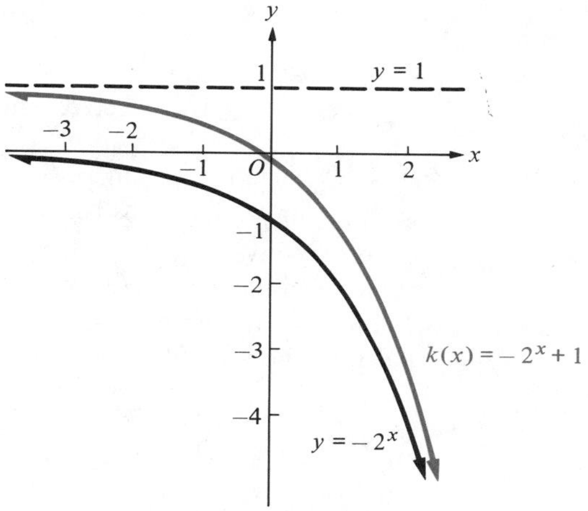

The following examples illustrate the use of exponential functions in the world of business, investment, and finance.

© **Example 7**   Bankers use the **compound interest** formula

$$S = P\left(1 + \frac{r}{n}\right)^{nt}$$

for the **final value** S dollars of a **principal** P dollars invested for a **term** of t years at a **nominal annual interest rate** r **compounded** n **times per year.** If you invest $P = \$500$ at a nominal annual interest rate of 8% (that is, $r = 0.08$) compounded quarterly ($n = 4$), what is the final value S of your investment after a term of $t = 3$ years?

Solution   Using a calculator, we find that

$$S = P\left(1 + \frac{r}{n}\right)^{nt} = 500\left(1 + \frac{0.08}{4}\right)^{4(3)} = 500(1.02)^{12} = \$634.12.$$ ■

Example 8   When a bank offers compound interest, it usually specifies not only the nominal annual interest rate r but also the **effective** simple annual interest rate R; that is, the rate of simple annual interest that would yield the same final value over a 1-year term as the compound interest. The formula

$$R = \left(1 + \frac{r}{n}\right)^n - 1$$

is used to calculate R in terms of r and n. Find the effective simple annual interest rate R corresponding to a nominal annual interest rate of 12% (that is, $r = 0.12$) compounded semiannually ($n = 2$).

Solution $R = \left(1 + \frac{r}{n}\right)^n - 1 = \left(1 + \frac{0.12}{2}\right)^2 - 1 = (1.06)^2 - 1 = 0.1236;$

in other words, the effective simple annual interest rate is 12.36%.   ∎

Money that you will receive in the future is worth *less* to you than the same amount of money received now, because you miss out on the interest you could collect by investing the money now. For this reason, we use the idea of the *present value* of money to be received in the future. If you have an opportunity to invest P dollars at a nominal annual interest rate r compounded n times a year, this principal plus the interest it earns will amount to S dollars after t years, as given by

$$S = P\left(1 + \frac{r}{n}\right)^{nt}.$$

Thus, P dollars in hand *right now* is worth S dollars to be received t years *in the future*. Solving the equation above for P in terms of S, we get an equation for the **present value** of an offer of S dollars to be received t years in the future:

$$P = S\left(1 + \frac{r}{n}\right)^{-nt}.$$

You could do just as well by investing P dollars now and collecting S dollars from your investment after t years.

© **Example 9**   Find the present value of $500 to be paid to you 2 years in the future, if investments during this period are earning a nominal annual interest rate of 10% compounded monthly.

Solution   Here $S = 500$, $r = 0.10$, $n = 12$, $t = 2$, and

$$P = S\left(1 + \frac{r}{n}\right)^{-nt} = 500\left(1 + \frac{0.10}{12}\right)^{-12(2)}$$

$$= 500\left(1 + \frac{1}{120}\right)^{-24} = \$409.70.$$ ∎

The idea of present value helps people to make intelligent investment choices. It allows the investor to translate various complex arrangements for future payments into single figures that are easy to compare.

Problem Set 5.1

In each problem set, problems with colored numbers constitute a good representation of the main ideas of the section.

1. By finding values of 4^x for $x = -2, -\frac{3}{2}, -1, -\frac{1}{2}, 0, \frac{1}{2}, 1, \frac{3}{2}$, and 2, plotting the resulting points $(x, 4^x)$, and drawing a smooth curve through these points, sketch the graph of $f(x) = 4^x$.

2. (a) Using the graph obtained in Problem 1 and approximating $\sqrt{2}$ as 1.4, find the approximate value of $4^{\sqrt{2}}$.

   C (b) Using a calculator with a y^x key, find the value of $4^{\sqrt{2}}$ to as many decimal places as you can.

3. Sketch the graph of $F(x) = (\frac{1}{4})^x$.

4. Sketch graphs of $g(x) = 5^x$ and $G(x) = (\frac{1}{5})^x$ on the same coordinate system.

C In Problems 5 and 6, use a calculator with a y^x key to find the value of each quantity to as many significant digits as you can.

5. (a) $2^{\sqrt{2}}$ (b) $2^{-\sqrt{2}}$ (c) 2^π (d) $2^{-\pi} - \pi^{-2}$

6. (a) $\sqrt{2}^{\sqrt{2}}$ (b) π^π
   (c) $\sqrt{3}^{-\sqrt{5}}$ (d) $3.0157^{2.7566}$

In Problems 7 to 20, sketch the graph of the given function; determine its domain, its range, and any horizontal or vertical asymptotes; and indicate whether the function is increasing or decreasing. (C Use a calculator if you wish.)

7. $f(x) = 2^x + 1$

8. $g(x) = (\frac{2}{3})^x$

9. $h(x) = 4^x - 1$

10. $F(x) = (0.2)^x$

11. $G(x) = 3 \cdot 2^x$

12. $H(x) = 3 \cdot 2^{-x}$

13. $f(x) = 2^{-x} - 3$

14. $g(x) = \frac{3}{2}(\frac{1}{3})^{-x}$

15. $h(x) = 2^{x-3}$

16. $F(x) = 3 - 2^{x-1}$

17. $g(x) = 2^{x-3} + 3$

18. $H(x) = (0.3)^x - 1$

19. $k(x) = (0.5)^x + 2$

20. $K(x) = 3^x - 2$

C In Problems 21 to 26, use a calculator with a y^x key to verify each equation for the indicated values of the variables.

21. $a^x a^y = a^{x+y}$ for $a = 3.074, x = 2.183, y = 1.075$

22. $a^{x+y} = a^x a^y$ for $a = 2.471, x = 5.507, y = 0.012$

23. $(a^x)^y = a^{xy}$ for $a = 1.777, x = -2.058, y = 3.333$

24. $a^{x-y} = \dfrac{a^x}{a^y}$ for $a = \sqrt{2}, x = \sqrt{5}, y = \sqrt{3}$

25. $(ab)^x = a^x b^x$ for $a = \sqrt{7}, b = \pi, x = \sqrt{\pi}$

26. $\left(\dfrac{a}{b}\right)^x = \dfrac{a^x}{b^x}$ for $a = 2 + \pi, b = \sqrt{2} - 1, x = \sqrt{5} - \sqrt{3}$

In Problems 27 and 28, use the properties of exponents to simplify each expression.

27. $(2^x + 2^{-x})^2 - (2^x - 2^{-x})^2$

28. $(b^x - b^{-x})(b^y + b^{-y}) + (b^y - b^{-y})(b^x + b^{-x})$

C In Problems 29 to 34, assume that you have invested a principal P dollars at a nominal annual interest rate r compounded n times per year for a term of t years. Calculate (a) the final value $S = P\left(1 + \dfrac{r}{n}\right)^{nt}$ dollars of your investment and (b) the effective simple annual interest rate $R = \left(1 + \dfrac{r}{n}\right)^n - 1$.

29. $P = \$1000, r = 0.07\ (7\%), n = 1, t = 13$ years

30. $P = \$1000, r = 0.07\ (7\%), n = 12, t = 13$ years

31. $P = \$1000, r = 0.12\ (12\%), n = 1, t = 13$ years

32. $P = \$1000$, $r = 0.12$ (12%), $n = 52$, $t = 13$ years

33. $P = \$50,000$, $r = 0.135$ (13.5%), $n = 12$, $t = \frac{1}{2}$ year

34. $P = \$25,000$, $r = 0.155$ (15.5%), $n = 52$, $t = \frac{1}{4}$ year

© **35.** Suppose that a bank offers to pay a nominal annual interest rate of 0.08 (8%) on money left on deposit for 2 years. Assume that a principal $P = \$1000$ is deposited. Find the final value S after the 2-year term if the interest is compounded (a) annually, (b) semiannually, (c) quarterly, (d) monthly, (e) weekly, (f) daily, and (g) hourly.

© **36.** Find out the nominal annual interest rate r offered by your local savings bank for regular savings accounts and the number of times n per year that the interest is compounded. Calculate the effective simple annual interest rate R.

© **37.** Find the present value of $1000 five years in the future at a nominal annual rate of 12% ($r = 0.12$) compounded weekly.

© **38.** A fund compounds interest quarterly. If a principal $P = \$14,000$ yields a final value $S = \$45,510$ after a 5-year term, (a) find the nominal annual interest rate r, and (b) find the effective simple annual interest rate R.

© **39.** Suppose that someone owes you money and that your local savings bank offers savings accounts at 8% nominal annual interest compounded monthly. Use this interest rate to determine the present value to you of money offered in the future. If your debtor offers to pay you $100 six months from now, what is the present value to you of this offer?

© **40.** On Leroy's 16th birthday, his father promises to give him $25,000 when he turns 21 to help set him up in business. Local banks are offering savings accounts at a nominal annual interest rate of 8% compounded quarterly. Leroy, who has studied the mathematics of finance, says, "Dad, I'll settle for _____ dollars right now!" Fill in the blank appropriately.

5.2 THE EXPONENTIAL FUNCTION WITH BASE e

If we increase the base b, the graph of the exponential function $f(x) = b^x$ rises more rapidly. This is illustrated in Figure 1 for $b = 2$ and $b = 3$. Since the graphs of all exponential functions contain the point $(0, 1)$, a good indication of how rapidly such a graph rises is its "steepness" at this point. The steepness of the graph of $f(x) = b^x$ at the point $(0, 1)$ can be measured by the slope m of its **tangent line**—that is, the straight line that just grazes the curve at this point (Figure 2). Although the tangent line is easily sketched by eye, its precise determination requires the use of calculus.

As we increase the base b of the exponential function $f(x) = b^x$, the graph of f in Figure 2 becomes steeper at the point $(0, 1)$ and the slope m of the tangent line

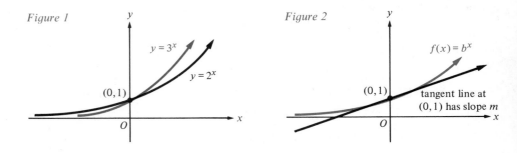

Table 1

b	m
0.5	-0.69
1	0
2	0.69
3	1.1
4	1.4

increases. If you sketch accurate graphs for the values of b in Table 1, draw the tangent lines at $(0, 1)$ by eye, and measure their slopes m, you will obtain approximately the values shown in the table (Problem 3).

From Table 1 we see that m is less than 1 when $b = 2$, and that m is greater than 1 when $b = 3$. As you might suspect, somewhere between 2 and 3 there is a value of b for which m is exactly 1. This particular value of the base is denoted by e, in honor of the great Swiss mathematician Leonhard Euler (1707–1783) (pronounced "oiler"), who was one of the first to recognize its immense importance. Like π, the value of e is an irrational number. By using advanced mathematical methods and high-speed computers, the numerical value of e has been calculated to thousands of decimal places. Rounded off to three decimal places,

$$e \approx 2.718$$

Figure 3

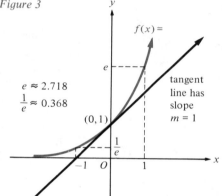

$e \approx 2.718$
$\dfrac{1}{e} \approx 0.368$

Ⓒ **Example 1**   Sketch the graph of the exponential function

$$f(x) = e^x.$$

Solution   The graph of the exponential function $f(x) = e^x$ rises at just the right rate so that the tangent line at $(0, 1)$ has slope $m = 1$. Using this fact, recalling the general shape of graphs of exponential functions with bases greater than 1, and plotting points corresponding to

$$f(1) = e^1 = e \approx 2.718 \quad \text{and} \quad f(-1) = e^{-1} = \frac{1}{e} \approx \frac{1}{2.718} \approx 0.368,$$

we can sketch the graph of $f(x) = e^x$ (Figure 3). As usual, accuracy is enhanced if a calculator is used to plot additional points. ∎

The exponential function $f(x) = e^x$ plays an important role in calculus, and it is essential in the applications of mathematics to many fields ranging from engineering to public health. You will find a variety of these applications in the remainder of this chapter. Indeed, the function $f(x) = e^x$ is used so often that people simply call it the **exponential function.** Whenever anyone uses this term without specifying the base, you can be certain that the function $f(x) = e^x$ is intended. On some scientific calculators* and in many of the standard computer languages (such as BASIC), the exponential function is denoted by exp (or by EXP). Thus,

$$\exp(x) = e^x.$$

When you deal with algebraic expressions that involve the exponential function, keep in mind that e^x is *always positive* (Figure 3).

* On certain calculators, you must press INV and then LN (inverse natural logarithm) for exponential function. Check the instruction manual furnished with your particular calculator.

Example 2   Find the zeros of $f(x) = xe^x - e^x$.

Solution   Factoring, we have

$$f(x) = (x - 1)e^x.$$

Since e^x cannot be zero, the only way that $f(x)$ can be equal to zero is to have

$$(x - 1) = 0.$$

Therefore, $x = 1$ is the only zero of $f(x)$.   ∎

ⓒ **Example 3**   Using a scientific calculator, evaluate

(a) e^1 **(b)** $e^{\sqrt{2}}$ **(c)** $e^{-5.0321}$

Solution   Using a 10-digit calculator, we find that

(a) $e^1 = e = 2.718281828$
(b) $e^{\sqrt{2}} = e^{1.414213562} = 4.113250377$
(c) $e^{-5.0321} = 6.525093476 \times 10^{-3} = 0.006525093476$ ∎

Because the tangent line to the graph of $y = e^x$ at the point $(0, 1)$ has slope $m = 1$, the slope-intercept equation of the tangent line is $y = 1 + x$ (Figure 4). Obviously,

Figure 4

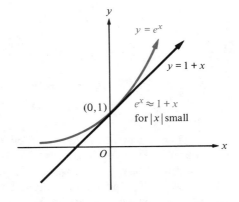

the curve $y = e^x$ and the tangent line $y = 1 + x$ are very close together near the point $(0, 1)$; hence, *for small values of* $|x|$, we have

$$\boxed{e^x \approx 1 + x,}$$

and this approximation becomes more and more accurate as $|x|$ gets smaller and smaller.

ⓒ **Example 4**   How accurate is the approximation $e^x \approx 1 + x$ if $x = 0.01$?

Solution   Using a calculator, we find that

$$e^{0.01} = 1.010050167.$$

Since $$1 + 0.01 = 1.010000000,$$

the discrepancy in the approximation $e^{0.01} \approx 1 + 0.01$ first occurs in the fifth decimal place. ∎

Some banks offer savings accounts with interest compounded not quarterly, not weekly, not daily, not hourly, but *continuously*. The formula for continuously compounded interest involves the exponential function. Although the derivation of this formula requires methods studied in calculus, we can derive it informally as follows. We begin with the formula

$$S = P\left(1 + \frac{r}{n}\right)^{nt}$$

for the final value S dollars of a principal P dollars invested for t years at a nominal annual interest rate r compounded n times a year. We're interested in what happens as n gets larger and larger. Let $x = r/n$ and notice that the larger n is, the smaller x is. Using the approximation $e^x \approx 1 + x$ for small x and the fact that $xn = r$, we have

$$S = P\left(1 + \frac{r}{n}\right)^{nt} = P(1 + x)^{nt} \approx P(e^x)^{nt} = Pe^{xnt} = Pe^{rt}.$$

As n becomes larger and larger, $x = r/n$ becomes smaller and smaller, and the approximation

$$S \approx Pe^{rt}$$

becomes more and more accurate. Therefore, for **continuously compounded** interest at a nominal annual rate r, bankers use the formula

$$S = Pe^{rt}$$

for the final value S dollars of a principal P dollars invested for a term of t years.

© **Example 5**   The New Mattoon Savings Bank offers a savings account with continuously compounded interest at a nominal annual rate of 7% (that is, $r = 0.07$).

(a) Sketch a graph showing the amount of money S dollars in such an account after t years, $0 \le t \le 20$, if a principal $P = \$100$ is deposited when $t = 0$.

(b) What is the final value S dollars of an investment of $P = \$100$ for a term of $t = 20$ years?

Solution

(a) Here $r = 0.07$, $P = 100$, and $S = Pe^{rt} = 100e^{0.07t}$.

The graph is sketched in Figure 5.

(b) When $t = 20$ years,

$$S = 100e^{0.07(20)}$$
$$= 100e^{1.4}$$
$$= \$405.52.$$ ∎

Figure 5

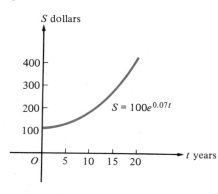

$S = 100e^{0.07t}$

Exponential Growth and Decay

If x and y are variable quantities, we say that y **increases** or **grows exponentially** as a function of x if there are positive constants y_0 and k such that

$$y = y_0 e^{kx}.$$

Similarly, if

$$y = y_0 e^{-kx},$$

we say that y **decreases** or **decays exponentially** as a function of x. Graphs of y as a function of x for exponential growth and exponential decay are shown in Figure 6. Notice that y_0 is the y intercept in both of the graphs; that is, y_0 is the value of y when $x = 0$. (Why?) The constant k, which determines how rapidly the growth or decay takes place, is called the **growth constant** or the **decay constant.**

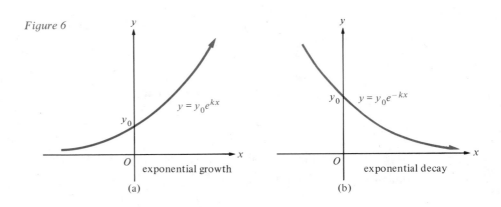

Figure 6

(a) exponential growth

(b) exponential decay

As we have seen, S dollars in a savings account with continuously compounded interest grows exponentially as a function of time. Since $S = Pe^{rt}$, the growth constant is equal to the nominal annual interest rate r. On the other hand, radioactive materials provide a good example of exponential decay.

© **Example 6**   Polonium, a radioactive element discovered by Marie Curie in 1898 and named after her native country Poland, decays exponentially. If y_0 grams of polonium are initially present, the number of grams y present after t days is given by

$$y = y_0 e^{-0.005t}.$$

(a) If $y_0 = 5$ grams, sketch a graph showing the amount y grams of polonium left after t days for $0 \le t \le 730$.

(b) Of a 5-gram sample of polonium, how much is left after 2 years (730 days)?

Solution   **(a)** The graph of $y = 5e^{-0.005t}$ for $0 \le t \le 730$ is sketched in Figure 7.

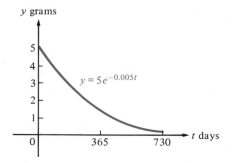

Figure 7

(b) When $t = 730$, we have

$$y = 5e^{-(0.005)(730)} = 5e^{-3.65} \approx 0.13 \text{ gram.}$$

Problem Set 5.2

Ⓒ In Problems 1 and 2, use a scientific calculator to evaluate the given quantity to as many significant digits as you can.

1. (a) e^{-1} (b) e^{-2} (c) e^3
   (d) e^{π} (e) $e^{\sqrt{5}}$

2. (a) e^e (b) $e^{-3.11}$ (c) e^{-e}
   (d) $\exp(0.5)$ (e) $\exp(1 - \sqrt{2})$

Ⓒ **3.** Using a calculator with a y^x key, sketch accurate graphs of $y = b^x$ for the values of b in Table 1 on page 271. Draw tangent lines to these graphs by eye at the point $(0, 1)$, measure the slopes of the tangent lines, and compare your slopes to the entries in the table.

Ⓒ **4.** Using a scientific calculator and a sheet of graph paper (available at your college bookstore), sketch the graph of the exponential function $\exp(x) = e^x$ as accurately as you can for $-2 \le x \le 2$.

In Problems 5 to 12, use the graph of $y = e^x$ (Figure 3) and the techniques of shifting, stretching, and reflecting to sketch the graph of each function. (Ⓒ Use a calculator if you wish.)

5. $f(x) = e^x + 1$ **6.** $g(x) = e^{x+1}$

7. $h(x) = -e^x$ **8.** $F(x) = -3e^{x+1}$

9. $G(x) = e^{-x}$ **10.** $H(x) = e^{1-x}$

11. $f(x) = e^{-x} + 1$ **12.** $g(x) = 2e^{1-x} + 2$

Ⓒ In Problems 13 to 16, use a scientific calculator to verify each equation for the indicated values of the variables.

13. $e^x e^y = e^{x+y}$ for $x = \sqrt{2}, y = \sqrt{3}$

14. $e^{x+y} = e^x e^y$ for $x = \sqrt{5}, y = -\pi$

15. $(e^x)^y = e^{xy}$ for $x = \dfrac{\pi}{2}, y = 1 - \sqrt{3}$

16. $e^{x-y} = \dfrac{e^x}{e^y}$ for $x = 3.9, y = 2.5$

In Problems 17 to 20, find the zeros of each function.

17. $f(x) = 2xe^{-x} - x^2 e^{-x}$ [*Hint:* e^{-x} is always positive.]

18. $g(x) = \dfrac{3(1 - x^2)e^{3x} + 2xe^{3x}}{(1 - x^2)^2}$

19. $h(x) = e^{2x} - 2e^x + 1$ [*Hint:* $e^{2x} = (e^x)^2$.]

20. $F(x) = e^{2x} - 2e^{x+1} + e^2$

In Problems 21 and 22, simplify each expression.

21. $\left(\dfrac{e^x + e^{-x}}{2}\right)^2 - \left(\dfrac{e^x - e^{-x}}{2}\right)^2$

22. $\left(\dfrac{2}{e^x + e^{-x}}\right)^2 + \left(\dfrac{e^x - e^{-x}}{e^x + e^{-x}}\right)^2$

© **23.** How accurate is the approximation $e^x \approx 1 + x$ if:

   (a) $x = 0.05$ (b) $x = 0.1$

   (c) $x = 0.5$ (d) $x = 1$

24. If n is a large number, show that $\left(1 + \dfrac{1}{n}\right)^n \approx e$.

   [*Hint:* Let $x = 1/n$ and use the approximation $e^x \approx 1 + x$.]

© **25.** When a flexible cord or chain is suspended from its ends, it hangs in a curve called a *catenary* whose equation has the form $y = \dfrac{a}{2}(e^{x/a} + e^{-x/a})$. Sketch the graph of the catenary for $a = 2$.

26. The famous Gateway Arch to the West in St. Louis has the shape of an *inverted catenary* (see Problem 25). Find an equation of such an inverted catenary if it is formed by reflecting the catenary $y = \dfrac{a}{2}(e^{x/a} + e^{-x/a})$ across the x axis and then

shifting it vertically so that its apex is on the y axis h units above the origin.

© **27.** Suppose that you invest a principal of $P = \$1000$ at a nominal annual interest rate of 10% ($r = 0.1$) for a period of $t = 5$ years. Calculate the final value S of your investment if the interest is compounded (a) monthly and (b) continuously.

© **28.** The concentration C of a drug in a person's circulatory system decreases as the drug is eliminated by the liver and kidneys or absorbed by other organs. Medical researchers often use the equation $C = C_0 e^{-kt}$ to predict the concentration C at a time t hours after the drug is administered, where C_0 is the initial concentration when $t = 0$ and k is a constant depending on the type of drug. If $C_0 = 3$ milligrams per liter and $k = 0.173$:

   (a) Sketch a graph of C as a function of t for $0 \leq t \leq 4$ hours.

   (b) Find C when $t = 4$ hours.

© **29.** Ecologists have determined that the approximate population N of bears in a certain protected forest area is given by $N = 225e^{0.02t}$, where t is the elapsed time in years since 1977.

   (a) Sketch a graph showing the bear population N as a function of t for $0 \leq t \leq 10$ years.

   (b) Estimate the number of bears that will inhabit the region in 1990.

30. If P dollars is invested for $t = 1$ year at a nominal annual interest rate r compounded continuously, the final value S dollars at the end of the year is given by $S = Pe^{r \cdot 1} = Pe^r$. Since P dollars invested for $t = 1$ year at a simple annual interest rate R yields a final value $S = P(1 + R)$ dollars, it follows that the effective simple annual interest rate R corresponding to the continuous nominal annual interest rate r satisfies the equation $P(1 + R) = Pe^r$.

(a) Solve for R in terms of r.

© (b) Find R if $r = 0.07$ (7%).

© 31. Carbon 14 decays exponentially according to the equation $y = y_0 e^{-0.0001212t}$ where y grams is the amount left after t years and y_0 grams is the initial amount.

(a) If $y_0 = 10$ grams, sketch a graph of y as a function of t for $0 \le t \le 10{,}000$ years.

(b) Of a 10-gram sample of carbon 14, how much will be left after 10,000 years?

© 32. The electric current I in amperes flowing in a series circuit having an inductance L henrys, a resistance R ohms, and a constant electromotive force E volts (Figure 8) satisfies the equation

$$I = \frac{E}{R} - \frac{E}{R}\exp\left(-\frac{Rt}{L}\right),$$

where t is the time in seconds after the current begins to flow. If $E = 12$ volts, $R = 5$ ohms, and $L = 0.03$ henry, sketch the graph of I as a function of t.

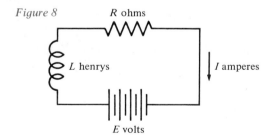

Figure 8   R ohms

L henrys   I amperes

E volts

© 33. The population N of a small country after t years is given by $N = 2{,}000{,}000e^{0.03t}$.

(a) What was the population when $t = 0$?

(b) What is the projected population when $t = 20$ years?

34. Find a formula for the present value P of S dollars t years in the future at a nominal annual interest rate r compounded continuously.

© 35. A biologist finds that the number N of bacteria in a culture after t hours is given by $N = 2000e^{0.7t}$.

(a) How many bacteria were present when $t = 0$?

(b) How many bacteria will be present after $t = 12$ hours?

(c) Sketch a graph of N as a function of t for $0 \le t \le 6$ hours.

(d) Using the graph in part (c), estimate the time at which $N = 32{,}000$ bacteria.

© 36. In statistics, the *normal probability density function* is defined by

$$f(x) = \frac{1}{\sigma\sqrt{2}}\exp\left[-\frac{(x - \mu)^2}{2\sigma^2}\right],$$

where σ and μ are certain constants called the *standard deviation* and the *mean*, respectively, and $\sigma > 0$. When students ask to be graded "on the curve," the curve in question is the graph of f. Sketch the graph of f for $\sigma = 1$ and $\mu = 0$.

© 37. Sketch the graph of $h(x) = \dfrac{2}{e^x - e^{-x}}$.

© 38. The function

$$g(x) = \begin{cases} 0 & \text{if } x = 0 \\ e^{-1/x^2} & \text{if } x \ne 0 \end{cases}$$

is useful in advanced mathematics because its graph is very "flat" near the origin. Sketch the graph of g.

© 39. In calculus it is shown that, for small values of x, $e^x \approx 1 + x + (x^2/2)$ gives an even better approximation than does $e^x \approx 1 + x$. Sketch graphs of $y = e^x$ and $y = 1 + x + (x^2/2)$ on the same coordinate system.

© 40. In 1973, it was projected that in t years the annual consumption of gasoline in the United States would be A billion barrels, where $A = 2.4e^{0.0381t}$. Use this equation to estimate the gasoline consumption of the United States in the year 1990.

5.3 LOGARITHMS AND THEIR PROPERTIES

If you put \$100 in a savings account at 8% nominal annual interest compounded quarterly, the final value S dollars of your investment after t years is given by

$$S = 100\left(1 + \frac{0.08}{4}\right)^{4t} = 100(1.02)^{4t}.$$

It's natural to ask how long you'll have to wait until your money doubles. To find out, you have to solve the equation

$$200 = 100(1.02)^{4t} \qquad \text{or} \qquad 1.02^{4t} = 2$$

for t. We'll explain later (in Section 5.5) how to solve this equation, but you can verify using a calculator that, the solution is (approximately) $t = 8.75$. In other words, you'll have to wait 8 years and 9 months for your money to double. An equation such as $1.02^{4t} = 2$, in which an unknown appears in an exponent, is called an **exponential equation.**

The simple exponential equations in Example 1 can be solved by using the fact that *if $b > 0$ and $b \neq 1$, then*

$$\boxed{b^x = b^y \qquad \text{if and only if} \qquad x = y}$$

(Problem 18).

Example 1 Solve each exponential equation.

(a) $2^x = 64$ **(b)** $36^t = 216^{2t-1}$

Solution **(a)** We begin by expressing 64 as a power of 2 so that both sides of the equation will have the same base. Since $64 = 2^6$, we can rewrite $2^x = 64$ as $2^x = 2^6$, from which it follows that $x = 6$.

(b) Since $36 = 6^2$ and $216 = 6^3$, we can rewrite the given equation as

$$(6^2)^t = (6^3)^{2t-1} \qquad \text{or} \qquad 6^{2t} = 6^{3(2t-1)},$$

from which it follows that

$$2t = 3(2t - 1), \qquad 2t = 6t - 3, \qquad 4t = 3, \qquad \text{and} \qquad t = \tfrac{3}{4}. \qquad ∎$$

If $b > 0$, $b \neq 1$, and $c > 0$, the solution x of the exponential equation

$$b^x = c$$

is denoted by

$$x = \log_b c,$$

which is read "x equals the **logarithm to the base b of c**." In other words,

$\log_b c$ is the power to which you must raise b to obtain c.

For instance, because $10^2 = 100$, it follows that $\log_{10} 100 = 2$. More generally, the fact that $\log_b c$ is the solution of $b^x = c$ is expressed by the equation

$$b^{\log_b c} = c.$$

As a useful memory device for this important identity, notice that *a logarithm is an exponent.*

Example 2   Find

(a) $\log_2 16$ **(b)** $\log_3 27$ **(c)** $\log_{10} \frac{1}{10}$ **(d)** $\log_e e^5$

Solution   **(a)** We ask ourselves, "To what power must we raise 2 to obtain 16?" Since $2^4 = 16$, the answer is 4. Therefore, $\log_2 16 = 4$.

(b) Since $3^3 = 27$, it follows that $\log_3 27 = 3$.

(c) Since $10^{-1} = \frac{1}{10}$, it follows that $\log_{10} \frac{1}{10} = -1$.

(d) We ask ourselves, "To what power must we raise e to obtain e^5?" The answer is 5, so $\log_e e^5 = 5$.   ■

To generalize part (d) of Example 2, notice that, if you ask yourself, "To what power must I raise b in order to obtain b^y," the obvious answer is y; hence,

$$\log_b b^y = y$$

holds for any real number y and any positive base $b \neq 1$. In particular, letting $y = 0$ and noting that $b^0 = 1$, you can see that

$$\log_b 1 = 0.$$

Similarly, letting $y = 1$ and noting that $b^1 = b$ you can see that

$$\log_b b = 1.$$

Using the fact that for $b > 0$, $b \neq 1$, and $c > 0$,

$$x = \log_b c \qquad \text{if and only if} \qquad b^x = c,$$

you can convert equations from logarithmic form to exponential form and vice versa. For instance:

Logarithmic Form	Exponential Form
$2 = \log_2 4$	$2^2 = 4$
$\log_{10} 10{,}000 = 4$	$10{,}000 = 10^4$
$-\frac{1}{2} = \log_{64} \frac{1}{8}$	$64^{-1/2} = \frac{1}{8}$
$\log_b x = y$	$b^y = x$ $\quad (b > 0, b \neq 1, x > 0)$
$k = \log_x d$	$x^k = d$ $\quad (x > 0, x \neq 1, d > 0)$

Notice that whenever you write $\log_b c$, you must make sure that c is positive and that b is positive and not equal to 1.

Example 3   Solve each equation.

(a) $\log_3 x^2 = 4$ (b) $\log_x 25 = 2$ (c) $\log_t(6t + 7) = 2$

Solution   (a) The equation $\log_3 x^2 = 4$ is equivalent to $3^4 = x^2$; that is, $x^2 = 81$. The solutions are $x = 9$ and $x = -9$.

(b) The equation $\log_x 25 = 2$ is equivalent to $x^2 = 25$, with the restriction that $x > 0$ and $x \neq 1$; hence, $x = 5$ is the solution.

(c) The equation $\log_t(6t + 7) = 2$ is equivalent to the equation $t^2 = 6t + 7$; that is, $t^2 - 6t - 7 = 0$, or $(t - 7)(t + 1) = 0$. Here again we have the restriction that $t > 0$ and $t \neq 1$, so $t = 7$ is the only solution ($t = -1$ is rejected). ∎

Using the connection between logarithms and exponents, we can translate properties of exponents into properties of logarithms. Some of these properties are as follows.

Properties of Logarithms

Let M, N, and b be positive numbers, $b \neq 1$, and let y be any real number. Then:

(i) $\log_b(MN) = \log_b M + \log_b N$

(ii) $\log_b \dfrac{M}{N} = \log_b M - \log_b N$

(iii) $\log_b N^y = y \log_b N$

(iv) $\log_b \dfrac{1}{N} = -\log_b N$

We verify Property (i) here and leave it to you to check Properties (ii), (iii), and (iv) (Problem 46). To prove Property (i), let

$$x = \log_b M \quad \text{and} \quad y = \log_b N,$$

so that
$$b^x = M \quad \text{and} \quad b^y = N.$$

Then, $\log_b(MN) = \log_b(b^x b^y) = \log_b(b^{x+y}) = x + y = \log_b M + \log_b N.$ ∎

Example 4   Suppose that $\log_b 2 = 0.48$ and $\log_b 3 = 0.76$. Find

(a) $\log_b 6$ **(b)** $\log_b \frac{3}{2}$ **(c)** $\log_b \sqrt[4]{2}$.

Solution   **(a)** By Property (i),

$$\log_b 6 = \log_b(2 \cdot 3) = \log_b 2 + \log_b 3 = 0.48 + 0.76 = 1.24.$$

(b) By Property (ii),

$$\log_b \tfrac{3}{2} = \log_b 3 - \log_b 2 = 0.76 - 0.48 = 0.28.$$

(c) By Property (iii),

$$\log_b \sqrt[4]{2} = \log_b 2^{1/4} = \tfrac{1}{4} \log_b 2 = \tfrac{1}{4}(0.48) = 0.12.$$ ∎

Example 5   Use the properties of logarithms to rewrite each expression as a sum or difference of multiples of logarithms.

(a) $\log_3 2x$ **(b)** $\log_2 \dfrac{1}{4}$ **(c)** $\log_b \dfrac{z}{uv}$ **(d)** $\log_2 \dfrac{(x^2 + 5)(2x + 5)^{3/2}}{\sqrt[4]{3x + 1}}$

Solution   **(a)** Assuming that x is positive, we have

$$\log_3 2x = \log_3 2 + \log_3 x \qquad \text{[Property (i)]}$$

(b) $\log_2 \tfrac{1}{4} = -\log_2 4 = -2$ \qquad\qquad\qquad\qquad [Property (iv)]

(c) Assuming that z, u, and v are positive, we have

$$\log_b \frac{z}{uv} = \log_b z - \log_b(uv) \qquad\qquad \text{[Property (ii)]}$$
$$= \log_b z - (\log_b u + \log_b v) \qquad \text{[Property (i)]}$$
$$= \log_b z - \log_b u - \log_b v.$$

(d) Assuming that $2x + 5$ and $3x + 1$ are positive, we have

$$\log_2 \frac{(x^2 + 5)(2x + 5)^{3/2}}{\sqrt[4]{3x + 1}} = \log_2[(x^2 + 5)(2x + 5)^{3/2}] - \log_2 \sqrt[4]{3x + 1}$$
$$= \log_2(x^2 + 5) + \log_2(2x + 5)^{3/2} - \log_2(3x + 1)^{1/4}$$
$$= \log_2(x^2 + 5) + \tfrac{3}{2} \log_2(2x + 5) - \tfrac{1}{4} \log_2(3x + 1),$$

where we applied Property (iii) twice in the last step. ∎

Example 6   Rewrite each expression as a single logarithm.

(a) $\log_5 7 + \log_5 x$

(b) $\log_4 t - \log_4 5$

(c) $2 \log_{10} x + 3 \log_{10}(x + 1)$

(d) $\log_b\left(x + \dfrac{x}{y}\right) - \log_b\left(z + \dfrac{z}{y}\right)$

Solution   Assuming that all quantities whose logarithms are taken are positive, we have the following:

(a) $\log_5 7 + \log_5 x = \log_5 7x$ [Property (i)]

(b) $\log_4 t - \log_4 5 = \log_4 \dfrac{t}{5}$ [Property (ii)]

(c) $2 \log_{10} x + 3 \log_{10}(x + 1) = \log_{10} x^2 + \log_{10}(x + 1)^3 = \log_{10}[x^2(x + 1)^3]$

(d) $\log_b\left(x + \dfrac{x}{y}\right) - \log_b\left(z + \dfrac{z}{y}\right) = \log_b \dfrac{x + \dfrac{x}{y}}{z + \dfrac{z}{y}} = \log_b \dfrac{\left(1 + \dfrac{1}{y}\right)x}{\left(1 + \dfrac{1}{y}\right)z} = \log_b \dfrac{x}{z}$ ■

Example 7   Solve each equation:

(a) $\log_{10} x + \log_{10}(x + 21) = 2$

(b) $\log_7(3t^2 - 5t - 2) - \log_7(t - 2) = 1$

Solution   (a) We begin by noticing that x must be positive for $\log_{10} x$ to be defined. If x is positive, so is $x + 21$, and $\log_{10}(x + 21)$ is also defined. Applying Property (i), we rewrite

$$\log_{10} x + \log_{10}(x + 21) = 2$$

as

$$\log_{10}[x(x + 21)] = 2.$$

The last equation can be rewritten in exponential form as

$$x(x + 21) = 10^2,$$

that is,

$$x^2 + 21x - 100 = 0.$$

Factoring, we have

$$(x + 25)(x - 4) = 0,$$

so $x = -25$ or $x = 4$. Since x must be positive, we can eliminate $x = -25$ as an extraneous root. Therefore, the solution is $x = 4$.

(b) Applying Property (ii), we can rewrite the given equation

$$\log_7(3t^2 - 5t - 2) - \log_7(t - 2) = 1$$

as

$$\log_7 \dfrac{3t^2 - 5t - 2}{t - 2} = 1,$$

provided that both $3t^2 - 5t - 2$ and $t - 2$ are positive. The last equation can be simplified by reducing the fraction,

$$\frac{3t^2 - 5t - 2}{t - 2} = \frac{(3t + 1)(t - 2)}{t - 2} = 3t + 1,$$

so

$$\log_7(3t + 1) = 1.$$

that is,

$$3t + 1 = 7^1 = 7.$$

The solution of this equation is

$$t = (7 - 1)/3 = 2.$$

We must check this answer against the original restrictions on the variables. However, if $t = 2$, then $t - 2 = 0$ and $\log_7(t - 2)$ is undefined. Thus, $t = 2$ is an extraneous root, and the original equation has no solution. ■

It is often useful to rewrite a logarithm in terms of logarithms to other bases. This is accomplished by using the following formula, which holds for $a > 0$, $b > 0$, $a \neq 1$, $b \neq 1$, and $c > 0$.

The Base-Changing Formula

$$\log_a c = \frac{\log_b c}{\log_b a}.$$

To prove the base-changing formula, let

$$x = \log_a c.$$

Then,

$$a^x = c,$$

and it follows that

$$\log_b a^x = \log_b c.$$

Using Property (iii), the last equation can be rewritten as

$$x \log_b a = \log_b c,$$

or

$$x = \frac{\log_b c}{\log_b a}.$$

Therefore, since $x = \log_a c$,

$$\log_a c = \frac{\log_b c}{\log_b a}.$$ ■

Example 8   Rewrite $\log_e 3$ in terms of logarithms to base 10.

Solution   By the base-changing formula with $a = e$, $c = 3$, and $b = 10$, we have

$$\log_e 3 = \frac{\log_{10} 3}{\log_{10} e}.$$

In working with logarithms, you must be careful not to confuse the fact that

$$\log_b\left(\frac{c}{a}\right) = \log_b c - \log_b a$$

(Property (ii) on page 280) and the fact that

$$\frac{\log_b c}{\log_b a} = \log_a c,$$

which is a consequence of the base-changing formula.

Problem Set 5.3

In Problems 1 to 12, solve each exponential equation.

1. $2^x = 8$

2. $3^{x^2} = 81$

3. $25^x = 5$

4. $2^{x^3} = 256$

5. $3^{2x+1} = 27$

6. $(1/10)^{4x} = 1000$

7. $3^{2-8x} = 9^{3x+1}$

8. $8^{3t} = 32^{4t-1}$

9. $5^{x^2+x} = 25$

10. $7^{x^2+x} = 1$

11. $3^{2t} - 3^t - 6 = 0$ [*Hint:* Let $x = 3^t$.]

12. $2^{2x+1} + 2^x = 10$

13. Find

(a) $\log_2 4$

(b) $\log_2 8$

(c) $\log_3 81$

(d) $\log_9 9^5$

(e) $\log_3 \frac{1}{9}$

(f) $\log_8 \frac{1}{64}$

(g) $\log_{10} 100,000$

14. Find

(a) $\log_2 \frac{1}{4}$

(b) $\log_3 \sqrt{3}$

(c) $\log_9 1$

(d) $\log_e e^\pi$

(e) $\log_2 4^3$

(f) $\log_3 9^{-0.5}$

(g) $\log_{10} \dfrac{1}{100,000}$

15. Rewrite each logarithmic equation as an equivalent exponential equation.

(a) $\log_2 32 = 5$

(b) $\log_{16} 2 = \frac{1}{4}$

(c) $\log_9 \frac{1}{3} = -\frac{1}{2}$

(d) $\log_e e = 1$

(e) $\log_{\sqrt{3}} 9 = 4$

(f) $\log_{10} 10^n = n$

(g) $\log_x x^5 = 5$

16. Give a geometric argument based on the graph of $f(x) = b^x$ to show that if $b > 0$, $b \neq 1$, and $c > 0$, then the exponential equation $b^x = c$ has exactly one solution.

17. Rewrite each exponential equation as an equivalent logarithmic equation.

(a) $8^0 = 1$

(b) $10^{-4} = 0.0001$

(c) $4^4 = 256$

(d) $27^{-1/3} = \frac{1}{3}$

(e) $8^{2/3} = 4$

(f) $a^c = y$

18. Using the result of Problem 16, show that if $b > 0$, $b \neq 1$, and x and y are real numbers such that $b^x = b^y$, it follows that $x = y$.

In Problems 19 to 44, solve each equation.

19. $x = \log_6 36$

20. $\log_5 x = 2$

21. $\log_x 125 = 3$

22. $x = \log_3 729$

23. $\log_7 x = 1$

24. $\log_3 x = 1$

25. $\log_x 16 = -\frac{4}{3}$

26. $\log_{\sqrt{3}} x = 6$

27. $\log_{\sqrt{2}} x = -6$

28. $\log_x \frac{27}{8} = -\frac{3}{2}$

29. $x = \log_{10} 10^{-7}$

30. $x = \log_e e^{-0.01}$

31. $x = \log_{27} \frac{1}{9}$

32. $x = \log_2(\log_4 256)$

33. $x = \log_5 \sqrt[4]{5}$

34. $x = \log_{3/4} \frac{4}{3}$

35. $\log_2(2x - 1) = 3$

36. $\log_5(2x - 3) = 2$

37. $\log_3(3x - 4) = 4$

38. $\log_7(2x - 7) = 0$

39. $\log_2(t^2 + 3t + 4) = 1$

40. $\log_5(y^2 - 4y) = 1$

41. $\log_4(9u^2 + 6u + 1) = 2$

42. $\log_3 |3 - 2t| = 2$

43. $\log_x(10 - 3x) = 2$

44. $\log_x(1 - x + x^2) = 3$

45. Suppose that $\log_b 2 = 0.53$, $\log_b 3 = 0.83$, $\log_b 5 = 1.22$, and $\log_b 7 = 1.48$. Find

(a) $\log_b 21$ (b) $\log_b 35$ (c) $\log_b \frac{2}{7}$

(d) $\log_b \frac{35}{3}$ (e) $\log_b \sqrt{7}$ (f) $\log_b \sqrt[3]{42}$

(g) $\log_b 3\sqrt{8}$

Round off all answers to two decimal places.

46. Verify Properties (ii), (iii), and (iv) on page 280.

In Problems 47 to 60, rewrite each expression as a sum or difference of multiples of logarithms. (Make the necessary assumptions about the values of the variables.)

47. $\log_3 7t$

48. $\log_2 \dfrac{3}{y}$

49. $\log_5 \sqrt{p}$

50. $\log_4 u^{-2/3}$

51. $\log_2 \dfrac{uv}{w}$

52. $\log_{0.1} \dfrac{p}{q}$

53. $\log_b[x(x + 1)]$

54. $\log_a(x^4 \sqrt{y})$

55. $\log_{10}[x^2(x + 1)]$

56. $\log_c \sqrt{\dfrac{x}{x + 7}}$

57. $\log_3 \dfrac{x^3 y^2}{z}$

58. $\log_e \dfrac{t(t + 1)}{(t + 2)^3}$

59. $\log_e \sqrt{x(x + 3)}$

60. $\log_b \sqrt[3]{(x + 1)^2 \sqrt{x + 7}}$

In Problems 61 to 68, rewrite each expression as a single logarithm. (Make the necessary assumptions about the values of the variables.)

61. $2 \log_3 x + 7 \log_3 x$

62. $\log_{10} \dfrac{a^3}{b} + \log_{10} \dfrac{b^2}{5a}$

63. $\dfrac{1}{2}[\log_5 4 - \log_5 9]$

64. $\log_x \dfrac{y^5}{z^4} - \log_x \dfrac{y^3}{z^2}$

65. $\log_e \dfrac{x}{x - 1} + \log_e \dfrac{x^2 - 1}{x}$

66. $\log_b \dfrac{x + y}{z} - \log_b \dfrac{1}{x + y}$

67. $\log_3 \dfrac{x^2 + 14x - 15}{x^2 + 4x - 5} - \log_3 \dfrac{x^2 + 12x - 45}{x^2 + 6x - 27}$

68. $\log_e \dfrac{m^2 - 2m - 24}{m^2 - m - 30} + \log_e \dfrac{(m + 5)^2}{m^2 - 16}$

In Problems 69 to 76, solve each equation.

69. $\log_4 x + \log_4(x + 6) = 2$

70. $\log_{10} x + \log_{10}(x + 3) = 1$

71. $\log_7 x + \log_7(18x + 61) = 1$

72. $\log_2 x + \log_2(x - 2) = \log_2(9 - 2x)$

73. $\log_3(x^2 + x) - \log_3(x^2 - x) = 1$

74. $\log_5(4x^2 - 1) = 2 + \log_5(2x + 1)$

75. $\log_8(x^2 - 9) - \log_8(x + 3) = 2$

76. $2 \log_2 x - \log_2(x - 1) = 2$

77. Use the base-changing formula to rewrite $\log_{10} 5$ in terms of logarithms to base e.

78. Using the base-changing formula, evaluate $\dfrac{\log_7 9}{\log_7 3}$.

79. Suppose that $a > 0$, $b > 0$, $a \neq 1$, and $b \neq 1$. Using the base-changing formula and the fact that $\log_b b = 1$, show that

$$\log_a b = \frac{1}{\log_b a}.$$

80. Derive the following alternative base-changing formula: If $a > 0$, $b > 0$, $a \neq 1$, $b \neq 1$, and $c > 0$, then

$$\log_a c = (\log_a b)(\log_b c).$$

5.4 LOGARITHMIC FUNCTIONS

A function F of the form

$$F(x) = \log_b x,$$

where $b > 0$ and $b \neq 1$, is called a **logarithmic function with base b.** The domain of F is the interval $(0, \infty)$ of all positive real numbers. If we let

$$f(x) = b^x,$$

then, since

$$b^{\log_b x} = x \quad \text{for } x > 0, \quad \text{and} \quad \log_b b^x = x \quad \text{for } x \text{ in } \mathbb{R},$$

it follows that

$$f[F(x)] = x \quad \text{for } x > 0, \quad \text{and} \quad F[f(x)] = x \quad \text{for } x \text{ in } \mathbb{R}.$$

In other words, the functions f and F are inverses of each other. (You may wish to review the idea of inverse functions in Section 3.8.) Therefore, *the graph of the logarithmic function $F(x) = \log_b x$ is the mirror image of the graph of the exponential function $f(x) = b^x$ across the line $y = x$.*

Example 1   Sketch the graphs of $F(x) = \log_3 x$ and $G(x) = \log_{1/3} x$.

Solution   Graphs of the functions $f(x) = 3^x$ and $g(x) = (\frac{1}{3})^x$ were shown in Figure 2 of Section 5.1 (page 263). By reflecting these graphs across the line $y = x$, we obtain the graphs of $F(x) = \log_3 x$ (Figure 1a) and $G(x) = \log_{1/3} x$ (Figure 1b). Notice that the graph of G can be obtained by reflecting the graph of F across the x axis. (For the reason why, see Problem 48.) ∎

Figure 1

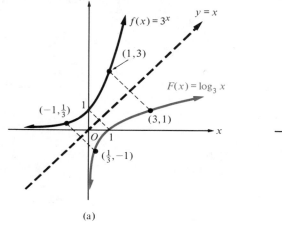

(a)

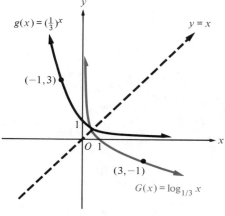

(b)

In general, graphs of logarithmic functions have the characteristic shapes shown in Figure 2. Thus, if $b > 1$, the graph of $F(x) = \log_b x$ rises to the right (Figure 2a), whereas if $0 < b < 1$, the graph falls to the right (Figure 2b).

Figure 2

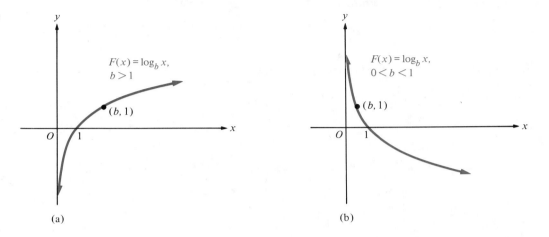

(a)   (b)

Notice that the graph of $F(x) = \log_b x$ always contains the point $(b, 1)$ (because $\log_b b = 1$) and that its x intercept is always 1 (because $\log_b 1 = 0$). There is no y intercept; in fact, the y axis is a vertical asymptote of the graph. The range of $F(x) = \log_b x$ is $\mathbb{R}$.

The Common and Natural Logarithm Functions

Before the development of electronic calculators and computers, logarithms were extensively used to facilitate numerical calculations. Because the usual positional system for writing numerals is based on 10, arithmetic calculation is easiest when logarithms with base 10 are used. Logarithms with base 10 are called **common logarithms,** and the symbol "log x"* (with no subscript) is often used as an abbreviation for $\log_{10} x$. Thus, the **common logarithm function** is defined by

$$\log x = \log_{10} x \qquad \text{for } x > 0.$$

These days, because of the wide availability of inexpensive and reliable electronic calculators, the common logarithm function is rarely used for purposes of numerical calculation. However, it still has many applications, ranging from the measurement of pH in chemistry to the measurement of sound pollution in the health sciences (see Section 5.5). For this reason, most scientific calculators have both a 10^x key and a log key.

* In some textbooks, and in the BASIC computer language, log or LOG is used as an abbreviation for the logarithm to base e.

© **Example 2**   Use a calculator with a log key to evaluate

(a) log 2110   (b) log 0.004326

Solution   On a 10-digit calculator, we obtain

(a) log 2110 = 3.324282455   (b) log 0.004326 = −2.363913485   ∎

© **Example 3**   Use a calculator to verify that $10^{\log x} = x$ for $x = \pi$.

Solution   Rounded off to 9 decimal places,

$$\pi = 3.141592654$$

and

$$\log \pi = 0.497149873.$$

Now,

$$10^{0.497149873} = 3.141592654 = \pi,$$

confirming that

$$10^{\log \pi} = \pi.$$ ∎

If a calculator isn't available, you can use Appendix Table IB to find common logarithms. Appendix I explains how.

In advanced mathematics and its applications, many otherwise cumbersome formulas become much simpler if logarithmic and exponential functions with base $e \approx 2.718$ are used (see Section 5.2). Logarithms with base e are called **natural logarithms** and the symbol "ln x"* is often used as an abbreviation for $\log_e x$. Thus, the **natural logarithm function** is defined by

$$\ln x = \log_e x \quad \text{for } x > 0.$$

In other words, for $x > 0$,

$$y = \ln x \quad \text{if and only if} \quad e^y = x$$

Of course, all scientific calculators have an ln key.

© **Example 4**   Use a calculator with an ln key to evaluate

(a) ln 7124   (b) ln 0.05319

Solution   On a 10-digit calculator, we obtain

(a) ln 7124 = 8.871224644   (b) ln 0.05319 = −2.933884870   ∎

© **Example 5**   Use a calculator to verify that $\ln e^x = x$ for $x = \sqrt{5}$.

---

* ln x is sometimes read as "ell en x."

Solution   Rounded off to 9 decimal places,

$$\sqrt{5} = 2.236067977$$

and $$e^{\sqrt{5}} = 9.356469012.$$

Now, $$\ln 9.356469012 = 2.236067977 = \sqrt{5},$$

confirming that $$\ln e^{\sqrt{5}} = \sqrt{5}.$$ ∎

If a calculator isn't available, you can use Appendix Table IA to find natural logarithms. Alternatively, you can use Appendix Table IB for common logarithms and the formula

$$\ln x = M \log x,$$

where $M = \ln 10 \approx 2.302585093$ (Problem 16).

Because *the natural logarithm function is the inverse of the exponential function*, the graph of

$$y = \ln x$$

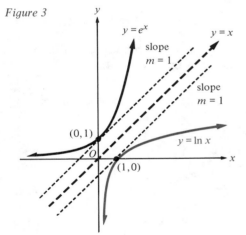

Figure 3

can be obtained by reflecting the graph of $y = e^x$ across the line $y = x$ (Figure 3). Recall that the tangent line to the graph of $y = e^x$ at $(0, 1)$ has slope $m = 1$ (Section 5.2, Figure 3). It follows that the tangent line to the graph of $y = \ln x$ at $(1, 0)$ also has slope $m = 1$. If you keep this fact in mind whenever you sketch the graph of the natural logarithm function, you will obtain a more accurate graph.

Although common and natural logarithm functions are sufficient for most purposes, there are situations in which logarithms with bases other than 10 and e are useful. For instance, in communications engineering and computer science, the bases 2 and 8 are often used. Since scientific calculators ordinarily have keys only for log and ln, you must use the *base-changing formula*

$$\log_a x = \frac{\log x}{\log a},$$

derived in Section 5.3, if you want to calculate logarithms with other bases.*

C **Example 6**   Use a calculator and the base-changing formula to find $\log_2 3$.

Solution

$$\log_2 3 = \frac{\log 3}{\log 2} = \frac{0.477121255}{0.301029996} = 1.584962500.$$ ∎

* In the BASIC computer language, where LOG refers to the natural logarithm, you must use the formula LOG A (X) = LOG (X)/LOG (A) to obtain the logarithm to the base A of X.

Graphs of Functions Involving Logarithms

In dealing with functions involving logarithms, you must keep in mind that logarithms of negative numbers and zero are undefined.

Example 7   Find the domain of

(a) $h(x) = \log_2(x + 1)$ **(b)** $g(x) = \log x^2$

Solution   **(a)** $\log_2(x + 1)$ is defined if and only if $x + 1 > 0$; that is, $x > -1$. Therefore, the domain of h is the interval $(-1, \infty)$.

(b) $\log x^2$ is defined if and only if $x^2 > 0$, that is, $x \neq 0$. Therefore, the domain of g is the set of all nonzero real numbers. ∎

Figure 4

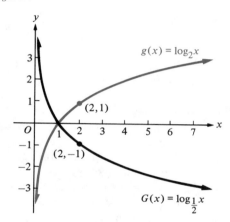

Example 8   Sketch the graphs of $g(x) = \log_2 x$ and $G(x) = \log_{1/2} x$ on the same coordinate system.

Solution   The graph of $g(x) = \log_2 x$ has the characteristic shape shown in Figure 2a, contains the point $(2, 1)$, and has x intercept 1 (Figure 4). The graph of $G(x) = \log_{1/2} x$ is obtained by reflecting the graph of $g(x) = \log_2 x$ across the x axis (Figure 4). ∎

The techniques of graph sketching presented in Section 3.6 can be applied to functions involving logarithms. As usual, accuracy is enhanced by plotting more points, and you may wish to use a calculator to find coordinates of such points quickly.

In Examples 9 and 10, determine the domain of each function, find any y or x intercepts of its graph, use the techniques of Section 3.6 to sketch the graph, determine the range of the function, indicate where it is increasing or decreasing, and find any horizontal or vertical asymptotes of the graph. (© If you wish, use a calculator for greater accuracy.)

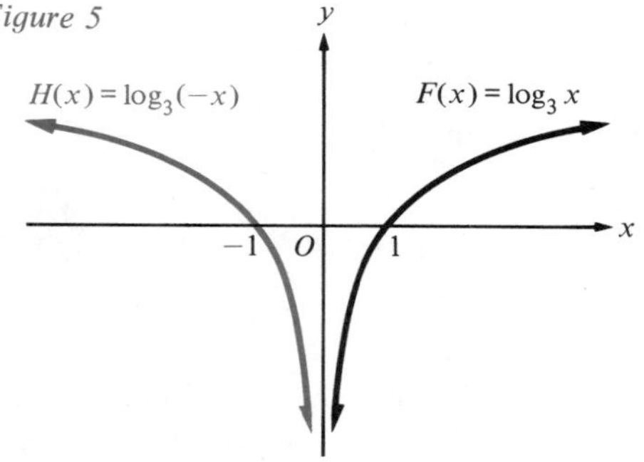

Example 9 $H(x) = \log_3(-x)$

Figure 5

Solution   The domain of H consists of all values of x for which $-x > 0$, that is, the interval $(-\infty, 0)$. Because $H(0)$ is undefined, there is no y intercept. The x intercept is the solution of the equation $H(x) = 0$; that is, $\log_3(-x) = 0$. Rewriting the last equation in exponential form, we obtain

$$-x = 3^0 = 1,$$

so the x intercept is $x = -1$. The graph of $H(x) = \log_3(-x)$ (Figure 5) is the mirror image of the graph of $F(x) = \log_3 x$ (Figure 1a) across the y axis. From the graph, we see that the range of H is the set $\mathbb{R}$

of all real numbers, and that the function H is decreasing over its entire domain. The y axis is a vertical asymptote, and there is no horizontal asymptote. ∎

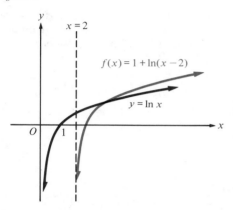

Figure 6

Example 10 $f(x) = 1 + \ln(x - 2)$

Solution   The domain of f is the interval $(2, \infty)$; hence, $f(0)$ is undefined and there is no y intercept. The x intercept is the solution of the equation $f(x) = 0$, that is,

$$1 + \ln(x - 2) = 0 \quad \text{or} \quad \ln(x - 2) = -1.$$

Rewriting the last equation in exponential form, we obtain

$$x - 2 = e^{-1},$$

so the x intercept is $x = 2 + e^{-1} \approx 2.37$. The graph of the function $f(x) = 1 + \ln(x - 2)$ (Figure 6) is obtained by shifting the graph of $y = \ln x$ (Figure 3) 1 unit upward and 2 units to the right. From the graph, we see that the range of f is the set $\mathbb{R}$ of all real numbers, and that the function f is increasing over its entire domain. Because the y axis is a vertical asymptote of the graph of $y = \ln x$, it follows that the line $x = 2$ is a vertical asymptote of the graph of f. There is no horizontal asymptote. ∎

Problem Set 5.4

In Problems 1 to 6, sketch the graph of each logarithmic function by reflecting the graph of its inverse exponential function across the line $y = x$, as in Example 1.

1. $f(x) = \log_4 x$

2. $g(x) = \log_5 x$

3. $F(x) = \log_{1/4} x$

4. $G(x) = \log_{1/5} x$

5. $h(x) = \log_6 x$

6. $H(x) = \log_{1/6} x$

C **7.** Use a calculator with a log key to evaluate

    (a) log 6.373

    (b) log 1230.4

    (c) log 0.03521

    (d) $\log(3.047 \times 10^{11})$

    (e) $\log(6.562 \times 10^{-9})$

8. Use Appendix Table IB to evaluate

    (a) log 2.74

    (b) log 0.00333

    (c) log 3470

    (d) $\log(9.09 \times 10^{21})$

    (e) $\log(3.11 \times 10^{-13})$

C **9.** Using a calculator, verify the property $\log(xy) = \log x + \log y$ for

    (a) $x = 31.27, y = 5.246$

    (b) $x = \pi, y = \sqrt{2}$

C **10.** Using a calculator, verify that

    (a) $\log 10^{\sqrt{7}} = \sqrt{7}$

    (b) $10^{\log \sqrt{7}} = \sqrt{7}$

© **11.** Use a calculator with an ln key to evaluate

(a) ln 4126   (b) ln 2.704

(c) ln 0.040404   (d) $\ln(7.321 \times 10^8)$

(e) $\ln(1.732 \times 10^{-7})$

12. Use Appendix Table IA to evaluate

(a) ln 2.68   (b) ln 3.33

(c) ln 25   [*Hint:* $25 = 5^2$.]

© **13.** Using a calculator, verify that

(a) $\ln e^\pi = \pi$ (b) $e^{\ln \pi} = \pi$

© **14.** Using a calculator, verify that $\ln y^x = x \ln y$ for $x = 77.01$ and $y = 3.352$.

© **15.** Using a calculator and the base-changing formula, evaluate

(a) $\log_2 25$ (b) $\log_3 2$ (c) $\log_8 e$

(d) $\log_\pi 5$ $\log \sqrt{2}/2$ 0.07301

16. Show that, for $x > 0$, $\ln x = M \log x$, where $M = \ln 10$.

17. Find the domain of each function.

(a) $f(x) = \log(x - 2)$ (b) $g(x) = \log_2 \sqrt{x}$

(c) $h(x) = \log_8(4 - x)$ (d) $H(x) = \log |x|$

(e) $K(x) = \dfrac{1}{\ln x}$

(f) $G(x) = \log(x^2 - 5x + 7)$

(g) $k(x) = \ln(x^2 + 1)$

18. Find any y and x intercepts of the graph of each function in Problem 17.

In Problems 19 to 36, determine the domain of each function, find any y or x intercepts of its graph, use the techniques of Section 3.6 to sketch the graph, determine the range of the function, indicate where it is increasing or decreasing, and find any horizontal or vertical asymptotes of the graph. (© If you wish, use a calculator for greater accuracy.)

19. $f(x) = 1 + \log_2 x$ **20.** $g(x) = \log_5(-x)$

21. $h(x) = \log_2 x^{-2}$ **22.** $F(x) = \log|x|$

23. $G(x) = \log \sqrt{x}$ **24.** $H(x) = \log_5 \sqrt[3]{x}$

25. $f(x) = \log_2(x - 1)$ **26.** $g(x) = (\ln x) - 1$

27. $h(x) = \log x^3$

28. $F(x) = |\log_2 x|$

29. $G(x) = -\log_3 x$

30. $H(x) = \ln(1 - x)$

31. $f(x) = 2 + \ln(x - 1)$

32. $g(x) = (\ln x)^{-1}$

33. $h(x) = \ln \dfrac{1}{x}$

34. $F(x) = 2 - \ln(1 - x)$

35. $G(x) = \ln(x + 2) - 2$ **36.** $H(x) = \log_{1/4} x^3$

In Problems 37 to 44, justify each property of the natural logarithm function by using either its definition or the properties of logarithms given in Section 5.3.

37. $\ln 1 = 0$

38. $\ln \dfrac{1}{x} = -\ln x, \qquad x > 0$

39. $\ln e = 1$

40. $\ln x = \dfrac{\log x}{\log e}, \qquad x > 0$

41. $\ln xy = \ln x + \ln y, \qquad x > 0, y > 0$

42. $\ln x = \dfrac{1}{\log_x e}, \qquad x > 0, x \neq 1$

43. $\ln \dfrac{x}{y} = \ln x - \ln y, \qquad x > 0, y > 0$

44. $\ln y^x = x \ln y, \qquad y > 0, x \text{ in } \mathbb{R}$

In Problems 45 and 46, find the inverse of each function.

45. $f(x) = e^{x+2}$ **46.** $g(x) = e^{2x-1} + 1$

47. Explain why every real number y can be written in the form $y = \ln x$ for a suitable value of x.

48. If $b > 0$ and $b \neq 1$, show that $\log_{1/b} x = -\log_b x$ holds for $x > 0$.

49. Sketch the graphs of $y = \log_2 x$, $y = \log_3 x$, and $y = \log_4 x$ on the same coordinate system. Describe the relationships among these graphs.

50. (a) Using the data in Table 1, Section 5.2, page 271, sketch a graph of m as a function of b.

(b) By looking at the graph in part (a), guess what function this is.

51. The function $F(x) = \ln|x|$ is used quite often in calculus. Sketch a graph of this function.

52. Solve the equation $\log_x(2x)^{3x} = 4^{\log_4 4x}$.

53. Find the value of b if the graph of $y = \log_b x$ contains the point $(\frac{1}{2}, -1)$.

54. If $a > 0$, $b > 0$, and $b \neq 1$, find a base c such that $a \log_b x = \log_c x$ holds for all $x > 0$.

55. Using the fact that the tangent line to the graph of $y = \ln x$ at $(1, 0)$ has slope $m = 1$, explain why $\ln x \approx x - 1$ for values of x close to 1, with the approximation becoming more and more accurate as x comes closer and closer to 1.

56. Show that $\ln(1 + x) \approx x$ if $|x|$ is small and that the approximation becomes more and more accurate as

$|x|$ gets smaller and smaller. [*Hint:* Use the result of Problem 55.]

57. The formula $y^x = e^{x \ln y}$ for $y > 0$ is used in calculus to write y^x in terms of the exponential and natural logarithm functions. Derive this formula. [*Hint:* $x \ln y = \ln y^x$.]

Ⓒ **58.** Using a calculator with y^x, e^x, and $\ln$ keys, check the formula in Problem 57 for $x = 1.59$ and $y = 7.47$.

59. Criticize the following statement: Since $\ln x^2 = 2 \ln x$, the graph of $y = \ln x^2$ can be found by doubling all ordinates on the graph of $y = \ln x$.

5.5 APPLICATIONS OF EXPONENTIAL AND LOGARITHMIC FUNCTIONS

In this and the next section, we present a small sample of the many and varied applications of exponential and logarithmic functions.

Solving Exponential Equations

You can often solve an exponential equation by taking the logarithm of both sides and using the properties of logarithms (page 280) to simplify the resulting equation. For this purpose, you can use either common or natural logarithms.

Ⓒ **Example 1**   Solve the exponential equation $7^{2x+1} = 3^{x-2}$.

Solution
$$7^{2x+1} = 3^{x-2}$$
$$\log 7^{2x+1} = \log 3^{x-2}.$$

Thus, by Property (iii),
$$(2x + 1) \log 7 = (x - 2) \log 3.$$

Distributing, we have
$$(2 \log 7)x + \log 7 = (\log 3)x - 2 \log 3$$
$$(2 \log 7 - \log 3)x = -\log 7 - 2 \log 3$$
$$(\log 7^2 - \log 3)x = -(\log 7 + \log 3^2)$$
$$(\log 49 - \log 3)x = -(\log 7 + \log 9)$$
$$(\log \tfrac{49}{3})x = -\log 63$$
$$x = \frac{-\log 63}{\log \tfrac{49}{3}}.$$

Therefore, using a 10-digit calculator, we find that

$$x = \frac{-\log 63}{\log \frac{49}{3}} = \frac{-1.799340549}{1.213074825} = -1.483289004.$$ ∎

The following example shows how to answer the question that we raised in the introduction to Section 5.3.

© **Example 2**   If you put \$100 in a savings account at 8% nominal annual interest compounded quarterly, how long will it take for your money to double?

Solution   The final value S dollars of your investment after t years is given by

$$S = 100\left(1 + \frac{0.08}{4}\right)^{4t} = 100(1.02)^{4t}.$$

If t is the time required to double your money, then

$$200 = 100(1.02)^{4t} \qquad \text{or} \qquad 1.02^{4t} = 2.$$

Taking the logarithm of both sides of the last equation, we get

$$4t \log 1.02 = \log 2$$

so that

$$t = \frac{\log 2}{4 \log 1.02} \approx 8.75 \text{ years.}$$ ∎

Applications in Chemistry, Earth Sciences, Psychophysics, and Physics

Measuring pH *in Chemistry*   In chemistry, the **pH** of a substance is defined by

$$pH = -\log[H^+],$$

where $[H^+]$ is the concentration of hydrogen ions in the substance, measured in moles per liter. The pH of distilled water is 7. A substance with a pH of less than 7 is known as an *acid*, whereas a substance with a pH of greater than 7 is called a *base*.

Environmentalists constantly monitor the pH of rain and snow because of the destructive effects of "acid rain" caused largely by sulfur dioxide emissions from factories and coal-burning power plants. Because of dissolved carbon dioxide from the atmosphere, rain and snow have a natural concentration of $[H^+] = 2.5 \times 10^{-6}$ mole per liter.

© **Example 3**   Find the natural pH of rain and snow.

Solution $$pH = -\log[H^+] = -\log(2.5 \times 10^{-6}) \approx -(-5.6) = 5.6.$$ ∎

Measuring Altitude: The Barometric Equation   The **barometric equation**

$$h = (30T + 8000) \ln \frac{P_0}{P}$$

relates the height h in meters above sea level, the air temperature T in degrees Celsius, the atmospheric pressure P_0 in centimeters of mercury at sea level, and the atmospheric pressure P in centimeters of mercury at height h. The altimeters most commonly used in aircraft measure the atmospheric pressure P and display the altitude by means of a scale calibrated according to the barometric equation.

ⓒ **Example 4**   Atmospheric pressure at the summit of Pike's Peak in Colorado on a certain day measures 44.7 centimeters of mercury. If the average air temperature is 5° Celsius and the atmospheric pressure at sea level is 76 centimeters of mercury, find the height of Pike's Peak.

Solution   We use the barometric equation with $T = 5$, $P_0 = 76$, and $P = 44.7$ to obtain

$$h = [30(5) + 8000] \ln \frac{76}{44.7} = 8150 \ln \frac{76}{44.7}$$

$$\approx 4330 \text{ meters.}$$ ∎

Measuring Sensation   In 1860, the German physicist Gustav Fechner (1801–1887) published a psychophysical law relating the intensity S of a sensation to the intensity P of the physical stimulus causing it. This law, which was based on experiments originally reported in 1829 by the German physiologist Ernst Weber (1795–1878), states that the change in S caused by a small change in P is proportional not to the change in P as one might suppose, but rather to the *percentage* of change in P. Using calculus, it can be shown, as a consequence, that S varies linearly as the natural logarithm of P, so that

$$S = A + B \ln P,$$

where A and B are suitable constants.

Suppose that P_0 denotes the **threshold intensity** of the physical stimulus; that is, the largest value of the intensity P for which there is no sensation. Then, substituting 0 for S and P_0 for P in the equation above, we find that

$$0 = A + B \ln P_0 \qquad \text{or} \qquad A = -B \ln P_0.$$

It follows that

$$S = -B \ln P_0 + B \ln P = B(\ln P - \ln P_0) = B \ln \frac{P}{P_0}.$$

If you prefer to write the relationship between S and P in terms of the common logarithm, use the equation

$$\ln \frac{P}{P_0} = M \log \frac{P}{P_0}, \qquad \text{where } M = \ln 10$$

(see Problem 16 on page 292) to rewrite $S = B \ln \dfrac{P}{P_0}$ as

$$S = MB \log \frac{P}{P_0}.$$

Finally, letting $C = MB$, you will obtain the **Weber–Fechner law** in the form

$$S = C \log \frac{P}{P_0}.$$

Choice of the constant C determines the units in which S is measured.

Early in the development of the telephone, it became necessary to have a unit to measure the loudness of telephone signals at various points in the network. The proposed unit, called the *bel* in honor of Alexander Graham Bell, the inventor of the telephone, is obtained by taking the constant $C = 1$ in the Weber–Fechner law. (The power P of the electrical signal is measured in watts.) It turns out that $\frac{1}{10}$ of a bel, called a **decibel** is roughly the smallest noticeable difference in the loudness of two sounds. For this reason, loudness is commonly measured in decibels rather than in bels. Thus, in electronics, an amplifier with an input of P_0 watts and an output of P watts is said to provide a *gain* of

$$S = 10 \log \frac{P}{P_0} \text{ decibels.}$$

Centennial model of Bell's telephone, 1976

Decibels are used to measure the loudness of sound produced by any source, not just a telephone receiver or an electronic amplifier. The intensity P of the air vibrations that produce the sensations of sound is commonly measured by the power (in watts) per square meter of wavefront. The threshold of human hearing is approximately

$$P_0 = 10^{-12} \text{ watt per square meter}$$

at the eardrum, and the formula

$$S = 10 \log \frac{P}{P_0}$$

thus gives the loudness in decibels produced by a sound wave of intensity P watts per square meter at the eardrum. The rustle of leaves in a gentle breeze corresponds to about 20 decibels, while loud thunder may reach 110 decibels or more. Sound in excess of 120 decibels can cause pain. In order to prevent "boilermaker's deafness," the American Academy of Ophthalmology and Otolaryngology recommends that no worker be exposed to a continuous sound level of 85 decibels or more for 5 hours or more per day without protective devices.

Ⓒ **Example 5**   Suppose that an employee in a factory must work in the vicinity of a machine producing 60 decibels of sound during an 8-hour shift. A second machine producing 70 decibels is moved into the same vicinity. Does the worker now need ear protection?

Solution   We can add sound intensities, but not decibels; thus, a combination of 60 decibels and 70 decibels does *not* amount to 130 decibels. In fact, suppose that P_1 and P_2 are the sound-wave intensities in watts per square meter produced by the two machines separately. Then the sound-wave intensity P produced by the machines operating together is given by $P = P_1 + P_2$. We know that

$$10 \log \frac{P_1}{P_0} = 60 \quad \text{and} \quad 10 \log \frac{P_2}{P_0} = 70,$$

so that

$$\log \frac{P_1}{P_0} = 6 \quad \text{and} \quad \log \frac{P_2}{P_0} = 7;$$

that is,

$$\frac{P_1}{P_0} = 10^6 \quad \text{and} \quad \frac{P_2}{P_0} = 10^7$$

or

$$P_1 = 10^6 P_0 \quad \text{and} \quad P_2 = 10^7 P_0.$$

It follows that

$$P = P_1 + P_2 = 10^6 P_0 + 10^7 P_0,$$

and the loudness of the sound produced by the machines operating together is

$$S = 10 \log \frac{P}{P_0} = 10 \log \frac{10^6 P_0 + 10^7 P_0}{P_0} = 10 \log(10^6 + 10^7)$$

$$= 10 \log 11{,}000{,}000 \approx 70.4 \text{ decibels,}$$

well below the 85-decibel limit. Ear protection will not be needed.   ■

Measuring the Rate of Radioactive Decay   As we mentioned in Section 5.2, radio-active materials decay according to the formula

$$y = y_0 e^{-kt},$$

where y is the mass of the material at time t and y_0 was the original mass when $t = 0$.

Let's consider how much the mass y changes when t changes by one unit of time. The change is

$$y_0 e^{-kt} - y_0 e^{-k(t+1)} = y_0(e^{-kt} - e^{-kt-k})$$
$$= y_0(e^{-kt} - e^{-kt}e^{-k})$$
$$= y_0 e^{-kt}(1 - e^{-k}).$$

The percentage of the change in mass in one unit of time is therefore given by

$$\frac{\text{change in mass}}{\text{original mass}} \times 100\% = \frac{y_0 e^{-kt}(1 - e^{-k})}{y_0 e^{-kt}} \times 100\%$$

$$= (1 - e^{-k}) \times 100\%.$$

In other words, if the percentage of the change in mass in one unit of time is expressed as a decimal K, we have

$$K = 1 - e^{-k} \quad \text{or} \quad e^{-k} = 1 - K,$$

so that

$$-k = \ln(1 - K) \quad \text{or} \quad k = -\ln(1 - K).$$

For small values of K, it can be shown (Problem 36) that $k \approx K$.

The rate of decay of a radioactive material can be measured not only in terms of k or K, but in terms of its **half-life,** which is defined to be the period of time T required for it to decay to half of its original mass. Thus,

$$\tfrac{1}{2} y_0 = y_0 e^{-kT} \quad \text{or} \quad \tfrac{1}{2} = e^{-kT};$$

that is,

$$e^{kT} = 2 \quad \text{or} \quad kT = \ln 2.$$

It follows that

$$T = \frac{\ln 2}{k}.$$

© **Example 6**   Potassium 42 is a radioactive element that is often used as a tracer in biological experiments. Its half-life is approximately 12.5 hours.

(a) If y_0 milligrams of potassium 42 are initially present, write a formula for the number of milligrams y present after t hours.

(b) What percent decrease in the amount of potassium 42 occurs in a sample during 1 hour?

(c) Of a 1-milligram sample of potassium 42, how much is left after 5 hours?

Solution   (a) $T = 12.5$; hence, from the equation $T = \dfrac{\ln 2}{k}$, we have

$$k = \frac{\ln 2}{T} = \frac{\ln 2}{12.5} = 0.05545.$$

Therefore, $y = y_0 e^{-0.05545t}$.

(b) Here $k = 0.05545$, so

$$K = 1 - e^{-k} = 1 - e^{-0.05545} \approx 0.0539.$$

Thus, the percentage of decrease per hour is about 5.39%.

(c) Putting $y_0 = 1$ milligram and $t = 5$ hours in the equation of part (a), we find that the amount left after 5 hours is

$$y = 1 \cdot e^{(-0.05545)(5)} = e^{-0.27725} \approx 0.758 \text{ milligram.} \qquad \blacksquare$$

Problem Set 5.5

C In Problems 1 to 18, use logarithms to solve each exponential equation.

1. $4^x = 3$

2. $2^{-x} = 5$

3. $7.07^x = 2001$

4. $8.97^x = 7.27^{x-1}$

5. $4^{2x+1} = 6^x$

6. $2^{3x} = 3^{2x-1}$

7. $10^{\sqrt{x}} = 100$

8. $3^{2x} = 46^{3x-2}$

9. $1000^{\sqrt[3]{x}} = 10$

10. $5^{3x-1} = 45^{-x}$

11. $3^{-3x+1} = e^{-x}$

12. $5^{x+1} = 5^{3x+1}$

13. $2^{2x+1} = 3^{2x+3}$

14. $e^{2x+1} = (2.3)^x$

15. $e^{2x+1} = 3^{x-1}$

16. $e^{1-3x} = 5^{2x}$

17. $(2.11)^{3x} = (1.77)^{x-1}$

18. $x^{\log x} = 1000x$

In Problems 19 to 22, solve for y in terms of x.

19. $x = e^{2y-1}$

20. $3e^{4y+1} = x - 2$

21. $e^{2y} - 2xe^y - 1 = 0$
    [Hint: $e^{2y} = (e^y)^2$; use the quadratic formula.]

22. $x = \frac{1}{2}(e^y + e^{-y})$ where $x \geq 1$.

C 23. If you invest $1000 in a savings account at 6% nominal annual interest compounded weekly, how long will it take for your money to double?

24. If a sum of money is invested at a nominal annual interest rate r compounded continuously, show that it doubles in $(\ln 2)/r$ years.

C 25. Find the pH of each substance (rounded off to two significant digits).

   (a) eggs: $[H^+] = 1.6 \times 10^{-8}$ mole/liter
   (b) tomatoes: $[H^+] = 6.3 \times 10^{-5}$ mole/liter
   (c) milk: $[H^+] = 4 \times 10^{-7}$ mole/liter

C 26. Find the hydrogen ion concentration $[H^+]$ in moles/liter of each substance (rounded off to two significant digits).

   (a) vinegar: pH = 3.1   (b) beer: pH = 4.3
   (c) lemon juice: pH = 2.3

C 27. Atmospheric pressure at the summit of Mt. Everest on a certain day measures 25.1 centimeters of mercury, and the air temperature is 0° Celsius. If the atmospheric pressure at sea level is 76 centimeters of mercury, use the barometric equation to find the approximate height of Mt. Everest.

C 28. Suppose that the pilot of a light aircraft has neglected to recalibrate the altimeter, which is set to a sea-level pressure of 76 centimeters of mercury, when the actual sea-level pressure has dropped to 75 centimeters because of weather conditions. Using $T = 0°$ Celsius in the barometric equation, find the amount of error in the pilot's altimeter reading.

C 29. Use the barometric equation with $T = 0°$ Celsius to find the elevation at which one-half of the atmosphere lies below and one-half lies above. [Hint: At this height, atmospheric pressure will have dropped to one-half its sea-level value.]

C 30. Using the barometric equation with $T = 0°$ Celsius and $P_0 = 76$ centimeters of mercury, sketch a graph of the height h as a function of atmospheric pressure P.

C 31. At takeoff, a certain supersonic jet produces a sound wave of intensity 0.2 watt per square meter. Taking P_0 to be 10^{-12} watt per square meter, find the loudness in decibels of the takeoff.

32. Show that a combination of two sounds with loudnesses S_1 and S_2 decibels produces a sound of loudness

$$S = 10 \log(10^{S_1/10} + 10^{S_2/10}) \text{ decibels.}$$

C 33. Find the ratio of the sound in watts per square meter at the threshold of pain (about 120 decibels) and at the threshold of hearing (0 decibels).

C 34. Suppose that the smallest weight you can perceive is $W_0 = 0.5$ gram, and that you can just barely notice the difference between 100 grams and 125 grams. Using the Weber–Fechner law, develop a scale of perceived heaviness. Call one unit on this scale a *heft* and write a formula for the number S of hefts corresponding to a weight of W grams.

c **35.** The brightness of a star perceived by the naked eye is measured in units called *magnitudes;* the brightest stars are of magnitude 1 and the dimmest are of magnitude 6. If I is the actual intensity of light from a star, and I_0 is the intensity of light from a just-visible star, the magnitude M corresponding to I is given by

$$M = 6 - 2.5 \log \frac{I}{I_0}.$$

Calculate the ratio of light intensities from a star of magnitude 1 and a star of magnitude 5.

36. If k is a small positive number, show that $k \approx 1 - e^{-k}$. [*Hint:* Use the approximation $e^x \approx 1 + x$ for small $|x|$.]

c **37.** The half-life of radium is 1656 years.

   (a) If y_0 grams of radium are initially present, write a formula for the number of grams y present after t years.

   (b) What percent decrease occurs in the amount of radium in a sample over 1 year?

   (c) How much of a 1-gram sample of radium is present after 20 years?

c **38.** A manufacturing plant estimates that the value V dollars of a machine is decreasing exponentially according to the equation

$$V = 76,000e^{-0.14t},$$

where t is the number of years since the machine was placed in service.

   (a) Find the value of the machine after 12 years.

   (b) Find the "half-life" of the machine.

c **39.** Sociologists in a certain country estimate that, of all couples married this year, 40% will be divorced within 15 years. Assuming that marriages fail exponentially, find the "half-life" of a marriage in that country.

c **40.** The intensity I of light below the surface of the ocean decreases exponentially with the depth d, the decay constant k depending on the amount of dissolved organic material, the pollution level, and other factors. In the surface layer of the ocean, called the *photic zone,* there is sufficient light for plant growth. Nearly all life in the ocean is dependent on microscopic plants called *phytoplankton,* which can live only in the photic zone. In coastal waters, the photic zone has a depth of as little as 5 meters; while in clear oceanic waters this zone may extend to a depth of 150 meters.

   (a) Find the ratio of the decay constant k for the intensity of light in coastal waters to its value in clear oceanic waters.

   (b) Assuming that the light intensity at the bottom of the photic zone is 1% of the surface light intensity, find the value of k for clear oceanic waters.

   (c) Find the depth at which the light intensity in clear oceanic waters has dropped to half its value at the surface.

c **41.** In psychological tests it is often found that if a group of people memorize a list of nonsense words, the fraction F of these people who remember all the words t hours later is given by $F = 1 - k \ln(t + 1)$, where k is a constant depending on the length of the list of words and other factors. A certain group was given such a memory test, and after 3 hours only half of the group's members could remember all the words.

   (a) Find the value of k for this experiment.

   (b) Predict the approximate fraction of group members who will remember all the words after 5 hours.

c **42.** In engineering thermodynamics, it is shown that when n moles of a gas are compressed isothermally (that is, at a constant temperature $T°$ Celsius) from volume V_0 to volume V, the resulting work W joules done on the gas is given by

$$W = n(8.314)(T + 273) \ln \frac{V_0}{V}.$$

Calculate the work done in compressing $n = 5$ moles of carbon dioxide isothermally at a temperature of $T = 100°$ Celsius from an initial volume $V_0 = 0.5$ cubic meter to a final volume of $V = 0.1$ cubic meter.

5.6 MATHEMATICAL MODELS AND POPULATION GROWTH

In the later years of his life, the Italian scientist Galileo Galilei (1564–1642) wrote about his experiments with motion in a treatise called *Dialogues Concerning Two New Sciences*. Here he described his wonderful discovery that distances covered in consecutive equal time intervals by balls rolling down inclined planes are proportional to the successive odd positive integers. Thus, if $d(n)$ denotes the distance covered during the nth time interval, then

$$d(n) = k(2n - 1),$$

where k is the proportionality constant. Galileo determined that the constant k depends only on the incline and not on the mass of the ball or the material of which it is composed. He reasoned that the same result should hold for freely falling bodies if air resistance is neglected.

Galileo's discovery of a mathematical model for uniformly accelerated motion is considered to have been the beginning of the science of dynamics. A **mathematical model** is an equation or set of equations in which variables represent real-world quantities—for instance, distance and time intervals in the equation $d(n) = k(2n - 1)$. A mathematical model is often an *idealization* of the real-world situation it supposedly describes; for instance, Galileo's mathematical model assumes a perfectly smooth inclined plane, no friction, and no air resistance.

In this section, we shall consider some of the mathematical models currently used in the life sciences to describe population growth and biological growth in general. The construction of realistic mathematical models nearly always requires the patient accumulation of experimental data, sometimes over a period of many years. In Figure 1, we have plotted points (t, N) showing the population N of the United States in the year $1790 + t$, according to the U.S. Census Bureau. These points seem to lie

Figure 1

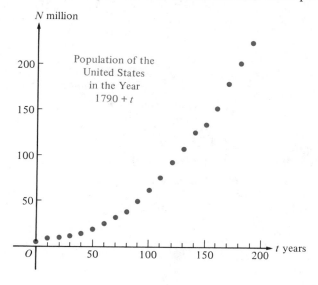

along a curve that is reminiscent of the graph of exponential growth (Figure 6a in Section 5.2) and they therefore suggest the mathematical model $N = N_0 e^{kt}$ for the growth of the population of the United States.

In 1798, the English economist Thomas Malthus made similar observations about the world population in his *Essay on the Principle of Population*. Because Malthus also proposed a linear model for the expansion of food resources, he forecast that the exponentially growing population would eventually be unable to feed itself. This dire prediction had such an impact on economic thought that the exponential model for population growth came to be known as the **Malthusian model.**

Consider a population growing according to the Malthusian model

$$N = N_0 e^{kt}.$$

Of course, N_0 is the population when $t = 0$; but what is the meaning of the growth constant k? To find out, let's consider how the population changes in 1 year. At the beginning of the $(t + 1)$st year, the population is $N_0 e^{kt}$, and at the end of this year it is $N_0 e^{k(t+1)}$. During the year, the population increase is

$$N_0 e^{k(t+1)} - N_0 e^{kt} = N_0(e^{kt+k} - e^{kt}) = N_0(e^{kt}e^k - e^{kt}) = N_0 e^{kt}(e^k - 1).$$

The percentage of the increase in population during the year is therefore given by

$$\frac{\text{increase during the year}}{\text{population at the beginning of the year}} \times 100\% = \frac{N_0 e^{kt}(e^k - 1)}{N_0 e^{kt}} \times 100\%$$

$$= (e^k - 1) \times 100\%.$$

In other words, if the yearly percentage increase in population is expressed as a decimal K, we have

$$K = e^k - 1 \qquad \text{or} \qquad e^k = 1 + K;$$

so that,

$$k = \ln(1 + K)$$

Example 1   According to the U.S. Census Bureau, the population of the United States in 1980 was $N_0 = 226$ million. Suppose that the population grows according to the Malthusian model at 1.1% per year.

(a) Write an equation for the population N million of the United States t years after 1980.

(b) Predict N in the year 2000.

(c) Predict N in the year 2020.

Solution   (a) 1.1% expressed as a decimal is $K = 0.011$. Therefore,

$$k = \ln(1 + K) = \ln 1.011 \approx 0.011,$$

and we have

$$N = N_0 e^{kt} = 226 e^{0.011t}.$$

(b) In the year 2000, $t = 20$ and

$$N = 226e^{0.011(20)} = 226e^{0.22} \approx 282 \text{ million.}$$

(c) In the year 2020, $t = 40$ and

$$N = 226e^{0.011(40)} = 226e^{0.44} \approx 351 \text{ million.}$$

     In Example 1, the growth constant k, rounded off to two significant digits, is the same as the yearly percentage of population increase expressed as a decimal K. This is no accident. Indeed, if K is small, then we can use the approximation $e^x \approx 1 + x$ (Section 5.2, page 272) with $x = K$ to get $e^K \approx 1 + K$, so that

$$k = \ln(1 + K) \approx \ln e^K = K.$$

If $K \leq 0.06$ (6% per year), the error in the approximation $k \approx K$ is less than 3%.

     Whenever exponential growth is involved as in the Malthusian model, it is interesting to ask when the growing quantity doubles. If T is the **doubling time,** then

$$N_0 e^{k(t + T)} = 2N_0 e^{kt} \quad \text{or} \quad e^{kt + kT} = 2e^{kt};$$

that is

$$e^{kt} e^{kT} = 2e^{kt} \quad \text{or} \quad e^{kT} = 2.$$

Thus, $$kT = \ln 2,$$

and we have the formula

$$T = \frac{\ln 2}{k} \approx \frac{\ln 2}{K}$$

for the doubling time.

© **Example 2**   If the population of the United States grows according to the Malthusian model at 1.1% per year, in approximately how many years will it double?

Solution   Here $K = 0.011$, so $T \approx \dfrac{\ln 2}{K} \approx \dfrac{\ln 2}{0.011} \approx 63$ years.

The Logistic Growth Model

When one-celled organisms reproduce by simple cell division in a culture containing an unlimited supply of nutrients, the Malthusian or exponential model $N = N_0 e^{kt}$ for the number N of organisms at time t is often quite accurate. In a natural environment, however, growth is often inhibited by various constraints that have a greater and greater effect as time goes on—depletion of the food supply, build-up of toxic wastes, physical crowding, and so on—and the Malthusian model may no longer apply.

If N_0 organisms are introduced into a habitat at time $t = 0$ and the population N of these organisms is plotted as a function of time t, the result is often an S-shaped curve (Figure 2). Typically, such a curve shows an increasing rate of growth up to a point P_I, called the **inflection point,** followed by a declining rate of growth as the population N levels off and approaches the maximum N_{max} that can be supported by the habitat. The horizontal line $N = N_{max}$ is an asymptote of the graph. Notice that the graph is bending upward to the left of the inflection point and bending downward to the right of it.

Figure 2

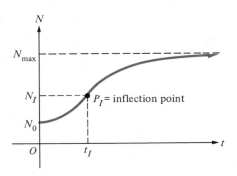

There are many equations that have graphs with the characteristic S-shape of Figure 2, and that are therefore used as mathematical models for population growth. Of these, one of the most popular is the **logistic model**

$$N = \frac{N_0 N_{max}}{N_0 + (N_{max} - N_0)e^{-kt}}$$

(Problem 8). Using calculus, it can be shown that the coordinates (t_I, N_I) of the inflection point P_I of the logistic model are

$$t_I = \frac{1}{k} \ln \frac{N_{max} - N_0}{N_0}, \qquad N_I = \frac{N_{max}}{2}.$$

Using logarithms, we can solve the logistic equation for t in terms of N; the result is

$$t = \frac{1}{k} \ln \frac{N(N_{max} - N_0)}{N_0(N_{max} - N)}$$

(Problem 10).

© **Example 3**   Suppose that the population N million of the United States grows according to the logistic model

$$N = \frac{N_0 N_{max}}{N_0 + (N_{max} - N_0)e^{-kt}},$$

where t is the time in years since 1780, and where $k = 0.03$. If the population in 1780 was 3 million and the population in 1880 was 50 million:

(a) Find $N_{\max}$.

(b) Find N in the year 1980.

(c) Find N in the year 2000.

(d) Determine the year in which the inflection occurred.

Solution **(a)** From the data given, $N_0 = 3$ million in 1780 when $t = 0$, while in 1880 when $t = 100$, $N = 50$ million; that is,

$$50 = \frac{N_0 N_{\max}}{N_0 + (N_{\max} - N_0)e^{-kt}} = \frac{3N_{\max}}{3 + (N_{\max} - 3)e^{-0.03(100)}}.$$

It follows that

$$50[3 + (N_{\max} - 3)e^{-3}] = 3N_{\max}$$

or

$$150 + 50e^{-3}N_{\max} - 150e^{-3} = 3N_{\max}.$$

Thus,

$$(3 - 50e^{-3})N_{\max} = 150(1 - e^{-3}),$$

so, rounding off to two significant digits, we have

$$N_{\max} = \frac{150(1 - e^{-3})}{3 - 50e^{-3}} \approx 280 \text{ million.}$$

(b) In 1980, we have $t = 200$ years and

$$N = \frac{N_0 N_{\max}}{N_0 + (N_{\max} - N_0)e^{-kt}} = \frac{3(280)}{3 + (280 - 3)e^{-0.03(200)}} \approx 230 \text{ million}$$

[*Note:* The correct value according to the 1980 census was 226 million, so our logistic model has predicted the correct value with an error of less than 2%.]

(c) In the year 2000, $t = 220$ years and

$$N = \frac{3(280)}{3 + (280 - 3)e^{-0.03(220)}} \approx 250 \text{ million.}$$

[Compare this with the prediction of 282 million according to the Malthusian model (Example 1). Notice that the "leveling off" built into the logistic model has apparently taken hold.]

(d) $t_I = \dfrac{1}{k} \ln \dfrac{N_{\max} - N_0}{N_0} = \dfrac{1}{0.03} \ln \dfrac{280 - 3}{3} = \dfrac{1}{0.03} \ln \dfrac{277}{3} \approx 150$ years.

Thus, according to the logistic model, the inflection would have taken place in the year $1780 + 150 = 1930$.

Problem Set 5.6

C **1.** The population of a small country was 10 million in 1980 and it is growing according to the Malthusian model 3% per year.

    (a) Write an equation for the population N million of the country t years after 1980.

    (b) Predict N in the year 2000.

    (c) Find the doubling time T for the population.

C **2.** The population of a certain city is growing according to the Malthusian model and it is expected to double in 35 years. Approximately what percent of growth will occur in this population over 1 year?

C **3.** The number of bacteria in an unrefrigerated chicken salad doubles in 3 hours. Assuming exponential growth, in how many hours will the number of bacteria be increased by a factor of 10?

4. Suppose that a population that is growing according to the Malthusian model increases by $100K\%$ per year. Write an *exact* formula (not an approximation) for the doubling time T in terms of K.

C **5.** The bacterium *Escherichia coli* is found in the human intestine. When *E. coli* is cultivated under ideal conditions in a biological laboratory, the population doubles in $T = 20$ minutes. Suppose that $N_0 = 10$ *E. coli* cells are placed in a nutrient broth medium extracted from yeast at time $t = 0$ minutes.

    (a) Write an equation for the number N of *E. coli* bacteria in the colony t minutes later.

    (b) Find N when $t = 60$ minutes.

6. Suppose that a population is growing according to the Malthusian model with doubling time T. If N_0 is the original size of the population when $t = 0$, show that the population t units of time later is given by $N = N_0 2^{t/T}$.

C **7.** The fruit fly *Drosophila melanogaster* is often used by biologists for genetic experiments because it breeds rapidly and has a short life cycle. Suppose a colony of *D. melanogaster* in a laboratory is observed to double in size in $T = 2$ days. Assuming a Malthusian model for growth of the colony, determine the approximate daily percentage increase in the size of the colony.

8. For the logistic model

$$N = \frac{N_0 N_{max}}{N_0 + (N_{max} - N_0)e^{-kt}},$$

show that:

    (a) $N = N_0$ when $t = 0$.

    (b) $N = N_{max}$ is a horizontal asymptote of the graph of N as a function of t. [*Hint:* As t gets larger and larger, e^{-kt} gets closer and closer to 0.]

C **9.** Suppose that a herd of 300 deer, newly introduced into a game preserve, grows according to the logistic model (Problem 8) with $k = 0.1$. Assume that the herd has grown to 387 deer after 5 years.

    (a) Find N_{max}, the maximum possible size of the herd in this habitat.

    (b) Find the population of the herd after 7 years.

    (c) When does the inflection occur in the population of the herd?

10. Solve the logistic equation (Problem 8) for t in terms of N.

C **11.** Sketch an accurate graph of the population N of the deer herd in Problem 9 as a function of the time t in years since the herd was introduced into the habitat.

12. If a certain population N is thought to be growing according to the logistic model (Problem 8), the value of the constant k is often found experimentally as follows: First, an estimate is made for N_{max}, the maximum possible size of the population in the given habitat. Then values of N are measured corresponding to several different values of t and

$$y = \ln \frac{N}{N_{max} - N}$$

is calculated for each value of N. The points (t, y) are plotted on a graph (Figure 3) and a straight line L that "best fits" these points is drawn. The constant k is taken to be the slope of the line L. Justify this procedure. [*Hint:* Use the result of Problem 10 to show that $kt = y - b$, where b is the value of y when $t = 0$.]

Figure 3

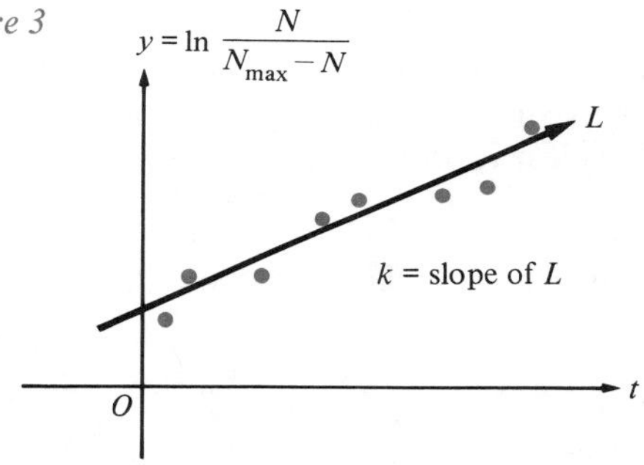

13. Consider the alternative growth model

$$N = N_{max}\left[1 - \left(1 - \frac{N_0}{N_{max}}\right)e^{-ct}\right]$$

for the deer herd in Problem 9, where $c = 0.05$, $N_0 = 300$, and $N_{max} = 700$. Sketch the graph of N as a function of t according to this model and compare it with the graph in Problem 11.

14. Outline a procedure for determining the constant c for the alternative growth model in Problem 13. The procedure should be similar to that in Problem 12, but with

$$y = \ln \frac{N_{max}}{N_{max} - N}.$$

15. The **Gompertz growth model** is $N = N_{max}\exp(-Be^{-Ct})$, where $B = \ln \frac{N_{max}}{N_0}$ and C is a positive constant. (Recall that $\exp x = e^x$.) Take $N_0 = 300$, $N_{max} = 700$, and $C = 0.07$; sketch the resulting graph of N as a function of t; and compare it with the graph in Problem 11.

16. For the Gompertz growth model (Problem 15), show that:

(a) N_0 is the value of N when $t = 0$.

(b) $N = N_{max}$ is a horizontal asymptote of the graph.

REVIEW PROBLEM SET, CHAPTER 5

1. Let $f(x) = 3^x$, $g(x) = (\frac{1}{3})^x$, and $h(x) = 2^{-x}$. Find:

(a) $f(3)$ (b) $g(0)$ (c) $g(-1)$
(d) $h(2)$ (e) $f[h(1)]$ (f) $g(\frac{1}{2})$

2. Using a calculator with a y^x key, find the value of each quantity to as many significant digits as you can.

(a) $0.47308^{7.7703}$ (b) $32.273^{-0.35742}$

In Problems 3 to 10, sketch the graph of the given function, determine its domain, its range, and any horizontal or vertical asymptotes, and indicate whether the function is increasing or decreasing. (Just make a rough sketch—do not use a calculator or tables.)

3. $f(x) = 2^{3+x}$
4. $g(x) = (\frac{1}{3})^{2-x}$
5. $F(x) = (\frac{1}{10})^{x+1}$
6. $G(x) = 6^{-x} - 1$
7. $H(x) = 2^{-x} + 3$
8. $h(x) = 3 + (\frac{1}{3})^{x-3}$
9. $F(x) = 1 - 2(6^{-x})$
10. $g(x) = 4(\frac{1}{10})^{1-x} + 2$

11. A savings account earns a nominal annual interest of 6% compounded quarterly. If there is $5000 in the account now, how much money will it contain after 5 years?

© 12. A certain bank advertises interest at a nominal annual rate of 6.5% compounded hourly. Find the effective simple annual interest rate.

© 13. Suppose that your local savings bank offers certificates of deposit paying nominal annual interest of 12% compounded semiannually. Use this interest rate to determine the present value to you of money in the future. If a debtor offers to pay you $1000 eighteen months from now, what is the present value to you of this offer?

14. True or false: If $f(x) = b^x$, then $f(x^2) = [f(x)]^2$. Justify your answer.

© 15. Use a calculator to find the value of each quantity to as many significant digits as you can.

(a) $e^{\sqrt{11}}$ (b) $e^{-\pi}$

© 16. Use a calculator to verify that $e^{x+y} = e^x e^y$ for $x = 7.7077$ and $y = 3.0965$.

In Problems 17 to 22, sketch the graph of the given function, determine its domain, its range and any horizontal or vertical asymptotes, and indicate whether the function is increasing or decreasing. (Just make a rough sketch—do not use a calculator or tables.)

17. $f(x) = 3 + e^x$

18. $g(x) = 1 - e^{-x}$

19. $h(x) = 4e^{x-4}$

20. $F(x) = 3 - e^{2-x}$

21. $G(x) = 3e^{x-1} + 2$

22. $H(x) = 2 + \left(\dfrac{1}{e}\right)^x$

© 23. A savings bank with $28,000,000 in regular savings accounts is paying interest at a nominal rate of 5.5% compounded quarterly. The bank is contemplating offering the same rate of interest, but compounding continuously rather than quarterly. How much more interest will the bank have to pay out per year if the new plan is adopted?

24. Infusion of a glucose solution into the bloodstream is a standard medical technique. Medical technicians use the formula $y = A + (B - A)e^{-kt}$, where A, B, and k are positive constants, to determine the concentration y of glucose in the blood t minutes after the beginning of the infusion.

(a) Interpret the meaning of the constant B.

(b) Interpret the meaning of the constant A.

(c) The constant k has to do with the rate at which the glucose is converted and removed from the bloodstream. If the glucose is being removed from the bloodstream rapidly, would this suggest that k is relatively large or small?

© 25. Assume that y grows or decays exponentially as a function of x, that y_0 is the value of y when $x = 0$, and that k is the growth or decay constant. Find the indicated quantity from the information given.

(a) Growth, $k = 5$, $y_0 = 3$. Find y when $x = 2$.

(b) Decay, $k = 0.12$, $y_0 = 5000$. Find y when $x = 10$.

(c) Growth, $y = 5$ when $x = 1$, $y = 7$ when $x = 3$. Find y_0.

(d) Decay, $k = 4$. Find the percent of change in y when x changes from 10 to 11.

© 26. Using a calculator, determine the percent error in approximating e by $[1 + (1/n)]^n$ for $n = 1000$. [See Problem 24 in Problem Set 5.2.]

In Problems 27 to 38, (a) rewrite each logarithmic equation as an exponential equation and (b) solve the exponential equation without using a calculator or tables.

27. $x = \log_5 \frac{1}{125}$

28. $\log_x 16 = -2$

29. $\log_3(2 + x) = 1$

30. $\log_4 |2x| = 1$

31. $\log_3 81 = 5x$

32. $x = \log_3 9\sqrt{3}$

33. $\log_x \frac{1}{49} = 2$

34. $\log_5 |3x - 5| = 0$

35. $\log_3 \frac{1}{81} = -2x$

36. $\ln(x^2 - 4) = 0$

37. $1 + \log_7 49 = |x|$

38. $\log |2x - 1| = 2$

39. Rewrite each exponential equation as an equivalent logarithmic equation.

(a) $3^6 = 729$ (b) $2^{-10} = \frac{1}{1024}$

(c) $64^{4/3} = 256$ (d) $x^a = w$

(e) $10^x = y$ (f) $e^x = y$

40. Solve each exponential equation without using a calculator or tables.

(a) $5^{2x} - 26(5^x) + 25 = 0$ (b) $2^x + 2^{-x} = 2$

(c) $7^{3x^2 - 2x} = 49(7^3)$

(d) $(2^{|x|+1} - 1)^{-1} = \frac{1}{3}$

41. Find each logarithm without using a calculator or tables.

(a) $\log_2 16$ (b) $\log_{16} 2$ (c) $\log_5 \sqrt{5}$

(d) $\log_7 1$ (e) $\log_3 \frac{1}{27}$ (f) $\ln e^{33}$

(g) $\log 100^{-0.7}$

42. Suppose that $\log_b 2 = 0.9345$, $\log_b 3 = 1.4812$, and $\log_b 7 = 2.6237$. Find each value.

(a) $\log_b 6$ (b) $\log_b \frac{7}{6}$ (c) $\log_b \frac{18}{7}$

(d) $\log_b 2^{64}$ (e) $\log_b 21$ (f) $\log_b 0.5$

(g) $\log_b 49$ (h) $\log_b \sqrt[3]{28}$ (i) $\log_b (36/\sqrt{56})$

(j) $\log_7 b$ ⓒ (k) b

ⓒ **43.** Use a calculator to find each value to as many significant digits as you can.

(a) $\log e$ (b) $\ln 100$ (c) $\ln 33.3$

(d) $2.07^{-15.32}$ (e) $\log(\log 14)$ (f) $\ln(\ln 200)$

(g) e^{e-1} (h) $\pi^{\pi^{\pi}}$ (i) $\log e^{e^e}$

ⓒ **44.** Use a calculator with y^x, e^x, and $\ln$ keys to verify the formula $y^x = e^{x \ln y}$ for the indicated values of x and y.

(a) $x = 3.22$, $y = 2.03$ (b) $x = 4.71$, $y = 0.83$

(c) $x = -4.01$, $y = 8.11$

In Problems 45 to 48, rewrite each expression as a sum or difference of multiples of logarithms. (Make the necessary assumptions about the values of the variables.)

45. $\log_b \left(\dfrac{\sqrt[n]{p}}{R^n} \right)$ **46.** $\ln(e^x q^n \sqrt{p})$

47. $\log \sqrt{\dfrac{4-x}{4+x}}$ **48.** $\log \dfrac{y^2 \sqrt{4-y^2}}{16(3y+7)^{3/2} y^4}$

In Problems 49 to 52, rewrite each expression as a single logarithm. (Make the necessary assumptions about the values of the variables.)

49. $a \log_b x + \dfrac{1}{c} \log_b y$

50. $4 \log(x^2 - 1) - 2 \log(x + 1)$

51. $\ln \dfrac{x^2 - 4}{x^2 - 3x - 4} - \ln \dfrac{x^2 - 3x - 10}{2x^2 + x - 1}$

52. $\ln \sqrt{\dfrac{x}{x+2}} + \ln \dfrac{\sqrt{x^2 - 4}}{x^2}$

In Problems 53 to 56, solve each equation.

53. $\log_3(x + 1) + \log_3(x + 3) = 1$

54. $\log_4(x + 3) - \log_4 x = 2$

55. $\log_3(7 - x) - \log_3(1 - x) = 1$

56. $\log_6(x + 3) + \log_6(x + 4) = 1$

ⓒ **57.** Use a calculator and the base-changing formula to evaluate

(a) $\log_2 11$ (b) $\log_8 10^{10}$

(c) $\log_{1/3} 0.707$ (d) $\log_{\sqrt{5}} \sqrt{2}$

58. Use Appendix Table IB (with linear interpolation when necessary) to find the common logarithm of each number rounded off to four decimal places.

(a) $\log 5.43$ (b) $\log 5430$

(c) $\log 0.00543$ (d) $\log 8.03$

(e) $\log 0.000803$ (f) $\log 903{,}200$

(g) $\log 7152$ (h) $\log 0.001347$

ⓒ In Problems 59 to 64, use a calculator to verify each equation for the indicated values of the variables.

59. $\log \sqrt{x} = \frac{1}{2} \log x$ for $x = 33.20477$

60. $\log(xy) = \log x + \log y$ for $x = 72.1355$, $y = 0.774211$

61. $\ln x = \dfrac{\log x}{\log e}$ for $x = 77.0809$

62. $\log_a b = \dfrac{1}{\log_b a}$ for $a = 10$, $b = e$

63. $\log y^x = x \log y$ for $y = 0.001507$, $x = 10.1333$

64. $\log \dfrac{x}{y} = \log x - \log y$ for $x = 5.3204 \times 10^{-7}$, $y = 3.2211 \times 10^3$

ⓒ In Problems 65 to 70, *an error has been made*. In each case, show that the equation is false by evaluating both sides with the aid of a calculator.

65. $\dfrac{\log \pi}{\log e} = \log \pi - \log e$?   **66.** $\log 6 = (\log 2)(\log 3)$?

67. $\dfrac{1}{\log \frac{2}{3}} = \log \frac{3}{2}$?   **68.** $-\log 5 = \log(-5)$?

69. $\log_2 3 = \log 8$?   **70.** $\dfrac{1}{\ln 2} + \dfrac{1}{\ln 3} = \dfrac{1}{\ln 5}$?

71. Find the domain of each function.

(a) $f(x) = \log_2(4x - 3)$ (b) $g(x) = \ln|x + 1|$
(c) $h(x) = \ln(x^2 - 4x - 4)$ (d) $F(x) = \log e^x$

72. Find the y and x intercepts of the graph of each function in Problem 71.

© In Problems 73 to 78, determine the domain and range of each function; find any y or x intercepts and any horizontal or vertical asymptotes of its graph; indicate where the function is increasing or decreasing; and sketch the graph.

73. $f(x) = 1 + \ln(x - 1)$ **74.** $g(x) = 1 - \log x^3$

75. $h(x) = \log \sqrt{4 - x}$ **76.** $F(x) = \dfrac{1}{\log_x 10}$

77. $G(x) = \ln|2 - x|$ **78.** $H(x) = \log_{1/e} x$

In Problems 79 to 82, simplify each expression.

79. $\dfrac{10^{\log(x^2 - 3x + 2)}}{x - 2}$ **80.** $e^{(1/2)\ln(x^2 - 2x + 1)}$

81. $\dfrac{\log 10^{x^2 - 4x - 5}}{x + 1}$ **82.** $\dfrac{\log y^x}{10^{\log x}}$

© In Problems 83 to 88, use logarithms and a calculator to solve each exponential equation.

83. $10^x = 20$ **84.** $e^{x+3} = 5^{-2}$

85. $10^{x+7} = 14^3$ **86.** $3^{x-1} = (15.4)^x$

87. $5^{2x} = 4(3^x)$ **88.** $3^{2x-1} = 4^{x+2}$

© **89.** As part of a psychological experiment, a group of students took an exam on material they had just studied in a physics course. At monthly intervals thereafter they were given equivalent exams. After t months, the average score S of the group was found to be given by $S = 78 - 55 \log(t + 1)$. What was the average score (a) when they took the original exam and (b) 5 months later? (c) When will the group have forgotten everything they learned?

© **90.** In 1935, the U.S. seismologist Charles Richter established the **Richter scale** for measuring the *magnitude M* of an earthquake in terms of the total energy E joules released by it. One form of Richter's equation is $M = \frac{2}{3}\log(E/E_0)$, where E_0

is a minimum amount of released energy used for comparison. An earthquake that measures 5.5 on the Richter scale, a magnitude corresponding to an energy release of 10^{13} joules, can cause local damage. An earthquake of magnitude 6 produces considerable destruction, and an earthquake of magnitude 7 or higher can result in catastrophic damage.

(a) Find E_0.

(b) The most powerful earthquake ever recorded occurred in Colombia on January 31, 1906, and measured 8.6 on the Richter scale. Approximately how many joules of energy were released?

© **91.** Suppose that a supersonic plane on takeoff produces 120 decibels of sound. How many decibels of sound would be produced by two such planes taking off side by side?

© **92.** An alternative method for measuring the loudness of sound is the *sone* scale. Whereas loudness S in decibels is given by $S = 10 \log(P/P_0)$, loudness s in sones is given by $s = 10^{2.4}P^{0.3}$, where P is the intensity of the sound wave in watts per square meter, and $P_0 = 10^{-12}$ watt per square meter. Show that a 10-decibel increase in loudness doubles the loudness on the sone scale.

© **93.** In 1921, President Warren G. Harding presented Marie Curie a gift of 1 gram of radium on behalf of the women of the United States. Using the fact that the half-life of radium is 1656 years, determine how much of the 1-gram gift was left in 1981.

94. A couple wishes to invest their life savings, P dollars, but finds that inflation is reducing the value of a dollar by $100K\%$ per year. They also find that they will have to pay state and local income taxes amounting to $100Q\%$ of their interest at the end of each year. Write a formula for the nominal annual interest rate r, compounded continuously, that the couple must receive on their investment just to break even at the end of the year.

© **95.** You have just won first prize in a state lottery and you have your choice of the following: (a) $30,000 will be placed in a savings account in your name, and the money will be compounded continuously at a nominal annual rate of 10% or (b) one penny will be placed in a fund in your name, and the amount in the fund will be doubled every 6 months over the next 12 years. Which plan do you choose and why?

96. Suppose a bank offers savings accounts at a nominal annual rate r compounded n times per year. By regularly depositing a fixed sum of p dollars in such an account every $\frac{1}{n}$th of a year, money can be accumulated for future needs. Such a plan is called a **sinking fund.** If it is required that such a sinking fund yield an amount S dollars after a term of t years, the periodic payment p dollars must be

$$p = \frac{Sr}{n\left[\left(1 + \frac{r}{n}\right)^{nt} - 1\right]}.$$

If a restaurant anticipates a capital expenditure of $80,000 for expanding in 5 years, how much money should be deposited quarterly in a sinking fund, at a nominal annual interest rate of 10% compounded quarterly, in order to provide the required amount for the expansion?

97. An interest-bearing debt is said to be **amortized** if the principal P dollars and the interest I dollars are paid over a term of t years by regular payments of p dollars every $\frac{1}{n}$th of a year. The amortization formulas are

$$p = \frac{Pr}{n\left[1 - \left(1 + \frac{r}{n}\right)^{-nt}\right]}$$

and
$$I = np\left[t - \frac{1 - \left(1 + \frac{r}{n}\right)^{-nt}}{r}\right].$$

Suppose you want to buy a car and need to borrow $P = 6000 from your local bank. The bank charges a nominal annual interest of 14% ($r = 0.14$) on automobile loans with monthly payments ($n = 12$) over a term of 36 months ($t = 3$ years). Use the amortization formulas to calculate (a) your monthly payment p dollars and (b) the total interest charge I dollars.

98. An automobile dealer's advertisement reads: "No money down, $300 per month for 36 months, 12

percent annual interest rate compounded monthly on the unpaid balance, puts you in the driver's seat of a new Wildebeest." Use the amortization formulas (Problem 97) to figure out how much of the $10,800 to be paid goes toward the car and how much is interest. [Take $p = 300$, $t = 3$, $r = 0.12$, and $n = 12$.]

99. The amortization formulas (Problem 97) with $n = 12$ apply if you borrow P dollars from a bank for a home mortgage at a nominal annual interest rate r payable in successive equal monthly payments of p dollars each over a term of t years. The balance due to the bank at the beginning of the kth month, P_k dollars, is customarily considered to be

$$P_k = P\frac{\left(1 + \frac{r}{12}\right)^{12t} - \left(1 + \frac{r}{12}\right)^{k-1}}{\left(1 + \frac{r}{12}\right)^{12t} - 1}.$$

Suppose that you purchase a new home for $65,000, paying $15,000 down, and taking out a 30-year mortgage on the remaining $50,000 at a nominal annual interest rate of 12%.

(a) What is your monthly payment on the mortgage?

(b) After 15 years, how much of the original $50,000 will be paid off?

100. In Problem 99, what part of your 180th payment goes toward interest and what part goes toward reduction of the principal?

101. Suppose that the population N of cancer cells in an experimental culture grows exponentially as a function of time. If $N_0 = 3000$ cells are present when $t = 0$ days, and if the number of cells is increasing at a rate of 0.35% per day:

(a) How many cells are present in the culture after 100 days?

(b) What is the approximate doubling time?

102. Suppose that y is growing or decaying exponentially as a function of x and that k is the growth or decay constant. Assume that $y = y_1$ when $x = x_1$ and that $y = y_2$ when $x = x_2$. If $x_1 \neq x_2$, prove that

$$k = \frac{1}{x_2 - x_1}\ln\frac{y_2}{y_1}.$$

103. In 1960, the American scientist Willard Libby received the Nobel Prize in Physical Chemistry for his discovery of the technique of radiocarbon dating. When a plant or animal dies, it receives no more of the naturally occurring radioactive carbon C^{14} from the atmosphere. Libby developed methods for determining the fraction F of the original C^{14} that is left in a fossil and made an experimental determination of the half-life T of C^{14}.

(a) If F is the fraction of the original C^{14} left in a fossil, show that the original plant or animal died t years ago, where $t = -T(\ln F/\ln 2)$.

© (b) An ancient scroll is unearthed, and it is determined that it contains only 76% of its original C^{14}. If $T = 5580$ years, how old is the scroll?

© **104.** According to **Newton's law of cooling,** if a hot object with initial temperature T_0 is placed at time $t = 0$ in a surrounding medium with a lower temperature T_1, the object cools in such a way that its temperature T at time t is given by

$$T = T_1 + (T_0 - T_1)e^{-kt},$$

where the constant k depends on the materials involved. Suppose an iron ball is heated to 300°F and allowed to cool in a room where the air temperature is 80°F. If after 10 minutes the temperature of the ball has dropped to 250°F, what is its temperature after 20 minutes?

© **105.** A new species of plant is introduced on a small island. Assume that 500 plants were introduced initially, and that the number N of plants increases according to the logistic model (Section 5.6, page 304) with $k = 0.62$. Suppose that there are 1700 plants on the island 2 years after they were introduced.

(a) Find N_{max}, the maximum possible number of plants the island will support.

(b) Find N after 10 years.

(c) When does the inflection occur?

(d) Sketch a graph of N as a function of time t in years.

© **106.** A rumor started by a single individual is spreading among students at a university. Assume that the number N of students who have heard the rumor t days after it was started grows according to the logistic model, with $N_0 = 1$ and $N_{max} = 20,000$ (the entire student body). If 10,000 students have heard the rumor after 2 days:

(a) Find the value of k.

(b) Determine how long it takes until 15,000 students have heard the rumor.

Chapter 6

Trigonometric Functions

Hipparchus

Hipparchus, a Greek astronomer and mathematician who lived in the second century B.C., is often regarded as the originator of the science of trigonometry. The word "trigonometry" is derived from the Greek words "trigon" for triangle and "metra" for measurement; it refers to "triangle measurement" based on the relationship between the vertex angles of a right triangle and ratios of sides of the triangle. The six *trigonometric functions*, which are defined in terms of these ratios, are used routinely in calculations made by surveyors, engineers, and navigators. Trigonometric functions also have applications in the physical and life sciences, where they provide mathematical models for periodic phenomena—for instance, the ebb and flow of the tides, electromagnetic waves, seasonal variations in an animal's food supply, or the alpha rhythms of the human brain.

We begin this chapter by considering angles and their measure. We define the trigonometric functions, first for acute angles in a right triangle and then for general angles. We study the evaluation of trigonometric functions, circular functions, graphs of trigonometric functions, and the simple harmonic model.

6.1 ANGLES AND THEIR MEASURE

In order to define the trigonometric functions so that they can be used not only for triangle measurement but also for modeling periodic phenomena, we must give a definition of an angle that is somewhat more general than a vertex angle of a triangle.

If A and B are distinct points, the portion of the straight line that starts at A and continues indefinitely through B is called a **ray** with **endpoint** A (Figure 1a). An **angle** is determined by rotating a ray about its endpoint. This rotation can be indicated by a curved arrow as in Figure 1b. The endpoint of the rotated ray is called the **vertex** of the angle. The position of the ray before the rotation is called

Figure 1

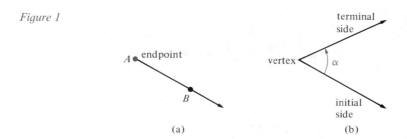

(a)   (b)

the **initial side** of the angle, and the position of the ray after the rotation is called the **terminal side.** Angles will often be denoted by small Greek letters, such as angle α in Figure 1b (α is the Greek letter *alpha*).

Angles determined by a *counterclockwise* rotation are said to be **positive** and angles determined by a *clockwise* rotation are said to be **negative.** In Figure 2a, angle β is positive (β is the Greek letter *beta*); in Figure 2b, angle γ is negative (γ is the Greek letter *gamma*).

Figure 2

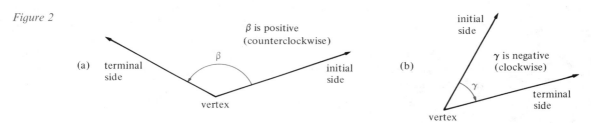

One full turn of a ray about its endpoint is called **one revolution** (Figure 3a); one-half of a revolution is called a **straight angle** (Figure 3b); and one-quarter of a revolution is called a **right angle** (Figure 3c). The initial and terminal sides of a right angle are perpendicular; hence, in drawing a right angle, we often replace the usual curved arrow by two perpendicular line segments (Figure 3c).

Figure 3

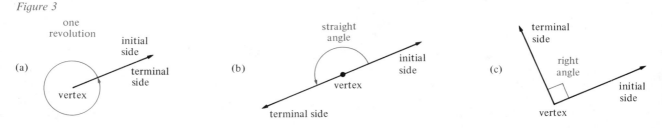

Angles that have the same initial sides and the same terminal sides are called **coterminal** angles. For instance, Figure 4 shows three coterminal angles α, β, and γ. Because different angles can be coterminal, an angle is not completely determined merely by specifying its initial and terminal sides. Nevertheless, angles are often named by using three letters such as ABC to denote a point A on the initial side, the vertex B, and a point C on the terminal side (Figure 5). Usually, the

Figure 4

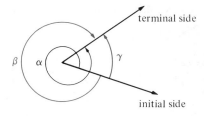

Figure 5

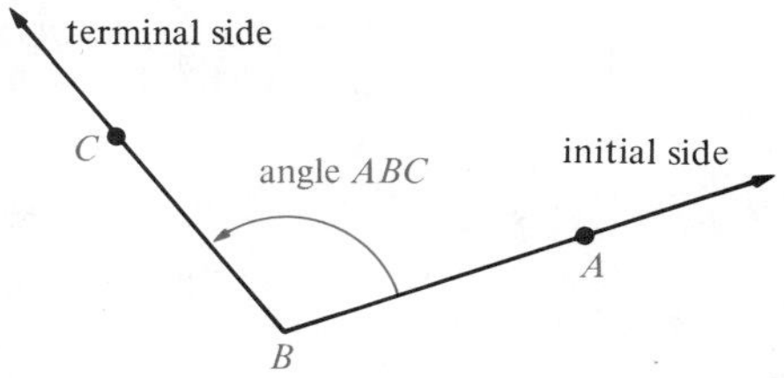

intended angle is the *smallest* rotation about the vertex that carries the initial side around to the terminal side. Sometimes, when it is perfectly clear which angle is intended, we refer to the angle by simply naming its vertex. For instance, the angle in Figure 5 could be called angle *B*.

There are various ways to assign numerical measures to angles. Of these, the most familiar is the *degree* measure: **One degree** (1°) is the measure of an angle formed by $\frac{1}{360}$ of a counterclockwise revolution. Negative angles are measured by a negative number of degrees; for instance, $-360°$ is the measure of one clockwise revolution. A positive right angle is $\frac{1}{4}$ of a counterclockwise revolution, and it therefore has a measure of $\frac{1}{4}(360°) = 90°$. Similarly, 180° represents 180°/360° or $\frac{1}{2}$ of a counterclockwise revolution.

Example 1   What is the degree measure of one-eighth of a clockwise revolution?

Solution $\frac{1}{8}(-360°) = -45°$ ∎

Example 2   What fraction of a revolution is an angle of 30°?

Solution $30°/360° = \frac{1}{12}$ of a counterclockwise revolution.   ∎

Although fractions of a degree can be expressed as decimals, such fractions are sometimes given in "minutes" and "seconds." **One minute** (1′) is defined to be 1/60 of a degree and **one second** (1″) is defined to be 1/60 of a minute. Hence, one second is 1/3600 of a degree. Using the relationships

$$1' = \left(\frac{1}{60}\right)°, \qquad 1'' = \left(\frac{1}{3600}\right)°, \qquad 1° = 60', \qquad \text{and} \qquad 1' = 60'',$$

you can convert from degrees, minutes, and seconds to decimals and vice versa.*

Ⓒ **Example 3**   Express the angle measure 72°13′59″ as a decimal. Round off your answer to four decimal places.

Solution $72°13'59'' = 72° + \left(\frac{13}{60}\right)° + \left(\frac{59}{3600}\right)° \approx 72.2331°$ ∎

* Because calculators work with angles expressed in decimal form, minutes and seconds are not as popular as they once were. Some scientific calculators can convert decimal degrees into degrees, minutes, and seconds, and vice versa.

Example 4   Express the angle measure $173.372°$ in degrees, minutes, and seconds. Round off to the nearest second.

Solution
$$0.372° = 0.372(1°) = 0.372(60') = 22.32'$$

and
$$0.32' = 0.32(1') = 0.32(60'') = 19.2'' \approx 19''.$$

Hence,

$$173.372° = 173° + 0.372° = 173° + 22.32' = 173° + 22' + 0.32'$$
$$\approx 173° + 22' + 19'' = 173°22'19''. \qquad \blacksquare$$

Figure 6

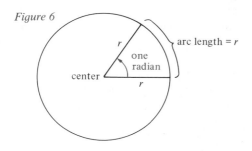

Although the degree measure of angles is used in most elementary applications of trigonometry, more advanced applications (especially those that involve calculus) require radian measure. **One radian** is the measure of an angle that has its vertex at the center of a circle (that is, a **central angle**) and intercepts an arc on the circle equal in length to the radius r (Figure 6).

A central angle of 2 radians in a circle of radius r intercepts an arc of length $2r$ on the circle, a central angle of $\frac{3}{4}$ radian intercepts an arc of length $\frac{3}{4}r$, and so forth. More generally, if a central angle AOB of θ radians (θ is the Greek letter *theta*) intercepts an arc $\overset{\frown}{AB}$ of length s on a circle of radius r (Figure 7), then we have

Figure 7

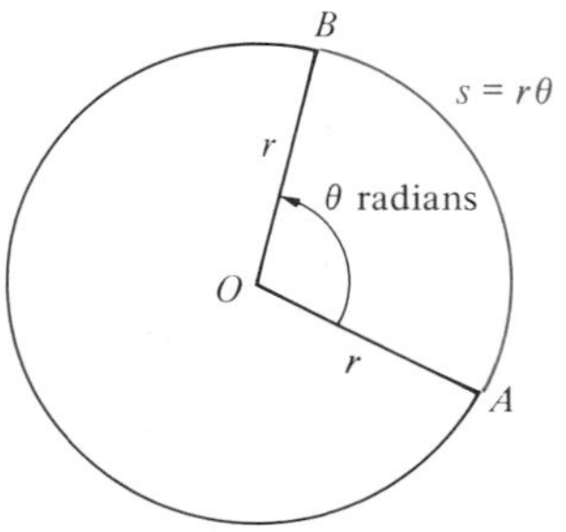

$$s = r\theta.$$

It follows that angle AOB has radian measure θ given by the formula

$$\theta = \frac{s}{r}.$$

Example 5   Find the length s of the arc intercepted on a circle of radius $r = 3$ meters by a central angle whose measure is $\theta = 4.75$ radians.

Solution $s = r\theta = 3(4.75) = 14.25$ meters. $\blacksquare$

Example 6   A central angle in a circle of radius 27 inches intercepts an arc of length 9 inches. Find the measure θ of the angle in radians.

Solution   Here $s = 9$ inches, $r = 27$ inches, and

$$\theta = \frac{s}{r} = \frac{9}{27} = \frac{1}{3} \text{ radian.} \qquad \blacksquare$$

Figure 8

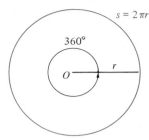

A central angle of $360°$ corresponds to one revolution; hence it intercepts an arc $s = 2\pi r$ equal to the entire circumference of the circle (Figure 8). Therefore, if θ is the radian measure of the $360°$ angle,

$$\theta = \frac{s}{r} = \frac{2\pi r}{r} = 2\pi \text{ radians};$$

that is, $360° = 2\pi$ radians, or

$$\boxed{180° = \pi \text{ radians.}}$$

You can use this relationship to convert degrees into radians and vice versa. In particular,

$$1° = \frac{\pi}{180} \text{ radian} \quad \text{and} \quad 1 \text{ radian} = \left(\frac{180}{\pi}\right)°.$$

Thus, we have the following *conversion rules:*

(i) Multiply degrees by $\dfrac{\pi}{180°}$ to convert to radians.

(ii) Multiply radians by $\dfrac{180°}{\pi}$ to convert to degrees.

Example 7   Convert each degree measure to radian measure.

(a) $60°$ (b) $-22.5°$

Solution   (a) $60° = \dfrac{\pi}{180°} (60°) \text{ radians} = \dfrac{\pi}{3} \text{ radians}$

(b) $-22.5° = \dfrac{\pi}{180°} (-22.5°) \text{ radian} = -\dfrac{\pi}{8} \text{ radian}$ ■

Example 8   Convert each radian measure to degree measure.

(a) $\dfrac{13\pi}{10} \text{ radians}$ (b) $-\dfrac{\pi}{4} \text{ radian}$

Solution   (a) $\dfrac{13\pi}{10} \text{ radians} = \dfrac{180°}{\pi} \left(\dfrac{13\pi}{10}\right) = 234°$ (b) $-\dfrac{\pi}{4} \text{ radian} = \dfrac{180°}{\pi} \left(-\dfrac{\pi}{4}\right) = -45°$ ■

As Examples 7 and 8 illustrate, when radian measures are expressed as rational multiples of π, we may leave them in that form rather than writing them as decimals.

Table 1

Degrees	Radians
0°	0
30°	$\pi/6$
45°	$\pi/4$
60°	$\pi/3$
90°	$\pi/2$
180°	π
360°	2π

Also, when angles are measured in radians, the word "radian" is often omitted. For instance, rather than writing $60° = \pi/3$ radians, we write $60° = \pi/3$.

> When no unit of angular measure is indicated, it is always understood that radian measure is intended.

Table 1 shows degree measures and the corresponding radian measures of several special angles.

Until now, we have been careful to distinguish between an *angle* (a rotation) and the *measure* of the angle in degrees or radians (a real number). However, continual use of such phrases as "the measure of the angle α is 30°" can become tedious, so people often write "$\alpha = 30°$" as an abbreviation for the correct statement. We shall follow this practice whenever it seems convenient and harmless.

Linear and Angular Speed

Figure 9

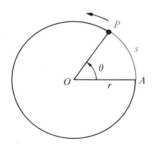

Angles measured in radians are especially useful in studying the motion of a particle moving at constant speed around a circle of radius r with center O (Figure 9). Suppose that the moving particle starts at the point A and that t units of time later it is at the point P. If the arc $\overset{\frown}{AP}$ has length s, then the particle has moved s units of distance in t units of time, and we say that its *speed* v is given by

$$v = \frac{s}{t} \text{ units of distance per unit of time.}$$

Here we assume that v is a constant.

Even though the particle in Figure 9 is moving around the circle, we speak of $v = s/t$ as its **linear speed** (since we could imagine the circular path straightened into a line). Now suppose that the radial line segment $\overline{OP}$ sweeps through an angle of θ radians in t units of time. Then we say that the **angular speed** of the particle is θ/t radians per unit of time. Angular speed is usually denoted by the Greek letter ω (*omega*), so we have

$$\omega = \frac{\theta}{t} \text{ radians per unit of time.}$$

To find the relationship between linear speed v and angular speed ω, we begin by recalling the formula

$$s = r\theta.$$

Dividing both sides of this equation by t, we obtain

$$\frac{s}{t} = r\frac{\theta}{t}$$

or

$$v = r\omega.$$

In words, *linear speed is the product of the radius and the angular speed.*

Example 9   A stone is whirled around in a circle at the end of a string 70 centimeters long. If it makes 5 revolutions in 2 seconds, find **(a)** its angular speed ω and **(b)** its linear speed v.

Solution   **(a)** Since the stone travels five times around the circle in 2 seconds, the radial line segment (the string) sweeps through $5(2\pi) = 10\pi$ radians in 2 seconds. Therefore, the angular speed of the stone is

$$\omega = \frac{\theta}{t} = \frac{10\pi}{2} = 5\pi \text{ radians per second.}$$

(b) The linear speed of the stone is

$$v = r\omega = 70(5\pi) = 350\pi$$
$$\approx 1100 \text{ centimeters per second.}$$ ■

Problem Set 6.1

In each problem set, problems with colored numbers constitute a good representation of the main ideas of the section.

In Problems 1 to 4, find the degree measure of each angle.

1. One-sixth of a counterclockwise revolution.

2. Two-thirds of a clockwise revolution.

3. Eleven-fifths of a clockwise revolution.

4. The angle through which the small hand on a clock turns in 24 hours.

In Problems 5 to 12, indicate the number of revolutions or the fraction of a revolution represented by the angle whose degree measure is given. Indicate whether the rotation is clockwise or counterclockwise.

5. $45°$

6. $-1080°$

7. $120°$

8. $800°$

9. $-330°$

10. $15°$

11. $-12°$

12. $-700°$

© In Problems 13 to 18, express each angle measure as a decimal.

13. $2'$

14. $5°6'0''$

15. $3''$

16. $20°45''$

17. $100°30'20''$

18. $-28°70'3''$

© In Problems 19 to 24, express each angle measure in degrees, minutes, and seconds.

19. $62.25°$

20. $-371.12°$

21. $-0.125°$

22. $880.23°$

23. $21.16°$

24. $1/17$ of a counterclockwise revolution

In Problems 25 to 30, s denotes the length of the arc intercepted on a circle of radius r by a central angle of θ radians. Find the missing quantity.

25. $r = 2$ meters, $\theta = 1.65$ radians, $s = ?$

26. $r = 1.8$ centimeters, $\theta = 8$ radians, $s = ?$

27. $r = 9$ feet, $s = 12$ feet, $\theta = ?$

28. $s = 4\pi$ kilometers, $\theta = \frac{\pi}{2}$ radians, $r = ?$

29. $r = 12$ inches, $\theta = \dfrac{5\pi}{18}$ radian, $s = ?$

30. $r = 5$ meters, $s = 13\pi$ meters, $\theta = ?$

31. Convert each degree measure to radians. Do not use a calculator. Write your answers as rational multiples of π.

   (a) $30°$ (b) $45°$ (c) $90°$

   (d) $120°$ (e) $-150°$ (f) $520°$

   (g) $72°$ (h) $67.5°$ (i) $-330°$

   (j) $450°$ (k) $21°$ (l) $-360°$

© 32. Use a calculator to convert each degree measure to an approximate radian measure expressed as a decimal. Round off all answers to five significant digits.

   (a) $7°$ (b) $33.333°$ (c) $-11.227°$

   (d) $571°$ (e) $1229°$ (f) $0.017320°$

33. Convert each radian measure to degrees. Do not use a calculator.

   (a) $\dfrac{\pi}{2}$ (b) $\dfrac{\pi}{3}$ (c) $\dfrac{\pi}{4}$ (d) $\dfrac{\pi}{6}$

   (e) $\dfrac{2\pi}{3}$ (f) $-\pi$ (g) $\dfrac{3\pi}{5}$ (h) $-\dfrac{5\pi}{2}$

   (i) $\dfrac{9\pi}{4}$ (j) $-\dfrac{3\pi}{8}$ (k) 7π (l) $-\dfrac{\pi}{14}$

© 34. Use a calculator to convert each radian measure to an approximate degree measure. Round off all answers to five significant digits.

   (a) $\dfrac{2}{3}$ (b) -2 (c) 200

   (d) $\dfrac{7\pi}{12}$ (e) 2.7333 (f) 1.5708

35. Indicate both the degree measure and the radian measure of the angle formed by (a) three-eighths of a clockwise revolution, (b) four and one-sixth counterclockwise revolutions, and (c) the hands of a clock at 4:00.

36. (a) Find a formula for the length s of the arc intercepted in a circle of radius r by a central angle of α *degrees.*

   (b) Use part (a) to find s if $r = 10$ meters and $\alpha = 35°$.

37. As a monkey swings on a 10-meter rope from one platform to another, the rope turns through an angle of $\pi/3$. How far does the monkey travel along the arc of the swing?

© 38. What radius should be used for a circular monorail track if the track is to change its direction by $7°$ in a distance of 24 meters (measured along the arc of the track)?

© 39. A motorcycle is moving at a speed of 88 kilometers per hour. If the radius of its wheels is 0.38 meter, find the angle in *degrees* through which a spoke of a wheel turns in 3 seconds.

40. A wheel of radius r is turning at a constant speed of n revolutions per second. Find a formula for (a) the angular speed ω and (b) the linear speed v of a point on the rim of the wheel.

© 41. A satellite is orbiting a certain planet in a perfectly circular orbit of radius 7680 kilometers. If it makes two-thirds of a revolution every hour, find (a) its angular speed ω and (b) its linear speed v.

© 42. **A nautical mile** may be defined as the arc length intercepted on the surface of the earth by a central angle of measure 1 minute. The radius of the earth is 2.09×10^7 feet. How many feet are there in a nautical mile?

© 43. A girl rides her bicycle to school at a speed of 12 miles per hour. If the wheel diameter is 26 inches, what is the angular speed of the wheels?

© 44. To determine the approximate speed of the water in a river, a water wheel of radius 1.5 meters is used. If the wheel makes 13 revolutions per minute, what is the speed of the river in kilometers per hour?

© 45. A $33\frac{1}{3}$-rpm (revolutions per minute) phonograph record has a radius of 14.6 centimeters. What is the linear speed (in centimeters per second) of a point on the rim of the record?

46. A first pulley of radius r_1 and a second pulley of radius r_2 are connected by a belt (Figure 10). The first pulley makes n_1 revolutions per minute and the second makes n_2 revolutions per minute. A point on the rim of either pulley will be moving at the same linear speed v as the belt.

   (a) Find formulas for n_1 and n_2 in terms of v, r_1, and r_2.

(b) Show that $n_1/n_2 = r_2/r_1$.

Figure 10

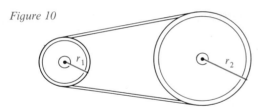

C 47. The earth, which is approximately 93,000,000 miles from the sun, revolves about the sun in a nearly circular orbit in approximately 365 days. Find the approximate linear speed (in miles per hour) of the earth in its orbit.

48. A **sector** *POQ* of a circle is the region inside the circle bounded by the arc $\overarc{PQ}$ and the radial segments $\overline{OP}$ and $\overline{OQ}$ (Figure 11). If the central angle *POQ* has a measure of θ radians and r is the radius of the circle, show that the area A of the sector is

given by $A = \frac{1}{2}r^2\theta$. [*Hint:* The area of the circle is πr^2. What fraction of this is the area of the sector?]

Figure 11

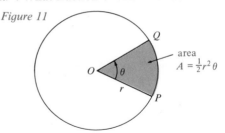

area $A = \frac{1}{2}r^2\theta$

49. Use the formula $A = \frac{1}{2}r^2\theta$ (Problem 48) to find the area A of a circular sector of radius r with a central angle of θ radians if (a) $r = 7$ centimeters and $\theta = 3\pi/14$ and (b) $r = 9$ inches and $\theta = 13\pi/9$.

50. A radar beam has an effective range of 70 miles and sweeps through an angle of $3\pi/4$ radians. What is the effective area in square miles swept by the radar beam?

6.2 TRIGONOMETRIC FUNCTIONS OF ACUTE ANGLES IN RIGHT TRIANGLES

In this section, we introduce the six functions—sine, cosine, tangent, cotangent, secant, and cosecant—upon which trigonometry is based. Here we define the values of these functions for *vertex* angles of right triangles; in Section 6.3 we extend the definitions to general angles.

An angle with a measure greater than 0° but less than 90° is called an **acute angle.** For instance, in a right triangle *ACB* (Figure 1a) the vertex angle *C* is a right angle (90°), and the other two vertex angles *A* and *B* are acute angles. Recall that the side $\overline{AB}$ opposite the right angle *C* is called the *hypotenuse.* If we denote one of the two acute angles by θ, we can abbreviate the lengths of the side **opposite** θ, the **hypotenuse,** and the side **adjacent** to θ as **opp, hyp,** and **adj,** respectively (Figure 1b).

Figure 1

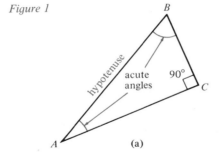

(a)

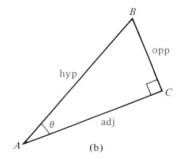

(b)

The six trigonometric functions of the acute angle θ are defined as follows:

Name of Function	Abbreviation	Value at θ
sine	sin	$\sin \theta = \dfrac{opp}{hyp}$
cosine	cos	$\cos \theta = \dfrac{adj}{hyp}$
tangent	tan	$\tan \theta = \dfrac{opp}{adj}$
cotangent	cot	$\cot \theta = \dfrac{adj}{opp}$
secant	sec	$\sec \theta = \dfrac{hyp}{adj}$
cosecant	csc	$\csc \theta = \dfrac{hyp}{opp}$

The values of the six trigonometric functions depend only on the angle θ and not on the size of the right triangle ACB in Figure 1. Indeed, two right triangles with the same acute vertex angle θ are similar (Figure 2); hence, ratios of corresponding sides are equal.*

Figure 2

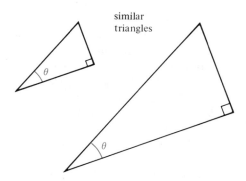

similar triangles

Figure 3

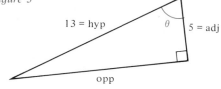

Example 1   Find the values of the six trigonometric functions of the acute angle θ in Figure 3.

Solution   For angle θ, we have

$$adj = 5 \qquad \text{and} \qquad hyp = 13,$$

but opp isn't given. However, by the Pythagorean theorem,*

$$(adj)^2 + (opp)^2 = (hyp)^2,$$

* See Appendix III for a review of the geometry of triangles.

so that

$$(\text{opp})^2 = (\text{hyp})^2 - (\text{adj})^2 = 13^2 - 5^2 = 144.$$

Therefore,
$$\text{opp} = \sqrt{144} = 12.$$

It follows that

$$\sin \theta = \frac{\text{opp}}{\text{hyp}} = \frac{12}{13} \qquad \csc \theta = \frac{\text{hyp}}{\text{opp}} = \frac{13}{12}$$

$$\cos \theta = \frac{\text{adj}}{\text{hyp}} = \frac{5}{13} \qquad \sec \theta = \frac{\text{hyp}}{\text{adj}} = \frac{13}{5}$$

$$\tan \theta = \frac{\text{opp}}{\text{adj}} = \frac{12}{5} \qquad \cot \theta = \frac{\text{adj}}{\text{opp}} = \frac{5}{12}.$$

If you look at the ratios that define the six trigonometric functions, you will see that

$$\tan \theta = \frac{\sin \theta}{\cos \theta} \qquad \cot \theta = \frac{\cos \theta}{\sin \theta}$$

$$\sec \theta = \frac{1}{\cos \theta} \qquad \csc \theta = \frac{1}{\sin \theta}.$$

For instance,

$$\tan \theta = \frac{\text{opp}}{\text{adj}} = \frac{\text{opp/hyp}}{\text{adj/hyp}} = \frac{\sin \theta}{\cos \theta},$$

and the remaining equations are derived similarly (Problem 42). Perhaps the easiest way for you to become familiar with all six trigonometric functions is to learn the two basic definitions

$$\sin \theta = \frac{\text{opp}}{\text{hyp}} \qquad \text{and} \qquad \cos \theta = \frac{\text{adj}}{\text{hyp}}$$

and use the equations obtained above to write the values of the other four functions in terms of sines and cosines. Thus, in Example 1, once you know that

$$\sin \theta = \frac{12}{13} \qquad \text{and} \qquad \cos \theta = \frac{5}{13},$$

you can immediately write

$$\tan \theta = \frac{\sin \theta}{\cos \theta} = \frac{12/13}{5/13} = \frac{12}{5} \qquad \cot \theta = \frac{\cos \theta}{\sin \theta} = \frac{5/13}{12/13} = \frac{5}{12}$$

$$\sec \theta = \frac{1}{\cos \theta} = \frac{1}{5/13} = \frac{13}{5} \qquad \csc \theta = \frac{1}{\sin \theta} = \frac{1}{12/13} = \frac{13}{12}.$$

Two angles are called **complementary** if their sum is a right angle. For instance, *the two acute angles of a right triangle are **complementary angles.*** (Why?) If θ is the degree measure of an angle, then the degree measure of the complementary angle is $90° - \theta$. For radian measure, the angle $(\pi/2) - \theta$ is complementary to θ.

Example 2   Find the angle complementary to θ if

(a) $\theta = 37°$ (b) $\theta = \pi/5$

Solution   (a) $90° - \theta = 90° - 37° = 53°$ (b) $\dfrac{\pi}{2} - \dfrac{\pi}{5} = \dfrac{5\pi}{10} - \dfrac{2\pi}{10} = \dfrac{3\pi}{10}$ ∎

Figure 4

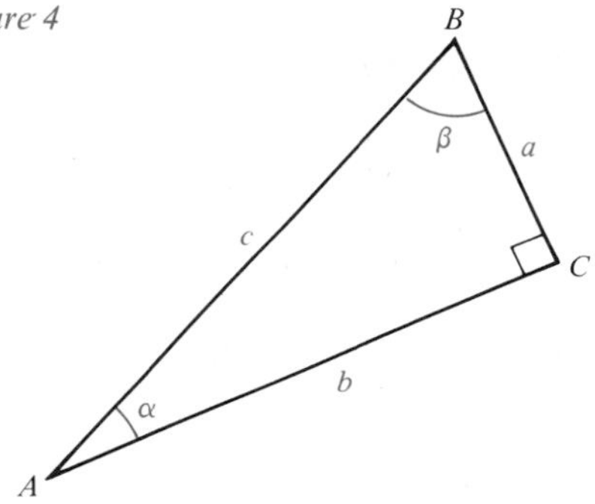

In Figure 4, the two acute angles α and β of right triangle ABC are complementary, a is the length of the side opposite α, b is the length of the side opposite β, and c is the length of the hypotenuse. Notice that the side opposite either one of the acute angles is adjacent to the other. Therefore,

$$\sin \alpha = \frac{a}{c} = \cos \beta \qquad \csc \alpha = \frac{c}{a} = \sec \beta$$

$$\cos \alpha = \frac{b}{c} = \sin \beta \qquad \sec \alpha = \frac{c}{b} = \csc \beta$$

$$\tan \alpha = \frac{a}{b} = \cot \beta \qquad \cot \alpha = \frac{b}{a} = \tan \beta.$$

By using the idea of *cofunctions,* you can remember these equations more easily. We say that sine and cosine are **cofunctions** of one another. Likewise, tangent and cotangent are said to be cofunctions, and so are secant and cosecant. Thus, *the value of any trigonometric function of an acute angle is the same as the value of the cofunction of the complementary angle:*

> "Cosine" is "sine of the complementary angle."
> "Cotangent" is "tangent of the complementary angle."
> "Cosecant" is "secant of the complementary angle."

Example 3   Express the value of each trigonometric function as the value of the cofunction of the complementary angle.

(a) $\cos 70°$ (b) $\cot \dfrac{\pi}{12}$ (c) $\sin 10°$ (d) $\sec \dfrac{\pi}{4}$

Solution   (a) $\cos 70° = \sin(90° - 70°) = \sin 20°$ (b) $\cot \dfrac{\pi}{12} = \tan\left(\dfrac{\pi}{2} - \dfrac{\pi}{12}\right) = \tan \dfrac{5\pi}{12}$

(c) $\sin 10° = \cos(90° - 10°) = \cos 80°$

(d) $\sec \dfrac{\pi}{4} = \csc\left(\dfrac{\pi}{2} - \dfrac{\pi}{4}\right) = \csc \dfrac{\pi}{4}$ ■

Figure 5

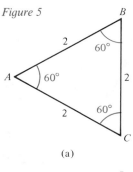

(a)

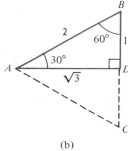

(b)

The special angles with measures of 30° ($\pi/6$ radian), 45° ($\pi/4$ radian), and 60° ($\pi/3$ radians) have a way of showing up over and over again in various applications of trigonometry. Notice that 30° and 60° angles are complementary, whereas a 45° angle is complementary to itself.

Using some elementary geometry, we can find the values of the six trigonometric functions of the special angles. To begin with, consider an equilateral triangle ABC having sides of length 2 units (Figure 5a). Recall from plane geometry that the vertex angles of an equilateral triangle all measure 60°, and that the bisector $\overline{AD}$ of vertex angle A is the perpendicular bisector of the opposite side $\overline{BC}$ (Figure 5b). Therefore,

$$|\overline{BD}| = \tfrac{1}{2}|\overline{BC}| = \tfrac{1}{2}(2) = 1 \quad \text{and} \quad \text{angle } DAB = \tfrac{1}{2}(60°) = 30°.$$

Applying the Pythagorean theorem to right triangle ADB, we have

$$|\overline{AD}|^2 + |\overline{BD}|^2 = |\overline{AB}|^2 \quad \text{or} \quad |\overline{AD}|^2 + 1^2 = 2^2,$$

so that

$$|\overline{AD}|^2 = 2^2 - 1^2 = 3 \quad \text{or} \quad |\overline{AD}| = \sqrt{3}.$$

Thus, from triangle ADB in Figure 5b, we have:

$$\sin 30° = \cos 60° = 1/2 \qquad\qquad \csc 30° = \sec 60° = 2$$
$$\cos 30° = \sin 60° = \sqrt{3}/2 \qquad\qquad \sec 30° = \csc 60° = 2/\sqrt{3} = 2\sqrt{3}/3$$
$$\tan 30° = \cot 60° = 1/\sqrt{3} = \sqrt{3}/3 \qquad \cot 30° = \tan 60° = \sqrt{3}.$$

Figure 6

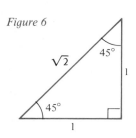

To find the values of the trigonometric functions of a 45° angle, we construct an isosceles right triangle with two equal sides of length 1 unit (Figure 6). We know from plane geometry that both of the acute angles in this right triangle measure 45°. Also, by the Pythagorean theorem,

$$1^2 + 1^2 = (\text{hyp})^2 \quad \text{or} \quad 2 = (\text{hyp})^2$$

so that

$$\text{hyp} = \sqrt{2}.$$

It follows that

$$\sin 45° = 1/\sqrt{2} = \sqrt{2}/2 \qquad \csc 45° = \sqrt{2}$$
$$\cos 45° = 1/\sqrt{2} = \sqrt{2}/2 \qquad \sec 45° = \sqrt{2}$$
$$\tan 45° = 1 \qquad\qquad\qquad \cot 45° = 1.$$

For convenience, the values of the trigonometric functions for the special angles are gathered in Table 1.*

Table 1

θ degrees	θ radians	$\sin \theta$	$\cos \theta$	$\tan \theta$	$\cot \theta$	$\sec \theta$	$\csc \theta$
30°	$\dfrac{\pi}{6}$	$\dfrac{1}{2}$	$\dfrac{\sqrt{3}}{2}$	$\dfrac{\sqrt{3}}{3}$	$\sqrt{3}$	$\dfrac{2\sqrt{3}}{3}$	2
45°	$\dfrac{\pi}{4}$	$\dfrac{\sqrt{2}}{2}$	$\dfrac{\sqrt{2}}{2}$	1	1	$\sqrt{2}$	$\sqrt{2}$
60°	$\dfrac{\pi}{3}$	$\dfrac{\sqrt{3}}{2}$	$\dfrac{1}{2}$	$\sqrt{3}$	$\dfrac{\sqrt{3}}{3}$	2	$\dfrac{2\sqrt{3}}{3}$

You can find values of the trigonometric functions for angles other than the special angles in Table 1 by using a scientific calculator. If a calculator isn't available, you can use Appendix Tables IIA and IIB. Instructions for using these tables are included in Appendix II.

> Before using a calculator, you must decide whether you are going to measure angles in *degrees* or *radians* and set the calculator accordingly.†

For most calculators, if angles are given in degrees, minutes, and seconds, you must first convert to decimal form before pressing a trigonometric key. For details refer to the instruction manual with your calculator.

© **Example 4** Use a calculator to find the approximate value of each trigonometric function.

(a) $\sin 41.3°$　　**(b)** $\cos 19°21'17''$　　**(c)** $\tan \dfrac{5\pi}{36}$　　**(d)** $\sin 0.8432$

Solution **(a)** We make sure the calculator is set in degree mode, enter 41.3, and press the sine (SIN) key. On a 10-digit calculator we get

$$\sin 41.3° = 0.660001668.$$

An 8-digit calculator gives this result rounded off as

$$\sin 41.3° = 0.6600017.$$

If a calculator isn't available, Appendix Table IIA can be used to obtain the result rounded off to four decimal places:

$$\sin 41.3° = 0.6600.$$

* A useful memory device for the sines and cosines of the special angles can be found in Problem 63 in Problem Set 6.4.
† See Problem 39.

(b) We begin by using the calculator to convert to decimal form:

$$19°21'17'' = 19° + \left(\frac{21}{60}\right)^° + \left(\frac{17}{3600}\right)^° = 19.35472222°.$$

Now, we make sure the calculator is set in *degree mode*, and press the cosine (COS) key to get

$$\cos 19°21'17'' = \cos 19.35472222° = 0.943484853.$$

(c) We set the calculator in *radian mode*, press the π key (or enter 3.141592654), multiply by 5, divide by 36, and press the tangent (TAN) key to obtain

$$\tan \frac{5\pi}{36} = 0.466307658.$$

(d) Because degrees are not indicated, we know that the angle is measured in radians. We set the calculator in radian mode, enter 0.8432, and press the sine (SIN) key to obtain

$$\sin 0.8432 = 0.746775185.$$

Some scientific calculators have keys only for sine, cosine, and tangent.* To evaluate the cotangent function on such a calculator, you can use the fact that

$$\cot \theta = \frac{1}{\tan \theta}$$

(Problem 43). Similarly, to evaluate the secant or cosecant functions, you can use the relations

$$\sec \theta = \frac{1}{\cos \theta} \quad \text{and} \quad \csc \theta = \frac{1}{\sin \theta}.$$

©️ **Example 5**   Use a calculator to find the approximate values of each trigonometric function.

(a) $\cot 50°$

(b) $\sec 1.098$

Solution   **(a)** With the calculator in degree mode, we enter 50, press the tangent (TAN) key, then we take the reciprocal of the result by pressing the 1/x key. (We're using the fact that $\cot 50° = 1/\tan 50°$.) The result is

$$\cot 50° = 0.839099631.$$

(b) With the calculator in radian mode, we enter 1.098, press the cosine (COS) key, and then press the 1/x key. The result is

$$\sec 1.098 = 2.195979642.$$

* This also applies to the BASIC computer language, where ordinarily only the functions SIN, COS, and TAN are built in.

Problem Set 6.2

In Problems 1 to 8, find the values of the six trigonometric functions of the acute angle θ in the indicated figure.

1.
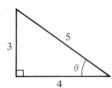

2.
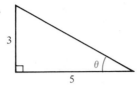

3.

4.

5.
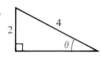

6.
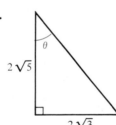

7.

8.
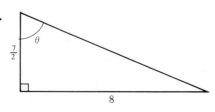

In Problems 9 to 16, find the angle complementary to each of the given angles.

9. 35°   **10.** 89°   **11.** $\dfrac{\pi}{5}$ **12.** $\dfrac{\pi}{7}$

13. $\dfrac{5\pi}{12}$ **14.** $\dfrac{7\pi}{33}$ **15.** 22.34°

16. one-sixth of a counterclockwise revolution

In Problems 17 to 24, express the value of each trigonometric function as the value of the cofunction of the complementary angle.

17. sin 31°   **18.** cos 13°

19. $\tan \dfrac{\pi}{9}$ **20.** $\csc \dfrac{\pi}{11}$

21. $\cot \dfrac{2\pi}{5}$ **22.** sin 9.03°

23. csc 77.03°   **24.** tan 0.774

© In Problems 25 to 34, use a calculator to find the approximate values of the six trigonometric functions of each angle.

25. 48°   **26.** 7°

27. 23°12′33″   **28.** 82°55′55″

29. 16.19°   **30.** $\dfrac{3\pi}{10}$

31. $\dfrac{2\pi}{7}$ **32.** $\dfrac{5\pi}{21}$

33. 0.7764   **34.** 1.33

© In Problems 35 to 38, use a calculator to verify that the given equation is true for the indicated value of the angle.

35. $\cos \theta = \sin(90° - \theta)$ for $\theta = 33°$

36. $\sin \theta = \cos(90° - \theta)$ for $\theta = 77.77°$

37. $\sin \theta = \cos\left(\dfrac{\pi}{2} - \theta\right)$ for $\theta = \dfrac{2\pi}{9}$

38. $\cos \theta = \sin\left(\dfrac{\pi}{2} - \theta\right)$ for $\theta = 1.0785$

39. How could you check to make sure that a calculator is set in degree mode? [*Hint:* We know that $\sin 30° = \frac{1}{2}$.]

40. Justify the following procedure: To convert from degrees, minutes, and seconds to degrees in decimal form, divide the number of seconds by 60, add the quotient to the number of minutes, divide this sum by 60, and add the result to the number of degrees.

Ⓒ **41.** Using a calculator, verify the entries in Table 1 on page 326.

42. Show that

(a) $\cot O = \dfrac{\cos \theta}{\sin \theta}$ (b) $\sec \theta = \dfrac{1}{\cos \theta}$ (c) $\csc \theta = \dfrac{1}{\sin \theta}$

43. Show that $\cot \theta = \dfrac{1}{\tan \theta}$.

Ⓒ **44.** Use a calculator to verify that $\tan \theta = \dfrac{\sin \theta}{\cos \theta}$ for

(a) $\theta = 10°$ (b) $\theta = \dfrac{3\pi}{13}$ (c) $\theta = 0.9474$

45. Using the Pythagorean theorem, show that, if θ is an acute angle in a right triangle, then

$$(\cos \theta)^2 + (\sin \theta)^2 = 1.$$

Ⓒ **46.** Use a calculator to verify that $(\cos \theta)^2 + (\sin \theta)^2 = 1$ for

(a) $\theta = 20°$ (b) $\theta = \dfrac{4\pi}{17}$ (c) $\theta = 0.5678$

6.3 TRIGONOMETRIC FUNCTIONS OF GENERAL ANGLES

In order to extend the definitions of the six trigonometric functions to *general angles*, we shall make use of the following ideas: In a Cartesian coordinate system, *an angle α is said to be in* **standard position** *if its vertex is at the origin O and its initial side coincides with the positive x axis* (Figure 1). *An angle is said to be* **in** *a certain quadrant if, when the angle is in standard position, the terminal side lies in that quadrant.* For instance, a 65° angle is in quadrant I (Figure 2a). Slight variations of this terminology are often used; for instance, we might also say that a 65° angle **lies in quadrant I** or simply that it is a **quadrant I angle**. As Figure 2b shows, an angle of −187° is a quadrant II angle.

If the terminal side of an angle in standard position lies along either the x axis or the y axis, then the angle is called **quadrantal**. For example, −360°, −270°, −180°, −90°, 0°, 90°, 180°, 270°, and 360° are quadrantal angles. Evidently, an angle is quadrantal if and only if its measure is an integer multiple of 90° (or, π/2 radians).

Figure 1

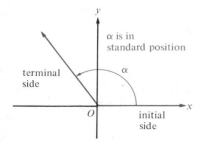

Figure 2

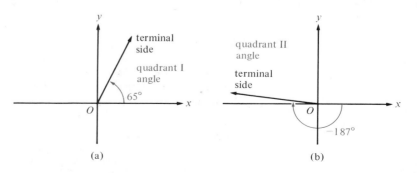

Example 1   For each point given, sketch two coterminal angles α and β in standard position whose terminal side contains the point. Arrange it so that α is positive, β is negative, and neither angle exceeds one revolution. In each case, name the quadrant in which the angle lies, or indicate that the angle is quadrantal.

(a) $(4, 3)$ (b) $(-4, 4)$ (c) $(-3, 0)$ (d) $(3, -4)$

Solution   (a) α and β are quadrant I angles (Figure 3a).

(b) α and β are quadrant II angles (Figure 3b).

(c) α and β are quadrantal angles (Figure 3c).

(d) α and β are quadrant IV angles (Figure 3d).   ∎

Figure 3

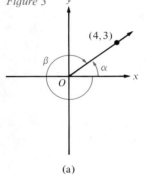

(a)

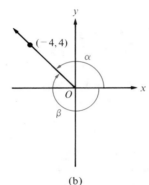

(b)

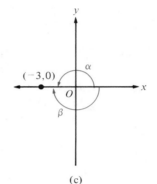

(c)

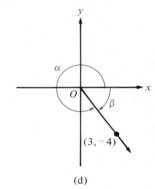

(d)

Figure 4

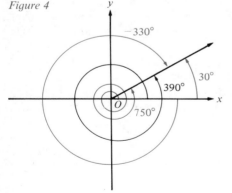

Example 2   Specify and sketch three angles that are coterminal with a 30° angle in standard position.

Solution   The three angles shown in Figure 4 are obtained as follows:

$$30° + 360° = 390°,$$

$$30° + 2(360°) = 750°,$$

$$30° - 360° = -330°.$$

Of course, there are other possibilities, such as

$$30° + 5(360°) = 1830°, \quad 30° - 13(360°) = -4650°,$$

and so on.   ∎

Now, let θ be one of the acute angles of a right triangle OQP placed on a Cartesian coordinate system with θ in standard position and the hypotenuse $\overline{OP}$ in quadrant I (Figure 5). Then $P = (x, y)$ is on the terminal side of θ, and we have

$$x = \text{adj} \quad \text{and} \quad y = \text{opp}.$$

If we let

$$r = \text{hyp} = \sqrt{x^2 + y^2},$$

then the six trigonometric functions of θ are given by the ratios shown in Figure 5.

Figure 5

$$\sin \theta = \frac{y}{r} \quad \csc \theta = \frac{r}{y}$$

$$\cos \theta = \frac{x}{r} \quad \sec \theta = \frac{r}{x}$$

$$\tan \theta = \frac{y}{x} \quad \cot \theta = \frac{x}{y}$$

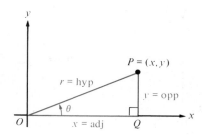

Figure 5 suggests the following definition.

Definition 1   **Trigonometric Functions of a General Angle**

Let θ be an angle in standard position and suppose that (x, y) is any point other than $(0,0)$ on the terminal side of θ (Figure 6). If $r = \sqrt{x^2 + y^2}$ is the distance between (x, y) and $(0,0)$, then the six trigonometric functions of θ are defined by

$$\sin \theta = \frac{y}{r} \qquad \csc \theta = \frac{r}{y}$$

$$\cos \theta = \frac{x}{r} \qquad \sec \theta = \frac{r}{x}$$

$$\tan \theta = \frac{y}{x} \qquad \cot \theta = \frac{x}{y},$$

Figure 6

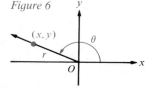

provided that the denominators are not zero.

By using similar triangles, you can see that the values of the six trigonometric functions in Definition 1 depend only on the angle θ and not on the choice of the point (x, y) on the terminal side of θ (Problem 30).

Example 3   Evaluate the six trigonometric functions of the angle θ in standard position if the terminal side of θ contains the point $(x, y) = (2, -1)$ (Figure 7).

Solution   Here $x = 2$, $y = -1$, and

$$r = \sqrt{x^2 + y^2} = \sqrt{2^2 + (-1)^2} = \sqrt{5}.$$

Figure 7

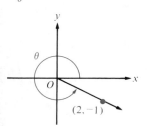

Thus,

$$\sin \theta = \frac{y}{r} = \frac{-1}{\sqrt{5}} = -\frac{\sqrt{5}}{5} \qquad \csc \theta = \frac{r}{y} = \frac{\sqrt{5}}{-1} = -\sqrt{5}$$

$$\cos \theta = \frac{x}{r} = \frac{2}{\sqrt{5}} = \frac{2\sqrt{5}}{5} \qquad \sec \theta = \frac{r}{x} = \frac{\sqrt{5}}{2}$$

$$\tan \theta = \frac{y}{x} = \frac{-1}{2} = -\frac{1}{2} \qquad \cot \theta = \frac{x}{y} = \frac{2}{-1} = -2.$$

∎

Figure 8

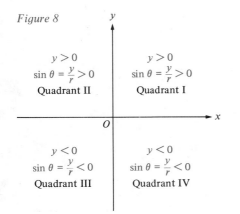

You can determine the algebraic signs of the trigonometric functions for angles in the various quadrants by recalling the algebraic signs of x and y in these quadrants and remembering that r *is always positive*. For instance, as Figure 8 shows, $\sin \theta = y/r$ is positive for θ in quadrants I and II (where both y and r are positive), and it is negative in quadrants III and IV (where y is negative and r is positive). By proceeding in a similar way, you can determine the algebraic signs of the remaining trigonometric functions in the various quadrants and thus confirm the results shown in Table 1.

Table 1

Quadrant Containing θ	Positive Functions	Negative Functions
I	All	None
II	$\sin \theta$, $\csc \theta$	$\cos \theta$, $\sec \theta$, $\tan \theta$, $\cot \theta$
III	$\tan \theta$, $\cot \theta$	$\sin \theta$, $\csc \theta$, $\cos \theta$, $\sec \theta$
IV	$\cos \theta$, $\sec \theta$	$\sin \theta$, $\csc \theta$, $\tan \theta$, $\cot \theta$

Example 4   Find the quadrant in which θ lies if $\tan \theta > 0$ and $\sin \theta < 0$.

Solution   This example can be worked by using Table 1; however, rather than relying on the table, we prefer to reason as follows: Let (x, y) be a point other than the origin on the terminal side of θ (in standard position). Because $\tan \theta = (y/x) > 0$, we see that x and y have the same algebraic sign. Furthermore, since $\sin \theta = (y/r) < 0$, it follows that $y < 0$; therefore, $x < 0$. Because $x < 0$ and $y < 0$, the angle θ is in quadrant III. ∎

The same basic relationships obtained in Section 6.2 for the trigonometric functions of acute angles continue to hold for general angles.

Theorem 1   **Reciprocal Identities**

> If θ is an angle for which the functions are defined, then
>
> **(i)** $\csc \theta = \dfrac{1}{\sin \theta}$ **(ii)** $\sec \theta = \dfrac{1}{\cos \theta}$ **(iii)** $\cot \theta = \dfrac{1}{\tan \theta}$

Figure 9

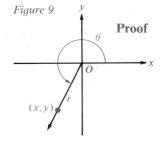

Proof   Let (x, y) be a point on the terminal side of θ in standard position and let $r = \sqrt{x^2 + y^2}$ (Figure 9). Then, provided that the denominators are not zero,

(i) $\csc \theta = \dfrac{r}{y} = \dfrac{1}{y/r} = \dfrac{1}{\sin \theta}$ **(ii)** $\sec \theta = \dfrac{r}{x} = \dfrac{1}{x/r} = \dfrac{1}{\cos \theta}$

(iii) $\cot \theta = \dfrac{x}{y} = \dfrac{1}{y/x} = \dfrac{1}{\tan \theta}$ ∎

In a similar way, you can prove the following theorem (Problem 35).

Theorem 2   **Quotient Identities**

> If θ is an angle for which the functions are defined, then:
>
> $$\tan \theta = \frac{\sin \theta}{\cos \theta} \quad \text{and} \quad \cot \theta = \frac{\cos \theta}{\sin \theta}.$$

Example 5   If $\sin \theta = \frac{1}{3}$ and $\cos \theta = -2\sqrt{2}/3$, find the values of the other four trigonometric functions of θ.

Solution   Using Theorems 1 and 2, we have

$$\tan \theta = \frac{\sin \theta}{\cos \theta} = \frac{1/3}{(-2\sqrt{2}/3)} = -\frac{1}{2\sqrt{2}} = -\frac{\sqrt{2}}{4}$$

$$\sec \theta = \frac{1}{\cos \theta} = \frac{1}{(-2\sqrt{2}/3)} = -\frac{3}{2\sqrt{2}} = -\frac{3\sqrt{2}}{4}$$

$$\cot \theta = \frac{1}{\tan \theta} = \frac{1}{(-\sqrt{2}/4)} = -\frac{4}{\sqrt{2}} = -2\sqrt{2}$$

$$\csc \theta = \frac{1}{\sin \theta} = \frac{1}{1/3} = 3. \qquad \blacksquare$$

By using the reciprocal and quotient identities, you can quickly recall the algebraic signs of the secant, cosecant, tangent, and cotangent in the four quadrants (Table 1), if you know the algebraic signs of the sine and cosine in these quadrants.

Another important identity is derived as follows: Again suppose that θ is an angle in standard position and that (x, y) is a point other than the origin on the terminal side of θ (Figure 9). Because $r = \sqrt{x^2 + y^2}$, we have $x^2 + y^2 = r^2$, so

$$(\cos \theta)^2 + (\sin \theta)^2 = \left(\frac{x}{r}\right)^2 + \left(\frac{y}{r}\right)^2 = \frac{x^2}{r^2} + \frac{y^2}{r^2} = \frac{x^2 + y^2}{r^2} = \frac{r^2}{r^2} = 1.$$

The relationship

$$(\cos \theta)^2 + (\sin \theta)^2 = 1$$

is called the **fundamental Pythagorean identity** because its derivation involves the fact that $x^2 + y^2 = r^2$, which is a consequence of the Pythagorean theorem.

The fundamental Pythagorean identity is used quite often, and it would be bothersome to write the parentheses each time for $(\cos \theta)^2$ and $(\sin \theta)^2$; yet, if the parentheses were simply omitted, the resulting expressions would be misunderstood. (For instance, $\cos \theta^2$ is usually understood to mean the cosine of the square of θ.) Therefore, it is customary to write $\cos^2 \theta$ and $\sin^2 \theta$ to mean $(\cos \theta)^2$ and $(\sin \theta)^2$.

Similar notation is used for the remaining trigonometric functions and for powers other than 2. Thus, $\cot^4 \theta$ means $(\cot \theta)^4$, $\sec^n \theta$ means $(\sec \theta)^n$, and so forth.* With this notation, the fundamental Pythagorean identity becomes

$$\cos^2 \theta + \sin^2 \theta = 1.$$

Actually, there are three Pythagorean identities—the fundamental identity and two others derived from it.

Theorem 3 **Pythagorean Identities**

> If θ is an angle for which the functions are defined, then:
>
> **(i)** $\cos^2 \theta + \sin^2 \theta = 1$ **(ii)** $1 + \tan^2 \theta = \sec^2 \theta$ **(iii)** $\cot^2 \theta + 1 = \csc^2 \theta$

Proof   We have already proved (i). To prove (ii), we divide both sides of (i) by $\cos^2 \theta$ to obtain

$$1 + \frac{\sin^2 \theta}{\cos^2 \theta} = \frac{1}{\cos^2 \theta}$$

or

$$1 + \left(\frac{\sin \theta}{\cos \theta}\right)^2 = \left(\frac{1}{\cos \theta}\right)^2,$$

provided that $\cos \theta \neq 0$. Since

$$\frac{\sin \theta}{\cos \theta} = \tan \theta \quad \text{and} \quad \frac{1}{\cos \theta} = \sec \theta,$$

we have

$$1 + \tan^2 \theta = \sec^2 \theta.$$

Identity (iii) is proved by dividing both sides of (i) by $\sin^2 \theta$ (Problem 34). ∎

In Examples 6 and 7, the value of one of the trigonometric functions of an angle θ is given along with information about the quadrant in which θ lies. Find the values of the other five trigonometric functions of θ.

Example 6 $\sin \theta = \dfrac{5}{13}$, θ in quadrant II.

Solution   By the fundamental Pythagorean identity, $\cos^2 \theta + \sin^2 \theta = 1$, so

$$\cos^2 \theta = 1 - \sin^2 \theta = 1 - \left(\frac{5}{13}\right)^2 = 1 - \frac{25}{169} = \frac{169 - 25}{169} = \frac{144}{169}.$$

* To avoid confusion with the inverse trigonometric functions (Section 7.6), this notation is not used for the exponent -1.

Therefore,
$$\cos \theta = \pm \sqrt{\frac{144}{169}} = \pm \frac{12}{13}.$$

Because θ is in quadrant II, we know that $\cos \theta$ is negative; hence,

$$\cos \theta = -\frac{12}{13}.$$

It follows that

$$\tan \theta = \frac{\sin \theta}{\cos \theta} = \frac{5/13}{(-12/13)} = -\frac{5}{12} \qquad \cot \theta = \frac{1}{\tan \theta} = \frac{1}{(-5/12)} = -\frac{12}{5}$$

$$\sec \theta = \frac{1}{\cos \theta} = \frac{1}{(-12/13)} = -\frac{13}{12} \qquad \csc \theta = \frac{1}{\sin \theta} = \frac{1}{5/13} = \frac{13}{5}.$$

Example 7 $\tan \theta = -8/7$ and $\sin \theta < 0$.

Solution   Because $\tan \theta < 0$ only in quadrants II and IV, and $\sin \theta < 0$ only in quadrants III and IV, it follows that θ must be in quadrant IV. By part (ii) of Theorem 3, $\sec^2 \theta = 1 + \tan^2 \theta$, so

$$\sec \theta = \pm \sqrt{1 + \tan^2 \theta} = \pm \sqrt{1 + \left(-\frac{8}{7}\right)^2} = \pm \sqrt{1 + \frac{64}{49}} = \pm \sqrt{\frac{113}{49}} = \pm \frac{\sqrt{113}}{7}.$$

Since θ is in quadrant IV, $\sec \theta > 0$; hence,

$$\sec \theta = \frac{\sqrt{113}}{7}.$$

Because
$$\sec \theta = \frac{1}{\cos \theta},$$

it follows that

$$\cos \theta = \frac{1}{\sec \theta} = \frac{1}{\sqrt{113}/7} = \frac{7}{\sqrt{113}} = \frac{7\sqrt{113}}{113}.$$

Now,
$$\tan \theta = \frac{\sin \theta}{\cos \theta}$$

so
$$\sin \theta = (\tan \theta)(\cos \theta) = \left(-\frac{8}{7}\right)\left(\frac{7\sqrt{113}}{113}\right) = -\frac{8\sqrt{113}}{113}.$$

Finally,
$$\csc \theta = \frac{1}{\sin \theta} = \frac{1}{(-8\sqrt{113}/113)} = -\frac{\sqrt{113}}{8}$$

and
$$\cot \theta = \frac{1}{\tan \theta} = \frac{1}{(-8/7)} = -\frac{7}{8}.$$

In the applications of trigonometry, and especially in calculus, it is often necessary to make trigonometric calculations, as we have done in this section, *without the use of calculators or tables*. In working the following problems, you should avoid using a calculator except in those few cases where its use is indicated.

Problem Set 6.3

In Problems 1 to 10, sketch two coterminal angles α and β in standard position whose terminal side contains the given point. Arrange it so that α is positive, β is negative, and neither angle exceeds one revolution. In each case, name the quadrant in which the angle lies, or indicate that the angle is quadrantal.

1. $(1, 2)$
2. $(2, -\frac{4}{3})$
3. $(-5, 0)$
4. $(-3, -4)$
5. $(5, -3)$
6. $(0, 4)$
7. $(-1, 1)$
8. $(\sqrt{2}/2, 0)$
9. $(-\frac{3}{4}, -\frac{3}{4})$
10. $(0, -3)$

In Problems 11 to 18, specify and sketch three angles that are coterminal with the given angle in standard position.

11. $60°$
12. $-15°$
13. $-\dfrac{\pi}{4}$
14. $\dfrac{4\pi}{3}$
15. $-612°$
16. $-\dfrac{7\pi}{4}$
17. $\dfrac{5\pi}{6}$
18. $1440°$

In Problems 19 to 28, evaluate the six trigonometric functions of the angle θ in standard position if the terminal side of θ contains the given point (x, y). [Do not use a calculator—leave all answers in the form of a fraction or an integer.] In each case, sketch one of the coterminal angles θ.

19. $(4, 3)$
20. $(2, 7)$
21. $(-5, 12)$
22. $(-2, 4)$
23. $(-3, -4)$
24. $(1, 1)$
25. $(7, 3)$
26. $(1, -3)$
27. $(-\sqrt{3}, -\sqrt{2})$
28. $(20, 21)$

29. Is there any angle θ for which $\sin \theta = \frac{5}{4}$? Explain.

30. Using similar triangles, show that the values of the six trigonometric functions in Definition 1 depend only on the angle θ and not on the choice of the point (x, y) on the terminal side of θ.

31. In each case, assume that θ is an angle in standard position and find the quadrant in which it lies.

   (a) $\tan \theta > 0$ and $\sec \theta > 0$
   (b) $\sin \theta > 0$ and $\sec \theta < 0$
   (c) $\sin \theta > 0$ and $\cos \theta < 0$
   (d) $\sec \theta > 0$ and $\tan \theta < 0$
   (e) $\tan \theta > 0$ and $\csc \theta < 0$
   (f) $\cos \theta < 0$ and $\csc \theta < 0$
   (g) $\sec \theta > 0$ and $\cot \theta < 0$
   (h) $\cot \theta > 0$ and $\sin \theta > 0$

32. Is there any angle θ for which $\sin \theta > 0$ and $\csc \theta < 0$? Explain.

33. Give the algebraic sign of each of the following.

   (a) $\cos 163°$
   (b) $\sin 211°$
   (c) $\sec(-355°)$
   (d) $\tan \dfrac{7\pi}{4}$
   (e) $\cot\left(-\dfrac{28\pi}{45}\right)$
   (f) $\csc\left(-\dfrac{5\pi}{12}\right)$
   (g) $\sec \dfrac{38\pi}{15}$

34. Prove part (iii) of Theorem 3.

35. Prove Theorem 2.

36. If θ is an angle for which the functions are defined, show that $\sec \theta - (\sin \theta)(\tan \theta) = \cos \theta$.

37. If $\sin \theta = -\frac{5}{13}$ and $\cos \theta = \frac{12}{13}$, use the reciprocal and quotient identities to find

   (a) $\sec \theta$ (b) $\csc \theta$ (c) $\tan \theta$ (d) $\cot \theta$

38. If $\sec \theta = -\frac{5}{4}$ and $\csc \theta = -\frac{5}{3}$, use the reciprocal and quotient identities to find

   (a) $\sin \theta$ (b) $\cos \theta$ (c) $\tan \theta$ (d) $\cot \theta$

In Problems 39 to 56, the value of one of the trigonometric functions of an angle θ is given along with information about the quadrant (Q) in which θ lies. Find the values of the other five trigonometric functions of θ.

39. $\sin \theta = \frac{4}{5}$, θ in Q_I

40. $\cos \theta = \frac{12}{13}$, θ in Q_{IV}

41. $\sin \theta = -\frac{3}{4}$, θ in Q_{III}

42. $\sin \theta = \frac{3}{4}$, θ not in Q_{I}

43. $\cos \theta = \frac{4}{7}$, $\sin \theta < 0$

44. $\cos \theta = \frac{1}{3}$, θ not in Q_{I}

45. $\csc \theta = \frac{3}{2}$, θ in Q_{I}

46. $\sec \theta = -\frac{5}{3}$, θ in Q_{III}

47. $\tan \theta = \frac{4}{3}$, θ in Q_{I}

48. $\tan \theta = 2$, $\sin \theta < 0$

49. $\cot \theta = -\frac{12}{5}$, $\csc \theta > 0$

50. $\csc \theta = -\frac{25}{7}$, $\sec \theta < 0$

51. $\sec \theta = -3$, $\csc \theta > 0$

52. $\cot \theta = -\sqrt{15}/7$, $\sin \theta > 0$

🄲 **53.** $\sin \theta = 0.4695$, θ in Q_{I}

🄲 **54.** $\tan \theta = -0.5095$, $\sin \theta < 0$

🄲 **55.** $\csc \theta = 2.613$, θ in Q_{II}

🄲 **56.** $\sec \theta = 2.924$, $\tan \theta > 0$

57. According to Theorem 1, the six trigonometric functions can be grouped in three reciprocal pairs. They can also be grouped in three pairs of cofunctions. Is any reciprocal pair also a pair of cofunctions?

6.4 EVALUATION OF TRIGONOMETRIC FUNCTIONS

In this section, we obtain values of the trigonometric functions for quadrantal angles, we introduce the idea of reference angles, and we discuss the use of a calculator to evaluate trigonometric functions of general angles.

In Definition 1, Section 6.3, *the domain of each trigonometric function consists of all angles θ for which the denominator in the corresponding ratio is not zero.* Because $r > 0$, it follows that $\sin \theta = y/r$ and $\cos \theta = x/r$ are defined for all angles θ. However, $\tan \theta = y/x$ and $\sec \theta = r/x$ are not defined when the terminal side of θ lies along the y axis (so that $x = 0$). Likewise, $\cot \theta = x/y$ and $\csc \theta = r/y$ are not defined when the terminal side of θ lies along the x axis (so that $y = 0$). Therefore, when you deal with a trigonometric function of a quadrantal angle, you must check to be sure that the function is actually defined for that angle.

Example 1   Find the values (if they are defined) of the six trigonometric functions for the quadrantal angle $\theta = 90°$ (or $\theta = \pi/2$).

Figure 1

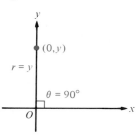

Solution   In order to use Definition 1 in Section 6.3, we begin by choosing any point $(0, y)$, with $y > 0$, on the terminal side of the $90°$ angle (Figure 1). Because $x = 0$, it follows that $\tan 90°$ and $\sec 90°$ are undefined. Since $y > 0$, we have

$$r = \sqrt{x^2 + y^2} = \sqrt{0^2 + y^2} = \sqrt{y^2} = |y| = y.$$

Therefore, $\sin 90° = \dfrac{y}{r} = \dfrac{y}{y} = 1$ $\cos 90° = \dfrac{x}{r} = \dfrac{0}{y} = 0$

$\csc 90° = \dfrac{r}{y} = \dfrac{y}{y} = 1$ $\cot 90° = \dfrac{x}{y} = \dfrac{0}{y} = 0.$ ∎

The values of the trigonometric functions for other quadrantal angles are found in a similar manner (Problem 1). The results appear in Table 1. Dashes in the table indicate that the function is undefined for that angle.

Table 1

θ degrees	θ radians	$\sin \theta$	$\cos \theta$	$\tan \theta$	$\cot \theta$	$\sec \theta$	$\csc \theta$
0°	0	0	1	0	—	1	—
90°	$\dfrac{\pi}{2}$	1	0	—	0	—	1
180°	π	0	−1	0	—	−1	—
270°	$\dfrac{3\pi}{2}$	−1	0	—	0	—	−1
360°	2π	0	1	0	—	1	—

It follows from Definition 1 in Section 6.3 that the values of each of the six trigonometric functions remain unchanged if the angle is replaced by a coterminal angle. If an angle exceeds one revolution or is negative, you can change it to a non-negative coterminal angle that is less than one revolution by adding or subtracting an integer multiple of 360° (or 2π radians). For instance,

$$\sin 450° = \sin(450° - 360°) = \sin 90° = 1$$

$$\sec 7\pi = \sec[7\pi - (3 \times 2\pi)] = \sec \pi = -1$$

$$\cos(-660°) = \cos[-660° + (2 \times 360°)] = \cos 60° = \tfrac{1}{2}.$$

In Examples 2 and 3, replace each angle by a nonnegative coterminal angle that is less than one revolution and then find the values of the six trigonometric functions (if they are defined).

Example 2   1110°

Solution   By dividing 1110 by 360, we find that the largest integer multiple of 360° that is less than 1110° is $3 \times 360° = 1080°$. Thus,

$$1110° - (3 \times 360°) = 1110° - 1080° = 30°.$$

(Or, we could have started with 1110° and repeatedly subtracted 360° until we obtained 30°.) It follows that

$$\sin 1110° = \sin 30° = 1/2 \qquad \csc 1110° = \csc 30° = 2$$

$$\cos 1110° = \cos 30° = \sqrt{3}/2 \qquad \sec 1110° = \sec 30° = 2\sqrt{3}/3$$

$$\tan 1110° = \tan 30° = \sqrt{3}/3 \qquad \cot 1110° = \cot 30° = \sqrt{3}.$$

Example 3 $-5\pi/2$

Solution   We repeatedly add 2π to $-5\pi/2$ until we obtain a nonnegative coterminal angle:

$$(-5\pi/2) + 2\pi = -\pi/2 \qquad \text{(still negative)}$$
$$(-\pi/2) + 2\pi = 3\pi/2.$$

Therefore, by Table 1 for quadrantal angles,

$$\sin(-5\pi/2) = \sin(3\pi/2) = -1 \qquad \cot(-5\pi/2) = \cot(3\pi/2) = 0$$
$$\cos(-5\pi/2) = \cos(3\pi/2) = 0 \qquad \csc(-5\pi/2) = \csc(3\pi/2) = -1,$$

and both $\tan(-5\pi/2)$ and $\sec(-5\pi/2)$ are undefined.   ■

In Section 6.2 we found exact values of the six trigonometric functions for the special acute angles $30°$, $45°$, and $60°$ (see Table 1 on page 326). This information can be applied to certain angles in quadrants II, III, and IV by using the idea of a "reference angle."

> If θ is an angle in standard position and if θ is not a quadrantal angle, then the **reference angle** θ_R for θ is defined to be the positive acute angle formed by the terminal side of θ and the x axis.

Figure 2 shows the reference angle θ_R for an angle θ in each of the four quadrants.

Figure 2

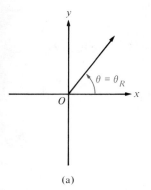

(a)

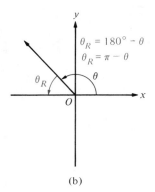

(b)

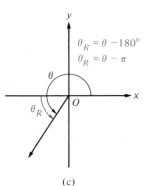

(c)

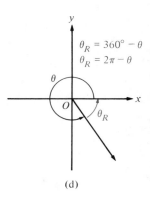

(d)

Example 4   Find the reference angle θ_R for each angle θ.

(a) $\theta = 60°$ **(b)** $\theta = \dfrac{3\pi}{4}$ **(c)** $\theta = 210°$ **(d)** $\theta = \dfrac{5\pi}{3}$

Solution   **(a)** By Figure 2a, $\theta_R = \theta = 60°$.

(b) By Figure 2b, $\theta_R = \pi - \theta = \pi - \dfrac{3\pi}{4} = \dfrac{\pi}{4}$.

(c) By Figure 2c, $\theta_R = \theta - 180° = 210° - 180° = 30°.$

(d) By Figure 2d, $\theta_R = 2\pi - \theta = 2\pi - \dfrac{5\pi}{3} = \dfrac{\pi}{3}.$ ■

The value of any trigonometric function of an angle θ is the same as the value of the function for the reference angle θ_R, except possibly for a change of algebraic sign.

Thus,

$$\sin \theta = \pm \sin \theta_R, \qquad \cos \theta = \pm \cos \theta_R,$$

and so forth. You can always determine the correct algebraic sign by considering in which quadrant θ lies.

We shall verify the statement above for θ in quadrant II, and leave the remaining cases as an exercise (Problem 16). Figure 3a shows a quadrant II angle θ in standard position; the corresponding reference angle θ_R is shown in standard position in Figure 3b. In the two figures, we have chosen points P and P' as shown, both r units from the origin on the terminal sides of θ and θ_R. Dropping perpendiculars $\overline{PQ}$ and $\overline{P'Q'}$ from P and P' to the x axis, we find that the right triangles PQO and $P'Q'O$ are congruent. (Why?) Therefore, if the coordinates of P' are (x, y), it follows that the coordinates of P are $(-x, y)$. Applying Definition 1 in Section 6.3 to Figures 3a and 3b, we have:

$$\sin \theta = \frac{y}{r} = \sin \theta_R \qquad\qquad \csc \theta = \frac{r}{y} = \csc \theta_R$$

$$\cos \theta = \frac{-x}{r} = -\frac{x}{r} = -\cos \theta_R \qquad \sec \theta = \frac{r}{-x} = -\frac{r}{x} = -\sec \theta_R$$

$$\tan \theta = \frac{y}{-x} = -\frac{y}{x} = -\tan \theta_R \qquad \cot \theta = \frac{-x}{y} = -\frac{x}{y} = -\cot \theta_R.$$

Figure 3

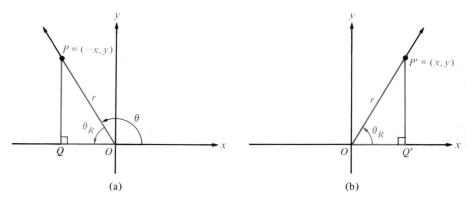

(a)   (b)

In Examples 5 to 7, find the reference angle θ_R for the given angle θ, and then use the information in Table 1 on page 326 to find the values of the six trigonometric functions of θ.

Example 5 $\theta = 135°$

Solution   The reference angle is $\theta_R = 180° - 135° = 45°$ (Figure 4). Since $135°$ is a quadrant II angle, we have

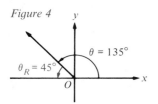

Figure 4

$$\sin 135° = \sin 45° = \frac{\sqrt{2}}{2} \qquad \csc 135° = \csc 45° = \sqrt{2}$$

$$\cos 135° = -\cos 45° = -\frac{\sqrt{2}}{2} \qquad \sec 135° = -\sec 45° = -\sqrt{2}$$

$$\tan 135° = -\tan 45° = -1 \qquad \cot 135° = -\cot 45° = -1.$$

Example 6 $\theta = 210°$

Solution   The reference angle is $\theta_R = 210° - 180° = 30°$ (Figure 5). Since $210°$ is a quadrant III angle, we have

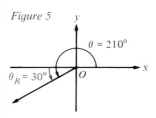

Figure 5

$$\sin 210° = -\sin 30° = -\frac{1}{2} \qquad \csc 210° = -\csc 30° = -2$$

$$\cos 210° = -\cos 30° = -\frac{\sqrt{3}}{2} \qquad \sec 210° = -\sec 30° = -\frac{2\sqrt{3}}{3}$$

$$\tan 210° = \tan 30° = \frac{\sqrt{3}}{3} \qquad \cot 210° = \cot 30° = \sqrt{3}.$$

Example 7 $\theta = \frac{5\pi}{3}$

Solution   The reference angle is $\theta_R = 2\pi - \frac{5\pi}{3} = \frac{\pi}{3}$ (Figure 6). Since $\frac{5\pi}{3}$ is a quadrant IV angle, we have

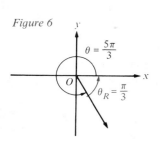

Figure 6

$$\sin \frac{5\pi}{3} = -\sin \frac{\pi}{3} = -\frac{\sqrt{3}}{2} \qquad \csc \frac{5\pi}{3} = -\csc \frac{\pi}{3} = -\frac{2\sqrt{3}}{3}$$

$$\cos \frac{5\pi}{3} = \cos \frac{\pi}{3} = \frac{1}{2} \qquad \sec \frac{5\pi}{3} = \sec \frac{\pi}{3} = 2$$

$$\tan \frac{5\pi}{3} = -\tan \frac{\pi}{3} = -\sqrt{3} \qquad \cot \frac{5\pi}{3} = -\cot \frac{\pi}{3} = -\frac{\sqrt{3}}{3}.$$

To find the reference angle θ_R for an angle θ in standard position when θ is negative or greater than one revolution, you can replace θ by a nonnegative coterminal angle that is less than one revolution, and then proceed as in Examples 5 to 7 above.

Example 8 Find the reference angle θ_R for the given angle θ, and then use the information in Table 1 on page 326 to find the values of the six trigonometric functions of θ.

(a) $\theta = \dfrac{21\pi}{4}$ **(b)** $\theta = -300°$

Figure 7

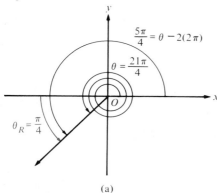

(a)

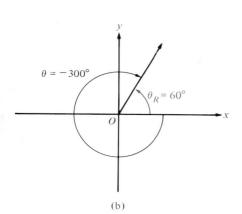

(b)

Solution

(a) Here $\theta = \dfrac{21\pi}{4}$ is coterminal with $\dfrac{21\pi}{4} - 2(2\pi) = \dfrac{5\pi}{4}$, a quadrant

III angle (Figure 7a). The corresponding reference angle is $\theta_R = \dfrac{5\pi}{4} - \pi = \dfrac{\pi}{4}$, so

$$\sin \frac{21\pi}{4} = -\sin \frac{\pi}{4} = -\frac{\sqrt{2}}{2} \qquad \csc \frac{21\pi}{4} = -\csc \frac{\pi}{4} = -\sqrt{2}$$

$$\cos \frac{21\pi}{4} = -\cos \frac{\pi}{4} = -\frac{\sqrt{2}}{2} \qquad \sec \frac{21\pi}{4} = -\sec \frac{\pi}{4} = -\sqrt{2}$$

$$\tan \frac{21\pi}{4} = \tan \frac{\pi}{4} = 1 \qquad \cot \frac{21\pi}{4} = \cot \frac{\pi}{4} = 1.$$

(b) The angle $\theta = -300°$ is coterminal with $-300° + 360° = 60°$, a quadrant I angle (Figure 7b). Here the reference angle is also $\theta_R = 60°$, and we have

$$\sin(-300°) = \sin 60° = \frac{\sqrt{3}}{2} \qquad \csc(-300°) = \csc 60° = \frac{2\sqrt{3}}{3}$$

$$\cos(-300°) = \cos 60° = \frac{1}{2} \qquad \sec(-300°) = \sec 60° = 2$$

$$\tan(-300°) = \tan 60° = \sqrt{3} \qquad \cot(-300°) = \cot 60° = \frac{\sqrt{3}}{3}.$$ ∎

In the examples above, we obtained *exact* values of the trigonometric functions for certain angles. Approximate values of these functions for all other angles can be found by using a scientific calculator. No special techniques are required beyond those already presented in Section 6.2 for evaluating the trigonometric functions of acute angles*—correct algebraic signs are automatically provided. (If a calculator isn't available, you can use Appendix Tables IIA and IIB.)

ⓒ **Example 9** Use a calculator to find the approximate value of each trigonometric function.

(a) $\cot 300.42°$ **(b)** $\sec \dfrac{40\pi}{7}$ **(c)** $\sin(-2.007)$

---

* Some scientific calculators have limitations on the magnitudes of the angles for which you can evaluate trigonometric functions. Check the instruction manual for your calculator.

Solution   **(a)** With the calculator in degree mode, we enter 300.42, press the tangent (TAN) key, then press the 1/x key. (We're using the fact that $\cot 300.42° = \dfrac{1}{\tan 300.42°}$.) On a 10-digit calculator, the result is

$$\cot 300.42° = -0.587165830.$$

(b) With the calculator in radian mode, we enter π, multiply by 40, divide by 7, press the cosine (COS) key, and then press the 1/x key. On a 10-digit calculator, we get

$$\sec \frac{40\pi}{7} = 1.603875452.$$

(c) Because degrees are not indicated, we know that the angle is measured in radians. We set the calculator in radian mode, enter -2.007, and press the sine (SIN) key to obtain

$$\sin(-2.007) = -0.906362145.$$ ∎

Problem Set 6.4

In Problems 1 and 2, find the values (if they are defined) of the six trigonometric functions of the given quadrantal angles. (Do not use a calculator.)

1. (a) $0°$ (b) $180°$ (c) $270°$ (d) $360°$

   [When you have finished, compare your answers with the results in Table 1.]

2. (a) 5π (b) 6π (c) -7π

   (d) $\dfrac{5\pi}{2}$ (e) $\dfrac{7\pi}{2}$

In Problems 3 to 14, replace each angle by a nonnegative coterminal angle that is less than one revolution, and then find the exact values of the six trigonometric functions (if they are defined) for the angle.

3. $1440°$

4. $810°$

5. $900°$

6. $-720°$

7. $750°$

8. $1845°$

9. $-675°$

10. $\dfrac{19\pi}{2}$

11. 5π

12. $\dfrac{25\pi}{6}$

13. $-\dfrac{17\pi}{3}$

14. $-\dfrac{31\pi}{4}$

©15. What happens when you try to evaluate $\tan 90°$ on a calculator? [Try it.]

16. Let θ be a quadrant III angle in standard position and let θ_R be its reference angle. Show that the value of any trigonometric function of θ is the same as the value of the function for θ_R, except possibly for a change of algebraic sign. Repeat for θ in quadrant IV.

In Problems 17 to 36, find the reference angle θ_R for each angle θ, and then find the exact values of the six trigonometric functions of θ.

17. $\theta = 150°$

18. $\theta = 120°$

19. $\theta = 240°$

20. $\theta = 225°$

21. $\theta = 315°$

22. $\theta = \dfrac{9\pi}{4}$

23. $\theta = -150°$

24. $\theta = 675°$

25. $\theta = -60°$

26. $\theta = -\dfrac{7\pi}{6}$

27. $\theta = -\dfrac{\pi}{4}$

28. $\theta = -\dfrac{13\pi}{6}$

29. $\theta = -\dfrac{2\pi}{3}$

30. $\theta = \dfrac{53\pi}{6}$

31. $\theta = \dfrac{7\pi}{4}$

32. $\theta = -\dfrac{50\pi}{3}$

33. $\theta = \dfrac{11\pi}{3}$

34. $\theta = -\dfrac{147\pi}{4}$

35. $\theta = -420°$

36. $\theta = -5370°$

37. Complete the following tables. (Do not use a calculator.)

θ degrees	θ radians	$\sin\theta$	$\cos\theta$	$\tan\theta$
210°	$7\pi/6$			
225°	$5\pi/4$			
240°	$4\pi/3$			
300°	$5\pi/3$			
315°	$7\pi/4$			
330°	$11\pi/6$			

θ degrees	θ radians	$\cot\theta$	$\sec\theta$	$\csc\theta$
210°	$7\pi/6$			
225°	$5\pi/4$			
240°	$4\pi/3$			
300°	$5\pi/3$			
315°	$7\pi/4$			
330°	$11\pi/6$			

38. A calculator is set in radian mode, π is entered, and the sine (SIN) key is pressed. The display shows -4.1×10^{-10}. But we know that $\sin\pi = 0$. Explain.

ⓒ In Problems 39 to 58, use a calculator to find the approximate values of the six trigonometric functions of each angle.

39. 46°

40. 162°

41. 143°

42. $-225.31°$

43. $-61.37°$

44. 852.48°

45. 61°35′

46. $-31°8′$

47. $-97°9′8″$

48. $\dfrac{12\pi}{5}$

49. $-\dfrac{2\pi}{7}$

50. $\dfrac{\sqrt{2}\pi}{3}$

51. 1.67

52. 3.725

53. -2.436

54. 16.57

55. -10.79

56. -6.203

57. 9.673

58. -3985.2

ⓒ In Problems 59 to 62, use a calculator to verify that the equation is true for the indicated value of the angle θ.

59. $\tan\theta = \sin\theta/\cos\theta$ for $\theta = 35°$

60. $(\cos\theta)(\tan\theta) = \sin\theta$ for $\theta = 5\pi/7$

61. $\sin^2\theta + \cos^2\theta = 1$ for $\theta = 3\pi/5$

62. $1 + \tan^2\theta = \sec^2\theta$ for $\theta = 17.75°$

63. Verify that for $\theta = 0°$, 30°, 45°, 60°, and 90°, we have $\sin\theta = \sqrt{0}/2$, $\sqrt{1}/2$, $\sqrt{2}/2$, $\sqrt{3}/2$, and $\sqrt{4}/2$, respectively. [Although there is no theoretical significance to this pattern, people often use it as a memory aid to help recall these values of $\sin\theta$.]

6.5 GRAPHS OF THE SINE AND COSINE FUNCTIONS

Until now, we have considered trigonometric functions of *angles*, measured either in degrees or radians. In advanced work, particularly in calculus and its applications, it is necessary to modify the trigonometric functions so that their domains consist of *real numbers*. For reasons that we shall soon give, the resulting functions are sometimes called the **circular functions.**

Trigonometric Functions of Real Numbers

If t is a real number, then the value of $\sin t$ is defined by

$$\sin t = \sin \theta,$$

where θ is an angle whose measure is t radians.

Similarly, the values of $\cos t$, $\tan t$, $\cot t$, $\sec t$, $\csc t$ *are defined to be the values of the corresponding trigonometric functions of an angle θ whose measure is t radians.* We shall now explain just what these so-called circular functions have to do with circles.

Consider the **unit circle;** that is, the circle with center at the origin and radius $r = 1$ (Figure 1a). Let t be any real number. Starting at the point $(1,0)$ on the unit circle, we move $|t|$ units around the circle, counterclockwise if $t > 0$ and clockwise if $t < 0$, to the point (x, y) (Figure 1a). Now let θ be the angle in standard position whose terminal side contains the point (x, y). Recall from Section 6.1 (page 316) that $t = r\theta$. Since $r = 1$, it follows that angle θ has radian measure t. Hence,

$$\sin t = \sin \theta = \frac{y}{r} = \frac{y}{1} = y \qquad \text{and} \qquad \cos t = \cos \theta = \frac{x}{r} = \frac{x}{1} = x.$$

Therefore, $$(x, y) = (\cos t, \sin t)$$

(Figure 1b).

In Figure 1b, the coordinates of $(\cos t, \sin t)$ repeat themselves each time the point goes completely around the circle; that is, each time t increases by 2π (the circumference of the unit circle). Thus, the sine and cosine functions are **periodic** in the sense that their values repeat themselves whenever the independent variable increases by 2π, that is,

$$\sin(t + 2\pi) = \sin t \qquad \text{and} \qquad \cos(t + 2\pi) = \cos t.$$

Figure 1

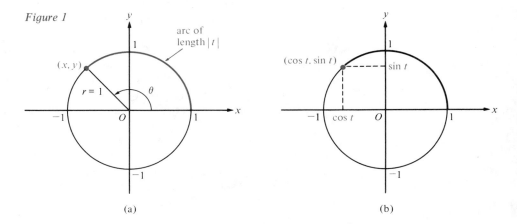

(a)   (b)

As t increases from 0 to 2π, the point $(\cos t, \sin t)$ (Figure 1b) moves once around the unit circle in a counterclockwise direction, and the coordinates $\cos t$ and $\sin t$ change as shown in Table 1.

Table 1

As t Increases from	sin t	cos t
0 to $\dfrac{\pi}{2}$	is positive and increases from 0 to 1	is positive and decreases from 1 to 0
$\dfrac{\pi}{2}$ to π	is positive and decreases from 1 to 0	is negative and decreases from 0 to -1
π to $\dfrac{3\pi}{2}$	is negative and decreases from 0 to -1	is negative and increases from -1 to 0
$\dfrac{3\pi}{2}$ to 2π	is negative and increases from -1 to 0	is positive and increases from 0 to 1

The idea of a circular function (that is, a trigonometric function of a real number rather than of an angle) is more a point of view than a new mathematical concept. For instance, put a calculator in radian mode, enter $\pi/3$, and press the sine key. The display will show the approximate value of $\sin(\pi/3)$, and it makes no difference whether you have in mind that $\pi/3$ is the measure of an angle in radians, the length of an arc on the unit circle, or simply a real number. In any case,

$$\sin \frac{\pi}{3} = \frac{\sqrt{3}}{2} \approx 0.866025404.$$

For this reason, people speak of "trigonometric functions" and "circular functions" more or less interchangeably.

On most scientific calculators, the number in the display register is denoted by x. Pressing the 1/x key gives you the reciprocal of x, pressing the x² key gives you the square of x, and, likewise, pressing the sine or cosine key gives you sin x or cos x. This notation is consistent with the custom of denoting the independent variable of a function by x (unless there are good reasons for doing otherwise). Therefore, in what follows we shall sketch graphs of $y = \sin x$ and of $y = \cos x$ on the usual xy coordinate system. Notice that x, used in this way as the independent variable for a circular function, must not be confused with the abscissa of the point (x, y) in Figure 1.

We can make a rough sketch of the graph of $y = \sin x$ for $0 \le x \le 2\pi$ by using only the information in Table 1, if we replace t by x. However, we can make a more accurate sketch by using the values of sin x shown in Table 2. If even more accuracy is desired, additional points can be found with the aid of a calculator. The graph of $y = \sin x$ for $0 \le x \le 2\pi$ is sketched in Figure 2. This curve is called **one cycle**

Table 2

x	$y = \sin x$
0	0
$\pi/6$	$1/2$
$\pi/4$	$\sqrt{2}/2$
$\pi/3$	$\sqrt{3}/2$
$\pi/2$	1
$2\pi/3$	$\sqrt{3}/2$
$3\pi/4$	$\sqrt{2}/2$
$5\pi/6$	$1/2$
π	0
$7\pi/6$	$-1/2$
$5\pi/4$	$-\sqrt{2}/2$
$4\pi/3$	$-\sqrt{3}/2$
$3\pi/2$	-1
$5\pi/3$	$-\sqrt{3}/2$
$7\pi/4$	$-\sqrt{2}/2$
$11\pi/6$	$-1/2$
2π	0

Figure 2

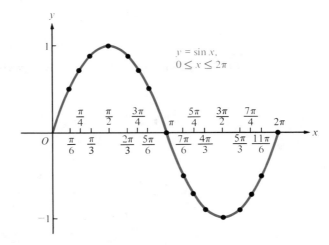

of the graph of $y = \sin x$. (To show more detail, we have used different scales on the x and y axes.)

Because the sine is a periodic function, the complete graph of $y = \sin x$ consists of an endless sequence of copies of the cycle in Figure 2, repeated to the right and left over successive intervals of length 2π (Figure 3). The graph shows that the range of the sine function is the closed interval $[-1, 1]$. In particular,

$$-1 \leq \sin x \leq 1$$

holds for all real numbers x. Other features of the sine function are indicated by its graph. For instance, the graph appears to be symmetric about the origin, which suggests that the sine is an odd function,

$$\sin(-x) = -\sin x.$$

Figure 3

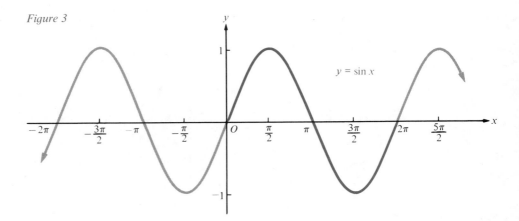

(We prove this in Section 7.1.) Another interesting feature suggested by the graph is that

$$\sin(x + \pi) = -\sin x$$

(Figure 4). This property is proved in Section 7.3.

Figure 4

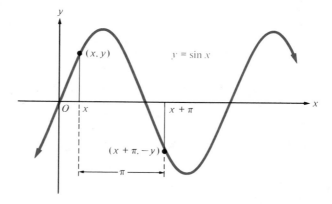

By using the information in Table 1 and the values of the cosine function for angles such as 0, $\pi/6$, $\pi/4$, $\pi/3$, $\pi/2$, and so on, you can sketch the graph of $y = \cos x$ for $0 \le x \le 2\pi$ (Problem 1). This curve (Figure 5) is called one *cycle* of the graph of $y = \cos x$. By extending the cycle to the right and to the left in a periodic fashion, we obtain the complete graph of $y = \cos x$ (Figure 6). Figure 6 also includes the graph of $y = \sin x$, which, as you can see, is the graph of $y = \cos x$ shifted $\pi/2$ units to the right. This indicates that

$$\sin x = \cos\left(x - \frac{\pi}{2}\right),$$

a fact which is proved in Section 7.3.

Figure 5

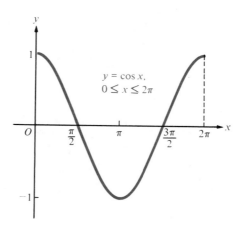

Figure 6

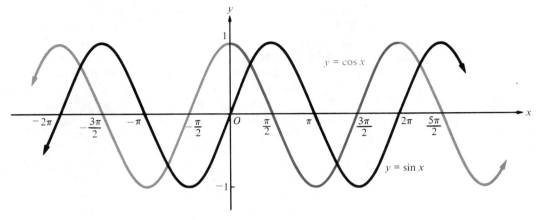

Like the sine function, the range of the cosine function is the interval $[-1, 1]$; hence,

$$-1 \le \cos x \le 1$$

holds for all real numbers x. The graph of cos x in Figure 6 appears to be symmetric about the y axis, indicating that the cosine is an even function,

$$\cos(-x) = \cos x.$$

(Again, this is proved in Section 7.1.) By drawing a figure for $y = \cos x$ similar to Figure 4, you can also see graphically that

$$\cos(x + \pi) = -\cos x$$

(Problem 2).

If the techniques of graphing discussed in Section 3.6 are used to stretch or shift the graphs of sine or cosine functions, the resulting curves always have the wavelike form shown in Figure 7. Such a curve is called a **simple harmonic curve**, a **sine wave**,

Figure 7

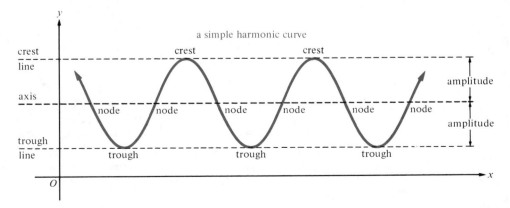

or a **sinusoidal curve.** The highest points on a simple harmonic curve are called **crests** and the lowest points are called **troughs.** The crests lie along one straight line, the troughs lie along another, and the two lines are parallel. The straight line midway between the crest line and the trough line is called the **axis** of the simple harmonic curve, and the points where the curve crosses the axis are called **nodes.** The distance between the axis and the crest line (or trough line) is called the **amplitude** of the simple harmonic curve. Thus, the amplitude is half the distance between the crest line and the trough line.

The portion of a simple harmonic curve from any one point P to the next point Q at which the curve starts to repeat itself is called a **cycle,** and the distance between P and Q is called the **period** (Figure 8). In sketching a cycle, we usually begin at a crest, node, or trough. If you are given (or have already sketched) one cycle, you can easily sketch as much of the rest of the curve as you please by repeating the cycle to the right and to the left over successive intervals, each of which has length equal to the period. It is often convenient to use different units of length along the x and y axes.

Figure 8

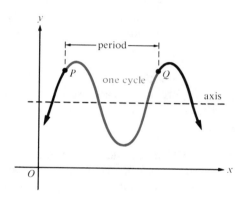

In Examples 1 to 3, sketch the graph of the function defined by each equation, find the amplitude and the period, and specify one cycle (starting at a node) on your graph.

Example 1   **(a)** $y = 3 \sin x$ **(b)** $y = -3 \sin x$

Solution   **(a)** We begin by sketching the graph of $y = \sin x$ (Figure 3); then we multiply each ordinate by 3 (Figure 9a). The x axis is the axis of the simple harmonic curve. The crests are 3 units above the axis, so the amplitude is 3. The period is 2π. One cycle, starting at $(0, 0)$ and ending at $(2\pi, 0)$, forms part of the graph.

(b) We obtain the graph of $y = -3 \sin x$ by reflecting the graph of $y = 3 \sin x$ across the x axis (Figure 9b). The amplitude of the reflected graph is still 3, and its period is still 2π. One cycle, starting at $(0, 0)$ and ending at $(2\pi, 0)$, forms part of the graph.   ■

Figure 9

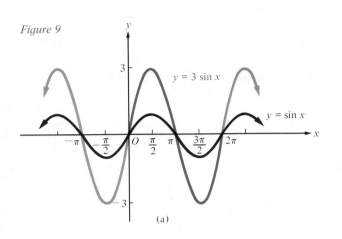

(a)

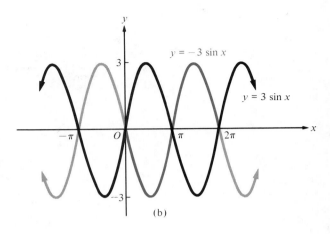

(b)

Example 2   **(a)** $y = \frac{1}{3}\cos x$ **(b)** $y = 1 + \frac{1}{3}\cos x$

Solution   **(a)** The graph of $y = \frac{1}{3}\cos x$ is obtained from the graph of $y = \cos x$ (Figure 6) by multiplying each ordinate by $\frac{1}{3}$ (Figure 10a). The amplitude is $\frac{1}{3}$, the period is 2π, and one cycle, starting at $x = \pi/2$ and ending at $x = (\pi/2) + 2\pi = 5\pi/2$, forms part of the graph.

(b) The graph of $y = 1 + \frac{1}{3}\cos x$ is obtained by shifting the graph of $y = \frac{1}{3}\cos x$ one unit upward (Figure 10b). This shift does not affect the amplitude, which is still $\frac{1}{3}$, or the period, which is still 2π. One cycle, starting at the point $(\pi/2, 1)$ and ending at the point $(5\pi/2, 1)$, forms part of the graph.   ∎

Figure 10

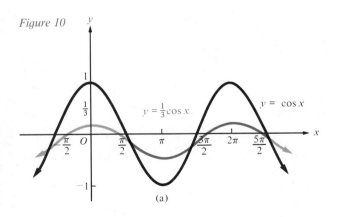

(a)

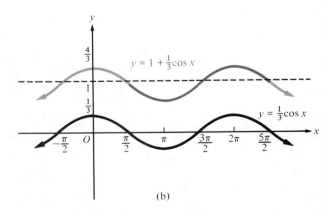

(b)

Example 3 $y = 3\sin 2x$

Solution   The graph of $y = 3\sin 2x$ will cover one cycle as $2x$ varies from 0 to 2π. When $2x = 0$, we have $x = 0$; when $2x = 2\pi$, we have $x = \pi$. Therefore, the graph covers one cycle as x varies from 0 to π, and the period is π. As x varies from 0 to π, sin $2x$ will take on maximum and minimum values of $+1$ and -1; hence, $3 \sin 2x$ will take on maximum and minimum values of $+3$ and -3. Thus, the amplitude is 3. We use this information to sketch one cycle, and then repeat the cycle on either side (Figure 11). Figure 11 includes the graph of $y = 3 \sin x$ for comparison. Notice that the graph of $y = 3 \sin 2x$ oscillates up and down twice as fast as the graph of $y = 3 \sin x$. ∎

Figure 11

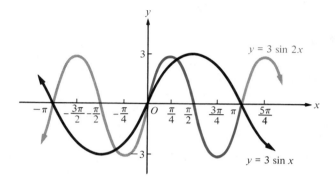

More generally, if a, b, c, and k are constants, $a \neq 0$, and $b > 0$, then the graph of

$$y = a \sin(bx - c) + k$$

is a sine wave with amplitude $|a|$ and horizontal axis $y = k$. This graph will cover one cycle as $bx - c$ varies from 0 to 2π. When $bx - c = 0$, we have $x = c/b$; when $bx - c = 2\pi$, we have $x = (2\pi + c)/b = (2\pi/b) + c/b$. Therefore, the graph completes one cycle, starting from a *node*, over the interval

$$\left[\frac{c}{b}, \frac{2\pi}{b} + \frac{c}{b} \right]$$

Because the node at the beginning of this cycle is shifted $|c/b|$ units from the origin (to the *right* if $c/b > 0$ and to the *left* if $c/b < 0$), the number c/b is called the **phase shift** of the graph. Since the length of the interval corresponding to one cycle is $2\pi/b$ units, the period of the sine wave is $2\pi/b$.

Similar remarks apply to the graph of

$$y = a \cos(bx - c) + k$$

except that the cycle over the interval from c/b to $(2\pi/b) + c/b$ starts from a *crest* or a *trough*, depending on whether $a > 0$ or $a < 0$, respectively.

Example 4   For the graph of $y = 3 \sin\left(2x - \dfrac{\pi}{2} \right)$, find the amplitude, the horizontal axis, the phase shift, and the period. Specify one cycle on the graph, starting from a node, and sketch the graph.

Solution   The amplitude is 3 and the horizontal axis is $y = 0$. The graph will cover one cycle as $2x - \dfrac{\pi}{2}$ varies from 0 to 2π. When $2x - \dfrac{\pi}{2} = 0$, we have $x = \dfrac{\pi}{4}$; when $2x - \dfrac{\pi}{2} = 2\pi$, we have $2x = 2\pi + \dfrac{\pi}{2} = \dfrac{5\pi}{2}$, so $x = \dfrac{5\pi}{4}$. Therefore, the graph covers one cycle as x

varies from $\dfrac{\pi}{4}$ to $\dfrac{5\pi}{4}$. It follows that the phase shift is $\dfrac{\pi}{4}$ and the period is $\dfrac{5\pi}{4} - \dfrac{\pi}{4} = \pi$.

Using this information we can sketch the graph (Figure 12). Notice that the graph

of $y = 3 \sin\left(2x - \dfrac{\pi}{2}\right) = 3 \sin 2\left(x - \dfrac{\pi}{4}\right)$ is the graph of $y = 3 \sin 2x$ (Figure 11)

shifted $\dfrac{\pi}{4}$ unit to the right, thus illustrating the idea of phase shift. ■

Figure 12

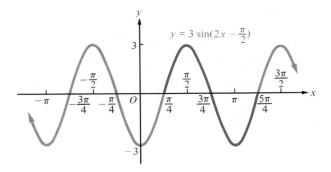

$y = 3 \sin(2x - \frac{\pi}{2})$

Problem Set 6.5

1. Use the information in Table 1 (page 326) and the values of the cosine function for the angles $0, \dfrac{\pi}{6}, \dfrac{\pi}{4}$, $\dfrac{\pi}{3}, \dfrac{\pi}{2}, \dfrac{2\pi}{3}, \dfrac{3\pi}{4}, \dfrac{5\pi}{6}, \pi, \dfrac{7\pi}{6}, \dfrac{5\pi}{4}, \dfrac{4\pi}{3}, \dfrac{3\pi}{2}, \dfrac{5\pi}{3}, \dfrac{7\pi}{4}, \dfrac{11\pi}{6}$, and 2π to sketch an accurate graph of $y = \cos x$ for $0 \le x \le 2\pi$. When you have finished your sketch compare it with Figure 5. ([c] For enhanced accuracy, you may wish to use a calculator to obtain additional points on the graph.)

2. By sketching a figure for $y = \cos x$ similar to Figure 4, show graphical evidence for the fact that $\cos(x + \pi) = -\cos x$.

[c] In Problems 3 to 6, use a calculator to verify that the equation is true for the indicated value of the variable.

3. $\sin(-x) = -\sin x$, $x = 4.203$

4. $\cos(-x) = \cos x$, $x = \dfrac{5\pi}{9}$

5. $\sin(x + \pi) = -\sin x$, $x = 2.771$

6. $\cos(x + \pi) = -\cos x$, $x = \dfrac{11\pi}{7}$

In Problems 7 to 12, use the graphs of $y = \sin x$ (Figure 3) and $y = \cos x$ (Figure 6) to help answer each question.

7. If $-2\pi \le x \le 2\pi$, for what values of x does $\sin x$ reach its maximum value and what is this maximum value?

8. If $-2\pi \le x \le 2\pi$, for what values of x is $\cos x = \frac{1}{2}$?

9. If $-2\pi \le x \le 2\pi$, for what values of x is $\cos x = 0$?

10. If h is a constant with $0 < h < 1$ and $-2\pi \le x \le 2\pi$, how many different values of x are there for which $\sin x = h$?

11. If $-2\pi \le x \le 2\pi$, for what values of x does $\sin x$ reach its minimum value and what is this minimum value?

12. Complete the following sentence: $\sin x$ reaches its maximum and minimum values at the same values of x for which the graph of $y = \cos x$ _____ .

13. Suppose a friend who is just beginning to study trigonometry says, "I understand how to take the sine of an *angle*, but how do you take the sine of a *number*?" Answer your friend's question in your own words.

14. Suppose you are going to measure an acute angle with a protractor and then use a calculator to find the sine of the angle. Of course, if there is an error in your measurement, there will be an error in the calculated value of the sine. Will the error in the sine value be more pronounced when the acute angle is small or when it is large? Why?

In Problems 15 to 41, for the graph of each equation, find the amplitude, the horizontal axis, the phase shift, and the period. Specify one cycle on the graph, starting at a node for the sine functions and a crest or trough for the cosine functions, and sketch the graph.

15. $y = 2 \sin x$

16. $y = -2 \sin x$

17. $y = \frac{1}{2} \sin x$

18. $y = 1 + \cos x$

19. $y = 1 - \frac{2}{3} \cos x$

20. $y = 3 \cos x - \frac{1}{2}$

21. $y = \sin 2x$

22. $y = 2 \sin \frac{x}{3}$

23. $y = \cos 6x$

24. $y = 1 + \cos \pi x$

25. $y = 1 + \cos \frac{x}{2}$

26. $y = \cos \frac{\pi x}{2} + 1$

27. $y = 2 - \pi \cos \pi x$

28. $y = \frac{1}{3} - \frac{1}{3} \cos \frac{3\pi x}{2}$

29. $y = \cos\left(x - \frac{\pi}{6}\right)$

30. $y = 1 + \sin\left(x + \frac{\pi}{4}\right)$

31. $y = 2 \cos\left(x - \frac{\pi}{3}\right)$

32. $y = -\cos\left(x - \frac{7\pi}{8}\right)$

33. $y = 1 - \cos\left(x - \frac{\pi}{2}\right)$

34. $y = -2 + 2 \cos\left(2x - \frac{\pi}{2}\right)$

35. $y = 3 \cos\left(\frac{x}{4} - \pi\right)$

36. $y = 2 \sin\left(\frac{x}{2} + \frac{\pi}{2}\right) + 1$

37. $y = 3 \sin(4x - \pi) + 1$

38. $y = -2 \cos\left(\frac{2\pi}{3} - \frac{x}{3}\right)$

39. $y = -\frac{1}{2} \sin\left(3x + \frac{3\pi}{4}\right)$

40. $y = -3 \cos(\pi - 2x)$

41. $y = 1 - \frac{1}{2} \cos\left(\frac{3x}{4} + \frac{\pi}{4}\right)$

6.6 GRAPHS OF THE TANGENT, COTANGENT, SECANT, AND COSECANT FUNCTIONS

In the previous section, we studied the graphs of the sine and cosine functions. In this section, we consider the graphs of the remaining four trigonometric functions.

Ⓒ **Example 1**　Sketch the graph of

$$y = \tan x \qquad \text{for} \qquad -\frac{\pi}{2} < x < \frac{\pi}{2}.$$

Solution　Table 1 shows values of $\tan x$ rounded off to two decimal places and obtained by using a calculator. Of course, $\tan x = \sin x/\cos x$ is undefined when $\cos x$ is zero; for instance, when $x = -\pi/2$ or when $x = \pi/2$. For values of x slightly smaller than $\pi/2$, the numerator $\sin x$ is close to 1, the denominator $\cos x$ is small, and thus,

$\tan x = \sin x/\cos x$ is very large. (Why?) Similarly, for values of x slightly larger than $-\pi/2$, $\tan x$ is negative with a very large absolute value. Thus the vertical lines

$$x = -\frac{\pi}{2} \quad \text{and} \quad x = \frac{\pi}{2}$$

are asymptotes of the graph. Plotting the points in Table 1, and using the information about the asymptotes, we obtain a sketch of the graph (Figure 1). ■

Table 1

x	$\tan x$
$-\pi/2$	undefined
$-5\pi/12$	-3.73
$-\pi/3$	-1.73
$-\pi/4$	-1
$-\pi/6$	-0.58
$-\pi/12$	-0.27
0	0
$\pi/12$	0.27
$\pi/6$	0.58
$\pi/4$	1
$\pi/3$	1.73
$5\pi/12$	3.73
$\pi/2$	undefined

Figure 1

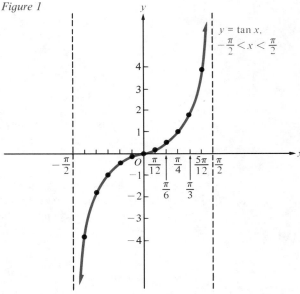

Like the sine and cosine functions, the tangent function is periodic; however, its values repeat themselves whenever the independent variable increases by π. In other words,

$$\boxed{\tan(x + \pi) = \tan x}$$

holds for every value of x in the domain of the tangent function. This can be seen by combining the identities

$$\sin(x + \pi) = -\sin x \quad \text{and} \quad \cos(x + \pi) = -\cos x,$$

which were mentioned in Section 6.5. Thus,

$$\tan(x + \pi) = \frac{\sin(x + \pi)}{\cos(x + \pi)} = \frac{-\sin x}{-\cos x} = \frac{\sin x}{\cos x} = \tan x.$$

Notice that the interval from $-\pi/2$ to $\pi/2$ in Figure 1 has length π units. Therefore, the complete graph of $y = \tan x$ consists of an endless sequence of copies of the curve in Figure 1, repeated to the right and left over successive intervals of length

Figure 2

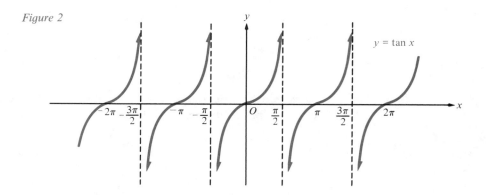

π (Figure 2). The domain of the tangent function is all real numbers *except odd multiples of* $\pi/2$. On each open interval from one odd multiple of $\pi/2$ to the next, the tangent function is increasing. As you can see, the range of the tangent function is $\mathbb{R}$. Also, the graph appears to be symmetric about the origin, suggesting that the tangent is an odd function,

$$\tan(-x) = -\tan x$$

for all values of x in the domain.

Like the tangent function, the values of the cotangent function repeat themselves periodically whenever the independent variable increases by π; that is,

$$\cot(x + \pi) = \cot x$$

holds for every value of x in the domain of the cotangent function (Problem 4). Because $\cot x = 1/\tan x$, the graph of the cotangent function has vertical asymptotes where the graph of the tangent function has x intercepts and vice versa. To sketch the graph of $y = \cot x$, start by sketching the graph for $0 < x < \pi$ (Problem 1), then draw copies of this curve to the left and right over successive intervals of length π (Problem 2). The resulting graph is shown in Figure 3. The domain of the cotangent function is all real numbers *except integer multiples of* π. On each open interval from one multiple of π to the next, the cotangent function is decreasing. As you can

Figure 3

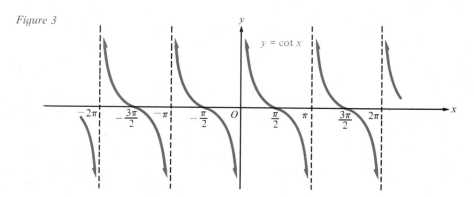

see, the range of the cotangent function is $\mathbb{R}$. Also, the graph appears to be symmetric about the origin, suggesting that the cotangent is an odd function,

$$\cot(-x) = -\cot x$$

for all values of x in the domain.

Ⓒ **Example 2**   Sketch the graph of $y = \sec x$.

Solution   Since $\sec x = 1/\cos x$, we obtain the graph of the secant function by using a calculator to find the reciprocals of nonzero ordinates of points on the graph of the cosine function (Figure 4). The graph of the secant function has vertical asymptotes where the graph of the cosine function has x intercepts.   ■

Figure 4

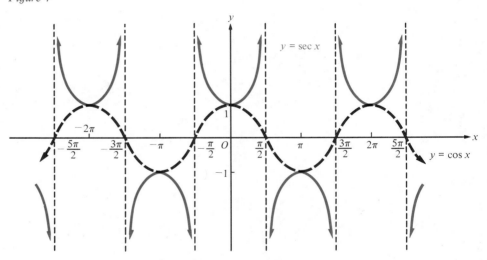

As you can see, the range of the secant function consists of the intervals $(-\infty, -1]$ and $[1, \infty)$. In particular,

$$|\sec x| \geq 1$$

holds for all values of x in the domain of the secant function.

You can sketch the graph of $y = \csc x$ (Figure 5) by using a calculator to find the reciprocals of nonzero ordinates of points on the graph of $y = \sin x$ (Problem 3). Like the secant function, the range of the cosecant function consists of the two intervals $(-\infty, -1]$ and $[1, \infty)$, and

$$|\csc x| \geq 1$$

holds for all values of x in the domain of the function.

You can use the techniques of graphing discussed in Section 3.6 to stretch, shift, or reflect the graphs of the tangent, cotangent, secant, and cosecant functions.

Figure 5

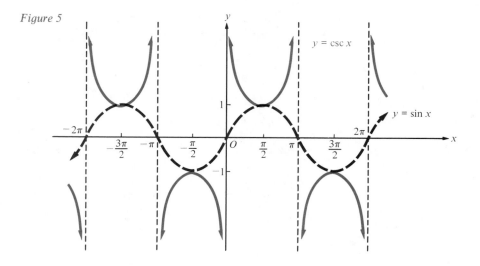

Example 3   Sketch the graph of $y = 3 \sec(x - \pi)$.

Solution   The graph is obtained by multiplying ordinates of points on the graph of $y = \sec x$ (Figure 4) by 3 and then shifting the resulting graph π units to the right (Figure 6). ∎

Figure 6

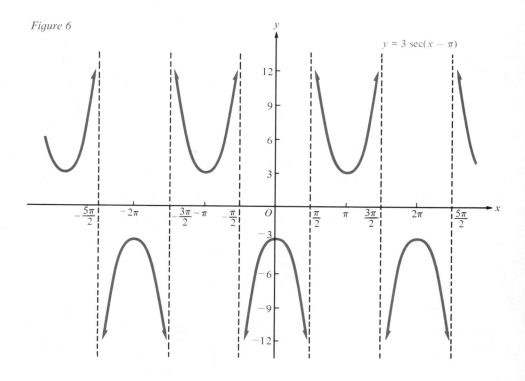

Problem Set 6.6

C **1.** (a) Make a table showing the values of $\cot x$, rounded off to two decimal places, for the following values of x: $\dfrac{\pi}{12}, \dfrac{\pi}{6}, \dfrac{\pi}{4}, \dfrac{\pi}{3}, \dfrac{5\pi}{12}, \dfrac{\pi}{2}, \dfrac{7\pi}{12}, \dfrac{2\pi}{3}$, $\dfrac{3\pi}{4}, \dfrac{5\pi}{6}$, and $\dfrac{11\pi}{12}$.

    (b) Use the table in part (a) to sketch an accurate graph of $y = \cot x$ for $0 < x < \pi$.

2. Using the result of Problem 1, sketch the graph of $y = \cot x$. When you have finished your sketch, compare it with Figure 3.

C **3.** With the aid of a calculator, sketch an accurate graph of $y = \csc x$. When you have finished, compare your graph with Figure 5.

4. Assuming that $\sin(x + \pi) = -\sin x$ and $\cos(x + \pi) = -\cos x$, show that $\cot(x + \pi) = \cot x$ holds for all values of x in the domain of the cotangent function.

C In Problems 5 to 10, use a calculator to verify that the equation is true for the indicated value of the variable.

5. $\tan(x + \pi) = \tan x$, $x = \dfrac{2\pi}{3}$

6. $\cot(x + \pi) = \cot x$, $x = -\dfrac{5\pi}{12}$

7. $\tan(-x) = -\tan x$, $x = 1.334$

8. $\cot(-x) = -\cot x$, $x = 0.7075$

9. $\sec(-x) = \sec x$, $x = \dfrac{5\pi}{12}$

10. $\csc(-x) = -\csc x$, $x = 0.3544$

In Problems 11 to 16, use the graphs of $y = \tan x$, $y = \cot x$, $y = \sec x$, and $y = \csc x$ to help answer each question.

11. If $-2\pi \le x \le 2\pi$, for what values of x is $\tan x = 0$?

12. If h is a positive constant and if $-2\pi \le x \le 2\pi$, how many different values of x are there for which $\tan x = h$?

13. If $-2\pi \le x \le 2\pi$, for what values of x does $\sec x$ reach its smallest positive value and what is this value?

14. How is the graph of $y = -\tan x$ related to the graph of $y = \cot x$?

15. For the graph of $y = \sec x$, what is the distance from one vertical asymptote to the next?

16. How is the graph of $y = \csc x$ related to the graph of $y = \sec x$?

17. Assuming that the cosine is an even function, show that the secant is also an even function.

18. The portion of the graph of
$$y = \sec x$$
for
$$-\pi/2 < x < \pi/2$$
looks vaguely like a parabola. Could it possibly be a parabola? Why?

In Problems 19 to 32, use the graphs of $y = \tan x$, $y = \cot x$, $y = \sec x$, and $y = \csc x$ and the techniques of shifting, stretching, and reflecting to sketch the graph of each equation.

19. $y = 2 \tan x$

20. $y = -2 \tan x$

21. $y = 1 - \frac{1}{2} \cot x$

22. $y = 3 - \sec x$

23. $y = 1 + \dfrac{1}{2} \sec x$

24. $y = \dfrac{2}{3} \csc\left(x - \dfrac{\pi}{3}\right)$

25. $y = \tan\left(x - \dfrac{\pi}{4}\right)$

26. $y = \tan\left(x + \dfrac{\pi}{4}\right)$

27. $y = \dfrac{2}{3} \cot\left(x - \dfrac{\pi}{2}\right)$

28. $y = 5 \sec\left(x - \dfrac{\pi}{2}\right)$

29. $y = \sec\left(x + \dfrac{\pi}{6}\right)$

30. $y = 1 - \csc\left(x + \dfrac{5\pi}{6}\right)$

31. $y = 2 \csc\left(x - \dfrac{\pi}{3}\right)$

32. $y = 2 + \dfrac{1}{2} \csc\left(x + \dfrac{\pi}{3}\right)$

6.7 THE SIMPLE HARMONIC MODEL

In the real world there are a number of quantities that oscillate or vibrate in a uniform manner, repeating themselves periodically in definite intervals of time. Examples include alternating electrical currents; sound waves; light waves, radio waves, and other electromagnetic waves; pendulums, mass-spring systems, and other mechanical oscillators; tides, geysers, seismic waves, the seasons, climatic cycles, and other periodic phenomena of interest in the earth sciences; and biological phenomena ranging from a human heartbeat to periodic variation of the population of a plant or animal species. Simple harmonic curves often provide useful mathematical models for such oscillations. It is only necessary to replace the variable x in the equation of a simple harmonic curve by the variable t representing units of time.

The mathematical model for a quantity y that is oscillating in a **simple harmonic** manner is the equation

$$y = a \cos(\omega t - \phi) + k$$

where a and ω are *positive* constants and ϕ and k are constants. Here a is the **amplitude,** ω (the Greek letter omega) is called the **angular frequency** (see Problem 36), ϕ (the Greek letter phi) is called the **phase angle,** k is the **vertical shift,** and the independent variable t is the **time.**

To sketch a graph of

$$y = a \cos(\omega t - \phi) + k,$$

we begin by noticing that y will go through one cycle as $\omega t - \phi$ goes from 0 to 2π. When $\omega t - \phi = 0$, we have $t = \phi/\omega$; when $\omega t - \phi = 2\pi$, we have $t = (\phi/\omega) + (2\pi/\omega)$. Therefore, one cycle of the graph of $y = a \cos(\omega t - \phi) + k$ covers the interval from $t = \phi/\omega$ to $t = (\phi/\omega) + (2\pi/\omega)$ (Figure 1). The vertical shift k moves the axis of the simple harmonic curve from $y = 0$ to $y = k$ (Figure 1). The **period** of the oscillation is the amount of time T required for y to proceed through one cycle. From

Figure 1

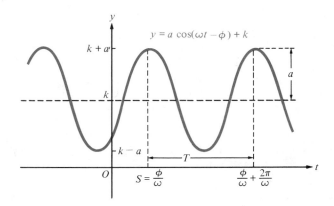

Figure 1, you can see that

$$T = \left(\frac{\phi}{\omega} + \frac{2\pi}{\omega}\right) - \frac{\phi}{\omega} = \frac{2\pi}{\omega}.$$

The equation

$$T = \frac{2\pi}{\omega}$$

is often written in the alternative form

$$\omega = \frac{2\pi}{T}.$$

Notice that one cycle (from crest to crest) of the graph $y = a\cos(\omega t - \phi)$ has the same general shape as one cycle (from crest to crest) of the cosine function (Figure 5 in Section 6.5), but shifted horizontally by the **phase shift** ϕ/ω as we saw in Section 6.5. In what follows, we denote the phase shift by S, so that

$$S = \frac{\phi}{\omega}.$$

The last equation is often written in the alternative form

$$\phi = \omega S.$$

Example 1   If y is oscillating according to the simple harmonic model

$$y = 2\cos\left(\frac{\pi}{4}t - \frac{\pi}{2}\right) - 1,$$

find **(a)** the amplitude a, **(b)** the angular frequency ω, **(c)** the period T, **(d)** the phase angle ϕ, **(e)** the phase shift S, and **(f)** the vertical shift k. Then **(g)** sketch the graph of one cycle.

Solution   **(a)** $a = 2$ **(b)** $\omega = \dfrac{\pi}{4}$

(c) $T = \dfrac{2\pi}{\omega} = \dfrac{2\pi}{\pi/4} = 8$ **(d)** $\phi = \dfrac{\pi}{2}$

(e) $S = \dfrac{\phi}{\omega} = \dfrac{\pi/2}{\pi/4} = 2$ **(f)** $k = -1$

(g) See Figure 2.

Figure 2

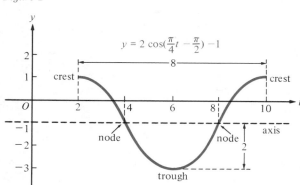

$$y = 2 \cos(\tfrac{\pi}{4}t - \tfrac{\pi}{2}) - 1$$

Figure 3

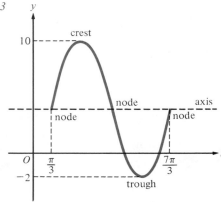

Example 2   The curve in Figure 3 is one cycle of the graph of a simple harmonic model. Determine **(a)** the amplitude a, **(b)** the period T, **(c)** the phase shift S, **(d)** the angular frequency ω, **(e)** the phase angle ϕ, and **(f)** the vertical shift k. Then **(g)** write an equation for y in terms of t.

Solution   **(a)** The vertical distance between the crest and the trough is $10 - (-2) = 12$ units, so $a = \tfrac{12}{2} = 6$.

(b) $T = \dfrac{7\pi}{3} - \dfrac{\pi}{3} = 2\pi$

(c) We need to calculate the distance S from the y axis to the crest. The distance from the y axis to the first node is $\pi/3$ units. The horizontal distance from this node to the crest is one-fourth of the period T, or $T/4 = 2\pi/4 = \pi/2$. Hence,

$$S = \frac{\pi}{3} + \frac{\pi}{2} = \frac{5\pi}{6}.$$

(d) $\omega = \dfrac{2\pi}{T} = \dfrac{2\pi}{2\pi} = 1$

(e) $\phi = \omega S = 1\left(\dfrac{5\pi}{6}\right) = \dfrac{5\pi}{6}$

(f) The axis is a units below the crest, so $k = 10 - a = 10 - 6 = 4$. (The axis is the horizontal line $y = 4$.)

(g) $y = a \cos(\omega t - \phi) + k = 6 \cos\left(t - \dfrac{5\pi}{6}\right) + 4$ ∎

In the simple harmonic model

$$y = a \cos(\omega t - \phi) + k,$$

one cycle requires $T = 2\pi/\omega$ units of time, so y **oscillates** *through* $1/T = \omega/(2\pi)$ *cycles in one unit of time.* We call $1/T$ the **frequency of the oscillation.** The Greek letter v (nu) is often used to denote frequency:

$$v = \frac{1}{T} = \frac{\omega}{2\pi}.$$

For instance, a quantity oscillating with a period of $T = 1/60$ second will oscillate through $v = 1/T = 60$ cycles per second.

One cycle per second is called a **hertz** (abbreviated Hz) in honor of the German physicist Heinrich Hertz (1857–1894), who discovered radio waves in the late 1880s. Broadcast AM radio waves have frequencies of thousands of hertz (kilohertz), whereas FM and TV waves have frequencies of millions of hertz (megahertz). Alternating current generated by electrical utilities in the United States has a standard frequency of 60 Hz (that is, 60 cycles per second).

Heinrich Hertz

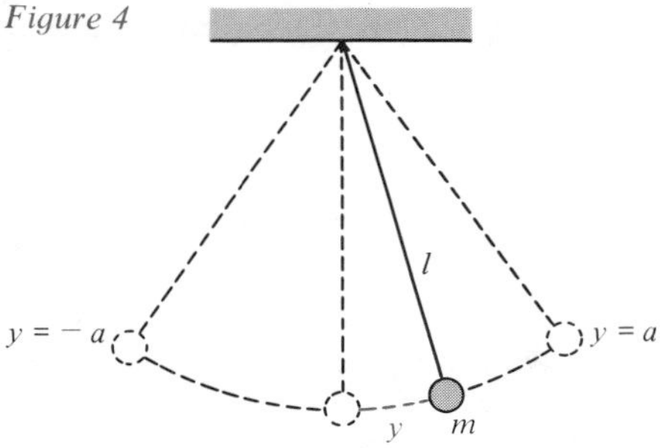

The idea of simple harmonic oscillation is nicely illustrated by a **simple pendulum,** which is an idealized object consisting of a point mass m suspended by a weightless string of length l (Figure 4). When pulled to one side of its vertical position and released, it moves periodically to and fro. Let y denote the displacement of the mass from its vertical position, measured along the arc of the swing (positive to the right and negative to the left) at time t. Suppose that $y = a$ when $t = 0$, the instant of release. Then, if a is not too large, the quantity y will oscillate (approximately) according to a simple harmonic model

$$y = a \cos \omega t,$$

with amplitude a, phase angle $\phi = 0$, vertical shift $k = 0$, and period

$$T = 2\pi \sqrt{\frac{l}{g}},$$

Figure 4

where g is the acceleration of gravity. (At the surface of the earth $g \approx 32$ feet/sec² or $g \approx 9.8$ meters/sec².)

Ⓒ **Example 3**   A pendulum of length $l = 1.2$ meters is pulled to the right through an arc of $a = 0.05$ meter and released at $t = 0$ seconds. Find **(a)** the period T, **(b)** the frequency v, **(c)** the angular frequency ω, and **(d)** the equation for y as a function of t. Round off your answers in parts (a), (b), and (c) to 2 significant digits.

Solution   **(a)** $T = 2\pi \sqrt{\dfrac{l}{g}} = 2\pi \sqrt{\dfrac{1.2}{9.8}} \approx 2.2$ seconds

(b) $v = \dfrac{1}{T} \approx \dfrac{1}{2.2} \approx 0.45$ Hz

(c) $\omega = \dfrac{2\pi}{T} \approx \dfrac{2\pi}{2.2} \approx 2.9$

(d) $y = a \cos \omega t = 0.05 \cos(2.9t)$

The simple harmonic model does not accurately represent the motion of a pendulum swinging with a large amplitude a. But if a is no larger than $l/4$, the formula $T = 2\pi\sqrt{l/g}$ gives the true period with an error of less than 1%.

There is an important relationship between simple harmonic motion and circular motion with constant speed (called **uniform circular motion**) that was discussed in Section 6.1.

> Simple harmonic motion is the perpendicular projection along a diameter of uniform circular motion.

Figure 5

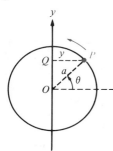

Indeed, in Figure 5, suppose that the point P is moving counterclockwise with constant angular speed ω around a circle of radius $r = a$ with center at the origin O of the y axis. Let Q be the perpendicular projection of P on the y axis, and let y denote the y coordinate of Q. Then y oscillates according to the simple harmonic model

$$y = a \cos(\omega t - \phi)$$

with amplitude a, angular frequency ω, and vertical shift $k = 0$ (Problem 36). If the central angle θ (Figure 5) has the value $\theta = \theta_0$ when $t = 0$, then the phase angle ϕ is given by

$$\phi = \frac{\pi}{2} - \theta_0$$

(Problem 36).

Problem Set 6.7

In Problems 1 to 10, suppose that y is oscillating according to the given simple harmonic model. Find (a) the amplitude a, (b) the angular frequency ω, (c) the period T, (d) the phase angle ϕ, (e) the phase shift S, and (f) the vertical shift k. Then (g) sketch the graph of one cycle.

1. $y = 2 \cos\left(t - \frac{\pi}{3}\right) + 1$

2. $y = 3 \cos\left(t + \frac{\pi}{6}\right) - 5$

3. $y = 4 \cos\left(t + \frac{\pi}{4}\right) + 2$

4. $y = 2 \cos(2t - \pi) + 2$

5. $y = 3 \cos\left(3t + \frac{5\pi}{2}\right)$

6. $y = \frac{1}{2} \cos\left(2t - \frac{\pi}{2}\right) + 1$

7. $y = \frac{3}{4} \cos(4t + 12) - \frac{3}{4}$

8. $y = \frac{1}{3} \cos\left(\frac{\pi t}{4} - \frac{7\pi}{8}\right) + \frac{2}{3}$

9. $y = 110 \cos\left(120\pi t + \frac{3\pi}{2}\right)$

10. $y = \sqrt{3} \cos(t - \sqrt{10}) - \frac{\sqrt{3}}{3}$

11. In the odd problems 1 to 9, find the frequency v.

12. For the simple harmonic model

$$y = a \cos(\omega t - \phi) + k,$$

show that $\phi = 2\pi S/T$, where S is the phase shift and T is the period.

In Problems 13 to 18, the figure shows one cycle of the graph of a simple harmonic model. Determine (a) the amplitude a, (b) the period T, (c) the phase shift S, (d) the angular frequency ω, (e) the phase angle ϕ, and (f) the vertical shift k. Then (g) write an equation for y in terms of t.

13.

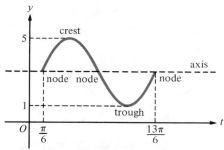

14.

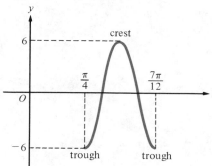

15.

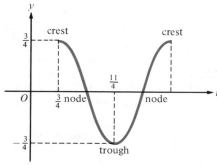

16.

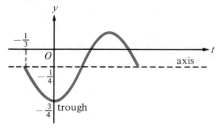

17.

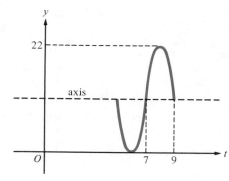

18.

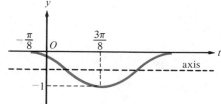

19. In Problems 13, 15, and 17, find the frequency v.

20. For the graph of a simple harmonic model with period T, complete the following sentences.

   (a) The distance between two successive nodes is _____ units.

   (b) The horizontal distance between a node and the next crest or trough is _____ units.

21. Figure 6 shows one cycle, from crest to crest, of the graph of a simple harmonic model

$$y = a\cos(\omega t - \phi) + k.$$

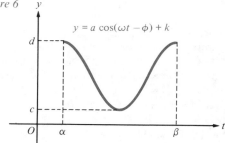

Figure 6

Show that

   (a) $a = \dfrac{d-c}{2}$ (b) $\omega = \dfrac{2\pi}{\beta - \alpha}$

(c) $\phi = \dfrac{2\pi\alpha}{\beta - \alpha}$ (d) $k = \dfrac{d + c}{2}$

22. Assuming that $\sin x = \cos[x - (\pi/2)]$, show that the graphs of $y = a\sin(\omega t - \mu) + k$ and the simple harmonic model $y = a\cos(\omega t - \phi) + k$ are the same where $\phi = \mu + (\pi/2)$. [μ is the Greek letter mu.]

23. Figure 7 shows one cycle, starting at a node, of the graph of a simple harmonic model

$$y = a\cos(\omega t - \phi) + k.$$

Show that

(a) $a = \dfrac{d - c}{2}$ (b) $\omega = \dfrac{2\pi}{\beta - \alpha}$

(c) $\phi = \dfrac{(3\alpha + \beta)\pi}{2(\beta - \alpha)}$ (d) $k = \dfrac{d + c}{2}$

Figure 7

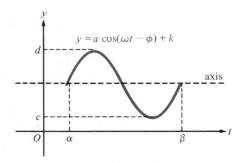

24. What does *amplitude modulation* (AM) mean? What does *frequency modulation* (FM) mean? [If you don't know, use a dictionary.]

In Problems 25 to 30, sketch one cycle of the graph of a quantity y oscillating according to a simple harmonic model $y = a\cos(\omega t - \phi) + k$, with amplitude a, frequency v, phase angle ϕ, and vertical shift k.

25. $a = 1$, $v = 50$ Hz, $\phi = \dfrac{\pi}{2}$, $k = 0$

26. $a = 220$, $v = 60$ Hz, $\phi = 0$, $k = 0$

27. $a = \dfrac{2}{3}$, $v = \dfrac{1}{4\pi}$ Hz, $\phi = \dfrac{\pi}{4}$, $k = \dfrac{2}{3}$

28. $a = 5$, $v = \dfrac{1}{28}$ cycle per day, $\phi = \dfrac{-3\pi}{14}$, $k = 0$

29. $a = 8.4$, $v = 0.18$ cycle per week, $\phi = 0$, $k = 0$

30. $a = 5 \times 10^{-7}$, $v = 88.5$ megahertz, $\phi = 0.8$, $k = 0$

31. A pendulum of length $l = 0.5$ foot is pulled to the right through an arc of $a = 0.1$ foot and released at $t = 0$ seconds. Find the period T, the frequency v, the angular frequency ω, and the equation giving the displacement y at time t.

C 32. An astronaut on the surface of the moon finds that a pendulum has a period 2.45 times as long as it does on the surface of the earth. From this information, she is able to calculate the acceleration of gravity on the moon. How does she do this, and what is her answer?

Earthrise seen from the moon

33. You are lost on a desert island and your watch is broken. You have managed to salvage a toolbox containing, among other things, a tape measure and a ball of string. Explain how you go about constructing a crude "clock" that will measure time in seconds.

C 34. In the northern sky, the constellation Cepheus includes a number of stars that pulsate in brightness or magnitude. One of these stars, Beta Cephei, has a magnitude y that varies from a minimum of 3.141 to a maximum of 3.159 with a period of 4.5 hours. Using a simple harmonic model with a phase angle of $\phi = 6.33$:

(a) Sketch a graph showing the variation of y over the time interval $t = 0$ to $t = 8$ hours.

(b) Determine the magnitude of Beta Cephei at $t = 4$ hours.

ⓒ 35. A typical mass-spring system consists of a mass m suspended on a spring (Figure 8). A vertical y axis is set up, with the equilibrium level of the mass at $y = k$. Suppose that the mass is raised to a level $y = k + a$ and released at time $t = 0$. It is shown in physics that, neglecting friction and air resistance, the mass will bob up and down according to a simple harmonic model $y = a \cos(\omega t - \phi) + k$ with phase angle $\phi = 0$ and frequency

$$v = \frac{1}{2\pi} \sqrt{\frac{K}{m}}.$$

Here K, a constant that measures the stiffness of the spring, is the amount of force required to stretch it by one unit of distance. If $m = 0.05$ kilogram, $a = 0.04$ meter, and $K = 1.25$ newtons per meter, sketch two cycles of the graph of y.

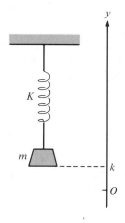

Figure 8

36. In Figure 5, suppose that the point P is moving counterclockwise around the circle with angular speed ω and that $\theta = \theta_0$ when $t = 0$. Show that $y = a \cos(\omega t - \phi)$, where $\phi = (\pi/2) - \theta_0$. [*Hint:* Show that $\theta = \omega t + \theta_0$, then use $\sin \theta = \cos[\theta - (\pi/2)]$.]

37. An electrical circuit consisting of a coil with inductance L henries and a capacitor with a capacitance of C farads is called an *LC-circuit* (Figure 9). If the capacitor is charged and the switch is closed at time $t = 0$ seconds, then the potential difference E volts across the plates of the capacitor will oscillate in a simple harmonic manner with amplitude E_0, angular frequency $\omega = 1/\sqrt{LC}$, and phase angle $\phi = 0$. If $E_0 = 100$ volts, $L = 10$ henries, and $C = 0.0005$

farad, find the period T, the frequency v, and the equation giving the voltage E at time t seconds.

Figure 9

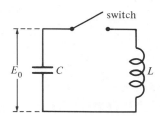

ⓒ 38. A **sound wave** in air consists of a continuous train of alternate compressions and rarefactions traveling at a speed of approximately 331 meters per second. The distance between any two successive compressions is called the **wavelength.** If the air pressure at any one fixed point varies in a simple harmonic manner, the sound heard by a human ear at that point is perceived as a **pure tone.** The amplitude of this variation determines the loudness of the tone, and its frequency determines the pitch. The pure tone middle C has a frequency of 256 Hz.

(a) Find the wavelength of a sound wave that produces the pure tone middle C.

(b) Suppose the sound wave in (a) causes an air pressure variation of amplitude 0.05 newton/meter^2 at a certain point. Using a phase angle $\phi = \pi/2$, write the equation for the air pressure y at this point at time t, if normal undisturbed atmospheric pressure is 10^5 newtons/meter^2.

ⓒ 39. Some people believe that we are all subject to periodic variations, called **biorhythms,** in our physical stamina, emotional well-being, and intellectual ability. The physical cycle is supposed to have a period of 23 days, the emotional cycle a period of 28 days, and the intellectual cycle a period of 33 days. Three graphs plotted on the same coordinate system (using, say, three different colors) showing the variations of these three factors for a particular individual over a period of a month or more make up a **biorhythm chart.** Most practitioners use simple harmonic curves all of the same (arbitrary) amplitude a. Taking $a = 1$, plot a biorhythm chart over a 62-day period for an individual with the following phase angles: physical phase angle = 5.804, emotional phase angle = 3.030, and intellectual phase angle = 6.045.

Ⓒ **40.** Believers in biorhythms (see Problem 39) refer to the nodes of any one of three biorhythm curves as **critical points,** and to the corresponding days as **critical days.** Contrary to what one might at first think, it is believed that the critical points are more significant than the crests or the troughs of the bio-rhythm curves. For the biorhythm chart plotted in Problem 39, locate the physical, emotional, and intellectual critical days. Are there any doubly critical days—that is, days on which two curves pass through a node simultaneously? What about triply critical days?

REVIEW PROBLEM SET, CHAPTER 6

In Problems 1 to 4, find the degree measure of the given angle.

1. One-tenth of a counterclockwise revolution.

2. Five-sixths of a counterclockwise revolution.

3. Four-fifths of a clockwise revolution.

4. The angle through which the minute hand on a clock turns in 5 hours.

In Problems 5 to 10, indicate the number of revolutions or the fraction of a revolution represented by the given angle. Indicate whether the rotation is clockwise or counterclockwise.

5. $60°$ **6.** $140°$

7. $-1440°$ **8.** $-310°$

9. $930°$ **10.** $-610°$

Ⓒ In Problems 11 to 14, express each angle measure as a decimal. Round off to four decimal places.

11. $2°3'$ **12.** $23°19'13''$

13. $55°45'35''$ **14.** $5''$

In Problems 15 to 18, express each angle measure in degrees, minutes, and seconds.

15. $87.35°$ **16.** $-62.45°$

Ⓒ **17.** $-24.53°$ Ⓒ **18.** $165.37°$

In Problems 19 to 24, s denotes the length of the arc intercepted on a circle of radius r by a central angle of θ radians. Find the missing quantity.

19. $r = 5$ meters, $\theta = 0.57$ radian, $s = ?$

20. $r = 40$ centimeters, $s = 4$ centimeters, $\theta = ?$

21. $s = 3\pi$ feet, $\theta = \pi$ radians, $r = ?$

22. $r = 13$ kilometers, $\theta = \dfrac{3\pi}{7}$ radians, $s = ?$

23. $r = 2$ meters, $s = \pi$ meters, $\theta = ?$

24. $s = 17\pi$ microns, $\theta = \dfrac{5\pi}{6}$ radians, $r = ?$

In Problems 25 to 30, convert each degree measure to radian measure. Do not use a calculator. Write your answer as a rational multiple of π.

25. $80°$ **26.** $570°$

27. $-355°$ **28.** $-810°$

29. $-310°$ **30.** $765°$

In Problems 31 to 36, convert each radian measure to degree measure. Do not use a calculator.

31. $\dfrac{2\pi}{5}$ **32.** $-\dfrac{13\pi}{4}$ **33.** $-\dfrac{7\pi}{8}$

34. $\dfrac{35\pi}{3}$ **35.** $\dfrac{51\pi}{4}$ **36.** $\dfrac{18\pi}{5}$

Ⓒ **37.** Use a calculator to convert each degree measure to an approximate radian measure rounded off to four decimal places.

(a) $5°$ (b) $27.7533°$

(c) $-17.173°$ (d) $35°16'55''$

Ⓒ **38.** Use a calculator to convert each radian measure to an approximate degree measure rounded off to four decimal places.

(a) 5 (b) 3.9 (c) -7.63 (d) -21.403

Ⓒ In Problems 39 to 42, s denotes the length of the arc intercepted on a circle of radius r by a central angle θ measured in *degrees*. Find the missing quantity. Round off your answers to three significant digits.

39. $r = 10$ feet, $\theta = 36°$, $s = ?$

40. $s = 111$ centimeters, $\theta = 135°$, $r = ?$

41. $r = 12$ meters, $s = 3\pi$ meters, $\theta = ?$

42. $r = 5$ kilometers, $\theta = 65°$, $s = ?$

Ⓒ In Problems 43 and 44, find the area $A = \frac{1}{2}r^2\theta$ of a sector of a circle of radius r with central angle θ. (See problem 48 in Problem Set 6.1.)

43. $r = 25$ centimeters, $\theta = \dfrac{\pi}{6}$

44. $r = 3.5$ meters, $\theta = 60°$

Ⓒ **45.** The minute hand on a tower clock is 0.6 meter long. How far does the tip of the hand travel in 4 minutes?

Ⓒ **46.** In Problem 45, what is the area of the sector swept out by the minute hand in 4 minutes?

Ⓒ **47.** A wheel with radius 2 feet makes 30 revolutions per minute. Find the linear speed of a point on the rim of the wheel.

Ⓒ **48.** Find the approximate diameter of the moon if its disk subtends an angle of 30′ at a point on the earth 240,000 miles away (Figure 1). [*Hint:* Approximate the diameter $|\overline{DE}|$ by the length of the arc $\overset{\frown}{BC}$.]

Figure 1

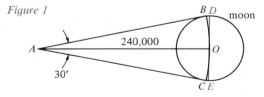

Ⓒ **49.** A satellite in a circular orbit above the earth is known to have a linear speed of 9.92 kilometers per second. In 10 seconds it moves along an arc that subtends an angle of 0.75° at the center of the earth. If the radius of the earth is 6371 kilometers, how high is the satellite above the surface of the earth?

Ⓒ **50.** A laser beam is projected from a point on the surface of the earth to a reflector on the surface of the moon 384,000 kilometers away. If the laser beam diverges at an angle of 1″, find the approximate diameter of the beam when it strikes the surface of the moon.

Ⓒ **51.** A phonograph record is rotating at 33.33 revolutions per minute. When the needle is in a groove 10 centimeters from the center, what is the linear speed of the groove with respect to the needle?

Ⓒ **52.** A small sprocket (toothed wheel) of radius 7 centimeters is connected by a chain to a large sprocket of radius 11 centimeters. The small sprocket is turning at 60 revolutions per minute.

(a) Find the linear speed of the chain connecting the sprockets.

(b) Find the rate of rotation of the large sprocket.

53. Find the values of the six trigonometric functions of the acute angle θ in Figure 2.

Figure 2

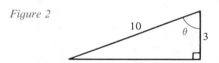

54. Given that $\sin 67° = 0.9205$, $\tan 67° = 2.356$, and $\sec 67° = 2.559$, find the values of

(a) $\cos 23°$ (b) $\cot 23°$ (c) $\csc 23°$

55. Express the value of each trigonometric function as the value of the cofunction of the complementary angle.

(a) $\sin 50°$ (b) $\cos \dfrac{\pi}{7}$

(c) $\sec 89°$ (d) $\cot \dfrac{3\pi}{28}$

Ⓒ **56.** Suppose that α and β are complementary acute angles and that $\sin \alpha = 0.9063$. Find

(a) $\cos \beta$ (b) $\csc \alpha$ (c) $\sec \beta$

57. Sketch two coterminal angles α and β in standard position whose terminal side contains each point. Arrange it so that α is positive, β is negative, and neither angle exceeds one revolution. In each case, name the quadrant in which the angle lies or in-dicate that it is quadrantal.

(a) $(5, 12)$ (b) $(-3, 5)$ (c) $(-7, -6)$

(d) $(0, -4)$ (e) $(\sqrt{2}, -\sqrt{3})$ (f) $(\sqrt{5}, 0)$

58. Sketch each angle in standard position and name the quadrant in which it lies or specify that it is quadrantal.

(a) $60°$ (b) $-210°$ (c) $110°$

(d) $-2160°$ (e) $-340°$ (f) $-750°$

59. In each case, specify and sketch three different angles that are coterminal with the given angle in standard position.

(a) $-15°$ (b) $460°$ (c) $170°$

(d) $-980°$ (e) $\dfrac{5\pi}{3}$

60. Indicate which of the six trigonometric functions are *not* defined for each angle.

(a) $1260°$ (b) 37π (c) 38π (d) $\dfrac{19\pi}{2}$

61 For each point (x, y), evaluate the six trigonometric functions of an angle θ in standard position if the terminal side of θ contains (x, y). In each case, sketch one of the coterminal angles θ thus formed. Do not use a calculator.

(a) $(-3, 5)$ (b) $(2, -3)$

(c) $(-6, -8)$ (d) $(\sqrt{3}, -1)$

62. Without using a calculator, give the algebraic sign of

(a) $\sin 183°$ (b) $\tan \dfrac{37\pi}{39}$

(c) $\cos(-269°)$ (d) $\cot\left(-\dfrac{27\pi}{17}\right)$.

63. Indicate the quadrant in which θ lies if

(a) $\sin \theta > 0$ and $\cos \theta < 0$

(b) $\tan \theta > 0$ and $\sec \theta < 0$

(c) $\csc \theta < 0$ and $\tan \theta < 0$

(d) $\sin \theta < 0$ and $\sec \theta > 0$.

64. If n is an integer, explain why $\cos(n\pi) = (-1)^n$.

65. If $\sin \theta = -\frac{4}{5}$, and $\cos \theta = \frac{3}{5}$, use the reciprocal and quotient identities to find the values of the other four trigonometric functions.

⌐C 66. If $\sin \theta = 0.3145$ and $\cos \theta = -0.9493$, use the reciprocal and quotient identities to find the values of the other four trigonometric functions.

In Problems 67 to 74, the value of one of the trigonometric functions is given along with information about the quadrant (Q) in which θ lies. Find the values of the other five trigonometric functions of θ.

67. $\sin \theta = -\frac{5}{13}$, θ in Q_{IV}

68. $\tan \theta = \frac{3}{2}$, θ not in Q_I

69. $\csc \theta = \frac{13}{12}$, θ in Q_{II}

70. $\sec \theta = -\frac{5}{4}$, θ in Q_{III}

71. $\sin \theta = \frac{3}{5}$, $\cos \theta < 0$

72. $\sec \theta = -5$, $\csc \theta < 0$

⌐C 73. $\cot \theta = 0.6249$, $\sin \theta < 0$

⌐C 74. $\tan \theta = 0.5543$, $\sec \theta < 0$

75. Without using a calculator, find the values (if they are defined) of the six trigonometric functions of

(a) $-180°$ (b) $540°$ (c) $990°$

(d) $-360°$ (e) $\dfrac{7\pi}{2}$ (f) -5π

(g) 17π (h) 18π

76. If n is an integer, explain why

$$\sin\left[\frac{(2n - 1)\pi}{2}\right] = (-1)^{n+1}.$$

77. Using reference angles and the values of the six trigonometric functions for $30°$, $45°$, and $60°$ (or $\pi/6$, $\pi/4$, and $\pi/3$), find the values of the six trigonometric functions of

(a) $-150°$ (b) $-315°$ (c) $780°$

(d) $\dfrac{13\pi}{3}$ (e) $-\dfrac{15\pi}{4}$

78. Trigonometric tables ordinarily give approximate values of the trigonometric functions for acute angles only. Explain in your own words how to use such a table to find values of these functions for angles that are not acute.

© In Problems 79 to 96, use a calculator to find the approximate values of the indicated trigonometric functions.

79. $\sin 27°20'$

80. $\sin 421°15'$

81. $\cos 53.47°$

82. $\cos(-113.81°)$

83. $\tan(-117°15'30'')$

84. $\tan(-281°31'25'')$

85. $\sec 16.43°$

86. $\sec(-248.2°)$

87. $\csc \dfrac{4\pi}{5}$

88. $\csc 5.132$

89. $\cot(-3.18)$

90. $\cot(-7.167)$

91. $\cos(-19.213)$

92. $\csc 18.113$

93. $\sin 5.015$

94. $\cot \sqrt{3}$

95. $\cos \dfrac{\sqrt{2}\pi}{4}$

96. $\sin[\tan(-71.32)]$

In Problems 97 to 106, use the graphs of the six trigonometric functions (Sections 6.5 and 6.6) and the techniques of graphing (Section 3.6) to sketch the graph of each equation.

97. $y = 1 + \frac{1}{3} \sin x$

98. $y = 1 - \frac{2}{5} \sin(x + \pi)$

99. $y = 2 + \dfrac{1}{2} \cos\left(x - \dfrac{\pi}{2}\right)$

100. $y = -0.1 \cos\left(x - \dfrac{\pi}{4}\right)$

101. $y = -\tan\left(x - \dfrac{\pi}{6}\right)$

102. $y = \sec\left(x - \dfrac{\pi}{4}\right) + 1$

103. $y = \sec(x - \pi)$

104. $y = 1 - 4 \csc\left(x - \dfrac{\pi}{3}\right)$

105. $y = \dfrac{2}{3} \cot\left(x - \dfrac{\pi}{2}\right)$

106. $y = 2 - \tan(x + \pi)$

107. If $-2\pi \le x \le 2\pi$, for what values of x does $\cos x$ reach its maximum and minimum values and what are these values?

108. If $-2\pi \le x \le 2\pi$, for what values of x is $\sin x = \sqrt{3}/2$?

109. Sketch the graphs of
$$y = \sin\left(\frac{3x}{2} - \frac{\pi}{4}\right) \quad \text{and} \quad y = \cos\left(\frac{3x}{2} - \frac{\pi}{4}\right)$$
on the same coordinate system.

110. Sketch the graph of $y = |\sin x|$.

In Problems 111 to 116, suppose that y is a quantity that is oscillating according to the given simple harmonic model. Find (a) the amplitude a, (b) the angular frequency ω, (c) the period T, (d) the phase angle ϕ, (e) the phase shift S, (f) the vertical shift k, and (g) the frequency v. Then (h) sketch the graph of one cycle.

111. $y = 2 \cos(\pi t + \pi) + 1$

112. $y = 3 \cos(0.4t - 1) + 3$

113. $y = 3 \cos\left(\dfrac{t}{2} - \dfrac{\pi}{2}\right) - 3$

114. $y = 5 \cos\left(\dfrac{t}{3} + \dfrac{5\pi}{6}\right) - 2$

115. $y = 0.2 \cos(0.25t - \pi)$

116. $y = 0.1 \cos\left[3\left(t - \dfrac{\pi}{3}\right)\right]$

© **117.** A pendulum of length $l = 5.2$ meters is pulled to the right through an arc of $a = 0.2$ meter and released at $t = 0$ seconds. Find the period T, the frequency v, the angular frequency ω, and the equation giving the displacement y at time t. Sketch a graph showing two cycles of the motion starting from $t = 0$.

© **118.** The long-period variable star T Cephei has a magnitude y that varies from a minimum of 5.13 to a maximum of 10.87 with a period of 390 days. Using a simple harmonic model with a phase angle of $\phi = 0.16$:

  (a) Sketch a graph showing the variation of y over the time interval $t = 0$ to $t = 600$ days.

  (b) Determine the magnitude of T Cephei at $t = 24$ days.

119. A sound wave in air with a frequency of 440 Hz is perceived as the pure tone "concert A." Suppose that this sound wave causes a sinusoidal air-pressure variation of amplitude 0.03 newton/meter^2 at a certain point. Using a phase angle of $\phi = \pi/2$, write the equation for the air pressure y at this point at time t if normal undisturbed atmospheric pressure is 10^5 newtons/meter.

[C] 120. An *LC*-circuit (Problem 37, Section 6.7) contains an inductance of $L = 5 \times 10^{-4}$ henry and a variable capacitor that allows the circuit to be tuned from a frequency of 5.5×10^5 Hz to a frequency of 1.6×10^6 Hz. Sketch a graph showing the capacitance C as a function of the frequency v over this range of values.

121. People who believe in biorhythms (see Problem 39, Section 6.7) claim that the physical, emotional, and intellectual curves start with a common node at birth, $t = 0$ days. Assuming this, determine the number of years that will elapse before the three biorhythms again reach a common node.

[C] 122. In engineering and physics, an oscillation with an amplitude that decreases exponentially with the passage of time is said to be *damped*. The mathematical model for **damped simple harmonic oscillation** is

$$y = ae^{-Kt} \cos(\omega t - \phi) + k,$$

where the positive constant K is called the *damping constant*. Sketch a graph of a damped harmonic oscillation for $K = \frac{1}{2}$, $a = 2$, $\omega = \pi/4$, $\phi = \pi/2$, and $k = 0$.

Chapter 7

Trigonometric Identities and Equations

In Section 6.3, we derived the fundamental *reciprocal*, *quotient*, and *Pythagorean* identities. In this chapter, we establish the remaining standard trigonometric identities, including the important formulas for *sums*, *differences*, and *multiples* of angles. (For convenience, all of the standard trigonometric identities are listed inside the back cover of this book.) In many applications, ranging from calculus, engineering, and physics to economics and the life sciences, these identities and formulas are routinely used to simplify complicated expressions and to help solve equations involving trigonometric functions.

7.1 THE FUNDAMENTAL TRIGONOMETRIC IDENTITIES

A **trigonometric equation** is, by definition, an equation that involves at least one trigonometric function of a variable. Such an equation is called a **trigonometric identity** if it is true for all values of the variable for which both sides of the equation are defined. An equation that is not an identity is called a **conditional equation.**

For instance, the trigonometric equation

$$\csc t = \frac{1}{\sin t}$$

is an identity, since it is true for all values of t (except, of course, for those values for which $\csc t$ or $1/\sin t$ is undefined). On the other hand, the trigonometric equation

$$\sin t = \cos t$$

is a conditional equation, since there are values of t (for instance, $t = 0$) for which it isn't true.

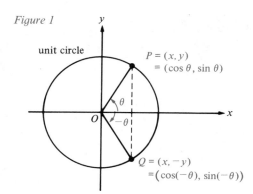

Figure 1

Now we are going to derive the trigonometric identities

$$\sin(-\theta) = -\sin\theta \quad \text{and} \quad \cos(-\theta) = \cos\theta$$

suggested by the graphs of the sine and cosine functions in Section 6.5. Figure 1 shows an angle θ and the corresponding angle $-\theta$, both in standard position. Evidently, the points P and Q, where the terminal sides of these angles intersect the unit circle, are mirror images of each other across the x axis. Therefore, if $P = (x, y)$, it follows that $Q = (x, -y)$. In Section 6.5, we showed that

$$P = (x, y) = (\cos\theta, \sin\theta).$$

Likewise,
$$Q = (x, -y) = (\cos(-\theta), \sin(-\theta)).$$

Therefore,
$$\sin(-\theta) = -y = -\sin\theta$$

and
$$\cos(-\theta) = x = \cos\theta.$$

If we now combine the identities obtained above with the quotient identity, $\tan\theta = \sin\theta/\cos\theta$, we find that

$$\tan(-\theta) = \frac{\sin(-\theta)}{\cos(-\theta)} = \frac{-\sin\theta}{\cos\theta} = -\tan\theta.$$

Similar arguments apply to $\cot(-\theta)$, $\sec(-\theta)$, and $\csc(-\theta)$ (Problem 51). The results are summarized in the following theorem.

Theorem 1   **Even–Odd Identities**

For all values of θ in the domains of the functions:

(i) $\sin(-\theta) = -\sin\theta$	**(ii)** $\cos(-\theta) = \cos\theta$	**(iii)** $\tan(-\theta) = -\tan\theta$
(iv) $\cot(-\theta) = -\cot\theta$	**(v)** $\sec(-\theta) = \sec\theta$	**(vi)** $\csc(-\theta) = -\csc\theta$

Notice that only the cosine and its reciprocal the secant are even functions—the remaining four trigonometric functions are odd. The even–odd identities are often used to simplify expressions, as in the following example:

Example 1   Use the even–odd identities to simplify each expression.

(a) $\dfrac{\sin(-\theta) + \cos(-\theta)}{\sin(-\theta) - \cos(-\theta)}$ **(b)** $1 + \tan^2(-t)$

Solution   **(a)** $\dfrac{\sin(-\theta) + \cos(-\theta)}{\sin(-\theta) - \cos(-\theta)} = \dfrac{-\sin\theta + \cos\theta}{-\sin\theta - \cos\theta} = \dfrac{-(\sin\theta - \cos\theta)}{-(\sin\theta + \cos\theta)} = \dfrac{\sin\theta - \cos\theta}{\sin\theta + \cos\theta}$

(b) $1 + \tan^2(-t) = 1 + [\tan(-t)]^2 = 1 + (-\tan t)^2 = 1 + \tan^2 t = \sec^2 t$ ∎

The reciprocal, quotient, and Pythagorean identities obtained in Section 6.3, together with the even–odd identities, are often called the *fundamental trigonometric identities.* For convenience, we summarize these identities here.

Fundamental Trigonometric Identities

1. $\csc \theta = \dfrac{1}{\sin \theta}$	**2.** $\sec \theta = \dfrac{1}{\cos \theta}$	**3.** $\cot \theta = \dfrac{1}{\tan \theta}$
4. $\tan \theta = \dfrac{\sin \theta}{\cos \theta}$	**5.** $\cot \theta = \dfrac{\cos \theta}{\sin \theta}$	**6.** $\cos^2 \theta + \sin^2 \theta = 1$
7. $1 + \tan^2 \theta = \sec^2 \theta$	**8.** $\cot^2 \theta + 1 = \csc^2 \theta$	**9.** $\sin(-\theta) = -\sin \theta$
10. $\cos(-\theta) = \cos \theta$	**11.** $\tan(-\theta) = -\tan \theta$	**12.** $\cot(-\theta) = -\cot \theta$
13. $\sec(-\theta) = \sec \theta$	**14.** $\csc(-\theta) = -\csc \theta$	

Not only should you memorize these fourteen fundamental identities, but they should become so familiar to you that you can recognize them quickly even when they are written in equivalent forms. For instance, $\csc \theta = 1/\sin \theta$ can also be written as

$$(\sin \theta)(\csc \theta) = 1 \qquad \text{or} \qquad \sin \theta = \frac{1}{\csc \theta}.$$

Incidentally, a product of values of trigonometric functions such as $(\sin \theta)(\csc \theta)$ is usually written simply as $\sin \theta \csc \theta$, unless the parentheses are necessary to prevent confusion.

In Examples 2 to 4, simplify each trigonometric expression by using the fundamental identities.

Example 2 $\csc \theta \cos \theta$

Solution $\csc \theta \cos \theta = \dfrac{1}{\sin \theta} \cos \theta = \dfrac{\cos \theta}{\sin \theta} = \cot \theta$ ■

Example 3 $\tan^2 t - \sec^2 t$

Solution Because $1 + \tan^2 t = \sec^2 t$, it follows that

$$\tan^2 t - \sec^2 t = -1.$$ ■

Example 4 $\csc^4 x - 2 \csc^2 x \cot^2 x + \cot^4 x$

Solution The given expression is the square of $\csc^2 x - \cot^2 x$. Because $\cot^2 x + 1 = \csc^2 x$, we have $\csc^2 x - \cot^2 x = 1$. Therefore,

$$\csc^4 x - 2 \csc^2 x \cot^2 x + \cot^4 x = (\csc^2 x - \cot^2 x)^2 = 1^2 = 1.$$ ■

The reciprocal and quotient identities enable us to write $\csc \theta$, $\sec \theta$, $\tan \theta$, and $\cot \theta$ in terms of $\sin \theta$ and $\cos \theta$. Therefore:

> Any trigonometric expression can be rewritten in terms of sines and cosines.

This fact and the Pythagorean identity $\cos^2 \theta + \sin^2 \theta = 1$ can often be used to simplify trigonometric expressions.

In Examples 5 and 6, rewrite each trigonometric expression in terms of sines and cosines, and then simplify the result.

Example 5 $\csc t - \dfrac{\cot t}{\sec t}$

Solution $\csc t - \dfrac{\cot t}{\sec t} = \dfrac{1}{\sin t} - \dfrac{\dfrac{\cos t}{\sin t}}{\dfrac{1}{\cos t}} = \dfrac{1}{\sin t} - \dfrac{\cos t}{\sin t} \cos t$

$$= \dfrac{1}{\sin t} - \dfrac{\cos^2 t}{\sin t} = \dfrac{1 - \cos^2 t}{\sin t} = \dfrac{\sin^2 t}{\sin t} = \sin t \qquad \blacksquare$$

Example 6 $\dfrac{\csc^2 x \sec^2 x}{\csc^2 x + \sec^2 x}$

Solution $\dfrac{\csc^2 x \sec^2 x}{\csc^2 x + \sec^2 x} = \dfrac{\left(\dfrac{1}{\sin^2 x}\right)\left(\dfrac{1}{\cos^2 x}\right)}{\dfrac{1}{\sin^2 x} + \dfrac{1}{\cos^2 x}} = \dfrac{\sin^2 x \cos^2 x \left(\dfrac{1}{\sin^2 x}\right)\left(\dfrac{1}{\cos^2 x}\right)}{\sin^2 x \cos^2 x \left(\dfrac{1}{\sin^2 x} + \dfrac{1}{\cos^2 x}\right)}$

$$= \dfrac{1}{\cos^2 x + \sin^2 x} = \dfrac{1}{1} = 1 \qquad \blacksquare$$

The Pythagorean identity $\cos^2 \theta + \sin^2 \theta = 1$ can be rewritten as

$$\sin^2 \theta = 1 - \cos^2 \theta \qquad \text{or} \qquad \cos^2 \theta = 1 - \sin^2 \theta.$$

Therefore, we have

> **(i)** $\sin \theta = \pm\sqrt{1 - \cos^2 \theta}$ **(ii)** $\cos \theta = \pm\sqrt{1 - \sin^2 \theta}.$

In either case, the correct algebraic sign is determined by the quadrant or coordinate axis containing the terminal side of the angle θ in standard position. After you have rewritten a trigonometric expression in terms of sines and cosines, you can use these equations to bring the expression into a form involving *only the sine* or *only the cosine*.

Example 7   Rewrite the expression $\cot \theta \csc^2 \theta$ in terms of $\sin \theta$ only.

Solution $$\cot \theta \csc^2 \theta = \frac{\cos \theta}{\sin \theta} \cdot \frac{1}{\sin^2 \theta} = \frac{\cos \theta}{\sin^3 \theta} = \frac{\pm\sqrt{1 - \sin^2 \theta}}{\sin^3 \theta}$$ ∎

Algebraic expressions not originally containing trigonometric functions can often be simplified by substituting trigonometric expressions for the variable. This technique, called **trigonometric substitution**, is routinely used in calculus to rewrite radical expressions as trigonometric expressions containing no radicals.

Example 8   If a is a positive constant, rewrite the radical expression $\sqrt{a^2 - u^2}$ as a trigonometric expression containing no radical by using the trigonometric substitution $u = a \sin \theta$. Assume that $-\dfrac{\pi}{2} \leq \theta \leq \dfrac{\pi}{2}$, so that $\cos \theta \geq 0$.

Solution $$\sqrt{a^2 - u^2} = \sqrt{a^2 - (a \sin \theta)^2} = \sqrt{a^2 - a^2 \sin^2 \theta}$$
$$= \sqrt{a^2(1 - \sin^2 \theta)} = \sqrt{a^2 \cos^2 \theta}$$
$$= a \cos \theta$$ ∎

Figure 2

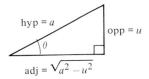

hyp = a
opp = u
θ
adj = $\sqrt{a^2 - u^2}$

The trigonometric substitution in Example 8 is illustrated geometrically for $0 < \theta < \pi/2$ by the right triangle in Figure 2. In this triangle, $\sin \theta = $ opp/hyp $= u/a$, so that $u = a \sin \theta$. By the Pythagorean theorem, adj $= \sqrt{a^2 - u^2}$; hence, $\cos \theta = $ adj/hyp $= \sqrt{a^2 - u^2}/a$, and it follows that $\sqrt{a^2 - u^2} = a \cos \theta$, as in Example 8.

Problem Set 7.1

In each problem set, problems with colored numbers constitute a good representation of the main ideas of the section.

In Problems 1 to 6, use the even–odd identities to simplify each expression.

1. $\sin(-\theta) \cos(-\theta)$

2. $\cot^2(-u) + 1$

3. $\tan t + \tan(-t)$

4. $\cos(-x) \sec x$

5. $\dfrac{1 + \csc(-\alpha)}{1 - \cot(-\beta)}$

6. $[1 + \sin \gamma][1 + \sin(-\gamma)]$

In Problems 7 to 28, use the fundamental identities to simplify each expression.

7. $\sec \theta \sin \theta$

8. $1 + \dfrac{\tan \alpha}{\cot \alpha}$

9. $\cot v \sec v$

10. $\dfrac{\csc^2 u}{1 + \tan^2 u}$

11. $\dfrac{\csc \beta}{\sec \beta}$

12. $\dfrac{\sin^2 \theta - 1}{\sec \theta}$

13. $\cot^2 \alpha - \csc^2 \alpha$

14. $\dfrac{\sec^2 t - 1}{\sec^2 t}$

15. $(\csc u - 1)(\csc u + 1)$

16. $\dfrac{1 + \cot^2 y}{1 + \tan^2 y}$

17. $\dfrac{1}{\sec^2 x} + \dfrac{1}{\csc^2 x}$

18. $\dfrac{(\sec \gamma - 1)(\sec \gamma + 1)}{\tan \gamma}$

19. $\sin^4 t + 2 \cos^2 t \sin^2 t + \cos^4 t$

20. $\sin^4 u + 2 \cos^2 u - \cos^4 u$

21. $\tan^4 \alpha - 2 \tan^2 \alpha \sec^2 \alpha + \sec^4 \alpha$

22. $(1 + \tan^2 \theta)(1 - \sin^2 \theta)$

23. $\cos x \sin^3 x + \sin x \cos^3 x$

24. $(1 - \cos^2 \beta)(1 + \cot^2 \beta)$

25. $\dfrac{1}{\sin t \cos t} - \dfrac{\cos t}{\sin t}$

26. $\dfrac{\cos \gamma}{1 - \sin \gamma} + \dfrac{\cos \gamma}{1 + \sin \gamma}$

27. $\dfrac{\sin t}{1 + \cos t} + \dfrac{1 + \cos t}{\sin t}$

28. $\dfrac{\sin \alpha + \sin \beta}{\cos \alpha + \cos \beta} + \dfrac{\cos \alpha - \cos \beta}{\sin \alpha - \sin \beta}$

In Problems 29 to 38, rewrite each trigonometric expression in terms of sines and cosines, and then simplify the result.

29. $\dfrac{\tan x}{\sec x}$

30. $(\cos \theta + \tan \theta \sin \theta) \cot \theta$

31. $\dfrac{\csc(-t)}{\sec(-t) \cot t}$

32. $\dfrac{\csc^2 x + \sec^2 x}{\csc^2 x \sec x}$

33. $\dfrac{\sec \alpha}{\csc \alpha (\tan \alpha + \cot \alpha)}$

34. $\dfrac{\sin y + \tan y}{1 + \sec y}$

35. $\dfrac{1 + \tan \theta}{\sec \theta}$

36. $\dfrac{\cot(-t) - 1}{1 - \tan(-t)}$

37. $\dfrac{\tan u + \sin u}{\cot u + \csc u}$

38. $\dfrac{\csc \beta}{\csc \beta + \tan \beta} + \dfrac{\csc \beta}{\csc \beta - \tan \beta}$

In Problems 39 to 44, rewrite each expression in terms of the indicated function only.

39. $\sec^2 \theta \tan \theta$ in terms of $\cos \theta$

40. $\dfrac{\sin t + \cot t \cos t}{\cot t}$ in terms of $\sec t$

41. $\dfrac{1 + \cot^2 x}{\cot^2 x}$ in terms of $\cos x$

42. $\dfrac{\csc^2 y + \sec^2 y}{\csc y \sec y}$ in terms of $\tan y$

43. $\dfrac{\sin(-\alpha) + \tan(-\alpha)}{1 + \sec(-\alpha)}$ in terms of $\sin \alpha$

44. $(\cot u + \csc u)(\tan u - \sin u)$ in terms of $\sec u$

In Problems 45 to 48, rewrite each radical expression as a trigonometric expression containing no radical, by making the indicated trigonometric substitution. Assume that a is a positive constant.

45. $\sqrt{a^2 + u^2}$, $u = a \tan \theta$, $-\pi/2 < \theta < \pi/2$

46. $\sqrt{u^2 - a^2}$, $u = a \sec \theta$, $0 \le \theta < \pi/2$

47. $\sqrt{(4 - x^2)^3}$, $x = 2 \cos t$, $0 \le t \le \pi$

48. $\sqrt{9 - 25x^2}$, $x = (3/5) \sin u$, $-\pi/2 \le u \le \pi/2$

49. Draw right triangles to illustrate the substitutions in Problems 45 and 47 for angles in quadrant I.

50. Let (x, y) be a point 1 unit from the origin (that is, on the unit circle) on the terminal side of an angle θ in standard position. Sketch figures to show that $(x, -y)$ is the point 1 unit from the origin on the terminal side of $-\theta$ for the following cases: (a) θ lies in quadrants II, III, or IV; (b) θ is a quadrantal angle; and (c) θ is a negative angle.

51. Prove parts (iv), (v), and (vi) of Theorem 1.

52. Show that every trigonometric expression can be rewritten in terms of each of the six trigonometric functions alone, provided that ambiguous $\pm$ signs are allowed.

7.2 VERIFYING TRIGONOMETRIC IDENTITIES

By combining the fundamental trigonometric identities, it is possible to derive a large number of related identities. It isn't necessary to memorize these derived identities—when you need one of them, you can obtain it by manipulating the fundamental identities.

Although there is no universal step-by-step procedure for proving that a trigonometric equation is an identity, there are some useful guidelines. Perhaps the most important one concerns what *not* to do:

> Do not treat a trigonometric equation as if it were an identity until after you have proved that it really is one.

The easiest way to avoid this pitfall is to *treat both sides of the equation separately.* To see the necessity for being careful, consider the following simple example. It's clear that

$$\sin x = -\sin x$$

is *not* an identity. Yet, if we square both sides, we get

$$\sin^2 x = \sin^2 x,$$

which *is* an identity.

Here are some additional suggestions.

Suggestion 1. *Take one side of the equation and try to reduce it to the other side by a sequence of manipulations. It usually pays to start with the more complicated side and try to reduce it to the simpler side.*

Suggestion 2. *If suggestion 1 doesn't seem to work, try to simplify each side of the equation separately. If you can reduce each side to the same expression, you can conclude that the original equation is an identity.*

Suggestion 3. *In carrying out suggestion 1 or 2, try the following:*

(a) *Perform indicated additions or subtractions of fractions. (First obtain common denominators, of course.)*

(b) *Perform indicated multiplications or divisions of expressions.*

(c) *Simplify fractions by canceling common factors in the numerator and denominator.*

(d) *See whether you can come closer to your goal by factoring combinations of terms.*

(e) *Try multiplying both numerator and denominator of a fraction by the same expression.*

(f) *Try rewriting all trigonometric expressions in terms of sines and cosines.*

In Examples 1 to 4, prove that each equation is an identity.

Example 1 $(\sin^2 \theta)(1 + \cot^2 \theta) = 1$

Solution   We follow suggestion 1, starting with the left side of the equation (the more complicated side):

$$(\sin^2 \theta)(1 + \cot^2 \theta) = \sin^2 \theta \csc^2 \theta = (\sin \theta \csc \theta)^2 = 1^2 = 1.$$ ∎

Example 2 $\cos^4 t - \sin^4 t = 1 - 2 \sin^2 t$

Solution   Again, we try suggestion 1. The left side of the equation involves fourth powers, so it is perhaps a bit more complicated than the right side. Therefore, we start with the left side. Trying parts (a) to (f) of suggestion 3, one at a time, we find that (d) is the first reasonable idea. So we begin by factoring:

$$\cos^4 t - \sin^4 t = (\cos^2 t)^2 - (\sin^2 t)^2 = (\cos^2 t + \sin^2 t)(\cos^2 t - \sin^2 t)$$
$$= 1(\cos^2 t - \sin^2 t) = \cos^2 t - \sin^2 t = (1 - \sin^2 t) - \sin^2 t$$
$$= 1 - 2 \sin^2 t.$$ ∎

Example 3 $2 \sec^2 x = \dfrac{1}{1 - \sin x} + \dfrac{1}{1 + \sin x}$

Solution   Here we start with the right side, which seems more complicated than the left side, and we apply part (a) of suggestion 3. Thus,

$$\frac{1}{1 - \sin x} + \frac{1}{1 + \sin x} = \frac{1(1 + \sin x)}{(1 - \sin x)(1 + \sin x)} + \frac{1(1 - \sin x)}{(1 + \sin x)(1 - \sin x)}$$
$$= \frac{1 + \sin x}{1 - \sin^2 x} + \frac{1 - \sin x}{1 - \sin^2 x} = \frac{1 + \sin x + 1 - \sin x}{1 - \sin^2 x}$$
$$= \frac{2}{1 - \sin^2 x} = \frac{2}{\cos^2 x} = 2 \sec^2 x.$$ ∎

Example 4 $\dfrac{1 + \sin(-t)}{\cos(-t)} = \dfrac{\cos(-t)}{1 - \sin(-t)}$

Solution   Since $\sin(-t) = -\sin t$ and $\cos(-t) = \cos t$, the left side of the equation is equal to $(1 - \sin t)/\cos t$ and the right side is equal to $\cos t/(1 + \sin t)$. Therefore, it will be enough to prove that

$$\frac{1 - \sin t}{\cos t} = \frac{\cos t}{1 + \sin t}.$$

Notice that it would *not* be correct to "cross multiply" here, since we don't know that the equation is true—indeed, our job is to show that it is true. Thus, we work with both sides separately. Neither side appears to be more complicated than the other, so let's start with the left side. We try part (e) of suggestion 3. Thus,

$$\frac{1 - \sin t}{\cos t} = \frac{(1 - \sin t)(1 + \sin t)}{\cos t\,(1 + \sin t)} = \frac{1 - \sin^2 t}{\cos t\,(1 + \sin t)} = \frac{\cos^2 t}{\cos t\,(1 + \sin t)}$$

$$= \frac{\cos t}{1 + \sin t}.$$ ∎

Example 5 Show that the trigonometric equation

$$\sin x + \cos x = \tan x + 1$$

is *not* an identity.

Solution If we let $x = 0$, we find that both sides of the equation are defined and that the equation becomes $1 = 1$, which is true. From this, however, we can't conclude that the equation is an identity, since there may be other values of x for which it is false. If we let $x = \pi/2$, the left side of the equation is $\sin(\pi/2) + \cos(\pi/2) = 1 + 0 = 1$, but the right side of the equation is undefined. Again, this is inconclusive, since an identity is only required to be true for values of the variable for which both sides are defined. Suppose we try $x = \pi/4$. Then the left side of the equation is $\sin(\pi/4) + \cos(\pi/4) = (\sqrt{2}/2) + (\sqrt{2}/2) = \sqrt{2}$, whereas the right side of the equation is $\tan(\pi/4) + 1 = 1 + 1 = 2$. Therefore, the equation is not an identity. ∎

Problem Set 7.2

In Problems 1 to 52, show that each trigonometric equation is an identity.

1. $\sin \theta \sec \theta = \tan \theta$

2. $\cos \alpha \tan \alpha \csc \alpha = 1$

3. $\tan x \cos x = \sin x$

4. $\sin \beta \cot \beta \sec \beta = 1$

5. $\csc(-t) \tan(-t) = \sec t$

6. $\sin(-u) = \sin^2 u \csc(-u)$

7. $\tan \alpha \sin \alpha + \cos \alpha = \sec \alpha$

8. $\dfrac{\sec x \csc x}{\tan x + \cot x} = 1$

9. $\dfrac{\sin \beta}{\csc \beta} + \dfrac{\cos \beta}{\sec \beta} = 1$

10. $2 - \sin^2 \theta = 1 + \cos^2 \theta$

11. $\cos^2 t(1 + \tan^2 t) = 1$

12. $\sec^2 v(1 - \sin^2 v) = 1$

13. $\sec^2 w \cot^2 w - \cos^2 w \csc^2 w = 1$

14. $\tan^4 u - \sec^4 u = 1 - 2 \sec^2 u$

15. $\sin^2 \theta \cot^2 \theta + \cos^2 \theta \tan^2 \theta = 1$

16. $\cot^2 \gamma - \cos^2 \gamma = \cot^2 \gamma \cos^2 \gamma$

17. $\sin^2 v + \tan^2 v + \cos^2 v = \sec^2 v$

18. $2 \csc \beta - \cot \beta \cos \beta = \sin \beta + \csc \beta$

19. $\sin^2 x + \cos^2 x(1 - \tan^2 x) = \cos^2 x$

20. $\sin^4 t - \cos^4 t + 2 \sin^2 t \cot^2 t = 1$

21. $\dfrac{\tan \theta}{1 + \tan^2 \theta} = \dfrac{\sin \theta}{\sec \theta}$

22. $\dfrac{\cos^2 s}{\sin s} + \dfrac{1}{\csc s} = \csc s$

23. $\dfrac{\sin^2 t}{\cos t} + \cos t = \sec t$

24. $\dfrac{1}{\tan \theta + \cot \theta} = \sin \theta \cos \theta$

25. $\dfrac{\sin^3 t}{\cos t} + \sin t \cos t = \tan t$

26. $\dfrac{\csc x - \sec x}{\csc x + \sec x} = \dfrac{\cot x - 1}{\cot x + 1}$

27. $\dfrac{\sin \beta + \cos \beta}{\sin \beta - \cos \beta} = \dfrac{\sec \beta + \csc \beta}{\sec \beta - \csc \beta}$

28. $\left(\dfrac{1 + \csc t}{\csc t}\right)^2 \sec^2 t = \dfrac{1 + \sin t}{1 - \sin t}$

29. $\dfrac{\sin x \cos x}{1 - 2 \sin^2 x} = \dfrac{\tan x}{1 - \tan^2 x}$

30. $\dfrac{(1 - \cot y)^2}{\csc^2 y} + 2 \sin y \cos y = 1$

31. $\dfrac{\tan u \sin u}{\tan u - \sin u} = \dfrac{\sin u}{1 - \cos u}$

32. $(\sec \gamma - \tan \gamma)^2 = \dfrac{1 - \sin \gamma}{1 + \sin \gamma}$

33. $\dfrac{1 - \cot(-\alpha)}{1 - \tan(-\alpha)} = \cot \alpha$

34. $\dfrac{1}{\csc x - \cot x} = \dfrac{2}{\sin x} - \dfrac{1}{\csc x + \cot x}$

35. $(\cot \beta + \csc \beta)^2 = \dfrac{\sec \beta + 1}{\sec \beta - 1}$

36. $\dfrac{\csc^2 t + \sec^2 t}{\csc t \sec t} = \cot t + \tan t$

37. $\dfrac{\cos \alpha}{1 + \cos \alpha \tan \alpha} = \dfrac{1 - \cos \alpha \tan \alpha}{\cos \alpha}$

38. $\dfrac{\cot \beta - \csc \beta + 1}{\cot \beta + \csc \beta - 1} = \dfrac{\sin \beta}{1 + \cos \beta}$

39. $\dfrac{\sin \theta}{\cot \theta + \csc \theta} - \dfrac{\sin \theta}{\cot \theta - \csc \theta} = 2$

40. $\dfrac{\tan x - \tan y}{1 + \tan x \tan y} = \dfrac{\cot y - \cot x}{1 + \cot x \cot y}$

41. $\dfrac{1}{\sin^2 t} + \dfrac{1}{\cos^2 t} = \dfrac{1}{\sin^2 t - \sin^4 t}$

42. $\cos^6 \theta - \sin^6 \theta = (2 \cos^2 \theta - 1)(1 - \sin^2 \theta \cos^2 \theta)$

43. $\dfrac{\cos(-\alpha)}{1 + \tan(-\alpha)} - \dfrac{\sin(-\alpha)}{1 + \cot(-\alpha)} = \sin \alpha + \cos \alpha$

44. $(1 + \tan \beta + \sec \beta)(1 + \cot \beta - \csc \beta) = 2$

45. $\dfrac{\sec u}{\csc u (1 + \sec u)} + \dfrac{1 + \cos u}{\sin u} = 2 \csc u$

46. $\dfrac{\sec^4 x + \tan^4 x}{\sec^2 x \tan^2 x} = \dfrac{\cos^4 x}{\sin^2 x} + 2$

47. $(1 + \tan \beta + \cot \beta)(\cos \beta - \sin \beta) = \dfrac{\csc \beta}{\sec^2 \beta} - \dfrac{\sec \beta}{\csc^2 \beta}$

48. $\sqrt{\dfrac{\sec \gamma - \tan \gamma}{\sec \gamma + \tan \gamma}} = \dfrac{|\cos \gamma|}{1 + \sin \gamma}$

49. $(1 - \cot w)^2 (1 + \cot w)^2 + 4 \cot^2 w = \csc^4 w$

50. $\dfrac{\cot \alpha + \csc \alpha}{\sin \alpha + \cot(-\alpha) + \csc(-\alpha)} + \sec \alpha = 0$

51. $(1 + \sin \omega t + \cos \omega t)^2 = 2(1 + \sin \omega t)(1 + \cos \omega t)$

52. $\dfrac{\cot x}{1 - \tan x} + \dfrac{\tan x}{1 - \cot x} = 1 + \dfrac{\sec x}{\sin x}$

In Problems 53 to 59, show that the given trigonometric equation is *not* a trigonometric identity.

53. $\sin \theta - \sec \theta = \tan \theta - 1$

54. $(\sin x + \cos x)^2 = 1$

55. $\dfrac{\sin t + \tan t}{\cos t + \tan t} = \tan t$

56. $\cos(\gamma + \pi) = \cos \gamma$

57. $\sqrt{1 + \sin^2 u} = 1 + \sin u$

58. $\boxed{c}$ $\ln(\sin x) = \sin(\ln x)$

59. $\boxed{c}$ $\sin(t^2) = \sin^2 t$

60. Give an example of a trigonometric equation that is true for three different values of the variable but isn't an identity.

7.3 TRIGONOMETRIC FORMULAS FOR SUMS AND DIFFERENCES

The fundamental trigonometric identities considered in Section 7.1 express relationships among trigonometric functions of a single variable. In this section we develop trigonometric identities involving sums or differences of *two* variables.

Since $75° = 30° + 45°$, you might ask whether $\cos 75°$ can be found by calculating $\cos 30° + \cos 45°$. The answer is *no*; $\cos 75° < 1$, whereas

$$\cos 30° + \cos 45° = \frac{\sqrt{3}}{2} + \frac{\sqrt{2}}{2} = \frac{\sqrt{3} + \sqrt{2}}{2} > 1.$$

The correct formula for the cosine of the sum of two angles is given by the following theorem.

Theorem 1   **Cosine of a Sum**

$$\cos(\alpha + \beta) = \cos \alpha \cos \beta - \sin \alpha \sin \beta$$

To see why the formula in Theorem 1 is correct, consider the unit circles in Figure 1 (a) and (b). In Figure 1(a), angle AOB is $\alpha + \beta$. Triangle COD in Figure 1(b) is obtained by rotating triangle AOB through the angle $-\alpha$. In Figure 1(a), we have $A = (\cos(\alpha + \beta), \sin(\alpha + \beta))$ and $B = (1, 0)$, so, by the distance formula,

$$\begin{aligned}
|\overline{AB}|^2 &= [\cos(\alpha + \beta) - 1]^2 + [\sin(\alpha + \beta) - 0]^2 \\
&= \cos^2(\alpha + \beta) - 2\cos(\alpha + \beta) + 1 + \sin^2(\alpha + \beta) \\
&= [\cos^2(\alpha + \beta) + \sin^2(\alpha + \beta)] - 2\cos(\alpha + \beta) + 1 \\
&= 1 - 2\cos(\alpha + \beta) + 1 = 2 - 2\cos(\alpha + \beta).
\end{aligned}$$

Figure 1

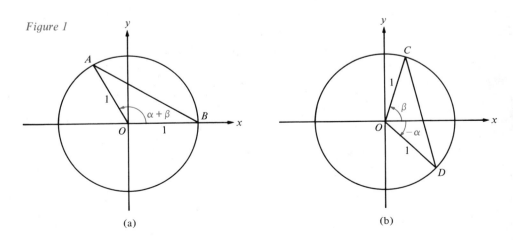

(a)   (b)

In Figure 1(b),

$$C = (\cos \beta, \sin \beta) \quad \text{and} \quad D = (\cos(-\alpha), \sin(-\alpha)) = (\cos \alpha, -\sin \alpha),$$

so, by the distance formula again,

$$
\begin{aligned}
|\overline{CD}|^2 &= (\cos \beta - \cos \alpha)^2 + [\sin \beta - (-\sin \alpha)]^2 = (\cos \beta - \cos \alpha)^2 + (\sin \beta + \sin \alpha)^2 \\
&= \cos^2 \beta - 2 \cos \beta \cos \alpha + \cos^2 \alpha + \sin^2 \beta + 2 \sin \beta \sin \alpha + \sin^2 \alpha \\
&= (\cos^2 \beta + \sin^2 \beta) + (\cos^2 \alpha + \sin^2 \alpha) - 2 \cos \beta \cos \alpha + 2 \sin \beta \sin \alpha \\
&= 1 + 1 - 2(\cos \beta \cos \alpha - \sin \beta \sin \alpha) \\
&= 2 - 2(\cos \alpha \cos \beta - \sin \alpha \sin \beta).
\end{aligned}
$$

Because triangle AOB is congruent to triangle COD, it follows that

$$|\overline{AB}|^2 = |\overline{CD}|^2.$$

Hence,
$$
\begin{aligned}
2 - 2\cos(\alpha + \beta) &= 2 - 2(\cos \alpha \cos \beta - \sin \alpha \sin \beta) \\
-2\cos(\alpha + \beta) &= -2(\cos \alpha \cos \beta - \sin \alpha \sin \beta) \\
\cos(\alpha + \beta) &= \cos \alpha \cos \beta - \sin \alpha \sin \beta,
\end{aligned}
$$

which is the formula given in Theorem 1.

Example 1   Find the exact value of $\cos 75°$.

Solution   By Theorem 1, $\quad \cos 75° = \cos(30° + 45°) = \cos 30° \cos 45° - \sin 30° \sin 45°$

$$= \frac{\sqrt{3}}{2} \cdot \frac{\sqrt{2}}{2} - \frac{1}{2} \cdot \frac{\sqrt{2}}{2} = \frac{(\sqrt{3} - 1)\sqrt{2}}{4}, \quad \text{or} \quad \frac{\sqrt{6} - \sqrt{2}}{4}. \quad \blacksquare$$

Theorem 2   **Cosine of a Difference**

$$\cos(\alpha - \beta) = \cos \alpha \cos \beta + \sin \alpha \sin \beta$$

Proof   In Theorem 1, replace β by $-\beta$ to obtain

$$
\begin{aligned}
\cos(\alpha - \beta) &= \cos[\alpha + (-\beta)] = \cos \alpha \cos(-\beta) - \sin \alpha \sin(-\beta) \\
&= \cos \alpha \cos \beta - \sin \alpha(-\sin \beta) = \cos \alpha \cos \beta + \sin \alpha \sin \beta. \quad \blacksquare
\end{aligned}
$$

Example 2   Find the exact value of $\cos 15°$.

Solution   By Theorem 2, $\quad \cos 15° = \cos(45° - 30°) = \cos 45° \cos 30° + \sin 45° \sin 30°$

$$= \frac{\sqrt{2}}{2} \cdot \frac{\sqrt{3}}{2} + \frac{\sqrt{2}}{2} \cdot \frac{1}{2} = \frac{\sqrt{2}(\sqrt{3} + 1)}{4}, \quad \text{or} \quad \frac{\sqrt{6} + \sqrt{2}}{4}. \quad \blacksquare$$

In Section 6.2, we discussed the function-cofunction relationships for trigonometric functions of *acute angles*. The following theorem shows that these relationships hold for arbitrary angles.

Theorem 3   **Cofunction Theorem**

If θ is a real number or an angle measured in radians, then

(i) $\cos\left(\dfrac{\pi}{2} - \theta\right) = \sin\theta$ **(ii)** $\sin\left(\dfrac{\pi}{2} - \theta\right) = \cos\theta$

(iii) $\cot\left(\dfrac{\pi}{2} - \theta\right) = \tan\theta$ **(iv)** $\tan\left(\dfrac{\pi}{2} - \theta\right) = \cot\theta$

(v) $\csc\left(\dfrac{\pi}{2} - \theta\right) = \sec\theta$ **(vi)** $\sec\left(\dfrac{\pi}{2} - \theta\right) = \csc\theta$

Proof   We prove (i), (ii), and (iii) here and leave the remaining proofs as an exercise (Problem 49).

(i) By Theorem 2,

$$\cos\left(\frac{\pi}{2} - \theta\right) = \cos\frac{\pi}{2}\cos\theta + \sin\frac{\pi}{2}\sin\theta = 0 \cdot \cos\theta + 1 \cdot \sin\theta = \sin\theta.$$

(ii) In part (i), we replace θ by $(\pi/2) - \theta$ to obtain

$$\cos\left[\frac{\pi}{2} - \left(\frac{\pi}{2} - \theta\right)\right] = \sin\left(\frac{\pi}{2} - \theta\right).$$

Thus, $$\cos\theta = \sin\left(\frac{\pi}{2} - \theta\right),$$

and therefore (ii) holds.

(iii) By parts (i) and (ii),

$$\cot\left(\frac{\pi}{2} - \theta\right) = \frac{\cos\left(\dfrac{\pi}{2} - \theta\right)}{\sin\left(\dfrac{\pi}{2} - \theta\right)} = \frac{\sin\theta}{\cos\theta} = \tan\theta. \qquad \blacksquare$$

Naturally, if we measure an angle α in degrees instead of radians, the relationships expressed in Theorem 3 are written as

$$\cos(90° - \alpha) = \sin\alpha, \qquad \sin(90° - \alpha) = \cos\alpha,$$

and so forth.

Now, by combining Theorem 2 and Theorem 3, we can prove the following important theorem.

Theorem 4   **Sine of a Sum**

$$\sin(\alpha + \beta) = \sin \alpha \cos \beta + \sin \beta \cos \alpha$$

Proof   By parts (i) and (ii) of Theorem 3,

$$\sin(\alpha + \beta) = \cos\left[\frac{\pi}{2} - (\alpha + \beta)\right] = \cos\left(\frac{\pi}{2} - \alpha - \beta\right) = \cos\left[\left(\frac{\pi}{2} - \alpha\right) - \beta\right]$$

$$= \cos\left(\frac{\pi}{2} - \alpha\right)\cos \beta + \sin\left(\frac{\pi}{2} - \alpha\right)\sin \beta = \sin \alpha \cos \beta + \cos \alpha \sin \beta$$

$$= \sin \alpha \cos \beta + \sin \beta \cos \alpha. \qquad \blacksquare$$

Example 3   Find the exact value of $\sin \dfrac{7\pi}{12}$.

Solution   By Theorem 4 and the fact that $\dfrac{\pi}{3} + \dfrac{\pi}{4} = \dfrac{7\pi}{12}$, we have

$$\sin \frac{7\pi}{12} = \sin\left(\frac{\pi}{3} + \frac{\pi}{4}\right) = \sin \frac{\pi}{3} \cos \frac{\pi}{4} + \sin \frac{\pi}{4} \cos \frac{\pi}{3}$$

$$= \frac{\sqrt{3}}{2} \cdot \frac{\sqrt{2}}{2} + \frac{\sqrt{2}}{2} \cdot \frac{1}{2} = \frac{\sqrt{2}(\sqrt{3} + 1)}{4}, \qquad \text{or} \qquad \frac{\sqrt{6} + \sqrt{2}}{4}. \qquad \blacksquare$$

The following theorem follows directly from Theorem 4 and the even–odd identities.

Theorem 5   **Sine of a Difference**

$$\sin(\alpha - \beta) = \sin \alpha \cos \beta - \sin \beta \cos \alpha$$

Proof   Replacing β by $-\beta$ in Theorem 4, we have

$$\sin(\alpha - \beta) = \sin[\alpha + (-\beta)] = \sin \alpha \cos(-\beta) + \sin(-\beta) \cos \alpha$$

$$= \sin \alpha \cos \beta - \sin \beta \cos \alpha. \qquad \blacksquare$$

The four identities

$$\sin(\alpha + \beta) = \sin \alpha \cos \beta + \sin \beta \cos \alpha \qquad \sin(\alpha - \beta) = \sin \alpha \cos \beta - \sin \beta \cos \alpha$$

$$\cos(\alpha + \beta) = \cos \alpha \cos \beta - \sin \alpha \sin \beta \qquad \cos(\alpha - \beta) = \cos \alpha \cos \beta + \sin \alpha \sin \beta$$

are used so often that they should be memorized. We recommend that you memorize the first identity in the form: *The sine of the sum of two angles is equal to the sine of the first times the cosine of the second plus the sine of the second times the cosine of the first.* The remaining three identities can be stated in a similar manner.

Example 4   Simplify each expression:

(a) $\sin(90° + \theta)$ (b) $\cos\left(\dfrac{\pi}{2} + x\right)$

Solution   (a) $\sin(90° + \theta) = \sin 90° \cos \theta + \sin \theta \cos 90° = 1 \cdot \cos \theta + (\sin \theta) \cdot 0 = \cos \theta$

(b) $\cos\left(\dfrac{\pi}{2} + x\right) = \cos \dfrac{\pi}{2} \cos x - \sin \dfrac{\pi}{2} \sin x = 0 \cdot \cos x - 1 \cdot \sin x = -\sin x$ ∎

Example 5   Suppose that α and β are angles in standard position; α is in quadrant I, $\sin \alpha = \frac{3}{5}$, β is in quadrant II, and $\cos \beta = -\frac{5}{13}$. Find:

(a) $\cos \alpha$ (b) $\sin \beta$ (c) $\sin(\alpha + \beta)$ (d) $\cos(\alpha + \beta)$

(e) the quadrant containing angle $\alpha + \beta$ in standard position

Solution   (a) Since α is in quadrant I, $\cos \alpha$ is positive, and

$$\cos \alpha = \sqrt{1 - \sin^2 \alpha} = \sqrt{1 - (\tfrac{3}{5})^2} = \sqrt{\tfrac{16}{25}} = \tfrac{4}{5}.$$

(b) Since β is in quadrant II, $\sin \beta$ is positive and

$$\sin \beta = \sqrt{1 - \cos^2 \beta} = \sqrt{1 - (-\tfrac{5}{13})^2} = \sqrt{\tfrac{144}{169}} = \tfrac{12}{13}.$$

(c) $\sin(\alpha + \beta) = \sin \alpha \cos \beta + \sin \beta \cos \alpha = \tfrac{3}{5}(-\tfrac{5}{13}) + \tfrac{12}{13} \cdot \tfrac{4}{5} = \tfrac{33}{65}$

(d) $\cos(\alpha + \beta) = \cos \alpha \cos \beta - \sin \alpha \sin \beta = \tfrac{4}{5}(-\tfrac{5}{13}) - \tfrac{3}{5} \cdot \tfrac{12}{13} = -\tfrac{56}{65}$

(e) Since $\sin(\alpha + \beta)$ is positive and $\cos(\alpha + \beta)$ is negative, it follows that $\alpha + \beta$ is a quadrant II angle.   ∎

Example 6   Simplify each expression without the use of a calculator or tables:

(a) $\cos 25° \cos 20° - \sin 25° \sin 20°$ (b) $\sin \dfrac{2\pi}{5} \cos \dfrac{3\pi}{5} + \sin \dfrac{3\pi}{5} \cos \dfrac{2\pi}{5}$

Solution   (a) Reading the formula for the cosine of a sum from right to left, we have

$$\cos 25° \cos 20° - \sin 25° \sin 20° = \cos(25° + 20°) = \cos 45° = \sqrt{2}/2.$$

(b) Reading the formula for the sine of a sum from right to left, we have

$$\sin \dfrac{2\pi}{5} \cos \dfrac{3\pi}{5} + \sin \dfrac{3\pi}{5} \cos \dfrac{2\pi}{5} = \sin\left(\dfrac{2\pi}{5} + \dfrac{3\pi}{5}\right) = \sin \pi = 0.$$ ∎

Example 7   As we noticed in Section 6.5, the graphs of the sine and cosine functions suggest that

$$\sin(x + \pi) = -\sin x \quad \text{and} \quad \cos(x + \pi) = -\cos x.$$

Prove these identities.

Solution $\sin(x + \pi) = \sin x \cos \pi + \sin \pi \cos x = (\sin x)(-1) + 0 \cdot \cos x$
$= -\sin x$

$\cos(x + \pi) = \cos x \cos \pi - \sin x \sin \pi = (\cos x)(-1) - (\sin x) \cdot 0$
$= -\cos x$ ∎

Example 8   Prove the trigonometric identity

$$\cos(x + y)\cos(x - y) = \cos^2 x - \sin^2 y.$$

Solution   Beginning with the left side and applying the formulas for the cosine of a sum and for the cosine of a difference, we have

$$
\begin{aligned}
\cos(x + y)\cos(x - y) &= (\cos x \cos y - \sin x \sin y)(\cos x \cos y + \sin x \sin y) \\
&= (\cos x \cos y)^2 - (\sin x \sin y)^2 \\
&= \cos^2 x \cos^2 y - \sin^2 x \sin^2 y \\
&= \cos^2 x(1 - \sin^2 y) - (1 - \cos^2 x)\sin^2 y \\
&= \cos^2 x - \cos^2 x \sin^2 y - \sin^2 y + \cos^2 x \sin^2 y \\
&= \cos^2 x - \sin^2 y.
\end{aligned}
$$

Do you see what gave us the idea to substitute $1 - \sin^2 y$ for $\cos^2 y$ and $1 - \cos^2 x$ for $\sin^2 x$ in the fourth step of the calculation above? [*Hint:* We wanted the expression $\cos^2 x - \sin^2 y$ to show up.] ∎

Using the formulas for the sine and cosine of a sum, we can derive a useful formula for the tangent of a sum.

Theorem 6   **Tangent of a Sum**

> If $\cos \alpha \neq 0$, $\cos \beta \neq 0$, and $\cos(\alpha + \beta) \neq 0$, then
> $$\tan(\alpha + \beta) = \frac{\tan \alpha + \tan \beta}{1 - \tan \alpha \tan \beta}.$$

Proof
$$\tan(\alpha + \beta) = \frac{\sin(\alpha + \beta)}{\cos(\alpha + \beta)} = \frac{\sin \alpha \cos \beta + \sin \beta \cos \alpha}{\cos \alpha \cos \beta - \sin \alpha \sin \beta}$$

Dividing numerator and denominator of the fraction on the right side of the equation above by $\cos \alpha \cos \beta$, we get

$$
\tan(\alpha + \beta) = \frac{\dfrac{\sin \alpha \cos \beta}{\cos \alpha \cos \beta} + \dfrac{\sin \beta \cos \alpha}{\cos \alpha \cos \beta}}{\dfrac{\cos \alpha \cos \beta}{\cos \alpha \cos \beta} - \dfrac{\sin \alpha \sin \beta}{\cos \alpha \cos \beta}} = \frac{\dfrac{\sin \alpha}{\cos \alpha} + \dfrac{\sin \beta}{\cos \beta}}{1 - \left(\dfrac{\sin \alpha}{\cos \alpha}\right)\left(\dfrac{\sin \beta}{\cos \beta}\right)} = \frac{\tan \alpha + \tan \beta}{1 - \tan \alpha \tan \beta}. \quad ∎
$$

Example 9   Assuming that $\cos \alpha \neq 0$, $\cos \beta \neq 0$, and $\cos(\alpha - \beta) \neq 0$, show that

$$\tan(\alpha - \beta) = \frac{\tan \alpha - \tan \beta}{1 + \tan \alpha \tan \beta}.$$

Solution   In Theorem 6, we replace β by $-\beta$ to obtain

$$\tan(\alpha - \beta) = \tan[\alpha + (-\beta)] = \frac{\tan \alpha + \tan(-\beta)}{1 - \tan \alpha \tan(-\beta)} = \frac{\tan \alpha - \tan \beta}{1 + \tan \alpha \tan \beta}.$$

■

Example 10   Simplify the expression

$$\frac{\tan 43° + \tan 17°}{1 - \tan 43° \tan 17°}$$

without the use of a calculator or tables.

Solution   Reading the formula in Theorem 6 from right to left, we have

$$\frac{\tan 43° + \tan 17°}{1 - \tan 43° \tan 17°} = \tan(43° + 17°) = \tan 60° = \sqrt{3}.$$

■

Problem Set 7.3

In Problems 1 to 12, find the exact value of the indicated trigonometric function. Do not use a calculator or tables.

1. $\sin 105°$ [*Hint:* $105° = 60° + 45°$.]

2. $\cos 105°$

3. $\sin 195°$ [*Hint:* $195° = 150° + 45°$.]

4. $\tan 195°$

5. $\cos \dfrac{17\pi}{12}$ $\left[Hint: \dfrac{17\pi}{12} = \dfrac{5\pi}{4} + \dfrac{\pi}{6}. \right]$

6. $\cot \dfrac{17\pi}{12}$

7. $\sin \dfrac{11\pi}{12}$ $\left[Hint: \dfrac{11\pi}{12} = \dfrac{7\pi}{6} - \dfrac{\pi}{4}. \right]$

8. $\sec \dfrac{11\pi}{12}$

9. $\tan \dfrac{13\pi}{12}$ $\left[Hint: \dfrac{13\pi}{12} = \dfrac{3\pi}{4} + \dfrac{\pi}{3}. \right]$

10. $\cot \dfrac{13\pi}{12}$

11. $\tan \dfrac{19\pi}{12}$ $\left[Hint: \dfrac{19\pi}{12} = \dfrac{11\pi}{6} - \dfrac{\pi}{4}. \right]$

12. $\csc \dfrac{19\pi}{12}$

In Problems 13 to 34, simplify each expression without the use of a calculator or tables.

13. $\sin(180° - \theta)$

14. $\cos(180° + t)$

15. $\cos(\pi - \alpha)$

16. $\sin(270° - x)$

17. $\sin(360° - s)$

18. $\cos(2\pi - \gamma)$

19. $\tan(180° - t)$

20. $\sec(\pi + u)$

21. $\csc\left(\dfrac{3\pi}{2} + \beta \right)$

22. $\tan(270° + \theta)$ [*Caution:* Theorem 6 doesn't work here.]

23. $\sin 17° \cos 13° + \sin 13° \cos 17°$

24. $\cos 43° \cos 17° - \sin 43° \sin 17°$

25. $\sin \dfrac{17\pi}{36} \cos \dfrac{11\pi}{36} - \sin \dfrac{11\pi}{36} \cos \dfrac{17\pi}{36}$

26. $\cos \dfrac{5\pi}{7} \sin \dfrac{2\pi}{7} + \cos \dfrac{2\pi}{7} \sin \dfrac{5\pi}{7}$

27. $\cos 2x \cos x + \sin 2x \sin x$

28. $\sin(\alpha - \beta) \cos \beta + \cos(\alpha - \beta) \sin \beta$

29. $\cos(\alpha + \beta) \cos \beta + \sin(\alpha + \beta) \sin \beta$

30. $\cos(x + y) - \cos(x - y)$

31. $\dfrac{\tan 42° + \tan 18°}{1 - \tan 42° \tan 18°}$ **32.** $\dfrac{\tan 50° - \tan 20°}{1 + \tan 50° \tan 20°}$

33. $\dfrac{\tan 8x - \tan 7x}{1 + \tan 8x \tan 7x}$ **34.** $\tan\left(\dfrac{\pi}{2} + \theta\right)$

35. Suppose that α and β are angles in standard position; α is in quadrant I, $\sin \alpha = \frac{4}{5}$, β is in quadrant II, and $\sin \beta = \frac{12}{13}$. Find:

(a) $\cos \alpha$ (b) $\cos \beta$

(c) $\sin(\alpha + \beta)$ (d) $\cos(\alpha + \beta)$

(e) $\sin(\alpha - \beta)$ (f) $\tan(\alpha + \beta)$

(g) The quadrant containing the angle $\alpha + \beta$ in standard position.

36. Suppose that $\dfrac{\pi}{2} < s < \pi$, $\dfrac{\pi}{2} < t < \pi$, $\sin s = \frac{5}{13}$, and $\tan t = -\frac{3}{4}$. Find:

(a) $\sin(s - t)$ (b) $\cos(s - t)$

(c) $\sin(s + t)$ (d) $\sec(s + t)$

(e) The quadrant containing the angle θ in standard position if the radian measure of θ is $s + t$.

37. If α and β are positive acute angles, $\sin(\alpha + \beta) = \frac{56}{65}$, and $\sin \beta = \frac{5}{13}$, find $\sin \alpha$. [*Hint:* $\sin \alpha = \sin((\alpha + \beta) - \beta)$.]

38. Derive a formula for $\tan(\alpha + \beta)$ for the case in which $\cos \alpha = 0$ and $\sin \beta \neq 0$.

In Problems 39 to 48, show that each equation is an identity.

39. $\dfrac{\sin(\alpha + \beta)}{\sin \alpha \cos \beta} = 1 + \cot \alpha \tan \beta$

40. $\tan(\theta + \pi) = \tan \theta$

41. $\sin\left(t - \dfrac{\pi}{3}\right) + \cos\left(t - \dfrac{\pi}{6}\right) = \sin t$

42. $\cos(x + y) \cos(x - y) = 1 - \sin^2 x - \sin^2 y$

43. $\tan\left(x + \dfrac{\pi}{2}\right) = \tan\left(x - \dfrac{\pi}{2}\right)$

44. $\tan \alpha - \tan \beta = \dfrac{\sin(\alpha - \beta)}{\cos \alpha \cos \beta}$

45. $\cot(x + y) = \dfrac{\cot x \cot y - 1}{\cot x + \cot y}$

46. $\dfrac{\sin(s + t)}{\sin(s - t)} = \dfrac{\tan s + \tan t}{\tan s - \tan t}$

47. $\dfrac{1 - \tan^2\left(\dfrac{\pi}{4} - \theta\right)}{1 + \tan^2\left(\dfrac{\pi}{4} - \theta\right)} = 2 \sin \theta \cos \theta$

48. $\tan^2\left(\dfrac{\pi}{2} - t\right) \sec^2 t - \sin^2\left(\dfrac{\pi}{2} - t\right) \csc^2 t = 1$

49. Prove parts (iv), (v), and (vi) of Theorem 3.

50. If $x + y = \pi/4$, show that

$$\tan y = \dfrac{1 - \tan x}{1 + \tan x}.$$

© In Problems 51 to 56, use a calculator to verify the equation for the indicated values of the variables.

51. $\cos(\alpha - \beta) = \cos \alpha \cos \beta + \sin \alpha \sin \beta$

for $\alpha = \dfrac{3\pi}{5}$, $\beta = \dfrac{\pi}{5}$

52. $\sin(x + y) + \sin x \cos y = \sin y \cos x$

for $x = 2.31$, $y = 1.07$

53. $\tan(\alpha + \beta) = \dfrac{\tan \alpha + \tan \beta}{1 - \tan \alpha \tan \beta}$

for $\alpha = 22.4°$, $\beta = 17.3°$

54. $\cos(s + t) = \cos s \cos t - \sin s \sin t$

for $s = \dfrac{5\pi}{11}$, $t = \dfrac{7\pi}{9}$

55. $\sin(s - t) = \sin s \cos t - \cos s \sin t$

for $s = 0.791$, $t = 1.234$

56. $\tan(\alpha - \beta) = \dfrac{\tan \alpha - \tan \beta}{1 + \tan \alpha \tan \beta}$

for $\alpha = 91.7°$, $\beta = 87.2°$

7.4 DOUBLE-ANGLE AND HALF-ANGLE FORMULAS

In this section we study the important double-angle and half-angle formulas. These identities, which are obtained as special cases of the formulas derived in the last section, enable us to evaluate trigonometric functions of 2θ or $\theta/2$ in terms of trigonometric functions of θ.

The sine of twice an angle is *not* in general the same as twice the sine of the angle. For instance, $\sin 30° = \frac{1}{2}$; $\sin 2(30°) = \sin 60° = \sqrt{3}/2 \neq 2 \sin 30°$. The correct formula for the sine of twice an angle is given by the following theorem.

Theorem 1   **Double-Angle Formula for Sine**

$$\sin 2\theta = 2 \sin \theta \cos \theta$$

Proof   In the formula

$$\sin(\alpha + \beta) = \sin \alpha \cos \beta + \sin \beta \cos \alpha,$$

let $\alpha = \beta = \theta$, so that

$$\sin(\theta + \theta) = \sin \theta \cos \theta + \sin \theta \cos \theta$$

or $\sin 2\theta = 2 \sin \theta \cos \theta.$ ∎

Example 1   Verify Theorem 1 for $\theta = 30°$.

Solution   We know that $\sin 60° = \sqrt{3}/2$. Using Theorem 1 with $\theta = 30°$, we also get

$$\sin 60° = \sin 2(30°) = 2 \sin 30° \cos 30° = 2\left(\frac{1}{2}\right)\left(\frac{\sqrt{3}}{2}\right) = \frac{\sqrt{3}}{2}.$$ ∎

Example 2   Write $\sin 100°$ in terms of trigonometric functions of $50°$.

Solution $\sin 100° = \sin 2(50°) = 2 \sin 50° \cos 50°$ ∎

Example 3   Reduce the expression $2 \sin 8x \cos 8x$ to the value of a function of $16x$.

Solution   Reading Theorem 1 from right to left with $\theta = 8x$, we have

$$2 \sin 8x \cos 8x = \sin[2(8x)] = \sin 16x.$$ ∎

Example 4   Suppose that α is a quadrant I angle and that $\sin \alpha = \frac{3}{5}$. Find $\sin 2\alpha$.

Solution   We are going to use the formula $\sin 2\alpha = 2 \sin \alpha \cos \alpha$, but first we must find $\cos \alpha$. Because α is a quadrant I angle, we know that $\cos \alpha$ is positive, so

$$\cos \alpha = \sqrt{1 - \sin^2 \alpha} = \sqrt{1 - (\tfrac{3}{5})^2} = \sqrt{\tfrac{16}{25}} = \tfrac{4}{5}.$$

Therefore, $\sin 2\alpha = 2 \sin \alpha \cos \alpha = 2(\tfrac{3}{5})(\tfrac{4}{5}) = \tfrac{24}{25}.$ ∎

There are three double-angle formulas for the cosine—a basic formula and two alternative versions of it.

Theorem 2   **Double-Angle Formulas for Cosine**

> (i) $\cos 2\theta = \cos^2 \theta - \sin^2 \theta$
>
> (ii) $\cos 2\theta = 2 \cos^2 \theta - 1$
>
> (iii) $\cos 2\theta = 1 - 2 \sin^2 \theta$

Proof   **(i)**   In the formula

$$\cos(\alpha + \beta) = \cos \alpha \cos \beta - \sin \alpha \sin \beta,$$

let $\alpha = \beta = \theta$, so that

$$\cos(\theta + \theta) = \cos \theta \cos \theta - \sin \theta \sin \theta$$

or $\cos 2\theta = \cos^2 \theta - \sin^2 \theta.$

(ii)   In the basic formula just derived, substitute $1 - \cos^2 \theta$ for $\sin^2 \theta$ to obtain

$$\cos 2\theta = \cos^2 \theta - (1 - \cos^2 \theta) = 2 \cos^2 \theta - 1.$$

(iii)   In the basic formula $\cos 2\theta = \cos^2 \theta - \sin^2 \theta$, substitute $1 - \sin^2 \theta$ for $\cos^2 \theta$ to obtain

$$\cos 2\theta = 1 - \sin^2 \theta - \sin^2 \theta = 1 - 2 \sin^2 \theta. \qquad \blacksquare$$

Example 5   Write $\cos(\pi/4)$ in terms of trigonometric functions of $\pi/8$.

Solution   By Theorem 2, with $\theta = \pi/8$,

$$\cos \frac{\pi}{4} = \cos 2\left(\frac{\pi}{8}\right) = \cos^2 \frac{\pi}{8} - \sin^2 \frac{\pi}{8}.$$

Alternatively, we could have written

$$\cos \frac{\pi}{4} = 2 \cos^2 \frac{\pi}{8} - 1 \qquad \text{or} \qquad \cos \frac{\pi}{4} = 1 - 2 \sin^2 \frac{\pi}{8}. \qquad \blacksquare$$

Example 6   Reduce the expression $1 - 2 \sin^2 5\theta$ to the value of a function of 10θ.

Solution   Reading part (iii) of Theorem 2 from right to left and replacing θ by 5θ, we have

$$1 - 2 \sin^2 5\theta = \cos 2(5\theta) = \cos 10\theta. \qquad \blacksquare$$

Example 7   If $\sin \beta = -\frac{7}{10}$, find $\cos 2\beta$.

Solution   By part (iii) of Theorem 2 with $\theta = \beta$, we have

$$\cos 2\beta = 1 - 2 \sin^2 \beta = 1 - 2(-\tfrac{7}{10})^2 = \tfrac{50}{50} - \tfrac{49}{50} = \tfrac{1}{50}. \qquad \blacksquare$$

Example 8   Express $\cos 4x$ in terms of $\sin x$.

Solution   By part (ii) of Theorem 2 with $\theta = 2x$,

$$\cos 4x = \cos 2(2x) = 2\cos^2 2x - 1 = 2(\cos 2x)^2 - 1.$$

Therefore, by part (iii) of Theorem 2 with $\theta = x$,

$$\cos 4x = 2(1 - 2\sin^2 x)^2 - 1 = 2(1 - 4\sin^2 x + 4\sin^4 x) - 1$$
$$= 1 - 8\sin^2 x + 8\sin^4 x.$$

■

If we solve the equation $\cos 2\theta = 2\cos^2\theta - 1$ in part (ii) of Theorem 2 for $\cos^2\theta$, we obtain the identity

$$\cos^2\theta = \frac{1 + \cos 2\theta}{2}.$$

Similarly, the equation $\cos 2\theta = 1 - 2\sin^2\theta$ in part (iii) of Theorem 2 can be solved for $\sin^2\theta$, and we have

$$\sin^2\theta = \frac{1 - \cos 2\theta}{2}.$$

The formulas above for $\cos^2\theta$ and $\sin^2\theta$ are used extensively in calculus, and they provide the basis for the half-angle formulas to be derived shortly. We record them for future use in the following theorem.

Theorem 3   **Formulas for $\cos^2\theta$ and $\sin^2\theta$**

$$\cos^2\theta = \frac{1 + \cos 2\theta}{2} \quad \text{and} \quad \sin^2\theta = \frac{1 - \cos 2\theta}{2}$$

Example 9   Prove the identity

$$\sin^4\theta = \tfrac{3}{8} - \tfrac{1}{2}\cos 2\theta + \tfrac{1}{8}\cos 4\theta.$$

Solution   Using the second equation of Theorem 3, we can write

$$\sin^4\theta = (\sin^2\theta)^2 = \left(\frac{1 - \cos 2\theta}{2}\right)^2 = \tfrac{1}{4}(1 - 2\cos 2\theta + \cos^2 2\theta).$$

Now, using the first equation of Theorem 3 with θ replaced by 2θ, we have

$$\sin^4\theta = \tfrac{1}{4}(1 - 2\cos 2\theta + \cos^2 2\theta) = \tfrac{1}{4}\left[1 - 2\cos 2\theta + \left(\frac{1 + \cos 4\theta}{2}\right)\right]$$

$$= \tfrac{1}{4} - \tfrac{1}{2}\cos 2\theta + \tfrac{1}{8} + \tfrac{1}{8}\cos 4\theta = \tfrac{3}{8} - \tfrac{1}{2}\cos 2\theta + \tfrac{1}{8}\cos 4\theta.$$

■

If we put $\alpha = \beta = \theta$ in Theorem 6 of Section 7.3, we obtain the double-angle formula for the tangent:

$$\tan 2\theta = \frac{2\tan\theta}{1 - \tan^2\theta}.$$

A formula for $\tan^2 \theta$ can be obtained as follows:

$$\tan^2 \theta = \frac{\sin^2 \theta}{\cos^2 \theta} = \frac{\left(\dfrac{1 - \cos 2\theta}{2}\right)}{\left(\dfrac{1 + \cos 2\theta}{2}\right)}$$

so that

$$\tan^2 \theta = \frac{1 - \cos 2\theta}{1 + \cos 2\theta}.$$

If you let $\theta = \alpha/2$ in Theorem 3 and in the preceding formula for $\tan^2 \theta$, you will obtain the formulas in the following theorem (Problem 34).

Theorem 4   **Half-Angle Formulas**

(i) $\cos^2 \dfrac{\alpha}{2} = \dfrac{1 + \cos \alpha}{2}$

(ii) $\sin^2 \dfrac{\alpha}{2} = \dfrac{1 - \cos \alpha}{2}$

(iii) $\tan^2 \dfrac{\alpha}{2} = \dfrac{1 - \cos \alpha}{1 + \cos \alpha}$,   for $\cos \alpha \neq -1$.

The equations in Theorem 4 can be written in the form

$$\cos \frac{\alpha}{2} = \pm\sqrt{\frac{1 + \cos \alpha}{2}}, \quad \sin \frac{\alpha}{2} = \pm\sqrt{\frac{1 - \cos \alpha}{2}}, \quad \text{and} \quad \tan \frac{\alpha}{2} = \pm\sqrt{\frac{1 - \cos \alpha}{1 + \cos \alpha}}.$$

In using these formulas, you can determine the algebraic sign ($+$ or $-$) by considering which quadrant contains the angle $\alpha/2$ in standard position.

Example 10   Find the exact values of

(a) $\sin 22.5°$ (b) $\cos(\pi/12)$

Solution   (a) Because 22.5° is a quadrant I angle, we have

$$\sin 22.5° = \sin \frac{45°}{2} = \sqrt{\frac{1 - \cos 45°}{2}} = \sqrt{\frac{1 - (\sqrt{2}/2)}{2}} = \sqrt{\frac{2 - \sqrt{2}}{4}} = \frac{\sqrt{2 - \sqrt{2}}}{2}$$

(b) Because $\pi/12$ is a quadrant I angle, we have

$$\cos \frac{\pi}{12} = \cos \frac{\pi/6}{2} = \sqrt{\frac{1 + \cos(\pi/6)}{2}} = \sqrt{\frac{1 + (\sqrt{3}/2)}{2}} = \sqrt{\frac{2 + \sqrt{3}}{4}} = \frac{\sqrt{2 + \sqrt{3}}}{2}$$

Example 11   Suppose that $\dfrac{3\pi}{2} < \theta < 2\pi$ and $\cos\theta = \dfrac{5}{13}$. Find:

(a) $\sin\dfrac{\theta}{2}$ (b) $\tan\dfrac{\theta}{2}$

Solution   Because $3\pi/2 < \theta < 2\pi$, we have $\frac{1}{2}(3\pi/2) < \frac{1}{2}\theta < \frac{1}{2}(2\pi)$, or $3\pi/4 < \theta/2 < \pi$. Therefore, $\theta/2$ is a quadrant II angle, and so $\sin(\theta/2) > 0$ and $\tan(\theta/2) < 0$.

(a) $\sin\dfrac{\theta}{2} = \sqrt{\dfrac{1-\cos\theta}{2}} = \sqrt{\dfrac{1-(5/13)}{2}} = \sqrt{\dfrac{1}{2}\left(\dfrac{8}{13}\right)} = \sqrt{\dfrac{4}{13}} = \dfrac{2\sqrt{13}}{13}$

(b) $\tan\dfrac{\theta}{2} = -\sqrt{\dfrac{1-\cos\theta}{1+\cos\theta}} = -\sqrt{\dfrac{1-(5/13)}{1+(5/13)}} = -\sqrt{\dfrac{13-5}{13+5}}$

$= -\sqrt{\dfrac{8}{18}} = -\sqrt{\dfrac{4}{9}} = -\dfrac{2}{3}$ ■

Example 12*   Let $z = \tan(\theta/2)$. Show that:

(a) $\cos\theta = \dfrac{1-z^2}{1+z^2}$ (b) $\sin\theta = \dfrac{2z}{1+z^2}$ (c) $\tan\dfrac{\theta}{2} = \dfrac{1-\cos\theta}{\sin\theta}$

Solution   (a) Using part (iii) of Theorem 4, we have

$$\frac{1-z^2}{1+z^2} = \frac{1-\tan^2\dfrac{\theta}{2}}{1+\tan^2\dfrac{\theta}{2}} = \frac{1-\left(\dfrac{1-\cos\theta}{1+\cos\theta}\right)}{1+\left(\dfrac{1-\cos\theta}{1+\cos\theta}\right)}$$

$$= \frac{1+\cos\theta-(1-\cos\theta)}{1+\cos\theta+(1-\cos\theta)} = \frac{2\cos\theta}{2} = \cos\theta.$$

(b) Using the fundamental identities, we have

$$\frac{2z}{1+z^2} = \frac{2\tan\dfrac{\theta}{2}}{1+\tan^2\dfrac{\theta}{2}} = \frac{2\tan\dfrac{\theta}{2}}{\sec^2\dfrac{\theta}{2}} = 2\tan\frac{\theta}{2}\cos^2\frac{\theta}{2}$$

$$= 2\frac{\sin\dfrac{\theta}{2}}{\cos\dfrac{\theta}{2}}\cos^2\frac{\theta}{2} = 2\sin\frac{\theta}{2}\cos\frac{\theta}{2}.$$

* The formulas in Example 12 provide the basis for the tangent-half-angle substitution in integral calculus.

But, by Theorem 1,

$$2 \sin \frac{\theta}{2} \cos \frac{\theta}{2} = \sin\left[2\left(\frac{\theta}{2}\right)\right] = \sin \theta.$$

Therefore,

$$\frac{2z}{1 + z^2} = \sin \theta.$$

(c) Using the results of (a) and (b), we have

$$\frac{1 - \cos \theta}{\sin \theta} = \frac{1 - \left(\dfrac{1 - z^2}{1 + z^2}\right)}{\left(\dfrac{2z}{1 + z^2}\right)} = \frac{(1 + z^2) - (1 - z^2)}{2z} = \frac{2z^2}{2z} = z = \tan \frac{\theta}{2}. \qquad \blacksquare$$

Example 13   Find the exact value of $\tan \dfrac{\pi}{8}$.

Solution   By part (c) of Example 12,

$$\tan \frac{\pi}{8} = \tan\left[\frac{1}{2}\left(\frac{\pi}{4}\right)\right] = \frac{1 - \cos \dfrac{\pi}{4}}{\sin \dfrac{\pi}{4}} = \frac{1 - \dfrac{\sqrt{2}}{2}}{\dfrac{\sqrt{2}}{2}} = \frac{2 - \sqrt{2}}{\sqrt{2}} = \frac{2}{\sqrt{2}} - 1 = \sqrt{2} - 1. \qquad \blacksquare$$

Example 14   If $z = \tan \dfrac{\theta}{2}$, write $\dfrac{1}{1 - \cos \theta}$ in terms of z.

Solution   By part (a) of Example 12,

$$\frac{1}{1 - \cos \theta} = \frac{1}{1 - \left(\dfrac{1 - z^2}{1 + z^2}\right)} = \frac{1 + z^2}{(1 + z^2) - (1 - z^2)} = \frac{1 + z^2}{2z^2} \qquad \blacksquare$$

Problem Set 7.4

In Problems 1 to 6, use the double-angle formulas to write the given expression in terms of the values of trigonometric functions of an angle half as large.

1. $\sin 76°$

2. $\tan 62°$

3. $\cos 144°$

4. $\cos 4x$

5. $\sin \dfrac{2\pi}{9}$

6. $\tan \dfrac{\pi}{7}$

In Problems 7 to 12, use the double-angle formulas to reduce the given expression to the value of a single trigonometric function of an angle twice as large.

7. $2 \sin 29° \cos 29°$

8. $\cos^2 4\alpha - \sin^2 4\alpha$

9. $2 \cos^2 \dfrac{5\theta}{2} - 1$

10. $\dfrac{\tan 5t}{1 - \tan^2 5t}$

11. $1 - 2 \sin^2 \dfrac{\pi}{17}$

12. $2 \sin\left(\dfrac{x + y}{2}\right) \cos\left(\dfrac{x + y}{2}\right)$

In Problems 13 to 18, use the given information to find (a) $\sin 2\theta$, (b) $\cos 2\theta$, and (c) $\tan 2\theta$.

13. $\sin \theta = \frac{4}{5}$, θ in quadrant I

14. $\csc \theta = \frac{5}{3}$, θ in quadrant II

15. $\cos \theta = -\frac{5}{13}$, θ in quadrant III

16. $\sec \theta = -\frac{17}{8}$, θ not in quadrant II

17. $\tan \theta = -\frac{12}{5}$, θ in quadrant II

18. $\cot \theta = \frac{1}{2}$, θ in quadrant III

In Problems 19 to 26, rewrite each expression in a form involving only a single trigonometric function.

19. $\dfrac{\sin 2x}{2 \sin x}$

20. $(\sin 2\alpha + \cos 2\alpha)^2 - 2 \sin 4\alpha$

21. $(\sin t - \cos t)^2 + 3 \sin 2t$

22. $\dfrac{1 - \cos 2u}{\sin 2u}$

23. $\sin x \cos^3 x + \sin^3 x \cos x$

24. $\dfrac{\sin 2\beta}{1 + \cos 2\beta}$

25. $\dfrac{\tan 2\theta - \sin 2\theta}{2 \tan 2\theta}$

26. $\dfrac{\sin 2x}{1 - \cos 2x}$

In Problems 27 to 33, (a) use the half-angle formulas to write the given expression in terms of the value of a trigonometric function of an angle twice as large, and (b) find the exact numerical value of each expression. Do not use a calculator or tables.

27. $\cos 15°$

28. $\sin 67.5°$

29. $\cos \dfrac{5\pi}{8}$

30. $\tan \dfrac{\pi}{12}$

31. $\tan 157.5°$

32. $\cos \dfrac{11\pi}{12}$

33. $\sin 202.5°$

34. Prove Theorem 4.

In Problems 35 to 38, use the given information to find (a) $\sin(\theta/2)$ (b) $\cos(\theta/2)$, and (c) $\tan(\theta/2)$.

35. $\cos \theta = \dfrac{7}{25}$, $0° < \theta < 90°$

36. $\cos \theta = -\dfrac{4}{5}$, $90° < \theta < 180°$

37. $\cos \theta = -\dfrac{5}{13}$, $\pi < \theta < \dfrac{3\pi}{2}$

38. $\cot \theta = -\dfrac{15}{8}$, $\dfrac{3\pi}{2} < \theta < 2\pi$

In Problems 39 to 44, reduce the given expression to the value of a single trigonometric function of an angle half as large.

39. $-\sqrt{\dfrac{1 + \cos 250°}{2}}$

40. $-\sqrt{\dfrac{1 - \cos 400°}{2}}$

41. $\dfrac{1 - \cos 6x}{\sin 6x}$

42. $\dfrac{\sin 10\theta}{1 - \cos 10\theta}$

43. $\sqrt{\dfrac{1 + \cos(2\pi/5)}{2}}$

44. $\sqrt{\dfrac{1 - \cos(11\pi/15)}{2}}$

In Problems 45 to 55, prove that each trigonometric equation is an identity.

45. $\cos^4 \theta - \sin^4 \theta = \cos 2\theta$

46. $2 \csc 2x = \sec x \csc x$

47. $\sin 4t = 2(\sin 2t)(1 - 2 \sin^2 t)$

48. $\dfrac{\sin^2 2u}{\sin^2 u} = 4 \cos^2 u$

49. $\sin 3\theta = (\sin \theta)(3 - 4 \sin^2 \theta)$ [*Hint:* $3\theta = \theta + 2\theta$.]

50. $\cos 3t = 4 \cos^3 t - 3 \cos t$

51. $\dfrac{\sec^2 x}{2 - \sec^2 x} = \sec 2x$

52. $\sin^2 t \cos^2 t = \frac{1}{8}(1 - \cos 4t)$

53. $\dfrac{\cot \alpha - 1}{\cot \alpha + 1} = \dfrac{\cos 2\alpha}{1 + \sin 2\alpha}$

54. $\cos^4 x = \frac{3}{8} + \frac{1}{2} \cos 2x + \frac{1}{8} \cos 4x$

55. $\csc \theta - \cot \theta = \tan \dfrac{\theta}{2}$

56. Show that $\cos 20°$ is a root of $8x^3 - 6x - 1 = 0$. [*Hint:* Use Problem 50.]

In Problems 57 to 60, write the given expression in terms of z, where $z = \tan(\theta/2)$. (Use the results of Example 12.)

57. $\sec \theta$

58. $\csc \theta$

59. $\cot \theta$

60. $\dfrac{\sin \theta}{2 - \cos \theta}$

61. Using Theorem 3, show that, if b and c are positive constants, the quantity $y = b \cos^2 ct$ oscillates in accordánce with the simple harmonic model $y = a \cos(\omega t - \phi) + k$, with $a = b/2$, $\omega = 2c$, $\phi = 0$, and $k = b/2$.

62. Let $x = (1 - z^2)/(1 + z^2)$ and $y = 2z/(1 + z^2)$. Using the results of Example 12, show that, as z runs through the real numbers, the point (x, y) runs over every point on the unit circle except for $(-1, 0)$.

7.5 PRODUCT FORMULAS, SUM FORMULAS, AND GRAPHS OF COMBINATIONS OF SINE AND COSINE FUNCTIONS

In applied mathematics, it is often necessary to rewrite a product of sine or cosine functions as a sum or difference of such functions, or vice versa. This can be done by using the identities in Theorems 1 and 2.

Theorem 1   **Product Formulas**

> **(i)** $\sin \alpha \cos \beta = \frac{1}{2}[\sin(\alpha + \beta) + \sin(\alpha - \beta)]$
>
> **(ii)** $\cos \alpha \sin \beta = \frac{1}{2}[\sin(\alpha + \beta) - \sin(\alpha - \beta)]$
>
> **(iii)** $\cos \alpha \cos \beta = \frac{1}{2}[\cos(\alpha + \beta) + \cos(\alpha - \beta)]$
>
> **(iv)** $\sin \alpha \sin \beta = \frac{1}{2}[\cos(\alpha - \beta) - \cos(\alpha + \beta)]$

Proof   We prove (i) and (iv) here and leave (ii) and (iii) as exercises (Problem 22).

(i)   Using the formulas

$$\sin(\alpha + \beta) = \sin \alpha \cos \beta + \sin \beta \cos \alpha$$

$$\sin(\alpha - \beta) = \sin \alpha \cos \beta - \sin \beta \cos \alpha,$$

and adding the second equation to the first, we obtain

$$\sin(\alpha + \beta) + \sin(\alpha - \beta) = 2 \sin \alpha \cos \beta.$$

It follows that

$$\sin \alpha \cos \beta = \frac{1}{2}[\sin(\alpha + \beta) + \sin(\alpha - \beta)].$$

(iv) Using the formulas

$$\cos(\alpha - \beta) = \cos \alpha \cos \beta + \sin \alpha \sin \beta$$

$$\cos(\alpha + \beta) = \cos \alpha \cos \beta - \sin \alpha \sin \beta,$$

and subtracting the second equation from the first, we obtain

$$\cos(\alpha - \beta) - \cos(\alpha + \beta) = 2 \sin \alpha \sin \beta.$$

It follows that

$$\sin \alpha \sin \beta = \tfrac{1}{2}[\cos(\alpha - \beta) - \cos(\alpha + \beta)].$$

■

Example 1   Express each product as a sum or difference:

(a) $\sin 5\theta \cos 3\theta$ **(b)** $\sin 3t \sin t$

Solution   **(a)** By part (i) of Theorem 1 with $\alpha = 5\theta$ and $\beta = 3\theta$,

$$\sin 5\theta \cos 3\theta = \tfrac{1}{2}[\sin(5\theta + 3\theta) + \sin(5\theta - 3\theta)] = \tfrac{1}{2} \sin 8\theta + \tfrac{1}{2} \sin 2\theta.$$

(b) By part (iv) of Theorem 1 with $\alpha = 3t$ and $\beta = t$,

$$\sin 3t \sin t = \tfrac{1}{2}[\cos(3t - t) - \cos(3t + t)] = \tfrac{1}{2} \cos 2t - \tfrac{1}{2} \cos 4t.$$

■

Theorem 2   **Sum Formulas**

> **(i)** $\sin \gamma + \sin \theta = 2 \sin\left(\dfrac{\gamma + \theta}{2}\right) \cos\left(\dfrac{\gamma - \theta}{2}\right)$
>
> **(ii)** $\sin \gamma - \sin \theta = 2 \cos\left(\dfrac{\gamma + \theta}{2}\right) \sin\left(\dfrac{\gamma - \theta}{2}\right)$
>
> **(iii)** $\cos \gamma + \cos \theta = 2 \cos\left(\dfrac{\gamma + \theta}{2}\right) \cos\left(\dfrac{\gamma - \theta}{2}\right)$
>
> **(iv)** $\cos \gamma - \cos \theta = -2 \sin\left(\dfrac{\gamma + \theta}{2}\right) \sin\left(\dfrac{\gamma - \theta}{2}\right)$

Proof   We prove (i) here and leave (ii), (iii), and (iv) as exercises (Problem 24). In part (i) of Theorem 1, let

$$\alpha = \frac{\gamma + \theta}{2} \quad \text{and} \quad \beta = \frac{\gamma - \theta}{2}.$$

Thus,

$$\alpha + \beta = \frac{\gamma + \theta}{2} + \frac{\gamma - \theta}{2} = \frac{2\gamma}{2} = \gamma$$

and

$$\alpha - \beta = \frac{\gamma + \theta}{2} - \frac{\gamma - \theta}{2} = \frac{2\theta}{2} = \theta.$$

It follows that

$$\sin\left(\frac{\gamma + \theta}{2}\right) \cos\left(\frac{\gamma - \theta}{2}\right) = \tfrac{1}{2}[\sin \gamma + \sin \theta],$$

or

$$\sin \gamma + \sin \theta = 2 \sin\left(\frac{\gamma + \theta}{2}\right) \cos\left(\frac{\gamma - \theta}{2}\right).$$

■

Example 2   Express $\sin 10x + \sin 4x$ as a product.

Solution   Using part (i) of Theorem 2 with $\gamma = 10x$ and $\theta = 4x$, we have

$$\sin 10x + \sin 4x = 2 \sin\left(\frac{10x + 4x}{2}\right) \cos\left(\frac{10x - 4x}{2}\right) = 2 \sin 7x \cos 3x. \qquad \blacksquare$$

Example 3*   Prove that the equation

$$\frac{\sin (x + h) - \sin x}{h} = \cos\left(x + \frac{h}{2}\right) \frac{\sin(h/2)}{h/2}$$

is an identity.

Solution   Using part (ii) of Theorem 2 with $\gamma = x + h$ and $\theta = x$, we have

$$\frac{\sin(x + h) - \sin x}{h} = \frac{2 \cos\left[\dfrac{(x + h) + x}{2}\right] \sin\left[\dfrac{(x + h) - x}{2}\right]}{h}$$

$$= \frac{2 \cos\left(\dfrac{2x + h}{2}\right) \sin \dfrac{h}{2}}{h}$$

$$= \cos\left(x + \frac{h}{2}\right) \frac{\sin(h/2)}{h/2}. \qquad \blacksquare$$

Graphs of Combinations of Sine and Cosine Functions

We have already graphed the sum of algebraic functions in Section 3.7 by adding ordinates. The same method works for graphing sums of trigonometric functions.

© **Example 4**   Sketch $y = \sin x + 2 \cos 2x$ by adding ordinates.

Solution   We begin by sketching the graphs of

$$y_1 = \sin x \qquad \text{and} \qquad y_2 = 2 \cos 2x$$

on the same coordinate system (Figure 1). Then we add the ordinates y_1 and y_2 for selected values of x to obtain values of

$$y = \sin x + 2 \cos 2x.$$

(A calculator is useful here.) By plotting the resulting points and connecting them with a smooth curve, we obtain the desired graph (Figure 1). $\blacksquare$

* The identity in Example 3 is important in the development of the differential calculus.

Figure 1

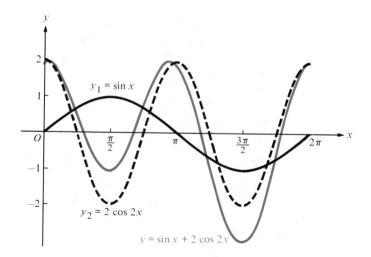

In sketching the graphs of certain combinations of sines and cosines, you will find the following theorem very helpful.

Theorem 3   **Reduction of $A \cos \theta + B \sin \theta$ to the Form $a \cos(\theta - \phi)$**

> Suppose that A and B are constants; let $a = \sqrt{A^2 + B^2}$, and let ϕ be an angle in standard position with the point (A, B) on its terminal side. Then $A = a \cos \phi$, $B = a \sin \phi$, and, for all values of θ,
>
> $$A \cos \theta + B \sin \theta = a \cos(\theta - \phi).$$

Figure 2

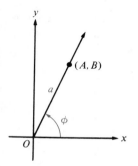

Proof   Notice that $a = \sqrt{A^2 + B^2}$ is the distance between (A, B) and $(0, 0)$ (Figure 2). Hence, by the definitions of $\sin \phi$ and $\cos \phi$,

$$\sin \phi = \frac{B}{a} \qquad \text{and} \qquad \cos \phi = \frac{A}{a}.$$

It follows that

$$B = a \sin \phi \qquad \text{and} \qquad A = a \cos \phi.$$

Therefore,

$$A \cos \theta + B \sin \theta = (a \cos \phi) \cos \theta + (a \sin \phi) \sin \theta$$
$$= a(\cos \theta \cos \phi + \sin \theta \sin \phi) = a \cos(\theta - \phi). \qquad \blacksquare$$

Example 5   Sketch the graph of $y = \sqrt{3} \cos \pi t + \sin \pi t$.

Solution   We could use the method of adding ordinates; but, since the same quantity πt appears in both terms, we prefer to use Theorem 3 with $\theta = \pi t$, $A = \sqrt{3}$, and $B = 1$.

Then $a = \sqrt{A^2 + B^2} = \sqrt{4} = 2$ and the equations $A = a \cos \phi$ and $B = a \sin \phi$ become $\sqrt{3} = 2 \cos \phi$ and $1 = 2 \sin \phi$, or

$$\cos \phi = \frac{\sqrt{3}}{2} \quad \text{and} \quad \sin \phi = \frac{1}{2}.$$

Because both $\cos \phi$ and $\sin \phi$ are positive, ϕ is a quadrant I angle. Evidently, $\phi = \pi/6$. Therefore, by Theorem 3, we can rewrite the original equation in the form $y = a \cos(\theta - \phi)$, that is,

$$y = 2 \cos\left(\pi t - \frac{\pi}{6}\right).$$

According to the discussion in Section 6.7, the graph is a simple harmonic curve with amplitude $a = 2$, angular frequency $\omega = \pi$, phase angle $\phi = \pi/6$, phase shift $S = \phi/\omega = (\pi/6)/\pi = \frac{1}{6}$, period $T = 2\pi/\omega = 2\pi/\pi = 2$, and vertical shift $k = 0$ (Figure 3). ∎

Figure 3

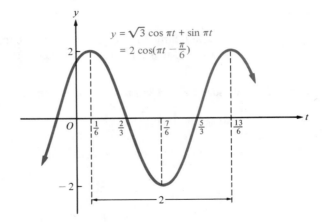

$y = \sqrt{3} \cos \pi t + \sin \pi t$
$= 2 \cos(\pi t - \frac{\pi}{6})$

To sketch the graph of a product of sine and cosine functions, you can use Theorem 1 to rewrite the product as a sum or difference, and then use the method of adding or subtracting ordinates.

© **Example 6**   Sketch the graph of

$$y = 2 \sin \frac{3x}{2} \sin \frac{x}{2}.$$

Solution   Using the product formula

$$\sin \alpha \sin \beta = \tfrac{1}{2}[\cos(\alpha - \beta) - \cos(\alpha + \beta)]$$

in part (iv) of Theorem 1 with $\alpha = 3x/2$ and $\beta = x/2$, we find that

$$2 \sin \frac{3x}{2} \sin \frac{x}{2} = \cos\left(\frac{3x}{2} - \frac{x}{2}\right) - \cos\left(\frac{3x}{2} + \frac{x}{2}\right)$$

$$= \cos x - \cos 2x.$$

Therefore, the original equation can be rewritten as

$$y = \cos x - \cos 2x.$$

In Figure 4, we have sketched graphs of $y_1 = \cos x$ and $y_2 = \cos 2x$ on the same coordinate system. Then, subtracting the ordinate y_2 from the ordinate y_1 for selected values of x, we obtain corresponding values of $y = \cos x - \cos 2x$. (Here a calculator proves to be quite useful.) By plotting the resulting points and connecting them with a smooth curve, we obtain the required graph (Figure 4).   ■

Figure 4

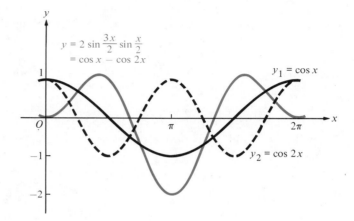

$$y = 2 \sin \frac{3x}{2} \sin \frac{x}{2}$$
$$= \cos x - \cos 2x$$

$y_1 = \cos x$

$y_2 = \cos 2x$

Many natural variations result from combinations of two or more simple harmonic oscillations with different frequencies, amplitudes, and phase angles. These variations may no longer be simple harmonic oscillations. For instance, pulsations in the brightness or magnitude of a variable star may be caused by a combination of factors, such as periodic expansion and contraction, variable temperature, and periodic eclipses by a companion star. The resulting pulsations, although they may occur in a regular pattern such as that shown in Figure 5, may not be simple harmonic oscillations.

Figure 5

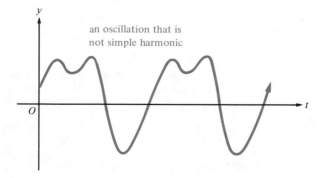

an oscillation that is
not simple harmonic

When two simple harmonic oscillations (for instance, musical tones) combine by addition, the resulting oscillation is called a **superposition** of the original oscillations. Thus, a superposition of the oscillations represented by

$$y_1 = a_1 \cos(\omega_1 t - \phi_1) + k_1 \qquad \text{and} \qquad y_2 = a_2 \cos(\omega_2 t - \phi_2) + k_2$$

is represented by

$$y = y_1 + y_2.$$

The graph of such a superposition can be sketched by adding ordinates. This graph is often **periodic,** in that it repeats itself over intervals of length T (called the **period**).

It can be shown that the superposition of two simple harmonic oscillations with periods T_1 and T_2 will be a periodic oscillation only if T_1/T_2 is a rational number. If

$$T_1/T_2 = n/m,$$

where n and m are positive integers and the fraction n/m is reduced to lowest terms, then the period T of the superposition is given by

$$T = mT_1 = nT_2.$$

Example 7   Find the period T of the oscillation represented by

$$y = 2 \cos\left(3\pi t - \frac{\pi}{3}\right) + \frac{1}{2}\cos\left(\frac{4}{5}\pi t + \frac{\pi}{12}\right).$$

Solution   The oscillation is a superposition of the simple harmonic oscillations

$$y_1 = 2 \cos\left(3\pi t - \frac{\pi}{3}\right) \qquad \text{and} \qquad y_2 = \frac{1}{2}\cos\left(\frac{4}{5}\pi t + \frac{\pi}{12}\right),$$

which have periods

$$T_1 = \frac{2\pi}{\omega_1} = \frac{2\pi}{3\pi} = \frac{2}{3} \qquad \text{and} \qquad T_2 = \frac{2\pi}{\omega_2} = \frac{2\pi}{(4/5)\pi} = \frac{5}{2}.$$

Here,

$$\frac{T_1}{T_2} = \frac{2/3}{5/2} = \frac{4}{15},$$

so the period of the superposition is given by

$$T = 15T_1 = 15(\tfrac{2}{3}) = 10$$

[and also by $T = 4T_2 = 4(\tfrac{5}{2}) = 10$].   ■

Even if the superposition of two simple harmonic oscillations is periodic, it usually isn't a simple harmonic oscillation (for instance, see Figure 4). However, *the superposition of simple harmonic oscillations*

$$y_1 = a_1 \cos(\omega t - \phi_1) + k_1 \qquad and \qquad y_2 = a_2 \cos(\omega t - \phi_2) + k_2,$$

which have the same period $T = 2\pi/\omega$, is also a simple harmonic oscillation of period T (for instance, see Figure 3). We leave the proof as an exercise (Problem 60).

Problem Set 7.5

In Problems 1 to 10, use Theorem 1 to express each product as a sum or difference.

1. $\sin 105° \cos 40°$

2. $\cos \dfrac{5\pi}{12} \sin \dfrac{\pi}{12}$

3. $2 \cos \dfrac{5\pi}{8} \cos \dfrac{\pi}{8}$

4. $10 \sin 160° \sin 42°$

5. $\sin 3\theta \cos 5\theta$

6. $\cos 7y \sin 3y$

7. $\sin 7s \sin 5t$

8. $\cos 3t \cos 5t$

9. $\sin x \cos 3x$

10. $a \cos(\omega_1 t + \phi_1) \cos(\omega_2 t + \phi_2)$

In Problems 11 to 20, use Theorem 2 to express each sum or difference as a product.

11. $\sin 80° + \sin 20°$

12. $\sin 50° - \sin 20°$

13. $\cos \dfrac{3\pi}{8} + \cos \dfrac{\pi}{8}$

14. $\cos \dfrac{7\pi}{12} - \cos \dfrac{5\pi}{12}$

15. $\sin 4\theta + \sin 2\theta$

16. $\sin 4x - \sin 6x$

17. $\cos 2t - \cos 4t$

18. $\cos 3\gamma + \cos 8\gamma$

19. $\sin(\alpha + \beta) - \sin \alpha$

20. $a \cos(\omega_1 t + \phi_1) + a \cos(\omega_2 t + \phi_2)$

21. Use Theorem 2 to simplify each expression; then find its exact numerical value. [*Hint:* Do not attempt to use a calculator.]

(a) $\dfrac{\cos 75° - \cos 15°}{\sin 75° + \sin 15°}$

(b) $\dfrac{\sin 80° + \sin 10°}{\cos 80° + \cos 10°}$

22. Prove parts (ii) and (iii) of Theorem 1.

23. Show that $\sin 50° + \sin 10° = \sin 70°$.

24. Prove parts (ii), (iii), and (iv) of Theorem 2.

In Problems 25 to 38, prove that each equation is a trigonometric identity.

25. $\dfrac{\sin 5\alpha + \sin \alpha}{\cos \alpha - \cos 5\alpha} = \cot 2\alpha$

26. $\dfrac{\cos \theta + \cos 9\theta}{\sin \theta + \sin 9\theta} = \cot 5\theta$

27. $2 \sin 3s \cos 2t = \sin(3s + 2t) + \sin(3s - 2t)$

28. $\sin(x + y) \sin(x - y) = \cos^2 y - \cos^2 x$

29. $2 \sin\left(\dfrac{\pi}{4} + x\right) \sin\left(\dfrac{\pi}{4} - x\right) = \cos 2x$

30. $(\sin \alpha)(\sin 3\alpha + \sin \alpha) = \cos^2 \alpha - \cos \alpha \cos 3\alpha$

31. $\dfrac{\sin 4t - \sin 2t}{\sin 4t + \sin 2t} = \dfrac{\tan t}{\tan 3t}$

32. $\dfrac{2 \sin 3x}{\cos 2x} - \dfrac{1}{\sin x} = -\cos 4x \sec 2x \csc x$

33. $\dfrac{\cos 5s + \cos 2s}{\sin 5s + \sin 2s} = \cot \dfrac{7s}{2}$

34. $\dfrac{\sin x + \sin y}{\sin x - \sin y} = \dfrac{\tan \frac{1}{2}(x + y)}{\tan \frac{1}{2}(x - y)}$

35. $\dfrac{\sin x - \sin y}{\cos x - \cos y} = -\cot\left(\dfrac{x + y}{2}\right)$

36. $\dfrac{\cos x + \cos y}{\cos y - \cos x} = \cot\left(\dfrac{x + y}{2}\right) \cot\left(\dfrac{x - y}{2}\right)$

37. $\dfrac{\cos(x + h) - \cos x}{h} = -\sin\left(x + \dfrac{h}{2}\right) \dfrac{\sin(h/2)}{h/2}$

38. $\dfrac{\cos 4t + \cos 3t + \cos 2t}{\sin 4t + \sin 3t + \sin 2t} = \cot 3t$

[C] In Problems 39 to 44, use addition or subtraction of ordinates to sketch each graph.

39. $y = \sin x + \cos x$

40. $y = \sin 2x + 2 \cos x$

41. $y = \sin x - \cos 2x$

42. $y = 3 \sin 2\pi x + 2 \sin 4\pi x$

43. $y = \sin 2t + \sin 3t$

44. $y = 2 \cos t - \sin \frac{1}{2}t$

In Problems 45 to 48, use Theorem 3 to rewrite each equation in the form $y = a \cos(\omega t - \phi)$, and then sketch the graph.

45. $y = \sin 2t + \cos 2t$

46. $y = -\sqrt{3} \cos\left(\dfrac{\pi}{2} t\right) + \sin\left(\dfrac{\pi}{2} t\right)$

47. $y = \cos 2\pi t - \sqrt{3} \sin 2\pi t$

48. $y = 3 \cos(2t - 1) - 3 \sin(2t - 1)$

C In Problems 49 to 54, use Theorem 1 to rewrite the product as a sum or difference, and then sketch the graph by adding or subtracting ordinates.

49. $y = \cos 4x \sin 2x$ **50.** $y = 3 \cos 3t \cos 7t$

51. $y = 2 \sin 3t \sin t$ **52.** $y = -\sin \dfrac{3x}{2} \cos \dfrac{x}{2}$

53. $y = -\cos \pi x \cos 2\pi x$ **54.** $y = 5 \sin \dfrac{4t}{3} \cos \dfrac{2t}{3}$

In Problems 55 to 58, find the period T of the oscillation represented by each equation.

55. $y = 3 \cos(2\pi t - \pi) + 2 \cos\left(\dfrac{\pi}{2} t - \dfrac{\pi}{2}\right) + 4$

56. $y = 2 \cos\left(\dfrac{3}{2} t + 1\right) + 4 \cos\left(\dfrac{2}{3} t - 2\right) - 5$

57. $y = \cos \dfrac{\sqrt{2}}{2} t + \cos(3\sqrt{2}t - 1)$

58. $y = \sqrt{3} \cos\left(\dfrac{4\pi}{3} t - \dfrac{\pi}{7}\right) + 2 \sin\left(\dfrac{3\pi}{2} t + \dfrac{\pi}{5}\right)$

C **59.** Suppose you sketch the graph of

$$y = 19 \sin x - 10 \sin 2x$$

on the interval $0 \le x \le \pi$ by plotting points corresponding to

$$x = 0, \frac{\pi}{6}, \frac{\pi}{4}, \frac{\pi}{3}, \frac{\pi}{2}, \frac{2\pi}{3}, \frac{3\pi}{4}, \frac{5\pi}{6}, \text{ and } \pi.$$

If you connect these points with a smooth curve, you will obtain the graph shown in Figure 6. However, this is *not* the correct graph. Find the error, correct it, and redraw the graph.

Figure 6

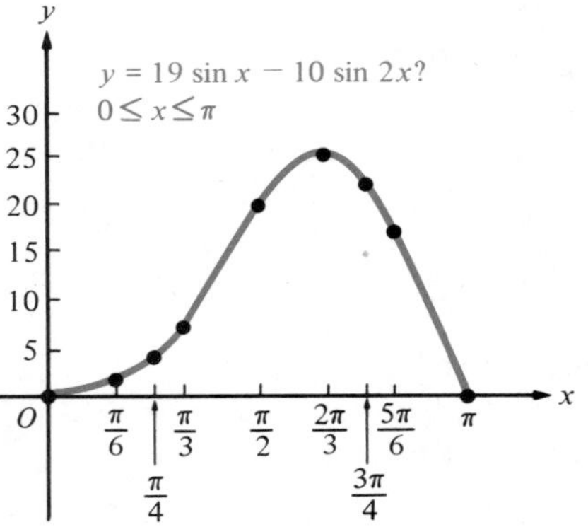

$y = 19 \sin x - 10 \sin 2x$?
$0 \le x \le \pi$

60. Suppose that $y_1 = a_1 \cos(\omega t - \phi_1) + k_1$ and $y_2 = a_2 \cos(\omega t - \phi_2) + k_2$. Let $\theta = \omega t - \phi_1$, $A = a_1 + a_2 \cos(\phi_2 - \phi_1)$, $B = a_2 \sin(\phi_2 - \phi_1)$, $k = k_1 + k_2$, and $y = y_1 + y_2$.

(a) Show that $\omega t - \phi_2 = \theta - (\phi_2 - \phi_1)$.

(b) Use the result in part (a) to show that

$$y_2 = a_2 \cos(\phi_2 - \phi_1) \cos \theta + a_2\sin(\phi_2 - \phi_1) \sin \theta + k_2.$$

(c) Use the result in part (b) to show that

$$y = A \cos \theta + B \sin \theta + k.$$

(d) Use the result in part (c) together with Theorem 3 to rewrite y in the form

$$y = a \cos(\theta - \phi) + k.$$

(e) Conclude that $y = a \cos(\omega t - \phi_0) + k$, where $\phi_0 = \phi_1 + \phi$.

C **61.** Even when sketched carefully and correctly, the graph of $y = 19 \sin x - 10 \sin 2x$ (see Problem 59) appears to have a maximum at $x = 2\pi/3$. Use a calculator to check values of y for values of x close to $2\pi/3$ to see if this is really so.

62. A 50 kilohertz radio frequency oscillation with the equation $y_1 = 10 \sin(\pi \times 10^5)t$ is amplitude modulated by an audio frequency tone with the equation $y_2 = \sin(2\pi \times 10^4)t$ to produce a signal with the equation $y = y_2 y_1$. Sketch three graphs over the interval $0 \le t \le \frac{1}{5000}$ to show:

(a) the radio frequency oscillation

(b) the audio frequency tone

(c) the amplitude modulated signal.

7.6 INVERSE TRIGONOMETRIC FUNCTIONS

In Sections 6.2, 6.3, and 6.4, we learned how to find the values (or the approximate values) of the trigonometric functions of known angles. In many applications of trigonometry, the problem is the other way around—to find unknown angles when information is given about the values of trigonometric functions of the angles. The **inverse trigonometric functions** can be used to solve such problems.

Recall from Section 3.8 that by reflecting the graph of an invertible function f across the line $y = x$ we obtain the graph of the inverse function f^{-1} (Figure 1). However, in order for f to be invertible, no horizontal line can intersect its graph more than once.

Figure 1

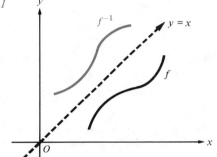

Figure 2

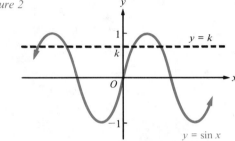

Because the sine function is periodic, any horizontal line $y = k$ with $-1 \leq k \leq 1$ intersects the graph of $y = \sin x$ repeatedly (Figure 2), and therefore the sine function is *not* invertible. But consider the portion of the graph of $y = \sin x$ between $x = -\pi/2$ and $x = \pi/2$ (Figure 3a). No horizontal line intersects this curve more than once, so its reflection across the line $y = x$ (Figure 3b) is the graph of a function. This function is often called the "inverse sine" and denoted by $\sin^{-1}$.

Of course, the terminology "inverse sine function" and the notation $\sin^{-1}$ aren't really correct because the sine function (defined on the set of all real numbers) has no inverse. Because of this, some people use the notation Sine (with a *capital* S) for

Figure 3

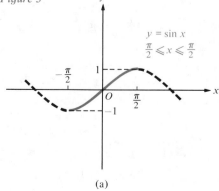

(a)   (b)

the function in Figure 3a and Sin^{-1} for its inverse in Figure 3b. Another source of difficulty is the possible confusion of sin^{-1} x with (sin x)$^{-1}$ = 1/sin x = csc x. To avoid such confusion, many people use the notation "arcsin x," meaning the arc (or angle) whose sine is x, rather than sin^{-1} x. Because both the notations sin^{-1} x and arcsin x are in common use in mathematics and its applications, we shall use them interchangeably. Thus, we make the following definition.

Definition 1   **The Inverse Sine or Arcsine**

> The **inverse sine** or **arcsine** function, denoted by sin^{-1} or arcsin, is defined by
>
> $$\sin^{-1} x = y \quad \text{if and only if} \quad x = \sin y \quad \text{and} \quad -\frac{\pi}{2} \le y \le \frac{\pi}{2}.$$

In other words:

> sin^{-1} x is the angle (or number) between $-\pi/2$ and $\pi/2$ whose sine is x.

Figure 4

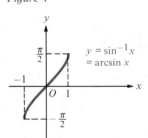

The graph of

$$y = \sin^{-1} x \quad \text{or} \quad y = \arcsin x$$

(Figure 4) shows that sin^{-1} is an increasing function with domain $[-1, 1]$ and range $[-\pi/2, \pi/2]$. The graph appears to be symmetric about the origin, indicating that sin^{-1} is an odd function; that is,

$$\sin^{-1}(-x) = -\sin^{-1} x.$$

A scientific calculator can be used to find approximate values of sin^{-1} x (or arcsin x). (If a calculator isn't available, you can find these values by reading Appendix Table IIB "backward.") Some calculators have a sin^{-1} key; some have a key marked arcsin; and some have an INV key, which must be pressed before the SIN key to give the inverse sine. Again, you must be careful to put the calculator in radian mode if you want sin^{-1} x in radians—otherwise, you will get the angle in *degrees* (between $-90°$ and $90°$) whose sine is x.

Example 1   Find each value:

(a) sin^{-1} $\frac{1}{2}$ (b) arcsin($-\sqrt{3}/2$)   © (c) sin^{-1} 0.7321

Solution   (a) Here a calculator isn't needed because we know that $\pi/6$ radian (30°) is an angle whose sine is $\frac{1}{2}$. Thus, because $-\pi/2 \le \pi/6 \le \pi/2$, we have sin^{-1} $\frac{1}{2}$ = $\pi/6$.

(b) Again we don't need a calculator because we know that $-\pi/3$ radians ($-60°$) is an angle whose sine is $-\sqrt{3}/2$. Thus, because $-\pi/2 \le -\pi/3 \le \pi/2$, we have arcsin($-\sqrt{3}/2$) = $-\pi/3$.

(c) Using a 10-digit calculator in radian mode, we obtain

$$\sin^{-1} 0.7321 = 0.821399673.$$

(In degree mode, we get sin^{-1} 0.7321 = 47.06273457°.)   ■

The remaining five trigonometric functions are also periodic, and therefore, like the sine function, they are not invertible. However, by restricting these functions to suitable intervals we can define corresponding inverses, just as we did for the sine function. The appropriate portions of the graphs of the cosine and tangent functions are shown in Figure 5.

Figure 5

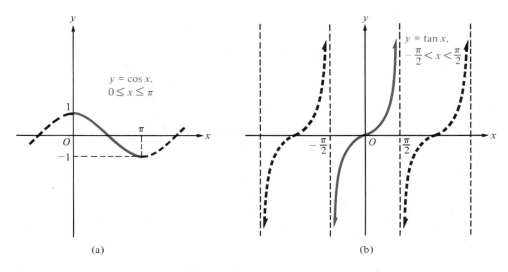

$y = \cos x$, $0 \le x \le \pi$

$y = \tan x$, $-\dfrac{\pi}{2} < x < \dfrac{\pi}{2}$

(a)   (b)

Thus, we make the following definitions.

Definition 2   **The Inverse Cosine or Arccosine**

> The **inverse cosine** or **arccosine** function, denoted by $\cos^{-1}$ or arccos, is defined by
>
> $$\cos^{-1} x = y \qquad \text{if and only if} \qquad x = \cos y \qquad \text{and} \qquad 0 \le y \le \pi.$$

Definition 3   **The Inverse Tangent or Arctangent**

> The **inverse tangent** or **arctangent** function, denoted by $\tan^{-1}$ or arctan, is defined by
>
> $$\tan^{-1} x = y \qquad \text{if and only if} \qquad x = \tan y \qquad \text{and} \qquad -\frac{\pi}{2} < y < \frac{\pi}{2}.$$

Figure 6

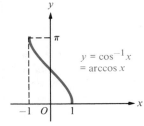

$y = \cos^{-1} x$ $= \arccos x$

In words, $\cos^{-1} x$ is the angle (or number) between 0 and π whose cosine is x. Likewise, $\tan^{-1} x$ is the angle (or number) between $-\pi/2$ and $\pi/2$ whose tangent is x.

The graph of $y = \cos^{-1} x$, obtained by reflecting the curve in Figure 5a across the line $y = x$, shows that $\cos^{-1}$ (or arccos) is a decreasing function with domain $[-1, 1]$ and range $[0, \pi]$ (Figure 6). Notice that the graph of $\cos^{-1}$ is *not* symmetric about the origin nor about the y axis. Thus, in spite of the fact that cosine is an even function, $\cos^{-1}$ is not.

Figure 7

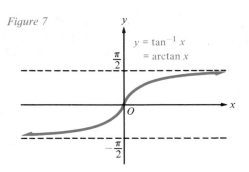

The graph of $y = \tan^{-1} x$, obtained by reflecting the curve in Figure 5b across the line $y = x$, shows that $\tan^{-1}$ (or arctan) is an increasing function with domain $\mathbb{R}$ and range $(-\pi/2, \pi/2)$ (Figure 7). Notice that the lines $y = -\pi/2$ and $y = \pi/2$ are horizontal asymptotes. The graph appears to be symmetric about the origin, indicating that $\tan^{-1}$ is an odd function; that is,

$$\tan^{-1}(-x) = -\tan^{-1} x$$

(Problem 39).

Example 2   Find each value:

(a) $\cos^{-1}(-\sqrt{2}/2)$ ⓒ (b) arccos 0.6675

(c) arctan $\sqrt{3}$ ⓒ (d) $\tan^{-1}(-2.498)$

Solution   (a) If $\cos^{-1}(-\sqrt{2}/2) = y$, then $-\sqrt{2}/2 = \cos y$ and $0 \le y \le \pi$. Therefore, since the cosine is negative in quadrant II, $y = 3\pi/4$.

(b) Using a 10-digit calculator in radian mode, we find that

$$\text{arccos } 0.6657 = 0.839950077.$$

(c) If arctan $\sqrt{3} = y$, then $\sqrt{3} = \tan y$ and $-\pi/2 < y < \pi/2$. Therefore, $y = \pi/3$.

(d) Using a 10-digit calculator in radian mode, we find that

$$\tan^{-1}(-2.498) = -1.190013897.$$

∎

If you enter a number x between -1 and 1 in a calculator, take $\sin^{-1} x$, and then take the sine of the result, you will get x back again. Thus

$$x \xrightarrow{\sin^{-1}} y \xrightarrow{\sin} x.$$

Do you see why? Therefore, we have the identity

$$\boxed{\sin(\sin^{-1} x) = x \qquad \text{for} \qquad -1 \le x \le 1.}$$

However, if you try the same thing the other way around, first taking the sine and then taking the inverse sine, you might not get your original number back. For instance, starting with $x = 2$, we get (on a 10-digit calculator in radian mode)

$$2 \xrightarrow{\sin} 0.909297427 \xrightarrow{\sin^{-1}} 1.141592654.$$

The reason is simply that $\sin^{-1} 0.909297427$ is the number *between* $-\pi/2$ *and* $\pi/2$ whose sine is 0.909297427, and 2 does not lie between $-\pi/2$ and $\pi/2$. However, if you start with a number *between* $-\pi/2$ *and* $\pi/2$, take the sine, and then take the inverse sine, you will get your original number back. In other words,

$$\sin^{-1}(\sin x) = x \qquad \text{for} \qquad -\pi/2 \le x \le \pi/2.$$

Similar rules apply to cos and $\cos^{-1}$, and to tan and $\tan^{-1}$ (Problems 45 and 47).

In calculus, it's sometimes necessary to find exact values of expressions such as $\sin(\tan^{-1}\frac{2}{3})$. The following example shows how this can be done.

Example 3   Find the exact value of $\sin(\tan^{-1}\frac{2}{3})$.

Figure 8

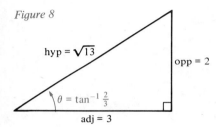

Solution   We begin by sketching an acute angle θ in a right triangle such that $\tan^{-1}\frac{2}{3} = \theta$; that is, $\frac{2}{3} = \tan\theta = \text{opp/adj}$. This is accomplished simply by letting opp $= 2$ units and adj $= 3$ units (Figure 8). By the Pythagorean theorem,

$$\text{hyp} = \sqrt{\text{opp}^2 + \text{adj}^2} = \sqrt{2^2 + 3^2} = \sqrt{13}.$$

Therefore, $\sin(\tan^{-1}\frac{2}{3}) = \sin\theta = \dfrac{\text{opp}}{\text{hyp}} = \dfrac{2}{\sqrt{13}} = \dfrac{2\sqrt{13}}{13}.$

Another method for finding exact values of expressions involving inverse trigonometric functions is to use identities such as the one developed in the following example.

Example 4   Show that $\cos(\tan^{-1} x) = \dfrac{1}{\sqrt{1 + x^2}}$.

Solution   Let $\tan^{-1} x = y$, so that $x = \tan y$ and $-\pi/2 < y < \pi/2$. Then

$$1 + x^2 = 1 + \tan^2 y = \sec^2 y.$$

Because $-\pi/2 < y < \pi/2$, it follows that $\sec y > 0$; hence,

$$\sqrt{1 + x^2} = \sec y.$$

Therefore, $\dfrac{1}{\sqrt{1 + x^2}} = \dfrac{1}{\sec y} = \cos y = \cos(\tan^{-1} x)$

The remaining inverse trigonometric functions $\cot^{-1}$, $\sec^{-1}$, and $\csc^{-1}$ (that is, arccot, arcsec, and arccsc, respectively) aren't used as often as $\sin^{-1}$, $\cos^{-1}$, and $\tan^{-1}$, so you won't find them on the keys of many scientific calculators. (See Problems 48 to 51.)

Problem Set 7.6

In Problems 1 to 12, evaluate each expression without using a calculator or tables.

1. $\sin^{-1} 1$

2. $\arcsin \dfrac{\sqrt{3}}{2}$

3. $\arcsin\left(-\dfrac{\sqrt{2}}{2}\right)$

4. $\cos^{-1}\left(-\dfrac{1}{2}\right)$

5. $\arccos 1$

6. $\cos^{-1}\dfrac{\sqrt{3}}{2}$

7. $\sin^{-1}\dfrac{\sqrt{2}}{2}$

8. $\cos^{-1} 0$

9. $\arccos \dfrac{1}{2}$

10. $\arctan 1$

11. $\tan^{-1}(-1)$

12. $\tan^{-1}\dfrac{\sqrt{3}}{3}$

[C] In Problems 13 to 24, use a calculator (or Appendix Table IIB) to evaluate each expression. Give answers in *radians*.

13. $\arcsin 0.6442$

14. $\arccos 0.6675$

15. $\cos^{-1} 0.9051$

16. $\tan^{-1} 0.2500$

17. $\arctan 2$

18. $\sin^{-1}(-0.5495)$

19. $\tan^{-1}(-3.224)$

20. $\cos^{-1}(-\tfrac{1}{8})$

21. $\arcsin(-0.5505)$

22. $\arccos\left(-\dfrac{5}{11}\right)$

23. $\sin^{-1}\dfrac{\sqrt{5}}{4}$

24. $\tan^{-1}\left(-\dfrac{\sqrt{7}}{3}\right)$

In Problems 25 to 36, find the exact value of each expression without using a calculator or tables.

25. $\sin\left(\sin^{-1}\dfrac{3}{4}\right)$

26. $\tan(\tan^{-1} 3)$

27. $\sin^{-1}\left(\sin\dfrac{\pi}{6}\right)$

28. $\tan^{-1}\left(\tan\dfrac{5\pi}{4}\right)$

29. $\sin\left(\tan^{-1}\dfrac{4}{3}\right)$

30. $\tan\left(\sin^{-1}\dfrac{\sqrt{5}}{5}\right)$

31. $\cos\left(\sin^{-1}\dfrac{\sqrt{10}}{10}\right)$

32. $\cos[\arctan(-2)]$

33. $\tan\left(\sin^{-1}\dfrac{4}{5}\right)$

34. $\sin\left(2 \arcsin \dfrac{2}{3}\right)$

35. $\sec(\cos^{-1}\tfrac{7}{10})$

36. $\cos(2 \sin^{-1}\tfrac{1}{8})$

In Problems 37 to 44, show that the given equation is an identity.

37. $\sin^{-1}(-x) = -\sin^{-1} x$

38. $\cos^{-1} x = \dfrac{\pi}{2} - \sin^{-1} x$

39. $\arctan(-x) = -\arctan x$

40. $\cos^{-1}(-x) = \pi - \cos^{-1} x$

41. $\tan(\sin^{-1} x) = \dfrac{x}{\sqrt{1 - x^2}}$

42. $\sin(\arctan x) = \dfrac{x}{\sqrt{1 + x^2}}$

43. $\sin(2 \arcsin x) = 2x\sqrt{1 - x^2}$

44. $\tan(\tfrac{1}{2} \arccos x) = \sqrt{\dfrac{1 - x}{1 + x}}$

45. For what values of x is it true that

  (a) $\cos(\cos^{-1} x) = x$?   (b) $\cos^{-1}(\cos x) = x$?

[C] **46.** To the nearest hundredth of a degree, find the two vertex angles α and β of a 3–4–5 right triangle (Figure 9).

Figure 9

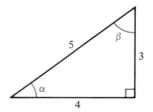

47. For what values of x is it true that

  (a) $\tan(\arctan x) = x$?   (b) $\arctan(\tan x) = x$?

48. The **inverse cotangent** function $\cot^{-1}$ or arccot, is defined by $\cot^{-1} x = y$ if and only if $x = \cot y$ and $0 < y < \pi$.

(a) Sketch the graph of $y = \cot^{-1} x$.

(b) Show that

$$\cot^{-1} x = \frac{\pi}{2} - \tan^{-1} x.$$

49. One popular definition of the **inverse secant** function, $\sec^{-1}$ or arcsec, is $\sec^{-1} x = y$ if and only if $x = \sec y$ and $0 \le y \le \pi$ with $y \ne \pi/2$.

(a) Sketch the graph of $y = \sec^{-1} x$.

(b) Show that

$$\sec^{-1} x = \cos^{-1}\left(\frac{1}{x}\right) \qquad \text{for } |x| \ge 1.$$

50. An alternative definition of the inverse secant function (see Problem 49) used in some calculus textbooks is $\sec^{-1} x = y$ if and only if $x = \sec y$ and either $0 \le y < \pi/2$ or else $\pi \le y < 3\pi/2$. Sketch the graph of $\sec^{-1}$ according to this definition.

51. One popular definition of the **inverse cosecant** function, $\csc^{-1}$ or arccsc, is $\csc^{-1} x = y$ if and only if $x = \csc y$ and $-\pi/2 \le y \le \pi/2$ with $y \ne 0$.

(a) Sketch the graph of $y = \csc^{-1} x$.

(b) Show that

$$\csc^{-1} x = \sin^{-1}\left(\frac{1}{x}\right) \qquad \text{for } |x| \ge 1.$$

7.7 TRIGONOMETRIC EQUATIONS

In Section 7.1 we mentioned that a trigonometric equation that is not an identity is called a **conditional** equation. To **solve** a conditional trigonometric equation means to find all values of the unknown for which the equation is true; these values are called the **solutions** of the equation. If a side condition such as $0° \le \theta < 360°$ is given along with a trigonometric equation, we understand that the solutions consist of all values of θ that satisfy *both* the equation and the side condition.

An equation, such as $\sin \theta = \sqrt{3}/2$, whose left side is a trigonometric function of the unknown and whose right side is a constant, is called a **simple** trigonometric equation. A side condition such as $0° \le \theta < 360°$ may or may not be involved. Even if there is no side condition, we usually *begin* by solving the equation for nonnegative values of the unknown that are less than $360°$ or 2π radians. The complete set of solutions is then obtained by adding all integer multiples of $360°$ or 2π radians to the values thus obtained. (Why?)

In Examples 1 to 4, solve each simple trigonometric equation. Use degree or radian measure as indicated.

Example 1 $\quad \sin \theta = \dfrac{\sqrt{3}}{2}, \qquad 0° \le \theta < 360°$

Figure 1

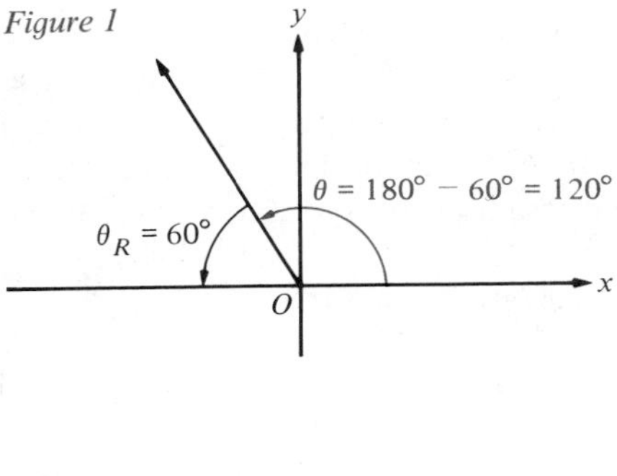

Solution Recall that $\sin 60° = \sqrt{3}/2$. Therefore, $\theta = 60°$ is the only solution in quadrant I. Because the values of $\sin \theta$ are negative in quadrants III and IV, the only possible remaining solution between $0°$ and $360°$ must lie in quadrant II, and its reference angle must be $\theta_R = 60°$ (Figure 1). Hence, $\theta = 180° - 60° = 120°$. Therefore, the solutions are

$$\theta = 60° \qquad \text{and} \qquad \theta = 120°.$$

Example 2 $\sec \alpha = -\sqrt{2}, \qquad 0° \le \alpha < 360°$

Solution   Recall that the values of sec α are negative only in quadrants II and III, and that $\sec 45° = \sqrt{2}$. Therefore, by constructing angles in quadrants II and III with $45°$ reference angles (Figure 2), we find the solutions

$$\alpha = 180° - 45° = 135° \qquad \text{and} \qquad \alpha = 180° + 45° = 225°. \qquad \blacksquare$$

Figure 2

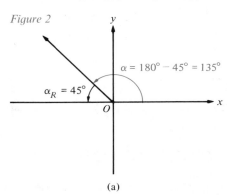

(a)

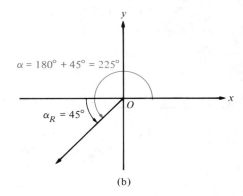

(b)

Example 3 $\sec t = -\sqrt{2}, \qquad t$ in radians

Solution   Here there is no side condition. Nevertheless, we *begin* by solving the equation with the side condition $0 \le t < 2\pi$. Proceeding as in Example 2, but measuring angles in radians, we find the two solutions

$$t = \pi - \frac{\pi}{4} = \frac{3\pi}{4} \qquad \text{and} \qquad t = \pi + \frac{\pi}{4} = \frac{5\pi}{4}.$$

Therefore, the complete set of solutions consists of all real numbers

$$t = \frac{3\pi}{4} + 2\pi k \qquad \text{and} \qquad t = \frac{5\pi}{4} + 2\pi k,$$

where k denotes an arbitrary integer. $\qquad \blacksquare$

Ⓒ **Example 4** $\cot x = -0.8333, \qquad 0 \le x < 2\pi$

Solution   We begin by rewriting the equation as

$$\tan x = 1/(-0.8333)$$

so that we can use the $\tan^{-1}$ (or arctan) key on a scientific calculator. If we simply calculate $\tan^{-1}[1/(-0.8333)]$, we get -0.876077723, which is *not* a solution because it doesn't satisfy the condition $0 \le x < 2\pi$. Recall that the values of tan x are negative in quadrants II and IV. With the calculator in radian mode, we enter 0.8333, take the reciprocal, and then take the inverse tangent to obtain the reference angle

$$\theta_R = \tan^{-1}(1/0.8333) = 0.876077723$$

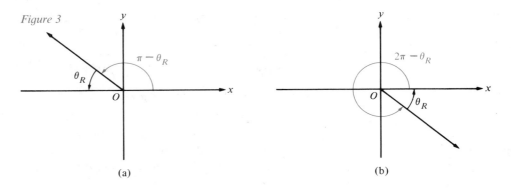

Figure 3

(a)

(b)

(Figure 3). Therefore, the solution in quadrant II is $x = \pi - \theta_R$ and the solution in quadrant IV is $x = 2\pi - \theta_R$; that is,

$$x = \pi - 0.876077723 = 2.265514931$$

and $$x = 2\pi - 0.876077723 = 5.407107585.$$

Note that both of these angles could have been obtained from $\tan^{-1}[1/(-0.8333)]$ by adding π and 2π, respectively. ■

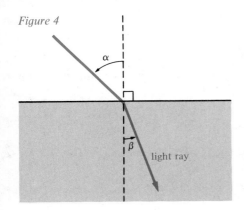

Figure 4

light ray

Snell's law of refraction, which was discovered around 1620 by the Dutch physicist Willebrord Snell (1591–1626), says that a light ray is bent (refracted) as it passes from a first medium into a second medium according to the equation

$$\frac{\sin \alpha}{\sin \beta} = \mu,$$

where α is the **angle of incidence** (Figure 4), β is the **angle of refraction** (Figure 4), and μ (the Greek letter mu) is a constant called the **index of refraction** of the second medium with respect to the first.

© **Example 5**   The index of refraction of flint glass with respect to air is $\mu = 1.650$. Determine the angle of refraction β of a ray of light that strikes a block of flint glass with an angle of incidence $\alpha = 35°$.

Solution   From Snell's law $\dfrac{\sin \alpha}{\sin \beta} = \mu$, we obtain the simple trigonometric equation

$$\sin \beta = \frac{\sin \alpha}{\mu} = \frac{\sin 35°}{1.650}.$$

From the geometry of the problem (Figure 4), we have the side condition $0 < \beta < 90°$. Using a calculator in degree mode, we find that

$$\beta = \sin^{-1}\left(\frac{\sin 35°}{1.650}\right) = 20.34°$$

to the nearest hundredth of a degree.

∎

Many of the techniques used for solving trigonometric equations are similar to those used for solving algebraic equations. The basic idea is to reduce the equation to one or more simple trigonometric equations, and then use the method illustrated above.

In Examples 6 to 10, solve each trigonometric equation subject to the side condition $0° \leq \theta < 360°$ *or* $0 \leq t < 2\pi$.

Example 6 $\cos 3t = \sqrt{2}/2$

Solution   Let $x = 3t$, and rewrite the equation as the simple trigonometric equation $\cos x = \sqrt{2}/2$. The side condition $0 \leq t < 2\pi$ is equivalent to $0 \leq 3t < 6\pi$ or $0 \leq x < 6\pi$. The solutions of $\cos x = \sqrt{2}/2$ for $0 \leq x < 2\pi$, obtained using the method described previously, are $x = \pi/4$ and $x = 7\pi/4$. Adding 2π to these, we obtain two more solutions,

$$x = \frac{\pi}{4} + 2\pi = \frac{9\pi}{4} \qquad \text{and} \qquad x = \frac{7\pi}{4} + 2\pi = \frac{15\pi}{4},$$

both of which satisfy the condition $2\pi \leq x < 4\pi$. Finally, adding 2π again, we obtain two more solutions,

$$x = \frac{9\pi}{4} + 2\pi = \frac{17\pi}{4} \qquad \text{and} \qquad x = \frac{15\pi}{4} + 2\pi = \frac{23\pi}{4},$$

both of which satisfy the condition $4\pi \leq x < 6\pi$. Thus, the solutions for $0 \leq x < 6\pi$ are

$$3t = x = \frac{\pi}{4}, \frac{7\pi}{4}, \frac{9\pi}{4}, \frac{15\pi}{4}, \frac{17\pi}{4}, \text{ and } \frac{23\pi}{4},$$

and so the values of t are

$$t = \frac{x}{3} = \frac{\pi}{12}, \frac{7\pi}{12}, \frac{3\pi}{4}, \frac{5\pi}{4}, \frac{17\pi}{12}, \text{ and } \frac{23\pi}{12}.$$

∎

Example 7 $\cos \theta \cot \theta = \cos \theta$

Solution   We rewrite the equation as

$$\cos \theta \cot \theta - \cos \theta = 0 \qquad \text{or} \qquad \cos \theta (\cot \theta - 1) = 0,$$

so that, with the side condition $0° \leq \theta < 360°$, we have:

$$\cos \theta = 0 \qquad\qquad\qquad \cot \theta - 1 = 0$$

$$\theta = 90°, 270° \qquad\qquad\qquad \cot \theta = 1$$

$$\theta = 45°, 225°$$

Therefore, the solutions are

$$\theta = 45°, 90°, 225°, \text{ and } 270°.$$

(*Note:* We were careful *not* to divide both sides of the original equation by the variable factor $\cos \theta$. Had we done so, we would have lost the solutions $\theta = 90°$ and $\theta = 270°$.) ∎

Example 8 $2 \cos^2 t - \sin t - 1 = 0$.

Solution   Using the Pythagorean identity $\cos^2 t + \sin^2 t = 1$, we rewrite the equation as

$$2(1 - \sin^2 t) - \sin t - 1 = 0 \qquad \text{or} \qquad -2 \sin^2 t - \sin t + 1 = 0,$$

that is,

$$2 \sin^2 t + \sin t - 1 = 0 \qquad \text{or} \qquad (2 \sin t - 1)(\sin t + 1) = 0.$$

Setting each factor equal to zero and keeping in mind the side condition $0 \leq t < 2\pi$, we have:

$$2 \sin t - 1 = 0 \qquad\qquad\qquad \sin t + 1 = 0$$

$$\sin t = \tfrac{1}{2} \qquad\qquad\qquad \sin t = -1$$

$$t = \frac{\pi}{6}, \frac{5\pi}{6} \qquad\qquad\qquad t = \frac{3\pi}{2}$$

Therefore, the solutions are

$$t = \frac{\pi}{6}, \frac{5\pi}{6}, \text{ and } \frac{3\pi}{2}.$$

∎

Example 9 $\tan \theta - \sec \theta = 1$

Solution   We rewrite the equation as $\sec \theta = \tan \theta - 1$, and square both sides. (*Note:* When proving a trigonometric *identity*, we are not permitted to square both sides. However, in solving a *conditional* equation, we may square both sides *provided that we later check for extraneous roots*.) We have

$$\sec^2 \theta = \tan^2 \theta - 2 \tan \theta + 1.$$

Using the Pythagorean identity $1 + \tan^2 \theta = \sec^2 \theta$, we can rewrite the last equation as

$$1 + \tan^2 \theta = \tan^2 \theta - 2 \tan \theta + 1 \qquad \text{or} \qquad 0 = -2 \tan \theta,$$

that is,

$$\tan \theta = 0.$$

Therefore, with the side condition $0° \leq \theta < 360°$, we have

$$\theta = 0° \quad \text{or} \quad \theta = 180°.$$

We must now check our solutions. Substituting $\theta = 0°$ in the original equation, we get

$$\tan 0° - \sec 0° = 1 \quad \text{or} \quad 0 - 1 = 1,$$

which is *false*. Therefore, $\theta = 0°$ isn't a solution. Substituting $\theta = 180°$ in the original equation, we get

$$\tan 180° - \sec 180° = 1 \quad \text{or} \quad 0 - (-1) = 1.$$

which is *true*. Thus, $\theta = 180°$ is the only solution.

$\boxed{\text{C}}$ **Example 10** $5 \tan^2 t + 2 \tan t - 7 = 0$ (Round off to four decimal places.)

Solution   Factoring the left side of the equation, we have

$$(5 \tan t + 7)(\tan t - 1) = 0.$$

Setting the first factor equal to zero, we obtain the simple trigonometric equation

$$\tan t = -\tfrac{7}{5} = -1.4.$$

Using a calculator (as in Example 4), we find the solutions

$$t \approx 2.1910 \quad \text{and} \quad t \approx 5.3326.$$

Setting the second factor equal to zero, we obtain the simple trigonometric equation

$$\tan t = 1,$$

which has the solutions

$$t = \frac{\pi}{4} \approx 0.7854 \quad \text{and} \quad t = \frac{5\pi}{4} \approx 3.9270.$$

Hence, rounded off to four decimal places, the solutions are

$$t = 0.7854, 2.1910, 3.9270, \text{ and } 5.3326.$$

Ecologists often use the following equations as a **simple predator–prey model:**

$$x = a_1 \cos \omega t + k_1$$

$$y = a_2 \cos(\omega t - \phi) + k_2,$$

where x is the number of prey at time t and y is the number of predators at time t.

Here ω, a_1, a_2, k_1, k_2, and ϕ are constants determined by the particular predator–prey relationship. Notice that when $t = 0$, the prey population has its maximum value $x = a_1 + k_1$.

That x and y oscillate in a simple harmonic fashion is explained as follows: As the predators begin consuming the prey, the predator population increases and the prey population decreases. After a while, there aren't enough prey left to support the increased number of predators, and the predators begin to die off. Fewer predators means that more prey survive, and the prey population begins to increase. These oscillations continue indefinitely, unless something disturbs the ecological balance.

Example 11   In a certain habitat, the simple predator–prey model for owls (the predator) and field mice (the prey) is determined by the values $\omega = 2$, $a_1 = 500$, $a_2 = 10$, $k_1 = 2000$, $k_2 = 50$, and $\phi = \pi/4$, where t is the time in years measured from an instant when the prey population had its maximum value $x = 2500$.

(a) Find the values of $t > 0$ for which the owl population is $y = 55$.

© **(b)** Find the field mouse population x when $y = 55$.

Solution   **(a)** Substitute $y = 55$, $a_2 = 10$, $\omega = 2$, $\phi = \pi/4$, and $k_2 = 50$ in the equation $y = a_2 \cos(\omega t - \phi) + k_2$ to obtain

$$55 = 10 \cos\left(2t - \frac{\pi}{4}\right) + 50;$$

that is,

$$\cos\left(2t - \frac{\pi}{4}\right) = \frac{1}{2}.$$

It follows that

$$2t - \frac{\pi}{4} = \frac{\pi}{3} + 2\pi n \qquad \text{or} \qquad 2t - \frac{\pi}{4} = \frac{5\pi}{3} + 2\pi n,$$

where n is an integer. Thus,

$$t = \frac{7\pi}{24} + \pi n \qquad \text{or} \qquad t = \frac{23\pi}{24} + \pi n.$$

Since $t > 0$, then n must be a *nonnegative* integer.

(b) When $t = \dfrac{7\pi}{24} + \pi n$,

$$x = 500 \cos 2\left(\frac{7\pi}{24} + \pi n\right) + 2000$$

$$= 500 \cos\left(\frac{7\pi}{12} + 2\pi n\right) + 2000$$

$$= 500 \cos\frac{7\pi}{12} + 2000 \approx 1871.$$

When $t = \dfrac{23\pi}{24} + \pi n$,

$$x = 500 \cos 2\left(\dfrac{23\pi}{24} + \pi n\right) + 2000$$

$$= 500 \cos\left(\dfrac{23\pi}{12} + 2\pi n\right) + 2000$$

$$= 500 \cos \dfrac{23\pi}{12} + 2000 \approx 2483.$$

Thus, when there are 55 owls in the habitat, the model predicts that there are either 1871 or 2483 field mice. ∎

Problem Set 7.7

In Problems 1 to 16, solve each simple trigonometric equation. Use degree or radian measure as indicated. Do not use a calculator or tables.

1. $\cos \theta = 0$, $\quad$ $0° \leq \theta < 360°$

2. $\cos x = 1$, $\quad$ $0 \leq x < 2\pi$

3. $\cot \theta = 1$, $\quad$ $0° \leq \theta < 360°$

4. $\tan s = 1$, $\quad$ $0 \leq s < 2\pi$

5. $\sin t = \frac{1}{2}$, $\quad$ $0 \leq t < 2\pi$

6. $\sec \alpha = 2$, $\quad$ $0° \leq \alpha < 360°$

7. $\tan x = -\sqrt{3}$, $\quad$ $0 \leq x < 2\pi$

8. $\csc t = -2$, $\quad$ $0 \leq t < 2\pi$

9. $\sin t = -1$, $\quad$ $0 \leq t < 2\pi$

10. $\sec \theta = -2\sqrt{3}/3$, $\quad$ $0° \leq \theta < 720°$

11. $\tan \theta = 0$, $\quad$ θ in degrees

12. $\sin \theta = 0$, $\quad$ $0° \leq \theta < 720°$

13. $\sec t = \sqrt{2}$, $\quad$ t in radians

14. $\sec x = 2$ in radians

15. $\tan x = \sqrt{3}$, $\quad$ x in radians

16. $\csc \beta = 2\sqrt{3}/3$, $\quad$ β in degrees

[C] In Problems 17 to 22, use a calculator to solve each simple trigonometric equation. Use degree or radian measure as indicated and round off all answers to four decimal places.

17. $\sin \theta = \frac{3}{4}$, $\quad$ $0° \leq \theta < 360°$

18. $\csc x = 4.5201$, $\quad$ $x > 0$ in radians

19. $\cos t = \frac{2}{3}$, $\quad$ $0 \leq t < 2\pi$

20. $\tan \beta = \frac{5}{4}$, $\quad$ $0° \leq \beta < 720°$

21. $\cot x = 0.2884$, $\quad$ x in radians

22. $\sec \alpha = 1.5763$, $\quad$ α in degrees

In Problems 23 to 50, solve each trigonometric equation with the side condition $0° \leq \theta < 360°$ or $0 \leq t < 2\pi$. Do not use a calculator or tables.

23. $\sin 2\theta = \frac{1}{2}$

24. $\sec 2t = 2$

25. $\tan 3\theta = \sqrt{3}$

26. $\csc 3t = -2$

27. $\sin \frac{1}{2}t = \sqrt{3}/2$

28. $\sec \frac{1}{2}\theta = 2\sqrt{3}/3$

29. $\cos \frac{1}{3}t = 0$

30. $\tan \frac{1}{3}\theta = 1$

31. $2 \sin \theta + 1 = 0$

32. $\sin t \cos t = 0$

33. $\sin^2 t = \frac{1}{2} \sin t$

34. $\sin(\theta + 40°) = 1$

35. $3 \tan^2 \theta - 1 = 0$

36. $2 \cos\left(t - \dfrac{\pi}{6}\right) = -\sqrt{2}$

37. $\sec^2 t - 2 = 0$

38. $4 \sin t \cos t = 1$

39. $\tan t \sin t = \sqrt{3} \sin t$

40. $\cos \theta + \sqrt{3} \sin \theta = -1$

41. $\tan^2 \theta - \tan \theta = 0$

42. $\sin \theta + \cos \theta = 1$

43. $\cot^2 t + \csc^2 t = 3$

44. $\tan^4 t - 2 \sec^2 t + 3 = 0$

45. $(\sin \theta - 1)(2 \cos \theta + \sqrt{3}) = 0$

46. $\sec^2 t + \csc^2 t = \sec^2 t \csc^2 t$

47. $2 \cos^2 t + \cos t - 1 = 0$

48. $\sin^2 t - 2 \sin t - 3 = 0$

49. $2 \sin^2 \theta - 3 \sin \theta + 1 = 0$

50. $2 \cos^3 t + \cos^2 t - 2 \cos t - 1 = 0$

[C] In Problems 51 to 54, solve each trigonometric equation with the side condition $0° \le \theta < 360°$ or $0 \le t < 2\pi$. Use a calculator and round off all answers to four decimal places.

51. $7 \sin^2 t - 10 \sin t + 3 = 0$

52. $\cot^2 \theta - 2 \cot \theta - 3 = 0$

53. $3 \sin^2 \theta + \sqrt{3} \cos \theta = 3$

54. $8 \sec^2 t + 2 \tan t - 9 = 0$

55. An oscillating signal voltage is given by the equation $E = 75 \cos(\omega t - \phi)$ millivolts, with angular frequency $\omega = 120\pi$, phase angle $\phi = \pi/2$, and time t in seconds. This voltage is applied to an oscilloscope, and the triggering circuit of the oscilloscope starts the sweep when E reaches the value 35 millivolts. Find the smallest positive value of t for which triggering occurs.

56. An irrigation ditch has a cross section in the shape of an isosceles trapezoid that is wider at the top than at the bottom (Figure 5). The bottom and the equal sides of the trapezoid are each 2 meters long. If θ is the acute angle between the horizontal and the side of the ditch, (a) show that the cross-

sectional area of the ditch is $A = 4 \sin \theta(1 + \cos \theta)$, and (b) solve for θ if $A = 3\sqrt{3}$ square meters.

Figure 5

57. The index of refraction of water with respect to air is $\mu = 1.333$. Use Snell's law (page 415) to determine the angle of refraction β of a ray of light that strikes the surface of a tank of water at an angle of incidence $\alpha = 15°$.

58. In optics, it is shown that if the index of refraction of a first medium with respect to a second is μ, then the index of refraction of the second medium with respect to the first is $1/\mu$. The index of refraction of flint glass with respect to air is $\mu = 1.650$. Figure 6 shows a light ray initiating in a block of flint glass and emerging into the air. The *critical angle* α_C for flint glass is the value of α for which $\beta = 90°$. (For $\alpha \ge \alpha_C$ the ray is totally reflected back into the glass when it strikes the surface.) Find the critical angle for flint glass.

Figure 6

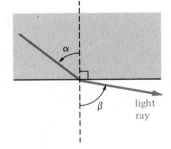

59. Suppose that in a certain tropical habitat, the simple predator–prey model (page 418) for boa constrictors (the predator) and wild pigs (the prey) is determined by the values $\omega = \frac{1}{2}$, $a_1 = 10$, $a_2 = 4$, $k_1 = 40$, $k_2 = 10$, and $\phi = \pi/3$, where t is the time in years measured from an instant when the prey population had its maximum value $x = 50$.

(a) Find the values of t for which the boa population is $y = 8$.

(b) Find the wild pig population x when $y = 8$.

REVIEW PROBLEM SET, CHAPTER 7

In Problems 1 to 10, simplify each expression.

1. $\dfrac{\sin(-\theta)}{\cos(-\theta)}$

2. $\dfrac{-\sin(-\alpha)}{-\cos(-\alpha)}$

3. $\csc x - \cos x \cot x$

4. $\sec \theta - \sin \theta \tan \theta$

5. $\csc^2 t \tan^2 t - 1$

6. $(\cot x + 1)^2 - \csc^2 x$

7. $\dfrac{\sec^2 u + 2 \tan u}{1 + \tan u}$

8. $\dfrac{\sec \beta}{\cot \beta + \tan \beta}$

9. $\dfrac{\sin^2 \theta + 2 \cos^2 \theta}{\sin \theta \cos \theta} - 2 \cot \theta$

10. $\dfrac{1}{\csc y - \cot y} - \dfrac{1}{\csc y + \cot y}$

In Problems 11 and 12, rewrite each radical expression as a trigonometric expression containing no radical by making the indicated trigonometric substitution.

11. $\sqrt{(25 - x^2)^3}$, $x = 5 \sin \theta$, $-\dfrac{\pi}{2} \le \theta \le \dfrac{\pi}{2}$

12. $\dfrac{x}{\sqrt{x^2 - 4}}$, $x = 2 \sec t$, $0 \le t < \dfrac{\pi}{2}$

In Problems 13 to 26, prove that each equation is an identity.

13. $\dfrac{\sin \theta}{\csc \theta} - 1 = \dfrac{-1}{\sec^2 \theta}$

14. $\dfrac{\csc \alpha}{\cot \alpha + \tan \alpha} = \cos \alpha$

15. $\dfrac{1 - \cot^2 t}{\tan^2 t - 1} = \cot^2 t$

16. $\dfrac{1}{\tan \beta} + \dfrac{\sin \beta}{1 + \cos \beta} = \csc \beta$

17. $\dfrac{1 + \tan x}{1 + \cot x} = \dfrac{\sec x}{\csc x}$

18. $\dfrac{\sin y + \tan y}{\cot y + \csc y} = \dfrac{\sin y}{\cot y}$

19. $\dfrac{\sin \beta + \cos \beta}{\sec \beta + \csc \beta} = \dfrac{\sin \beta}{\sec \beta}$

20. $\dfrac{1 - (\sin t - \cos t)^2}{\sin t} = 2 \cos t$

21. $\dfrac{1 - \tan \theta}{1 + \tan \theta} = \dfrac{\cot \theta - 1}{\cot \theta + 1}$

22. $(\sec u - \tan u)^2 = \dfrac{1 - \sin u}{1 + \sin u}$

23. $\dfrac{\cos s}{\sec s - \tan s} = 1 + \sin s$

24. $\sqrt{\dfrac{\sec x - 1}{\sec x + 1}} = \dfrac{|\sin x|}{1 + \cos x}$

25. $\dfrac{\sin t}{\cot t} - \dfrac{\cos t}{\sec t} = \dfrac{\tan t - \cos t \cot t}{\csc t}$

26. $(\csc \omega + \sec \omega)^2 = \dfrac{\sec^2 \omega + 2 \tan \omega}{\sin^2 \omega}$

In Problems 27 to 36, simplify each expression.

27. $\cos(360° - \theta)$

28. $\tan(2\pi - \beta)$

29. $\sin(270° + \alpha)$

30. $\cos(270° - \phi)$

31. $\sin(2\pi + t)$

32. $\cot\left(\dfrac{3\pi}{2} + x\right)$

33. $\sin 37° \cos 23° + \cos 37° \sin 23°$

34. $\dfrac{\tan(\pi/5) + \tan(\pi/20)}{1 - \tan(\pi/5)\tan(\pi/20)}$

35. $\sin x \cos y - \sin\left(x + \dfrac{\pi}{2}\right)\sin(-y)$

36. $\cos(\pi - t) - \tan t \cos\left(\dfrac{\pi}{2} - t\right)$

37. Use the fact that $\dfrac{7\pi}{12} = \dfrac{\pi}{4} + \dfrac{\pi}{3}$ to find the exact numerical value of

    (a) $\sin \dfrac{7\pi}{12}$ (b) $\cos \dfrac{7\pi}{12}$ (c) $\tan \dfrac{7\pi}{12}$

38. Use the fact that $75° = 120° - 45°$ to find the exact numerical value of

    (a) $\sin 75°$ (b) $\cos 75°$ (c) $\sec 75°$

In Problems 39 to 48, assume that α is in quadrant IV, $\cos \alpha = \frac{3}{5}$, β is in quadrant I, $\sin \beta = \frac{8}{17}$, γ is in quadrant II, $\cos \gamma = -\frac{24}{25}$, θ is in quadrant II, and $\sin \theta = \frac{5}{13}$. Find the exact numerical value of each expression.

39. $\sin(\alpha + \beta)$

40. $\cos(\gamma + \theta)$

41. $\sin(\beta + \theta)$

42. $\sin(\alpha - \gamma)$

43. $\cos(\beta - \gamma)$

44. $\sin(\beta - \gamma)$

45. $\tan(\beta - \gamma)$

46. $\sec(\beta - \gamma)$

47. $\sin(\theta - \gamma)$

48. $\cos(\beta - \theta)$

In Problems 49 to 56, prove that each equation is an identity.

49. $\sin(90° + \theta) = \sin(90° - \theta)$

50. $\sin\left(t + \dfrac{\pi}{6}\right) + \cos\left(t + \dfrac{\pi}{3}\right) = \cos t$

51. $\dfrac{\sin(s + t)}{\cos s \cos t} = \tan s + \tan t$

52. $\sin(\beta + 30°) + \cos(60° - \beta) = 2\sin(\beta + 30°)$

53. $\dfrac{\cos(x - y)}{\cos x \sin y} = \tan x + \cot y$

54. $\tan(\alpha + 135°) = \dfrac{\tan \alpha - 1}{\tan \alpha + 1}$

55. $\sin(\alpha - \beta) \cos \beta + \cos(\alpha - \beta) \sin \beta = \sin \alpha$

56. $\tan(\alpha + \beta) \tan(\alpha - \beta) = \dfrac{\tan^2 \alpha - \tan^2 \beta}{1 - \tan^2 \alpha \tan^2 \beta}$

57. If $\sec x = \frac{25}{7}$ and $0 < x < \pi/2$, find the exact numerical value of

 (a) $\sin 2x$ (b) $\cos 2x$

 (c) $\cos(x/2)$ (d) $\tan(x/2)$

58. If $\tan \theta = \frac{5}{12}$ and $180° < \theta < 270°$, find the exact numerical value of

 (a) $\sin(\theta/2)$ (b) $\cos(\theta/2)$

 (c) $\cos 2\theta$ (d) $\tan 2\theta$

In Problems 59 to 68, simplify each expression.

59. $\cos^2 2x - \sin^2 2x$

60. $1 - 2\sin^2 \dfrac{t}{2}$

61. $2 \sin \dfrac{t}{2} \cos \dfrac{t}{2}$

62. $\cos^4 2\theta - \sin^4 2\theta$

63. $2 \sin^2 \dfrac{\theta}{2} + \cos \theta$

64. $\dfrac{\sin 4\pi t}{4 \sin \pi t \cos \pi t}$

65. $\dfrac{\tan \omega t}{1 - \tan^2 \omega t}$

66. $\dfrac{2 \tan(t/2)}{1 - \tan^2(t/2)}$

67. $\dfrac{\cos^2(v/2) - \cos v}{\sin^2(v/2)}$

68. $2 \sin 2x \cos^3 2x + 2 \sin^3 2x \cos 2x$

In Problems 69 to 84, prove that each equation is an identity.

69. $\dfrac{\sin 2\theta + \sin \theta}{\cos 2\theta + \cos \theta + 1} = \tan \theta$

70. $\dfrac{\sec x - 1}{2 \sec x} = \sin^2 \dfrac{x}{2}$ **71.** $\tan t + \cot t = 2 \csc 2t$

72. $\csc w - \cot w = \tan \dfrac{w}{2}$

73. $\dfrac{\cos^2 \dfrac{x}{2} - \cos x}{\sin^2 \dfrac{x}{2}} = 1$

74. $\dfrac{\tan \theta - \sin \theta}{\sin \theta \sec \theta} = \tan \dfrac{\theta}{2} \sin \theta$

75. $8 \sin^2 \dfrac{\theta}{2} \cos^2 \dfrac{\theta}{2} = 1 - \cos 2\theta$

76. $\cos^4 \dfrac{x}{2} + \sin^4 \dfrac{x}{2} = 1 - \dfrac{1}{2} \sin^2 x$

77. $\dfrac{\tan x + \sin x}{2 \tan x} = \cos^2 \dfrac{x}{2}$

78. $\dfrac{\tan \alpha - \sin \alpha}{2 \tan \alpha} = \sin^2 \dfrac{\alpha}{2}$

79. $\sin 4t = 4 \sin t \cos t (1 - 2 \sin^2 t)$

80. $\cos 4w = 8 \cos^4 w - 8 \cos^2 w + 1$

81. $4 \sin \beta \cos^2 \dfrac{\beta}{2} = \sin 2\beta + 2 \sin \beta$

82. $\dfrac{1 - \cos x - \tan^2 \dfrac{x}{2}}{\sin^2 \dfrac{x}{2}} = \dfrac{2 \cos x}{1 + \cos x}$

83. $\dfrac{\sin 3x}{\sin x} + \dfrac{\cos 3x}{\cos x} = \sin 4x \sec x \csc x$

84. $\dfrac{\cos 3u}{\sin u} - \dfrac{\sin 3u}{\cos u} = \dfrac{\cos 4u}{\sin u \cos u}$

In Problems 85 to 88, express each product as a sum or difference.

85. $\sin \dfrac{3x}{2} \cos \dfrac{x}{2}$

86. $\cos 4\beta \sin 2\beta$

87. $\sin 37.5° \sin 7.5°$

88. $\sin 75° \cos 15°$

In Problems 89 to 92, rewrite each expression as a product.

89. $\sin 55° + \sin 5°$

90. $\cos \dfrac{5\pi}{3} + \cos \dfrac{\pi}{12}$

91. $\sin 4\beta - \sin \beta$

92. $\sin 11t + \sin 5t$

In Problems 93 to 98, prove that each equation is an identity.

93. $\dfrac{\cos 5x + \cos 3x}{\sin 5x - \sin 3x} = \cot x$

94. $\dfrac{2 \sin 2\theta \cos \theta - \sin \theta}{\cos \theta - 2 \sin 2\theta \sin \theta} = \tan 3\theta$

95. $4 \sin 3t \cos 3t \sin t = \cos 5t - \cos 7t$

96. $4 \sin 4\alpha \cos 2\alpha \sin \alpha =$
$\qquad \cos \alpha - \cos 3\alpha + \cos 5\alpha - \cos 7\alpha$

97. $\dfrac{\sin \theta + \sin 3\theta + \sin 5\theta}{\cos \theta + \cos 3\theta + \cos 5\theta} = \tan 3\theta$

98. $2 \sin\left(\dfrac{\alpha - \beta}{2} + \dfrac{\pi}{4}\right) \cos\left(\dfrac{\alpha + \beta}{2} - \dfrac{\pi}{4}\right) = \sin \alpha + \cos \beta$

© In Problems 99 to 102, use addition or subtraction of ordinates to sketch each graph. (A calculator will be helpful.)

99. $y = 2 \sin 3x + 3 \cos 2x$

100. $y = 0.12 \sin \pi t + 0.6 \sin 2\pi t$

101. $y = \sin t - \sin \dfrac{t}{3}$

102. $y = \sin x + \tan x$

© **103.** The variations in air pressure caused at a certain point by simultaneously playing the pure tone "concert A" and the pure tone one octave higher are given by

$$y = 0.12 \sin 880\pi t + 0.06 \sin 1760\pi t,$$

where t is the time in seconds. Sketch the graph showing these air-pressure variations.

© **104.** Believers in biorhythms sometimes plot a *composite* biorhythm curve by adding all three ordinates (physical, emotional, and intellectual) on a biorhythm chart. Using the data in Problem 39 on page 367, sketch a composite biorhythm curve over a 62-day period.

© **105.** Use the product formula to rewrite the quantity y given by $y = 4 \sin t \cos(4t/3)$ as a superposition of two simple harmonic oscillations. Then plot a graph of y as a function of t using the method of addition of ordinates.

106. The equation for a radio frequency carrier wave with frequency v_r, which is amplitude modulated by a pure audio tone with frequency v_a, is

$$y = a(1 + m \sin 2\pi v_a t) \sin 2\pi v_r t,$$

where a and m are constants. Rewrite this equation to show that y is a superposition of simple harmonic oscillations.

In Problems 107 to 110, use Theorem 3 on page 401 to rewrite the given equation in the form $y = a \cos(\omega t - \phi)$ and then sketch the graph.

107. $y = 2 \sin\left(\dfrac{\pi}{3} t - \pi\right) - 2 \cos\left(\dfrac{\pi}{3} t - \pi\right)$

© **108.** $y = 4 \cos\left(\dfrac{\pi}{2} t - \dfrac{\pi}{3}\right) - 3 \sin\left(\dfrac{\pi}{2} t - \dfrac{\pi}{3}\right)$

109. $y = \sqrt{3} \cos(2t - 3) - \sin(2t - 3)$

© **110.** $y = 2 \cos(3t - 1) - 3 \sin(3t - 1)$

In Problems 111 and 112, find the period T of the oscillation represented by each equation.

111. $y = 2 \cos\left(\dfrac{\pi}{3} t - \dfrac{\pi}{4}\right) + \sqrt{3} \cos\left(\dfrac{\pi}{5} t - \dfrac{\pi}{3}\right) + 2$

112. $y = \sqrt{2} \cos\left(\dfrac{\sqrt{3}}{2} t - \sqrt{5}\right) - \sin\left(\dfrac{\sqrt{3}}{3} t - \sqrt{3}\right) - 1$

113. Write the oscillation represented by

$$y = 2 \cos \pi t \cos 3\pi t$$

as a superposition of two simple harmonic oscillations and determine its period T.

114. Figure 1 illustrates the phenomenon of *beats* or *heterodyning* that occurs when two sounds of slightly different frequencies v_1 and v_2 are superposed. If the sounds are represented by $y_1 = a\cos(2\pi v_1 t)$ and $y_2 = a\cos(2\pi v_2 t)$, show that the superposition $y = y_1 + y_2$ can be regarded as a sound of frequency $(v_1 + v_2)/2$ amplitude modulated by an oscillation of frequency $|v_1 - v_2|/2$. Explain why one hears "beats" with a frequency of $|v_1 - v_2|$.

Figure 1

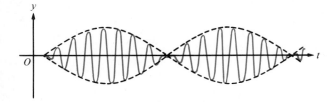

In Problems 115 to 118, evaluate each expression without the use of a calculator or tables.

115. $\sin^{-1}(-\frac{1}{2})$

116. $\arccos(-\sqrt{3}/2)$

117. $\arctan\sqrt{3}$

118. $\operatorname{arcsec}\sqrt{2}$

© In Problems 119 to 128, use a calculator (or Appendix Table IIB) to evaluate each expression. Give answers in radians.

119. $\sin^{-1} 0.3750$

120. $\arccos(-0.3901)$

121. $\cos^{-1} 0.9273$

122. $\tan^{-1} 57.29$

123. $\arctan 1.425$

124. $\cos^{-1}\frac{1}{8}$

125. $\arcsin(-\frac{3}{4})$

126. $\tan^{-1} 8$

127. $\cos^{-1}(-\frac{10}{11})$

128. $\arctan 1.007$

In Problems 129 to 134, find the exact value of each expression without using a calculator or tables.

129. $\cos(\tan^{-1}\frac{4}{3})$

130. $\sin[\arctan(-\frac{5}{12})]$

131. $\sin[\sin^{-1}(-\frac{12}{13})]$

132. $\tan[\arccos(-\frac{3}{5})]$

133. $\arcsin[\sin(19\pi/14)]$

134. $\sin(\sin^{-1}\frac{2}{3} + \sin^{-1}\frac{3}{4})$

In Problems 135 to 138, show that the given equation is an identity.

135. $\sin(\cos^{-1} x) = \sqrt{1 - x^2}$

136. $\sin(\frac{1}{2}\arccos x) = \sqrt{\dfrac{1 - x}{2}}$

137. $\tan(\arccos x) = \dfrac{\sqrt{1 - x^2}}{x}$

138. $\tan(\tan^{-1} x + \tan^{-1} y) = \dfrac{x + y}{1 - xy}$

139. A picture a meters high hangs on a wall so that its bottom is b meters above the eye level of an observer. If the observer stands x meters from the wall (Figure 2), show that the angle θ subtended by the picture is given by

$$\theta = \arctan\frac{a + b}{x} - \arctan\frac{b}{x}.$$

Figure 2

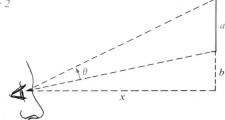

140. Two parallel lines, each at a distance a from the center of a circle of radius r, cut off the region of area A shown in Figure 3. Show that

$$A = 2a\sqrt{r^2 - a^2} + 2r^2 \arcsin\frac{a}{r}.$$

Figure 3

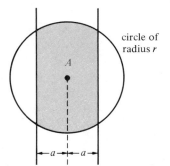

circle of radius r

In Problems 141 to 152, solve each trigonometric equation with the side condition $0° \le \theta < 360°$ or $0 \le t < 2\pi$. Do not use a calculator or tables.

141. $\tan\theta = \dfrac{\sqrt{3}}{3}$

142. $4\sin^2 t - 3 = 0$

143. $4 \cos^2 t - 3 = 0$

144. $\tan^2 \theta + (1 - \sqrt{3}) \tan \theta - \sqrt{3} = 0$

145. $2 \sin^2 \theta + \sqrt{2} \sin \theta = 0$

146. $\tan t = \sec t + 1$

147. $\cot^2 t + 3 \csc t + 3 = 0$

148. $\cot \theta = \csc \theta - 1$ **149.** $\tan 4\theta = \sqrt{3}$

150. $\sin 3t \cos t = \sin t \cos 3t + 1$

151. $\sin 3\theta + \sin \theta = 0$ **152.** $\cos \dfrac{t}{4} = \dfrac{\sqrt{3}}{2}$

© In Problems 153 to 156, solve each trigonometric equation with the side condition $0° \le \theta < 360°$ or $0 \le t < 2\pi$. Use a calculator and round off all answers to four decimal places.

153. $\cos 2\theta = \dfrac{3}{5}$ **154.** $\cot \dfrac{\theta}{2} = -2.017$

155. $6 \sin^2 t + \sin t - 2 = 0$

156. $2 \sec^2 t + \tan t - 3 = 0$

157. The index of refraction of a diamond with respect to air is $\mu = 2.417$. Use Snell's law (page 415) to determine the angle of refraction β of a ray of light

that strikes the surface of a diamond at an angle of incidence $\alpha = 20°$.

158. A ray of light passing through a plate of material with parallel faces is displaced, but not deviated (Figure 4); that is, the emerging ray is parallel to the ingoing ray. If d is the amount of displacement, t the thickness of the plate, α the angle of incidence, and β the angle of refraction, show that

$$\beta = \arctan\left(\tan \alpha - \frac{d}{t} \sec \alpha\right).$$

Figure 4

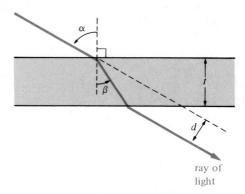

ray of light

Chapter 8

Applications of Trigonometry

In this chapter, we study some of the applications of the trigonometric functions to geometry. We begin by "solving triangles," that is, finding certain parts of triangles when other parts are known. To do this, we derive and apply the *law of sines* and the *law of cosines*. The chapter also includes an introduction to vectors and a discussion of polar coordinates.

8.1 RIGHT TRIANGLES

Figure 1

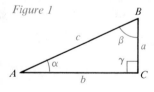

In the right triangle ACB shown in Figure 1, the angles are denoted by α at vertex A, β at vertex B, and γ at vertex C. The lengths of the sides opposite angles α, β, and γ are denoted by a, b, and c. Note that angles α and β are complementary acute angles, that angle γ is a right angle, and that c is the hypotenuse of right triangle ACB. Therefore,

$$\sin \alpha = \cos \beta = \frac{a}{c} \qquad \csc \alpha = \sec \beta = \frac{c}{a}$$

$$\cos \alpha = \sin \beta = \frac{b}{c} \qquad \sec \alpha = \csc \beta = \frac{c}{b}$$

$$\tan \alpha = \cot \beta = \frac{a}{b} \qquad \cot \alpha = \tan \beta = \frac{b}{a}.$$

If the lengths of two sides of a right triangle are given, or if one side and an acute angle are given, then these formulas can be used to solve for the remaining angles and sides of the triangle. This procedure is called **solving the right triangle.**

Example 1   Solve a right triangle ACB labeled as in Figure 1 if $\beta = 30°$ and $a = 24$. Sketch the resulting triangle.

Solution   We must find b, c, and α. Because α and β are complementary,

$$\alpha = 90° - \beta = 90° - 30° = 60°.$$

To find b, we use the fact that

$$\tan \beta = \frac{b}{a}, \qquad \text{so } b = a \tan \beta.$$

Therefore, $b = a \tan \beta = 24 \tan 30° = 24\left(\dfrac{\sqrt{3}}{3}\right) = 8\sqrt{3}.$

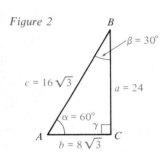

Figure 2

To find c, we use the fact that

$$\sec \beta = \frac{c}{a}, \qquad \text{so } c = a \sec \beta.$$

Therefore, $c = a \sec \beta = 24 \sec 30° = 24\left(\dfrac{2\sqrt{3}}{3}\right) = 16\sqrt{3}$

(Figure 2). (Of course, we could also have used the Pythagorean theorem to find $c = \sqrt{a^2 + b^2}$). ∎

Unless the special angles $30°$, $45°$, or $60°$ are involved, it is necessary to use a calculator or a table of trigonometric functions to solve a right triangle. You should always keep in mind that solutions obtained using a calculator or tables are *approximations*. In this section, whenever such approximations are involved, we round off all angles to the nearest hundredth of a degree, and all side lengths to four significant digits.

⒞ *In Examples 2 and 3, assume that right triangle ACB is labeled as in Figure 1. In each case, solve the triangle and sketch it.*

Example 2 $b = 31$, $\alpha = 43.33°$

Solution   We must find a, c, and β. Because α and β are complementary,

$$\beta = 90° - \alpha = 90° - 43.33° = 46.67°.$$

To find a, we notice that

$$\tan \alpha = \frac{a}{b}, \qquad \text{so } a = b \tan \alpha.$$

Using a calculator, we find that

$$\tan \alpha = \tan 43.33° = 0.943341386.$$

Therefore, rounded off to four significant digits,

$$a = b \tan \alpha = 31 \tan 43.33° = 29.24.$$

(Note that it isn't necessary to write down the intermediate result $\tan 43.33° = 0.943341386$. We have done it here so you can check your calculator work.) To find c, we notice that

$$\cos \alpha = \frac{b}{c}, \qquad \text{so } c = \frac{b}{\cos \alpha}.$$

Therefore, rounded off to four significant digits,

$$c = \frac{b}{\cos \alpha} = \frac{31}{\cos 43.33°} = \frac{31}{0.727413564} = 42.62.$$

Figure 3

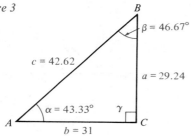

(Again, if you are using your calculator efficiently, it won't be necessary to write down the fraction $31/0.727413564$. You should be able to calculate $31/\cos 43.33°$ directly with only a few key strokes.) Notice that we could have used the Pythagorean theorem $c^2 = a^2 + b^2$ to solve for c:

$$c = \sqrt{a^2 + b^2} = \sqrt{(29.24)^2 + 31^2}$$
$$= 42.61.$$

(The discrepancy in the last decimal place was caused by using the rounded off value for a. The result $c = 42.62$ is actually more accurate.) The triangle is shown in Figure 3. ∎

Example 3 $a = 4, \qquad b = 3$

Solution   By the Pythagorean theorem,

$$c = \sqrt{a^2 + b^2} = \sqrt{16 + 9} = 5.$$

Figure 4

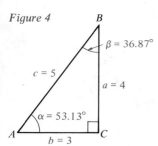

Now we have

$$\sin \alpha = \frac{a}{c} = \frac{4}{5} = 0.8,$$

so

$$\alpha = \sin^{-1} 0.8 = 53.13°.$$

It follows that

$$\beta = 90° - \alpha = 90° - 53.13° = 36.87°.$$

The triangle is shown in Figure 4. ∎

In many applications of trigonometry, we represent the relevant features of a real-world situation with triangles, and then solve the triangles for their unknown parts.

© **Example 4**   A railway track rises 300 feet per mile (5280 feet). Find the angle α at which the track is inclined from the horizontal (Figure 5).

Figure 5

Solution   From Figure 5,

$$\tan \alpha = \tfrac{300}{5280},$$

so

$$\alpha = \tan^{-1} \tfrac{300}{5280} = 3.25°.$$

The acute angle formed by a horizontal line and an observer's line of sight to any object above the horizontal is called an **angle of elevation** (Figure 6a). Similarly, the acute angle formed by a horizontal line and an observer's line of sight to any object below the horizontal is called an **angle of depression** (Figure 6b).

Figure 6

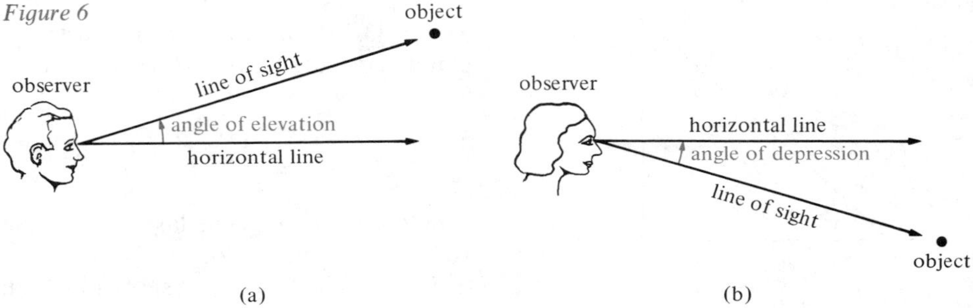

object

observer
line of sight
angle of elevation
horizontal line

observer
horizontal line
angle of depression
line of sight
object

(a)   (b)

ⓒ **Example 5**   From a point on level ground 75 meters from the base of a television transmitting tower, the angle of elevation of the top of the tower is 68.17°. Find the height h of the tower (Figure 7).

Solution   In Figure 7, $\tan 68.17° = \dfrac{h}{75}$,

so $h = 75 \tan 68.17° = 187.2$ meters.

Figure 7

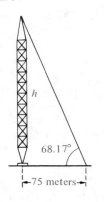

h

68.17°

|←75 meters→|

Figure 8

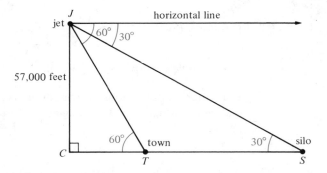

jet J horizontal line
60° 30°

57,000 feet

60° town   30°   silo
C T S

ⓒ **Example 6**   A high-altitude military-reconnaissance jet photographs a missile silo under construction near a small town. The jet is at an altitude of 57,000 feet, and the angles of depression of the town and silo are 60° and 30°, respectively. Assuming that the jet, the silo, and the town lie in the same vertical plane, find the distance between the town and the silo (Figure 8).

Solution

In Figure 8, we want to find $|\overline{TS}|$, which is $|\overline{CS}| - |\overline{CT}|$. From elementary geometry, we know that alternate interior angles are equal,* so angle $CTJ = 60°$ and angle $CSJ = 30°$. In right triangle TCJ,

$$\tan 60° = \frac{|\overline{CJ}|}{|\overline{CT}|} = \frac{57,000}{|\overline{CT}|},$$

so

$$|\overline{CT}| = \frac{57,000}{\tan 60°} = \frac{57,000}{\sqrt{3}} = \frac{57,000\sqrt{3}}{3} = 19,000\sqrt{3}.$$

In right triangle SCJ,

$$\tan 30° = \frac{|\overline{CJ}|}{|\overline{CS}|} = \frac{57,000}{|\overline{CS}|},$$

so

$$|\overline{CS}| = \frac{57,000}{\tan 30°} = \frac{57,000}{\sqrt{3}/3} = 57,000\sqrt{3}.$$

Hence,

$$|\overline{TS}| = |\overline{CS}| - |\overline{CT}| = 57,000\sqrt{3} - 19,000\sqrt{3} = 38,000\sqrt{3} \approx 65,820 \text{ feet.} \quad \blacksquare$$

In some applications of trigonometry, especially in surveying and navigation, the **direction** or **bearing** of a point Q as viewed from a point P is defined to be the positive angle the ray from P through Q makes with a north–south line through P. Such an angle is specified as being measured east or west from north or south. For instance, in Figure 9, the bearing from P to Q is 61° east of north, or N61°E. Similarly, the bearing from P to R is N28°W; the bearing from P to T is S15°W; and the bearing from P to V is S40°E.

Figure 9

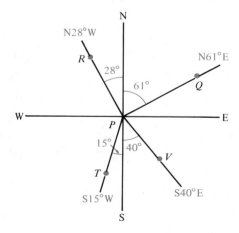

* See the review of geometry in Appendix III for a discussion of alternate interior angles.

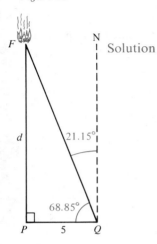

Figure 10

© **Example 7**   From an observation point P, a forest ranger sights a fire F directly to the north. Another ranger at point Q, 5 kilometers due east of P, sights the same fire at a bearing N21.15°W (Figure 10). Find the distance d between P and the fire.

Solution   In Figure 10, angle PQF is complementary to the 21.15° angle; therefore,

$$\text{angle } PQF = 90° - 21.15°$$
$$= 68.85°.$$

Also,
$$\tan 68.85° = \frac{d}{5},$$

so
$$d = 5 \tan 68.85° = 12.92.$$

Therefore, the fire is approximately 12.92 kilometers due north of observation point P.   ■

© **Problem Set 8.1**

In each problem set, problems with colored numbers constitute a good representation of the main ideas of the section.

In Problems 1 to 18, assume that the right triangle ACB is labeled as in Figure 11. In each case, solve the triangle and sketch it. When approximations are involved, round off angles to the nearest hundredth of a degree and side lengths to four significant digits.

Figure 11

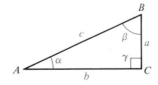

1. $a = 5$, $\alpha = 60°$

2. $b = 91$, $\beta = 30°$

3. $a = 10$, $c = 10\sqrt{2}$

4. $c = 100$, $\beta = 41°$

5. $b = 1700$, $\alpha = 31.23°$

6. $c = 10^4$, $\beta = 17.81°$

7. $a = 7.132$, $c = 9.209$

8. $b = 0.01523$, $\beta = 11.3°$

9. $a = 3.32$, $b = 4331$

10. $a = 113.5$, $c = 217.0$

11. $b = 2570$, $\beta = 15.45°$

12. $a = 50$, $b = 120$

13. $b = 5.673 \times 10^{-3}$ meter, $\alpha = 67°45'$

14. $a = 8.141 \times 10^5$ meters, $\alpha = 1.10°$

15. $b = 30.73$ miles, $c = 77.17$ miles

16. $a = 9.200 \times 10^7$ kilometers, $\alpha = 51°33'$

17. $a = 4.932 \times 10^{-6}$ meter,
    $b = 4.101 \times 10^{-6}$ meter

18. $c = 1.410 \times 10^{15}$ kilometers, $\alpha = 62°13'$

19. A 30-foot ladder leaning against a vertical wall just reaches a window sill. If the ladder makes an angle of 47° with the level ground, how high is the window sill? (Round off your answer to the nearest foot.)

20. A guy wire 8 meters long helps support a CB base antenna mounted on top of a flat roof. If the wire makes an angle of 49.5° with the horizontal roof, how far above the roof is it attached to the antenna? (Round off your answer to two decimal places.)

21. A kite string makes an angle of 28° with the level ground, and 73 meters of string is out. How high is the kite? (Round off your answer to the nearest meter.)

22. An engineer is designing an access ramp for an elevated expressway that is 40 feet above level ground. The ramp must be straight and cannot be inclined more than 15° from the horizontal. Because of existing structures near the expressway, the horizontal distance between the beginning of the ramp and the expressway cannot exceed 150 feet. Can the engineer design such a ramp?

23. A jetliner is climbing so that its path is a straight line that makes an angle of 8.5° with the horizontal. How many meters does the jetliner rise while traveling 300 meters along its path? (Round off your answer to the nearest meter.)

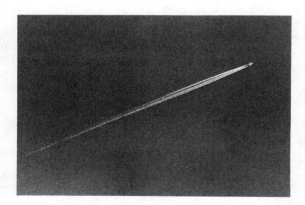

24. A portion of a tunnel under a river is straight for 150 meters and descends 10 meters in this distance.

   (a) To the nearest hundredth of a degree, what angle does this part of the tunnel make with the horizontal?

   (b) What is the *horizontal* distance, to the nearest 0.1 meter, between the ends of this part of the tunnel?

25. A monument 22 meters high casts a shadow 31 meters long. Find the angle of elevation of the sun to the nearest tenth of a degree.

26. To measure the height of a cloud cover at night, a spotlight is aimed straight upward from the ground. The resulting spot of light on the clouds is viewed from a point on the level ground 850 meters from the spotlight, and the angle of elevation is measured at 61.8°. Find the height of the cloud cover to the nearest meter.

27. A wrecking company has contracted to knock down an old brick chimney. In order to determine in advance just where the chimney will fall, it is necessary to find its height. The top of the chimney is viewed from a point 150 meters from its base, and the angle of elevation is measured at 33.25°. Find the height of the chimney, rounded off to the nearest 0.1 meter.

28. A rectangular panel to collect solar energy rests on flat ground and is tilted toward the sun. The edge resting on the ground is 3.217 meters long, and the upper edge is 1.574 meters above the ground. The panel is located near Chicago and its latitude is 41°50′. Solar engineers recommend that the angle between the panel and the ground be equal to the latitude of its location. Assuming that this recommendation has been followed, find the surface area of the panel. (Round off your answer to three decimal places.)

29. A lifeguard is seated on a high platform so that her eyes are 7 meters above sea level. Suddenly she spots the dorsal fin of a great white shark at a 4° angle of depression. Estimate, to the nearest meter, the horizontal distance between the platform and the shark.

30. A customs officer located on a straight shoreline observes a smuggler's motorboat making directly for the closest point on shore, an abandoned lighthouse. The angle between the shoreline and the officer's line of sight to the motorboat is 34.6°, the motorboat is traveling at 18 knots (nautical miles per hour), and it is 6 nautical miles from the lighthouse. The officer immediately departs by car on a straight road along the shoreline, hoping to reach the lighthouse 10 minutes before the smugglers in order to apprehend them as they land. How fast must the officer drive? (One nautical mile is 6076

feet, whereas one statute, or ordinary, mile is 5280 feet.)

31. Biologists studying the migration of birds are following a migrating flock in a light plane. The birds are flying at a constant altitude of 1200 feet and the plane is following at a constant altitude of 1700 feet. The biologists must maintain a distance of at least 600 feet between the plane and the flock in order to avoid disturbing the birds; therefore, they must monitor the angle of depression of the flock from the plane. Find the maximum allowable angle of depression, rounded off to the nearest degree.

32. An observer on a bluff 100 meters above the surface of a Scottish loch sees the head of an aquatic monster at an angle of depression of 18.45°. The monster, swimming directly away from the observer, immediately submerges. Five minutes later, the monster's head reappears, now at an angle of depression of 14.05°. How fast is the monster swimming?

33. A nature photographer using a telephoto lens photographs a rare bird roosting on a high branch of a tree at an angle of elevation of 22.5°. The distance between the lens and the bird is 330 feet. In order to obtain a more detailed photograph of the bird, the photographer cautiously moves closer to the base of the tree. The angle of elevation of the bird is now 51.25°. Find the new distance between the photographer's lens and the bird.

34. The angle of elevation of the top of a tower from a point 100 meters from the top is measured to be α degrees. Thus, the height h of the tower is calculated to be $h = 100 \sin \alpha$ meters. Suppose, however, that a 0.1° error has been made, and that the true angle of elevation is $\alpha + 0.1°$. Then the true height

of the tower is $100 \sin(\alpha + 0.1°)$, and the error E in the calculated value of h is given by $E = 100 \sin(\alpha + 0.1°) - 100 \sin \alpha$, the true value minus the calculated value. Using a calculator, find E if (a) $\alpha = 20°$, (b) $\alpha = 40°$, and (c) $\alpha = 60°$. (Round off your answer to two decimal places.)

35. A lifeguard at station A sights a swimmer directly south of him. Another lifeguard at station B, 230 feet directly east of A, sights the same swimmer at a bearing of S49.5°W. How far is the swimmer from station A?

36. Refer to Problem 34.

(a) Sketch a graph of E as a function of α for $0 < \alpha < 90°$.

(b) Does the error E increase or decrease as α increases?

(c) If you know that your measurement of α is subject to an error of as much as 0.1°, to how many decimal places should you round off your calculated value of h?

37. An airplane A is in the air at a position due west of an airport, and another airplane B is 15 miles south of A. From B, the bearing of the airport is N58.5°E. How far is airplane A from the airport?

38. To find the width $|\overline{PQ}|$ of a lake, a surveyor measures 3000 feet from P in the direction of N41°40′W to locate point R. The surveyor then determines that the bearing of Q from R is N71°30′E. Find the width of the lake if the point Q is located so that angle QPR is a right angle.

39. At 1:00 P.M. a ship is 24 nautical miles directly east of a lighthouse. The ship is sailing due north at 32 knots (nautical miles per hour). What is the bearing of the lighthouse from the ship at 3:00 P.M.?

8.2 THE LAW OF SINES

In this section and the next, we shall be studying relationships among the three angles α, β, and γ and the opposite sides a, b, and c of a *general* triangle ABC (Figure 1). The following theorem, called the *law of sines*, relates the lengths of the three sides to the sines of the three vertex angles.

Figure 1

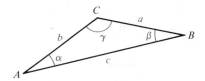

Theorem 1   **The Law of Sines**

In the general triangle *ABC* in Figure 1,

$$\frac{\sin \alpha}{a} = \frac{\sin \beta}{b} = \frac{\sin \gamma}{c}.$$

Proof   We show that $\dfrac{\sin \alpha}{a} = \dfrac{\sin \beta}{b}$ and leave the similar proof that $\dfrac{\sin \beta}{b} = \dfrac{\sin \gamma}{c}$ as an exercise (Problem 38). Drop a perpendicular $\overline{CD}$ from vertex *C* to the straight line *l* containing the vertices *A* and *B*. Figure 2 shows the case in which both angles α and β are acute, so that *D* lies *between* *A* and *B* on *l*. [If either α or β is not acute, *D* falls outside of the segment $\overline{AB}$ on *l*, and the following argument has to be modified slightly (Problem 37).] In right triangles *ADC* and *BDC*, we have

$$\sin \alpha = \frac{|\overline{CD}|}{b} \qquad \text{and} \qquad \sin \beta = \frac{|\overline{CD}|}{a},$$

so $$b \sin \alpha = |\overline{CD}| \qquad \text{and} \qquad a \sin \beta = |\overline{CD}|.$$

Consequently, $$b \sin \alpha = a \sin \beta.$$

Dividing both sides of the last equation by *ab*, we obtain

$$\frac{\sin \alpha}{a} = \frac{\sin \beta}{b}.$$

■

Figure 2

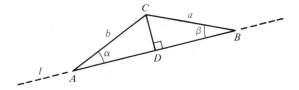

Notice that the law of sines (Theorem 1) can be written in the alternative form

$$\frac{a}{\sin \alpha} = \frac{b}{\sin \beta} = \frac{c}{\sin \gamma}.$$

If two angles of triangle ABC are given, then the third angle can be found by using the relationship

$$\alpha + \beta + \gamma = 180°;$$

hence, the three denominators $\sin \alpha$, $\sin \beta$, and $\sin \gamma$ can be found by using a calculator (or a table of sines). Now, if any one of the sides a, b, or c is also given, then the equations

$$\frac{a}{\sin \alpha} = \frac{b}{\sin \beta} = \frac{c}{\sin \gamma}$$

can be solved for the remaining two sides.

The following examples illustrate the procedure for solving a triangle when *two angles and one side* are given (or can be determined from information provided). Unless otherwise indicated, we shall round off all angles to the nearest hundredth of a degree, and all side lengths to four significant digits.

Figure 3

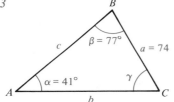

© **Example 1**   In triangle ABC (Figure 3), suppose that $\alpha = 41°$, $\beta = 77°$, and $a = 74$. Solve for γ, b, and c.

Solution $\gamma = 180° - \alpha - \beta = 180° - 41° - 77° = 62°$

By the law of sines,

$$\frac{a}{\sin 41°} = \frac{b}{\sin 77°} = \frac{c}{\sin 62°};$$

hence, since $a = 74$,

$$b = \frac{a \sin 77°}{\sin 41°} = \frac{74 \sin 77°}{\sin 41°} = 109.9$$

and

$$c = \frac{a \sin 62°}{\sin 41°} = \frac{74 \sin 62°}{\sin 41°} = 99.59.$$

■

© **Example 2**   A statue of height 70 meters ($\overline{BC}$ in Figure 4) stands atop a hill of height h. From a point A at ground level, the angle of elevation of the base B of the statue is $20.75°$ and the angle of elevation of the top C of the statue is $28.30°$. Find the height h.

Figure 4

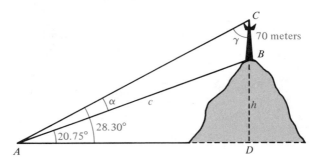

Solution   In right triangle ADB, we have

$$\sin 20.75° = h/c$$

so $$h = c \sin 20.75°.$$

Our plan is to find c and then use the last equation to find h. We know the length of one side $\overline{BC}$ in triangle ABC, so we can use the law of sines to find $c = |\overline{AB}|$, provided that we can find two of the vertex angles in this triangle. Angle α is easily found:

$$\alpha = 28.30° - 20.75° = 7.55°.$$

In right triangle ADC, angle γ is complementary to angle DAC; hence,

$$\gamma = 90° - \text{angle } DAC = 90° - 28.30° = 61.70°.$$

Now we can apply the law of sines to triangle ABC to obtain

$$\frac{c}{\sin \gamma} = \frac{70}{\sin \alpha}$$

so that $$c = \frac{70 \sin \gamma}{\sin \alpha} = \frac{70 \sin 61.70°}{\sin 7.55°},$$

and it follows that

$$h = c \sin 20.75° = \left(\frac{70 \sin 61.70°}{\sin 7.55°}\right) \sin 20.75° = 166.2 \text{ meters.} \quad \blacksquare$$

Example 3   Solve the triangle ABC if $\alpha = 82.17°$, $\gamma = 103.50°$, and $b = 615$.

Solution   We have

$$\beta = 180° - (\alpha + \gamma) = 180° - 185.67° = -5.67°,$$

which is impossible because we cannot have a triangle with a negative vertex angle. We conclude that there is no triangle satisfying the given conditions. $\quad \blacksquare$

More generally, if the specifications for a triangle require that the sum of two vertex angles exceeds $180°$, then no such triangle will exist.

The Ambiguous Case

Figure 5

Because there are several possibilities, the situation in which you are given the *lengths of two sides* of a triangle and the *angle opposite one of them* is called the **ambiguous case.** For instance, suppose you are given side a, side b, and angle α in triangle ABC. You might try to construct triangle ABC from this information by drawing a line segment $\overline{AC}$ of length b and a ray l that starts at A and makes an angle α with $\overline{AC}$ (Figure 5). To find the remaining vertex B, you could use a compass to draw an arc of a circle of radius a with center C. If the arc intersects the ray l at point B, then ABC is the desired triangle. (Why?)

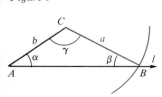

As Figure 6 illustrates, there are actually *three* possibilities if you try to construct triangle *ABC* by the method above: The circle does not intersect the ray *l* at all and there is **no triangle *ABC*** (Figure 6a); it intersects the ray *l* in exactly one point *B* and there is just **one triangle *ABC*** (Figures 5 and 6b); or it intersects the ray *l* in two points B_1 and B_2 and there are **two triangles AB_1C and AB_2C** that satisfy the given conditions (Figure 6c).

Figure 6

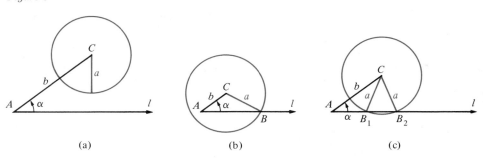

(a)   (b)   (c)

In the ambiguous case, you can use a calculator (or Appendix Table IIA) to solve the triangle *ABC* without bothering to draw any diagrams at all. Just use the law of sines,

$$\frac{\sin \alpha}{a} = \frac{\sin \beta}{b},$$

to evaluate $\sin \beta$:

$$\sin \beta = b\,\frac{\sin \alpha}{a}, \qquad 0 < \beta < 180°.$$

Recall that the sine of an angle is never greater than 1; hence, if $b\,\dfrac{\sin \alpha}{a} > 1$, this trigonometric equation has no solution, and *no triangle satisfies the given conditions.* If $b\,\dfrac{\sin \alpha}{a} = 1$, the equation has only one solution, $\beta = 90°$. If $b\,\dfrac{\sin \alpha}{a} < 1$, the trigonometric equation has two solutions (Problem 39), namely

$$\beta_1 = \sin^{-1}\left(b\,\frac{\sin \alpha}{a}\right) \qquad \text{and} \qquad \beta_2 = 180° - \beta_1.$$

Once you have determined β (or β_1 and β_2), you know two angles and two sides of the triangle (or triangles), and you can solve the triangle by using the methods previously explained. Even if there are two solutions β_1 and β_2 of the trigonometric equation for β, it is possible that only one of these solutions corresponds to an actual triangle satisfying the given conditions (see Figure 6b).

In Examples 4 to 7, solve each triangle by using the method for the ambiguous case.

© **Example 4** $a = 95,$ $b = 117,$ $\alpha = 65°$

Solution   Here,

$$b\,\frac{\sin \alpha}{a} = 117\,\frac{\sin 65°}{95} = 1.116 > 1,$$

so there is *no triangle* satisfying the given conditions.   ■

Example 5 $a = 10,$ $b = 20,$ $\alpha = 30°$

Solution   Because

$$\sin \beta = b\,\frac{\sin \alpha}{a} = 20\,\frac{\sin 30°}{10} = 20\,\frac{1/2}{10} = 1,$$

it follows that $\beta = 90°$, and the triangle is a right triangle with hypotenuse $b = 20$ (Figure 7). Therefore,

$$\gamma = 90° - \alpha = 90° - 30° = 60°.$$

Figure 7

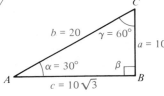

Also, by the Pythagorean theorem,

$$c^2 + a^2 = b^2,$$

so

$$c = \sqrt{b^2 - a^2}$$
$$= \sqrt{20^2 - 10^2}$$
$$= \sqrt{300} = 10\sqrt{3}.$$ ■

© **Example 6** $a = 10,$ $b = 11,$ $\alpha = 57°$

Solution   Here

$$b\,\frac{\sin \alpha}{a} = 11\,\frac{\sin 57°}{10} = 0.9225 < 1.$$

Thus, $$\beta_1 = \sin^{-1}\left(b\,\frac{\sin \alpha}{a}\right) = \sin^{-1}\left(11\,\frac{\sin 57°}{10}\right) = 67.30°$$

and $$\beta_2 = 180° - \beta_1 = 180° - 67.30° = 112.70°.$$

The solution $\beta_1 = 67.30°$ leads to a triangle AB_1C with

$$\gamma_1 = 180° - (\alpha + \beta_1)$$
$$= 180° - (57° + 67.30°) = 55.70°$$

and because $\dfrac{c_1}{\sin \gamma_1} = \dfrac{a}{\sin \alpha}$

$$c_1 = \frac{a \sin \gamma_1}{\sin \alpha} = \frac{10 \sin 55.70°}{\sin 57°} = 9.850$$

(Figure 8a). The solution $\beta_2 = 112.70°$ leads to a second triangle AB_2C with

$$\gamma_2 = 180° - (\alpha + \beta_2)$$
$$= 180° - (57° + 112.70°) = 10.30°$$

and because $\dfrac{c_2}{\sin \gamma_2} = \dfrac{a}{\sin \alpha}$

$$c_2 = \frac{a \sin \gamma_2}{\sin \alpha} = \frac{10 \sin 10.30°}{\sin 57°} = 2.132$$

(Figure 8b). ∎

Figure 8

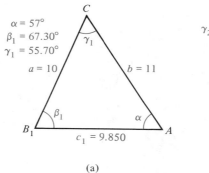

$\alpha = 57°$
$\beta_1 = 67.30°$
$\gamma_1 = 55.70°$
$a = 10$
$b = 11$
$c_1 = 9.850$

(a)

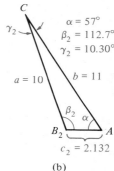

$\alpha = 57°$
$\beta_2 = 112.7°$
$\gamma_2 = 10.30°$
$a = 10$
$b = 11$
$c_2 = 2.132$

(b)

© **Example 7** $a = 70, \qquad b = 65, \qquad \alpha = 40°$

Solution   Here

$$b \frac{\sin \alpha}{a} = 65 \frac{\sin 40°}{70} = 0.5969 < 1.$$

Thus, $$\beta_1 = \sin^{-1}\left(b \frac{\sin \alpha}{a}\right) = \sin^{-1}\left(65 \frac{\sin 40°}{70}\right) = 36.65°$$

and $$\beta_2 = 180° - \beta_1 = 143.35°.$$

Because it is impossible to have an angle $\beta_2 = 143.35°$ in a triangle that already has an angle of $40°$ ($143.35° + 40° > 180°$), the solution $\beta_2 = 143.35°$ does not lead to an actual triangle. However, the solution $\beta = \beta_1 = 36.65°$ leads to

$$\gamma = 180° - (\alpha + \beta) = 180° - (40° + 36.65°) = 103.35°$$

and because $\dfrac{c}{\sin \gamma} = \dfrac{a}{\sin \alpha}$

$$c = \frac{a \sin \gamma}{\sin \alpha} = \frac{70 \sin 103.35°}{\sin 40°} = 106.0$$

Figure 9

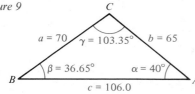

$a = 70$ $\gamma = 103.35°$ $b = 65$
$\beta = 36.65°$ $\alpha = 40°$
$c = 106.0$

(Figure 9). ∎

Problem Set 8.2

Ⓒ In Problems 1 to 8, use the law of sines to solve each triangle ABC. Round off angles to the nearest hundredth of a degree and side lengths to four significant digits.

1. $a = 32$, $\beta = 57°$, $\gamma = 38°$

2. $b = 16$, $\alpha = 120°$, $\gamma = 30°$

3. $b = 24$, $\beta = 38°$, $\gamma = 21°$

4. $a = 42.9$, $\alpha = 32°$, $\beta = 81.5°$

5. $b = 47.3$, $\alpha = 48.31°$, $\gamma = 57.82°$

6. $a = 120$, $\alpha = 132°$, $\beta = 61°$

7. $a = 95$, $\alpha = 132°$, $\beta = 24.45°$

8. $b = 67.5$, $\alpha = 61.25°$, $\beta = 65.75°$

Ⓒ In Problems 9 to 16, solve each triangle ABC by using the method for the ambiguous case. Round off angles to the nearest hundredth of a degree and side lengths to four significant digits. Be certain to find all possible triangles that satsify the given conditions.

9. $a = 50$, $b = 30$, $\alpha = 45°$

10. $a = 40$, $b = 70$, $\alpha = 30°$

11. $a = 31$, $b = 33$, $\alpha = 60°$

12. $a = 10$, $b = 26$, $\beta = 110°$

13. $a = 27$, $b = 52$, $\alpha = 70°$

14. $b = 4.5$, $c = 9$, $\gamma = 60°$

15. $a = 65.52$, $b = 55.51$, $\alpha = 111.5°$

16. $a = 12.41$, $b = 81.69$, $\beta = 36.67°$

Ⓒ In Problems 17 to 24, solve each triangle ABC. Round off angles to the nearest hundredth of a degree and side lengths to four significant digits. In the ambiguous cases, be certain to find all possible triangles that satisfy the given conditions.

17. $b = 52.05$ feet, $\alpha = 42.85°$, $\beta = 61.50°$

18. $b = 194.5$ meters, $\alpha = 82.25°$, $\beta = 69.45°$

19. $a = 110$ kilometers, $b = 100$ kilometers, $\alpha = 59.33°$

20. $a = 8.143 \times 10^6$ meters, $b = 1.271 \times 10^7$ meters, $\alpha = 34.65°$

21. $a = 2090$ meters, $b = 5579$ meters, $\alpha = 22°$

22. $a = 2.011 \times 10^{-10}$ meter, $b = 1.509 \times 10^{-10}$ meter, $\beta = 30.2°$

23. $a = 10.07$ kilometers, $b = 15.15$ kilometers, $\alpha = 67.67°$

24. $c = 0.4351$ meter, $\beta = 48.50°$, $\gamma = 96.25°$

Ⓒ 25. A surveyor lays out a baseline segment $\overline{AB}$ of length 346 meters. An inaccessible point C forms the third vertex of a triangle ABC. If the surveyor determines that angle $CAB = 62°40'$ and that angle $CBA = 54°30'$, find $|\overline{AC}|$ and $|\overline{BC}|$.

Ⓒ 26. The crankshaft $\overline{OA}$ of an engine is 2 inches long and the connecting rod $\overline{AP}$ is 8 inches long (Figure 10). If angle APO is $10°$ at a certain instant, find angle AOP at this instant. [*Caution:* There may be more than one solution.]

Figure 10

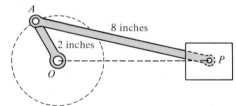

Ⓒ 27. An observation tower stands on the edge of a vertical cliff rising directly above a river (Figure 11). From the top of the tower, the angle of depression of a point on the opposite shore of the river is $24°$; from the base of the tower, the angle of depression of the same point is $18°$. If the top of the tower is 11 meters above its base, how wide is the river?

Figure 11

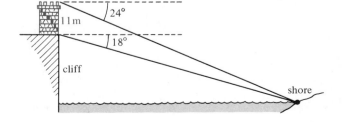

c 28. Plans for an oil pipeline require that it be diverted, as shown in Figure 12, around the mating grounds of a caribou herd. Suppose that angle $BAC = 22°$, angle $ABC = 48°$, and $|AB| = 7$ kilometers. If the pipeline costs $200,000 per kilometer, find the additional cost of the pipeline because of the diversion.

Figure 12

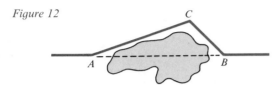

c 29. A nature photographer spots a rare northern shrike perched in a tree at an angle of elevation of 14°. Using the range finder on her camera, she determines that the shrike is 103 meters from her lens. She creeps toward the bird until its angle of elevation is 20°. Find the distance between the lens and the bird in the photographer's new position.

c 30. A commercial jet is flying at constant speed on a level course that will carry it directly over Albuquerque, New Mexico. Through a hole in the cloud cover, the captain observes the lights of the city at an angle of depression of 7°. Three minutes later, the captain catches a second glimpse of the city, now at an angle of depression of 13°. In how many *more* minutes will the jet be directly over the city?

c 31. A sports parachutist is sighted simultaneously by two observers who are 5 kilometers apart and on opposite sides of the parachutist. The observed angles of elevation are 26.5° and 18.2°. Assuming that the parachutist and the two observers lie in the same vertical plane, how high above the ground is the parachutist?

c 32. Riding at constant speed across the desert on his faithful camel Clyde, Lawrence sights the top of a tall palm tree. He assumes that the palm tree is located at an oasis, and heads directly toward it. When first sighted, the top of the palm tree was at an angle of elevation of 4°. Twenty minutes later, the angle of elevation has increased to 9°. In how many *more* minutes will Lawrence reach the palm tree?

c 33. A CB transmitter is illegally transmitting at excessive power from point C, and it is under surveillance by two FCC agents situated 3.8 kilometers apart at points A and B. Using direction finders,

the agents find that angle $CAB = 40.1°$ and angle $CBA = 31.8°$. How far is the illegal transmitter from the agent at point A?

c 34. Two Coast Guard stations A and B are located along a straight coastline. Station A is 20 miles due north of station B. A supertanker in distress is observed from station A at a bearing of S9°10′E, and from station B at a bearing of S37°40′E. Find the distance from the supertanker to the nearest point on the coastline.

c 35. Two joggers running at the same constant speed set out from the same point C. They run along separate straight paths toward a straight highway. The first jogger arrives at point A along the highway after running for 70 minutes. Some time later, the second jogger arrives at point B along the highway. If angle CAB is 105° and angle CBA is 50°, how many minutes were required by the second jogger to run from C to B?

36. In triangle ABC, use the law of sines and the fact that $\sin \gamma = \sin[180° - (\alpha + \beta)]$ to prove that $c = b \cos \alpha + a \cos \beta$.

37. In the proof of Theorem 1, suppose that angle α is not acute. Redraw Figure 2 to illustrate this situation and then show that $\dfrac{\sin \alpha}{a} = \dfrac{\sin \beta}{b}$ is still true. [*Hint:* $\sin(180° - \alpha) = \sin \alpha$.]

38. Complete the proof of Theorem 1 by showing that $\dfrac{\sin \beta}{b} = \dfrac{\sin \gamma}{c}$.

39. If $0 < \beta_1 < 90°$ and $\beta_2 = 180° - \beta_1$, show that $90° < \beta_2 < 180°$ and $\sin \beta_1 = \sin \beta_2$.

40. Use the law of sines to establish the **law of tangents:** For any triangle ABC,

$$\frac{a - b}{a + b} = \frac{\tan\left(\dfrac{\alpha - \beta}{2}\right)}{\tan\left(\dfrac{\alpha + \beta}{2}\right)}.$$

Begin by proving each of the following:

(a) $\dfrac{a - b}{b} = \dfrac{\sin \alpha - \sin \beta}{\sin \beta}$

(b) $\dfrac{a + b}{b} = \dfrac{\sin \alpha + \sin \beta}{\sin \beta}$

(c) $\dfrac{a-b}{a+b} = \dfrac{\sin\alpha - \sin\beta}{\sin\alpha + \sin\beta}$

ⓒ In Problems 41 to 43, use the law of tangents in Problem 40 to solve triangle ABC.

41. $a = 9,\ \beta = 73°,\ \alpha = 69°$

42. $b = 27,\ \beta = 33°,\ \gamma = 77°$

43. $a = 15,\ \gamma = 67°,\ \alpha = 81°$

44. The identity $(a-b)\cos\dfrac{\gamma}{2} = c\sin\left(\dfrac{\alpha-\beta}{2}\right)$ is called **Mollweide's formula.** It is often used to check the

solution of a triangle ABC because it contains all six parts of the triangle. If, when values are substituted in Mollweide's formula, the result is not a true statement, then an error has been made in solving the triangle. Prove Mollweide's formula.

ⓒ **45.** Use Mollweide's formula to check your solutions of the triangles in Problems 41 and 43.

46. (a) Show that the area $\mathscr{A}$ of triangle ABC is given by the formula $\mathscr{A} = \frac{1}{2}ab\sin\gamma$.

ⓒ (b) Use the formula in part (a) to find the area of triangle ABC if $a = 12.7$ meters, $b = 8.91$ meters, and $\gamma = 46.5°$.

8.3 THE LAW OF COSINES

The law of sines is not effective for solving a triangle when *two sides and the angle between them are given*, or when *all three sides are given*. However, the following *law of cosines* can be used in these cases.

Figure 1

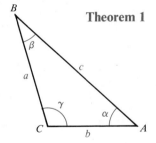

Theorem 1

The Law of Cosines

In triangle ABC (Figure 1):

(i) $c^2 = a^2 + b^2 - 2ab\cos\gamma$

(ii) $a^2 = b^2 + c^2 - 2bc\cos\alpha$

(iii) $b^2 = a^2 + c^2 - 2ac\cos\beta$

Figure 2

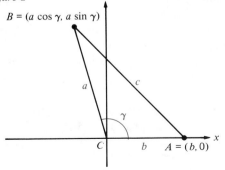

$B = (a\cos\gamma,\ a\sin\gamma)$

$A = (b, 0)$

Proof We prove (i) here; (ii) and (iii) are proved in exactly the same way. To begin with, we place triangle ABC on a Cartesian coordinate system with vertex C at the origin and vertex A on the positive x axis (Figure 2). Then vertex B has coordinates $B = (a\cos\gamma, a\sin\gamma)$. (Why?) Because $A = (b, 0)$, it follows from the distance formula that

$$c^2 = (a\cos\gamma - b)^2 + (a\sin\gamma - 0)^2$$
$$= a^2\cos^2\gamma - 2ab\cos\gamma + b^2 + a^2\sin^2\gamma$$
$$= a^2(\cos^2\gamma + \sin^2\gamma) + b^2 - 2ab\cos\gamma$$
$$= a^2 + b^2 - 2ab\cos\gamma.$$

∎

Rather than memorizing the three formulas in Theorem 1, we recommend that you learn the following statement, which summarizes all three formulas.

The Law of Cosines

> The square of the length of any side of a triangle is equal to the sum of the squares of the lengths of the other two sides minus twice the product of the lengths of these two sides and the cosine of the angle between them.*

Figure 3

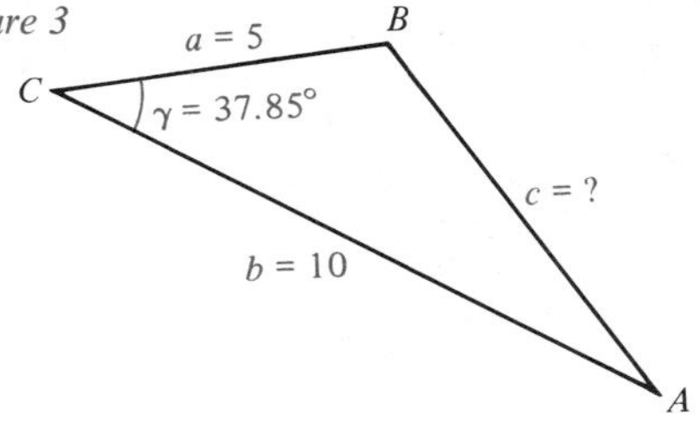

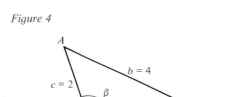

Ⓒ **Example 1**   In triangle ABC (Figure 3), find c to four significant digits if $a = 5$, $b = 10$, and $\gamma = 37.85°$.

Solution   By the law of cosines,

$$c^2 = a^2 + b^2 - 2ab \cos \gamma = 5^2 + 10^2 - 2(5)(10) \cos 37.85°$$
$$= 25 + 100 - 100 \cos 37.85° = 46.04.$$

Therefore, $c = \sqrt{46.04} = 6.875.$ ■

Figure 4

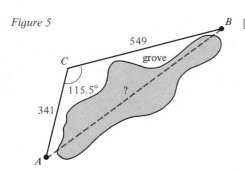

Ⓒ **Example 2**   In triangle ABC (Figure 4), find β to the nearest hundredth of a degree if $a = 3$, $b = 4$, and $c = 2$.

Solution   We begin by writing the law of cosines in the form that contains $\cos \beta$:

$$b^2 = a^2 + c^2 - 2ac \cos \beta.$$

Solving this equation for $\cos \beta$, we obtain

$$\cos \beta = \frac{a^2 + c^2 - b^2}{2ac}.$$

Therefore, $\beta = \cos^{-1} \dfrac{a^2 + c^2 - b^2}{2ac} = \cos^{-1} \dfrac{3^2 + 2^2 - 4^2}{2(3)(2)}$

$$= \cos^{-1}\left(\frac{-3}{12}\right) = \cos^{-1}\left(-\frac{1}{4}\right) = 104.48°.$$ ■

Figure 5

Ⓒ **Example 3**   In order to find the horizontal distance between two towers A and B that are separated by a grove of trees, a point C that is readily accessible from both A and B is chosen (Figure 5), and the following measurements are made: $|\overline{CA}| = 341$ meters, $|\overline{CB}| = 549$ meters, angle $ACB = 115.5°$. Find $|\overline{AB}|$ to four significant digits.

Solution   By the law of cosines,

$$|\overline{AB}|^2 = |\overline{CA}|^2 + |\overline{CB}|^2 - 2|\overline{CA}||\overline{CB}| \cos(\text{angle } ACB)$$
$$= 341^2 + 549^2 - 2(341)(549) \cos 115.5°;$$

---

* It's interesting to notice that, when the angle is 90°, the law of cosines reduces to the Pythagorean theorem (Problem 35).

hence,

$$|\overline{AB}| = \sqrt{341^2 + 549^2 - 2(341)(549)\cos 115.5°} = 760.8 \text{ meters.}$$ ■

The law of cosines can be used to establish a number of interesting formulas involving the geometry of triangles (see Problems 36 to 42). One of the most useful of these is **Hero's** (or Heron's) **area formula:**

$$\mathscr{A} = \sqrt{s(s-a)(s-b)(s-c)} \qquad \text{where } s = \tfrac{1}{2}(a+b+c),$$

which gives the area $\mathscr{A}$ of a triangle with sides of lengths a, b, and c (Problem 40). The quantity s, which is one-half of the perimeter $a + b + c$ of the triangle, is called the *semiperimeter*.

Figure 6

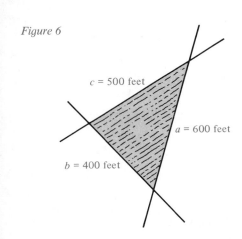

c = 500 feet

a = 600 feet

b = 400 feet

© **Example 4**   A cornfield that is bounded by three straight roads has side lengths $a = 600$ feet, $b = 400$ feet, and $c = 500$ feet (Figure 6). Find the area $\mathscr{A}$ of the cornfield.

Solution   We use Hero's formula. The semiperimeter of the cornfield is

$$s = \tfrac{1}{2}(a + b + c)$$
$$= \tfrac{1}{2}(600 + 400 + 500)$$
$$= 750.$$

Therefore,

$$\mathscr{A} = \sqrt{s(s-a)(s-b)(s-c)}$$
$$= \sqrt{750(750 - 600)(750 - 400)(750 - 500)}$$
$$= \sqrt{750(150)(350)(250)} = \sqrt{9{,}843{,}750{,}000}$$
$$\approx 99{,}216 \text{ square feet.}$$ ■

Problem Set 8.3

© In Problems 1 to 12, use the law of cosines to find the specified unknown part of the triangle ABC (Figure 1). Round off angles to the nearest hundredth of a degree and side lengths to four significant digits.

1. Find c if $a = 3$, $b = 10$, and $\gamma = 60°$.

2. Find a if $b = 68$, $c = 14$, and $\alpha = 24.5°$.

3. Find b if $a = 23$, $c = 31$, and $\beta = 112°$.

4. Find a if $b = 3.2$, $c = 2.4$, and $\alpha = 117°$.

5. Find a if $c = 5.78$, $b = 4.78$, and $\alpha = 35.25°$.

6. Find β if $a = 200$, $b = 50$, and $c = 177$.

7. Find γ if $a = 2$, $b = 3$, and $c = 4$.

8. Find α if $a = 2\sqrt{61}$, $b = 8$, and $c = 10$.

9. Find β if $a = 1240$, $b = 876$, and $c = 918$.

10. Find γ if $a = 0.64$, $b = 0.27$, and $c = 0.49$.

11. Find α if $a = 189$, $b = 214$, and $c = 325$.

12. Find β if $a = 189$, $b = 214$, and $c = 325$.

C In Problems 13 to 18, use the law of cosines or the law of sines (or both) to solve each triangle ABC. Round off angles to the nearest hundredth of a degree and side lengths to four significant digits. In the ambiguous cases, be certain to find all possible triangles that satisfy the given conditions.

13. $a = 327$, $b = 251$, $\gamma = 72.45°$

14. $a = 100$, $b = 150$, $c = 300$

15. $a = 312$, $b = 490$, $\alpha = 33.75°$

16. $a = 87$, $c = 124$, $\alpha = 35.33°$

17. $a = 321$, $b = 456$, $c = 654$

18. $a = 59$, $b = 22$, $\beta = 117°20'$

C **19.** To find the distance between two points A and B, a surveyor chooses a point C that is 52 meters from A and 64 meters from B. If angle ACB is $72°20'$, find the distance between A and B. Round off your answer to four significant digits.

C **20.** A tracking antenna A on the surface of the earth is following a spacecraft B (Figure 7). The tracking antenna is aimed 36° above the horizon, and the distance from the antenna to the spacecraft is 6521 miles. If the radius of the earth is 3959 miles, how high is the spacecraft above the surface of the earth?

Figure 7

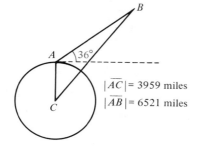

$|\overline{AC}| = 3959$ miles
$|\overline{AB}| = 6521$ miles

C **21.** Engineers are planning to construct a straight tunnel from point A on one side of a hill to point B on the other side. From a third point C, the distances $|\overline{CA}| = 837$ meters and $|\overline{CB}| = 1164$ meters are measured. If angle ACB is measured as 44.5°, find the length of the tunnel when it is completed. Round off your answer to four significant digits.

C **22.** A molecule of ammonia (NH_3) consists of three hydrogen atoms and one nitrogen atom, which form the vertices of a pyramid with four triangular faces. In one of the faces containing the nitrogen atom (N) and two hydrogen atoms (H), the distance between the two H atoms is 1.628 angstroms and the distance from the N atom to either H atom is 1.014 angstroms (Figure 8). [One angstrom (abbreviated Å) is 10^{-10} meter.] Find the angle formed by the two line segments joining the H atoms to the N atom.

Figure 8

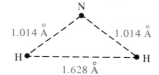

C **23.** Two jet fighters leave an air base at the same time and fly at the same speed along straight courses, forming an angle of 135.65° with each other. After the jets have each flown 402 kilometers, how far apart are they?

C **24.** A ship leaves a harbor at noon and sails S52°W at 22 knots (nautical miles per hour) until 2:30 P.M. At that time it changes its course and sails N22°W at a reduced speed of 18 knots until 5:00 P.M. At 5:00 P.M., how far is the ship from the harbor and what is its bearing from the harbor?

C **25.** Two straight roads cross at an angle of 75°. A bus on one road is 5 miles from the intersection and moving away from it at 50 miles per hour. At the same instant, a truck on the other road is 10 miles from the intersection and moving away from it at the rate of 55 miles per hour (Figure 9). If the bus and truck maintain constant speeds, what is the distance between them after 45 minutes?

Figure 9

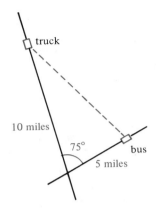

that $|\overline{AB}| = 320$ meters, $|\overline{BC}| = 200$ meters, $|\overline{CD}| = 360$ meters, angle $ABC = 146°$, and angle $BCD = 77°$. Find (a) the area of the bird sanctuary and (b) the distance $|\overline{AD}|$ along the shoreline. [*Hint:* Break the quadrilateral $ABCD$ into two triangles—for instance, ABC and ACD—then solve these triangles.]

Figure 10

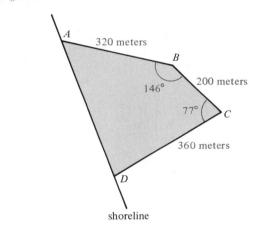

shoreline

© **26.** The navigator of an oil tanker plots a straight-line course from point A to point B. Because of an error, the tanker proceeds from point A along a slightly different straight-line course. After sailing for 5 hours on the wrong course, the error is discovered and the tanker is turned through an angle of $23°$ so that it is now heading directly toward point B. After 3 more hours, the tanker reaches point B. Assuming that the tanker was proceeding at a constant speed, calculate the time lost because of the error.

© In Problems 27 to 30, use Hero's area formula to find the area of a triangle with the given side lengths a, b, and c. Round off all answers to four significant digits.

27. $a = 5$ centimeters, $b = 8$ centimeters, $c = 7$ centimeters

28. $a = 7.301$ kilometers, $b = 11.15$ kilometers, $c = 15.22$ kilometers

29. $a = 13.03$ feet, $b = 15.77$ feet, $c = 17.20$ feet

30. $a = 1.110 \times 10^5$ meters, $b = 9.812 \times 10^4$ meters, $c = 1.728 \times 10^5$ meters

© **31.** Offshore drilling rights are granted to RIPCO oil company in a triangular region bounded by sides of lengths 33 kilometers, 28 kilometers, and 7 kilometers. Find the area of this region.

© **32.** A bird sanctuary along the straight shoreline of a lake is bounded by straight line segments $\overline{AB}$, $\overline{BC}$, and $\overline{CD}$ as shown in Figure 10. Suppose

© **33.** The Great Pyramid of Cheops at Al Giza, Egypt, has four congruent triangular faces. Each face is an isosceles triangle with equal sides of length 219 meters and base 230 meters (Figure 11). Find the *total* surface area of the pyramid.

Figure 11

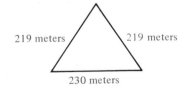

© **34.** In Problem 33, find (a) the height h of the pyramid and (b) the volume V of the pyramid. [*Hint:* The volume of a pyramid is one-third the height times the area of the base.]

35. Show that the Pythagorean theorem is a special case of the law of cosines.

36. Let $s = \frac{1}{2}(a + b + c)$ be the semiperimeter of triangle ABC (Figure 12). Combine the law of cosines

$$a^2 = b^2 + c^2 - 2bc \cos \alpha$$

and the half-angle formula

$$\cos^2 \frac{\alpha}{2} = \frac{1 + \cos \alpha}{2}$$

to obtain the formula

$$\cos \frac{\alpha}{2} = \sqrt{\frac{s(s - a)}{bc}}.$$

Figure 12

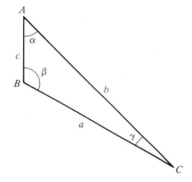

37. For every triangle ABC (Figure 12), show that

$$\frac{\cos \alpha}{a} + \frac{\cos \beta}{b} + \frac{\cos \gamma}{c} = \frac{a^2 + b^2 + c^2}{2abc}.$$

38. Proceeding as in Problem 36, but using the half-angle formula

$$\sin^2 \frac{\alpha}{2} = \frac{1 - \cos \alpha}{2},$$

show that

$$\sin \frac{\alpha}{2} = \sqrt{\frac{(s - b)(s - c)}{bc}}.$$

39. For every triangle ABC (Figure 12), show that γ is an acute angle if and only if $a^2 + b^2 > c^2$.

40. In Figure 13, the area $\mathscr{A}$ of triangle ABC is given by $\mathscr{A} = \frac{1}{2}hb$.

(a) Show that $\mathscr{A} = \frac{1}{2}bc \sin \alpha$.

(b) Using the double-angle formula, show that

$$\sin \alpha = 2 \sin \frac{\alpha}{2} \cos \frac{\alpha}{2}.$$

(c) Combine (a) and (b) to obtain

$$\mathscr{A} = bc \sin \frac{\alpha}{2} \cos \frac{\alpha}{2}.$$

(d) Combine (c), Problem 36, and Problem 38 to obtain Hero's formula

$$\mathscr{A} = \sqrt{s(s - a)(s - b)(s - c)}.$$

Figure 13

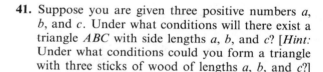

41. Suppose you are given three positive numbers a, b, and c. Under what conditions will there exist a triangle ABC with side lengths a, b, and c? [*Hint:* Under what conditions could you form a triangle with three sticks of wood of lengths a, b, and c?]

42. The center O of the circle inscribed in a triangle ABC always lies at the intersection of the bisectors of the three vertex angles (Figure 14). If the semiperimeter of triangle ABC is $s = \frac{1}{2}(a + b + c)$, show that the radius r of the inscribed circle is given by

$$r = \sqrt{\frac{(s - a)(s - b)(s - c)}{s}}.$$

[*Hint:* Use Hero's formula and the fact that the area of triangle ABC is the sum of the areas of triangles AOB, BOC, and COA.]

Figure 14

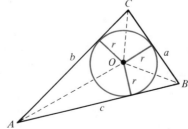

8.4 VECTORS

Many applications of mathematics involve force, velocity, acceleration, or other quantities that have both a *magnitude* and a *direction*. For instance, if you push on a stone, the force you exert has both a magnitude (say, 5 pounds) and a direction (say, due east). You can represent this force by an arrow that has a length (5 units) corresponding to the magnitude and that points in the appropriate direction (due east) (Figure 1).

Figure 1

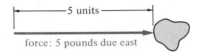

force: 5 pounds due east

Figure 2

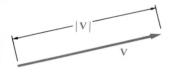

More generally, a **vector** can be thought of as an arrow, that is, a line segment with a direction specified by an arrowhead. We use boldface letters, such as **A**, **B**, **C**, and so forth, to denote vectors. If **V** is a vector, we denote the **length** or **magnitude** of **V** by $|\mathbf{V}|$ (Figure 2). A vector **V** with tail end, or **initial point,** A and head end, or **terminal point,** B may be written as

$$\mathbf{V} = \overrightarrow{AB}$$

Figure 3

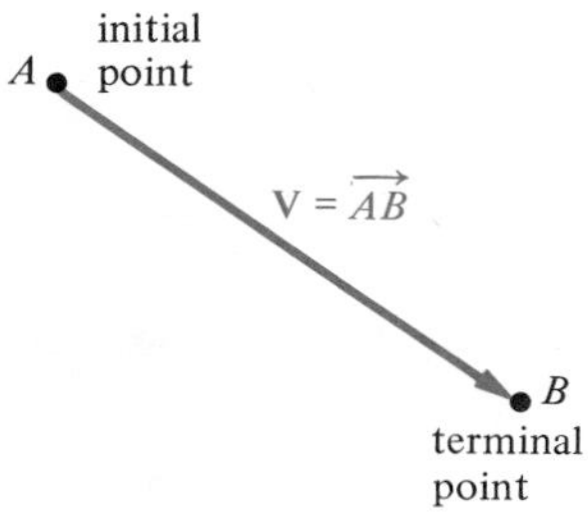

initial point
A •

$\mathbf{V} = \overrightarrow{AB}$

• B
terminal point

(Figure 3). This notation tells us that we are dealing with a **vector** $\overrightarrow{AB}$ rather than a line segment $\overline{AB}$, and it shows the direction of the vector—from the initial point A to the terminal point B. Notice that the magnitude

$$|\mathbf{V}| = |\overrightarrow{AB}| = |\overline{AB}|$$

is just the distance between A and B.

Let's agree to say that two vectors $\overrightarrow{AB}$ and $\overrightarrow{CD}$ are **equal** and to write

$$\overrightarrow{AB} = \overrightarrow{CD}$$

if they are parallel and have the same direction and magnitude (Figure 4). In other words, you may move a vector around freely if you keep it parallel to its original position. In particular, you can always move a vector **R** in the xy plane so that its tail end (initial point) is at the origin O (Figure 5). Then **R** will extend from O to its head end (terminal point) $P = (x, y)$, and we have

$$\mathbf{R} = \overrightarrow{OP}.$$

Figure 4

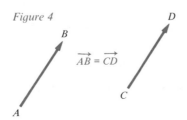

$\overrightarrow{AB} = \overrightarrow{CD}$

Figure 5

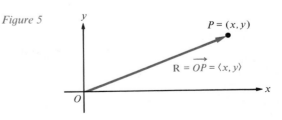

$\mathbf{R} = \overrightarrow{OP} = \langle x, y \rangle$

$P = (x, y)$

We say that $\mathbf{R} = \overrightarrow{OP}$ is the **position vector** of the point $P = (x, y)$; we call $\mathbf{x}$ and $\mathbf{y}$ the **components** of the vector $\mathbf{R}$; and we write

$$\mathbf{R} = \langle x, y \rangle.$$

The special symbol $\langle x, y \rangle$ is used to avoid confusing the *vector* $\langle x, y \rangle$ and the *point* $P = (x, y)$.

Figure 6

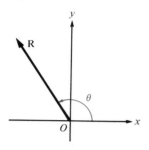

The **zero vector,** denoted by $\mathbf{0}$, is defined by $\mathbf{0} = \langle 0, 0 \rangle$. You can think of $\mathbf{0}$ as an arrow whose length has shrunk to 0, so that its initial and terminal points coincide. Not only is $\mathbf{0}$ the only vector whose magnitude is 0, but it is the only vector whose direction is indeterminate.

Let $\mathbf{R}$ be a nonzero vector in the xy plane. Move $\mathbf{R}$, if necessary, so that its initial point is at the origin O. If $\mathbf{R}$ falls along the terminal side of an angle θ in standard position, we say that θ is a **direction angle** for $\mathbf{R}$ (Figure 6). Notice that any angle coterminal with θ is also a direction angle for $\mathbf{R}$. The vector $\mathbf{R}$ can be specified either by giving its components $\mathbf{R} = \langle x, y \rangle$ or by giving its magnitude (length) $|\mathbf{R}|$ and its direction angle θ (Figure 7). The quantities x, y, $|\mathbf{R}|$, and θ are related as follows:

Figure 7

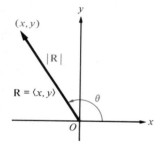

(i) $|\mathbf{R}| = \sqrt{x^2 + y^2}$

(ii) $\cos \theta = \dfrac{x}{|\mathbf{R}|}$ or $x = |\mathbf{R}| \cos \theta$

(iii) $\sin \theta = \dfrac{y}{|\mathbf{R}|}$ or $y = |\mathbf{R}| \sin \theta$

(iv) $\tan \theta = \dfrac{y}{x}$ if $x \neq 0$.

(See Definition 1 on page 331.) These equations can be used to find $|\mathbf{R}|$ and θ in terms of x and y or vice versa.

© **Example 1**   Sketch each vector, find its magnitude, and find its smallest positive direction angle θ.

(a) $\mathbf{A} = \langle 4, 3 \rangle$ (b) $\mathbf{B} = \langle 2\sqrt{3}, -1 \rangle$

Figure 8

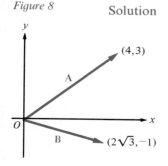

Solution   The two vectors, which extend from the origin to the points $(4, 3)$ and $(2\sqrt{3}, -1)$, are shown in Figure 8.

(a) By equation (i) above,

$$|\mathbf{A}| = \sqrt{4^2 + 3^2} = \sqrt{25} = 5 \text{ units.}$$

By equation (iv) above,

$$\tan \theta = \tfrac{3}{4} = 0.75.$$

Because the point $(4, 3)$ is in quadrant I, it follows that θ is an acute angle; hence,

$$\theta = \tan^{-1} 0.75 \approx 36.87°.$$

(b) $|\mathbf{B}| = \sqrt{(2\sqrt{3})^2 + (-1)^2} = \sqrt{13}$ units. Because the point $(2\sqrt{3}, -1)$ is in quadrant IV, it follows that $270° < \theta < 360°$. Also,

$$\tan \theta = \frac{-1}{2\sqrt{3}} = -\frac{\sqrt{3}}{6};$$

so the reference angle θ_R corresponding to θ is given by

$$\theta_R = \tan^{-1}(\sqrt{3}/6) \approx 16.10°.$$

Therefore $\qquad \theta = 360° - \theta_R \approx 360° - 16.10° = 343.90°.$

(Alternatively, we could have added $360°$ to $\tan^{-1}(-\sqrt{3}/6)$ to obtain θ.) ∎

Example 2 Find the components of the vector $\mathbf{V}$ if $|\mathbf{V}| = 3$ and a direction angle for $\mathbf{V}$ is $\theta = 150°$.

Solution We use equations (ii) and (iii) in the forms $x = |\mathbf{V}| \cos \theta$ and $y = |\mathbf{V}| \sin \theta$. Thus, we have

$$x = |\mathbf{V}| \cos \theta = 3 \cos 150° = 3\left(\frac{-\sqrt{3}}{2}\right) = -\frac{3\sqrt{3}}{2}$$

and $\qquad y = |\mathbf{V}| \sin \theta = 3 \sin 150° = 3\left(\frac{1}{2}\right) = \frac{3}{2}$

so $\qquad \mathbf{V} = \left\langle -\frac{3\sqrt{3}}{2}, \frac{3}{2} \right\rangle.$ ∎

Figure 9

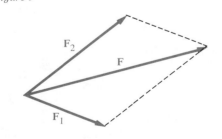

As we mentioned earlier, forces can be represented by vectors: the length $|\mathbf{F}|$ of a force vector $\mathbf{F}$ is the magnitude of the force, while the direction of $\mathbf{F}$ indicates the direction of the force. Laboratory experiments have revealed the following: If two forces $\mathbf{F}_1$ and $\mathbf{F}_2$ act simultaneously on a particle, the effect is the same as if a single force $\mathbf{F}$, called the **resultant,** were acting on the particle. The resultant $\mathbf{F}$ forms a diagonal of the parallelogram in which $\mathbf{F}_1$ and $\mathbf{F}_2$ are adjacent edges (Figure 9). The resultant vector $\mathbf{F}$ is also called the **sum** of the two vectors $\mathbf{F}_1$ and $\mathbf{F}_2$, and we write

$$\mathbf{F} = \mathbf{F}_1 + \mathbf{F}_2.$$

Figure 10

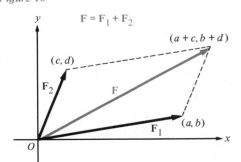

By applying the **parallelogram rule,** illustrated in Figure 9, you can form the sum $\mathbf{F} = \mathbf{F}_1 + \mathbf{F}_2$ of any two vectors, whether they represent forces or not. (Also, see Problem 31.) By examining Figure 10, you can see that *the components of the sum of two vectors are the sums of their corresponding components*, that is,

$$\langle a, b \rangle + \langle c, d \rangle = \langle a + c, b + d \rangle.$$

(See Problem 35).

Example 3   If $\mathbf{A} = \langle 2, -7 \rangle$ and $\mathbf{B} = \langle \sqrt{3}, 5 \rangle$, find the components of $\mathbf{A} + \mathbf{B}$.

Solution $\mathbf{A} + \mathbf{B} = \langle 2, -7 \rangle + \langle \sqrt{3}, 5 \rangle = \langle 2 + \sqrt{3}, -7 + 5 \rangle = \langle 2 + \sqrt{3}, -2 \rangle.$ ∎

Figure 11

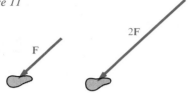

To *double* a force $\mathbf{F}$ means to double its magnitude and leave its direction the same. Thus, $2\mathbf{F}$ is a vector twice as long as $\mathbf{F}$, and pointing in the same direction (Figure 11).

More generally, if s is a real number and $\mathbf{F}$ is a vector, we define $s\mathbf{F}$ to be a vector $|s|$ times as long as $\mathbf{F}$, and in the same direction if s is positive, or in the opposite direction if s is negative. Naturally, if either $s = 0$ or $\mathbf{F} = \mathbf{0}$ (or both), we define $s\mathbf{F} = \mathbf{0}$. In a product $s\mathbf{F}$, the real number s is often called a **scalar** (because it corresponds to a point on a number scale). Thus, the vector $s\mathbf{F}$ is referred to as a **scalar multiple** of the vector $\mathbf{F}$. Notice that

$$|s\mathbf{F}| = |s|\,|\mathbf{F}|.$$

When a vector is multiplied by a scalar, the components of the vector are multiplied by the scalar. Thus, if $\mathbf{F} = \langle a, b \rangle$, then $s\mathbf{F} = \langle sa, sb \rangle$, that is,

$$s\langle a, b \rangle = \langle sa, sb \rangle.$$

You can confirm this for the case in which s is positive and $\mathbf{F} = \langle a, b \rangle$ lies in quadrant I by considering the similar triangles in Figures 12a and b. The remaining cases are left as an exercise (Problem 34).

Figure 12

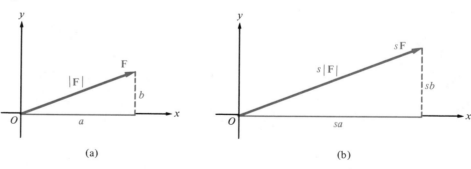

(a)   (b)

Example 4   Let $\mathbf{A} = \langle -2, \sqrt{3} \rangle$ and $\mathbf{B} = \langle 1, 4 \rangle$. Find the components of

(a) $5\mathbf{A}$ **(b)** $-3\mathbf{B}$ **(c)** $2\mathbf{A} + 3\mathbf{B}$.

Solution   **(a)** $5\mathbf{A} = 5\langle -2, \sqrt{3} \rangle = \langle 5(-2), 5\sqrt{3} \rangle = \langle -10, 5\sqrt{3} \rangle$

(b) $-3\mathbf{B} = -3\langle 1, 4 \rangle = \langle -3(1), -3(4) \rangle = \langle -3, -12 \rangle$

(c) $2\mathbf{A} + 3\mathbf{B} = 2\langle -2, \sqrt{3} \rangle + 3\langle 1, 4 \rangle = \langle 2(-2), 2\sqrt{3} \rangle + \langle 3(1), 3(4) \rangle$
$\qquad = \langle -4, 2\sqrt{3} \rangle + \langle 3, 12 \rangle = \langle -4 + 3, 2\sqrt{3} + 12 \rangle$
$\qquad = \langle -1, 2\sqrt{3} + 12 \rangle$ ∎

Figure 13

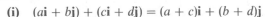

By the definition of multiplication of a vector by a scalar, the vector $(-1)\mathbf{F}$ has the same magnitude as $\mathbf{F}$, but points in the opposite direction. For simplicity, $(-1)\mathbf{F}$ is often written as $-\mathbf{F}$ (Figure 13). If $\mathbf{F} = \langle a, b \rangle$, then $-\mathbf{F} = \langle -a, -b \rangle$, that is,

$$-\langle a, b \rangle = \langle -a, -b \rangle$$

(Problem 36). The **difference** of two vectors $\mathbf{A}$ and $\mathbf{B}$ is defined by

$$\mathbf{A} - \mathbf{B} = \mathbf{A} + (-\mathbf{B}).$$

In other words, to **subtract** a vector $\mathbf{B}$ from a vector $\mathbf{A}$, you add $-\mathbf{B}$ to $\mathbf{A}$ (see Problem 33.) Evidently, if $\mathbf{A} = \langle a, b \rangle$ and $\mathbf{B} = \langle c, d \rangle$, then $\mathbf{A} - \mathbf{B} = \langle a - c, b - d \rangle$, that is,

$$\langle a, b \rangle - \langle c, d \rangle = \langle a - c, b - d \rangle$$

(Problem 37).

Example 5   If $\mathbf{A} = \langle 4, -1 \rangle$ and $\mathbf{B} = \langle 7, \frac{2}{3} \rangle$, find the components of

(a) $-\mathbf{A}$ **(b)** $\mathbf{A} - \mathbf{B}$ **(c)** $2\mathbf{A} - 3\mathbf{B}$

Solution   **(a)** $-\mathbf{A} = -\langle 4, -1 \rangle = \langle -4, -(-1) \rangle = \langle -4, 1 \rangle$

(b) $\mathbf{A} - \mathbf{B} = \langle 4, -1 \rangle - \langle 7, \frac{2}{3} \rangle = \langle 4 - 7, -1 - \frac{2}{3} \rangle = \langle -3, -\frac{5}{3} \rangle$

(c) $2\mathbf{A} - 3\mathbf{B} = 2\langle 4, -1 \rangle - 3\langle 7, \frac{2}{3} \rangle = \langle 8, -2 \rangle - \langle 21, 2 \rangle$
$= \langle 8 - 21, -2 - 2 \rangle = \langle -13, -4 \rangle$ ∎

Figure 14

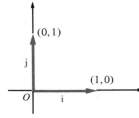

A popular alternative way to represent vectors in the xy plane is based on the two special vectors $\mathbf{i} = \langle 1, 0 \rangle$ and $\mathbf{j} = \langle 0, 1 \rangle$ (Figure 14). Notice that $\mathbf{i}$ and $\mathbf{j}$, which are called the **standard unit basis vectors,** have length 1 and point in the directions of the positive x and y axes, respectively. If $\mathbf{R} = \langle x, y \rangle$, then

$$\mathbf{R} = \langle x, y \rangle = \langle x, 0 \rangle + \langle 0, y \rangle = x\langle 1, 0 \rangle + y\langle 0, 1 \rangle = x\mathbf{i} + y\mathbf{j}.$$

In other words, $\mathbf{R} = \langle x, y \rangle$ and $\mathbf{R} = x\mathbf{i} + y\mathbf{j}$ are just two different ways of saying that $\mathbf{R}$ is the vector with components x and y. Using the standard unit basis vectors, we can write the following versions of the rules for addition, subtraction, and multiplication of vectors by scalars:

(i) $(a\mathbf{i} + b\mathbf{j}) + (c\mathbf{i} + d\mathbf{j}) = (a + c)\mathbf{i} + (b + d)\mathbf{j}$

(ii) $(a\mathbf{i} + b\mathbf{j}) - (c\mathbf{i} + d\mathbf{j}) = (a - c)\mathbf{i} + (b - d)\mathbf{j}$

(iii) $s(a\mathbf{i} + b\mathbf{j}) = (sa)\mathbf{i} + (sb)\mathbf{j}.$

Example 6   If $\mathbf{A} = 2\mathbf{i} + 5\mathbf{j}$ and $\mathbf{B} = 3\mathbf{i} - \mathbf{j}$, find

(a) $\mathbf{A} + \mathbf{B}$ (b) $\mathbf{A} - \mathbf{B}$ (c) $\frac{4}{5}\mathbf{A}$ (d) $3\mathbf{A} - 2\mathbf{B}$

Solution   (a) $\mathbf{A} + \mathbf{B} = (2\mathbf{i} + 5\mathbf{j}) + (3\mathbf{i} - \mathbf{j}) = (2 + 3)\mathbf{i} + (5 - 1)\mathbf{j} = 5\mathbf{i} + 4\mathbf{j}$

(b) $\mathbf{A} - \mathbf{B} = (2\mathbf{i} + 5\mathbf{j}) - (3\mathbf{i} - \mathbf{j}) = (2 - 3)\mathbf{i} + (5 + 1)\mathbf{j} = -\mathbf{i} + 6\mathbf{j}$

(c) $\frac{4}{5}\mathbf{A} = \frac{4}{5}(2\mathbf{i} + 5\mathbf{j}) = [\frac{4}{5}(2)]\mathbf{i} + [\frac{4}{5}(5)]\mathbf{j} = \frac{8}{5}\mathbf{i} + 4\mathbf{j}$

(d) $3\mathbf{A} - 2\mathbf{B} = 3(2\mathbf{i} + 5\mathbf{j}) - 2(3\mathbf{i} - \mathbf{j}) = (6\mathbf{i} + 15\mathbf{j}) - (6\mathbf{i} - 2\mathbf{j})$
     $= (6 - 6)\mathbf{i} + [15 - (-2)]\mathbf{j} = 17\mathbf{j}$

■

Problem Set 8.4

In Problems 1 to 12, sketch each vector, find its magnitude, and find its smallest nonnegative direction angle θ.

1. $\mathbf{R} = \langle -4, 4 \rangle$ Ⓒ 2. $\mathbf{A} = \langle 6, 8 \rangle$

Ⓒ 3. $\mathbf{B} = \langle 4, 3 \rangle$ 4. $\mathbf{D} = \langle 5, 0 \rangle$

5. $\mathbf{C} = \langle 1, -\sqrt{3} \rangle$ 6. $\mathbf{F} = \langle 0, 5 \rangle$

Ⓒ 7. $\mathbf{H} = \langle \frac{3}{8}, \frac{1}{2} \rangle$ Ⓒ 8. $\mathbf{K} = \langle 3.71, -4.88 \rangle$

Ⓒ 9. $\mathbf{M} = \langle -\sqrt{5}, \sqrt{11} \rangle$ 10. $\mathbf{N} = \langle \pi, \pi\sqrt{3} \rangle$

Ⓒ 11. $\mathbf{P} = 3\mathbf{i} + 4\mathbf{j}$ Ⓒ 12. $\mathbf{Q} = 2\sqrt{5}\mathbf{i} - 4\mathbf{j}$

In Problems 13 to 20, the magnitude $|\mathbf{V}|$ and a direction angle θ of a vector $\mathbf{V}$ are given. In each case, find the components of $\mathbf{V}$.

13. $|\mathbf{V}| = 5, \theta = 45°$ 14. $|\mathbf{V}| = 3, \theta = 180°$

15. $|\mathbf{V}| = 4\sqrt{2}, \theta = 135°$ 16. $|\mathbf{V}| = 9, \theta = -\dfrac{5\pi}{3}$

17. $|\mathbf{V}| = 2, \theta = -\dfrac{5\pi}{6}$ Ⓒ 18. $|\mathbf{V}| = 4, \theta = 82°$

Ⓒ 19. $|\mathbf{V}| = 7.03, \theta = 75.25°$

Ⓒ 20. $|\mathbf{V}| = 8.12, \theta = 247°10'$

In Problems 21 to 30, find (a) $\mathbf{A} + \mathbf{B}$, (b) $\mathbf{A} - \mathbf{B}$, (c) $3\mathbf{A} + 4\mathbf{B}$, and (d) $2\mathbf{A} - 3\mathbf{B}$.

21. $\mathbf{A} = \langle 3, 4 \rangle, \mathbf{B} = \langle -1, 4 \rangle$

22. $\mathbf{A} = \langle -2, 6 \rangle, \mathbf{B} = \langle 3, -5 \rangle$

23. $\mathbf{A} = \langle -3, 2 \rangle, \mathbf{B} = \langle \frac{5}{2}, \frac{7}{3} \rangle$

24. $\mathbf{A} = \langle 3, -3 \rangle, \mathbf{B} = \langle -5, 0 \rangle$

25. $\mathbf{A} = 2\langle 1, -3 \rangle, \mathbf{B} = 3\langle 3, -5 \rangle$

26. $\mathbf{A} = -\mathbf{i} - \mathbf{j}, \mathbf{B} = 9\mathbf{j}$

27. $\mathbf{A} = 2\mathbf{i} + 5\mathbf{j}, \mathbf{B} = -3\mathbf{i} + 7\mathbf{j}$

28. $\mathbf{A} = \frac{7}{4}\mathbf{i} - \mathbf{j}, \mathbf{B} = \mathbf{i} - \frac{11}{3}\mathbf{j}$

29. $\mathbf{A} = 6\mathbf{i} - 5\mathbf{j}, \mathbf{B} = 2(2\mathbf{i} - 3\mathbf{j})$

30. $\mathbf{A} = 3.07\mathbf{i} + 0.33\mathbf{j}, \mathbf{B} = 7.65\mathbf{i} - 3.44\mathbf{j}$

31. The sum $\mathbf{A} + \mathbf{B}$ of two vectors can be obtained geometrically by applying the following **head-to-tail rule:** Move $\mathbf{B}$ if necessary so that its tail end coincides with the head end of $\mathbf{A}$ (Figure 15). Then $\mathbf{A} + \mathbf{B}$ is the vector whose tail end is the tail end of $\mathbf{A}$ and whose head end is the head end of $\mathbf{B}$. Show that the head-to-tail rule follows from the parallelogram rule.

Figure 15

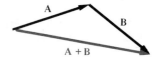

32. Using the head-to-tail rule (Problem 31), give a geometric justification of the following equation, which is known as the **associative law of vector addition:**

$$\mathbf{A} + (\mathbf{B} + \mathbf{C}) = (\mathbf{A} + \mathbf{B}) + \mathbf{C}.$$

33. Justify the **triangle rule** for subtraction of vectors: Move **B** if necessary so that its tail end coincides with the tail end of **A** (Figure 16). Then **A** − **B** is the vector whose tail end is the head end of **B** and whose head end is the head end of **A**.

Figure 16

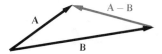

34. Complete the argument begun in the text (Figure 12) to show that if $\mathbf{F} = \langle a, b \rangle$, then $s\mathbf{F} = \langle sa, sb \rangle$. Consider the cases in which **F** lies in quadrants II, III, and IV as well as the situations in which s is 0 or negative.

35. Figure 10 shows a special situation in which both vectors $\mathbf{F}_1$ and $\mathbf{F}_2$ lie in quadrant I. Check several other configurations to convince yourself that the components of $\mathbf{F}_1 + \mathbf{F}_2$ are always equal to the sum of the corresponding components of $\mathbf{F}_1$ and $\mathbf{F}_2$.

36. Prove that if $\mathbf{F} = \langle a, b \rangle$, then $-\mathbf{F} = \langle -a, -b \rangle$.

37. Prove that if $\mathbf{A} = \langle a, b \rangle$ and $\mathbf{B} = \langle c, d \rangle$, then $\mathbf{A} - \mathbf{B} = \langle a - c, b - d \rangle$.

38. If $\mathbf{A} + \mathbf{X} = \mathbf{B}$, prove that $\mathbf{X} = \mathbf{B} - \mathbf{A}$.

39. If $\mathbf{A} + \mathbf{X} = \mathbf{A}$, prove that $\mathbf{X} = \mathbf{0}$.

40. If $s\mathbf{F} = \mathbf{0}$, prove that either $s = 0$ or $\mathbf{F} = \mathbf{0}$.

41. Prove that $s(\mathbf{A} + \mathbf{B}) = s\mathbf{A} + s\mathbf{B}$.

42. If **A** and **B** are two vectors and α is the angle between them, then the **dot product** of **A** and **B** is defined to be the number

$$\mathbf{A} \cdot \mathbf{B} = |\mathbf{A}|\,|\mathbf{B}|\cos\alpha.$$

Use the law of cosines to show that if $\mathbf{A} = \langle a, b \rangle$ and $\mathbf{B} = \langle c, d \rangle$, then

$$\mathbf{A} \cdot \mathbf{B} = ac + bd.$$

43. Use the formula obtained in Problem 42 to calculate $\mathbf{A} \cdot \mathbf{B}$ if

(a) $\mathbf{A} = \langle 5, 1 \rangle$ and $\mathbf{B} = \langle 4, -7 \rangle$
(b) $\mathbf{A} = \langle -3, 2 \rangle$ and $\mathbf{B} = \langle -4, 3 \rangle$
(c) $\mathbf{A} = \mathbf{i} + 2\mathbf{j}$ and $\mathbf{B} = 3\mathbf{i} - 5\mathbf{j}$
(d) $\mathbf{A} = \mathbf{i}$ and $\mathbf{B} = \mathbf{j}$
(e) $\mathbf{A} = \mathbf{B} = \mathbf{i}$

44. Use the result of Problem 42 to show that for any three vectors **A**, **B**, and **C**,

$$\mathbf{A} \cdot (\mathbf{B} + \mathbf{C}) = \mathbf{A} \cdot \mathbf{B} + \mathbf{A} \cdot \mathbf{C}.$$

8.5 POLAR COORDINATES

Figure 1

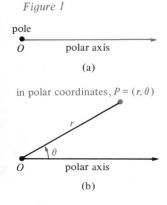

pole
O polar axis

(a)

in polar coordinates, $P = (r, \theta)$

r

θ

O polar axis

(b)

Until now we have specified the position of points in the plane by means of Cartesian coordinates; however, in some situations it is simpler and more natural to use a different system called *polar coordinates*. To set up a polar coordinate system in the plane, we choose a fixed point *O* called the **pole** and a fixed ray with endpoint *O*. The ray is called the **polar axis** (Figure 1a).

Now let *P* be any point in the plane and denote by *r* the distance between *P* and the pole *O*, so that $r = |\overline{OP}|$ (Figure 1b). If $P \neq O$, then the ray containing *P* with endpoint *O* forms the terminal side of an angle θ with the polar axis as its initial side (Figure 1b). We shall refer to the ordered pair (r, θ) as the **polar coordinates** of the point *P*, and we write

$$P = (r, \theta).$$

The angle θ can be measured either in degrees or radians.

The polar coordinates (r, θ) locate the point *P* on a grid formed by concentric circles with center *O* and rays with endpoint *O* (Figure 2). The value of *r* determines

Figure 2

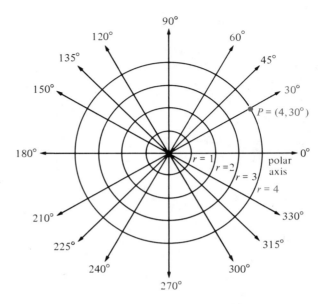

a particular circle of radius r; the value of θ designates a specific ray making an angle θ with the polar axis; and P lies at the intersection of the circle and the ray. For instance, the point

$$P = (4, 30°)$$

lies at the intersection of the circle $r = 4$ and the ray $\theta = 30°$.

If $r = 0$ in the polar coordinate sytem, we understand that the point $(r, \theta) = (0, \theta)$ is at the pole O no matter what the angle θ may be. Thus,

$$O = (0, \theta).$$

Also, it is convenient to allow r to be negative by making the following definition: The point $(-r, \theta°)$ is located $|r|$ units from the pole O, but on the ray opposite to the $\theta°$ ray, that is, on the $\theta° + 180°$ ray (Figure 3). Therefore,

$$(-r, \theta°) = (r, \theta° + 180°),$$

or if θ is measured in radians,

$$(-r, \theta) = (r, \theta + \pi).$$

Figure 3

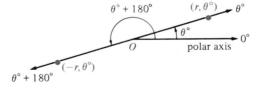

To "plot the polar point (r, θ)," means to draw a diagram showing the pole O, the polar axis, and the point P whose polar coordinates are (r, θ). You will find it easier to plot polar points if you use polar graph paper.

Example 1   Plot each point in the polar coordinate system.

    (a) $(2, 60°)$ **(b)** $(4, 135°)$ **(c)** $(-\frac{5}{2}, 30°)$ **(d)** $\left(2, -\frac{\pi}{4}\right)$

    (e) $(2, 0°)$ **(f)** $(0, 0°)$ **(g)** $(-4, -45°)$

Solution   **(a)** To plot the polar point $(2, 60°)$, we measure an angle of $60°$ from the polar axis and then locate the point 2 units from the pole on the terminal side of this angle.

    (b) to **(g)** The polar points in parts (b) through (g) are plotted in a similar way (Figure 4). ■

Figure 4

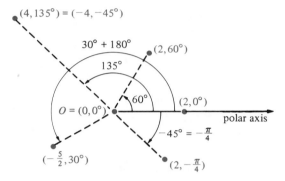

Unlike the Cartesian coordinate system, in which a point P has only one representation, a point P has many different representations in the polar coordinate system. Not only do we have $(r, θ°) = (-r, θ° + 180°)$ but, because $\pm 360°$ corresponds to one revolution about the pole, we also have

$$(r, θ°) = (r, θ° + 360°) = (r, θ° - 360°).$$

Indeed, if n is any integer,

$$(r, θ°) = (r, θ + 360° \cdot n),$$

or, if angles are measured in radians,

$$(r, θ) = (r, θ + 2πn).$$

Example 2   Give other polar representations $P = (r, θ°)$ of the polar point $P = (2, 30°)$ that satisfy each of the following conditions:

    (a) $r < 0$ and $0° \le θ° < 360°$ **(b)** $r > 0$ and $-360° < θ° \le 0°$
    (c) $r < 0$ and $-360° < θ° \le 0°$

Solution   **(a)** $P = (2, 30°) = (-2, 30° + 180°) = (-2, 210°)$

    (b) $P = (2, 30°) = (2, 30° - 360°) = (2, -330°)$

    (c) $P = (-2, 210°) = (-2, 210° - 360°) = (-2, -150°)$ ■

It is sometimes necessary to *convert from Cartesian to polar coordinates or vice versa.** To do this, we place the pole for the polar coordinate system at the origin of the Cartesian coordinate system, and position the polar axis along the positive x axis (Figure 5). Now suppose that the point P has Cartesian coordinates (x, y) and polar coordinates (r, θ) with $r > 0$. Then $\cos \theta = x/r$ and $\sin \theta = y/r$; hence,

$$x = r \cos \theta \quad \text{and} \quad y = r \sin \theta.$$

It can be shown (Problem 51) that these equations hold in all cases, even when $r \le 0$.

Figure 5

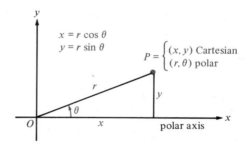

Example 3   Convert the following polar coordinates to Cartesian coordinates:

(a) $(4, 30°)$ **(b)** $\left(6, -\dfrac{3\pi}{4}\right)$ Ⓒ **(c)** $(-2, 133.5°)$

Solution   We use the conversion equations $x = r \cos \theta$ and $y = r \sin \theta$.

(a) $x = r \cos \theta = 4 \cos 30° = 4\left(\dfrac{\sqrt{3}}{2}\right) = 2\sqrt{3}$

$y = r \sin \theta = 4 \sin 30° = 4(\tfrac{1}{2}) = 2$

Hence, the Cartesian coordinates are $(2\sqrt{3}, 2)$.

(b) $x = r \cos \theta = 6 \cos\left(-\dfrac{3\pi}{4}\right) = 6\left(-\dfrac{\sqrt{2}}{2}\right) = -3\sqrt{2}$

$y = r \sin \theta = 6 \sin\left(-\dfrac{3\pi}{4}\right) = 6\left(-\dfrac{\sqrt{2}}{2}\right) = -3\sqrt{2}$

Hence, the Cartesian coordinates are $(-3\sqrt{2}, -3\sqrt{2})$.

(c) Using a calculator, and rounding off to four significant digits, we have

$$x = r \cos \theta = -2 \cos 133.5° = 1.377$$

$$y = r \sin \theta = -2 \sin 133.5° = -1.451.$$

Hence, the approximate Cartesian coordinates are $(1.377, -1.451)$. ∎

* Some scientific calculators will convert Cartesian coordinates to polar coordinates and vice versa. Consult the instruction manual for your particular calculator to see if it has this feature.

From the conversion equations $x = r \cos \theta$ and $y = r \sin \theta$, it follows that

$$x^2 + y^2 = r^2 \cos^2 \theta + r^2 \sin^2 \theta$$
$$= r^2(\cos^2 \theta + \sin^2 \theta)$$
$$= r^2,$$

so

$$r = \pm\sqrt{x^2 + y^2}.$$

Also, if $x \neq 0$,

$$\frac{y}{x} = \frac{r \sin \theta}{r \cos \theta} = \frac{\sin \theta}{\cos \theta} = \tan \theta,$$

so

$$\tan \theta = \frac{y}{x} \qquad \text{for} \qquad x \neq 0.$$

The equations $r = \pm\sqrt{x^2 + y^2}$ and $\tan \theta = y/x$ do not determine r and θ uniquely in terms of x and y, simply because the point P whose Cartesian coordinates are (x, y) has an unlimited number of different representations in the polar coordinate system. Thus, in using these equations to find the polar coordinates of P, you must pay attention to the quadrant in which P lies, because this will help you to determine an appropriate value of θ.

Example 4   Convert the following Cartesian coordinates to polar coordinates with $r \geq 0$ and $-180° < \theta° \leq 180°$:

(a) $(2, 2)$ ©**(b)** $(5.031, -2.758)$

Solution   **(a)** $r = \sqrt{2^2 + 2^2} = \sqrt{8} = 2\sqrt{2}$, and $\tan \theta = 2/2 = 1$. Because the point lies in quadrant I, it follows that $0° < \theta° < 90°$, and so $\theta = 45°$. Hence, the polar coordinates are $(2\sqrt{2}, 45°)$.

(b) Using a calculator, and rounding off to four significant digits, we have

$$r = \sqrt{(5.031)^2 + (-2.758)^2} = 5.737.$$

Here $\tan \theta = -2.758/5.031$, and, because the point lies in quadrant IV, we have $-90° < \theta° < 0°$. Therefore, rounded off to the nearest hundredth of a degree,

$$\theta = \tan^{-1} \frac{-2.758}{5.031} = -28.73°.$$

Hence, the approximate polar coordinates are $(5.737, -28.73°)$. ∎

Example 5   Convert the following Cartesian coordinates to polar coordinates with $r \geq 0$ and $-\pi < \theta \leq \pi$.

(a) $(0, -7)$ ©**(b)** $(-33.01, 29.77)$

Solution   **(a)** $r = \sqrt{0^2 + (-7)^2} = \sqrt{49} = 7$. Because the point lies on the negative y axis, $\theta = -\pi/2$, and the polar coordinates are $(7, -\pi/2)$.

(b) Using a calculator, and rounding off to four significant digits, we have

$$r = \sqrt{(-33.01)^2 + (29.77)^2} = 44.45.$$

Because the point lies in quadrant II, we have $\pi/2 < \theta < \pi$, and because $\tan \theta = 29.77/(-33.01)$, the corresponding reference angle is

$$\theta_R = \tan^{-1} \frac{29.77}{33.01} = 0.7338$$

(rounded off to four significant digits). Hence,

$$\theta = \pi - \theta_R = 2.408,$$

and the approximate polar coordinates are $(44.45, 2.408)$.   ■

Problem Set 8.5

In Problems 1 to 20, plot each point in the polar co-ordinate system.

1. $(6, 30°)$

2. $\left(10, \dfrac{\pi}{3}\right)$

3. $(3, 120°)$

4. $(0, 25°)$

5. $\left(4, -\dfrac{\pi}{6}\right)$

6. $(-4, 90°)$

7. $(-5, 135°)$

8. $\left(\dfrac{5}{2}, -270°\right)$

9. $(0, -45°)$

10. $\left(-\dfrac{7}{3}, \dfrac{\pi}{6}\right)$

11. $\left(\dfrac{17}{2}, -45°\right)$

12. $(5, -315°)$

13. $\left(\dfrac{5}{3}, -210°\right)$

14. $(3.75, -330°)$

15. $\left(-4.5, -\dfrac{\pi}{3}\right)$

16. $(-4.75, -30°)$

17. $\left(-2, -\dfrac{3\pi}{4}\right)$

18. $\left(-\dfrac{1}{2}, -\dfrac{3\pi}{2}\right)$

19. $(4.8, 0°)$

20. $(-3.25, 0)$

21. Plot the point $(5, 45°)$ in the polar coordinate system, and then give other polar representations $(r, \theta°)$

of the same point that satisfy each of the following conditions:

(a) $r < 0$ and $0° \le \theta° < 360°$
(b) $r > 0$ and $-360° < \theta° \le 0°$
(c) $r < 0$ and $-360° < \theta° \le 0°$

22. Plot the point $\left(-3, \dfrac{13\pi}{6}\right)$ in the polar coordinate system, and then give other polar representations (r, θ) of the same point that satisfy each of the following conditions:

(a) $r < 0$ and $0 \le \theta < 2\pi$
(b) $r > 0$ and $-2\pi < \theta \le 0$
(c) $r < 0$ and $-2\pi < \theta \le 0$

In Problems 23 to 34, convert the polar coordinates to Cartesian coordinates. If use of a calculator is indicated, round off your answers to four significant digits.

23. $(4, 45°)$

24. $(3, 60°)$

25. $\left(-5, \dfrac{\pi}{4}\right)$

26. $\left(0, \dfrac{\pi}{7}\right)$

27. $\left(1, -\dfrac{\pi}{4}\right)$

28. $\left(-5, \dfrac{5\pi}{6}\right)$

29. $(-4, 150°)$

30. $(-3, 25°)$

31. $(2.765, -122.73°)$

32. $(-6.772, 133.33°)$

Ⓒ **33.** $(17.76, -37.07°)$ Ⓒ **34.** $(3.271 \times 10^7, 44.33°)$

In Problems 35 to 44, convert the Cartesian coordinates to polar coordinates with $r \geq 0$ and $-180° < \theta° \leq 180°$. If use of a calculator is indicated, round off r to four significant digits, and θ to the nearest hundredth of a degree.

35. $(\sqrt{2}, \sqrt{2})$ **36.** $(-1, \sqrt{3})$

37. $(1, \sqrt{3})$ **38.** $(-2, 2)$

39. $(-3, 0)$ Ⓒ **40.** $(4, 3)$

Ⓒ **41.** $(-5, 12)$ Ⓒ **42.** $(-2, -3)$

Ⓒ **43.** $(13.01, -15.57)$

Ⓒ **44.** $(3.001 \times 10^{-5}, 2.774 \times 10^{-5})$

In Problems 45 to 50, convert the Cartesian coordinates to polar coordinates with $r \geq 0$ and $-\pi < \theta \leq \pi$. If use of a calculator is indicated, round off r and θ to four significant digits.

45. $(2\sqrt{3}, -2)$ **46.** $(-3, -3\sqrt{3})$

47. $(0, -2)$ **48.** $(4\sqrt{3}, -4)$

Ⓒ **49.** $(-72.02, -91.33)$

Ⓒ **50.** $(-2.102 \times 10^7, 9.901 \times 10^6)$

51. Show that the equations

$$x = r \cos \theta \quad \text{and} \quad y = r \sin \theta$$

work in all cases, even when $r \leq 0$, to convert the polar coordinates (r, θ) of a point P into the Cartesian coordinates (x, y) of P.

8.6 APPLICATIONS OF VECTORS AND POLAR COORDINATES

In this section, we present a few of the many applications of vectors and polar coordinates. Notice that there is a particularly close connection between vectors and polar coordinates: If **R** is the position vector of a point P, and if θ is a direction angle for **R** (Figure 1), then $(|\mathbf{R}|, \theta)$ are polar coordinates for P and

$$\mathbf{R} = \langle |\mathbf{R}| \cos \theta, |\mathbf{R}| \sin \theta \rangle.$$

Figure 1

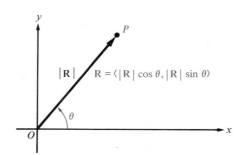

Applications of Vectors

The following example illustrates how vectors may be used to solve problems involving forces.

Ⓒ **Example 1** Two forces $\mathbf{F}_1$ and $\mathbf{F}_2$ act on a point. If $|\mathbf{F}_1| = 30$ pounds, $|\mathbf{F}_2| = 50$ pounds, and the angle α from $\mathbf{F}_1$ to $\mathbf{F}_2$ is $60°$, find:

(a) the magnitude $|\mathbf{F}|$ of the resultant force **F** **(b)** the angle θ from $\mathbf{F}_1$ to **F**

Figure 2

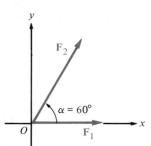

Solution

We begin by setting up an xy coordinate system with the origin O at the point on which both $\mathbf{F}_1$ and $\mathbf{F}_2$ act, and with the positive x axis in the direction of $\mathbf{F}_1$ (Figure 2). Then $\alpha = 60°$ is a direction angle for $\mathbf{F}_2$; hence, the components of $\mathbf{F}_2$ are given by

$$\mathbf{F}_2 = \langle |\mathbf{F}_2| \cos \alpha, |\mathbf{F}_1| \sin \alpha \rangle$$
$$= \langle 50 \cos 60°, 50 \sin 60° \rangle$$
$$= \left\langle \frac{50}{2}, \frac{50\sqrt{3}}{2} \right\rangle = \langle 25, 25\sqrt{3} \rangle.$$

Since $\mathbf{F}_1$ lies along the positive x axis and $|\mathbf{F}_1| = 30$, $\mathbf{F}_1 = \langle 30, 0 \rangle$. As a consequence,

$$\mathbf{F} = \mathbf{F}_1 + \mathbf{F}_2 = \langle 30, 0 \rangle + \langle 25, 25\sqrt{3} \rangle$$
$$= \langle 30 + 25, 0 + 25\sqrt{3} \rangle = \langle 55, 25\sqrt{3} \rangle.$$

(a) $|\mathbf{F}| = \sqrt{55^2 + (25\sqrt{3})^2} = \sqrt{3025 + 1875} = \sqrt{4900} = 70$ pounds

(b) The angle θ from $\mathbf{F}_1$ to $\mathbf{F}$ is the direction angle of $\mathbf{F}$ (Figure 3). Because the components of $\mathbf{F}$ are both positive, $\mathbf{F}$ lies in quadrant I, so θ is a quadrant I angle. Therefore, because

$$\tan \theta = \frac{25\sqrt{3}}{55} = \frac{5\sqrt{3}}{11},$$

it follows that

$$\theta = \tan^{-1} \frac{5\sqrt{3}}{11} \approx 38.21°.$$

Figure 3

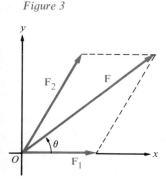

Another important application of vectors is in navigation, where the following terminology is commonly used:

1. The **heading** of an airplane is the direction in which it is pointed, and its **air speed** is its speed relative to the air. The vector $\mathbf{V}_1$ whose magnitude is the air speed and whose direction is the heading represents the **velocity** of the airplane *relative to the air* (Figure 4).

2. The **course** or **track** of an airplane is the direction in which it is actually moving over the ground, and its **ground speed** is its speed relative to the ground. The vector $\mathbf{V}$ whose magnitude is the ground speed and whose direction is the course represents the actual **velocity** of the airplane *relative to the ground* (Figure 4).

3. The difference $\mathbf{V}_2 = \mathbf{V} - \mathbf{V}_1$ is called the **drift vector.** It represents the *velocity of the air relative to the ground* (Figure 4). Thus,

$$\mathbf{V} = \mathbf{V}_1 + \mathbf{V}_2.$$

The angle α between $\mathbf{V}$ and $\mathbf{V}_1$ is called the **drift angle.**

(Similar terminology is used in ship navigation.)

Figure 4

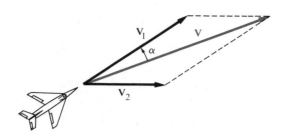

©️ **Example 2**   A commercial jet is headed N30°E with an air speed of 550 miles per hour. The wind is blowing N155°E at a speed of 50 miles per hour. Find:

(a) the ground speed   **(b)** the drift angle   **(c)** the course of the jet

Solution   Let $\mathbf{V}_1$ represent the velocity of the jet relative to the air; let $\mathbf{V}_2$ represent the velocity of the air relative to the ground; and let $\mathbf{V}$ represent the velocity of the jet relative to the ground (Figure 5).

Figure 5

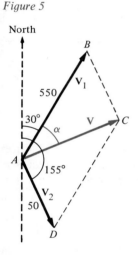

There are two ways to solve this problem: We can set up an xy coordinate system, find the components of $\mathbf{V}_1$ and $\mathbf{V}_2$, and then find the components of $\mathbf{V} = \mathbf{V}_1 + \mathbf{V}_2$; or we can use trigonometry to solve the triangles in Figure 5. Because the first method was illustrated in Example 1, we elect to use the second method here. Our plan is to find angle ABC so that we can apply the law of cosines to triangle ABC.

The four interior angles of a parallelogram add up to 360°, and the interior angles at opposite vertices are equal. Hence, in parallelogram $ABCD$,

$$2(\text{angle } ABC) + 2(\text{angle } DAB) = 360°.$$

Therefore, $$\text{angle } ABC = 180° - (\text{angle } DAB)$$
$$= 180° - (155° - 30°)$$
$$= 55°.$$

(a) Applying the law of cosines to triangle ABC, and noting that opposite sides of a parallelogram have the same length, we have

$$|\mathbf{V}|^2 = |\overline{AC}|^2 = |\overline{AB}|^2 + |\overline{BC}|^2 - 2|\overline{AB}||\overline{BC}| \cos 55°$$
$$= 550^2 + 50^2 - 2(550)(50) \cos 55°$$

so $$|\mathbf{V}| = \sqrt{550^2 + 50^2 - 2(550)(50) \cos 55°}$$
$$\approx 522.9 \text{ miles per hour.}$$

(b) Applying the law of sines to triangle ABC, we have

$$\frac{\sin \alpha}{50} = \frac{\sin 55°}{|\mathbf{V}|}$$

so $$\sin \alpha = \frac{50 \sin 55°}{|\mathbf{V}|} = \frac{50 \sin 55°}{522.9}.$$

Because α is a quadrant I angle, it follows that

$$\alpha = \sin^{-1}\frac{50 \sin 55^\circ}{522.9} \approx 4.49^\circ.$$

Hence, rounded off to the nearest hundredth of a degree, the drift angle is 4.49°.

(c) From Figure 5, the vector **V** is $30^\circ + \alpha = 34.49^\circ$ east of due north; hence, the course of the jet is N34.49°E. ∎

Graphs of Polar Equations

In calculus, polar coordinates are often used to write equations of curves in the plane. An equation, such as $r^2 = 9 \cos 2\theta$, that relates the polar coordinates r and θ is called a **polar equation.** Because each point in the plane has a multitude of different polar representations, it is necessary to define the graph of a polar equation with some care:

> The **graph** of a polar equation consists of all points P in the plane that have at least one polar representation (r, θ) that satisfies the equation.

Example 3 Sketch the graph of each polar equation:

(a) $r = 4$ **(b)** $r^2 = 16$ **(c)** $\theta = \dfrac{\pi}{6}$

Figure 6

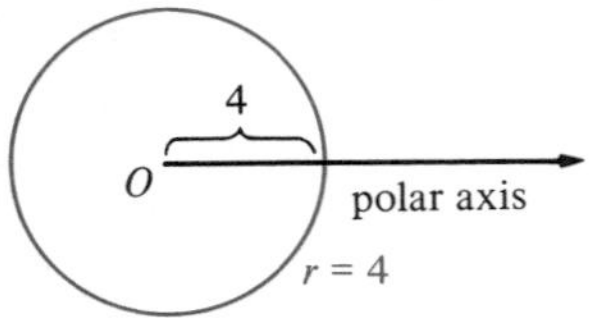

Solution

(a) The graph of $r = 4$ is a circle of radius 4 with center at the pole (Figure 6). Notice, for instance, that the point $P = (4, -\pi)$ belongs to the graph in spite of the fact that not all of its representations, such as $P = (-4, 0)$ or $P = (-4, 2\pi)$, satisfy the equation $r = 4$.

(b) The equation $r^2 = 16$ is equivalent to $|r| = 4$, and its graph is the same as the graph of $r = 4$ (Figure 6).

(c) The graph of $\theta = \pi/6$ consists of the entire line through O, making an angle of $\pi/6$ radian (30°) with the polar axis—not just the *ray* as one might at first think (Figure 7). Points of the form $P = \left(r, \dfrac{\pi}{6} + \pi\right)$ belong to the graph because they can be rewritten as $P = (-r, \pi/6)$. ∎

Figure 7

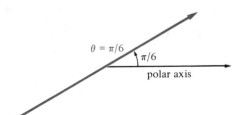

By making the substitutions $x = r \cos \theta$ and $y = r \sin \theta$, you can convert a Cartesian equation into a polar equation.

Example 4   Find a polar equation corresponding to the Cartesian equation $x^2 + (y + 4)^2 = 16$, and sketch the graph.

Figure 8

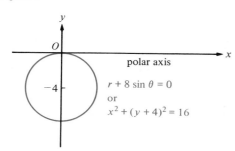

Solution   In Section 3.1, we learned that the graph of an equation having the form $(x - h)^2 + (y - k)^2 = r^2$ is a circle with center (h, k) and radius r. Hence, the graph of $x^2 + (y + 4)^2 = 16$ is a circle of radius 4 with center at the point with Cartesian coordinates $(0, -4)$ (Figure 8). Rewriting the equation as

$$x^2 + y^2 + 8y + 16 = 16 \quad \text{or} \quad x^2 + y^2 + 8y = 0,$$

substituting $x = r \cos \theta$ and $y = r \sin \theta$, and using the fact that $x^2 + y^2 = r^2$, we obtain the polar equation

$$r^2 + 8r \sin \theta = 0$$

or

$$r + 8 \sin \theta = 0.$$

In the last step of Example 4, we have divided by r; hence, we may have lost the solution $r = 0$. In this particular case, we have not lost the solution $r = 0$ because, if $r + 8 \sin \theta = 0$, then $r = 0$ when $\theta = 0$. More generally, if you multiply a polar equation by r, you *may* introduce an extraneous solution $r = 0$ that doesn't really belong and, conversely, if you divide a polar equation by r, you *may* lose a solution $r = 0$ that really does belong. You must check each case.

Example 5   Find a Cartesian equation corresponding to the polar equation $r = 4 \tan \theta \sec \theta$, and sketch the graph.

Solution   We have

$$r = 4 \tan \theta \sec \theta = 4 \frac{\sin \theta}{\cos \theta} \cdot \frac{1}{\cos \theta}$$

or

$$r \cos^2 \theta = 4 \sin \theta.$$

Multiplication by r gives

$$r^2 \cos^2 \theta = 4r \sin \theta$$

or

$$(r \cos \theta)^2 = 4r \sin \theta.$$

Because $x = r \cos \theta$ and $y = r \sin \theta$, the last equation can be rewritten as

$$x^2 = 4y \quad \text{or} \quad y = \tfrac{1}{4}x^2.$$

By Section 4.1, the graph of $y = \frac{1}{4}x^2$ is a parabola with a vertical axis of symmetry and vertex at $(0,0)$ (Figure 9). [Notice that multiplication by r did not introduce an extraneous solution $r = 0$ because, if $r\cos^2\theta = 4\sin\theta$, then $r = 0$ when $\theta = 0$.] ∎

Figure 9

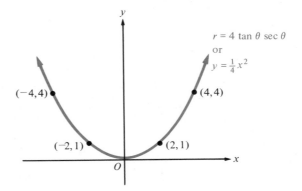

You can also sketch the graph of a polar equation by plotting polar points and connecting them with a smooth curve, in much the same way that you sketch the graph of a Cartesian equation.

Ⓒ **Example 6**   Sketch the graph of $r = 2(1 - \cos\theta)$.

Solution   Because $\cos(-\theta) = \cos\theta$, it follows that if the polar point $P = (r, \theta)$ belongs to the graph, so does the polar point $Q = (r, -\theta)$. Therefore, the graph is **symmetric about the polar axis.** We sketch the top half of the graph for $0 \le \theta \le \pi$ using the data in the table below, and then we reflect the top half across the polar axis to obtain the bottom half (Figure 10).

θ	$r = 2(1 - \cos\theta)$
0	0
$\pi/6$	0.27 (approximately)
$\pi/4$	0.59 (approximately)
$\pi/3$	1
$\pi/2$	2
$2\pi/3$	3
$5\pi/6$	3.73 (approximately)
π	4

Figure 10

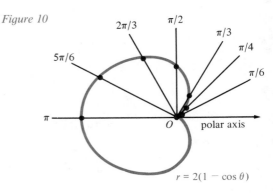

The curve in Figure 10 is called a **cardioid.** ∎

Problem Set 8.6

ⓒ In Problems 1 to 4, assume that the two forces $\mathbf{F}_1$ and $\mathbf{F}_2$ act on a point and that α is the angle from $\mathbf{F}_1$ to $\mathbf{F}_2$. If $\mathbf{F}$ is the resultant force, find (a) $|\mathbf{F}|$ and (b) the angle θ from $\mathbf{F}_1$ to $\mathbf{F}$.

1. $|\mathbf{F}_1| = 25$ pounds, $|\mathbf{F}_2| = 40$ pounds, $\alpha = 45°$

2. $|\mathbf{F}_1| = 41$ newtons, $|\mathbf{F}_2| = 7$ newtons, $\alpha = 27°$

3. $|\mathbf{F}_1| = 22$ newtons, $|\mathbf{F}_2| = 33$ newtons, $\alpha = 120°$

4. $|\mathbf{F}_1| = 3.4$ pounds, $|\mathbf{F}_2| = 7.3$ pounds, $\alpha = 133°$

ⓒ 5. Two children are pulling a third child across the ice on a sled. The first child pulls on her rope with a force of 10 pounds, and the second child pulls on his rope with a force of 7 pounds. If the angle between the ropes is 30°:

(a) What is the magnitude of the resultant force?

(b) What angle does the resultant force vector make with the first child's rope?

ⓒ 6. Two tugboats are pulling an ocean freighter (Figure 11). The first tugboat exerts a force of 3600 pounds on a cable making an angle of 25° with the axis of the ship. The second tugboat pulls on a cable making an angle of 32° with the axis of the ship. If the resultant force vector lies directly along the axis of the ship:

(a) What force is the second tugboat exerting on the ship?

(b) What is the magnitude of the resultant force vector?

Figure 11

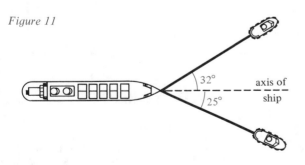

ⓒ 7. An airplane is headed N95°E with an air speed of 800 kilometers per hour. The wind is blowing

N25°E at a speed of 45 kilometers per hour. Find:

(a) the ground speed

(b) the drift angle

(c) the course of the plane

ⓒ 8. A long-distance swimmer can swim 3 kilometers per hour in still water. She is swimming away from and at right angles to a straight shoreline. If a current of 5 kilometers per hour is flowing parallel to the shoreline, find her speed relative to the land and the angle θ between the shoreline and her actual velocity vector.

ⓒ 9. A balloon is rising 3 meters per second while a wind is blowing horizontally at 2 meters per second. Find:

(a) the speed of the balloon

(b) the angle its track makes with the horizontal

10. Figure 12 shows the sail $\overline{AB}$ of an iceboat and a particle P of air moving with velocity $\mathbf{V}_1$ perpendicular to the iceboat's track. Suppose that $\mathbf{V}_2$ is the velocity of the iceboat and that the sail $\overline{AB}$ makes an angle θ with the iceboat's track. Assume that P is moving at such a speed that it just maintains contact with the sail.

(a) Show that $|\mathbf{V}_1| = |\mathbf{V}_2| \tan \theta$.

(b) Explain why an iceboat can be propelled by a wind at a speed greater than the speed of the wind.

Figure 12

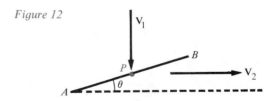

ⓒ 11. Three children located at points A, B, and C tug on ropes attached to a ring at point O (Figure 13). The child at A pulls with a force of 32 newtons in the direction S71°W; the child at B pulls with a force of 96 newtons in the direction S21°E; and the child at C pulls with a force of 56 newtons in the direction N82°E. In what direction and with what force should a fourth child pull on a rope attached to the ring in order to counterbalance the other three forces?

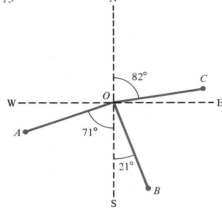

Figure 13

12. Suppose that two forces $\mathbf{F}_1$ and $\mathbf{F}_2$ act on a point and that α is the angle from $\mathbf{F}_1$ to $\mathbf{F}_2$. Let $\mathbf{F}$ be the resultant of $\mathbf{F}_1$ and $\mathbf{F}_2$ and let θ be the angle from $\mathbf{F}_1$ to $\mathbf{F}$. Show that:

(a) $|\mathbf{F}| = (|\mathbf{F}_1|^2 + |\mathbf{F}_2|^2 + 2|\mathbf{F}_1||\mathbf{F}_2|\cos \alpha)^{1/2}$

(b) $\theta = \sin^{-1} \dfrac{|\mathbf{F}_2| \sin \alpha}{|\mathbf{F}|}$

In Problems 13 to 18, sketch the graph of each polar equation.

13. $r = 1$

14. $r^2 = 9$

15. $\theta = \dfrac{\pi}{3}$

16. $\theta^2 = \dfrac{\pi^2}{16}$

17. $\theta + \dfrac{\pi}{2} = 0$

18. $\theta^2 - \dfrac{4\pi}{3}\theta + \dfrac{7\pi^2}{36} = 0$

In Problems 19 to 24, find a polar equation corresponding to each Cartesian equation and sketch the graph. Simplify your answer if possible.

19. $x^2 + y^2 = 25$

20. $y = 6x^2$

21. $3x - 2y = 6$

22. $xy = 1$

23. $(x - 1)^2 + y^2 = 1$

24. $y = \sin x$

In Problems 25 to 30, find a Cartesian equation corresponding to each polar equation and sketch the graph. Simplify your answer if possible.

25. $r = 1$

26. $r = 3 \cos \theta$

27. $r = \dfrac{1}{\sin \theta - 2 \cos \theta}$

28. $\theta = \dfrac{\pi}{12}$

29. $r \cos \theta = \tan \theta$

30. $r + \dfrac{1}{r} = 2(\cos \theta + \sin \theta)$

In Problems 31 to 34, use the point-plotting method to sketch the graph of each polar equation.

31. $r = 2(1 - \sin \theta)$ (This is called a *cardioid*.)

32. $r = 3 \sin 3\theta$ (This is called a *three-leaved rose*.)

33. $r = \dfrac{60}{\pi}$ (This is called an *Archimedean spiral*.)

34. $r = 3 - 2 \cos \theta$ (This is called a *limaçon*.)

8.7 TRIGONOMETRIC FORM AND *n*th ROOTS OF COMPLEX NUMBERS

We begin this section by giving an answer to the question, temporarily put aside in Section 1.8, of just what complex numbers are. Although there are several possible representations of the complex numbers, their interpretation as geometric points in the plane has the most intuitive appeal. According to this scheme, if a and b are

Figure 1

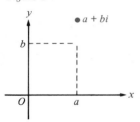

real numbers, then the complex number $a + bi$ is represented by the point (a, b) in the xy plane (Figure 1). When each point in a coordinate plane is made to correspond to a complex number in this way, we refer to the plane as the **complex plane.** This correspondence is so compelling that it is natural to *identify* a complex number with its corresponding geometric point, and we shall do so in what follows. Under this identification, there's nothing at all imaginary about i or any other complex number; nevertheless, the word "imaginary" continues to be used for historical reasons.

In the complex plane, the x axis is called the **real axis** because its points correspond to real numbers, and the y axis is called the **imaginary axis** because its points correspond to multiples of i by real numbers. Figure 2 shows the complex number $z = x + yi$ with real part x and imaginary part y. The distance r between the point $x + yi$ and the origin is called the **absolute value*** of z and denoted by

Figure 2

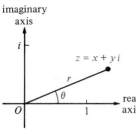

$$|z| = |x + yi| = \sqrt{x^2 + y^2}.$$

An angle θ in standard position that contains the point $x + yi$ on its terminal side is called an **argument** of z. Thus, the absolute value $r = |x + yi|$ and an argument θ of $x + yi$ form polar coordinates of the point $x + yi$. It follows that

$$x = r\cos\theta \quad \text{and} \quad y = r\sin\theta.$$

(Section 8.5, page 458).

If a complex number z is written as

$$z = x + yi,$$

we say that it is expressed in **Cartesian form.** If $r = |z|$ and θ is an argument of z, we can also write z in the **trigonometric** or **polar form**

$$z = r\cos\theta + (r\sin\theta)i = r(\cos\theta + i\sin\theta).$$

Figure 3

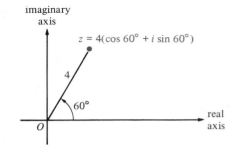

Example 1   Plot the point $z = 4(\cos 60° + i\sin 60°)$ in the complex plane and rewrite z in Cartesian form.

Solution   Here $r = 4$ and $\theta = 60°$. The point z is plotted in Figure 3. In Cartesian form,

$$z = 4\cos 60° + (4\sin 60°)i$$

$$= 4\left(\frac{1}{2}\right) + \left[4\left(\frac{\sqrt{3}}{2}\right)\right]i$$

$$= 2 + 2\sqrt{3}i.$$

---

* Note that $z\bar{z} = x^2 + y^2$, where $\bar{z} = x - yi$ is the complex conjugate of z; hence, we can also write $|z| = \sqrt{z\bar{z}}$. Also note that, if $y = 0$, so that $z = x$ is a real number, then $|z| = \sqrt{x^2}$ is just the usual absolute value of z.

Example 2   Plot the point $z = \sqrt{2} - \sqrt{2}i$ in the complex plane and rewrite z in polar form.

Solution   The point z is plotted in Figure 4. Because the real and imaginary parts of z are $x = \sqrt{2}$ and $y = -\sqrt{2}$, it follows that

$$r = \sqrt{x^2 + y^2}$$
$$= \sqrt{(\sqrt{2})^2 + (-\sqrt{2})^2}$$
$$= \sqrt{2 + 2} = 2.$$

Figure 4

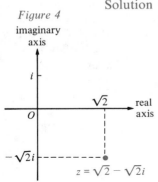

Also,

$$\cos \theta = \frac{x}{r} = \frac{\sqrt{2}}{2} \quad \text{and} \quad \sin \theta = \frac{y}{r} = \frac{-\sqrt{2}}{2}, \quad \text{so} \quad \theta = \frac{7\pi}{4}.$$

Thus,

$$z = 2\left(\cos \frac{7\pi}{4} + i \sin \frac{7\pi}{4}\right).$$

■

Although complex numbers are most easily added or subtracted when they are written in Cartesian form, multiplication and division are more easily carried out when the numbers are expressed in polar form. The following theorem shows why.

Theorem 1   **Multiplication and Division in Polar Form**

> Let $z_1 = r_1(\cos \theta_1 + i \sin \theta_1)$ and $z_2 = r_2(\cos \theta_2 + i \sin \theta_2)$. Then
>
> **(i)** $z_1 z_2 = r_1 r_2 [\cos(\theta_1 + \theta_2) + i \sin(\theta_1 + \theta_2)]$ and, provided that $z_2 \neq 0$,
>
> **(ii)** $\dfrac{z_1}{z_2} = \dfrac{r_1}{r_2} [\cos(\theta_1 - \theta_2) + i \sin(\theta_1 - \theta_2)]$.

Proof   We prove (i) and leave the proof of (ii) as an exercise (Problem 58).

$$z_1 z_2 = r_1(\cos \theta_1 + i \sin \theta_1) r_2(\cos \theta_2 + i \sin \theta_2)$$
$$= r_1 r_2 (\cos \theta_1 + i \sin \theta_1)(\cos \theta_2 + i \sin \theta_2)$$
$$= r_1 r_2 [(\cos \theta_1 \cos \theta_2 - \sin \theta_1 \sin \theta_2) + i(\sin \theta_1 \cos \theta_2 + \sin \theta_2 \cos \theta_1)]$$
$$= r_1 r_2 [\cos(\theta_1 + \theta_2) + i \sin(\theta_1 + \theta_2)].$$

■

In words, part (i) of Theorem 1 says:

> The absolute value of the product of two complex numbers is the product of their absolute values; whereas, an argument of the product is given by the sum of their arguments.

Part (ii) can be expressed in a similar way (Problem 55).

Example 3   If $z = 6\left(\cos\dfrac{\pi}{2} + i\sin\dfrac{\pi}{2}\right)$ and $w = 2\left(\cos\dfrac{\pi}{6} + i\sin\dfrac{\pi}{6}\right)$, find:

(a) zw **(b)** $\dfrac{z}{w}$.

Solution   Using Theorem 1, we have

(a) $zw = (6)(2)\left[\cos\left(\dfrac{\pi}{2} + \dfrac{\pi}{6}\right) + i\sin\left(\dfrac{\pi}{2} + \dfrac{\pi}{6}\right)\right] = 12\left(\cos\dfrac{2\pi}{3} + i\sin\dfrac{2\pi}{3}\right)$

(b) $\dfrac{z}{w} = \dfrac{6}{2}\left[\cos\left(\dfrac{\pi}{2} - \dfrac{\pi}{6}\right) + i\sin\left(\dfrac{\pi}{2} - \dfrac{\pi}{6}\right)\right] = 3\left(\cos\dfrac{\pi}{3} + i\sin\dfrac{\pi}{3}\right)$ ∎

Powers and Roots of Complex Numbers

Abraham De Moivre

If $z = r(\cos\theta + i\sin\theta)$, then, by part (i) of Theorem 1, $z^2 = r^2(\cos 2\theta + i\sin 2\theta)$. Therefore,

$$\begin{aligned}
z^3 &= z^2 z \\
&= [r^2(\cos 2\theta + i\sin 2\theta)][r(\cos\theta + i\sin\theta)] \\
&= r^3(\cos 3\theta + i\sin 3\theta).
\end{aligned}$$

A similar computation (Problem 57) shows that

$$z^4 = r^4(\cos 4\theta + i\sin 4\theta).$$

The pattern emerging here is quite clear. It is expressed formally in the following theorem, which is attributed to Abraham De Moivre (1667–1754), a Frenchman turned Englishman.

Theorem 3   **De Moivre's Theorem**

If $z = r(\cos\theta + i\sin\theta)$ and n is any positive integer, then

$$z^n = r^n(\cos n\theta + i\sin n\theta).$$

In words, De Moivre's theorem can be stated as follows:

To raise a complex number to a positive integer power n, you raise its absolute value to the power n and multiply its argument by n.

A formal proof of De Moivre's theorem can be made by using the principle of mathematical induction discussed in Section 11.1.

In Examples 4 and 5, use De Moivre's theorem to evaluate each expression.

Example 4 $[2(\cos 10° + i \sin 10°)]^3$

Solution $[2(\cos 10° + i \sin 10°)]^3 = 2^3[\cos 3(10°) + i \sin 3(10°)]$
$$= 8(\cos 30° + i \sin 30°)$$
$$= 8\left[\frac{\sqrt{3}}{2} + i\left(\frac{1}{2}\right)\right] = 4\sqrt{3} + 4i.$$ ■

Example 5 $(-1 + \sqrt{3}i)^5$

Solution   We begin by rewriting $-1 + \sqrt{3}i$ in polar form:
$$-1 + \sqrt{3}i = 2\left(\cos\frac{2\pi}{3} + i \sin\frac{2\pi}{3}\right).$$

Therefore,

$$(-1 + \sqrt{3}i)^5 = \left[2\left(\cos\frac{2\pi}{3} + i \sin\frac{2\pi}{3}\right)\right]^5 = 2^5\left(\cos\frac{10\pi}{3} + i \sin\frac{10\pi}{3}\right)$$

$$= 32\left[\cos\left(\frac{4\pi}{3} + 2\pi\right) + i \sin\left(\frac{4\pi}{3} + 2\pi\right)\right] = 32\left(\cos\frac{4\pi}{3} + i \sin\frac{4\pi}{3}\right)$$

$$= 32\left[-\frac{1}{2} + i\left(-\frac{\sqrt{3}}{2}\right)\right] = -16 - 16\sqrt{3}i.$$ ■

If w is a complex number and n is a positive integer, then any complex number z such that

$$z^n = w$$

is called a **complex nth root** of w. For $n = 2$ and for $n = 3$, the complex nth roots of w are called the **complex square roots** and the **complex cube roots** of w. By using the following theorem, you can find all complex nth roots of a nonzero complex number w.

Theorem 3   **Complex nth Roots**

Let $w = R(\cos \phi + i \sin \phi)$ be a nonzero complex number in polar form, and let n be a positive integer. Then there are exactly n different complex roots of w:

$$z_0, z_1, z_2, \ldots, z_{n-1},$$

where $$z_k = \sqrt[n]{R}\left[\cos\left(\frac{\phi}{n} + \frac{2\pi k}{n}\right) + i \sin\left(\frac{\phi}{n} + \frac{2\pi k}{n}\right)\right]$$

for $k = 0, 1, 2, \ldots, n - 1$.

The fact that each of the complex numbers $z_0, z_1, \ldots, z_{n-1}$ in Theorem 3 satisfies the equation $z^n = w$ is a consequence of De Moivre's theorem (Problem 56). We omit the proof that these complex numbers are different from each other and are *all* of the *n*th roots of *w*.

Example 6 Find the complex fourth roots of $w = -\dfrac{9}{2} - \dfrac{9\sqrt{3}}{2} i$.

Solution We begin by rewriting *w* in polar form:

$$w = 9\left(\cos\frac{4\pi}{3} + i\sin\frac{4\pi}{3}\right).$$

Then we use Theorem 3 with $R = 9$, $\phi = 4\pi/3$, and $n = 4$ to obtain the four complex fourth roots:

$$z_0 = \sqrt[4]{9}\left[\cos\left(\frac{4\pi/3}{3} + 0\right) + i\sin\left(\frac{4\pi/3}{3} + 0\right)\right] = \sqrt{3}\left(\cos\frac{\pi}{3} + i\sin\frac{\pi}{3}\right)$$

$$z_1 = \sqrt[4]{9}\left[\cos\left(\frac{4\pi/3}{4} + \frac{2\pi}{4}\right) + i\sin\left(\frac{4\pi/3}{4} + \frac{2\pi}{4}\right)\right] = \sqrt{3}\left(\cos\frac{5\pi}{6} + i\sin\frac{5\pi}{6}\right)$$

$$z_2 = \sqrt[4]{9}\left[\cos\left(\frac{4\pi/3}{4} + \frac{4\pi}{4}\right) + i\sin\left(\frac{4\pi/3}{4} + \frac{4\pi}{4}\right)\right] = \sqrt{3}\left(\cos\frac{4\pi}{3} + i\sin\frac{4\pi}{3}\right)$$

$$z_3 = \sqrt[4]{9}\left[\cos\left(\frac{4\pi/3}{4} + \frac{6\pi}{4}\right) + i\sin\left(\frac{4\pi/3}{4} + \frac{6\pi}{4}\right)\right] = \sqrt{3}\left(\cos\frac{11\pi}{6} + i\sin\frac{11\pi}{6}\right).$$

These roots can be rewritten in Cartesian form as

$$z_0 = \frac{\sqrt{3}}{2} + \frac{3}{2}i, \qquad z_1 = -\frac{3}{2} + \frac{\sqrt{3}}{2}i, \qquad z_2 = -\frac{\sqrt{3}}{2} - \frac{3}{2}i, \qquad z_3 = \frac{3}{2} - \frac{\sqrt{3}}{2}i. \qquad \blacksquare$$

If you plot the four complex roots z_0, z_1, z_2, and z_3 of

$$w = -\frac{9}{2} - \frac{9\sqrt{3}}{2}i$$

obtained in Example 6, you will find that they are equally spaced around a circle of radius $\sqrt{3}$ with center at the origin (Figure 5, page 474). More generally:

> The complex *n*th roots of a nonzero complex number *w* are equally spaced around a circle of radius $\sqrt[n]{|w|}$ with center at the origin.

We leave the proof of this fact as an exercise (Problem 54).

Figure 5

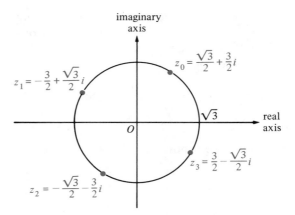

Problem Set 8.7

In Problems 1 to 8, plot each point z in the complex plane and rewrite z in Cartesian form.

1. $z = 4(\cos 30° + i \sin 30°)$

2. $z = 6\left(\cos \dfrac{\pi}{4} + i \sin \dfrac{\pi}{4}\right)$

3. $z = 7\left(\cos \dfrac{3\pi}{4} + i \sin \dfrac{3\pi}{4}\right)$

4. $z = 8\,(\cos 270° + i \sin 270°)$

© **5.** $z = 6\,(\cos 370° + i \sin 370°)$

© **6.** $z = 6\left(\cos \dfrac{7\pi}{10} + i \sin \dfrac{7\pi}{10}\right)$

7. $z = 2\left[\cos\left(-\dfrac{11\pi}{6}\right) + i \sin\left(-\dfrac{11\pi}{6}\right)\right]$

8. $z = 12\,(\cos 240° + i \sin 240°)$

In Problems 9 to 18, plot each point z in the complex plane and rewrite z in polar form.

9. $z = -1 + i$

10. $z = 3 - 3i$

11. $z = 3i$

12. $z = 4$

© **13.** $z = 5 + \sqrt{3}\,i$

© **14.** $z = -2 - \sqrt{3}\,i$

15. $z = -\sqrt{3} - i$

16. $z = \sqrt{3} - i$

17. $z = -5$

18. $z = -i$

In Problems 19 to 26, use Theorem 1 to find (a) zw and (b) z/w. Leave the results in polar form.

19. $z = 4\,(\cos 70° + i \sin 70°)$, $w = 2\,(\cos 40° + i \sin 40°)$

20. $z = 8\,(\cos 80° + i \sin 80°)$, $w = 4\,(\cos 20° + i \sin 20°)$

21. $z = 14\left(\cos \dfrac{3\pi}{2} + i \sin \dfrac{3\pi}{2}\right)$,

$w = 7\left(\cos \dfrac{5\pi}{4} + i \sin \dfrac{5\pi}{4}\right)$

22. $z = 15\left(\cos \dfrac{4\pi}{3} + i \sin \dfrac{4\pi}{3}\right)$,

$w = 5\left[\cos\left(-\dfrac{\pi}{3}\right) + i \sin\left(-\dfrac{\pi}{3}\right)\right]$

23. $z = 6(\cos 90° + i \sin 90°)$, $w = 3(\cos 45° + i \sin 45°)$

24. $z = 8(\cos 85° + i \sin 85°)$, $w = 4(\cos 55° + i \sin 55°)$

25. $z = 1 + i$, $w = 1 + \sqrt{3}\,i$ [Change to polar form.]

26. $z = \sqrt{3} + i$, $w = 1 - i$ [Change to polar form.]

In Problems 27 to 38, use De Moivre's theorem to evaluate each expression. Write the result in Cartesian form.

27. $[2(\cos 9° + i \sin 9°)]^5$

28. $[\sqrt{2}(\cos 15° + i \sin 15°)]^{10}$

29. $\left[\sqrt{3}\left(\cos \dfrac{\pi}{10} + i \sin \dfrac{\pi}{10}\right)\right]^{10}$

30. $\left[2\left(\cos \dfrac{2\pi}{9} + i \sin \dfrac{2\pi}{9}\right)\right]^6$

31. $\left[2\left(\cos \dfrac{\pi}{12} + i \sin \dfrac{\pi}{12}\right)\right]^9$

32. $\left(\cos \dfrac{2\pi}{5} + i \sin \dfrac{2\pi}{5}\right)^{15}$

33. $(\sqrt{3} - i)^5$

34. $(-1 - \sqrt{3}i)^8$

© **35.** $(-1 + 2i)^{12}$ **36.** $(1 + i)^{16}$

37. $\left(\dfrac{1 + i}{\sqrt{2}}\right)^7$ **©** **38.** $(2 + 3i)^{11}$

In Problems 39 to 48, use Theorem 3 to find all of the indicated complex *n*th roots.

39. The square roots of $\dfrac{9}{2} - \dfrac{9\sqrt{3}}{2} i$.

40. The cube roots of $-8i$.

41. The cube roots of $-1 + i$.

42. The fourth roots of $-2 + 2\sqrt{3}i$.

43. The fourth roots of $8 - 8\sqrt{3}i$.

44. The fifth roots of 1.

45. The fourth roots of -1.

46. The sixth roots of $-64i$.

47. The fifth roots of $-16\sqrt{2} - 16\sqrt{2}i$.

48. The fifth roots of $-\sqrt{3} - i$.

49. (a) Show that De Moivre's theorem (Theorem 2) holds for $n = 0$ and for $n = -1$, provided that $z \neq 0$.

(b) Using the fact that $z^{-n} = 1/z^n$ for $z \neq 0$, show that De Moivre's theorem holds for all integer values of n, provided that $z \neq 0$.

50. Suppose that $z = a + bi$, where a and b are real numbers, and let $r = |z|$.

(a) Show that $r + a \geq 0$.

(b) Show that $r - a \geq 0$.

(c) Let $x = \sqrt{\dfrac{r + a}{2}}$ and $y = \pm\sqrt{\dfrac{r - a}{2}}$, where the algebraic sign is chosen so that xy and b have the same algebraic sign. Show that $x + yi$ is a square root of z.

(d) Show that $-x - yi$ is a square root of z.

51. If n is a positive integer, the complex nth roots of 1 are called the **nth roots of unity.** Use Theorem 3 to find all nth roots of unity for

(a) $n = 2$ (b) $n = 3$ (c) $n = 4$.

52. If n is a positive integer, show that:

(a) The product of two nth roots of unity (Problem 51) is also an nth root of unity.

(b) The reciprocal of an nth root of unity is also an nth root of unity.

(c) All nth roots of unity lie on the unit circle in the complex plane.

53. If z and w are points in the complex plane, give a geometric interpretation of

(a) $\bar{z}$ (b) $-z$ (c) $-\bar{z}$
(d) iz (e) $|z - w|$.

54. Let w be a fixed nonzero complex number and let n be a positive integer. Show that the complex nth roots of w are equally spaced around a circle of radius $\sqrt[n]{|w|}$ with center at the origin.

55. Express part (ii) of Theorem 1 in words.

56. Using De Moivre's theorem, show that $z_0, z_1, \ldots,$ z_{n-1} in Theorem 3 are nth roots of w.

57. If $z = r(\cos \theta + i \sin \theta)$, show that $z^4 = r^4 (\cos 4\theta + i \sin 4\theta)$ by direct use of Theorem 1.

58. Prove part (ii) of Theorem 1.

REVIEW PROBLEM SET, CHAPTER 8

© In Problems 1 to 4, assume that the right triangle ACB is labeled as in Figure 1. In each case, solve the triangle and sketch it. Round off all angles to the nearest hundredth of a degree and all side lengths to four significant digits.

Figure 1

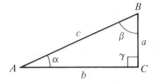

1. $c = 200$ meters, $\alpha = 31.25°$

2. $a = 92.70$ kilometers, $\alpha = 34.45°$

3. $b = 47.33$ microns, $\beta = 56.15°$

4. $b = 1.151 \times 10^4$ meters, $c = 4.703 \times 10^4$ meters

© **5.** A straight sidewalk is inclined at an angle of $4.5°$ from the horizontal. How far must a person walk along this sidewalk to change his or her elevation by 2 meters?

© **6.** A ship in distress at night fires a signal rocket straight upward. At an altitude of 0.2 kilometer, the rocket explodes with a brilliant flash that is observed at an angle of elevation of $15.75°$ by an approaching rescue ship. How far is the rescue ship from the ship in distress?

© **7.** A broadcasting antenna tower 200 meters high is to be held vertical by three cables running from a point 10 meters below the top of the tower to concrete anchors sunk in the ground (Figure 2). If the

Figure 2

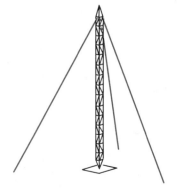

cables are to make angles of $60°$ with the horizontal, how many meters of cable will be required?

8. Let θ be the angle of depression of the horizon as seen by an astronaut A in a space vehicle h meters above the surface of a planet of radius r meters (Figure 3). Show that $h = r(\sec \theta - 1)$.

Figure 3

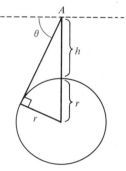

© **9.** A certain species of bird is known to fly at an average altitude of 260 meters during migration. An ornithologist observes a migrating flock of these birds flying directly away from her at an angle of elevation of $30°$. One minute later, she observes that the angle of elevation of the flock has changed to $20°$. Approximately how fast are the birds flying?

© **10.** In dealing with alternating current circuits, electrical engineers measure the opposition to current flow by a quantity called **impedance,** whose magnitude Z is measured in ohms. In a circuit driven by an alternating electromotive force (emf) and containing only a resistor and an inductor connected in series, the resistor and inductor both contribute to the impedance. If the resistance is R ohms, the inductance is L henrys, and the driving emf has a frequency of v hertz, the relationships among Z, R, L, and v is shown by an **impedance triangle** as in Figure 4. The angle ϕ is called the

Figure 4

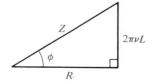

phase and the quantity $2\pi vL$ is called the **inductive reactance.** If $R = 4000$ ohms, $L = 2$ henrys, and $v = 60$ hertz, find Z and ϕ.

© **11.** An electronic echo locator on a commercial fishing boat indicates a school of fish at a slant distance of 575 meters from the boat with an angle of depression of 33.60°. What is the depth of the school of fish?

© **12.** A jetliner is flying at an altitude of 10 kilometers. The captain is preparing for a descent to an airport along a straight flight path making an angle of 5° with the horizontal. If the descent will take 12 minutes, what will be the average airspeed of the plane during the descent?

© **13.** A ship leaves port and sails S48°10′E for 41 nautical miles. At this point, it turns and sails S41°50′W for 75 nautical miles. Find the bearing of the ship from the port and its distance from the port.

© **14.** Two buildings A and B are 2175 feet apart; the angle of depression from the top of B to the top of A is 12.17°; and the angle of depression from the top of B to the bottom of A is 53.35°. Find the heights of the buildings.

© In Problems 15 to 20, use the law of sines to find the specified part of each triangle ABC (Figure 5). Round off angles to the nearest hundredth of a degree and side lengths to four significant digits.

Figure 5

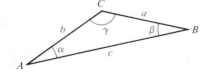

15. Find a if $b = 13$ feet, $\alpha = 47°$, and $\gamma = 118°$.

16. Find b if $a = 8.12$ centimeters, $\alpha = 62.45°$, and $\beta = 79.30°$.

17. Find c if $a = 23$ miles, $\alpha = 65°$, and $\beta = 70°$.

18. Find a if $c = 13.70$ kilometers, $\alpha = 62°50′$, and $\beta = 57°30′$.

19. Find γ if $b = 24$ meters, $c = 14$ meters, and $\beta = 38°$.

20. Find α if $a = 5.88$ microns, $c = 12.35$ microns, and $\gamma = 106.55°$.

© In Problems 21 to 26, solve each triangle ABC (Figure 5) by using the method for the ambiguous case. Round off angles to the nearest hundredth of a degree and side lengths to four significant digits. Be certain to find all possible triangles that satisfy the given conditions.

21. $b = 50\sqrt{3}$ feet, $c = 150$ feet, $\beta = 30°$

22. $a = 4.50$ kilometers, $b = 5.30$ kilometers, $\alpha = 60.33°$

23. $b = 50$ centimeters, $c = 58$ centimeters, $\gamma = 57.25°$

24. $b = 5.94$ angstroms, $c = 7.23$ angstroms, $\beta = 38°$

25. $a = 8.00$ miles, $b = 10.00$ miles, $\alpha = 54°$

26. $a = 40.33$ nautical miles, $b = 42.01$ nautical miles, $\alpha = 110.05°$

© In Problems 27 to 32, use the law of cosines to find the specified part of each triangle ABC (Figure 5). Round off angles to the nearest hundredth of a degree and side lengths to four significant digits.

27. Find α if $a = 7$ feet, $b = 8$ feet, and $c = 3$ feet.

28. Find a if $b = 11$ meters, $c = 12$ meters, and $\alpha = 81°$.

29. Find c if $a = 14$ kilometers, $b = 8$ kilometers, and $\gamma = 37°$.

30. Find b if $a = 7$ centimeters, $c = 11$ centimeters, and $\beta = 53.35°$.

31. Find a if $b = 12.30$ inches, $c = 4.85$ inches, and $\alpha = 161.15°$.

32. Find γ if $a = 48.31$ meters, $b = 35.11$ meters, and $c = 63.27$ meters.

© In Problems 33 to 38, use an appropriate method to solve each triangle ABC (Figure 5). Round off angles to the nearest hundredth of a degree and side lengths to four significant digits.

33. $a = 49.7$ kilometers, $b = 111$ kilometers, $\gamma = 41.05°$

34. $\alpha = 37.17°$, $\beta = 82.25°$, $a = 7777$ meters

35. $a = 1.12$ microns, $c = 0.98$ micron, $\gamma = 55°$

36. $a = 33.39$ centimeters, $b = 72.27$ centimeters, $c = 106.2$ centimeters

37. $\alpha = 42.45°$, $\beta = 32.15°$, $b = 1.41$ nautical miles

38. $a = 1.80$ millimeters, $b = 1.20$ millimeters, $\alpha = 47°$

© **39.** The three sides of a triangle have lengths $a = 27$ meters, $b = 44$ meters, and $c = 65$ meters. Use Hero's formula (page 445) to find the area of the triangle.

© **40.** A section of West Germany is called the brown-coal triangle because it contains vast reserves of lignite (brown coal). It is a triangular area with vertices at the industrial cities of Aachen, Cologne, and Düsseldorf. The distance from Aachen to Cologne is 67 kilometers; the distance from Cologne to Düsseldorf is 37 kilometers; and the distance from Düsseldorf to Aachen is 75 kilometers. Find the approximate area of the brown-coal triangle in square kilometers.

© **41.** In the western North Atlantic there is a mysterious region, shaped roughly like a triangle with sides of length 925 miles, 850 miles, and 1300 miles, that has been the scene of a number of unexplained disappearances and disasters involving both ships and aircraft. Find the approximate area of this region in square miles.

© **42.** Engineers excavating a new subway tunnel in a large city encounter a region of very hard rock (Figure 6). They can either drill the tunnel straight through the rock along path $\overline{AC}$ at a cost of $5500 per foot, or they can divert the tunnel along path $\overline{AB}$ and then path $\overline{BC}$ at a cost of $4700 per foot. If angle $BAC = 20°$ and angle $ACB = 10°$, which path would be the cheaper?

Figure 6

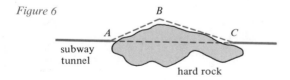

© **43.** Two observers are situated on level ground on opposite sides of a tall building. The top of the building is 1600 meters from the first observer at an angle of elevation of 15°. The top of the building is 650 meters from the second observer. How far apart are the observers?

© **44.** A vertical utility pole is supported on an embankment by a guy wire from 1 meter below the top of the pole to a point 20 meters up the embankment from the foot of the pole (Figure 7). If the embankment makes an angle of 17° with the horizontal

and if the guy wire makes an angle of 21° with the horizontal, how tall is the utility pole?

Figure 7

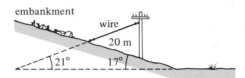

© **45.** A CB radio in a person's car can communicate with a base station located at the person's home over a maximum distance of 12 miles. The car leaves the home and travels due west for 5 miles to an interstate highway. The car moves onto the straight highway at precisely 1:00 P.M. traveling N30°E at 55 miles per hour. At what time (to the nearest minute) is communication lost between the car and the base station?

© **46.** Apollo objects in the solar system are asteriodlike masses whose orbits intersect the orbit of the earth. Geologists believe that several craterlike formations in Canada and elsewhere were caused by collisions of such objects with the earth. An astronomer observes through a telescope that the angle formed by the lines of sight to two Apollo objects is 43.33°, and that the distances to the two objects are 15,000,000 and 43,000,000 kilometers. How far apart are the two Apollo objects?

© **47.** A natural-gas pipeline is to be constructed through a swamp. An engineer obtains the distance across the swamp by establishing points A and B at both ends of it along the path the pipeline is to follow, and finding a third point C outside the swamp at a location that can be seen from both A and B. If $|\overline{CA}| = 1100$ meters, $|\overline{CB}| = 990$ meters, and angle ACB is 75°, what is the distance $|\overline{AB}|$ across the swamp?

© **48.** A group of antinuclear activists is conducting a protest march to the construction site of a nuclear power plant. They are marching at constant speed along a flat plane directly toward a huge concrete cooling tower that has already been finished. At 7:00 A.M. the leader of the march observes that the top of the cooling tower is at an angle of elevation of 2°. At 7:30 A.M. the angle of elevation has increased to 5°. Estimate to the nearest minute the time of arrival of the protest group at the construction site.

C 49. A subatomic particle that has been created at point C in a bubble chamber travels in a straight line for 3.5 cm to point B, where it collides with another particle and its path is deflected. It then travels in a straight line from B for 2.3 cm to point A, where it is annihilated. If $|\overline{CA}| = 4.4$ cm, find the deflection angle $\theta = 180° -$ angle CBA.

50. Two small ships are traveling at constant speeds on straight-line courses that intersect. The captain of the first ship measures the bearing of the second ship at the beginning and at the end of a 2-minute interval. The bearing has changed during this interval, so the captain concludes that the ships will not collide. Is this reasoning correct? Why or why not?

In Problems 51 to 62, let $\mathbf{A} = \langle 2, -1 \rangle$, $\mathbf{B} = \langle 3, 3 \rangle$, and $\mathbf{C} = \langle -5, 4 \rangle$. Evaluate each expression.

51. $5\mathbf{A}$

52. $-4\mathbf{B}$

53. $\mathbf{A} + \mathbf{B}$

54. $\mathbf{A} - \mathbf{B}$

55. $2\mathbf{A} + 3\mathbf{B}$

56. $3\mathbf{A} + \mathbf{C}$

57. $2\mathbf{C} - \mathbf{B}$

58. $3\mathbf{B} + 4\mathbf{C} - \mathbf{A}$

59. $|4\mathbf{A}|$

60. $|\mathbf{A} - \mathbf{B}|$

61. $|\mathbf{A} + \mathbf{B} - \mathbf{C}|$

62. $|2\mathbf{C} - 3\mathbf{B} + \mathbf{A}|$

63. Find the components of the vector $\mathbf{A}$ from the given information about its length $|\mathbf{A}|$ and its direction angle θ:

(a) $|\mathbf{A}| = 50, \theta = 30°$

C (b) $|\mathbf{A}| = 70, \theta = 61°$

C (c) $|\mathbf{A}| = 25, \theta = 310°$

(d) $|\mathbf{A}| = 250, \theta = -\dfrac{\pi}{6}$

64. Give an example to show that in general $|\mathbf{A} + \mathbf{B}| \neq |\mathbf{A}| + |\mathbf{B}|$. Are there any cases in which equality does hold?

65. Find the magnitude and the smallest nonnegative direction angle of each vector:

(a) $\langle 1, \sqrt{3} \rangle$ **C** (b) $\langle 12, 15 \rangle$

C (c) $\langle -30, 40 \rangle$ **C** (d) $\langle \frac{3}{16}, \frac{1}{4} \rangle$

C (e) $3\mathbf{i} + \sqrt{7}\mathbf{j}$

66. Give an example to show that the direction angle of the sum of two vectors is not necessarily the sum of their direction angles.

C 67. Suppose that $\mathbf{F}_1$ and $\mathbf{F}_2$ represent forces acting on a point and that α is the angle from $\mathbf{F}_1$ to $\mathbf{F}_2$. If $\mathbf{F}$ denotes the resultant force, find the magnitude $|\mathbf{F}|$ and the angle θ from $\mathbf{F}_1$ to $\mathbf{F}$ in each of the following cases:

(a) $|\mathbf{F}_1| = 20$ pounds, $|\mathbf{F}_2| = 18$ pounds, $\alpha = 71°$

(b) $|\mathbf{F}_1| = 139$ newtons, $|\mathbf{F}_2| = 156$ newtons, $\alpha = 83°$

(c) $|\mathbf{F}_1| = 7$ tonnes, $|\mathbf{F}_2| = 9$ tonnes, $\alpha = 125°$

(d) $|\mathbf{F}_1| = 300$ dynes, $|\mathbf{F}_2| = 450$ dynes, $\alpha = 270°$

C 68. Find the angle θ from the position vector $\mathbf{R}_1 = \langle 5, 3 \rangle$ to the position vector $\mathbf{R}_2 = \langle -7, 2 \rangle$.

C 69. A cargo jet is headed N45°W with an air speed of 500 knots. The wind is blowing N60°E at a speed of 70 knots. Find:

(a) the ground speed   (b) the drift angle

(c) the course of the jet

C 70. Using ropes, two workers are gently lowering a stone weighing 60 kilograms as shown in Figure 8. The ropes make angles of 60° and 45° with the horizontal. Make a vector diagram showing all three forces acting on the stone, including the force of gravity. Using the fact that all three forces are in equilibrium—that is, their vector sum is $\mathbf{0}$— find the magnitude of the force exerted by each of the workers on their ropes.

Figure 8

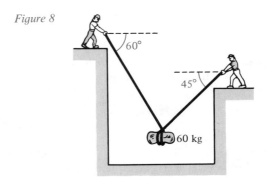

C 71. An oceanographic research vessel leaves Miami on a course for a small island whose bearing is S43°20′E. The vessel is moving through the water at a constant speed of 20 knots, and the captain has charted the course to compensate for the fact that a Gulf Stream current of 7 knots is running due north. Determine the heading of the vessel.

72. If a force represented by a vector **F** acts on a particle and produces a displacement represented by a vector **D**, the *work* done is given by $|\mathbf{F}|\,|\mathbf{D}|$ cos α, where α is the angle between **F** and **D**. Calculate the work done by the force $\mathbf{F} = \langle -3, 4 \rangle$ in producing the displacement $\mathbf{D} = \langle 0, 7 \rangle$. [If you use the law of cosines, you won't need a calculator to solve this problem.]

73. Convert the polar coordinates to Cartesian coordinates:

(a) $(2, 45°)$
(b) $\left(-2, \dfrac{\pi}{4}\right)$
© (c) $(-1, 280°)$
(d) $(1.5, 0°)$
(e) $(-4, -210°)$
(f) $(6, 315°)$

74. Plot the point $P = (-5, \pi/6)$ in the polar coordinate system and give other polar representations $P = (r, \theta)$ of P that satisfy each of the following conditions:

(a) $r < 0$ and $0 \le \theta < 2\pi$
(b) $r > 0$ and $-2\pi < \theta \le 0$
(c) $r < 0$ and $-2\pi < \theta \le 0$

75. Convert the Cartesian coordinates to polar coordinates with $r \ge 0$ and $-\pi < \theta \le \pi$. If an approximation is necessary, round off the angle θ to the nearest hundredth of a radian.

(a) $(-1, 0)$
(b) $(-7, -7)$
© (c) $(-5, 12)$
© (d) $(3, -4)$
© (e) $(-8, 15)$
© (f) $(-30, -16)$

76. Show that the distance d between the points (r_1, θ_1) and (r_2, θ_2) in the polar coordinate system is given by the formula

$$d = \sqrt{r_1^2 - 2r_1 r_2 \cos(\theta_1 - \theta_2) + r_2^2}.$$

In Problems 77 to 84, convert each Cartesian equation into a corresponding polar equation by making the substitutions $x = r \cos\theta$, $y = r \sin\theta$, and $x^2 + y^2 = r^2$. Simplify your answer if possible.

77. $y = 3x + 1$

78. $(x - 1)^2 + (y - 2)^2 = 4$

79. $x^2 + y^2 = 4y$

80. $y^2(2 - x) = x^3$

81. $y^2 = 6x$

82. $(x^2 + y^2)^2 = x^2 - y^2$

83. $4x^2 + y^2 = 4$

84. $x^2 - 4y^2 = 4$

In Problems 85 to 88, convert each polar equation into a corresponding Cartesian equation.

85. $r = \cos\theta + \sin\theta$ 86. $r = \sin 2\theta$

87. $r^2 \cos 2\theta = 2$ 88. $\theta = \dfrac{5\pi}{4}$

89. Using the result of Problem 87, sketch the graph of $r^2 \cos 2\theta = 2$.

90. Using the result in Problem 76, write the polar form of the equation of a circle of radius a with center at (r_0, θ_0).

© In Problems 91 to 94, sketch the graph of each polar equation by the point-plotting method.

91. $r = 4(1 + \cos\theta)$ (This is called a *cardioid*.)

92. $r = 2 - 3\sin\theta$ (This is called a *limaçon with a loop*.)

93. $r = \cos 2\theta$ (This is called a *four-leaved rose*.)

94. $r^2 = \cos 2\theta$ (This is called a *lemniscate*.)

In Problems 95 to 98, rewrite each complex number in Cartesian form.

95. $3(\cos 45° + i \sin 45°)$

96. $6\left[\cos\left(-\dfrac{\pi}{6}\right) + i \sin\left(-\dfrac{\pi}{6}\right)\right]$

97. $16\left(\cos\dfrac{5\pi}{3} + i \sin\dfrac{5\pi}{3}\right)$

98. $\overline{\left(\cos\dfrac{71\pi}{4} + i \sin\dfrac{71\pi}{4}\right)}$

In Problems 99 to 104, find each absolute value.

99. $|-4 - 3i|$ 100. $|6 - 8i|$

101. $|8i^7|$ 102. $|i^{17}|$

103. $\left|\dfrac{3+2i}{3-4i}\right|$ **104.** $\left|\dfrac{-3-2i}{-6+8i}\right|$

In Problems 105 to 108, express each complex number in polar form.

105. $-4+4i$ **106.** $3+0i$

107. $-1+\sqrt{3}i$ **108.** $7-7i$

In Problems 109 to 116, find (a) zw and (b) z/w. Express the answers in polar form.

109. $z=6(\cos 22° + i \sin 22°)$, $w=4(\cos 8° + i \sin 8°)$

110. $z=8(\cos 85° + i \sin 85°)$,
$w=2(\cos 95° + i \sin 95°)$

111. $z=12(\cos 235° + i \sin 235°)$,
$w=4(\cos 125° + i \sin 125°)$

112. $z=14(\cos 78° + i \sin 78°)$,
$w=7(\cos 18° + i \sin 18°)$

113. $z=10\left(\cos\dfrac{13\pi}{36} + i \sin\dfrac{13\pi}{36}\right)$,
$w=5\left(\cos\dfrac{2\pi}{9} + i \sin\dfrac{2\pi}{9}\right)$

114. $z=20\left(\cos\dfrac{143\pi}{180} + i \sin\dfrac{143\pi}{180}\right)$,
$w=4\left(\cos\dfrac{2\pi}{45} + i \sin\dfrac{2\pi}{45}\right)$

115. $z=3+\sqrt{3}i$, $w=\sqrt{3}+i$

116. $z=-3-\sqrt{3}i$, $w=2+2\sqrt{3}i$

In Problems 117 to 122, use De Moivre's theorem to evaluate each expression. Write the answer in Cartesian form.

117. $[2(\cos 300° + i \sin 300°)]^5$

118. $\left[2\left(\cos\dfrac{\pi}{3} + i \sin\dfrac{\pi}{3}\right)\right]^4$

119. $\left[2\left(\cos\dfrac{5\pi}{6} + i \sin\dfrac{5\pi}{6}\right)\right]^6$

120. $(1-\sqrt{3}i)^9$

121. $(-1+i)^6$ **122.** $(1-i)^{12}$

In Problems 123 to 126, find all of the indicated complex nth roots:

123. the cube roots of i

124. the fifth roots of $-\sqrt{3}-i$

125. the fourth roots of $\dfrac{1}{2} - \dfrac{\sqrt{3}}{2}i$

126. the seventh roots of 1

127. Find all complex roots of the equation $x^4 - 1 = 0$.

128. Find all complex roots of the equation $x^3 - i = 0$.

Chapter 9

Systems of Equations and Inequalities

In Chapter 2, we studied methods for solving equations and inequalities in *one unknown*. Now we turn our attention to equations and inequalities containing *two or more unknowns*. Because applications of mathematics frequently involve many unknown quantities, it is very important to be able to deal with more than one unknown. In this chapter, we study systems of equations, determinants, systems of linear inequalities, and linear programming.

9.1 SYSTEMS OF LINEAR EQUATIONS

A collection of two or more equations is called a **system** of equations. The equations in such a system are customarily written in a column enclosed by a brace on the left; for instance,

$$\begin{cases} 2x + y = -4 \\ x + 2y = 1 \end{cases}$$

is a system of two linear equations in the two unknowns x and y. Such a system is usually written as shown, with first-degree polynomials on the left and constants on the right of the equal signs.

If every equation in a system is true when we substitute particular numbers for the unknowns, we say that the substitution is a **solution** of the system. For instance, the substitution $x = -3$, $y = 2$ is a solution of the system

$$\begin{cases} 2x + y = -4 \\ x + 2y = 1. \end{cases}$$

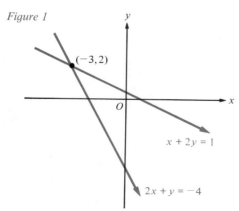

Figure 1

(Why?) Sometimes we write such a solution as an ordered pair $(-3, 2)$, with the value of x first and the value of y second. Then the solution $(-3, 2)$ can be interpreted geometrically as the point where the graph of $2x + y = -4$ intersects the graph of $x + 2y = 1$ (Figure 1).

In this section, we concentrate on solving systems of linear equations—later (in Section 9.6) we study the more general case in which nonlinear equations may be involved. The most general system of two linear equations in x and y can be written as

$$\begin{cases} ax + by = h \\ cx + dy = k, \end{cases}$$

where a, b, c, d, h, and k are constants. Unless both a and b, or both c and d, are zero, the graphs of the equations in this system are straight lines. If you draw these lines on the same coordinate system, one of the following cases will occur:

Case 1. *The two lines intersect at exactly one point and there is exactly one solution (corresponding to this point). In this case, we say that the equations in the system are* **consistent.**

Case 2. *The two lines are parallel, and therefore do not intersect. In this case, there is no solution, and we say that the equations in the system are* **inconsistent.**

Case 3. *The two lines coincide. In this case, every point on the common line corresponds to a solution, and we say that the equations in the system are* **dependent.**

Example 1   Use graphs to determine whether the equations in each system are consistent, inconsistent, or dependent, and indicate the solutions (if any) on the graphs.

(a) $\begin{cases} x - 2y = 2 \\ x + y = 5 \end{cases}$ **(b)** $\begin{cases} x + 2y = 4 \\ 2x + 4y = -3 \end{cases}$ **(c)** $\begin{cases} 2x + 4y = 6 \\ x + 2y = 3 \end{cases}$

Solution   The graphs of the equations in systems (a), (b), and (c) are shown in Figure 2.

(a) In Figure 2a, the graphs intersect at a point, so there is one solution, and the equations are consistent.

(b) In Figure 2b, the two lines have the same slope ($m = -\frac{1}{2}$), so they are parallel, there is no solution, and the equations are inconsistent.

(c) In Figure 2c, the two equations have the same graph, so there are infinitely many solutions (one for each point on the graph) and the equations are dependent. ∎

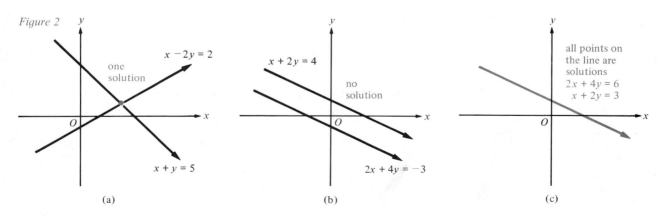

Figure 2

The Substitution Method

There are various algebraic methods for solving systems of equations. The **substitution method** works as follows:

> Choose one of the equations and solve it for *one* of the unknowns in terms of the remaining ones. Then substitute this solution into the remaining equation or equations. You will then have a system involving one fewer equation and one fewer unknown.

If necessary, you can repeat this procedure until the resulting system becomes simple enough to be solved.

In Examples 2 and 3, use the substitution method to solve each system of equations.

Example 2
$$\begin{cases} 2x + 3y = 1 \\ 3x - y = 7 \end{cases}$$

Solution   The second equation is easily solved for y in terms of x to obtain

$$y = 3x - 7.$$

We now substitute $3x - 7$ for y in the first equation $2x + 3y = 1$ to get

$$2x + 3(3x - 7) = 1$$

or

$$2x + 9x - 21 = 1,$$

that is,

$$11x = 22 \quad \text{or} \quad x = 2.$$

To obtain the corresponding value of y, we go back to the equation $y = 3x - 7$ and substitute $x = 2v$; hence,

$$y = 3(2) - 7 = -1.$$

Therefore, the solution is $x = 2$ and $y = -1$, or $(2, -1)$.   ■

Example 3
$$\begin{cases} 2x + y - z = 3 \\ 2x - 2y + 8z = -24 \\ x + 3y + 5z = -2 \end{cases}$$

Solution   We begin by solving the first equation for z in terms of x and y:

$$z = 2x + y - 3.$$

Now, we substitute $2x + y - 3$ for z in the second and third equations to obtain

$$\begin{cases} 2x - 2y + 8(2x + y - 3) = -24 \\ x + 3y + 5(2x + y - 3) = -2. \end{cases}$$

Simplifying these two equations, we have

$$\begin{cases} 18x + 6y = 0 \\ 11x + 8y = 13. \end{cases}$$

This simpler system can now be solved by another use of the substitution procedure. Solving the equation $18x + 6y = 0$ for y in terms of x, we get

$$y = -3x.$$

Substituting $y = -3x$ in the equation $11x + 8y = 13$, we have

$$11x + 8(-3x) = 13 \qquad \text{or} \qquad -13x = 13;$$

hence, $$x = -1.$$

Now, we substitute $x = -1$ in the previous equation $y = -3x$ to get

$$y = -3(-1) = 3.$$

Having found that $x = -1$ and $y = 3$, we need only substitute these values back into the equation $z = 2x + y - 3$ to find that

$$z = 2(-1) + 3 - 3 = -2.$$

Hence, our solution is $x = -1$, $y = 3$, $z = -2$. This solution can also be written as an ordered *triple* $(-1, 3, -2)$. ∎

The Elimination Method

Although the method of substitution can be quite efficient for solving a system of two or three equations, it tends to become cumbersome when more than three equations are involved. A more serious drawback of the substitution method is that it does not lend itself to being programmed on a computer. An alternative method, called the **method of elimination,** is similar in spirit to the method of substitution, but is much more systematic. It leads directly to **matrix methods** of solution, which are easily performed by computers.

The method of elimination is based on the idea of *equivalent* systems of equations. Two systems of equations (linear or not) are said to be **equivalent** if they have exactly the same solutions. You can change a system of equations into an equivalent system by any of the following three **elementary operations:**

1. Interchange the position of two equations in the system.

2. Multiply or divide an equation by a nonzero constant.

3. Add a constant multiple of one equation to another equation.

To solve a system of linear equations by the method of elimination, use a sequence of elementary operations to reduce the given system to a simple equivalent system whose solution is obvious. You do this by using the elementary operations to eliminate unknowns from the equations (which accounts for the name of the method). The following examples illustrate the method of elimination.

Example 4   Use the method of elimination to solve the system of linear equations

$$\begin{cases} \frac{1}{3}x + y = 3 \\ -2x + 5y = 4. \end{cases}$$

Solution   We begin by multiplying the first equation by 3 in order to remove the fraction $\frac{1}{3}$. The result is the equivalent system

$$\begin{cases} x + 3y = 9 \\ -2x + 5y = 4. \end{cases}$$

Next we multiply the first equation by 2 and add it to the second equation in order to eliminate x from the latter. [The actual addition of

$$2x + 6y = 18 \qquad \text{to} \qquad -2x + 5y = 4 \qquad \text{to obtain} \qquad 11y = 22$$

is best done separately to avoid messing up the system with arithmetic calculations.] The resulting equivalent system is

$$\begin{cases} x + 3y = 9 \\ 11y = 22. \end{cases}$$

[Note that the first equation was *not changed* by this operation—it was used to eliminate x from the second equation, but *only* the second equation was changed.] Now we divide the second equation by 11 to obtain the equivalent system

$$\begin{cases} x + 3y = 9 \\ y = 2. \end{cases}$$

The resulting system is so simple that its solution is at hand. We just substitute $y = 2$ from the second equation back into the first equation to get

$$x + 3(2) = 9 \qquad \text{or} \qquad x = 3.$$

Thus, the solution is

$$x = 3 \text{ and } y = 2 \qquad \text{or} \qquad (3, 2). \qquad \blacksquare$$

In the solution above, the final step in which the value of y is substituted into a previous equation is called *back substitution*.

Example 5   Solve the system of linear equations

$$\begin{cases} x + 2y = 3 \\ 3x + 6y = 10. \end{cases}$$

Solution   By adding -3 times the first equation to the second, we obtain

$$\begin{cases} x + 2y = 3 \\ 0 = 1. \end{cases}$$

The last equation cannot be true, so the system has *no* solution. Thus, the equations in the system are inconsistent. $\blacksquare$

Problem Set 9.1

In each problem set, problems with colored numbers constitute a good representation of the main ideas of the section.

In Problems 1 to 10, use graphs to determine whether the equations in each system are consistent, inconsistent, or dependent, and indicate the solutions (if any) on the graphs.

1. $\begin{cases} 4x + y = 5 \\ 3x - y = 2 \end{cases}$

2. $\begin{cases} y = 5x - 2 \\ y = 2x + 1 \end{cases}$

3. $\begin{cases} 3x - y = 4 \\ -6x + 2y = -8 \end{cases}$

4. $\begin{cases} x - y = 4 \\ -3x - 3y = -12 \end{cases}$

5. $\begin{cases} 2x + y = 3 \\ 4x + 2y = 7 \end{cases}$

6. $\begin{cases} 2u - v = 5 \\ u + 3v = -1 \end{cases}$

7. $\begin{cases} x + \frac{1}{3}y = 2 \\ 3x + y = -2 \end{cases}$

8. $\begin{cases} x = 2 \\ y = x \end{cases}$

9. $\begin{cases} \frac{1}{6}x + \frac{1}{4}y = \frac{1}{6} \\ \frac{1}{4}x - \frac{1}{2}y = 2 \end{cases}$

10. $\begin{cases} 0.5x - 1.2y = 0.3 \\ 0.7x + 1.5y = 3.6 \end{cases}$

11. Solve the systems in Problems 1, 3, and 5 by the substitution method.

12. Solve the systems in Problems 2, 4, and 6 by the substitution method.

In Problems 13 to 20, use the substitution method to solve each system of linear equations.

13. $\begin{cases} x - y = -2 \\ 2x - 3y = -7 \end{cases}$

14. $\begin{cases} x = 3 - y \\ 5y = 12 - 2x \end{cases}$

15. $\begin{cases} 2u + 3v = 5 \\ u - 2v = 6 \end{cases}$

16. $\begin{cases} 6s + 5t = 7 \\ 3s - 7t = 13 \end{cases}$

17. $\begin{cases} \frac{1}{2}x - \frac{3}{4}y = 1 \\ 3x + y = 1 \end{cases}$

18. $\begin{cases} 13x + 11y = 21 \\ 7x + 6y = -3 \end{cases}$

19. $\begin{cases} x - 2y + 3z = -3 \\ 2x - 3y - z = 7 \\ 3x + y - 2z = 6 \end{cases}$

20. $\begin{cases} s - 5t + 4u = 8 \\ 3s + t - 2u = 7 \\ -9s - 3t + 6u = 5 \end{cases}$

In Problems 21 to 28, use the method of elimination to solve each system of linear equations.

21. $\begin{cases} 2x + y = 10 \\ 3x - y = 5 \end{cases}$

22. $\begin{cases} 5u + v = 14 \\ 2u + v = 5 \end{cases}$

23. $\begin{cases} 2x + 3y = 18 \\ -7x + 9y = 15 \end{cases}$

24. $\begin{cases} x + \frac{1}{2}y = 2 \\ 3x - y = 1 \end{cases}$

25. $\begin{cases} x + y + z = 6 \\ 2x - y + z = 3 \\ x + 2y - 3z = -4 \end{cases}$

26. $\begin{cases} x - 5y + 4z = 8 \\ 3x + y - 2z = 7 \\ 9x + 3y - 6z = -5 \end{cases}$

27. $\begin{cases} 2x + y + z = 20 \\ x + 2y + 2z = 16 \\ x + y + 2z = 12 \end{cases}$

28. $\begin{cases} 2x_1 + 3x_2 - 2x_3 = 3 \\ 8x_1 + x_2 + x_3 = 2 \\ 2x_1 + 2x_2 + x_3 = 1 \end{cases}$

29. An appliance store sells dryers for $280 each and washing machines for $315 each. On a certain day, the store sells a total of 39 washers and dryers, and its total receipts for them are $11,375. Let x denote the number of dryers sold on this day and let y denote the number of washing machines sold.

   (a) Write two linear equations that must be satisfied by x and y.

   (b) Solve the resulting system.

30. Solve the system

$$\begin{cases} \dfrac{1}{s} + \dfrac{4}{t} - \dfrac{3}{u} = 4 \\ \dfrac{2}{s} - \dfrac{3}{t} + \dfrac{1}{u} = 1 \\ -\dfrac{3}{s} + \dfrac{2}{t} + \dfrac{2}{u} = -3 \end{cases}$$

for s, t, and u. $\left[\textit{Hint: Let } x = \dfrac{1}{s}, \ y = \dfrac{1}{t}, \text{ and } z = \dfrac{1}{u}. \right]$

9.2 THE ELIMINATION METHOD USING MATRICES

When you solve a system of linear equations by the elimination method, your arithmetic involves only the numerical coefficients and constants in the equations—the unknowns just "go along for the ride." This suggests abbreviating a system of linear equations by writing only the coefficients and constants. For instance, to abbreviate the system

$$\begin{cases} \frac{1}{3}x + y = 3 \\ -2x + 5y = 4, \end{cases}$$

we write only the coefficients on the left and the constants on the right:

$$\begin{array}{cc|c} \frac{1}{3} & 1 & 3 \\ -2 & 5 & 4. \end{array}$$

(The dashed line separating the coefficients and the constants is optional.) It is customary to enclose the resulting array of numbers in square brackets,

$$\begin{bmatrix} \frac{1}{3} & 1 & 3 \\ -2 & 5 & 4 \end{bmatrix},$$

and to refer to it as the **matrix*** of the system.

Example 1 Write the matrix of the system

$$\begin{cases} y - 2x + \frac{2}{3}z = 7 \\ z + 5x = 3 \\ -y - z = 4. \end{cases}$$

Solution We begin by rewriting the equations so that the unknowns on the left appear in the order x, y, z. Missing unknowns are written with zero coefficients:

$$\begin{cases} -2x + y + \frac{2}{3}z = 7 \\ 5x + 0y + z = 3 \\ 0x - y - z = 4. \end{cases}$$

The corresponding matrix is

$$\begin{bmatrix} -2 & 1 & \frac{2}{3} & 7 \\ 5 & 0 & 1 & 3 \\ 0 & -1 & -1 & 4 \end{bmatrix}.$$

∎

Example 2 Write the system of linear equations in $x, y,$ and z, corresponding to the matrix

$$\begin{bmatrix} 0 & 5 & -1 & \frac{2}{3} \\ 1 & 2 & 0 & 0 \end{bmatrix}.$$

* Some authors call this the **augmented** matrix of the system.

Solution   Since there are only two horizontal rows in the matrix, there are only two equations in the corresponding system:

$$\begin{cases} 0x + 5y - z = \frac{2}{3} \\ x + 2y + 0z = 0 \end{cases} \quad \text{or} \quad \begin{cases} 5y - z = \frac{2}{3} \\ x + 2y = 0. \end{cases}$$ ■

The horizontal rows in a matrix are called simply the **rows** and the vertical columns are called simply the **columns.** The numbers that appear in the matrix are called the **entries** or **elements** of the matrix. To specify a particular entry in a matrix, you can give its "address" by indicating the row and column to which it belongs. For instance, in the matrix

$$\begin{bmatrix} 0 & 8 & -5 & \frac{2}{3} \\ -5 & \frac{3}{7} & 4 & 11 \end{bmatrix},$$

the entry in the first row and third column is -5; the entry in the second row and second column is $\frac{3}{7}$; and so forth.

The three elementary operations on the equations of a system (Section 9.1) are represented by the following **elementary row operations** on the corresponding matrix:

---

1. Interchange two rows.

2. Multiply or divide all elements of a row by a nonzero constant.

3. Add a constant multiple of the elements of one row to the corresponding elements of another row.

We denote an interchange of the ith and jth rows of a matrix by

$$R_i \rightleftarrows R_j.$$

The operation of replacing the ith row by a nonzero constant multiple of the ith row is symbolized by

$$cR_i \rightarrow R_i,$$

where c is the constant multiplier. Similarly, the operation of replacing the jth row by the result of adding c times the ith row to the jth row is denoted by

$$cR_i + R_j \rightarrow R_j.$$

For instance, $2R_1 + R_2 \rightarrow R_2$ is read as "2 times R_1 added to R_2 replaces R_2." By performing these elementary row operations on a matrix, you can solve the corresponding system of linear equations by the elimination method. When an unknown is eliminated from a particular equation, a zero will appear in the corresponding row and column of the matrix.

For instance, the solution of Example 4 in Section 9.1 is abbreviated as follows:

$$\begin{bmatrix} \frac{1}{3} & 1 & \vdots & 3 \\ -2 & 5 & \vdots & 4 \end{bmatrix} \xrightarrow{3R_1 \rightarrow R_1} \begin{bmatrix} 1 & 3 & \vdots & 9 \\ -2 & 5 & \vdots & 4 \end{bmatrix} \xrightarrow{2R_1 + R_2 \rightarrow R_2} \begin{bmatrix} 1 & 3 & \vdots & 9 \\ 0 & 11 & \vdots & 22 \end{bmatrix}$$

$$\xrightarrow{\frac{1}{11}R_2 \rightarrow R_2} \begin{bmatrix} 1 & 3 & \vdots & 9 \\ 0 & 1 & \vdots & 2 \end{bmatrix};$$

that is,

$$\begin{cases} x + 3y = 9 \\ y = 2, \end{cases}$$

from which the solution $x = 3$, $y = 2$ is found, as before, by back substitution. In this calculation, the last matrix is in **echelon form** in the sense that the following conditions are satisfied:

> **1.** The **leading entry**—that is, the first nonzero entry, reading from left to right—in each row is 1.
>
> **2.** The leading entry in each row after the first is to the right of the leading entry in the previous row.
>
> **3.** If there are any rows with no leading entry—that is, rows consisting entirely of zeros—they are placed at the bottom of the matrix.

By a sequence of elementary row operations, the matrix corresponding to any system of linear equations can be brought into echelon form, and then the solution can be obtained by back substitution.

Example 3   Solve the system of equations

$$\begin{cases} 2x + y - z = 5 \\ 2x - 2y + 8z = -10 \\ 4y + z = 7. \end{cases}$$

Solution   We write the matrix of the system and then reduce it to echelon form by a sequence of elementary row operations as follows:

$$\begin{bmatrix} 2 & 1 & -1 & | & 5 \\ 2 & -2 & 8 & | & -10 \\ 0 & 4 & 1 & | & 7 \end{bmatrix} \xrightarrow{(-1)R_1 + R_2 \to R_2} \begin{bmatrix} 2 & 1 & -1 & | & 5 \\ 0 & -3 & 9 & | & -15 \\ 0 & 4 & 1 & | & 7 \end{bmatrix} \xrightarrow{(-\frac{1}{3})R_2 \to R_2}$$

$$\begin{bmatrix} 2 & 1 & -1 & | & 5 \\ 0 & 1 & -3 & | & 5 \\ 0 & 4 & 1 & | & 7 \end{bmatrix} \xrightarrow{(-4)R_2 + R_3 \to R_3} \begin{bmatrix} 2 & 1 & -1 & | & 5 \\ 0 & 1 & -3 & | & 5 \\ 0 & 0 & 13 & | & -13 \end{bmatrix} \xrightarrow{\frac{1}{13}R_3 \to R_3}$$

$$\begin{bmatrix} 2 & 1 & -1 & | & 5 \\ 0 & 1 & -3 & | & 5 \\ 0 & 0 & 1 & | & -1 \end{bmatrix} \xrightarrow{\frac{1}{2}R_1 \to R_1} \begin{bmatrix} 1 & \frac{1}{2} & -\frac{1}{2} & | & \frac{5}{2} \\ 0 & 1 & -3 & | & 5 \\ 0 & 0 & 1 & | & -1 \end{bmatrix}.$$

The last matrix is in echelon form and corresponds to the system

$$\begin{cases} x + \frac{1}{2}y - \frac{1}{2}z = \frac{5}{2} \\ y - 3z = 5 \\ z = -1. \end{cases}$$

Now we back substitute $z = -1$ from the third equation into the second equation $y - 3z = 5$ to obtain

$$y - 3(-1) = 5 \qquad \text{or} \qquad y = 2.$$

Finally, we back substitute $z = -1$ and $y = 2$ into the first equation $x + \frac{1}{2}y - \frac{1}{2}z = \frac{5}{2}$ to obtain

$$x + \tfrac{1}{2}(2) - \tfrac{1}{2}(-1) = \tfrac{5}{2} \qquad \text{or} \qquad x = 1.$$

Therefore, the solution is

$$x = 1, \; y = 2, \text{ and } z = -1 \qquad \text{or} \qquad (1, 2, -1). \qquad \blacksquare$$

For the relatively simple systems of linear equations considered here and in the problems at the end of this section, you can find, by trial and error, a suitable sequence of elementary row operations that reduces the matrix to echelon form. More advanced textbooks on linear algebra describe step-by-step procedures (eliminating all guesswork) for doing this. In these textbooks, you can also find a proof that, for any system of linear equations, there are just three possibilities:

Case 1. *There is exactly one solution, in which case we say that the equations in the system are* **consistent.**

Case 2. *There is no solution, in which case we say that the equations in the system are* **inconsistent.**

Case 3. *There is an infinite number of solutions, in which case we say that the equations in the system are* **dependent.**

In working with systems of linear equations, you can use whatever letters you please to denote the unknowns. Sometimes it is convenient to use just one letter with different subscripts (for instance, $x_1, x_2, x_3, \ldots, x_m$) to represent the different unknowns. In any case, before you can form the corresponding matrix, you must decide in what order the unknowns are to be written in the equations. When subscripts are used, the unknowns are usually written in the numerical order of the subscripts.

In solving a system of linear equations by the matrix method, it is often convenient to bring the matrix of the system into **row reduced** echelon form; that is, to continue applying the elementary row operations until the entries *above* each leading entry (as well as below) are zero.

Example 4   Solve the system $\begin{cases} x_1 + 3x_2 - 5x_3 = 1 \\ 2x_1 + 5x_2 - 2x_3 = 4. \end{cases}$

Solution   We begin by writing the matrix of the system and bringing it into echelon form:

$$\begin{bmatrix} 1 & 3 & -5 & | & 1 \\ 2 & 5 & -2 & | & 4 \end{bmatrix} \xrightarrow{(-2)R_1 + R_2 \to R_2} \begin{bmatrix} 1 & 3 & -5 & | & 1 \\ 0 & -1 & 8 & | & 2 \end{bmatrix}$$

$$\xrightarrow{(-1)R_2 \to R_2} \begin{bmatrix} 1 & 3 & -5 & | & 1 \\ 0 & 1 & -8 & | & -2 \end{bmatrix}.$$

The matrix is now in echelon form, but it is not in reduced echelon form because the entry above the leading entry in the second row is not zero. To bring it into reduced echelon form, we continue as follows:

$$\begin{bmatrix} 1 & 3 & -5 & | & 1 \\ 0 & 1 & -8 & | & -2 \end{bmatrix} \xrightarrow{(-3)R_2 + R_1 \rightarrow R_1} \begin{bmatrix} 1 & 0 & 19 & | & 7 \\ 0 & 1 & -8 & | & -2 \end{bmatrix}.$$

The last matrix is the matrix of the system

$$\begin{cases} x_1 + 19x_3 = 7 \\ x_2 - 8x_3 = -2, \end{cases}$$

or, equivalently,

$$\begin{cases} x_1 = -19x_3 + 7 \\ x_2 = 8x_3 - 2. \end{cases}$$

By assigning x_3 any value, say $x_3 = t$, we express x_1, x_2, and x_3 in terms of t:

$$x_1 = -19t + 7$$
$$x_2 = 8t - 2$$
$$x_3 = t.$$

Thus, the equations in the system are dependent, since there are infinitely many solutions $(-19t + 7, 8t - 2, t)$. For instance, $t = 0$ gives the solution $(7, -2, 0)$, $t = \frac{1}{2}$ gives the solution $(-\frac{5}{2}, 2, \frac{1}{2})$, and $t = \pi$ gives the solution $(-19\pi + 7, 8\pi - 2, \pi)$. ∎

Problem Set 9.2

In Problems 1 to 4, consider the matrix $\begin{bmatrix} 0 & 5 & -3 \\ 2 & -1 & 7 \end{bmatrix}$.

1. What are the elements in the first row?

2. What are the elements in the second column?

3. What is the element in the second row and first column?

4. What is the element in the first row and third column?

In Problems 5 to 8, write the matrix of each system of linear equations.

5. $\begin{cases} \frac{3}{4}x - \frac{2}{3}y = \frac{1}{7} \\ -x + 5y = 6 \end{cases}$

6. $\begin{cases} 0.5x_1 + 3.2x_2 = 7.1 \\ 5.3x_1 - 3.0x_2 = -6.5 \end{cases}$

7. $\begin{cases} 40x - z + 22y = -17 \\ y + z = 0 \\ 17y - 13x + 12z = 5 \end{cases}$

8. $\begin{cases} 3x_3 - 2x_2 = x_1 \\ 2x_1 - 5x_3 = -x_2 \\ x_2 + x_3 = 6 \end{cases}$

In Problems 9 to 12, write the system of linear equations in x and y or in x, y, and z corresponding to each matrix.

9. $\begin{bmatrix} 1 & 3 & | & 0 \\ 2 & -4 & | & 1 \end{bmatrix}$

10. $\begin{bmatrix} 0.1 & 3.2 & | & -1.7 \\ -4.4 & 0 & | & 0 \end{bmatrix}$

11. $\begin{bmatrix} 2 & 5 & 3 & | & 1 \\ -3 & 7 & \frac{1}{2} & | & \frac{3}{4} \\ 0 & \frac{2}{3} & 0 & | & -\frac{4}{5} \end{bmatrix}$

12. $\begin{bmatrix} 3 & 2 & | & 3 \\ 1 & \frac{2}{3} & | & 1 \\ 5 & 0 & | & -2 \end{bmatrix}$

In Problems 13 to 38, write the matrix of the system of linear equations and solve the system by the elimination method using matrices.

13. $\begin{cases} x + y = 4 \\ x - 4y = 8 \end{cases}$

14. $\begin{cases} x + 6y = 7 \\ 11x - 7y = -10 \end{cases}$

15. $\begin{cases} 3x + y = 15 \\ 3x - 7y = 15 \end{cases}$

16. $\begin{cases} -6x_1 + 2x_2 = 3 \\ 2x_1 + 5x_2 = 3 \end{cases}$

17. $\begin{cases} 4s + 3t = 17 \\ 2s + 3t = 13 \end{cases}$

18. $\begin{cases} \frac{1}{2}x + \frac{2}{3}y = 6 \\ -\frac{3}{2}x + \frac{1}{2}y = -3 \end{cases}$

19. $\begin{cases} 2x + y = 6 \\ 3x + 4y = 4 \end{cases}$

20. $\begin{cases} \frac{1}{7}u + \frac{3}{7}v = 1 \\ -\frac{1}{7}u + \frac{4}{7}v = 1 \end{cases}$

21. $\begin{cases} x + y = 1 \\ 2x + 2y = 0 \end{cases}$

22. $\begin{cases} 5x - 2y = y - 1 \\ 4x - 5y = 3 - 2x \end{cases}$

23. $\begin{cases} \frac{1}{3}x_1 + \frac{1}{6}x_2 = 1 \\ x_1 - x_2 = 3 \end{cases}$

24. $\begin{cases} x + by = 2 \\ bx + y = 3 \end{cases}$

25. $\begin{cases} x + 5y - z = -7 \\ 3x + 4y - 2z = 2 \\ 2x - 3y + 5z = 19 \end{cases}$

26. $\begin{cases} 3x + 2y + 5z = 7 \\ 2x - 3y - 2z = -3 \\ x + 2y + 3z = 5 \end{cases}$

27. $\begin{cases} 2x + y - 3z = 11 \\ x - 2y + 4z = -3 \\ 3x + y - 2z = 12 \end{cases}$

28. $\begin{cases} u + 5v - w = 2 \\ 2u + v + w = 7 \\ u - v + 2w = 11 \end{cases}$

29. $\begin{cases} 8r + 3s - 18t = 1 \\ 16r + 6s - 6t = 7 \\ 4r + 9s + 12t = 9 \end{cases}$

30. $\begin{cases} \frac{2}{3}x_1 + \frac{1}{4}x_2 - \frac{1}{3}x_3 = 3 \\ -\frac{3}{2}x_1 + \frac{1}{8}x_2 + x_3 = 1 \\ \frac{1}{2}x_1 - x_2 + x_3 = 4 \end{cases}$

31. $\begin{cases} x + 3y + z = 0 \\ 2y + 4z = 1 \\ -x + 3z = 2 \end{cases}$

32. $\begin{cases} x + 2y - z = 0 \\ 2x - y + 3z = 1 \\ 3x - 2y = -1 \end{cases}$

33. $\begin{cases} x_1 + 2x_2 - x_3 = 0 \\ 2x_1 - 2x_2 + x_3 = 0 \\ 6x_1 + 4x_2 + 3x_3 = 0 \end{cases}$

34. $\begin{cases} 3y - 2z = 2 \\ 4x + 5z = -1 \\ 5x + y = 0 \end{cases}$

35. $\begin{cases} 2x + 3z = 1 \\ x - y + 2z = 0 \\ 3x - y + 5z = -1 \end{cases}$

36. $\begin{cases} 3x_1 + 5x_2 - x_3 = 2 \\ x_2 + x_3 = 3 \\ 2x_1 + 3x_2 - x_3 = 4 \end{cases}$

37. $\begin{cases} 3x + 2y + z = 6 \\ x - 3y + 5z = 3 \end{cases}$

38. $\begin{cases} 2x_1 - x_2 - x_3 = 4 \\ x_1 + 2x_2 + 3x_3 = 8 \end{cases}$

9.3 MATRIX ALGEBRA

In Section 9.2, we used matrices merely as abbreviations for simultaneous systems of linear equations. Matrices, however, have a life of their own and, since their invention in 1858 by the English mathematician Arthur Cayley (1821–1895), they have played an ever increasing role in applications ranging from economics to quantum mechanics.

Under suitable conditions, matrices can be added, subtracted, multiplied, and (in a sense) divided. In this section, we take a brief look at the resulting "algebra" of matrices. Here we make no attempt to be complete, and we recommend that you look into a textbook on linear algebra or matrix theory for more details.

We use capital letters A, B, C, and so on to denote matrices. Here we consider only matrices whose elements are real numbers. These elements are arranged in a rectangular pattern of horizontal rows and vertical columns; for instance, the matrix

$$A = \begin{bmatrix} -1 & 0 & 5 & 7 & 2 \\ 3 & -2 & \frac{2}{3} & 0 & \frac{1}{2} \\ \frac{5}{4} & 0 & -4 & 5 & 1 \end{bmatrix}$$

has 3 rows and 5 columns. If a matrix has n rows and m columns, we call it an **n by m matrix.** Thus, A is a 3 by 5 matrix. By a **square matrix,** we mean a matrix with the same number of rows as columns. For instance, the matrix

$$\begin{bmatrix} 1 & -5 \\ -3 & 2 \end{bmatrix}$$

is a square 2 by 2 matrix. Notice that an n by m matrix has nm elements.

If A and B are n by m matrices and if each element of A is equal to the corresponding element of B, we say that the matrices A and B are **equal** and we write

$$A = B.$$

Such a **matrix equation** represents nm ordinary equations in a highly compact form.

If C and D are two n by m matrices, then their **sum** $C + D$ is defined to be the n by m matrix obtained by adding the corresponding elements of C and D. Likewise, the **difference** $C - D$ is defined to be the n by m matrix obtained by subtracting the elements of D from the corresponding elements of C.

Example 1   Let $C = \begin{bmatrix} 2 & 1 & 4 \\ 3 & -5 & 8 \end{bmatrix}$ and $D = \begin{bmatrix} -3 & 2 & -1 \\ 4 & -1 & 5 \end{bmatrix}$.

(a) Find $C + D$.   **(b)** Find $C - D$.

Solution   **(a)** $C + D = \begin{bmatrix} 2 & 1 & 4 \\ 3 & -5 & 8 \end{bmatrix} + \begin{bmatrix} -3 & 2 & -1 \\ 4 & -1 & 5 \end{bmatrix}$

$$= \begin{bmatrix} 2 + (-3) & 1 + 2 & 4 + (-1) \\ 3 + 4 & -5 + (-1) & 8 + 5 \end{bmatrix}$$

$$= \begin{bmatrix} -1 & 3 & 3 \\ 7 & -6 & 13 \end{bmatrix}$$

(b) $C - D = \begin{bmatrix} 2 & 1 & 4 \\ 3 & -5 & 8 \end{bmatrix} - \begin{bmatrix} -3 & 2 & -1 \\ 4 & -1 & 5 \end{bmatrix}$

$$= \begin{bmatrix} 2 - (-3) & 1 - 2 & 4 - (-1) \\ 3 - 4 & -5 - (-1) & 8 - 5 \end{bmatrix} = \begin{bmatrix} 5 & -1 & 5 \\ -1 & -4 & 3 \end{bmatrix} \quad \blacksquare$$

Notice that you can add or subtract matrices only if they have the same shape—that is, only if they have the same number of rows and the same number of columns. Because matrices are added by adding their corresponding elements, it follows from the commutative and associative properties of real numbers (Section 1.1) that matrix addition is also commutative and associative. Thus, if A, B, and C are matrices of the same shape, we have

$$A + B = B + A \quad \text{(commutative property of addition)}$$

and $$A + (B + C) = (A + B) + C \quad \text{(associative property of addition)}.$$

A matrix all of whose elements are zero is called a **zero matrix.** The zero matrix with n rows and m columns is denoted by $0_{n,m}$. For instance,

$$0_{1,3} = \begin{bmatrix} 0 & 0 & 0 \end{bmatrix} \qquad 0_{2,2} = \begin{bmatrix} 0 & 0 \\ 0 & 0 \end{bmatrix} \qquad 0_{4,3} = \begin{bmatrix} 0 & 0 & 0 \\ 0 & 0 & 0 \\ 0 & 0 & 0 \\ 0 & 0 & 0 \end{bmatrix}$$

and so forth. The subscripts indicating the shape of a zero matrix are often omitted because you can tell from the context how many rows and how many columns are involved. For instance, we have the property

$$A + 0 = A \qquad \text{(additive identity property),}$$

where it is understood that 0 denotes the zero matrix with the same shape as the matrix A.

If A is a matrix, we define $-A$ to be the matrix obtained by multiplying each element of A by -1. For instance,

$$-\begin{bmatrix} 2 & -\frac{1}{2} & 4 & 0 \\ -3 & 2 & 0 & -5 \end{bmatrix} = \begin{bmatrix} -2 & \frac{1}{2} & -4 & 0 \\ 3 & -2 & 0 & 5 \end{bmatrix}.$$

More generally, if k is any real number, we define kA to be the matrix obtained by multiplying each element of A by k. In particular,

$$(-1)A = -A.$$

Notice that $-A$ is an **additive inverse** of A in the sense that

$$A + (-A) = 0 \qquad \text{(additive inverse property).}$$

Example 2 If $A = \begin{bmatrix} 3 & -1 & 2 \\ 0 & 2 & -4 \end{bmatrix}$, find: **(a)** $-A$ **(b)** $\frac{1}{2}A$

Solution **(a)** $-A = -\begin{bmatrix} 3 & -1 & 2 \\ 0 & 2 & -4 \end{bmatrix}$

$$= \begin{bmatrix} -3 & 1 & -2 \\ 0 & -2 & 4 \end{bmatrix}$$

(b) $\frac{1}{2}A = \frac{1}{2}\begin{bmatrix} 3 & -1 & 2 \\ 0 & 2 & -4 \end{bmatrix}$

$$= \begin{bmatrix} \frac{3}{2} & -\frac{1}{2} & 1 \\ 0 & 1 & -2 \end{bmatrix}$$

The product of matrices is defined in a somewhat unexpected way. In order to introduce matrix multiplication, we begin by considering special matrices having only one row or column. A 1 by n matrix

$$R = \begin{bmatrix} x_1 & x_2 & x_3 & \cdots & x_n \end{bmatrix}$$

is called a **row vector,** and an n by 1 matrix

$$C = \begin{bmatrix} y_1 \\ y_2 \\ y_3 \\ \vdots \\ y_n \end{bmatrix}$$

is called a **column vector.**

If the number of elements in R is the same as the number of entries in C, we define the **product** RC of R and C to be the number obtained by pairing each element of R with the corresponding element of C, multiplying these pairs, and adding the resulting products. Thus,

$$RC = x_1y_1 + x_2y_2 + x_3y_3 + \cdots + x_ny_n.$$

Example 3   Find RC if $R = \begin{bmatrix} 2 & 8 & 3 \end{bmatrix}$ and $C = \begin{bmatrix} 60 \\ 20 \\ 300 \end{bmatrix}$.

Solution $RC = \begin{bmatrix} 2 & 8 & 3 \end{bmatrix} \begin{bmatrix} 60 \\ 20 \\ 300 \end{bmatrix} = 2(60) + 8(20) + 3(300) = 1180$ ∎

Although the "row by column" product may seem contrived at first, it has many practical uses. For instance, if a furniture store sells 2 tables, 8 chairs, and 3 sofas for $60 per table, $20 per chair, and $300 per sofa, the product of

the **demand vector** $\begin{bmatrix} 2 & 8 & 3 \end{bmatrix}$

and

the **revenue vector** $\begin{bmatrix} 60 \\ 20 \\ 300 \end{bmatrix}$

gives the **total revenue**

$$\begin{bmatrix} 2 & 8 & 3 \end{bmatrix} \begin{bmatrix} 60 \\ 20 \\ 300 \end{bmatrix} = 2(60) + 8(20) + 3(300) = 1180 \text{ dollars.}$$

As another example, notice that the linear equation

$$a_1x_1 + a_2x_2 + a_3x_3 + \cdots + a_nx_n = k$$

can be written in matrix form as

$$\begin{bmatrix} a_1 & a_2 & a_3 & \cdots & a_n \end{bmatrix} \begin{bmatrix} x_1 \\ x_2 \\ x_3 \\ \vdots \\ x_n \end{bmatrix} = k.$$

The rows of an n by m matrix can be regarded as row vectors, and its columns as column vectors. This permits us to give the following definition.

Definition 1   **The Product of Matrices**

Let A be an n by m matrix and let B be an m by p matrix. We define the **product** AB to be the n by p matrix determined by the following procedure: To find the element in the ith row and jth column of AB, we multiply the ith row of A by the jth column of B.

Example 4   Find AB if $A = \begin{bmatrix} 3 & -2 & 4 \\ 5 & 1 & -3 \end{bmatrix}$ and $B = \begin{bmatrix} 3 & 6 \\ -4 & 5 \\ 2 & -2 \end{bmatrix}$.

Solution   Here A is a 2 by 3 matrix and B is a 3 by 2 matrix, so AB will be a 2 by 2 matrix.

$$AB = \begin{bmatrix} 3 & -2 & 4 \\ 5 & 1 & -3 \end{bmatrix} \begin{bmatrix} 3 & 6 \\ -4 & 5 \\ 2 & -2 \end{bmatrix} = \begin{bmatrix} \text{1st row of } A \text{ times} & \text{1st row of } A \text{ times} \\ \text{1st column of } B & \text{2nd column of } B \\ \\ \text{2nd row of } A \text{ times} & \text{2nd row of } A \text{ times} \\ \text{1st column of } B & \text{2nd column of } B \end{bmatrix}$$

$$= \begin{bmatrix} 3(3) + (-2)(-4) + 4(2) & 3(6) + (-2)(5) + 4(-2) \\ 5(3) + 1(-4) + (-3)(2) & 5(6) + 1(5) + (-3)(-2) \end{bmatrix} = \begin{bmatrix} 25 & 0 \\ 5 & 41 \end{bmatrix}.$$ ∎

Notice that two matrices can be multiplied only if they fit together in the sense that the first matrix has as many columns as the second matrix has rows. If they do fit together in this way, the product has as many rows as the first matrix and as many columns as the second. If A is an n by m matrix, B is an m by p matrix, and C is a p by q matrix, it can be shown that

$$A(BC) = (AB)C \qquad \text{(associative property of multiplication).}$$

Distributive properties can also be proved, so that, for matrices of the appropriate shapes

$$A(B + C) = AB + AC \qquad \text{(distributive property)}$$

and $$(D + E)F = DF + EF \qquad \text{(distributive property).}$$

If A and B are square matrices of the same size, we can form the product AB and also the product BA; however, in general, $AB \neq BA$, so

the commutative property of multiplication fails for matrices.

Example 5   Let $A = \begin{bmatrix} 1 & -1 \\ 2 & 3 \end{bmatrix}$ and $B = \begin{bmatrix} 5 & -2 \\ 4 & 1 \end{bmatrix}$. Find:

(a) AB **(b)** BA

Solution **(a)** $AB = \begin{bmatrix} 1 & -1 \\ 2 & 3 \end{bmatrix} \begin{bmatrix} 5 & -2 \\ 4 & 1 \end{bmatrix} = \begin{bmatrix} 1(5) + (-1)4 & 1(-2) + (-1)1 \\ 2(5) + 3(4) & 2(-2) + 3(1) \end{bmatrix} = \begin{bmatrix} 1 & -3 \\ 22 & -1 \end{bmatrix}$

(b) $BA = \begin{bmatrix} 5 & -2 \\ 4 & 1 \end{bmatrix} \begin{bmatrix} 1 & -1 \\ 2 & 3 \end{bmatrix} = \begin{bmatrix} 5(1) + (-2)2 & 5(-1) + (-2)3 \\ 4(1) + 1(2) & 4(-1) + 1(3) \end{bmatrix} = \begin{bmatrix} 1 & -11 \\ 6 & -1 \end{bmatrix}$

Notice that $AB \neq BA$. ∎

Using the idea of matrix multiplication, you can write a system of linear equations such as

$$\begin{cases} a_1 x_1 + a_2 x_2 + a_3 x_3 = k_1 \\ b_1 x_1 + b_2 x_2 + b_3 x_3 = k_2 \\ c_1 x_1 + c_2 x_2 + c_3 x_3 = k_3 \end{cases}$$

in the matrix form

$$\begin{bmatrix} a_1 & a_2 & a_3 \\ b_1 & b_2 & b_3 \\ c_1 & c_2 & c_3 \end{bmatrix} \begin{bmatrix} x_1 \\ x_2 \\ x_3 \end{bmatrix} = \begin{bmatrix} k_1 \\ k_2 \\ k_3 \end{bmatrix}.$$

Thus,
$$AX = K,$$

where A represents the **matrix of coefficients**

$$A = \begin{bmatrix} a_1 & a_2 & a_3 \\ b_1 & b_2 & b_3 \\ c_1 & c_2 & c_3 \end{bmatrix},$$

X represents the column vector of unknowns, and K represents the column vector of constants:

$$X = \begin{bmatrix} x_1 \\ x_2 \\ x_3 \end{bmatrix}, \qquad K = \begin{bmatrix} k_1 \\ k_2 \\ k_3 \end{bmatrix}.$$

If $AX = K$ were an ordinary equation, you could multiply both sides by the reciprocal of A to obtain the solution

$$X = A^{-1}K.$$

As we shall see, it is often possible to solve the matrix equation in much the same way by using the **multiplicative inverse** A^{-1} of the matrix A.

A square n by n matrix with 1 in each position on the diagonal running from upper left to lower right, and zeros elsewhere, is called the **n by n identity matrix** and is denoted by I_n. For instance,

$$I_2 = \begin{bmatrix} 1 & 0 \\ 0 & 1 \end{bmatrix} \qquad \text{and} \qquad I_3 = \begin{bmatrix} 1 & 0 & 0 \\ 0 & 1 & 0 \\ 0 & 0 & 1 \end{bmatrix}.$$

The subscript indicating the size of the identity matrix is often omitted because you can tell from the context how many rows and columns are involved. An identity matrix I plays a role in matrix algebra similar to the role played by 1 in ordinary algebra. In particular, for matrices C and D of the appropriate shape,

$$IC = C \quad \text{and} \quad DI = D \qquad \text{(multiplicative identity property).*}$$

By analogy, with the reciprocal a^{-1} of a nonzero number in ordinary algebra, we have the following definition in matrix algebra.

Definition 2   **The Inverse of a Square Matrix**

> Let A be a square n by n matrix. We say that A is **nonsingular** if there exists an n by n matrix A^{-1} such that
>
> $$AA^{-1} = A^{-1}A = I.$$
>
> If A^{-1} exists, it is called the **inverse** of the matrix A.

It can be shown that a nonsingular matrix A has a unique inverse A^{-1}, which you can find by carrying out the following procedure:

> Form the n by $2n$ matrix
>
> $$[A \mid I]$$
>
> in which the first n columns are the columns of A and the last n columns are the columns of the identity matrix I_n. Then, using the elementary row operations (Section 9.2), reduce this matrix to the form
>
> $$[I \mid B],$$
>
> so that the first n columns are the columns of I_n. Then the matrix B formed by the last n columns is the inverse of A, that is,
>
> $$B = A^{-1}.$$

Example 6   Find A^{-1} if $A = \begin{bmatrix} 1 & -1 & 1 \\ 0 & 2 & -1 \\ 2 & 3 & 0 \end{bmatrix}$.

* See Problems 17, 19, and 27.

Solution　We start by forming the matrix $[A \mid I]$:

$$\begin{bmatrix} 1 & -1 & 1 & | & 1 & 0 & 0 \\ 0 & 2 & -1 & | & 0 & 1 & 0 \\ 2 & 3 & 0 & | & 0 & 0 & 1 \end{bmatrix}.$$

Now, we execute elementary row operations on the entire matrix until the left half is transformed into the identity matrix:

$$\begin{bmatrix} 1 & -1 & 1 & | & 1 & 0 & 0 \\ 0 & 2 & -1 & | & 0 & 1 & 0 \\ 2 & 3 & 0 & | & 0 & 0 & 1 \end{bmatrix} \xrightarrow{-2R_1 + R_3 \to R_3} \begin{bmatrix} 1 & -1 & 1 & | & 1 & 0 & 0 \\ 0 & 2 & -1 & | & 0 & 1 & 0 \\ 0 & 5 & -2 & | & -2 & 0 & 1 \end{bmatrix}$$

$$\xrightarrow{\frac{1}{2}R_2 \to R_2} \begin{bmatrix} 1 & -1 & 1 & | & 1 & 0 & 0 \\ 0 & 1 & -\frac{1}{2} & | & 0 & \frac{1}{2} & 0 \\ 0 & 5 & -2 & | & -2 & 0 & 1 \end{bmatrix}$$

$$\xrightarrow{R_2 + R_1 \to R_1} \begin{bmatrix} 1 & 0 & \frac{1}{2} & | & 1 & \frac{1}{2} & 0 \\ 0 & 1 & -\frac{1}{2} & | & 0 & \frac{1}{2} & 0 \\ 0 & 5 & -2 & | & -2 & 0 & 1 \end{bmatrix}$$

$$\xrightarrow{-5R_2 + R_3 \to R_3} \begin{bmatrix} 1 & 0 & \frac{1}{2} & | & 1 & \frac{1}{2} & 0 \\ 0 & 1 & -\frac{1}{2} & | & 0 & \frac{1}{2} & 0 \\ 0 & 0 & \frac{1}{2} & | & -2 & -\frac{5}{2} & 1 \end{bmatrix}$$

$$\xrightarrow{R_3 + R_2 \to R_2} \begin{bmatrix} 1 & 0 & \frac{1}{2} & | & 1 & \frac{1}{2} & 0 \\ 0 & 1 & 0 & | & -2 & -2 & 1 \\ 0 & 0 & \frac{1}{2} & | & -2 & -\frac{5}{2} & 1 \end{bmatrix}$$

$$\xrightarrow{-1R_3 + R_1 \to R_1} \begin{bmatrix} 1 & 0 & 0 & | & 3 & 3 & -1 \\ 0 & 1 & 0 & | & -2 & -2 & 1 \\ 0 & 0 & \frac{1}{2} & | & -2 & -\frac{5}{2} & 1 \end{bmatrix}$$

$$\xrightarrow{2R_3 \to R_3} \begin{bmatrix} 1 & 0 & 0 & | & 3 & 3 & -1 \\ 0 & 1 & 0 & | & -2 & -2 & 1 \\ 0 & 0 & 1 & | & -4 & -5 & 2 \end{bmatrix}.$$

Therefore,

$$A^{-1} = \begin{bmatrix} 3 & 3 & -1 \\ -2 & -2 & 1 \\ -4 & -5 & 2 \end{bmatrix}.$$

By using matrix multiplication, you can check that $AA^{-1} = I$ and $A^{-1}A = I$ (Problem 43). ▪

The following example illustrates the use of the inverse of a matrix to solve a system of linear equations.

Example 7   Use the inverse of the matrix of coefficients to solve the system of linear equations

$$\begin{cases} x - y + z = 8 \\ 2y - z = -7 \\ 2x + 3y = 1. \end{cases}$$

Solution   The matrix of coefficients is

$$A = \begin{bmatrix} 1 & -1 & 1 \\ 0 & 2 & -1 \\ 2 & 3 & 0 \end{bmatrix}.$$

If we let

$$X = \begin{bmatrix} x \\ y \\ z \end{bmatrix} \quad \text{and} \quad K = \begin{bmatrix} 8 \\ -7 \\ 1 \end{bmatrix},$$

we can write the system of linear equations as the matrix equation

$$AX = K.$$

In the previous example, we showed that A is nonsingular with

$$A^{-1} = \begin{bmatrix} 3 & 3 & -1 \\ -2 & -2 & 1 \\ -4 & -5 & 2 \end{bmatrix}.$$

If we multiply $AX = K$ on the left by A^{-1}, we obtain

$$A^{-1}(AX) = A^{-1}K$$
$$(A^{-1}A)X = A^{-1}K$$
$$IX = A^{-1}K$$
$$X = A^{-1}K.$$

Therefore,

$$\begin{bmatrix} x \\ y \\ z \end{bmatrix} = X = A^{-1}K$$

$$= \begin{bmatrix} 3 & 3 & -1 \\ -2 & -2 & 1 \\ -4 & -5 & 2 \end{bmatrix} \begin{bmatrix} 8 \\ -7 \\ 1 \end{bmatrix}$$

$$= \begin{bmatrix} 3(8) + 3(-7) + (-1)1 \\ (-2)8 + (-2)(-7) + 1(1) \\ (-4)8 + (-5)(-7) + 2(1) \end{bmatrix} = \begin{bmatrix} 2 \\ -1 \\ 5 \end{bmatrix}.$$

It follows that $x = 2$, $y = -1$, and $z = 5$.   ∎

Problem Set 9.3

In Problems 1 to 8, find (a) $A + B$, (b) $A - B$, (c) $-3A$, and (d) $-3A + 2B$.

1. $A = \begin{bmatrix} 2 & -3 \\ 5 & 1 \end{bmatrix}$, $B = \begin{bmatrix} 6 & 4 \\ 3 & -2 \end{bmatrix}$

2. $A = \begin{bmatrix} 3 & 2 & -4 & 1 \\ 0 & 3 & -5 & 6 \end{bmatrix}$, $B = \begin{bmatrix} -7 & 3 & 0 & 4 \\ 1 & 0 & -1 & -3 \end{bmatrix}$

3. $A = \begin{bmatrix} 3 & 2 \\ -2 & 5 \\ 2 & 1 \\ -4 & 4 \end{bmatrix}$, $B = \begin{bmatrix} -2 & 3 \\ 3 & 1 \\ 4 & -2 \\ 1 & 0 \end{bmatrix}$

4. $A = \begin{bmatrix} 1 & \frac{1}{3} & 3 \\ 2 & 0 & -\frac{4}{3} \\ 1 & \sqrt{3} & -2 \end{bmatrix}$, $B = \begin{bmatrix} 1 & \frac{1}{2} & -1 \\ 3 & -1 & 0 \\ 2 & 0 & -\frac{3}{2} \end{bmatrix}$

5. $A = \begin{bmatrix} 2 & -3 & 2 & -3 \\ -3 & 2 & 1 & 1 \\ 4 & 1 & -3 & 4 \end{bmatrix}$, $B = \begin{bmatrix} 2 & -3 & 0 & 2 \\ 3 & 2 & -1 & 5 \\ 0 & -2 & 1 & 0 \end{bmatrix}$

6. $A = \begin{bmatrix} 0 \\ 1 \\ 2 \\ 3 \\ 4 \end{bmatrix}$, $B = \begin{bmatrix} -1 \\ 3 \\ -4 \\ 2 \\ 0 \end{bmatrix}$

7. $A = \begin{bmatrix} 1 & \frac{1}{6} & 0 \\ \frac{4}{3} & \pi & -2 \\ 1 & 0 & \frac{5}{3} \end{bmatrix}$, $B = \begin{bmatrix} 1 & -\frac{5}{6} & \sqrt{2} \\ \frac{3}{2} & 1 & 0 \\ 3 & \frac{5}{2} & 0 \end{bmatrix}$

8. $A = \begin{bmatrix} 3.1 & 2.5 \\ 6.8 & 1.1 \\ 4.7 & -8.2 \end{bmatrix}$, $B = \begin{bmatrix} 1.9 & 0 \\ 7.4 & 1 \\ -1 & 2 \end{bmatrix}$

In Problems 9 to 20, let

$$A = \begin{bmatrix} a_1 & a_2 \\ a_3 & a_4 \end{bmatrix}, \quad B = \begin{bmatrix} b_1 & b_2 \\ b_3 & b_4 \end{bmatrix}, \text{ and } C = \begin{bmatrix} c_1 & c_2 \\ c_3 & c_4 \end{bmatrix},$$

and let p and q denote arbitrary numbers. Verify each equation by direct calculation.

9. $A + (B + C) = (A + B) + C$

10. $p(B + C) = pB + pC$ **11.** $(p + q)A = pA + qA$

12. $(pq)A = p(qA)$ **13.** $A + (-A) = 0$

14. $A(BC) = (AB)C$ **15.** $A + 0 = A$

16. $A(B + C) = AB + AC$ **17.** $AI = A$

18. $(A + B)C = AC + BC$ **19.** $IA = A$

20. $0A = 0$

In Problems 21 to 42, let

$$A = \begin{bmatrix} 1 & -1 & 3 \\ 2 & 0 & 4 \\ 2 & -3 & 6 \end{bmatrix}, \quad B = \begin{bmatrix} -1 & 2 & -1 \\ -3 & 4 & 3 \\ 0 & -1 & 2 \end{bmatrix},$$

$$C = \begin{bmatrix} 2 & -1 \\ 0 & 2 \\ -3 & 1 \end{bmatrix}, \quad D = \begin{bmatrix} 1 & 3 & 2 \\ 4 & -1 & -2 \end{bmatrix},$$

$$E = \begin{bmatrix} 1 & 3 \\ -1 & 2 \end{bmatrix}, \quad F = \begin{bmatrix} 2 & 0 \\ 4 & -1 \end{bmatrix}.$$

Find each product, if it exists.

21. EF **22.** CE **23.** FE **24.** EC

25. EE **26.** BC **27.** AI **28.** CB

29. AB **30.** BA **31.** CD **32.** AC

33. AA **34.** ABC **35.** EFE **36.** ACB

37. $(A + B)C$ **38.** $A(A + B)$

39. $E(F - I)$ **40.** $EDCF$

41. DE **42.** $(A - B)(A + B)$

43. Check the solution to Example 6 by verifying that $AA^{-1} = I$ and that $A^{-1}A = I$.

44. Let

$$A = \begin{bmatrix} a & b \\ c & d \end{bmatrix}$$

and suppose that $ad - bc \neq 0$. Prove that A is nonsingular with

$$A^{-1} = (ad - bc)^{-1} \begin{bmatrix} d & -b \\ -c & a \end{bmatrix}.$$

In Problems 45 to 58, find the inverse of each matrix if it exists. Check your answers (see Problem 43).

45. $\begin{bmatrix} 1 & 1 \\ 1 & -4 \end{bmatrix}$ **46.** $\begin{bmatrix} 1 & 6 \\ 11 & -7 \end{bmatrix}$ **47.** $\begin{bmatrix} -6 & 2 \\ 2 & 5 \end{bmatrix}$ **48.** $\begin{bmatrix} 3 & 6 \\ 1 & 2 \end{bmatrix}$

49. $\begin{bmatrix} 1 & -1 \\ 9 & 3 \end{bmatrix}$ **50.** $\begin{bmatrix} 1 & b \\ b & 1 \end{bmatrix}$ **51.** $\begin{bmatrix} 1 & 1 \\ 1 & 1 \end{bmatrix}$ **52.** $\begin{bmatrix} 0 & 1 \\ 1 & 0 \end{bmatrix}$

53. $\begin{bmatrix} 1 & 2 & -1 \\ 2 & -1 & 3 \\ 3 & -2 & 3 \end{bmatrix}$ **54.** $\begin{bmatrix} 1 & 5 & -1 \\ 2 & 1 & 1 \\ 1 & -1 & 2 \end{bmatrix}$

55. $\begin{bmatrix} 1 & 2 & -1 \\ 2 & -2 & 1 \\ 6 & 4 & 3 \end{bmatrix}$ **56.** $\begin{bmatrix} 1 & 2 & -1 \\ 1 & 3 & 2 \\ 2 & 5 & 1 \end{bmatrix}$

57. $\begin{bmatrix} 1 & 3 & 1 \\ 0 & 2 & 4 \\ -1 & 0 & 3 \end{bmatrix}$ **58.** $\begin{bmatrix} 0 & 0 & 1 \\ 0 & 1 & 0 \\ 1 & 0 & 0 \end{bmatrix}$

In Problems 59 to 68, use the inverse of the matrix of coefficients to solve each system of linear equations.

59. $\begin{cases} x + y = 4 \\ x - 4y = 8 \end{cases}$ (See Problem 45.)

60. $\begin{cases} x + 6y = 7 \\ 11x - 7y = -10 \end{cases}$ (See Problem 46.)

61. $\begin{cases} -6x + 2y = 3 \\ 2x + 5y = 3 \end{cases}$ (See Problem 47.)

62. $\begin{cases} x_1 + 6x_2 = a \\ 11x_1 - 7x_2 = b \end{cases}$ (See Problem 46.)

63. $\begin{cases} r - s = 6 \\ 9r + 3s = 14 \end{cases}$ (See Problem 49.)

64. $\begin{cases} x + by = 2 \\ bx + y = 3 \end{cases}$ (See Problem 50.)

65. $\begin{cases} x + 2y - z = 6 \\ 2x - y + 3z = -13 \\ 3x - 2y + 3z = -16 \end{cases}$ (See Problem 53.)

66. $\begin{cases} u + 5v - w = 2 \\ 2u + v + w = 7 \\ u - v + 2w = 11 \end{cases}$ (See Problem 54.)

67. $\begin{cases} x + 3y + z = 0 \\ 2y + 4z = 1 \\ -x + 3z = 2 \end{cases}$ (See Problem 57.)

68. $\begin{cases} x + 2y - z = 0 \\ 2x - 2y + z = 0 \\ 6x + 4y + 3z = 0 \end{cases}$ (See Problem 55.)

69. In economics, a square matrix in which the element in the ith row and jth column indicates the number of units of commodity number i used to produce one unit of commodity number j is called a **technology matrix.** Let $T = \begin{bmatrix} a & b \\ c & d \end{bmatrix}$ be the technology matrix for a simple economic model involving only two commodities. The column vector $X = \begin{bmatrix} x_1 \\ x_2 \end{bmatrix}$ in which x_1 (respectively, x_2) represents the number of units of commodity number 1 (respectively, commodity number 2) produced in unit time is called the **intensity vector.**

(a) Give the economic interpretation of the product TX.

(b) Give the economic interpretation of $X - TX$.

70. In Problem 69, let d_1 (respectively, d_2) denote the surplus number of units of commodity number 1 (respectively, commodity number 2) required per unit time for export. If $I - T$ is a nonsingular matrix and $D = \begin{bmatrix} d_1 \\ d_2 \end{bmatrix}$, give the economic interpretation of $(I - T)^{-1}D$.

9.4 DETERMINANTS AND CRAMER'S RULE

In this section, we consider an alternative method, called Cramer's rule, for solving systems of linear equations. Although Cramer's rule applies to systems of n linear equations in n unknowns for any positive integer n, its practical use is usually limited to the cases $n = 2$ or $n = 3$. For larger values of n, the elimination method using matrices (Section 9.2) is usually more efficient.

Cramer's rule is based on the idea of a *determinant*. If a, b, c, and d are any four numbers, the symbol

$$\begin{vmatrix} a & b \\ c & d \end{vmatrix}$$

is called a 2 by 2 **determinant** with *entries* or *elements* a, b, c, and d. Its **value** is defined to be the number $ad - cb$, that is,

$$\begin{vmatrix} a & b \\ c & d \end{vmatrix} = ad - cb.$$

The memory aid

$$\begin{vmatrix} a & b \\ c & d \end{vmatrix} = ad - cb$$

is often helpful.

Example 1   Evaluate the determinant $\begin{vmatrix} 4 & -3 \\ 2 & 1 \end{vmatrix}$.

Solution $\begin{vmatrix} 4 & -3 \\ 2 & 1 \end{vmatrix} = 4(1) - 2(-3) = 10$ ∎

Now, let's see how determinants can be used to solve systems of linear equations. Consider the system

$$\begin{cases} ax + by = h \\ cx + dy = k. \end{cases}$$

If we multiply the first equation by d and the second equation by b, we obtain

$$\begin{cases} adx + bdy = hd \\ bcx + bdy = bk. \end{cases}$$

So, subtracting the second equation from the first, we eliminate the terms involving y and get

$$adx - bcx = hd - bk$$

or

$$(ad - bc)x = hd - bk.$$

Using determinants, we can rewrite the last equation as

$$\begin{vmatrix} a & b \\ c & d \end{vmatrix} x = \begin{vmatrix} h & b \\ k & d \end{vmatrix}.$$

Therefore,

$$x = \dfrac{\begin{vmatrix} h & b \\ k & d \end{vmatrix}}{\begin{vmatrix} a & b \\ c & d \end{vmatrix}},$$

provided that $\begin{vmatrix} a & b \\ c & d \end{vmatrix} \neq 0$. A similar calculation (Problem 51) yields

$$y = \dfrac{\begin{vmatrix} a & h \\ c & k \end{vmatrix}}{\begin{vmatrix} a & b \\ c & d \end{vmatrix}}.$$

Thus, if

$$D = \begin{vmatrix} a & b \\ c & d \end{vmatrix}, \qquad D_x = \begin{vmatrix} h & b \\ k & d \end{vmatrix}, \qquad D_y = \begin{vmatrix} a & h \\ c & k \end{vmatrix},$$

and $D \neq 0$, then the system

$$\begin{cases} ax + by = h \\ cx + dy = k \end{cases}$$

has one and only one solution:

$$x = \frac{D_x}{D}, \qquad y = \frac{D_y}{D}.$$

This is **Cramer's rule** for two linear equations in two unknowns.

In Cramer's rule, D is called the **coefficient determinant** because its entries are the coefficients of the unknowns in the system:

$$\begin{cases} ax + by = h \\ cx + dy = k, \end{cases} \qquad D = \begin{vmatrix} a & b \\ c & d \end{vmatrix}.$$

Notice that D_x is obtained by replacing the *first* column of D (the coefficients of x), and that D_y is obtained by replacing the *second* column of D (the coefficients of y) by the constants on the right in the system of equations:

$$\begin{cases} ax + by = h \\ cx + dy = k, \end{cases} \qquad D_x = \begin{vmatrix} h & b \\ k & d \end{vmatrix}, \qquad D_y = \begin{vmatrix} a & h \\ c & k \end{vmatrix}.$$

Keep in mind that Cramer's rule can be applied only when $D \neq 0$. If $D = 0$, it can be shown (Problem 53) that the equations in the system are either inconsistent or dependent.

Example 2   Use Cramer's rule (if applicable) to solve each system:

(a) $\begin{cases} 2x - y = 7 \\ x + 3y = 14 \end{cases}$ **(b)** $\begin{cases} 2x - y = 7 \\ 4x - 2y = 3 \end{cases}$

Solution **(a)** $D = \begin{vmatrix} 2 & -1 \\ 1 & 3 \end{vmatrix} = 2(3) - 1(-1) = 7, \quad D_x = \begin{vmatrix} 7 & -1 \\ 14 & 3 \end{vmatrix} = 7(3) - 14(-1) = 35,$

$D_y = \begin{vmatrix} 2 & 7 \\ 1 & 14 \end{vmatrix} = 2(14) - 1(7) = 21;$

hence, by Cramer's rule

$$x = \frac{D_x}{D} = \frac{35}{7} = 5 \quad \text{and} \quad y = \frac{D_y}{D} = \frac{21}{7} = 3.$$

(b) Here,

$$D = \begin{vmatrix} 2 & -1 \\ 4 & -2 \end{vmatrix} = 2(-2) - 4(-1) = 0,$$

so Cramer's rule does not apply. [Actually, the system of equations is inconsistent—it has no solution.]

To extend Cramer's rule to systems of three linear equations in three unknowns, we begin by extending the definition of a determinant. The **value** of a 3 by 3 determinant is defined in terms of 2 by 2 determinants as follows:

$$\begin{vmatrix} a_1 & a_2 & a_3 \\ b_1 & b_2 & b_3 \\ c_1 & c_2 & c_3 \end{vmatrix} = a_1 \begin{vmatrix} b_2 & b_3 \\ c_2 & c_3 \end{vmatrix} - a_2 \begin{vmatrix} b_1 & b_3 \\ c_1 & c_3 \end{vmatrix} + a_3 \begin{vmatrix} b_1 & b_2 \\ c_1 & c_2 \end{vmatrix}.$$

We refer to this as the **expansion formula** for 3 by 3 determinants. Notice the *negative sign* on the middle term.

In the expansion formula, notice that each entry in the first row is multiplied by the 2 by 2 determinant that remains when the row and column containing the multiplier are (mentally) crossed out. Thus:

a_1 is multiplied by $\begin{vmatrix} a_1 & a_2 & a_3 \\ b_1 & b_2 & b_3 \\ c_1 & c_2 & c_3 \end{vmatrix} = \begin{vmatrix} b_2 & b_3 \\ c_2 & c_3 \end{vmatrix}$

a_2 is multiplied by $\begin{vmatrix} a_1 & a_2 & a_3 \\ b_1 & b_2 & b_3 \\ c_1 & c_2 & c_3 \end{vmatrix} = \begin{vmatrix} b_1 & b_3 \\ c_1 & c_3 \end{vmatrix}$, and

a_3 is multiplied by $\begin{vmatrix} a_1 & a_2 & a_3 \\ b_1 & b_2 & b_3 \\ c_1 & c_2 & c_3 \end{vmatrix} = \begin{vmatrix} b_1 & b_2 \\ c_1 & c_2 \end{vmatrix}.$

To **expand** a 3 by 3 determinant means to find its value by using the expansion formula. Again, we emphasize: When expanding a 3 by 3 determinant, *don't forget the negative sign on the middle term.*

Example 3   Expand the determinant $\begin{vmatrix} 3 & 1 & 2 \\ -4 & 2 & 4 \\ 1 & 0 & 5 \end{vmatrix}$.

Solution

$$\begin{vmatrix} 3 & 1 & 2 \\ -4 & 2 & 4 \\ 1 & 0 & 5 \end{vmatrix} = 3\begin{vmatrix} 2 & 4 \\ 0 & 5 \end{vmatrix} - 1\begin{vmatrix} -4 & 4 \\ 1 & 5 \end{vmatrix} + 2\begin{vmatrix} -4 & 2 \\ 1 & 0 \end{vmatrix}$$

$$= 3[2(5) - 0(4)] - 1[(-4)5 - 1(4)] + 2[(-4)0 - 1(2)]$$

$$= 3(10) - 1(-24) + 2(-2) = 50$$

Now we can state **Cramer's rule** for solving a system

$$\begin{cases} a_1 x + a_2 y + a_3 z = k_1 \\ b_1 x + b_2 y + b_3 z = k_2 \\ c_1 x + c_2 y + c_3 z = k_3 \end{cases}$$

of three linear equations in three unknowns: Form the **coefficient determinant**

$$D = \begin{vmatrix} a_1 & a_2 & a_3 \\ b_1 & b_2 & b_3 \\ c_1 & c_2 & c_3 \end{vmatrix}.$$

If $D \neq 0$, form the determinants

$$D_x = \begin{vmatrix} k_1 & a_2 & a_3 \\ k_2 & b_2 & b_3 \\ k_3 & c_2 & c_3 \end{vmatrix}, \qquad D_y = \begin{vmatrix} a_1 & k_1 & a_3 \\ b_1 & k_2 & b_3 \\ c_1 & k_3 & c_3 \end{vmatrix}, \qquad D_z = \begin{vmatrix} a_1 & a_2 & k_1 \\ b_1 & b_2 & k_2 \\ c_1 & c_2 & k_3 \end{vmatrix}.$$

Then the solution of the system of linear equations is

$$x = \frac{D_x}{D}, \qquad y = \frac{D_y}{D}, \qquad z = \frac{D_z}{D}.$$

If $D \neq 0$, this is the *only* solution of the system; that is, the equations in the system are *consistent* (see page 491). If $D = 0$, the equations are either *inconsistent* or *dependent*, and Cramer's rule is not applicable. You can find a proof of Cramer's rule in a textbook on linear algebra.

Example 4   Use Cramer's rule to solve the system

$$\begin{cases} 3x - 2y + z = -9 \\ x + 2y - z = 5 \\ 2x - y + 3z = -10. \end{cases}$$

Solution

$$D = \begin{vmatrix} 3 & -2 & 1 \\ 1 & 2 & -1 \\ 2 & -1 & 3 \end{vmatrix} = 3\begin{vmatrix} 2 & -1 \\ -1 & 3 \end{vmatrix} - (-2)\begin{vmatrix} 1 & -1 \\ 2 & 3 \end{vmatrix} + 1\begin{vmatrix} 1 & 2 \\ 2 & -1 \end{vmatrix}$$

$$= 3[2(3) - (-1)(-1)] + 2[1(3) - 2(-1)] + 1[1(-1) - 2(2)]$$

$$= 3(5) + 2(5) + 1(-5) = 20$$

Because $D \neq 0$, the system is consistent, and we can solve it by applying Cramer's rule:

$$D_x = \begin{vmatrix} -9 & -2 & 1 \\ 5 & 2 & -1 \\ -10 & -1 & 3 \end{vmatrix} = -9\begin{vmatrix} 2 & -1 \\ -1 & 3 \end{vmatrix} - (-2)\begin{vmatrix} 5 & -1 \\ -10 & 3 \end{vmatrix} + 1\begin{vmatrix} 5 & 2 \\ -10 & -1 \end{vmatrix} = -20$$

$$D_y = \begin{vmatrix} 3 & -9 & 1 \\ 1 & 5 & -1 \\ 2 & -10 & 3 \end{vmatrix} = 3\begin{vmatrix} 5 & -1 \\ -10 & 3 \end{vmatrix} - (-9)\begin{vmatrix} 1 & -1 \\ 2 & 3 \end{vmatrix} + 1\begin{vmatrix} 1 & 5 \\ 2 & -10 \end{vmatrix} = 40$$

$$D_z = \begin{vmatrix} 3 & -2 & -9 \\ 1 & 2 & 5 \\ 2 & -1 & -10 \end{vmatrix} = 3\begin{vmatrix} 2 & 5 \\ -1 & -10 \end{vmatrix} - (-2)\begin{vmatrix} 1 & 5 \\ 2 & -10 \end{vmatrix} + (-9)\begin{vmatrix} 1 & 2 \\ 2 & -1 \end{vmatrix} = -40;$$

hence,

$$x = \frac{D_x}{D} = \frac{-20}{20} = -1, \qquad y = \frac{D_y}{D} = \frac{40}{20} = 2, \qquad z = \frac{D_z}{D} = \frac{-40}{20} = -2.$$

■

Properties of Determinants

Determinants have a number of useful properties, some of which we now state (without proof).

> **Property 1.** If you interchange any two rows or any two columns of a determinant, you change its algebraic sign.

For instance,

$$\begin{vmatrix} 2 & 3 \\ 5 & 6 \end{vmatrix} = -\begin{vmatrix} 5 & 6 \\ 2 & 3 \end{vmatrix}.$$

(Check this yourself.)

> **Property 2.** If you multiply every entry in one row or one column of a determinant by a constant k, the effect is to multiply the value of the determinant by k.

For instance,

$$\begin{vmatrix} 2k & 3 \\ 5k & 6 \end{vmatrix} = k\begin{vmatrix} 2 & 3 \\ 5 & 6 \end{vmatrix}.$$

(Check this yourself.)

Property 2 allows you to "factor out" a common factor of all the elements of a single row or column of a determinant. For instance,

$$\begin{vmatrix} 8 & 28 & 3 \\ 3 & -14 & 2 \\ 5 & 35 & 4 \end{vmatrix} = 7 \begin{vmatrix} 8 & 4 & 3 \\ 3 & -2 & 2 \\ 5 & 5 & 4 \end{vmatrix}.$$

> **Property 3.** If you add a constant multiple of the entries in any one row of a determinant to the corresponding entries in any other row, the value of the determinant will not change. Likewise for columns.

For instance, in the determinant

$$\begin{vmatrix} -3 & 5 \\ 6 & -4 \end{vmatrix},$$

if you add 2 times the first row to the second row, the value of the determinant won't change, that is,

$$\begin{vmatrix} -3 & 5 \\ 6 & -4 \end{vmatrix} = \begin{vmatrix} -3 & 5 \\ 6 + 2(-3) & -4 + 2(5) \end{vmatrix} = \begin{vmatrix} -3 & 5 \\ 0 & 6 \end{vmatrix}.$$

(Check this yourself.)

> **Property 4.** If any two rows or any two columns of a determinant are the same, its value is zero. More generally, if the corresponding entries in any two rows or in any two columns are proportional, the value of the determinant is zero.

For instance,

$$\begin{vmatrix} 1 & 5 & -7 \\ 2 & 3 & 1 \\ 1 & 5 & -7 \end{vmatrix} = 0 \quad \text{and} \quad \begin{vmatrix} 2 & -1 & 10 \\ 3 & 7 & 15 \\ 4 & 2 & 20 \end{vmatrix} = 0.$$

(Check this yourself.)

The **main diagonal** of a determinant is the diagonal running from upper left to lower right. For instance, in the determinant

$$\begin{vmatrix} a & b & c \\ u & v & w \\ x & y & z \end{vmatrix}$$

the entries on the main diagonal are a, v, and z. A determinant is said to be in **triangular form** if all entries below the main diagonal are zero. For instance,

$$\begin{vmatrix} 3 & 5 & -7 \\ 0 & 2 & 4 \\ 0 & 0 & 6 \end{vmatrix}$$

is in triangular form.

> **Property 5.** If a determinant is in triangular form, its value is the product of the entries on its main diagonal.

For instance,

$$\begin{vmatrix} 3 & 5 & -7 \\ 0 & 2 & 4 \\ 0 & 0 & 6 \end{vmatrix} = 3(2)(6) = 36.$$

(Check this yourself.)

By using Properties 1 to 5, you can often evaluate a determinant more easily than by applying the expansion formula. The usual idea is to use Properties 1 to 3 to bring the determinant into triangular form, and then to apply Property 5.

Example 5   Use the properties of determinants to evaluate:

(a) $\begin{vmatrix} 3 & -1 & 2 \\ 6 & -2 & 4 \\ 7 & 0 & 3 \end{vmatrix}$ (b) $\begin{vmatrix} 4 & 3 & 3 \\ 1 & 0 & 2 \\ 6 & 6 & 7 \end{vmatrix}$

Solution   (a) The second row is proportional to the first, so, by Property 4,

$$\begin{vmatrix} 3 & -1 & 2 \\ 6 & -2 & 4 \\ 7 & 0 & 3 \end{vmatrix} = 0.$$

(b) $\begin{vmatrix} 4 & 3 & 3 \\ 1 & 0 & 2 \\ 6 & 6 & 7 \end{vmatrix} = 3\begin{vmatrix} 4 & 1 & 3 \\ 1 & 0 & 2 \\ 6 & 2 & 7 \end{vmatrix}$ (We factored out 3 from the second column—Property 2.)

$$= -3\begin{vmatrix} 1 & 4 & 3 \\ 0 & 1 & 2 \\ 2 & 6 & 7 \end{vmatrix}$$ (We interchanged the first and second columns—Property 1.)

$$= -3\begin{vmatrix} 1 & 4 & 3 \\ 0 & 1 & 2 \\ 0 & -2 & 1 \end{vmatrix}$$ (We added -2 times the first row to the third row—Property 3.)

$$= -3\begin{vmatrix} 1 & 4 & 3 \\ 0 & 1 & 2 \\ 0 & 0 & 5 \end{vmatrix}$$ (We added 2 times the second row to the third row—Property 3.)

$$= -3(1)(1)(5)$$ (We used Property 5.)
$$= -15.$$

It should come as no surprise to you to learn that 4 by 4, 5 by 5, and, in general, n by n determinants can be defined, and Properties 1 to 5 continue to hold for them as well. You can learn more about determinants from a textbook on linear algebra. Determinants have a wide variety of uses (other than for solving systems of linear equations) in linear algebra, calculus, and other branches of mathematics.

Problem Set 9.4

In Problems 1 to 8, evaluate each determinant.

1. $\begin{vmatrix} 2 & 3 \\ 1 & 4 \end{vmatrix}$

2. $\begin{vmatrix} e & \pi \\ \sqrt{3} & \sqrt{2} \end{vmatrix}$

3. $\begin{vmatrix} 6 & -4 \\ 3 & 7 \end{vmatrix}$

4. $\begin{vmatrix} x & -y \\ y & x \end{vmatrix}$

5. $\begin{vmatrix} \sqrt{6} & -2\sqrt{5} \\ 3\sqrt{5} & 4\sqrt{6} \end{vmatrix}$

6. $\begin{vmatrix} x+y & x+y \\ x-y & x+y \end{vmatrix}$

7. $\begin{vmatrix} \log 100 & 2 \\ \log 10 & 3 \end{vmatrix}$

8. $\begin{vmatrix} x & -x \\ y & -y \end{vmatrix}$

In Problems 9 to 14, use Cramer's rule (when applicable) to solve each system of linear equations.

9. $\begin{cases} 5x + 7y = -2 \\ 3x + 4y = -1 \end{cases}$

10. $\begin{cases} \frac{1}{2}x - \frac{2}{3}y = \frac{3}{4} \\ \frac{1}{3}x + 2y = \frac{5}{6} \end{cases}$

11. $\begin{cases} 2x_1 + x_2 = 5 \\ x_1 - 2x_2 = 0 \end{cases}$

12. $\begin{cases} 3u - 4v = 1 \\ -4u + \frac{16}{3}v = 2 \end{cases}$

13. $\begin{cases} 8u + 3v = 9 \\ 4u - 6v = 7 \end{cases}$

14. $\begin{cases} ax + y = 0 \\ x + ay = 0 \end{cases}$

In Problems 15 to 22, expand each determinant.

15. $\begin{vmatrix} 2 & 3 & -1 \\ 5 & 7 & 0 \\ 2 & -3 & 1 \end{vmatrix}$

16. $\begin{vmatrix} 1 & 5 & -7 \\ 3 & 0 & 2 \\ -1 & 4 & 1 \end{vmatrix}$

17. $\begin{vmatrix} 2 & 0 & 4 \\ 1 & 5 & 0 \\ 0 & 7 & 1 \end{vmatrix}$

18. $\begin{vmatrix} a & b & c \\ x & y & z \\ 1 & 1 & 1 \end{vmatrix}$

19. $\begin{vmatrix} 2 & -3 & 1 \\ 1 & 2 & 3 \\ 0 & 1 & 2 \end{vmatrix}$

20. $\begin{vmatrix} e & \sqrt{2} & \sqrt{3} \\ \pi & 0 & 1 \\ -1 & 2 & 0 \end{vmatrix}$

21. $\begin{vmatrix} \frac{1}{2} & 1 & -\frac{2}{3} \\ \frac{5}{2} & -\frac{4}{3} & 1 \\ \frac{3}{2} & 0 & \frac{3}{4} \end{vmatrix}$

22. $\begin{vmatrix} 0 & a & b \\ a & 0 & c \\ b & c & 0 \end{vmatrix}$

In Problems 23 to 28, use Cramer's rule (when applicable) to solve each system of linear equations.

23. $\begin{cases} x + 2y - z = -3 \\ 2x - y + z = 5 \\ 3x + 2y - 2z = -3 \end{cases}$

24. $\begin{cases} -u + 2v + w = -1 \\ 4u - 2v - w = 3 \\ 4u + 2v - w = 5 \end{cases}$

25. $\begin{cases} -3x + 4y + 6z = 30 \\ x + 2z = 6 \\ -x - 2y + 3z = 8 \end{cases}$

26. $\begin{cases} 2y - 3x = 1 \\ 3z - 2y = 5 \\ x + z = 4 \end{cases}$

27. $\begin{cases} 2x + 5y - z = 3 \\ -3x - 2y + 7z = 4 \\ -x + 3y + 6z = 0 \end{cases}$

28. $\begin{cases} 3x + y - z = 14 \\ x + 3y - z = 16 \\ x + y - 3z = -10 \end{cases}$

In Problems 29 to 36, use Properties 1 to 5 to evaluate each determinant.

29. $\begin{vmatrix} 1 & -2 & 3 \\ 2 & 3 & -2 \\ 3 & 1 & -1 \end{vmatrix}$

30. $\begin{vmatrix} 5 & 2 & 3 \\ 4 & -5 & -6 \\ 7 & -8 & -9 \end{vmatrix}$

31. $\begin{vmatrix} 1 & -5 & 2 \\ -4 & -1 & 5 \\ 3 & -4 & 3 \end{vmatrix}$

32. $\begin{vmatrix} 2 & 4 & 3 \\ -6 & 0 & 4 \\ -1 & 1 & 3 \end{vmatrix}$

33. $\begin{vmatrix} -1 & 1 & 1 \\ 4 & 2 & 3 \\ 1 & 3 & 0 \end{vmatrix}$

34. $\begin{vmatrix} 4 & 5 & 6 \\ 2 & 2 & 2 \\ 7 & 2 & -7 \end{vmatrix}$

35. $\begin{vmatrix} 3 & 1 & 4 \\ 1 & 7 & 3 \\ 5 & -10 & 5 \end{vmatrix}$

36. $\begin{vmatrix} 9 & 3 & 3 \\ -2 & 0 & 6 \\ 2 & 1 & 1 \end{vmatrix}$

© In Problems 37 and 38, evaluate each determinant with the aid of a calculator.

37. $\begin{vmatrix} 2.03 & -7.07 & 1.55 \\ 3.71 & 2.22 & 5.77 \\ 6.65 & -8.56 & 3.65 \end{vmatrix}$

38. $\begin{vmatrix} 0.071 & 0.029 & -0.095 \\ 0.101 & 0.210 & 0.055 \\ 0.077 & -0.101 & 0.039 \end{vmatrix}$

In Problems 39 to 44, evaluate each determinant mentally. Indicate which property or properties you use.

39. $\begin{vmatrix} 3 & \sqrt{2} & 19 \\ 0 & 1 & \frac{5}{2} \\ 0 & 0 & -4 \end{vmatrix}$

40. $\begin{vmatrix} 3 & 1 & 4 \\ -6 & -2 & -8 \\ 1 & 5 & -7 \end{vmatrix}$

41. $\begin{vmatrix} 1 & 1 & 1 \\ -1 & -1 & -1 \\ a & b & c \end{vmatrix}$

42. $\begin{vmatrix} a & b & c \\ 0 & d & h \\ 0 & 0 & \sqrt{5} \end{vmatrix}$

43. $\begin{vmatrix} 0 & 1 & 0 \\ 1 & 0 & 0 \\ 0 & 0 & 1 \end{vmatrix}$

44. $\begin{vmatrix} 1 & 2 & 3 \\ 3 & 2 & 1 \\ 4 & 4 & 4 \end{vmatrix}$

In Problem 45 to 50, assume that $\begin{vmatrix} a & b & c \\ u & v & w \\ x & y & z \end{vmatrix} = -3.$

Find the value of each determinant.

45. $\begin{vmatrix} u & v & w \\ a & b & c \\ x & y & z \end{vmatrix}$

46. $\begin{vmatrix} u & v & w \\ x & y & z \\ a & b & c \end{vmatrix}$

47. $\begin{vmatrix} c & a & b \\ w & u & v \\ z & x & y \end{vmatrix}$

48. $\begin{vmatrix} -a & -b & -c \\ 3u & 3v & 3w \\ 4x & 4y & 4z \end{vmatrix}$

49. $\begin{vmatrix} a+u & b+v & c+w \\ u & v & w \\ x+u & y+v & z+w \end{vmatrix}$

50. $\begin{vmatrix} a & b & c \\ u-2a & v-2b & w-2c \\ 3x & 3y & 3z \end{vmatrix}$

51. Complete the proof of Cramer's rule for two linear equations in two unknowns (page 505) by showing that if $D \neq 0$, then $y = D_y/D$. [*Hint:* Multiply the first equation by c and the second equation by a; then subtract the first equation from the second.]

52. Show that Property 4 can be derived from Property 2 and Property 3.

53. If $\begin{vmatrix} a & b \\ c & d \end{vmatrix} = 0$, show that the equations in the system
$$\begin{cases} ax + by = h \\ cx + dy = k \end{cases}$$
are either inconsistent or dependent.
[*Hint:* The equations $Dx = D_x$ and $Dy = D_y$ are true even if $D = 0$.]

54. Solve each equation for x:

(a) $\begin{vmatrix} 4 & -1 \\ x & 3 \end{vmatrix} = 2$ (b) $\begin{vmatrix} -1 & 5 & -2 \\ 2 & -2 & x \\ 3 & 1 & 0 \end{vmatrix} = -3.$

55. Show that in the xy plane, an equation of the line that contains the two points (a, b) and (c, d) is
$$\begin{vmatrix} x & y & 1 \\ a & b & 1 \\ c & d & 1 \end{vmatrix} = 0.$$

56. Show that
$$\begin{vmatrix} a_1 + a_2 & b_1 + b_2 & c_1 + c_2 \\ u & v & w \\ x & y & z \end{vmatrix} = \begin{vmatrix} a_1 & b_1 & c_1 \\ u & v & w \\ x & y & z \end{vmatrix} + \begin{vmatrix} a_2 & b_2 & c_2 \\ u & v & w \\ x & y & z \end{vmatrix}.$$

57. Solve the equation
$$\begin{vmatrix} 1 & x-2 & 2 \\ -2 & x-1 & 3 \\ 1 & 2 & x \end{vmatrix} = 0.$$

58. Show that $x - a$ and $x - b$ are factors of
$$\begin{vmatrix} 1 & 1 & 1 \\ x & a & b \\ x^2 & a^2 & b^2 \end{vmatrix}.$$

59. Show that the equation
$$\begin{vmatrix} a-x & b \\ b & c-x \end{vmatrix} = 0,$$
in which a, b, and c are real numbers and x is the unknown, always has real roots.

9.5 APPLICATIONS OF SYSTEMS OF LINEAR EQUATIONS

In this section, we present several problems that can be worked by setting up and solving systems of linear equations.

Partial Fractions

If you add the two fractions $\dfrac{2}{x-2}$ and $\dfrac{3}{x+1}$, you obtain

$$\frac{2}{x-2} + \frac{3}{x+1} = \frac{2(x+1) + 3(x-2)}{(x-2)(x+1)} = \frac{5x-4}{(x-2)(x+1)}.$$

The reverse process of "taking the fraction $\dfrac{5x-4}{(x-2)(x+1)}$ apart" into the sum of simpler fractions,

$$\frac{5x-4}{(x-2)(x+1)} = \frac{2}{x-2} + \frac{3}{x+1},$$

is called **decomposing** $\dfrac{5x-4}{(x-2)(x+1)}$ into the **partial fractions** $\dfrac{2}{x-2}$ and $\dfrac{3}{x+1}$. The process of decomposing fractions into simpler partial fractions is used routinely in calculus and other branches of mathematics. Here we shall give you a brief introduction to the subject—you can find further details in calculus textbooks.

It is easiest to decompose a rational expression into partial fractions if the degree of the numerator is less than the degree of the denominator, and if the denominator is (or can be) factored into linear factors* that are all different from one another. In this case, you simply provide a partial fraction of the form

$$\frac{\text{constant}}{\text{linear factor}}$$

for each linear factor in the denominator.

Example 1   Decompose $\dfrac{6x^2 + 2x + 2}{x(x-2)(2x+1)}$ into partial fractions.

Solution   The linear factors in the denominator are x, $x-2$, and $2x+1$. Since these factors are different from each other, we must provide partial fractions

$$\frac{\text{constant}}{x}, \qquad \frac{\text{constant}}{x-2}, \qquad \text{and} \qquad \frac{\text{constant}}{2x+1}.$$

---

* A linear factor has the form $ax + b$, where a and b are constants.

We denote the three constants by A, B, and C, so that

$$\frac{6x^2 + 2x + 2}{x(x-2)(2x+1)} = \frac{A}{x} + \frac{B}{x-2} + \frac{C}{2x+1}.$$

To determine the values of A, B, and C, we begin by multiplying both sides of this equation by $x(x-2)(2x+1)$ to clear the fractions. Thus, we have

$$6x^2 + 2x + 2 = A(x-2)(2x+1) + Bx(2x+1) + Cx(x-2)$$

or $6x^2 + 2x + 2 = A(2x^2 - 3x - 2) + B(2x^2 + x) + C(x^2 - 2x).$

Collecting like powers on the right, we obtain

$$6x^2 + 2x + 2 = (2A + 2B + C)x^2 + (-3A + B - 2C)x - 2A.$$

Now we equate the coefficients of like powers of x on both sides of the equation:

$$6 = 2A + 2B + C, \qquad 2 = -3A + B - 2C, \qquad \text{and} \qquad 2 = -2A.$$

In other words, the constants A, B, and C satisfy the system of linear equations

$$\begin{cases} 2A + 2B + C = 6 \\ -3A + B - 2C = 2 \\ -2A = 2. \end{cases}$$

Solving this system by one of the methods given in Sections 9.1, 9.2, and 9.4, we find that

$$A = -1, \qquad B = 3, \qquad \text{and} \qquad C = 2.$$

Therefore, $\dfrac{6x^2 + 2x + 2}{x(x-2)(2x+1)} = \dfrac{-1}{x} + \dfrac{3}{x-2} + \dfrac{2}{2x+1}.$ ∎

A factor of the form $(ax + b)^2$ in the denominator requires two partial fractions:

$$\frac{B}{ax+b} + \frac{C}{(ax+b)^2}.$$

Example 2   Decompose $\dfrac{2x^3 + 7x^2 + 6x + 2}{x^3 + 2x^2 + x}$ into partial fractions.

Solution   When the degree of the numerator is greater than or equal to the degree of the denominator, we first perform a long division, then decompose into partial fractions. By long division, we have

$$\frac{2x^3 + 7x^2 + 6x + 2}{x^3 + 2x^2 + x} = 2 + \frac{3x^2 + 4x + 2}{x^3 + 2x^2 + x} = 2 + \frac{3x^2 + 4x + 2}{x(x+1)^2}.$$

Notice that we have factored the denominator, in preparation for the method of partial fractions. Now, we write

$$\frac{3x^2 + 4x + 2}{x(x+1)^2} = \frac{A}{x} + \frac{B}{x+1} + \frac{C}{(x+1)^2}.$$

Multiplying both sides of the equation by $x(x + 1)^2$, we have

$$3x^2 + 4x + 2 = A(x + 1)^2 + Bx(x + 1) + Cx$$
$$= A(x^2 + 2x + 1) + B(x^2 + x) + Cx$$
$$= (A + B)x^2 + (2A + B + C)x + A.$$

Equating coefficients of like powers of x on both sides of the last equation, we obtain the system

$$\begin{cases} A + B &= 3 \\ 2A + B + C &= 4 \\ A &= 2. \end{cases}$$

Solving this system, we find that $A = 2$, $B = 1$, and $C = -1$. Hence,

$$\frac{3x^2 + 4x + 2}{x(x + 1)^2} = \frac{2}{x} + \frac{1}{x + 1} + \frac{-1}{(x + 1)^2}.$$

Therefore, $$\frac{2x^3 + 7x^2 + 6x + 2}{x^3 + 2x^2 + x} = 2 + \frac{2}{x} + \frac{1}{x + 1} + \frac{-1}{(x + 1)^2}.$$

A prime quadratic factor of the form $ax^2 + bx + c$ in the denominator requires a partial fraction

$$\frac{Bx + C}{ax^2 + bx + c}.$$

Example 3   Decompose $\dfrac{8x^2 + 3x + 20}{(x + 1)(x^2 + 4)}$ into partial fractions.

Solution   We begin by writing

$$\frac{8x^2 + 3x + 20}{(x + 1)(x^2 + 4)} = \frac{A}{x + 1} + \frac{Bx + C}{x^2 + 4}.$$

Multiplying both sides by $(x + 1)(x^2 + 4)$, we obtain

$$8x^2 + 3x + 20 = A(x^2 + 4) + (Bx + C)(x + 1)$$
$$= Ax^2 + 4A + Bx^2 + Bx + Cx + C$$
$$= (A + B)x^2 + (B + C)x + 4A + C.$$

Equating coefficients, we get the system

$$\begin{cases} A + B &= 8 \\ B + C &= 3 \\ 4A + C &= 20. \end{cases}$$

The solution of this system is $A = 5$, $B = 3$, and $C = 0$; hence,

$$\frac{8x^2 + 3x + 20}{(x + 1)(x^2 + 4)} = \frac{5}{x + 1} + \frac{3x}{x^2 + 4}.$$

Other Applications of Systems of Linear Equations

Word problems that lead to a system of linear equations can be solved by using a slight variation of the procedure given in Section 2.2 (page 79). We only need to modify step 2 of this procedure by introducing as many letters as may be necessary to represent all of the unknown quantities in the problem. Step 3 will then produce a *system* of equations to be solved for these unknowns.

Example 4   The price of admission to a play was $2 for adults, $1 for senior citizens, and $0.50 for children. Altogether, 270 tickets were sold and the total revenue was $360. Twice as many children as senior citizens attended the play. How many adults, how many children, and how many senior citizens attended the play?

Solution   Let x = the number of adults, y = the number of children, and z = the number of senior citizens who attended the play. From the fact that 270 tickets were sold, we have

$$x + y + z = 270.$$

Because the total revenue was $360, it follows that

$$2x + \tfrac{1}{2}y + z = 360.$$

Since twice as many children as senior citizens attended,

$$2z = y \quad \text{or} \quad y - 2z = 0.$$

Thus, we have the system of linear equations

$$\begin{cases} x + y + z = 270 \\ 2x + \tfrac{1}{2}y + z = 360 \\ y - 2z = 0. \end{cases}$$

Solving this system (say, by the elimination method) we find that $x = 135$, $y = 90$, and $z = 45$.   ■

Problem Set 9.5

In Problems 1 to 26, decompose each fraction into partial fractions.

1. $\dfrac{3}{(x-3)(x-2)}$

2. $\dfrac{x}{(x-1)(x-4)}$

3. $\dfrac{x+2}{(x+5)(x-1)}$

4. $\dfrac{3x+7}{x^2-2x-3}$

5. $\dfrac{x^2-5x-3}{x(x-2)(x+2)}$

6. $\dfrac{x+12}{x^3-x^2-6x}$

7. $\dfrac{8x+2}{x^3-x}$

8. $\dfrac{x^2-16x-12}{x^3-3x^2-4x}$

9. $\dfrac{3x^3+4x^2-17x-1}{x^3+x^2-6x}$

10. $\dfrac{2x^2+5x-4}{x^3+x^2-2x}$

11. $\dfrac{-2x^2+x-1}{(x-3)(x-1)^2}$

12. $\dfrac{13x-12}{x^2(x-3)}$

13. $\dfrac{x^3-x^2+1}{x^2(x-1)}$

14. $\dfrac{3y+4}{(y+2)^2(y-6)}$

15. $\dfrac{3x^2 + 18x + 15}{(x - 1)(x + 2)^2}$

16. $\dfrac{s + 4}{(s + 1)^2(s - 1)^2}$

17. $\dfrac{4x^2 - 7x + 10}{(x + 2)(3x - 2)^2}$

18. $\dfrac{1}{x^4 - 2x^3 + x^2}$

19. $\dfrac{t + 10}{(t + 1)(t^2 + 1)}$

20. $\dfrac{x^4 - x^2 - 2x + 3}{(x - 2)(x^2 + 2x + 2)}$

21. $\dfrac{x^5 + 9x^3 + 1}{x^3 + 9x}$

22. $\dfrac{4x + 3}{(x^2 + 1)(x^2 + 2)}$

23. $\dfrac{t + 3}{t(t^2 + 1)}$

24. $\dfrac{x}{(x + 1)^2(x^2 + 1)}$

25. $\dfrac{u}{u^4 - 1}$

26. $\dfrac{3s + 1}{s^2(s^2 + 1)}$

In Problems 27 to 30, find the values of the constants in each decomposition into partial fractions.

27. $\dfrac{3x^2 - 2x - 4}{x^3(x + 2)} = \dfrac{A}{x} + \dfrac{B}{x^2} + \dfrac{C}{x^3} + \dfrac{D}{x + 2}$

28. $\dfrac{x^3 + 3x^2 + 1}{(x^2 + 1)^2} = \dfrac{Ax + B}{x^2 + 1} + \dfrac{Cx + D}{(x^2 + 1)^2}$

29. $\dfrac{t^3 - t^2}{(t^2 + 3)^2} = \dfrac{At + B}{t^2 + 3} + \dfrac{Ct + D}{(t^2 + 3)^2}$

30. $\dfrac{x^5 - 2x^4 + 2x^3 + x - 2}{x^2(x^2 + 1)^2} = \dfrac{A}{x} + \dfrac{B}{x^2} + \dfrac{Cx + D}{x^2 + 1} + \dfrac{Ex + G}{(x^2 + 1)^2}$

31. The price of admission for a sporting event was $2 for adults and $1 for children. Altogether, 925 tickets were sold, and the resulting revenue was $1150. How many adults and how many children attended the game?

32. A veterinarian has put certain animals on a diet. Each animal receives, among other things, exactly 25 grams of protein and 9.5 grams of fat for each feeding. If the veterinarian buys two food mixes, the first containing 10% protein and 8% fat, and the second containing 20% protein and 2% fat, how many grams of each should be combined to provide the right diet for a single feeding of 10 animals?

33. One angle x of a triangle is $10°$ greater than a second angle y, and $40°$ less than the third angle z. Find x, y, and z.

34. A certain three-digit number is 56 times the sum of its digits. The unit's digit is 4 more than the ten's digit. If the unit's digit and the hundred's digit were interchanged, the resulting number would be 99 less than the original number. Find the number.

35. A department store has sold 80 men's suits of three different types at a discount. If the suits had been sold at their original prices—type I suits for $80, type II suits for $90, and type III suits for $95—the total receipts would have been $6825. However, the suits were sold for $75, $80, and $85, respectively, and the total receipts amounted to $6250. Determine the number of suits of each type sold during the sale.

36. A collection of dimes and quarters amounts to $2.70. If the total number of coins is 15, how many coins of each type are in the collection?

37. Suppose that the demand and supply equations for coal in a certain marketing area are $q = -2p + 150$ and $q = 3p$, respectively, where p is the price per ton in dollars and q is the quantity of coal in thousands of tons. In economics, *market equilibrium* is said to occur when these equations hold simultaneously. Solve the system for market equilibrium.

38. A chemist has two solutions, the first containing 20% acid and the second containing 50% acid. She wishes to mix the two solutions to obtain 9 liters of a 30% acid solution. How many liters of each solution should she use?

39. Suppose that x dollars is invested at a simple annual interest rate of 8.5%, and that y dollars is invested at 9.5%. If the total amount invested is $17,000 and the total interest from the two investments at the end of the year is $1535, find x and y.

40. A company has two hydraulic presses, an old one and an improved model. With both presses working together, a certain job is done in 2 hours and 24 minutes. On another job of the same kind, the old press is operated alone for 3 hours, then the new press is also put into operation and the two presses together finish the job in an additional 1 hour and 12 minutes. How long would it take each press operating alone to do this job?

41. On a certain date, 3 pounds of coffee, 4 quarts of milk, and 2 cans of tuna fish cost $13.20. One can of tuna fish cost twice as much as 1 quart of milk. Six months later, because of inflation, the price of the coffee had increased by 15%, the price of milk by 5%, and the price of tuna fish by 10%; so the same grocery order cost $14.82. Find the original prices of a pound of coffee, a quart of milk, and a can of tuna fish.

42. Professor Grumbles, who teaches morning and afternoon statistics classes of equal size, is accused of male chauvinism because in the two classes taken together, 80% of the male students passed, but only 20% of the female students did. However, the professor contends that the accusation is false because in the morning class, 10% of the men and 10% of the women passed; whereas in the afternoon class, 90% of the men and 90% of the women passed. Furthermore, the total number of men in the two classes is the same as the total number of women. Is this possible, and if so, how?

9.6 SYSTEMS CONTAINING NONLINEAR EQUATIONS

In this section, we consider the solution of systems containing nonlinear equations. To avoid technical difficulties, we shall study only the relatively simple case of two equations in two unknowns. The solution of such a system can be found (at least approximately) by sketching graphs of the two equations on the same coordinate system, and determining the points where the two graphs intersect.

The substitution and elimination methods, introduced for systems of linear equations in Section 9.1, can often be used (with minor modifications) for systems containing nonlinear equations. Even then, graphs can be sketched to determine the number of solutions and as a rough check on the calculations.

In Examples 1 and 2, sketch graphs to determine the number of solutions of each system, and then solve the system.

Example 1 $\begin{cases} x^2 - 2y = 0 \\ x + 2y = 6 \end{cases}$

Solution   The graph of $x^2 - 2y = 0$, or $y = \frac{1}{2}x^2$, is a parabola opening upward with vertex at the origin; the graph of $x + 2y = 6$ is a line that intersects the parabola at two points (Figure 1). Thus, there are two solutions of the system. To find these solutions algebraically, we use the method of elimination. Adding the second equation to the first (to eliminate y), we obtain the equivalent system

$$\begin{cases} x^2 + x = 6 \\ x + 2y = 6. \end{cases}$$

Now the first equation is quadratic in x and can be solved by factoring:

$$x^2 + x - 6 = 0 \qquad \text{or} \qquad (x + 3)(x - 2) = 0,$$

so $x = -3$ or $x = 2$. Substituting these values, one at a time, into

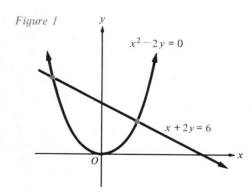

Figure 1

the second equation $x + 2y = 6$, we find that

$$y = \tfrac{9}{2} \quad \text{when} \quad x = -3 \quad \text{and} \quad y = 2 \quad \text{when} \quad x = 2.$$

Hence, the two solutions are $(-3, \tfrac{9}{2})$ and $(2, 2)$. ∎

Figure 2

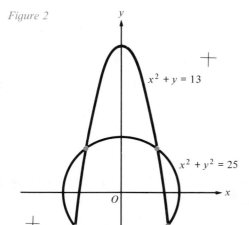

Example 2 $\begin{cases} x^2 + y^2 = 25 \\ x^2 + y\ = 13 \end{cases}$

Solution   Sketching the graphs of the two equations (a circle and a parabola) on the same coordinate system, we see four points of intersection (Figure 2). Thus, there are four solutions of the system. Again, we can use the method of elimination. Subtracting the second equation from the first (to eliminate x^2), we obtain the equivalent system

$$\begin{cases} y^2 - y = 12 \\ x^2 + y = 13. \end{cases}$$

Now the first equation is quadratic in y and can be solved by factoring:

$$y^2 - y - 12 = 0 \quad \text{or} \quad (y + 3)(y - 4) = 0,$$

so $y = -3$ or $y = 4$. Substituting $y = -3$ into the second equation, $x^2 + y = 13$, we obtain

$$x^2 - 3 = 13 \quad \text{or} \quad x^2 = 16;$$

hence, $x = 4$ or $x = -4$. Similarly, substituting $y = 4$ into the second equation, we obtain

$$x^2 + 4 = 13 \quad \text{or} \quad x^2 = 9;$$

hence, $x = 3$ or $x = -3$. Therefore, the solutions are $(-4, -3)$, $(4, -3)$, $(-3, 4)$, and $(3, 4)$. ∎

When the elimination method can be used, as in the examples above, it is usually the most efficient way to solve the system. Otherwise, try the substitution method.

Example 3   Solve the system $\begin{cases} x^2 + 2y^2 = 18 \\ xy = 4. \end{cases}$

Solution   Here there doesn't seem to be a simple way to eliminate variables, so we try the substitution method. Because the second equation is easily solved for y in terms of x, we begin there. If $xy = 4$, then $x \neq 0$ and

$$y = \frac{4}{x}$$

Substituting $4/x$ for y in the first equation, we obtain

$$x^2 + 2\left(\frac{4}{x}\right)^2 = 18 \quad \text{or} \quad x^2 + \frac{32}{x^2} = 18.$$

Multiplying both sides of the last equation by x^2, we get

$$x^4 + 32 = 18x^2 \quad \text{or} \quad x^4 - 18x^2 + 32 = 0.$$

Factoring, we have

$$(x^2 - 16)(x^2 - 2) = 0 \quad \text{or} \quad (x - 4)(x + 4)(x - \sqrt{2})(x + \sqrt{2}) = 0.$$

Setting each factor equal to zero gives us

$$x = 4, \quad x = -4, \quad x = \sqrt{2}, \quad \text{or} \quad x = -\sqrt{2}.$$

For each of these values of x, there is a corresponding value of y given by $y = 4/x$. Therefore, the solutions are

$$(4, 1), \quad (-4, -1), \quad (\sqrt{2}, 2\sqrt{2}), \quad \text{and} \quad (-\sqrt{2}, -2\sqrt{2}). \qquad \blacksquare$$

If a system of equations contains exponential or logarithmic functions, the properties of these functions may be used to help find the solutions.

Example 4   Solve the system $\begin{cases} y - \log_4(6x + 10) = 1 \\ y + \log_4 x = 2. \end{cases}$

Solution   We use the method of elimination. Subtracting the first equation from the second, we obtain

$$\log_4 x + \log_4(6x + 10) = 1.$$

Now we recall a basic property of logarithms [Property (iii), page 280], and the definition of logarithms to rewrite the last equation as

$$\log_4[x(6x + 10)] = 1 \quad \text{or} \quad x(6x + 10) = 4^1;$$

that is,

$$6x^2 + 10x - 4 = 0 \quad \text{or} \quad 3x^2 + 5x - 2 = 0.$$

Factoring, we get

$$(3x - 1)(x + 2) = 0; \quad \text{hence, } x = \tfrac{1}{3} \quad \text{or} \quad x = -2.$$

Because $\log_4 x$ is undefined when x is negative, $x = -2$ is an extraneous root, and we must reject it. Substituting $x = \tfrac{1}{3}$ into the equation $y = 2 - \log_4 x$, we find that

$$y = 2 - \log_4 \tfrac{1}{3} = 2 - \log_4 3^{-1} = 2 - (-1)\log_4 3 = 2 + \log_4 3.$$

Hence, the solution is $(\tfrac{1}{3}, 2 + \log_4 3)$. $\qquad \blacksquare$

We close this section with a brief indication of the way in which systems of equations, not all of which need be linear, arise in practical situations. We choose economics as our area of application. If p denotes the price per unit of a commodity, and q denotes the number of units of the commodity demanded in the marketplace at price p, a graph of q as a function of p produces a **demand curve** (Figure 3a). Note that q will ordinarily be a decreasing function of p. (Why?) On the other hand, if p represents the price per unit of a commodity in the marketplace, and q denotes the number of units that manufacturers are willing to supply at that price, a graph of q as a function of p produces a **supply curve** (Figure 3b). In this case, q will ordinarily be an increasing function of p. (Why?) If the supply and demand curves are plotted on the same coordinate system, the point where they intersect is called the **market equilibrium point** (Figure 3c). At the equilibrium price (the p coordinate of the market equilibrium point), the quantity supplied will be equal to the quantity demanded. Under the usual interpretations of price, supply, and demand, only the portions of the supply and demand curves that fall in the first quadrant are economically meaningful.

Figure 3

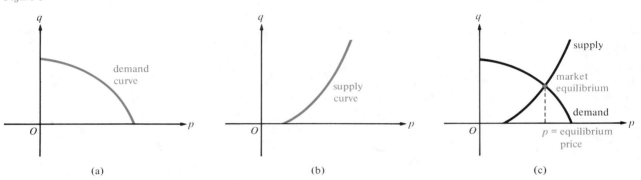

(a)   (b)   (c)

Example 5   Suppose that the weekly demand q million gallons of synthetic fuel made from oil shale is related to the price p in dollars per gallon by the demand equation $8p^2 + 5q = 100$. If the market price is p dollars per gallon, assume that producers are willing to supply q million gallons of the synthetic fuel per week according to the supply equation $6p^2 - p - 3q = 5$. Find:

(a) The market equilibrium point   **(b)** The equilibrium price

Solution   **(a)** The market equilibrium point is found by solving the system

$$\begin{cases} 8p^2 + 5q = 100 \\ 6p^2 - p - 3q = 5. \end{cases}$$

Multiplying the second equation by 5 and adding 3 times the first equation to the result, we obtain the equivalent system

$$\begin{cases} 8p^2 + 5q = 100 \\ 54p^2 - 5p = 325. \end{cases}$$

The second equation can be solved by factoring:

$$54p^2 - 5p - 325 = 0 \quad \text{or} \quad (2p - 5)(27p + 65) = 0;$$

hence, $p = \frac{5}{2}$ or $p = -\frac{65}{27}$. Because of the economic interpretation, p cannot be negative, so we reject the second solution and retain only the solution $p = \frac{5}{2}$. Substituting $p = \frac{5}{2}$ into the first equation $8p^2 + 5q = 100$, we obtain

$$8(\tfrac{5}{2})^2 + 5q = 100 \quad \text{or} \quad 5q = 50,$$

so

$$q = 10.$$

Therefore, the market equilibrium point is $(p, q) = (\frac{5}{2}, 10)$.

(b) The equilibrium price is

$$p = \tfrac{5}{2} = \$2.50 \text{ per gallon.} \qquad \blacksquare$$

Problem Set 9.6

In Problems 1 to 6, sketch graphs to determine the number of solutions of each system of equations, and then solve the system.

1. $\begin{cases} x^2 - 2y = 0 \\ 3x + 2y = 10 \end{cases}$

2. $\begin{cases} x^2 - 2y = 3 \\ x - y = 1 \end{cases}$

3. $\begin{cases} y^2 - 3x = 0 \\ 2x - y = 3 \end{cases}$

4. $\begin{cases} x^2 + y^2 = 25 \\ x^2 + y = 19 \end{cases}$

5. $\begin{cases} x^2 + y^2 = 4 \\ x - 2y = 4 \end{cases}$

6. $\begin{cases} 2x^2 + y = 9 \\ y - x^2 - 5x = 1 \end{cases}$

In Problems 7 to 26, solve each system of equations by the elimination or substitution methods.

7. $\begin{cases} 2x^2 + y = 4 \\ 2x - y = 1 \end{cases}$

8. $\begin{cases} x^2 + y^2 = 1 \\ x^2 - y = 3 \end{cases}$

9. $\begin{cases} x^2 - y^2 = 3 \\ -2x + y = 1 \end{cases}$

10. $\begin{cases} x^2 + y^2 = 1 \\ -x + 2y = -2 \end{cases}$

11. $\begin{cases} x^2 + 2y^2 = 22 \\ 2x^2 + y^2 = 17 \end{cases}$

12. $\begin{cases} x^2 + y^2 = 625 \\ x + y = 35 \end{cases}$

13. $\begin{cases} 2s^2 - 4t^2 = 8 \\ s^2 + 2t^2 = 10 \end{cases}$

14. $\begin{cases} 4r^2 + 7s^2 = 23 \\ -3r^2 + 11s^2 = -1 \end{cases}$

15. $\begin{cases} 4h^2 + 7k^2 = 32 \\ -3h^2 + 11k^2 = 41 \end{cases}$

16. $\begin{cases} 2a^2 - 5b^2 + 8 = 0 \\ a^2 - 7b^2 + 4 = 0 \end{cases}$

17. $\begin{cases} x^2 - y = 2 \\ 2x^2 + y = 6x + 7 \end{cases}$

18. $\begin{cases} x^2 + y^2 - 8y = -7 \\ y - x^2 = 1 \end{cases}$

19. $\begin{cases} x^2 - xy + 2y^2 = 8 \\ xy = 4 \end{cases}$

20. $\begin{cases} 4x^2 - 6xy + 9y^2 = 63 \\ 2x - 3y + 3 = 0 \end{cases}$

21. $\begin{cases} x - y = 21 \\ \sqrt{x} + \sqrt{y} = 7 \end{cases}$

22. $\begin{cases} \dfrac{3}{x^2} - \dfrac{2}{y^2} = 1 \\[2mm] \dfrac{7}{x^2} - \dfrac{6}{y^2} = 2 \end{cases}$

23. $\begin{cases} y - \sqrt[4]{x} = 0 \\ y^2 - \sqrt[4]{x} = 2 \end{cases}$

24. $\begin{cases} y = \sqrt[4]{x - 2} \\ y^2 = \sqrt[4]{x - 2} + 12 \end{cases}$

25. $\begin{cases} (x - y)^2 + (x + y)^2 = 17 \\ (x - y)^2 - 4(x + y)^2 = 12 \end{cases}$

26. $\begin{cases} \dfrac{2}{(x + 1)^2} - \dfrac{5}{(y - 1)^2} = 3 \\[2mm] \dfrac{1}{(x + 1)^2} + \dfrac{3}{(y - 1)^2} = 7 \end{cases}$

In Problems 27 to 32, use appropriate properties of the exponential and logarithmic functions to solve each system.

27. $\begin{cases} y - \log_6(x + 3) = 1 \\ y + \log_6(x + 4) = 2 \end{cases}$

28. $\begin{cases} y + \log_3(x + 1) = 2 \\ y - \log_3(x + 3) = 1 \end{cases}$

29. $\begin{cases} y = 5 + \log_{10} x \\ y - \log_{10}(x + 6) = 7 \end{cases}$

30. $\begin{cases} \log_4(x^2 + y^2) = 2 \\ 2y - x = 4 \end{cases}$

31. $\begin{cases} x - 5^y = 0 \\ x - 25^y = -20 \end{cases}$

32. $\begin{cases} y + 15 = 10^x \\ y - 10^{2-x} = 0 \end{cases}$

33. The ratio of two positive numbers is 3 to 8, and their product is 864. Find the numbers.

34. A sporting goods store sells two circular targets for archery. The radius of the larger target is 10 centimeters more than the radius of the smaller target, and the difference between the areas of the two targets is 2300 square centimeters. Find the radii of the targets.

35. A commercial artist is using a triangular template whose altitude exceeds its base by 4 inches, and whose area is 30 square inches. Find the altitude and base of the template.

36. The sum of the squares of the digits of a certain two-digit number is 61. The product of the number and the number with the digits reversed is 3640. What is the number?

37. A company determines that its total monthly production cost C in thousands of dollars satisfies the equation $C^2 = 8x + 4$ and that its monthly revenue R in thousands of dollars satisfies the equation $8R - 3x^2 = 0$, where x is the number of thousands of units of its product manufactured and sold per month. When the cost of manufacturing the product equals the revenue obtained from selling it, the company *breaks even*. How many units must the company produce in order to break even?

38. A commercial jet was delayed for 15 minutes before takeoff because of a problem with baggage loading. To make up for the delay, the pilot increased the jet's air speed and took advantage of a favorable tail wind. This increased the jet's average speed by 36 miles per hour, with the result that it landed on time, 3 hours and 45 minutes after takeoff. Find the usual average ground speed and the distance between the two airports.

39. Suppose that the annual demand, q million cars, for a front-wheel-drive economy car is related to its sticker price, p thousand dollars, by the demand equation $2q^2 = 9 - p$. At a sticker price of p thousand dollars, the manufacturers are willing to build q million cars a year according to the supply equation $q^2 + 5q = p - 1$. Find:

    (a) The market equilibrium point

    (b) The equilibrium price of a car

40. Find formulas for the length l and the width w of a rectangle in terms of its area A and its perimeter P.

9.7 SYSTEMS OF INEQUALITIES AND LINEAR PROGRAMMING

In this section, we consider problems involving systems of inequalities in *two* unknowns. The solution of such a system can be represented by a set of points in the xy plane, and the problem can then be solved geometrically. Many real-world applications of mathematics—especially those pertaining to the allocation of limited resources—give rise to systems of *linear* inequalities and thus to problems in *linear programming*. Therefore, before considering more general types of inequalities, we begin by considering linear inequalities.

The Graph of an Inequality

By a **linear inequality** in the two unknowns x and y, we mean an inequality having one of the forms

$$ax + by + c > 0, \quad ax + by + c < 0, \quad ax + by + c \geq 0, \quad \text{or} \quad ax + by + c \leq 0,$$

where a, b, and c are constants and a and b are not both zero. The first two inequalities are called **strict;** the second two, **nonstrict.** The **graph** of such an inequality is defined to be the set of all points (x, y) in the xy plane whose coordinates satisfy the inequality.

In order to study the graph of a linear inequality in x and y, we begin by considering the graph of the linear *equation*

$$ax + by + c = 0$$

Figure 1

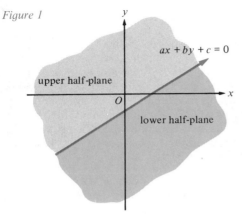

obtained by (temporarily) replacing the inequality sign with an equal sign. This graph is a straight line which divides the xy plane into two regions called **half-planes,** one on each side of the line (Figure 1). A half-plane is called **closed** if the points on the boundary line belong to it; a half-plane is said to be **open** if the points on the boundary line do not belong to it.

In Figure 1, the expression $ax + by + c$ is zero only for points on the line separating the two half-planes, so it is either positive or negative for points (x, y) in the open half-planes. Actually, $ax + by + c$ is positive on one of the open half-planes and negative on the other (Problems 50 and 51). Thus, we have the following procedure.

Procedure for Sketching the Graph of a Linear Inequality

Step 1. Draw the graph of the linear equation obtained by (temporarily) replacing the inequality sign with an equal sign. If the inequality is strict ($>$ or $<$), draw a dashed line; if the inequality is nonstrict ($\geq$ or $\leq$), draw a solid line.

Step 2. Select any convenient test point in one of the two open half-planes determined by the line in step 1. If the coordinates of the test point satisfy the inequality, shade the half-plane that contains it; otherwise, shade the other half-plane.

The shaded half-plane is the graph of the inequality. A dashed boundary line indicates an *open* half-plane, and a solid boundary line indicates a *closed* one.

Figure 2

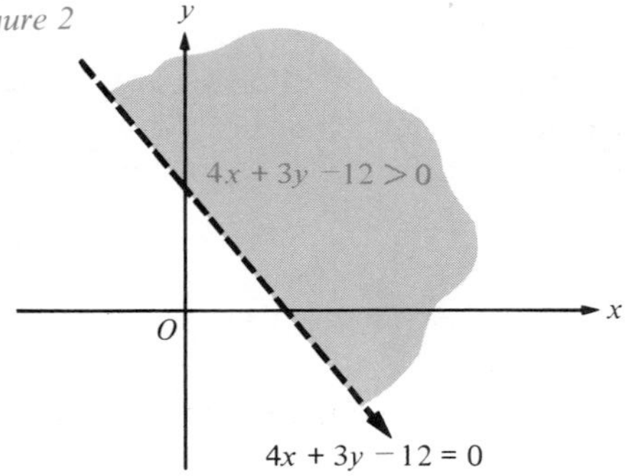

In Examples 1 and 2, sketch the graph of each linear inequality.

Example 1 $4x + 3y - 12 > 0$.

Solution   We follow the procedure above.

1. Since the inequality is strict, we draw the graph of the equation $4x + 3y - 12 = 0$ as a dashed line (Figure 2).

2. We test the inequality at the origin $O = (0, 0)$ by substituting $x = 0$ and $y = 0$ to obtain $-12 > 0$, which is *false*. Therefore, we shade the open half-plane *not* containing the origin (Figure 2). ∎

Figure 3

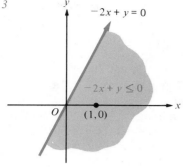

Example 2 $-2x + y \le 0$.

Solution   Again, we follow the procedure.

1. Since the inequality is nonstrict, we draw the graph of the equation $-2x + y = 0$ as a solid line (Figure 3).

2. We test the inequality at the point $(1, 0)$ by substituting $x = 1$ and $y = 0$ to obtain $-2 \le 0$, which is *true*. Therefore, we shade the closed half-plane containing the point $(1, 0)$ (Figure 3).   ∎

The method for sketching the graph of a nonlinear inequality is similar to the procedure illustrated above, except that the graph of the equation obtained in step 1 might divide the plane into more than two regions. (In special cases, it might not divide the plane at all.) In carrying out step 2 for nonlinear inequalities, it is necessary to select a test point in *each* of these regions, and to shade only those regions (if any) for which the chosen test point satisfies the given inequality.

Figure 4

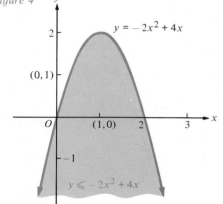

Example 3   Sketch the graph of the inequality $y \le -2x^2 + 4x$.

Solution   Since the inequality is nonstrict, we begin by sketching the graph of $y = -2x^2 + 4x$ as a solid curve (Figure 4). The result is a parabola that divides the plane into two regions, one region above the parabola and one region below it. We select (say) the test point $(0, 1)$ in the upper region and the test point $(1, 0)$ in the lower one. Substituting $x = 0$ and $y = 1$ in the inequality $y \le -2x^2 + 4x$, we obtain the *false* statement $1 \le 0$, so we do not shade the upper region. Substituting $x = 1$ and $y = 0$, we obtain the *true* statement $0 \le 2$, so we shade the region below the parabola.   ∎

The Graph of a System of Inequalities

The **graph** of a *system* of inequalities is defined to be the set of all points (x, y) in the xy plane whose coordinates satisfy every inequality in the system. Such a graph is obtained by sketching the graphs of all the inequalities on the same coordinate system. The region where all of these graphs overlap is the graph of the system of inequalities.

In Examples 4 and 5, sketch the graph of each system of linear inequalities.

Example 4
$$\begin{cases} x + y \le 2 \\ -x + 3y \ge 4 \end{cases}$$

Solution Using the procedure given above, we sketch the graphs of $x + y \le 2$ and $-x + 3y \ge 4$ on the same coordinate system (Figure 5). The graph of $x + y \le 2$ is the closed half-plane below the line $x + y = 2$, and the graph of $-x + 3y \ge 4$ is the closed half-plane above the line $-x + 3y = 4$. These two half-planes overlap in the region shaded in Figure 5; hence, this shaded region is the graph of the system of equations. ■

Figure 5

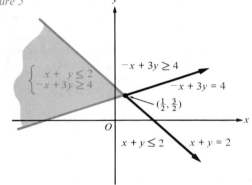

Figure 6

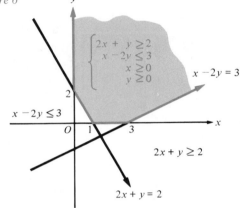

Example 5
$$\begin{cases} 2x + y \ge 2 \\ x - 2y \le 3 \\ x \ge 0 \\ y \ge 0 \end{cases}$$

Solution Again, we begin by sketching the graphs of the four linear inequalities on the same coordinate system (Figure 6). Notice that the graph of $x \ge 0$ is the closed half-plane to the right of the y axis, and that the graph of $y \ge 0$ is the closed half-plane above the x axis. These two closed half-planes overlap in the region consisting of the first quadrant together with the positive x and y axes and the origin. We intersect this region with the closed half-plane above the line $2x + y = 2$ [the graph of $2x + y \ge 2$] and the closed half-plane above the line $x - 2y = 3$ [the graph of $x - 2y \le 3$] to obtain the graph of the system of inequalities (Figure 6). ■

If the graph of a system of linear inequalities in x and y is a nonempty set of points, it will be bounded by straight line segments or rays meeting at "corner points" called **vertices.** After you have sketched the graph, you can see these vertices and you can find their coordinates by solving appropriate pairs of linear equations.

Example 6   Find the vertices of the graphs in

(a) Figure 5   **(b)** Figure 6.

Solution   **(a)** In Figure 5 on page 526, the vertex, found by solving the system

$$\begin{cases} x + y = 2 \\ -x + 3y = 4, \end{cases}$$

is $(\frac{1}{2}, \frac{3}{2})$.

(b) The graph in Figure 6 has three vertices, $(0, 2)$, $(1, 0)$, and $(3, 0)$. ▪

Because the graphs in Figures 5 and 6 extend indefinitely in certain directions, we say that they are **unbounded.** The graph in the next example is **bounded** in the sense that it is cut off by boundary curves in every direction.

Example 7   Sketch the graph of the system of inequalities:

$$\begin{cases} -x^2 + y \geq 1 \\ x + y < 3. \end{cases}$$

Solution   We begin by sketching the graph of the first inequality. Using test points above and below the graph of the parabola $-x^2 + y = 1$, that is, $y = x^2 + 1$, we find that the inequality $-x^2 + y \geq 1$ is satisfied for points on or above the parabola (Figure 7a). Next, we sketch the graph of the second inequality, and find that it consists of all points below the line $x + y = 3$ (Figure 7b). Finally, we sketch both the parabola and the line on the same coordinate system (Figure 7c) and shade the region above (or on) the parabola *and* below the line. ▪

Figure 7

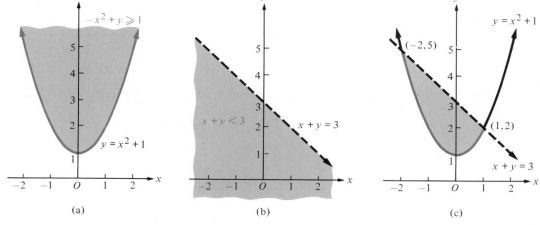

(a)   (b)   (c)

Linear Programming

Suppose that an oil company's profit depends on what portion of a limited allocation of crude oil the company refines into gasoline and what portion it converts into heating oil. Assume that there is a governmental restriction requiring that at least a certain fraction of the crude oil be converted into heating oil. In order to achieve the greatest possible profit under the governmental constraint, the company will have to plan or "program" its activities.

A problem in which it is necessary to find the maximum (largest) or minimum (smallest) value of a certain quantity (such as profit, cost, revenue, distance, or time) under a given set of constraints (restrictions) is sometimes called a **programming** problem. Here the word "programming" is intended to suggest planning. The quantity whose maximum or minimum value is desired is called the **objective function.** The objective function depends on one or more variables, and the constraints are expressed as conditions on the possible values of these variables. When the objective function is a linear (first-degree) polynomial in two or more variables and the constraints are expressed as a system of nonstrict linear inequalities, we have a **linear programming** problem.

Here we consider only linear programming problems in which the objective function F depends on two variables x and y, and the graph G of the system of linear inequalities that express the constraints is a *bounded* region in the xy plane. Thus, F has the form

$$F = ax + by + c,$$

where a, b, and c are constants, and G is bounded by a finite number of line segments meeting at a finite number of vertices. Under these circumstances, it can be shown that F *has a maximum and a minimum value on G, and these values occur at certain vertices of* G. Thus:

> To find the maximum and minimum values, you merely list all the vertices of G and calculate the value of F at each one.

The largest and smallest of these values are the maximum and minimum values of F on the region G.

Example 8   Find the maximum and minimum values of the objective function

$$F = 3x + 4y + 1,$$

subject to the constraints

$$\begin{cases} x + 2y \le 8 \\ x + y \le 5 \\ x \ge 0 \\ y \ge 0. \end{cases}$$

Solution   We begin by using the method described above to sketch the graph G of the system representing the constraints (Figure 8). By solving appropriate pairs of linear equations, we find the coordinates of the vertices of G. These coordinates are listed in Table 1 along with the corresponding values of $F = 3x + 4y + 1$ at each vertex. From Table 1, the minimum value of F, which is 1, occurs at $(0,0)$; and the maximum value of F, which is 19, occurs at $(2,3)$.

Figure 8

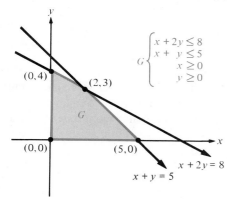

$$G \begin{cases} x + 2y \le 8 \\ x + y \le 5 \\ x \ge 0 \\ y \ge 0 \end{cases}$$

Table 1

Vertex	Value of $F = 3x + 4y + 1$	
$(0,0)$	1	minimum
$(5,0)$	16	
$(2,3)$	19	maximum
$(0,4)$	17	

The following example illustrates a typical application of linear programming.

Example 9   A large school system wants to design a lunch menu containing two food items X and Y. Each ounce of X supplies 1 unit of protein, 2 units of carbohydrates, and 1 unit of fat. Each ounce of Y supplies 1 unit of protein, 1 unit of carbohydrates, and 1 unit of fat. The two items together must provide at least 7 units of protein, at least 10 units of carbohydrates, and no more than 8 units of fat per serving. If each ounce of X costs 12 cents and each ounce of Y costs 8 cents, how many ounces of each item should each serving contain to meet the dietary requirements at the lowest cost?

Solution   Let x and y denote the number of ounces per serving of X and Y, respectively. Since each ounce of X costs 12 cents and each ounce of Y costs 8 cents, the cost per serving is

$$F = 12x + 8y \text{ cents.}$$

Since each ounce of X or of Y supplies 1 unit of protein, each serving will supply $x + y$ units of protein. To meet the protein requirement, we must have

$$x + y \ge 7.$$

Similarly, to meet the carbohydrate requirement, we must have

$$2x + y \ge 10,$$

and to meet the fat requirement,

$$x + y \le 8.$$

Because the number of ounces per serving cannot be negative, we also have the conditions $x \ge 0$ and $y \ge 0$. Therefore, the problem is to minimize the objective function

$$F = 12x + 8y$$

Figure 9

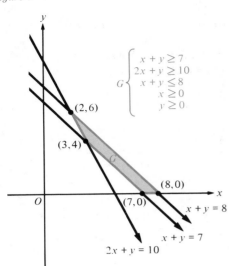

subject to the constraints

$$\begin{cases} x + y \geq 7 \\ 2x + y \geq 10 \\ x + y \leq 8 \\ x \geq 0 \\ y \geq 0. \end{cases}$$

The graph G of the system representing these constraints is sketched in Figure 9, from which we find the vertices listed in Table 2. Thus, if each serving contains 3 ounces of item X and 4 ounces of item Y, all dietary requirements will be met at the minimum cost of 68 cents per serving.

Table 2

Vertex	Value of $F = 12x + 8y$	
$(7,0)$	84	
$(8,0)$	96	
$(2,6)$	72	
$(3,4)$	68	minimum

Problem Set 9.7

In Problems 1 to 10, sketch the graph of each linear inequality.

1. $x \geq 0$

2. $y < 1$

3. $2x + y - 3 \leq 0$

4. $5x - 2y \leq 2$

5. $3x + 2y \geq 6$

6. $2x \geq 2 - y$

7. $3x + 5y < 15$

8. $4x > 3 - 2y$

9. $x < y$

10. $1 < x - y \leq 5$

In Problems 11 to 22, (a) sketch the graph of the system of linear inequalities, (b) determine whether the graph is bounded or unbounded, and (c) find all vertices of the graph.

11. $\begin{cases} 2x + 3y \leq 7 \\ 3x - y \leq 5 \end{cases}$

12. $\begin{cases} 2x + y < 3 \\ x + 3y > 4 \end{cases}$

13. $\begin{cases} x + y \leq 4 \\ y > 2x - 4 \end{cases}$

14. $\begin{cases} x + y < 1 \\ -x + 2y \geq 4 \\ y > 0 \end{cases}$

15. $\begin{cases} y + 1 < 3x \\ y - x > 3 \end{cases}$

16. $\begin{cases} 3x + y < 4 \\ y - 2x \geq -1 \\ x \leq 0 \end{cases}$

17. $\begin{cases} 4x + 7y \leq 28 \\ 2x - 3y \geq -6 \\ y \geq -2 \end{cases}$

18. $\begin{cases} x + 3y \leq 6 \\ 3x - 2y \leq 4 \\ y \geq 0 \end{cases}$

19. $\begin{cases} -2x + y \leq 2 \\ x + 2y \leq 8 \\ x \geq 0 \\ y \geq 0 \end{cases}$

20. $\begin{cases} 2x - 3y \geq 2 \\ y \geq 6 - x \\ x \geq 4 \end{cases}$

21. $\begin{cases} x + 2y \leq 6 \\ 3x + y \leq 9 \\ x \geq 0 \\ y \geq 0 \end{cases}$

22. $\begin{cases} \dfrac{x}{2} - 1 \leq y \leq 3 + 3x \\ 0 \leq 4x \leq 12 - 3y \end{cases}$

In Problems 23 to 30, sketch the graph of each system of inequalities.

23. $\begin{cases} -x^2 + y \geq 0 \\ x + y \leq 2 \end{cases}$

24. $\begin{cases} -x^2 + y \geq 0 \\ x^2 + y^2 < 1 \end{cases}$

25. $\begin{cases} x^2 + y^2 \le 1 \\ (x+1)^2 + y^2 \le 1 \end{cases}$

26. $\begin{cases} x^2 + y^2 \le 4 \\ (x-2)^2 + (y+2)^2 \ge 4 \end{cases}$

27. $\begin{cases} \sqrt{x} - y > 0 \\ x - 4y \le 0 \end{cases}$

28. $\begin{cases} x^2 + y \le 13 \\ x^2 + y^2 \ge 25 \\ x \ge 0 \end{cases}$

29. $\begin{cases} |x + y| \ge 1 \\ x^2 + y^2 \le 1 \end{cases}$

30. $\begin{cases} |x| + |y| \le 1 \\ 2x^2 + 2y^2 \ge 1 \\ xy \ge 0 \end{cases}$

31. Rewrite the system $\begin{cases} 0 \le x \le 7 - 2y \\ 0 \le 8y \le 5x + 3 \end{cases}$ as an equivalent system of four inequalities and sketch the graph of the resulting system.

32. A hardware store has display space for at most 40 spray cans of rustproofing paint, x cans of red and y cans of gray. If there are to be at least 10 cans of each type in the display, (a) sketch a graph showing the possible numbers of red and gray cans in the display, and (b) find all vertices of the graph.

In Problems 33 to 42, find the maximum and minimum values of the objective function F subject to the indicated constraints.

33. $F = 2x - y + 3$

$\begin{cases} x + y \le 1 \\ x \ge 0 \\ y \ge 0 \end{cases}$

34. $F = 3x + 2y - 1$

$\begin{cases} x + 2y \le 7 \\ 5x - 8y \le -3 \\ x \ge 0 \\ y \ge 0 \end{cases}$

35. $F = 5x + 4y$

$\begin{cases} x + 2y \ge 3 \\ x + 2y \le 5 \\ x \ge 0 \\ y \ge 0 \end{cases}$

36. $F = 4x - y + 7$

$\begin{cases} 3x + 8y \le 120 \\ 3x + y \le 36 \\ x \ge 0 \\ y \ge 0 \end{cases}$

37. $F = 3x - 5y + 2$

$\begin{cases} x + y \le 10 \\ x - 3y \ge -18 \\ x \ge 0 \\ y \ge 0 \end{cases}$

38. $F = 10x + 12y$

$\begin{cases} 0.2x + 0.4y \le 30 \\ 0.2x + 0.2y \le 20 \\ x \ge 0 \\ y \ge 0 \end{cases}$

39. $F = \frac{3}{2}x + y$

$\begin{cases} \frac{1}{2}x + y \ge 2 \\ \frac{1}{2}x - y \ge 1 \\ x \le 6 \\ y \ge 0 \end{cases}$

40. $F = -10x + 5y + 3$

$\begin{cases} x + \frac{3}{2}y \ge -60 \\ x + y \ge -50 \\ x \le 0 \\ y \le 0 \end{cases}$

41. $F = \frac{1}{2}x + \frac{3}{4}y$

$\begin{cases} \frac{1}{3}x + y \le 30 \\ x + \frac{1}{2}y \le 40 \\ x + y \ge 10 \end{cases}$

42. $F = 0.15x + 0.1y$

$\begin{cases} 4x + 5y \le 2000 \\ 12x + 5y \le 3000 \\ x + y \ge 100 \\ x \ge 0 \\ y \ge 0 \end{cases}$

43. An electronics company manufactures two models of household smoke detectors. Model A requires 1 unit of labor and 4 units of parts; model B requires 1 unit of labor and 3 units of parts. If 90 units of labor and 320 units of parts are available, and if the company makes a profit of \$5 on each model A detector and \$4 on each model B detector, how many of each model should it manufacture to maximize its profit?

[C] **44.** A family owns and operates a 312-acre farm on which it grows cotton and peanuts. The task of planting, picking, ginning, and baling the cotton requires 35 person-hours of labor per acre; the task of planting, harvesting, and bagging the peanuts requires 27 person-hours of labor per acre. The family is able to devote 9500 person-hours to these activities. If the profit for each acre of cotton grown is \$173 and the profit for each acre of peanuts grown is \$152, how many acres should be planted in cotton and how many in peanuts to maximize the family's profit?

45. In Problem 43, suppose that the company raises its prices so that its profit on each model A detector is \$7 and on each model B detector is \$5. Now how many detectors of each type should it manufacture to maximize its profit?

46. A supplier has 105 pounds of leftover beef which must be sold before it spoils. Of this beef, 15 pounds is prime grade, 40 pounds is grade A, and the rest is utility grade. A local restaurant will buy ground beef consisting of 20% prime grade, 40% grade A, and 40% utility grade for 75 cents per pound. A hamburger stand will buy ground beef consisting of 40% grade A and 60% utility grade for 55 cents per pound. How much ground beef should the supplier sell to the restaurant and how much to the hamburger stand in order to maximize its revenue?

47. A town operates two recycling centers. Each day that Center I is open, 300 pounds of glass, 200 pounds of paper, and 100 pounds of aluminum are deposited there. Each day that Center II is open,

100 pounds of glass, 600 pounds of paper, and 100 pounds of aluminum are deposited there. The town has contracted to supply at least 1200 pounds of glass, 2400 pounds of paper, and 800 pounds of aluminum per week to a salvage company. Supervision and maintenance at Center I costs the town $40 each day it is open, and Center II costs $50 each day it is open. How many days a week should each center remain open so as to minimize the total weekly cost for supervision and maintenance, yet allow the town to fulfill its contract with the salvage company?

48. In Problem 46, suppose that the owner of the restaurant learns that the meat is in danger of spoiling, and decides to pay only 50 cents per pound for it. If the hamburger stand will still pay 55 cents per pound, how much ground beef should the supplier now sell to the restaurant and how much to the hamburger stand?

49. Find a condition on the positive constants a and b such that the objective function

$$F = ax + by,$$

subject to the constraints $2x + 3y \le 9$, $x - y \le 2$, $x \ge 0$, and $y \ge 0$, will take on the *same* maximum value at the vertex $(0, 3)$ and the vertex $(3, 1)$.

50. Suppose that a, b, and c are constants and $b \ne 0$. Let L be the line

$$ax + by + c = 0.$$

(a) If $b > 0$, show that $ax + by + c > 0$ for (x, y) above L, and that $ax + by + c < 0$ for (x, y) below L.

(b) If $b < 0$, show that $ax + by + c < 0$ for (x, y) above L, and that $ax + by + c > 0$ for (x, y) below L.

51. Suppose that a and c are constants and $a \ne 0$. Let L be the vertical line

$$ax + c = 0.$$

(a) If $a > 0$, show that $ax + c > 0$ for (x, y) to the right of L, and that $ax + c < 0$ for (x, y) to the left of L.

(b) If $a < 0$, show that $ax + c < 0$ for (x, y) to the right of L, and that $ax + c > 0$ for (x, y) to the left of L.

REVIEW PROBLEM SET, CHAPTER 9

In Problems 1 to 6, use graphs to determine whether the equations in each system are consistent, inconsistent, or dependent, and indicate the solutions (if any) on the graphs.

1. $\begin{cases} y = 2x + 7 \\ 2x - y = 5 \end{cases}$

2. $\begin{cases} 4x + 2y = 3 \\ 2x + y = 5 \end{cases}$

3. $\begin{cases} 4x - y = 3 \\ -2x + 3y = 1 \end{cases}$

4. $\begin{cases} -2x + 6y = -8 \\ x - 3y = 4 \end{cases}$

5. $\begin{cases} 5x + 5y = 10 \\ -3x - 3y = -6 \end{cases}$

6. $\begin{cases} x - 3y = 4 \\ 2x + y = 15 \end{cases}$

In Problems 7 to 12, use the substitution method to solve each system of linear equations.

7. $\begin{cases} x - y = 1 \\ 2x + y = 5 \end{cases}$

8. $\begin{cases} s - t = 1 \\ 2s + t = -4 \end{cases}$

9. $\begin{cases} 3u + 2v = 7 \\ -u + 4v = 3 \end{cases}$

10. $\begin{cases} 6x_1 = 15 + 9x_2 \\ x_1 = 7 - \frac{3}{2}x_2 \end{cases}$

11. $\begin{cases} 2x - y - z = 0 \\ 2x + 3y \quad\quad = 1 \\ 8x \quad\quad - 3z = 4 \end{cases}$

12. $\begin{cases} x_1 + x_2 - x_3 = 0 \\ x_1 - x_2 + x_3 = 2 \\ 2x_1 + x_2 - 4x_3 = -8 \end{cases}$

In Problems 13 to 16, use the method of elimination (without matrices) to find the solution of each system of linear equations.

13. $\begin{cases} 6x - 9y = -3 \\ 2x + y = 3 \end{cases}$

14. $\begin{cases} -4x_1 + 5x_2 = 1 \\ x_1 - 2x_2 = -1 \end{cases}$

15. $\begin{cases} 2x \quad\quad - z = 12 \\ x + y \quad\quad = 7 \\ 5x \quad\quad + 4z = -9 \end{cases}$

16. $\begin{cases} x_1 + x_2 + x_3 = 6 \\ 2x_1 - x_2 + x_3 = 3 \\ 3x_1 + x_2 - x_3 = 2 \end{cases}$

In Problems 17 to 24, solve the system of linear equations by the elimination method using matrices.

17. $\begin{cases} 5x + 2y = 3 \\ 2x - 3y = 5 \end{cases}$

18. $\begin{cases} 2x_1 + 3y_1 = 4 \\ 3x_1 + 5y_1 = 5 \end{cases}$

19. $\begin{cases} 5u - v = 19 \\ -2u + 3v = 8 \end{cases}$

20. $\begin{cases} s + 26 = -4t \\ 3s + 8 = 2t \end{cases}$

21. $\begin{cases} 2x + y - z = -3 \\ 3x - 2y + 2z = 13 \\ x + 3y + 4z = 0 \end{cases}$

22. $\begin{cases} 2x + y = 11 \\ x + 2y - 4z = 17 \\ 3x - y + 3z = 1 \end{cases}$

23. $\begin{cases} 2x_1 - x_2 + 2x_3 = -8 \\ 3x_1 - x_2 - 4x_3 = 3 \\ x_1 + 2x_2 - 3x_3 = 9 \end{cases}$

24. $\begin{cases} \frac{1}{2}x + \frac{1}{3}y - \frac{1}{4}z = 2 \\ \frac{1}{3}x + \frac{1}{4}y - \frac{1}{2}z = \frac{1}{6} \\ -\frac{3}{2}x + 3y + 2z = 23 \end{cases}$

In Problems 25 to 36, perform the indicated matrix operations.

25. $A + 3B$ and $-2A + 4B$ if:

$A = \begin{bmatrix} -1 & 1 \\ 2 & 3 \end{bmatrix}, B = \begin{bmatrix} -5 & 1 \\ 2 & 4 \end{bmatrix}$

26. $2A - B$ and $3A + 3B$ if:

$A = \begin{bmatrix} 2 & -1 & 3 \\ 1 & 2 & 4 \end{bmatrix}, B = \begin{bmatrix} -3 & 2 & 1 \\ 5 & -2 & 4 \end{bmatrix}$

27. $4A + 2B$ and $B - 2A$ if

$A = \begin{bmatrix} 2 & 4 \\ 5 & -2 \\ -4 & 3 \end{bmatrix}$ and $B = \begin{bmatrix} -1 & 0 \\ 3 & 4 \\ 0 & -5 \end{bmatrix}$

28. $\frac{1}{2}A - 2B + C$ if $A = \begin{bmatrix} 3 & 1 \\ 1 & 0 \\ 4 & -1 \\ 0 & 3 \end{bmatrix}$,

$B = \begin{bmatrix} -2 & 2 \\ 1 & 4 \\ 3 & 1 \\ 1 & -2 \end{bmatrix}$,   and $C = \begin{bmatrix} 3 & -1 \\ 5 & -1 \\ 0 & 2 \\ 8 & 0 \end{bmatrix}$

29. AB and BA if

$A = \begin{bmatrix} 2 & 1 & -1 \\ 3 & 1 & 2 \end{bmatrix}$ and $B = \begin{bmatrix} -1 & 2 \\ 2 & 4 \\ 0 & 5 \end{bmatrix}$

30. A^2, B^2, and A^2B^2 if

$A = \begin{bmatrix} 1 & -1 \\ 2 & 3 \end{bmatrix}$ and $B = \begin{bmatrix} 1 & 3 \\ -2 & 1 \end{bmatrix}$

31. AB if

$A = \begin{bmatrix} 1 & -1 & 2 \\ 3 & 0 & -2 \\ 1 & 4 & 0 \end{bmatrix}$ and $B = \begin{bmatrix} 1 & -2 \\ 2 & 3 \\ 4 & -1 \end{bmatrix}$

32. $(A - 2B)B$ if

$A = \begin{bmatrix} -2 & 1 \\ 3 & 4 \end{bmatrix}$ and $B = \begin{bmatrix} 4 & 2 \\ 3 & -1 \end{bmatrix}$

33. A^{-1}, AA^{-1}, and $A^{-1}A$ if $A = \begin{bmatrix} 2 & -4 \\ 3 & 7 \end{bmatrix}$

34. $(AB)^{-1} - A^{-1}B^{-1}$ if

$A = \begin{bmatrix} 1 & 2 \\ 3 & -1 \end{bmatrix}$ and $B = \begin{bmatrix} -2 & 4 \\ 1 & 3 \end{bmatrix}$

35. A^{-1} if $A = \begin{bmatrix} 2 & -1 & 0 \\ 1 & 0 & 1 \\ 1 & -2 & 0 \end{bmatrix}$

36. A^{-1}, B^{-1}, and $(AB)^{-1} - B^{-1}A^{-1}$

if $A = \begin{bmatrix} 1 & 3 \\ 4 & -1 \end{bmatrix}$ and $B = \begin{bmatrix} 2 & 3 \\ 4 & 1 \end{bmatrix}$

37. Find a value of x for which $AB = BA$

if $A = \begin{bmatrix} 1 & x \\ 0 & -1 \end{bmatrix}$ and $B = \begin{bmatrix} 2 & 3 \\ 0 & 1 \end{bmatrix}$

38. If $A = \begin{bmatrix} a & b \\ c & d \end{bmatrix}$, the *determinant* of A, in symbols, $\det A$, is defined by $\det A = \begin{vmatrix} a & b \\ c & d \end{vmatrix}$.

If A and B are 2 by 2 matrices, prove that $\det(AB) = (\det A)(\det B)$.

39. If A and B are square n by n nonsingular matrices, show that AB is nonsingular.

40. Let A be a 2 by 2 matrix and let $d = \det A$ (see Problem 38). If A is nonsingular, prove that $d \neq 0$ and that $\det(AI) = d^{-1}$.

In Problems 41 to 46, (a) write the system of linear equations in the form $AX = K$ for suitable matrices A, X, and K; and (b) solve each system with the aid of A^{-1}.

41. $\begin{cases} 3x + 2y = 11 \\ 4x - 3y = 9 \end{cases}$

42. $\begin{cases} 7x_1 + 10x_2 = -3 \\ 5x_1 + 2x_2 = 3 \end{cases}$

43. $\begin{cases} 3x + 2y + z = 7 \\ 2x + 3z = 10 \\ 5x - y = -8 \end{cases}$

44. $\begin{cases} 7x - 4z = k_1 \\ 5y + 3z = k_2 \\ -3x + 7y = k_3 \end{cases}$

45. $\begin{cases} 2x + z = a \\ x + 6y + 4z = b \\ -x - y = c \end{cases}$

46. $\begin{cases} 3x + 2y = 8 - 2z \\ x + 6z = 8 + 5y \\ 6x = 4 + 8z \end{cases}$

In Problems 47 and 48, evaluate each determinant.

47. $\begin{vmatrix} 1/2 & 4/3 \\ -5/2 & 2/3 \end{vmatrix}$

Ⓒ **48.** $\begin{vmatrix} 5.007 & 13.142 \\ -3.733 & 2.501 \end{vmatrix}$

In Problems 49 and 50, use the expansion formula to evaluate each determinant.

49. $\begin{vmatrix} 7 & 4 & 5 \\ 6 & -5 & 1 \\ 3 & 2 & 0 \end{vmatrix}$

Ⓒ **50.** $\begin{vmatrix} 4.21 & 3.72 & -2.02 \\ -1.59 & 7.07 & -8.83 \\ 2.22 & 3.14 & 2.03 \end{vmatrix}$

In Problems 51 to 58, evaluate each determinant by any method you wish.

51. $\begin{vmatrix} 1 & -1 & 2 \\ 2 & -2 & 4 \\ \sqrt{2} & -\sqrt{2} & 2\sqrt{2} \end{vmatrix}$

52. $\begin{vmatrix} 1 & -1 & 2 \\ 2 & -2 & 1 \\ \sqrt{2} & -\sqrt{2} & 2\sqrt{2} \end{vmatrix}$

53. $\begin{vmatrix} x & 0 & 0 \\ 2 & x & 1 \\ 0 & 3 & 2 \end{vmatrix}$

54. $\begin{vmatrix} 25 & -6 & 0 \\ 0 & 3 & -4 \\ 35 & 9 & 2 \end{vmatrix}$

55. $\begin{vmatrix} 4 & \sqrt{2} & \pi \\ 0 & 5 & e \\ 0 & 0 & -1 \end{vmatrix}$

56. $\begin{vmatrix} 1 & 1 & 1 \\ x & a & b \\ x^2 & a^2 & b^2 \end{vmatrix}$

57. $\begin{vmatrix} 2 & 1 & -3 \\ 4 & 3 & -2 \\ 3 & -1 & 4 \end{vmatrix}$

58. $\begin{vmatrix} 1 & 3 & 4 \\ 2 & 2 & 4 \\ 3 & 1 & 4 \end{vmatrix}$

In Problems 59 to 64, assume that

$$\begin{vmatrix} a & b & c \\ u & v & w \\ x & y & z \end{vmatrix} = 2.$$

Find the value of each determinant.

59. $\begin{vmatrix} c & a & b \\ w & u & v \\ z & x & y \end{vmatrix}$

60. $\begin{vmatrix} -a & -b & -c \\ -u & -v & -w \\ -x & -y & -z \end{vmatrix}$

61. $\begin{vmatrix} 2a & 2b & 2c \\ 2u & 2v & 2w \\ 2x & 2y & 2z \end{vmatrix}$

62. $\begin{vmatrix} a & b+a & 2c \\ u & v+u & 2w \\ x & y+x & 2z \end{vmatrix}$

63. $\begin{vmatrix} a+u & b+v & c+w \\ u+x & v+y & w+z \\ -x & -y & -z \end{vmatrix}$

64. $\begin{vmatrix} a & u & x \\ b & v & y \\ c & w & z \end{vmatrix}$

In Problems 65 to 70, use Cramer's rule to solve each system.

65. $\begin{cases} 3x - 4y = -2 \\ 4x + y = 7 \end{cases}$

66. $\begin{cases} 3x_1 + 10x_2 = 3 \\ 6x_1 - 5x_2 = 16 \end{cases}$

67. $\begin{cases} x + y + 2z = 10 \\ 5x + 3y - z = 1 \\ 3x - y - 3z = -3 \end{cases}$

68. $\begin{cases} 3r + 6s - 5t = -1 \\ r - 2s + 4t = 4 \\ 5r + 6s - 7t = -5 \end{cases}$

69. $\begin{cases} x_1 + x_2 \qquad = \quad 0 \\ 2x_1 - 3x_2 + x_3 = \quad 7 \\ \qquad 3x_2 - 7x_3 = -17 \end{cases}$

70. $\begin{cases} s + t - u = 0 \\ 2s \qquad + 2u = 1 \\ \qquad 3t + u = 3 \end{cases}$

71. If $P = (a, b)$ and $Q = (c, d)$ are points in the xy plane, and $P \neq (0, 0)$, show that the perpendicular distance from Q to the line through $O = (0, 0)$ and P is the absolute value of

$$\begin{vmatrix} a & b \\ c & d \end{vmatrix} (a^2 + b^2)^{-1/2}.$$

72. For the points P and Q in Problem 71, show that the absolute value of $\begin{vmatrix} a & b \\ c & d \end{vmatrix}$ is the area of the parallelogram that has $\overline{OP}$ and $\overline{OQ}$ as two adjacent sides.

In Problems 73 to 78, decompose each fraction into partial fractions.

73. $\dfrac{2x}{(x - 1)(x + 1)}$

74. $\dfrac{4x + 7}{x^2 + 5x + 4}$

75. $\dfrac{4x^2 + 13x - 9}{x(x + 3)(x - 1)}$

76. $\dfrac{-x^2 + 13x - 26}{(x + 1)^2(x - 4)}$

77. $\dfrac{x^2 + 4}{x(x^2 + 1)}$

78. $\dfrac{6x^3 + 5x^2 + 21x + 12}{x(x + 1)(x^2 + 4)}$

79. If the supermarket price of a certain cut of beef is p dollars per pound, then q million pounds will be sold according to the demand equation $p + q = 4$. When the supermarket price is p dollars per pound, a large packing company will supply q million pounds of this meat according to the supply equation $p - 4q + 2 = 0$. Find the point of market equilibrium.

80. Two cars start from points 400 kilometers apart and travel toward each other at constant speeds. If the first car travels 20 kilometers per hour faster than the second, and if they both meet after 4 hours, find the speeds of both vehicles.

C 81. A person invested a total of $40,000, part of it in a conservative investment at 8.5% simple annual interest, and the rest in a riskier investment at 11.2% simple annual interest. At the end of the year, both investments paid off at the stipulated rates for a total interest of $4129. How much was invested at each rate of interest?

82. An industrial chemist, in preparing a batch of fertilizer, mixes 10 tons of nitrogen compounds with 12 tons of phosphates. The total cost of the ingredients is $5400. A second batch of fertilizer, prepared for special soil conditions, is mixed from 15 tons of nitrogen compounds and 8 tons of phosphates, for a total cost of $6100. Find the cost per ton of the nitrogen compounds and the cost per ton of the phosphates.

83. Byron, Jason, and Adrian receive a total weekly allowance of $12, which is split three ways. If Jason's allowance plus twice the sum of Byron's and Adrian's is $20, and if Adrian's allowance plus twice the sum of Byron's and Jason's is $22, find each boy's weekly allowance.

84. Joe is 4 years younger than twice Gus's age, and Jamal is 3 years older than Gus. Six years from now, Joe will be $\frac{4}{3}$ times as old as Gus will be. Find their present ages.

In Problems 85 to 94, solve each system of equations.

85. $\begin{cases} y^2 - 12x = 0 \\ y^2 + 9x^2 = 9 \end{cases}$

86. $\begin{cases} x^2 - y = 0 \\ x^2 + 4y^2 = 4 \end{cases}$

87. $\begin{cases} 2x + y = 1 \\ 4x^2 + y^2 = 13 \end{cases}$

88. $\begin{cases} 4x^2 - y^2 = 7 \\ 2x^2 + 5y^2 = 8 \end{cases}$

89. $\begin{cases} y^2 - x^2 = 3 \\ xy = 2 \end{cases}$

90. $\begin{cases} 9x^2 - 4y^2 = 36 \\ x^2 + y^2 = 43 \end{cases}$

91. $\begin{cases} \log_4(x^2 - 4y^2) = 1 \\ x - y = 1 \end{cases}$

92. $\begin{cases} x - y = 2 \\ 2^{x^2 - y^2} = 256 \end{cases}$

93. $\begin{cases} \log_2(2x^2 + y) = 2 \\ \log_2(2x - y) = 0 \end{cases}$

94. $\begin{cases} \dfrac{3}{x^2} - \dfrac{2}{y^2} = \dfrac{1}{2} \\ \dfrac{6}{x^2} + \dfrac{1}{y} = \dfrac{5}{2} \end{cases}$

95. The length of the hypotenuse of a right triangle is 25 centimeters, and the perimeter is 56 centimeters. Find the lengths of the legs.

96. A manufacturer of drafting supplies makes right triangles from plastic sheets. The perimeters of the triangles are 60 centimeters, and each triangle has a hypotenuse of 25 centimeters. Find the lengths of the sides of the triangles.

97. A town parking lot has an area of 7500 square meters. The town engineer suggests enlarging the lot by increasing both its length and its width by 10 meters. She points out that this will increase the area of the lot by 1850 square meters. Find the present dimensions of the lot.

98. In three years, Joshua will be twice as old as Miriam. At present, twice his age is the same as the product of Rebecca's and Miriam's ages. If Rebecca is now twice as old as Miriam, find the ages of all three children.

In Problems 99 to 105, (a) sketch the graph of each system of linear inequalities, (b) determine whether the graph is bounded or unbounded, and (c) find all of its vertices.

99. $\begin{cases} 2y - x > 10 \\ y + x \geq 5 \end{cases}$

100. $\begin{cases} 4y > x - 16 \\ y + x < 0 \end{cases}$

101. $\begin{cases} 3y + x \leq 2 \\ y > x + 1 \end{cases}$

102. $\begin{cases} y \leq x + 1 \\ x + y \leq 4 \\ x \geq 0 \end{cases}$

103. $\begin{cases} y - 3x \leq 2 \\ y + 2x \leq 4 \\ y \geq 0 \end{cases}$

104. $\begin{cases} x + y \leq 500 \\ 3y \leq x \\ x \leq 400 \\ y \geq 60 \end{cases}$

105. $\begin{cases} x + y \geq 5 \\ 2y \geq 8 - x \\ y \leq 5 \\ x \leq 10 \end{cases}$

106. A psychology professor is planning a 50-minute class. The professor will devote x minutes to a review of old material, and y minutes to a lecture on new material. The remaining time will be used for classroom discussion of the new material. At least 20 minutes must be reserved for the lecture on new material; at least 10 minutes will be needed for the classroom discussion; and no more than 15 minutes will be required for the review of old material. Sketch a graph in the xy plane illustrating the professor's options, and find all vertices of the graph.

In Problems 107 to 110, sketch the graph of each system of inequalities.

107. $\begin{cases} x^2 + 2y \geq 0 \\ x + 2y \leq 6 \end{cases}$

108. $\begin{cases} x^2 - y \leq 2 \\ 2x^2 + y < 6x + 7 \end{cases}$

109. $\begin{cases} x^2 + y^2 \leq 1 \\ x^2 - y > 3 \end{cases}$

110. $\begin{cases} |x| + |y| \leq 1 \\ x^2 - y \geq 0 \end{cases}$

In Problems 111 to 114, find the maximum and minimum values of the objective function F subject to the indicated constraints.

111. $F = 4x + 7y + 1$
$\begin{cases} y - x + 1 \geq 0 \\ 2y + x - 10 \geq 0 \\ x \geq 0 \\ y \leq 5 \end{cases}$

112. $F = 3x + 5y$
$\begin{cases} y \leq 4 - x \\ y \geq \frac{1}{2}x - 2 \\ x \geq 0 \\ y \geq 0 \end{cases}$

113. $F = x + 5y$
$\begin{cases} x + y \leq 4 \\ 4x + y \leq 7 \\ x \geq 0 \\ y \geq 0 \end{cases}$

114. $F = x$
$\begin{cases} 2y + x \leq 16 \\ x - y \leq 10 \\ x \geq 0 \end{cases}$

115. A firm manufactures two products, A and B. For each product, two different machines, M_1 and M_2, are used. Each machine can operate up to 20 hours per day. Product A requires 1 hour of time on machine M_1 and 3 hours of time on machine M_2 per 100 units. Product B requires 2 hours of time on machine M_1 and 1 hour of time on machine M_2 per 100 units. The firm makes a profit of $20 per unit on product A and $10 per unit on product B. Determine how many units of each product should be manufactured per day in order to maximize the profit.

© **116.** A woman has $20,000 to invest and two investment opportunities, one conservative and one somewhat riskier. The conservative investment pays 8.5% simple annual interest, and the risky investment pays 9.5% simple annual interest. She wants to invest at most three times as much in the risky investment as in the conservative one. Find her maximum possible dividend after 1 year if she decides to invest no more than $16,000 in the conservative investment and at least $2400 in the risky investment.

Chapter 10

Analytic Geometry and the Conics

In this chapter we study the **conic sections** (or **conics,** for short). These graceful curves were well known to the ancients; however, their study is immensely enhanced by the use of analytic geometry and calculus. They are obtained by sectioning, or cutting, a right circular cone of two nappes with a suitable plane (Figure 1).

Figure 1

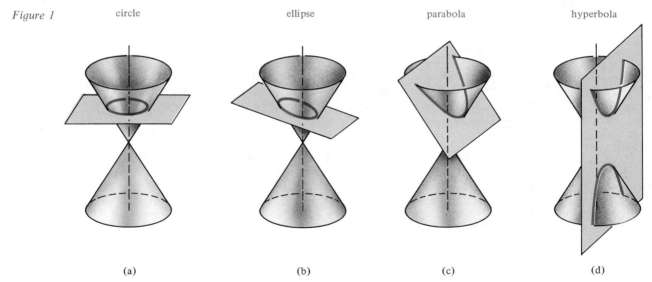

| circle | ellipse | parabola | hyperbola |

(a)   (b)   (c)   (d)

By shining a flashlight onto a white wall, you can see examples of the conic sections. If the axis of the flashlight is perpendicular to the wall, then the illuminated region is *circular* (Figure 2a); if the flashlight is tilted slightly upward, the illuminated region elongates and its boundary becomes an *ellipse* (Figure 2b). As the flashlight is tilted further, the ellipse becomes more and more elongated until it changes into a *parabola* (Figure 2c). Finally, if the flashlight is tilted still further, the edges of the parabola become straighter and it changes into a portion of a *hyperbola* (Figure 2d).

Figure 2

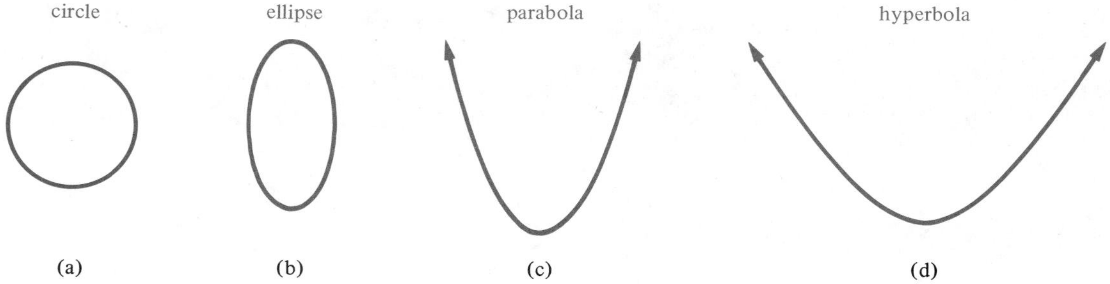

circle	ellipse	parabola	hyperbola

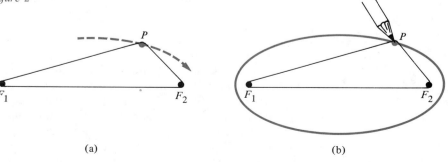

(a)   (b)   (c)   (d)

In Sections 3.1 and 8.6, we obtained equations for circles in both Cartesian and polar forms. Here we derive the standard equations for ellipses, parabolas, and hyperbolas, and we use these equations to illustrate the ideas of translation and rotation of coordinate axes.

Figure 1

10.1 THE ELLIPSE

Ellipses are of practical importance in fields ranging from art to astronomy. For instance, a circular object viewed in perspective forms an ellipse (Figure 1), and an orbiting satellite (natural or artificial) moves in an elliptical path. The geometric definition of an ellipse is as follows.

Definition 1   **Ellipse**

> An **ellipse** is the set of all points P in the plane such that the sum of the distances from P to two fixed points F_1 and F_2 is constant. Here F_1 and F_2 are called the **focal points,** or the **foci,** of the ellipse. The midpoint C of the line segment $\overline{F_1F_2}$ is called the **center** of the ellipse.

Figure 2a shows two fixed pins F_1 and F_2 and a loop of string of length l stretched tightly about them to the point P. Here we have

$$|\overline{PF_1}| + |\overline{PF_2}| + |\overline{F_1F_2}| = l \qquad \text{or} \qquad |\overline{PF_1}| + |\overline{PF_2}| = l - |\overline{F_1F_2}|.$$

Figure 2

(a)   (b)

Hence, as P is moved about, $|\overline{PF_1}| + |\overline{PF_2}|$ always has the constant value $l - |\overline{F_1F_2}|$. Thus, if a pencil point P is inserted into the loop of string and moved so as to keep the string tight, it traces out an ellipse (Figure 2b).

Consider the ellipse in Figure 3 with foci F_1 and F_2 and center C. We denote by c the distance between the center C and either focus F_1 or F_2. Notice that the ellipse is symmetric about the line through F_1 and F_2. Let V_1 and V_2 be the points where this line intersects the ellipse. The center C bisects the line segment $\overline{V_1V_2}$, and the ellipse is symmetric about the line through C perpendicular to $\overline{V_1V_2}$. Let V_3 and V_4 be the points where this perpendicular intersects the ellipse. The four points V_1, V_2, V_3, and V_4 are called the **vertices** of the ellipse. The line segment $\overline{V_1V_2}$ is called the **major axis,** and the line segment $\overline{V_3V_4}$ is called the **minor axis** of the ellipse. Let $2a$ denote the length of the major axis, and let $2b$ denote the length of the minor axis (Figure 3). The numbers a and b are called the **semimajor axis** and the **semiminor axis** of the ellipse.

Figure 3

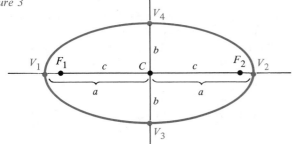

Figure 4

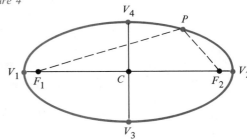

If a point P moves along the ellipse in Figure 4, then, by definition, the sum

$$|\overline{PF_1}| + |\overline{PF_2}|$$

does not change. Therefore, its value when P reaches V_1 is the same as its value when P reaches V_4; that is,

$$|\overline{V_1F_1}| + |\overline{V_1F_2}| = |\overline{V_4F_1}| + |\overline{V_4F_2}|.$$

By symmetry

$$|\overline{V_4F_1}| = |\overline{V_4F_2}|,$$

so the equation above can be rewritten

$$|\overline{V_1F_1}| + |\overline{V_1F_2}| = 2|\overline{V_4F_2}|.$$

But, by symmetry again,

$$|\overline{V_1F_2}| = |\overline{V_2F_1}|.$$

Hence, $2|\overline{V_4F_2}| = |\overline{V_1F_1}| + |\overline{V_1F_2}| = |\overline{V_1F_1}| + |\overline{V_2F_1}| = |\overline{V_1V_2}| = 2a,$

from which it follows that

$$|\overline{V_4F_2}| = a.$$

Figure 5

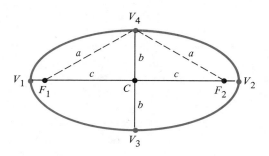

Therefore, applying the Pythagorean theorem to the right triangle V_4CF_2 (Figure 5), we find that

$$a^2 = b^2 + c^2.$$

If we place the ellipse of Figure 5 in the xy plane so that its center C is at the origin O and the foci F_1 and F_2 lie on the negative and positive portions of the x axis, respectively, then we can derive an equation of the ellipse as follows.

Theorem 1   **Ellipse Equation in Cartesian Form**

An equation of the ellipse with foci at $F_1 = (-c, 0)$ and $F_2 = (c, 0)$ is

$$\frac{x^2}{a^2} + \frac{y^2}{b^2} = 1,$$

where a is the semimajor axis, b is the semiminor axis, and $a^2 = b^2 + c^2$.

Proof   Let $P = (x, y)$ be any point on the ellipse (Figure 6). As Figure 5 shows, when $P = (0, b)$, we have

$$|\overline{PF_1}| + |\overline{PF_2}| = 2a.$$

Figure 6

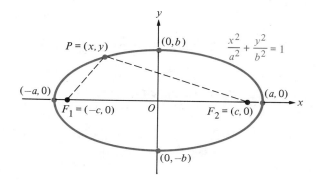

Therefore, by Definition 1, the equation

$$|\overline{PF_1}| + |\overline{PF_2}| = 2a$$

holds for every point $P = (x, y)$ on the ellipse. Using the distance formula, we can rewrite this as

$$\sqrt{(x + c)^2 + y^2} + \sqrt{(x - c)^2 + y^2} = 2a$$

or

$$\sqrt{(x + c)^2 + y^2} = 2a - \sqrt{(x - c)^2 + y^2}.$$

Squaring both sides of the last equation, we have

$$x^2 + 2cx + c^2 + y^2 = 4a^2 - 4a\sqrt{(x - c)^2 + y^2} + x^2 - 2cx + c^2 + y^2,$$

so that

$$4cx - 4a^2 = -4a\sqrt{(x - c)^2 + y^2} \qquad \text{or} \qquad cx - a^2 = -a\sqrt{(x - c)^2 + y^2}.$$

Squaring both sides of the last equation, we obtain

$$c^2x^2 - 2a^2cx + a^4 = a^2(x^2 - 2cx + c^2 + y^2),$$

so that

$$a^4 - a^2c^2 = (a^2 - c^2)x^2 + a^2y^2 \qquad \text{or} \qquad a^2(a^2 - c^2) = (a^2 - c^2)x^2 + a^2y^2.$$

Since $a^2 = b^2 + c^2$, we have $a^2 - c^2 = b^2$, and the equation above can be rewritten as

$$a^2b^2 = b^2x^2 + a^2y^2.$$

If both sides of the last equation are divided by a^2b^2, the result is

$$1 = \frac{x^2}{a^2} + \frac{y^2}{b^2}$$

as desired. By reversing the argument above, it can be shown that, conversely, if the equation $(x^2/a^2) + (y^2/b^2) = 1$ holds, then the point $P = (x, y)$ is on the ellipse (Problem 22). ∎

If a and b are positive constants and $a > b$, the Cartesian equation

$$\boxed{\frac{x^2}{a^2} + \frac{y^2}{b^2} = 1}$$

is called the **standard form** for the equation of an ellipse with center at the origin O and with a horizontal major axis. It is not difficult to derive an equation of an ellipse with center at the origin and with a vertical major axis (Figure 7). In this case, the ellipse has foci $F_1 = (0, -c)$ and $F_2 = (0, c)$ on the y axis and vertices $V_1 = (0, -a)$, $V_2 = (0, a)$, $V_3 = (-b, 0)$, and $V_4 = (b, 0)$. The semimajor axis is a, and the semiminor axis is b. The equation can be derived as in Theorem 1, the argument

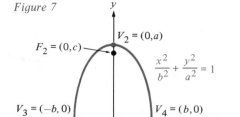

Figure 7

$$\frac{x^2}{b^2} + \frac{y^2}{a^2} = 1$$

being word for word the same except that the variables x and y interchange their roles. Therefore, the equation is

$$\frac{x^2}{b^2} + \frac{y^2}{a^2} = 1 \qquad \text{where } a > b.$$

This equation is also called the **standard form** for the equation of an ellipse.

The results obtained above are summarized in Table 1. In this table, we assume that $a > b$ and $c^2 = a^2 - b^2$.

Table 1 **Ellipses**

Major Axis	Center	Foci	Vertices On Major Axis	Vertices On Minor Axis	Standard Form Equation
Horizontal (on the x axis)	$(0,0)$	$(-c,0), (c,0)$	$(-a,0), (a,0)$	$(0,-b), (0,b)$	$\dfrac{x^2}{a^2} + \dfrac{y^2}{b^2} = 1$
Vertical (on the y axis)	$(0,0)$	$(0,-c), (0,c)$	$(0,-a), (0,a)$	$(-b,0), (b,0)$	$\dfrac{x^2}{b^2} + \dfrac{y^2}{a^2} = 1$

Example 1 Find the coordinates of the four vertices and the two foci of each ellipse and sketch the graph.

(a) $4x^2 + 9y^2 = 36$ (b) $4x^2 + y^2 = 4$.

Solution   (a) We divide both sides of the equation by 36 to obtain

$$\frac{x^2}{9} + \frac{y^2}{4} = 1,$$

that is,

$$\frac{x^2}{a^2} + \frac{y^2}{b^2} = 1$$

with $a = 3$ and $b = 2$. By Table 1, this is the equation of an ellipse, centered at the origin, with major axis on the x axis, and with vertices $(-3,0)$, $(3,0)$, $(0,-2)$, and $(0,2)$ (Figure 8a). Also, the foci are $(-c,0)$ and $(c,0)$, where

$$c^2 = a^2 - b^2 = 9 - 4 = 5,$$

that is, $c = \sqrt{5}$. Thus, $F_1 = (-\sqrt{5},0)$ and $F_2 = (\sqrt{5},0)$.

(b) We divide both sides of the equation by 4 to obtain

$$\frac{x^2}{1} + \frac{y^2}{4} = 1,$$

that is,

$$\frac{x^2}{b^2} + \frac{y^2}{a^2} = 1$$

Figure 8

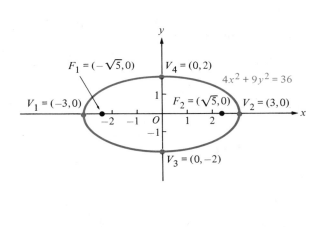

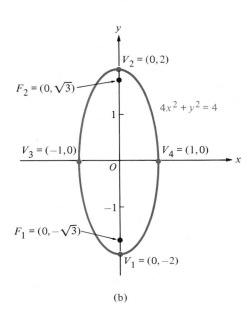

(a)   (b)

with $b = 1$ and $a = 2$. By Table 1, this is the equation of an ellipse, centered at the origin, with major axis on the y axis, and with vertices $(0, -2)$, $(0, 2)$, $(-1, 0)$, and $(1, 0)$ (Figure 8b). Also, the foci are $(0, -c)$ and $(0, c)$, where

$$c^2 = a^2 - b^2 = 4 - 1 = 3,$$

that is, $c = \sqrt{3}$. Thus, $F_1 = (0, -\sqrt{3})$ and $F_2 = (0, \sqrt{3})$.   ∎

Example 2   Find the equation in the standard form of the ellipse with foci $F_1 = (-\sqrt{3}, 0)$ and $F_2 = (\sqrt{3}, 0)$ and vertices $V_1 = (-2, 0)$ and $V_2 = (2, 0)$. Also, find the coordinates of the remaining two vertices, V_3 and V_4, and sketch the graph.

Solution   Here $c = \sqrt{3}$, $a = 2$; hence, $b = \sqrt{a^2 - c^2} = \sqrt{4 - 3} = 1$, and the equation is

$$\frac{x^2}{4} + \frac{y^2}{1} = 1.$$

Figure 9

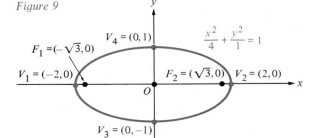

Also, $V_3 = (0, -1)$ and $V_4 = (0, 1)$. The graph appears in Figure 9.   ∎

By reasoning as in Section 3.6, you can see that, if $a > b$, and if an ellipse with center at the origin, semimajor axis a, and semiminor axis b is *shifted* so that its center is at the point $C = (h, k)$, then the equation of the shifted ellipse will have one of the following **standard forms:**

Standard Forms for Ellipses Centered at (h, k)

(i) $\dfrac{(x - h)^2}{a^2} + \dfrac{(y - k)^2}{b^2} = 1$, if the major axis is horizontal (Figure 10a)

(ii) $\dfrac{(x - h)^2}{b^2} + \dfrac{(y - k)^2}{a^2} = 1$, if the major axis is vertical (Figure 10b)

Figure 10

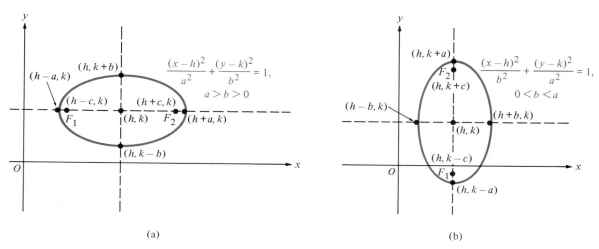

(a)

(b)

Example 3 Show that the graph of $25x^2 + 9y^2 - 100x - 54y = 44$ is an ellipse. Find the coordinates of the center, the vertices, and the foci, and sketch the graph.

Solution Here we have

$$25(x^2 - 4x \qquad) + 9(y^2 - 6y \qquad) = 44.$$

Completing the squares in the last equation, we obtain

$$25(x^2 - 4x + 4) + 9(y^2 - 6y + 9) = 44 + 25(4) + 9(9)$$

or

$$25(x - 2)^2 + 9(y - 3)^2 = 225.$$

Dividing the last equation by 225, we have

$$\frac{(x - 2)^2}{9} + \frac{(y - 3)^2}{25} = 1,$$

that is,

$$\frac{(x - h)^2}{b^2} + \frac{(y - k)^2}{a^2} = 1$$

with $h = 2$, $b = 3$, $k = 3$, and $a = 5$. This is the standard form for an ellipse with a vertical major axis with center $C = (h, k) = (2, 3)$. The distance c between the center and the foci is given by

$$c = \sqrt{a^2 - b^2} = \sqrt{25 - 9} = 4 \text{ units.}$$

Figure 11

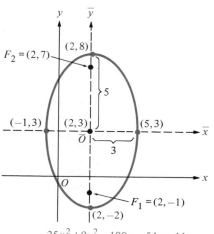

$25x^2 + 9y^2 - 100x - 54y = 44$

Figure 12

National Statuary Hall, Washington, D.C.

Thus, the foci F_1 and F_2 have xy coordinates

$$F_1 = (2, 3 - 4) = (2, -1) \quad \text{and} \quad F_2 = (2, 3 + 4) = (2, 7)$$

and the xy coordinates of the four vertices are

$$(2, -2), \quad (2, 8), \quad (-1, 3), \quad \text{and} \quad (5, 3).$$

The graph is sketched in Figure 11.

Elliptically shaped surfaces have the interesting property that light or sound produced at one focus will be reflected to the other focus. This **reflecting property** of the ellipse is useful in the design of optical apparatus, and it accounts for the strange **"whispering gallery"** effect sometimes noticed in buildings with elliptical ceilings. In such a building, words whispered softly at one focus can be heard clearly at the other focus. Among the more famous whispering galleries are the Mormon Tabernacle in Salt Lake City, St. Paul's Cathedral in London, and the National Statuary Hall of the Capitol in Washington, D.C. (Figure 12).

Problem Set 10.1

In each problem set, problems with colored numbers constitute a good representation of the main ideas of the section.

In Problems 1 to 8, find the coordinates of the vertices and foci of each ellipse, and sketch its graph.

1. $\dfrac{x^2}{16} + \dfrac{y^2}{4} = 1$

2. $\dfrac{x^2}{9} + y^2 = 1$

3. $4x^2 + y^2 = 16$

4. $36x^2 + 9y^2 = 144$

5. $x^2 + 16y^2 = 16$

6. $16x^2 + 25y^2 = 400$

7. $9x^2 + 36y^2 = 4$

8. $x^2 + 4y^2 = 1$

In Problems 9 to 12, find the equation in standard form of the ellipse that satisfies the conditions given.

9. Foci $F_1 = (-4, 0)$ and $F_2 = (4, 0)$; vertices $V_3 = (0, -3)$ and $V_4 = (0, 3)$

10. Vertices $V_1 = (-5, 0)$ and $V_2 = (5, 0)$; horizontal major axis; $c = 3$ units

11. Foci $F_1 = (0, -12)$ and $F_2 = (0, 12)$; vertices $V_1 = (0, -13)$ and $V_2 = (0, 13)$

12. Foci $F_1 = (0, -8)$ and $F_2 = (0, 8)$; semiminor axis $b = 6$ units

In Problems 13 to 20, find the coordinates of the center, the vertices, and the foci of each ellipse, and sketch its graph.

13. $\dfrac{(x - 1)^2}{9} + \dfrac{(y + 2)^2}{4} = 1$

14. $\dfrac{(x + 2)^2}{16} + \dfrac{(y - 1)^2}{4} = 1$

15. $4(x + 3)^2 + y^2 = 36$

16. $25(x + 1)^2 + 16(y - 2)^2 = 400$

17. $x^2 + 2y^2 + 6x + 7 = 0$

18. $4x^2 + y^2 - 8x + 4y - 8 = 0$

19. $2x^2 + 5y^2 + 20x - 30y + 75 = 0$

20. $9x^2 + 4y^2 + 18x - 16y - 11 = 0$

21. The segment cut by an ellipse from a line containing a focus and perpendicular to the major axis is called a **focal chord** of the ellipse (Figure 13).

Figure 13

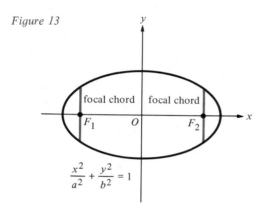

$$\frac{x^2}{a^2} + \frac{y^2}{b^2} = 1$$

(a) Show that $2b^2/a$ is the length of a focal chord of the ellipse $(x^2/a^2) + (y^2/b^2) = 1$.

(b) Find the length of a focal chord of the ellipse $9x^2 + 16y^2 = 144$.

22. Suppose that $a > b > 0$, where a and b are constants and let $c = \sqrt{a^2 - b^2}$. Assume that the numbers x and y satisfy the equation $(x^2/a^2) + (y^2/b^2) = 1$, and let $P = (x, y)$, $F_1 = (-c, 0)$, and $F_2 = (c, 0)$. Prove the following without reference to geometric diagrams:

(a) $c|x| < a^2$ (b) $\sqrt{(x - c)^2 + y^2} < 2a$
(c) $|\overline{PF_1}| + |\overline{PF_2}| = 2a$

23. How long a loop of rope should be used to lay out an elliptical flower bed 7 meters wide and 20 meters long? How far apart should the two stakes (foci) be? (See Figure 2.)

24. In Figure 2b, show that the semiminor axis b of the ellipse is given by

$$b = \tfrac{1}{2}\sqrt{l^2 - 4lc}, \qquad \text{where } c = \tfrac{1}{2}|F_1 F_2|.$$

25. A mathematician has accepted a position at a new university situated 6 kilometers from the straight shoreline of a large lake (Figure 14). The professor wishes to build a home that is half as far from the

Figure 14

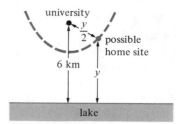

university as it is from the shore of the lake. The possible homesites satisfying this condition lie along a curve. Describe this curve, and find its equation with respect to a coordinate system having the shoreline as the x axis and the university at the point $(0, 6)$ on the y axis.

26. Find a formula for the semimajor axis of the ellipse in Figure 2b in terms of l and $c = \tfrac{1}{2}|F_1 F_2|$ (See Problem 24).

27. The *Ellipse* in Washington, D.C., is a park located between the White House and the Washington Monument. It is bounded by an elliptical path with a semimajor axis of 229 meters and a semiminor axis of 195 meters. What is the distance between the foci of this ellipse?

28. Except for minor perturbations, a satellite orbiting the earth moves in an ellipse with the center of the earth at one focus. Suppose that a satellite at perigee (nearest point to center of earth) is 400 kilometers from the surface of the earth and at apogee (farthest point to center of earth) is 600 kilometers from the surface of the earth. Assume that the earth is a sphere of radius 6371 kilometers. Find the semiminor axis b of the elliptical orbit.

29. The earth moves in an elliptical orbit with the sun at one focus. In early July, the earth is farthest from the sun (aphelion), a distance of 94,448,000 miles; in early January, it is closest to the sun (perihelion), a distance of 91,341,000 miles. Find the distance between the sun and the other focus of the earth's orbit.

30. An arch in the shape of the upper half of an ellipse with a horizontal major axis is to support a bridge over a river 100 meters wide. The center of the arch is to be 25 meters above the surface of the river. Find the equation in standard form for the ellipse.

10.2 THE PARABOLA

In Section 4.1, we mentioned that the graphs of equations of the form $y = ax^2$ are "examples of curves called *parabolas*." In this section we give a proof of this fact based on the following geometric definition.

Definition 1   **Parabola**

> A **parabola** is the set of all points P in the plane such that the distance from P to a fixed point F (the **focus**) is equal to the distance from P to a fixed line D (the **directrix**).

Parabolas appear often in the real world. A ball thrown up at an angle travels along a parabolic arc (Figure 1a), a main cable in a suspension bridge forms an arc of a parabola (Figure 1b), and the familiar "dish antennas" have parabolic cross sections (Figure 1c).

Figure 1

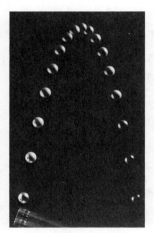

(a) *Multiflash photo of a ball thrown into the air*

(b) *Bay Bridge, San Francisco, showing paracables*

(c) *Parabolic dish antenna*

If the focus F of a parabola is placed on the positive y axis at the point $(0, p)$ and if the directrix D is placed parallel to the x axis and p units below it, the resulting parabola appears as in Figure 2. Its Cartesian equation is derived in the following theorem.

Theorem 1   **Parabola Equation in Cartesian Form**

> An equation of the parabola with focus $F = (0, p)$ and with directrix $y = -p$ is
> $$x^2 = 4py, \text{ or } y = \frac{1}{4p} x^2.$$

Figure 2

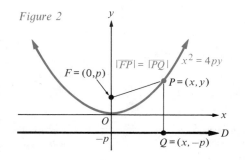

Proof Let $P = (x, y)$ be any point and let $Q = (x, -p)$ be the point at the foot of the perpendicular from P to the directrix D (Figure 2). The requirement for P to be on the parabola is $|\overline{FP}| = |\overline{PQ}|$; that is,

$$\sqrt{x^2 + (y - p)^2} = \sqrt{(y + p)^2}$$

The last equation is equivalent to

$$x^2 + (y - p)^2 = (y + p)^2$$

that is,

$$x^2 + y^2 - 2py + p^2 = y^2 + 2py + p^2 \qquad \text{or} \qquad x^2 = 4py \qquad ∎$$

Obvious modifications of the argument in Theorem 1 provide Cartesian equations for parabolas opening to the *right* (Figure 3a), to the *left* (Figure 3b), and *downward* (Figure 3c). These cases are summarized in Table 1. We assume that $p > 0$ for each situation.

Table 1 **Standard Forms for Parabolas with Vertex at the Origin**

Parabola Opens		Axis of Symmetry	Vertex	Focus	Directrix	Standard Form Equation
Right	(Figure 3a)	x axis	$(0,0)$	$(p,0)$	$x = -p$	$y^2 = 4px$
Left	(Figure 3b)	x axis	$(0,0)$	$(-p,0)$	$x = p$	$y^2 = -4px$
Upward	(Figure 2)	y axis	$(0,0)$	$(0,p)$	$y = -p$	$x^2 = 4py$
Downward	(Figure 3c)	y axis	$(0,0)$	$(0,-p)$	$y = p$	$x^2 = -4py$

Figure 3

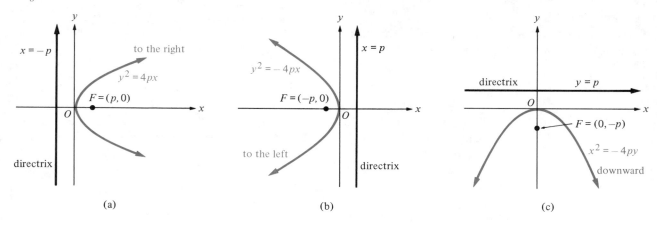

(a) (b) (c)

Example 1 Find the coordinates of the focus and an equation of the directrix of the parabola $y^2 = -8x$. Also determine its direction of opening and sketch the graph.

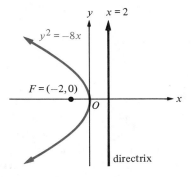

Figure 4

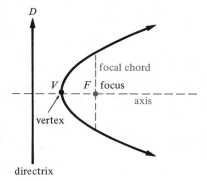

Figure 5

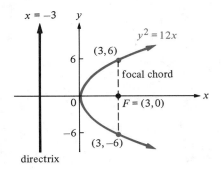

Figure 6

Solution   The equation has the form $y^2 = -4px$ with $p = 2$; hence, it corresponds to Figure 3b. Therefore, the graph is a parabola opening to the left with focus given by

$$F = (-p, 0) = (-2, 0)$$

and directrix $x = p$; that is, $x = 2$ (Figure 4).   ∎

A parabola with focus F and directrix D is evidently symmetric about the line through F perpendicular to D. This line is called the **axis** of the parabola (Figure 5). The axis intersects the parabola at the **vertex** V, which is located midway between the focus and the directrix. The segment cut by the parabola on the line through the focus and perpendicular to its axis is called the **focal chord** of the parabola (Figure 5). The following example shows how to find the length of the focal chord.

Example 2   A parabola opens to the right, has its vertex at the origin, and contains the point $(3, 6)$. Find its equation, sketch the parabola, and find the length of its focal chord.

Solution   The equation must have the form $y^2 = 4px$. Since the point $(3, 6)$ belongs to the graph, we put $x = 3$ and $y = 6$ in the equation to obtain $36 = 12p$, and we conclude that $p = 3$. Thus, an equation of the parabola is $y^2 = 12x$ (Figure 6). The focus is given by $F = (p, 0) = (3, 0)$. The focal chord lies along the line $x = 3$. Putting $x = 3$ in the equation $y^2 = 12x$ and solving for y, we obtain $y^2 = 36$, $y = \pm 6$. Thus, the points $(3, 6)$ and $(3, -6)$ are the endpoints of the focal chord, and therefore its length is 12 units.   ∎

A parabola that opens upward, to the right, downward, or to the left is said to be **in standard position.** If a parabola is in standard position, then its axis of symmetry is either horizontal or vertical. The parabolas in Table 1 are in standard position with vertices at the *origin*. By shifting these parabolas horizontally and vertically (as in Section 4.1), you can obtain all parabolas in standard position with vertices at a point (h, k). The results are summarized in Table 2. Again we assume that $p > 0$.

Table 2   **Standard Forms for Parabolas with Vertex (h, k)**

Vertex	Opening	Equation
(h, k)	To the right	$(y - k)^2 = 4p(x - h)$
(h, k)	To the left	$(y - k)^2 = -4p(x - h)$
(h, k)	Upward*	$(x - h)^2 = 4p(y - k)$
(h, k)	Downward*	$(x - h)^2 = -4p(y - k)$

* Note that a parabola with a vertical axis has an equation of the form $y = a(x - h)^2 + k$, where $a = \pm 1/(4p)$, in conformity with Theorem 1 in Section 4.1.

Example 3   Find the coordinates of the vertex V and the focus F of each parabola, determine the direction in which it opens, find an equation for its directrix, determine the length of the focal chord, and sketch the graph.

(a) $(y + 1)^2 = -12(x - 2)$ **(b)** $x^2 + 4x - 10y + 34 = 0$

Solution   **(a)** The equation can be written as

$$(y - k)^2 = -4p(x - h)$$

with $p = 3$, $h = 2$, and $k = -1$. By Table 2, the parabola opens to the left, $V = (2, -1)$, $F = (2 - 3, -1) = (-1, -1)$, and the directrix is $x = 5$. Since the x coordinate of the focus is -1, the focal chord lies along the vertical line $x = -1$. Putting $x = -1$ in the equation of the parabola, we obtain $(y + 1)^2 = 36$, so that $y + 1 = \pm 6$; that is, $y = 5$ or $y = -7$. Therefore, the endpoints of the focal chord are $(-1, 5)$ and $(-1, -7)$, and its length is 12 units (Figure 7).

(b) Completing the square, we obtain

$$x^2 + 4x + 4 - 10y + 34 = 4$$

or

$$(x + 2)^2 = 10y - 30$$

that is, $(x + 2)^2 = 10(y - 3)$. Thus, $p = \frac{10}{4} = \frac{5}{2}$, and the graph is parabola opening upward with vertex $V = (-2, 3)$, focus $F = (-2, \frac{11}{2})$,

Figure 7

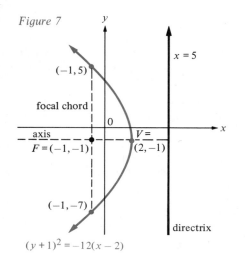

$(y + 1)^2 = -12(x - 2)$

Figure 8

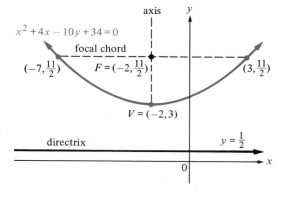

and directrix $y = \frac{1}{2}$ (Figure 8). Here, the focal chord lies along the horizontal line $y = \frac{11}{2}$. Putting $y = \frac{11}{2}$ in the equation of the parabola, we obtain $(x + 2)^2 = 10(\frac{11}{2}) - 30 = 25$, so $x = 3$ or $x = -7$. Therefore, the endpoints of the focal chord are $(-7, \frac{11}{2})$ and $(3, \frac{11}{2})$, and its length is 10 units. ∎

Parabolas have a **reflecting property** analogous to the reflecting property of an ellipse. Sound, light, or electromagnetic radiation emanating from the focus of a parabolic reflector is always reflected parallel to the axis (Figure 9). Thus, if an intense source of light such as a carbon arc or an incandescent filament is placed

Figure 9

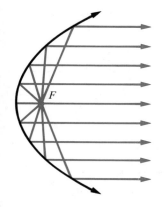

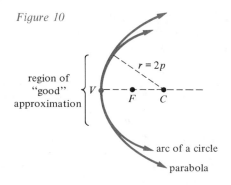

Figure 10

region of "good" approximation

at the focus of a parabolic mirror, the light is reflected and projected in a parallel beam. The same principle is used in reverse in a reflecting telescope—parallel rays of light from a distant object are brought together at the focus of a parabolic mirror.

In practice, it is very difficult to manufacture large parabolic mirrors, so it is often necessary to make do with mirrors whose cross section is a portion of a circle approximating the appropriate parabola (Figure 10). It can be shown that the circle that "best approximates" the parabola near its vertex V has its center C located on the axis of the parabola twice as far from the vertex V as the focus F of the parabola and so has radius $r = 2|\overline{VF}| = 2p$.

Problem Set 10.2

In Problems 1 to 6, find the coordinates of the vertex and the focus of the parabola. Also find an equation of the directrix and the length of the focal chord. Sketch the graph.

1. $y^2 = 4x$

2. $y^2 = -9x$

3. $x^2 - y = 0$

4. $x^2 - 4y = 0$

5. $x^2 + 9y = 0$

6. $3x^2 - 4y = 0$

7. Find an equation of the parabola whose focus is the point $(0, 3)$ and whose directrix is the line $y = -3$.

8. Find the vertex of the parabola $y = Ax^2 + Bx + C$, where A, B, and C are constants and $A \neq 0$.

In Problems 9 to 16, find the coordinates of the vertex and the focus of the parabola. Also find an equation of the directrix and the length of the focal chord. Sketch the graph.

9. $(y - 2)^2 = 8(x + 3)$

10. $(y + 1)^2 = -4(x - 1)$

11. $(x - 4)^2 = 12(y + 7)$

12. $(x + 1)^2 = -8y$

13. $y^2 - 8y - 6x - 2 = 0$

14. $2x^2 + 8x - 3y + 4 = 0$

15. $x^2 - 6x - 8y + 1 = 0$

16. $y^2 + 10y - x + 21 = 0$

In Problems 17 to 22, find an equation of the parabola that satisfies the conditions given.

17. Focus at $(4, 2)$ and directrix $x = 6$.

18. Focus at $(3, -1)$ and directrix $y = 5$.

19. Vertex at $(-6, -5)$ and focus at $(2, -5)$.

20. Vertex at $(2, -3)$ and directrix $x = -8$.

21. Axis is parallel to the x axis, vertex $(-\frac{1}{2}, -1)$, and contains the point $(\frac{5}{8}, 2)$.

22. Axis coincides with the y axis and parabola contains the points $(2, 3)$ and $(-1, -2)$.

23. A roadway 400 meters long is held up by a parabolic main cable. The main cable is 100 meters above the roadway at the ends and 4 meters above the roadway at the center. Vertical supporting cables run at 50-meter intervals along the roadway. Find the lengths of these vertical cables. [*Hint:* Set up an xy coordinate system with vertical y axis and having the vertex of the parabola 4 units above the origin.]

24. The surface of a roadway over a stone bridge follows a parabolic curve with the vertex in the middle of the bridge. The span of the bridge is 60 meters, and the road surface is 1 meter higher in the middle than at the ends. How much higher than the ends is a point on the roadway 15 meters from an end?

25. Show that, if p is the distance between the vertex and the focus of a parabola, then the focal chord of the parabola has length $4p$.

26. Let A, B, and C be constants with $A > 0$. Show that $y = Ax^2 + Bx + C$ is an equation of a parabola with a vertical axis of symmetry, opening upward. Find the coordinates of the vertex V and the focus F. Find p and the length of the focal chord. Find conditions for the graph to intersect the x axis.

27. Figure 11 shows a cross section of a parabolic dish antenna. Show that the focus F of the antenna is p units above the vertex V, where $p = a^2/(16b)$.

Figure 11

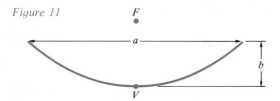

28. A parabola may be thought of as an enormous ellipse with one vertex infinitely far away. To see this, consider the ellipse in Figure 12. Show that, if we hold the lower vertex fixed at O and hold the lower

focus fixed at $(0, p)$, but allow the upper vertex to approach $+\infty$ along the y axis, then the ellipse approaches the parabola $y = (1/4p)x^2$ as a limiting curve.

Figure 12

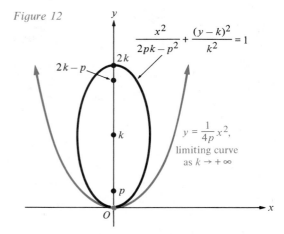

29. A student says, "If you've seen one parabola, you've seen them all,"

(a) Explain why this is essentially true.

(b) Explain why a similar statement cannot be made for ellipses.

10.3 THE HYPERBOLA

Hyperbolas are of practical importance in fields ranging from engineering to navigation. The natural-draft evaporative cooling towers used at large electric power stations have hyperbolic cross sections (Figure 1a); a comet or other object moving with more than enough kinetic energy to escape the sun's gravitational pull traces out one branch of a hyperbola (Figure 1b); and the long-range radio navigation system known as LORAN locates a ship or plane at the intersection of two hyperbolas (Figure 1c). The geometric definition of a hyperbola is as follows.

Definition 1   **Hyperbola**

> A **hyperbola** is the set of all points P in the plane such that the absolute value of the difference of the distances from P to two fixed points F_1 and F_2 is a constant positive number. Here F_1 and F_2 are called the **focal points**, or the **foci**, of the hyperbola. The midpoint C of the line segment $\overline{F_1 F_2}$ is called the **center** of the hyperbola.

Figure 1

(a) *Hyperbolic cooling towers for an electric power station*

(b) *Comet photographed through a telescope*

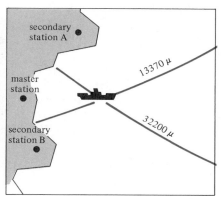

(c) *Synchronized pulses transmitted from three stations locate a ship at the intersection of two hyperbolas plotted on a LORAN chart.*

Figure 2

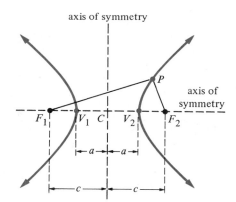

Figure 3

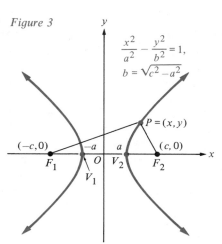

Figure 2 shows a hyperbola with foci F_1 and F_2. Notice that the line through the two foci is an axis of symmetry for the hyperbola, and so is the perpendicular bisector of the line segment $\overline{F_1F_2}$. The two points V_1 and V_2 where the two branches of the hyperbola intersect the line through F_1 and F_2 are called the **vertices**, and the line segment $\overline{V_1V_2}$ is called the **transverse axis** of the hyperbola. The distance from the center C to either focus is denoted by c, and the distance from the center C to either vertex is denoted by a. Thus, the length of the transverse axis is $2a$, and the distance between the two foci is $2c$.

As the point P in Figure 2 moves along the right-hand branch toward V_2, the difference

$$\left|\overline{PF_1}\right| - \left|\overline{PF_2}\right|$$

maintains a constant value (Definition 1). When P reaches V_2,

$$\left|\overline{PF_1}\right| - \left|\overline{PF_2}\right| = \left|\overline{V_2F_1}\right| - \left|\overline{V_2F_2}\right| = (c + a) - (c - a) = 2a$$

Therefore, for any point P on the hyperbola,

$$\left|\left|\overline{PF_1}\right| - \left|\overline{PF_2}\right|\right| = 2a$$

by Definition 1.

If we place the hyperbola in Figure 2 in the xy plane so that its center C is at the origin O and the foci F_1 and F_2 lie on the x axis (Figure 3), we can derive its Cartesian equation as in the following theorem.

Theorem 1 **Hyperbola Equation in Cartesian Form**

An equation of the hyperbola with foci $F_1 = (-c, 0)$, $F_2 = (c, 0)$ and vertices $V_1 = (-a, 0)$, $V_2 = (a, 0)$ is

$$\frac{x^2}{a^2} - \frac{y^2}{b^2} = 1 \qquad \text{where } b = \sqrt{c^2 - a^2}.$$

The proof of Theorem 1 is quite similar to the proof of Theorem 1 in Section 10.1; hence, it is left as an exercise (Problem 26). The equation

$$\frac{x^2}{a^2} - \frac{y^2}{b^2} = 1$$

is called the **standard form** for the equation of a hyperbola (Figure 3). We solve this equation for y in terms of x as follows:

$$\frac{y^2}{b^2} = \frac{x^2}{a^2} - 1$$

so

$$y^2 = b^2 \left(\frac{x^2}{a^2} - 1 \right) = \left(\frac{b^2 x^2}{a^2} \right) \left(1 - \frac{a^2}{x^2} \right).$$

Hence,

$$y = \pm \left(\frac{bx}{a} \right) \sqrt{1 - \frac{a^2}{x^2}}$$

provided that $x \neq 0$. Note that, as $|x|$ gets very large, a^2/x^2 will get very small and the expression under the square root will come closer and closer to 1. In other words, for large values of $|x|$, points on the hyperbola come very close to the lines

$$y = \frac{b}{a} x \qquad \text{and} \qquad y = -\frac{b}{a} x.$$

Thus, these lines are **asymptotes** of the hyperbola (see Section 4.6). They are good approximations to the hyperbola itself at large distances from the origin.

Although the asymptotes of the hyperbola are *not* part of the hyperbola itself, they are useful in sketching it. For instance, if we wish to sketch the graph of the equation $(x^2/a^2) - (y^2/b^2) = 1$, we begin by sketching the rectangle with height $2b$ and horizontal base $2a$ whose center is at the origin (Figure 4). The asymptotes are then drawn through the two diagonals of this rectangle. If we keep in mind that the vertices of the hyperbola are located at the midpoints of the left and right sides of the rectangle, and that the hyperbola approaches the asymptotes as it moves out away from the vertices, then it becomes an easy matter to sketch the graph (Figure 4).

Figure 4

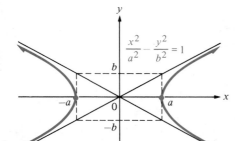

Figure 5

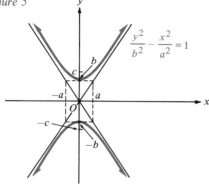

Suppose that we wish to find an equation of the hyperbola in Figure 5, which has a *vertical* transverse axis, center at the origin, vertices $V_1 = (0, -b)$ and $V_2 = (0, b)$, and foci $F_1 = (0, -c)$ and $F_2 = (0, c)$. Using Theorem 1 but interchanging x with y and interchanging a with b, we obtain the equation

$$\frac{y^2}{b^2} - \frac{x^2}{a^2} = 1 \qquad \text{where } a = \sqrt{c^2 - b^2}.$$

This equation is also called the **standard form** for the equation of a hyperbola. The asymptotes are still given by

$$y = \frac{b}{a}x \qquad \text{and} \qquad y = -\frac{b}{a}x.$$

(Why?)

Unlike the ellipse equation $(x^2/a^2) + (y^2/b^2) = 1$ and the equation $(x^2/b^2) + (y^2/a^2) = 1$, *there is no requirement that $a > b$ in the hyperbola equation $(x^2/a^2) - (y^2/b^2) = 1$ or in the hyperbola equation $(y^2/b^2) - (x^2/a^2) = 1$.*

Example 1   Find the coordinates of the foci and the vertices and find equations of the asymptotes of each hyperbola. Also, sketch the graph.

(a) $\dfrac{x^2}{4} - \dfrac{y^2}{1} = 1$

(b) $\dfrac{y^2}{16} - \dfrac{x^2}{9} = 1$

Figure 6

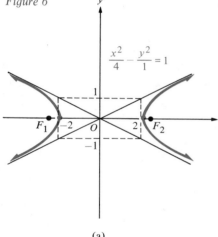

(a)

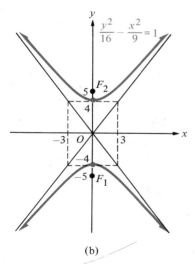

(b)

Solution   **(a)** The equation has the form

$$\frac{x^2}{a^2} - \frac{y^2}{b^2} = 1$$

with $a = 2$, $b = 1$. Hence, $c = \sqrt{a^2 + b^2} = \sqrt{5}$. Thus, the foci are $F_1 = (-\sqrt{5}, 0)$, $F_2 = (\sqrt{5}, 0)$, and the vertices are $V_1 = (-2, 0)$, $V_2 = (2, 0)$. The asymptotes are given by $y = \frac{1}{2}x$ and $y = -\frac{1}{2}x$ (Figure 6a).

(b) The equation has the form

$$\frac{y^2}{b^2} - \frac{x^2}{a^2} = 1$$

with $b = 4$, $a = 3$. Hence $c = \sqrt{a^2 + b^2} = \sqrt{25} = 5$. Thus, the foci are $F_1 = (0, -5)$, $F_2 = (0, 5)$, and the vertices are $V_1 = (0, -4)$, $V_2 = (0, 4)$. The asymptotes are given by $y = \frac{4}{3}x$ and $y = -\frac{4}{3}x$ (Figure 6b). ■

As with the ellipse in Section 10.1, if a hyperbola with either horizontal or vertical transverse axis is shifted so that its center is at the point $C = (h, k)$, then the equation of the shifted hyperbola will have one of the following standard forms:

Standard Forms for Hyperbolas Centered at (h, k)

(i) $\dfrac{(x - h)^2}{a^2} - \dfrac{(y - k)^2}{b^2} = 1$ if the transverse axis is horizontal (Figure 7a).

(ii) $\dfrac{(y - k)^2}{b^2} - \dfrac{(x - h)^2}{a^2} = 1$ if the transverse axis is vertical (Figure 7b).

Figure 7

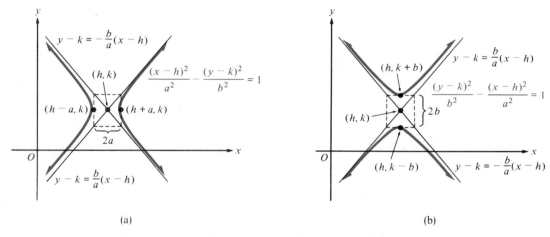

(a)   (b)

In either case, the asymptotes have the equations

$$y - k = \frac{b}{a}(x - h) \quad \text{and} \quad y - k = -\frac{b}{a}(x - h)$$

and the distance from the center (h, k) to either focus is given by

$$c = \sqrt{a^2 + b^2}.$$

Example 2   Find the coordinates of the center, the foci, and the vertices of the hyperbola $y^2 - 4x^2 - 8x - 4y - 4 = 0$. Also find equations of its asymptotes, and sketch its graph.

Solution   Completing the squares, we have

$$y^2 - 4y + 4 - 4(x^2 + 2x + 1) = 4$$

or

$$(y - 2)^2 - 4(x + 1)^2 = 4.$$

Figure 8 $y^2 - 4x^2 - 8x - 4y - 4 = 0$

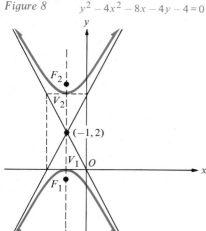

Dividing by 4, we obtain

$$\frac{(y-2)^2}{4} - \frac{(x+1)^2}{1} = 1,$$

the equation of a hyperbola with center $(-1, 2)$ and with a vertical transverse axis. Since $a = 1$ and $b = 2$, the equations of the asymptotes are

$$y - 2 = \frac{2}{1}(x + 1) \qquad \text{and} \qquad y - 2 = \frac{-2}{1}(x + 1)$$

that is,

$$y = 2x + 4 \qquad \text{and} \qquad y = -2x.$$

Also, $c = \sqrt{a^2 + b^2} = \sqrt{5}$, so $F_1 = (-1, 2 - \sqrt{5})$, $F_2 = (-1, 2 + \sqrt{5})$, $V_1 = (-1, 0)$, and $V_2 = (-1, 4)$ (Figure 8). ∎

Example 3   Two microphones are located at the points $(-c, 0)$ and $(c, 0)$ on the x axis (Figure 9). An explosion occurs at an unknown point P to the right of the y axis. The sound of the explosion is detected by the microphone at $(c, 0)$ exactly T seconds before it is detected by the microphone at $(-c, 0)$. Assuming that sound travels in air at the constant speed of v feet per second, show that the point P must have been located on the right-hand branch of the hyperbola whose equation is $(x^2/a^2) - (y^2/b^2) = 1$, where

$$a = \frac{vT}{2} \qquad \text{and} \qquad b = \frac{\sqrt{4c^2 - v^2 T^2}}{2}$$

Solution   Let d_1 and d_2 denote the distances from P to $(-c, 0)$ and $(c, 0)$, respectively. The sound of the explosion reaches $(-c, 0)$ in d_1/v seconds, and it reaches $(c, 0)$ in d_2/v seconds; hence, $d_1/v - d_2/v = T$, so $d_1 - d_2 = vT$. Putting $a = vT/2$, we notice that the condition $d_1 - d_2 = 2a$ requires that P belong to a hyperbola with foci $F_1 = (-c, 0)$ and $F_2 = (c, 0)$. By Theorem 1 we can write an equation of this hyperbola as $(x^2/a^2) - (y^2/b^2) = 1$, where

$$a = \frac{vT}{2} \qquad \text{and} \qquad b = \sqrt{c^2 - a^2} = \sqrt{c^2 - \left(\frac{vT}{2}\right)^2} = \frac{\sqrt{4c^2 - v^2 T^2}}{2} \qquad ∎$$

Figure 9

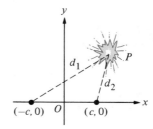

Problem Set 10.3

In Problems 1 to 8, find the coordinates of the vertices and the foci of each hyperbola. Also find equations of the asymptotes, and sketch the graph.

1. $\dfrac{x^2}{9} - \dfrac{y^2}{4} = 1$

2. $\dfrac{x^2}{1} - \dfrac{y^2}{9} = 1$

3. $\dfrac{y^2}{16} - \dfrac{x^2}{4} = 1$

4. $\dfrac{y^2}{4} - \dfrac{x^2}{1} = 1$

5. $4x^2 - 16y^2 = 64$

6. $49x^2 - 16y^2 = 196$

7. $36y^2 - 10x^2 = 360$

8. $y^2 - 4x^2 = 1$

In Problems 9 to 11, find an equation of the hyperbola that satisfies the conditions given.

9. Vertices at $(-4, 0)$ and $(4, 0)$, foci at $(-6, 0)$ and $(6, 0)$.

10. Vertices at $(0, -\frac{1}{2})$ and $(0, \frac{1}{2})$, foci at $(0, -1)$ and $(0, 1)$.

11. Vertices at $(-4, 0)$ and $(4, 0)$, the equations of the asymptotes are $y = -\frac{5}{4}x$ and $y = \frac{5}{4}x$.

12. Determine the values of a^2 and b^2 so that the graph of the equation $b^2x^2 - a^2y^2 = a^2b^2$ contains the pair of points (a) $(2, 5)$ and $(3, -10)$ and (b) $(4, 3)$ and $(-7, 6)$.

In Problems 13 to 20, find the coordinates of the center, the vertices, and the foci of each hyperbola. Also find equations of the asymptotes, and sketch the graph.

13. $\dfrac{(x - 1)^2}{9} - \dfrac{(y + 2)^2}{4} = 1$

14. $\dfrac{(x + 3)^2}{1} - \dfrac{(y - 1)^2}{9} = 1$

15. $\dfrac{(y + 1)^2}{16} - \dfrac{(x + 2)^2}{25} = 1$

16. $4x^2 - y^2 - 8x + 2y + 7 = 0$

17. $x^2 - 4y^2 - 4x - 8y - 4 = 0$

18. $16x^2 - 9y^2 + 180y = 612$

19. $9x^2 - 25y^2 + 72x - 100y + 269 = 0$

20. $9x^2 - 16y^2 - 90x - 256y = 223$

21. Find an equation of the hyperbola that satisfies the conditions given.

   (a) Foci at $(1, -1)$ and $(7, -1)$, length of transverse axis is 2.

   (b) Vertices at $(-4, 3)$ and $(0, 3)$, foci at $(-\frac{9}{2}, 3)$ and $(\frac{1}{2}, 3)$.

   (c) Center at $(2, 3)$, one vertex at $(2, 8)$, and one focus at $(2, -3)$.

22. The segment cut by a hyperbola from a line containing a focus and perpendicular to the transverse axis is called a **focal chord** of the hyperbola (Figure 10).

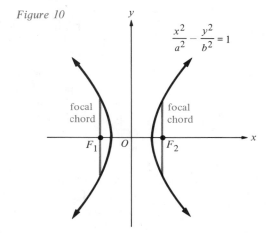

Figure 10

   (a) Show that the length of a focal chord of the hyperbola $(x^2/a^2) - (y^2/b^2) = 1$ is $2b^2/a$.

   (b) Find the length of a focal chord of the hyperbola $x^2 - 8y^2 = 16$.

23. A hyperbola is said to be **equilateral** if its two asymptotes are perpendicular. Find an equation of an equilateral hyperbola with horizontal transverse axis and center at the origin. (Denote the distance from the center to a vertex by a.)

24. Sketch the graph of the hyperbola

$$\dfrac{(y + b)^2}{b^2} - \dfrac{x^2}{2bp + p^2} = 1.$$

Show that as $b \to +\infty$, the upper branch of this hyperbola approaches the parabola $y = (1/4p)x^2$.

25. In Figure 11, hold the center C and the asymptotes fixed, but allow the foci F_1 and F_2 to move inward the center. What happens to the hyperbola?

Figure 11

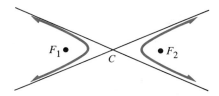

Figure 12

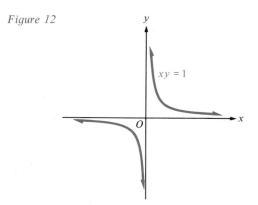

26. Give a proof of Theorem 1.

27. Sound travels with speed s in air, and a bullet travels with speed b from a gun at $(-h, 0)$ to a target at $(h, 0)$ in the xy plane. At what points (x, y) can the boom of the gun and the ping of the bullet hitting the target be heard simultaneously?

28. It can be shown that the graph of the equation $xy = 1$ is a hyperbola with center at the origin and with the x and y axes as asymptotes (Figure 12). Find the coordinates of the foci of this hyperbola. [*Hint:* The transverse axis makes a 45° angle with the x axis.]

Figure 13

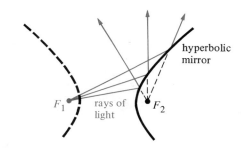

29. A point moves so that the product of the slopes of the line segments that join it to two given points is 9. Describe the path of the point.

30. The **reflection property** for hyperbolas is illustrated in Figure 13. State this property in words.

10.4 TRANSLATION AND ROTATION OF THE COORDINATE AXES

An equation of a curve in the plane can often be simplified by changing to a new pair of coordinate axes. In practice, this is usually done by choosing one or both of the new coordinate axes to coincide with an axis of symmetry of the curve.

Translation of the Coordinate Axes

If two Cartesian coordinate systems have corresponding axes that are parallel and have the same positive directions, then we say that these coordinate systems are obtained from one another by **translation.** Figure 1 shows a translation of an "old" xy coordinate system to a "new" $\bar{x}\bar{y}$ coordinate system whose origin $\bar{O}$ has the "old" coordinates (h, k). Consider the point P in Figure 1 having old coordinates (x, y), but having new coordinates $(\bar{x}, \bar{y})$. Evidently,

$$\begin{cases} x = h + \bar{x} \\ y = k + \bar{y} \end{cases} \quad \text{or} \quad \begin{cases} \bar{x} = x - h \\ \bar{y} = y - k. \end{cases}$$

Figure 1

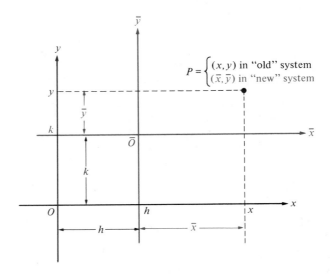

These equations, which allow us to change from the new to the old coordinates of a point P or vice versa, are called the **translation equations.** Notice that a translation of the coordinate axes does not change the position of a point P in the plane—it only changes the numerical "address" of the point.

Example 1   Let the $\bar{x}\bar{y}$ axes be obtained from the xy axes by a translation so that the new origin $\bar{O}$ has coordinates $(h, k) = (-3, 4)$ in the old xy coordinate system. Let P be the point whose old coordinates are $(x, y) = (2, 1)$. Find the new coordinates $(\bar{x}, \bar{y})$ of P.

Solution   By the translation equations,

$$\bar{x} = x - h = 2 - (-3) = 5$$

and

$$\bar{y} = y - k = 1 - 4 = -3,$$

so, in the new coordinate system, $P = (\bar{x}, \bar{y}) = (5, -3)$.   ∎

An equation of a curve in the plane depends not only on the set of points comprising it but also on our choice of the coordinate system. For instance, the circle in Figure 2 has the equation

$$(x + 1)^2 + (y - 2)^2 = 9$$

Figure 2

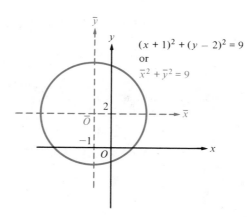

$(x + 1)^2 + (y - 2)^2 = 9$
or
$\bar{x}^2 + \bar{y}^2 = 9$

with respect to the old xy coordinate system. However, the very same circle has the simpler equation

$$\bar{x}^2 + \bar{y}^2 = 9$$

with respect to the new $\bar{x}\bar{y}$ coordinate system whose origin $\bar{O}$ is at the center of the circle. Notice that a translation of the coordinate axes does not change the position or the shape of a geometric curve in the plane—it only changes the *equation* of the curve.

Example 2   Find a translation of axes that will reduce the equation

$$4x^2 + 9y^2 - 8x + 36y + 4 = 0$$

to an equation involving no first-degree terms in $\bar{x}$ or $\bar{y}$. Identify the graph.

Solution   The simplest procedure is to complete the squares, so that the equation becomes

$$4(x^2 - 2x + 1) + 9(y^2 + 4y + 4) + 4 = 4(1) + 9(4)$$

or $4(x - 1)^2 + 9(y + 2)^2 = 36.$

Then we let $\bar{x} = x - 1$ and $\bar{y} = y + 2$ to obtain

$$4\bar{x}^2 + 9\bar{y}^2 = 36 \qquad \text{or} \qquad \frac{\bar{x}^2}{9} + \frac{\bar{y}^2}{4} = 1,$$

the equation in standard form of an ellipse with center at the origin $\bar{O}$ of the new coordinate system. An alternative solution is obtained by letting

$$x = \bar{x} + h \qquad \text{and} \qquad y = \bar{y} + k,$$

where h and k are yet to be determined, and then substituting into the equation $4x^2 + 9y^2 - 8x + 36y + 4 = 0$. Routine algebraic simplification then yields

$$4\bar{x}^2 + 9\bar{y}^2 + 8(h - 1)\bar{x} + 18(k + 2)\bar{y} + 4h^2 + 9k^2 - 8h + 36k + 4 = 0.$$

The first-degree terms in $\bar{x}$ and $\bar{y}$ drop out if we let $h = 1$ and $k = -2$, and the equation becomes

$$4\bar{x}^2 + 9\bar{y}^2 - 36 = 0 \qquad \text{or} \qquad \frac{\bar{x}^2}{9} + \frac{\bar{y}^2}{4} = 1.$$

■

Rotation of the Coordinate Axes

In some situations we can simplify an equation of a curve by **rotating** the coordinate system rather than translating it. Of course, the polar coordinate system (Section 8.5) is naturally adapted to rotation about the pole. Figure 3 shows an "old" polar axis and a "new" polar axis obtained by rotating the old polar axis counterclockwise about the pole $O = \bar{O}$ through the angle ϕ. Consider the point P in Figure 3 having old polar coordinates (r, θ), but having new polar coordinates $(\bar{r}, \bar{\theta})$. Evidently,

$$\begin{cases} r = \bar{r} \\ \theta = \phi + \bar{\theta} \end{cases} \quad \text{or} \quad \begin{cases} \bar{r} = r \\ \bar{\theta} = \theta - \phi. \end{cases}$$

These equations are called the **rotation equations for polar coordinates.**

Figure 3

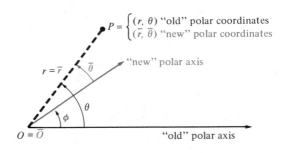

$$P = \begin{cases} (r, \theta) \text{ "old" polar coordinates} \\ (\bar{r}, \bar{\theta}) \text{ "new" polar coordinates} \end{cases}$$

"new" polar axis

$r = \bar{r}$ $\bar{\theta}$

θ

ϕ

$O = \bar{O}$ "old" polar axis

The rotation equations for Cartesian coordinates are

$$x = \bar{x} \cos \phi - \bar{y} \sin \phi$$
$$y = \bar{x} \sin \phi + \bar{y} \cos \phi,$$

where the "new" $\bar{x}\bar{y}$ coordinate system is obtained by rotating the "old" xy coordinate system counterclockwise about the origin through the angle ϕ (Figure 4). These Cartesian rotation equations can be derived by converting from Cartesian to polar coordinates, rotating the polar coordinate system through angle ϕ by using the rotation equations in polar coordinates, and then converting back to Cartesian coordinates. We leave the derivation of these Cartesian rotation equations as an exercise (Problem 36).

Figure 4

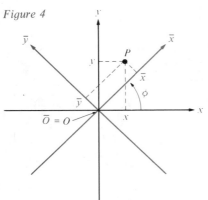

The Cartesian rotation equations given above can be solved for $\bar{x}$ and $\bar{y}$ in terms of x and y to obtain

$$\begin{cases} \bar{x} = x \cos \phi + y \sin \phi \\ \bar{y} = -x \sin \phi + y \cos \phi. \end{cases}$$

(See Problem 37.)

Example 3   The old xy coordinate system is rotated through $\pi/6$ radian to obtain a new $\bar{x}\bar{y}$ coordinate system (Figure 5). Find the new $\bar{x}\bar{y}$ coordinates of the point P whose old xy coordinates are $(1, 2)$.

Solution   We use the Cartesian rotation equations, solved for $\bar{x}$ and $\bar{y}$ in terms of x and y, with $\phi = \pi/6$, $x = 1$, and $y = 2$ to obtain

$$\bar{x} = x \cos \phi + y \sin \phi = 1 \cos \frac{\pi}{6} + 2 \sin \frac{\pi}{6} = \frac{\sqrt{3}}{2} + 2\left(\frac{1}{2}\right) = \frac{\sqrt{3}}{2} + 1$$

$$\bar{y} = -x \sin \phi + y \cos \phi = -1 \sin \frac{\pi}{6} + 2 \cos \frac{\pi}{6} = -\frac{1}{2} + 2\left(\frac{\sqrt{3}}{2}\right) = -\frac{1}{2} + \sqrt{3}.$$

Hence, the new coordinates of P are $(\bar{x}, \bar{y}) = \left(1 + \dfrac{\sqrt{3}}{2}, \sqrt{3} - \dfrac{1}{2}\right)$.   ∎

By a suitable rotation of the coordinate axes, an equation of a curve can often be "simplified" or brought into recognizable form.

Figure 5

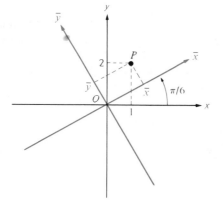

Figure 6

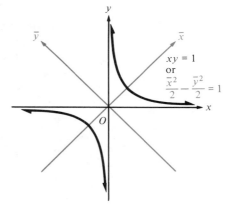

Example 4   Suppose that the old xy coordinate system is rotated through $\phi = 45°$ to obtain a new $\bar{x}\bar{y}$ coordinate system (Figure 6). Find an equation of the curve $xy = 1$ in the new $\bar{x}\bar{y}$ coordinate system and identify the curve.

Solution   The graph of $xy = 1$ looks suspiciously like a hyperbola whose transverse axis makes an angle of $45°$ with the x axis. Our calculations will confirm this. We substitute the Cartesian rotation equations with $\phi = 45°$, that is,

$$x = \bar{x} \cos 45° - \bar{y} \sin 45° = \frac{\sqrt{2}}{2}(\bar{x} - \bar{y})$$

$$y = \bar{x} \sin 45° + \bar{y} \cos 45° = \frac{\sqrt{2}}{2}(\bar{x} + \bar{y}),$$

into the equation $xy = 1$ to obtain

$$\frac{\sqrt{2}}{2}(\bar{x} - \bar{y})\frac{\sqrt{2}}{2}(\bar{x} + \bar{y}) = 1$$

or

$$\frac{\bar{x}^2}{2} - \frac{\bar{y}^2}{2} = 1.$$

Thus, just as we suspected, the curve is a hyperbola, since its equation has the standard form in the new $\bar{x}\bar{y}$ coordinate system. ∎

In Example 4 we were able to identify the curve $xy = 1$ as a hyperbola because rotation of the coordinate axis through the $45°$ angle brought the equation into a familiar form. This occurred because the new $\bar{x}\bar{y}$ coordinate axes were aligned with the axes of symmetry of the hyperbola. Essentially the same procedure can be applied to any equation of the form

$$\boxed{Ax^2 + Bxy + Cy^2 + Dx + Ey + F = 0}$$

containing a **mixed term Bxy.** The latter equation, in which the coefficients denote constant real numbers, is called the **general second-degree equation in x and y.**

The equation in standard form of a circle, an ellipse, a parabola, or a hyperbola can always be rewritten in the form of a general second-degree equation *containing no mixed term* (see Problem 38). For instance, the equation in standard form

$$\frac{(x - 3)^2}{4} - \frac{(y + 1)^2}{9} = 1$$

of a hyperbola can be rewritten as

$$\frac{x^2 - 6x + 9}{4} - \frac{y^2 + 2y + 1}{9} = 1$$

or

$$9(x^2 - 6x + 9) - 4(y^2 + 2y + 1) = 36;$$

that is,

$$9x^2 - 54x + 81 - 4y^2 - 8y - 4 - 36 = 0$$

or

$$9x^2 - 4y^2 - 54x - 8y + 41 = 0.$$

The last equation has the general second-degree form, with $A = 9$, $B = 0$, $C = -4$, $D = -54$, $E = -8$, and $F = 41$.

Now consider a curve in the plane whose equation has the general second-degree form, but with a mixed term Bxy, with $B \neq 0$. The following theorem shows that the mixed term can always be removed by a suitable rotation of the coordinate axes (see Problems 40 and 42).

Theorem 1   **Removal of the Mixed Term by Rotation**

If the old xy coordinate system is rotated about the origin through the angle ϕ to obtain a new $\bar{x}\bar{y}$ coordinate system, then the curve whose old equation was

$$Ax^2 + Bxy + Cy^2 + Dx + Ey + F = 0$$

will have a new equation of the form

$$\bar{A}\bar{x}^2 + \bar{B}\bar{x}\bar{y} + \bar{C}\bar{y}^2 + \bar{D}\bar{x} + \bar{E}\bar{y} + \bar{F} = 0.$$

If $B \neq 0$ and if ϕ is chosen so that $0 < \phi < \dfrac{\pi}{2}$ and

$$\cot 2\phi = \frac{A - C}{B},$$

then $\bar{B} = 0$, and the mixed term will not appear in the new equation.

Example 5   Rotate the old xy coordinates axes to remove the mixed term from the equation $x^2 - 4xy + y^2 - 6 = 0$ and sketch the graph showing both the old xy coordinate system and the new $\bar{x}\bar{y}$ coordinate system.

Solution   The equation

$$x^2 - 4xy + y^2 - 6 = 0$$

has the form

$$Ax^2 + Bxy + Cy^2 + Dx + Ey + F = 0,$$

with $A = 1$, $B = -4$, $C = 1$, $D = 0$, $E = 0$, and $F = -6$. According to Theorem 1, the mixed term can be removed by rotating the coordinate system through the angle ϕ, where $0 < \phi < \dfrac{\pi}{2}$ and $\cot 2\phi = \dfrac{A - C}{B} = \dfrac{1 - 1}{-4} = 0$. Therefore, we take $2\phi = \pi/2$ or $\phi = \pi/4$. The rotation equations with $\phi = \pi/4$ are

$$x = \bar{x} \cos \phi - \bar{y} \sin \phi = \frac{\sqrt{2}}{2}(\bar{x} - \bar{y})$$

and

$$y = \bar{x} \sin \phi + \bar{y} \cos \phi = \frac{\sqrt{2}}{2}(\bar{x} + \bar{y}).$$

Substituting these into the equation $x^2 - 4xy + y^2 - 6 = 0$, we obtain

$$\left[\frac{\sqrt{2}}{2}(\bar{x} - \bar{y})\right]^2 - 4\left[\frac{\sqrt{2}}{2}(\bar{x} - \bar{y})\right]\left[\frac{\sqrt{2}}{2}(\bar{x} + \bar{y})\right] + \left[\frac{\sqrt{2}}{2}(\bar{x} + \bar{y})\right]^2 - 6 = 0,$$

which simplifies to

$$\tfrac{1}{2}(\bar{x} - \bar{y})^2 - 2(\bar{x} - \bar{y})(\bar{x} + \bar{y}) + \tfrac{1}{2}(\bar{x} + \bar{y})^2 = 6$$

or

$$\tfrac{1}{2}(\bar{x}^2 - 2\bar{x}\bar{y} + \bar{y}^2) - 2(\bar{x}^2 - \bar{y}^2) + \tfrac{1}{2}(\bar{x}^2 + 2\bar{x}\bar{y} + \bar{y}^2) = 6.$$

Figure 7

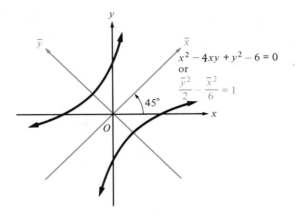

Collecting terms in the last equation, we get

$$-\bar{x}^2 + 3\bar{y}^2 = 6 \qquad \text{or} \qquad \frac{\bar{y}^2}{2} - \frac{\bar{x}^2}{6} = 1,$$

an equation of a hyperbola (Figure 7). ∎

If $0 < \phi < \dfrac{\pi}{2}$, then the trigonometric identities

$$\cos 2\phi = \frac{\cot 2\phi}{\sqrt{\cot^2 2\phi + 1}}, \qquad \cos \phi = \sqrt{\frac{1 + \cos 2\phi}{2}}, \qquad \sin \phi = \sqrt{\frac{1 - \cos 2\phi}{2}}$$

permit us to find $\cos \phi$ and $\sin \phi$ algebraically in terms of the value of $\cot 2\phi$. This is useful in applying Theorem 1, as the following example shows.

Ⓒ **Example 6**   Rotate the old xy coordinate axes to remove the mixed term from the equation
$$8x^2 - 4xy + 5y^2 = 144$$
and sketch the graph showing both the old and the new coordinate systems.

Solution   Here $A = 8$, $B = -4$, $C = 5$, $D = 0$, $E = 0$, and $F = -144$. Hence,

$$\cot 2\phi = \frac{A - C}{B} = \frac{8 - 5}{-4} = -\frac{3}{4}.$$

Thus,

$$\cos 2\phi = \frac{\cot 2\phi}{\sqrt{\cot^2 2\phi + 1}} = \frac{-3/4}{\sqrt{(-3/4)^2 + 1}} = \frac{-3/4}{\sqrt{(9/16) + 1}} = \frac{-3/4}{5/4} = -\frac{3}{5},$$

so that

$$\cos\phi = \sqrt{\frac{1 + \cos 2\phi}{2}} = \sqrt{\frac{1 - (3/5)}{2}} = \sqrt{\frac{2}{10}} = \sqrt{\frac{1}{5}} = \frac{\sqrt{5}}{5},$$

and

$$\sin\phi = \sqrt{\frac{1 - \cos 2\phi}{2}} = \sqrt{\frac{1 + (3/5)}{2}} = \sqrt{\frac{8}{10}} = \sqrt{\frac{4}{5}} = \frac{2\sqrt{5}}{5}.$$

Now, substituting the rotation equations

$$x = \bar{x}\cos\phi - \bar{y}\sin\phi = \frac{\sqrt{5}}{5}\bar{x} - \frac{2\sqrt{5}}{5}\bar{y}$$

$$y = \bar{x}\sin\phi + \bar{y}\cos\phi = \frac{2\sqrt{5}}{5}\bar{x} + \frac{\sqrt{5}}{5}\bar{y}$$

into the given equation $8x^2 - 4xy + 5y^2 = 144$, we obtain

$$8\left(\frac{\sqrt{5}}{5}\bar{x} - \frac{2\sqrt{5}}{5}\bar{y}\right)^2 - 4\left(\frac{\sqrt{5}}{5}\bar{x} - \frac{2\sqrt{5}}{5}\bar{y}\right)\left(\frac{2\sqrt{5}}{5}\bar{x} + \frac{\sqrt{5}}{5}\bar{y}\right)$$

$$+ 5\left(\frac{2\sqrt{5}}{5}\bar{x} + \frac{\sqrt{5}}{5}\bar{y}\right)^2 = 144.$$

This equation simplifies to

$$4\bar{x}^2 + 9\bar{y}^2 = 144 \qquad \text{or} \qquad \frac{\bar{x}^2}{36} + \frac{\bar{y}^2}{16} = 1,$$

an equation of an ellipse. Since $\sin\phi = 2\sqrt{5}/5$, it follows (using a calculator or Appendix Table IIa) that

$$\phi = \sin^{-1}\frac{2\sqrt{5}}{5} \approx 63.43°$$

(Figure 8). ■

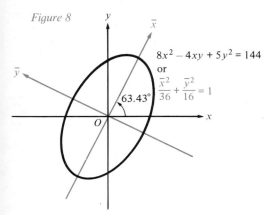

Figure 8

$8x^2 - 4xy + 5y^2 = 144$
or
$\dfrac{\bar{x}^2}{36} + \dfrac{\bar{y}^2}{16} = 1$

63.43°

We now have the means to sketch the graph of any second-degree equation in x and y. If the equation contains a mixed term, a suitable rotation of the coordinate axes removes it (Theorem 1). Then, by completing the squares as in Section 4.1, we *usually* obtain the equation in standard form for a circle, an ellipse, a parabola, or a hyperbola. The rotation of axes that removes the mixed term lines up the new coordinate system with the axes of symmetry of the conic section (see Figures 7 and 8).

There are some exceptional cases in which, even after completing the squares, an equation cannot be brought into any of the standard forms mentioned above. In these cases, we say that the graph is a **degenerate conic.** The only possible degenerate conics are (i) the whole plane, (ii) the empty set, (iii) a pair of straight lines, (iv) a straight line, or (v) one point. (See Problems 39, 41, 43, and 45.)

Problem Set 10.4

1. A new $\bar{x}\bar{y}$ coordinate system is obtained by translating the old xy coordinate system so that the origin $\bar{O}$ of the new system has old xy coordinates $(-1, 2)$. Find the new $\bar{x}\bar{y}$ coordinates of the points whose old xy coordinates are:

(a) $(0, 0)$ (b) $(-2, 1)$ (c) $(3, -3)$

(d) $(-3, -2)$ (e) $(5, 5)$ (f) $(6, 0)$.

2. A new $\bar{x}\bar{y}$ coordinate system is obtained by translating the old xy coordinate system so that the origin $\bar{O}$ of the new system has old xy coordinates $(2, -3)$. Find the old xy coordinates of the points whose new $\bar{x}\bar{y}$ coordinates are:

(a) $(0, 0)$ (b) $(3, 2)$ (c) $(-3, 4)$

(d) $(-2, 3)$ (e) $(\pi, \sqrt{2})$ (f) $(0, \pi)$.

In Problems 3 to 16, find a translation of axes $x = \bar{x} + h$ and $y = \bar{y} + k$ that will reduce each equation to an equation involving no first-degree terms in $\bar{x}$ or $\bar{y}$. Identify the graph.

3. $x^2 + y^2 + 4x - 2y + 1 = 0$

4. $3x^2 + 3y^2 + 7x - 5y + 3 = 0$

5. $x^2 + 4y^2 + 2x - 8y + 1 = 0$

6. $9x^2 + y^2 - 18x + 2y + 9 = 0$

7. $6x^2 + 9y^2 - 24x - 54y + 51 = 0$

8. $9x^2 + 4y^2 - 18x + 16y - 11 = 0$

9. $x^2 - 2x - y^2 + 6 = 0$

10. $4x^2 + 24x + 39 - 3y^2 = 0$

11. $x^2 - 10y - 4x + 21 = 0$

12. $3x^2 - y^2 + 12x + 8y = 7$

13. $4x^2 - 25y^2 + 24x + 50y + 22 = 0$

14. $5y^2 - 9x^2 + 10y + 54x = 112$

15. $x^2 - 4y^2 - 4x - 8y - 4 = 0$

16. $9x^2 - y^2 - 18x - 4y + 5 = 0$

In Problems 17 to 24, the old xy axis has been rotated counterclockwise about the origin through the angle ϕ to form a new $\bar{x}\bar{y}$ coordinate system. The point P has coordinates (x, y) in the old system and coordinates $(\bar{x}, \bar{y})$ in the new system. Supply the missing information.

17. $(x, y) = (4, -7)$, $\phi = 90°$, $(\bar{x}, \bar{y}) = ?$

18. $(x, y) = (2, 0)$, $(\bar{x}, \bar{y}) = (1, \sqrt{3})$,
$0 < \phi < 90°$, $\phi = ?$

19. $(\bar{x}, \bar{y}) = (-3, -3)$, $\phi = \pi/3$, $(x, y) = ?$

20. $(x, y) = (5\sqrt{2}, \sqrt{2})$, $\phi = 45°$, $(\bar{x}, \bar{y}) = ?$

21. $(\bar{x}, \bar{y}) = (-4, -2)$, $\phi = 30°$, $(x, y) = ?$

22. $(\bar{x}, \bar{y}) = (-3, \sqrt{2})$, $\phi = 3\pi/4$, $(x, y) = ?$

23. $(x, y) = (-4, 0)$, $\phi = \pi$, $(\bar{x}, \bar{y}) = ?$

24. $(x, y) = (1, -7)$, $\phi = 240°$, $(\bar{x}, \bar{y}) = ?$

25. The xy coordinate system is rotated $30°$ about the origin to form the $\bar{x}\bar{y}$ coordinate system. Rewrite each equation as an equation in the other coordinate system.

(a) $y^2 = 3x$ (b) $\bar{y} = 3\bar{x}$

(c) $\bar{x}^2 + \bar{y}^2 = 1$ (d) $5x - y = 4$

(e) $x^2 + y^2 = 1$ (f) $x^2 = 25$

26. Find an angle ϕ (if one exists) for which the Cartesian rotation equations give each of the following.

(a) $\bar{x} = y$ and $\bar{y} = -x$

(b) $\bar{x} = -x$ and $\bar{y} = -y$

(c) $\bar{x} = y$ and $\bar{y} = x$

(d) $\bar{x} = -y$ and $\bar{y} = -x$

In Problems 27 to 35, (a) use Theorem 1 to find an angle ϕ through which to rotate the coordinate system so as to remove the mixed term from the equation, (b) find x and y in terms of $\bar{x}$ and $\bar{y}$, (c) find the new equation in terms of $\bar{x}$ and $\bar{y}$, and (d) sketch the graph showing both the old xy axes and the new $\bar{x}\bar{y}$ axes.

Ⓒ **27.** $x^2 + 4xy - 2y^2 = 12$

28. $x^2 + 2xy + y^2 + x + y = 0$

29. $x^2 + 2xy + y^2 = 1$

30. $x^2 + 2xy + y^2 - 4\sqrt{2}x + 4\sqrt{2}y = 0$

Ⓒ **31.** $9x^2 - 24xy + 16y^2 = 144$

32. $2x^2 + 4\sqrt{3}xy - 2y^2 - 4 = 0$

Ⓒ **33.** $6x^2 - 6xy + 14y^2 = 45$

© **34.** $17x^2 - 12xy + 8y^2 - 68x + 24y - 12 = 0$

© **35.** $2x^2 + 6xy - 6y^2 + 2\sqrt{10}x + 3\sqrt{10}y - 16 = 0$

36. Derive the Cartesian rotation equations by converting from Cartesian to polar coordinates, rotating the polar coordinate system through the angle ϕ, and then converting back to Cartesian coordinates.

37. Solve the simultaneous equations

$$\begin{cases} \bar{x}\cos\phi - \bar{y}\sin\phi = x \\ \bar{x}\sin\phi + \bar{y}\cos\phi = y \end{cases}$$

for $\bar{x}$ and $\bar{y}$ in terms of x and y.

38. Show that an equation in standard form of a circle, an ellipse, a parabola, or a hyperbola can be rewritten in the form of a general second-degree equation *with no mixed term*.

39. The following second-degree equations in x and y have graphs that are degenerate conics. In each case, verfiy that the graph is as described.

(a) $0x^2 + 0yx + 0y^2 + 0x + 0y + 0 = 0$
(The whole xy plane)

(b) $x^2 + y^2 + 1 = 0$ (The empty set)

(c) $2x^2 - 4xy + 2y^2 = 0$ (A straight line)

(d) $4x^2 - y^2 + 16x + 2y + 15 = 0$
(Two intersecting straight lines)

(e) $x^2 - 2xy + y^2 - 18 = 0$
(Two parallel straight lines)

(f) $x^2 + y^2 - 6x + 4y + 13 = 0$
(A single point)

40. In Theorem 1, show by direct calculation that:

(a) $\bar{A} = A\cos^2\phi + B\cos\phi\sin\phi + C\sin^2\phi$

(b) $\bar{B} = 2(C - A)\cos\phi\sin\phi + B(\cos^2\phi - \sin^2\phi)$
$= (C - A)\sin 2\phi + B\cos 2\phi$

(c) $\bar{C} = A\sin^2\phi - B\cos\phi\sin\phi + C\cos^2\phi$

(d) $\bar{D} = D\cos\phi + E\sin\phi$

(e) $\bar{E} = -D\sin\phi + E\cos\phi$

(f) $\bar{F} = F$.

41. If $AC > 0$, show that the graph of the equation $Ax^2 + Cy^2 + Dx + Ey + F = 0$ is an ellipse, a circle, a single point, or the empty set.

42. Using the results of part (b) of Problem 40, prove Theorem 1.

43. If $AC < 0$, show that the graph of the equation $Ax^2 + Cy^2 + Dx + Ey + F = 0$ is a hyperbola or a pair of intersecting straight lines.

44. Using the results of Problem 40, show:
$A + C = \bar{A} + \bar{C}$.

45. If $AC = 0$, show that the graph of the equation $Ax^2 + Cy^2 + Dx + Ey + F = 0$ is a parabola, a pair of parallel straight lines, a single straight line, the whole plane, or the empty set.

46. Using the results of Problem 40, show that $B^2 - 4AC = \bar{B}^2 - 4\bar{A}\bar{C}$.

10.5 ECCENTRICITY AND CONICS IN POLAR FORM

In Sections 10.1 to 10.3, we derived equations in Cartesian form for the ellipse, the parabola, and the hyperbola from three special geometric definitions—one for each type of curve. By using the idea of *eccentricity*, it is possible to give a unified geometric definition for these curves. In this definition, F is a fixed point in the plane, D is a fixed line in the plane, and F does not belong to D.

Definition 1   **Conics in Terms of Eccentricity, Focus, and Directrix**

> Let e be a fixed positive number. A **conic** with **eccentricity** e, **focus** F, and **directrix** D is the set of all points P in the plane such that the distance from P to F, divided by the distance from P to D, is equal to e.

Figure 1

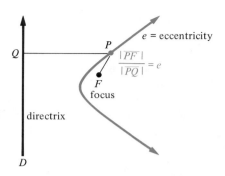

Thus, the point P belongs to the conic if and only if

$$\frac{|\overline{PF}|}{|\overline{PQ}|} = e$$

where Q is the foot of the perpendicular from P to D (Figure 1). Although we use the same symbol e for both the eccentricity of a conic and the base of the natural logarithm, no confusion should result—you can always tell from the context what is intended.

Example 1   Find the equation in Cartesian form of the conic with eccentricity $e = 2$ whose focus is at the origin and whose directrix is given by $x = -3$.

Solution   From Figure 2, the point $P = (x, y)$ belongs to the given conic if and only if $|\overline{PF}|/|\overline{PQ}| = 2$; that is,

Figure 2

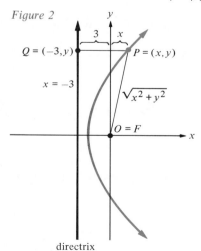

$$\frac{\sqrt{x^2 + y^2}}{3 + x} = 2 \qquad \text{or} \qquad \sqrt{x^2 + y^2} = 6 + 2x.$$

Squaring both sides of the last equation, we have

$$x^2 + y^2 = 36 + 24x + 4x^2$$

or

$$3x^2 + 24x - y^2 = -36.$$

Completing the square, we obtain

$$3(x^2 + 8x + 16) - y^2 = 3(16) - 36$$

or

$$3(x + 4)^2 - y^2 = 12,$$

that is,

$$\frac{(x + 4)^2}{4} - \frac{y^2}{12} = 1.$$

Therefore, the conic is a hyperbola with center at $(-4, 0)$. ▪

By an argument similar to that in the example above, it can be shown that a conic, defined as in Definition 1, is an ellipse, a parabola, or a hyperbola. In fact, we have the following theorem, whose proof is left as an exercise (Problem 42).

Theorem 1   **Conic Theorem**

Suppose that a conic with focus F at the origin and directrix $x = -d$ has eccentricity e, where e and d are positive. Then exactly one of the following holds:

Case (i) $e < 1$, and the conic is an ellipse with the equation

$$\frac{(x - c)^2}{a^2} + \frac{y^2}{b^2} = 1$$

where $a = \dfrac{ed}{1 - e^2}$, $b = \dfrac{ed}{\sqrt{1 - e^2}}$, and $c = \sqrt{a^2 - b^2} = ae$.

Case (ii) $e = 1$, and the conic is a parabola with the equation

$$4p(x + p) = y^2 \qquad \text{where } p = \frac{d}{2}.$$

Case (iii) $e > 1$, and the conic is a hyperbola with the equation

$$\frac{(x + c)^2}{a^2} - \frac{y^2}{b^2} = 1$$

where $a = \dfrac{ed}{e^2 - 1}$, $b = \dfrac{ed}{\sqrt{e^2 - 1}}$, and $c = \sqrt{a^2 + b^2} = ae$.

By translation and rotation (if necessary), the equation of any ellipse, parabola, or hyperbola can be brought into the form indicated in Theorem 1. Therefore, an ellipse or hyperbola is a conic with eccentricity

$$e = c/a$$

and a parabola is a conic with eccentricity 1.

Figure 3

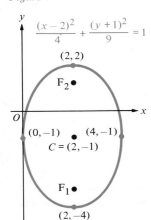

$$\frac{(x - 2)^2}{4} + \frac{(y + 1)^2}{9} = 1$$

$(2, 2)$

F_2

O

$(0, -1)$ $(4, -1)$

$C = (2, -1)$

F_1

$(2, -4)$

Example 2   Find the eccentricity of the ellipse

$$\frac{(x - 2)^2}{4} + \frac{(y + 1)^2}{9} = 1$$

and sketch the graph.

Solution   The major axis is vertical and the center is $C = (2, -1)$. The semimajor axis is $a = 3$, the semiminor axis is $b = 2$, and the distance from the center to the foci is $c = \sqrt{9 - 4} = \sqrt{5}$. Hence, the eccentricity is $e = c/a = \sqrt{5}/3$ (Figure 3). ∎

Conics in Polar Form

To find a polar equation for a conic of eccentricity e, we place the focus F at the pole O, and we place the directrix D perpendicular to the polar axis and d units to the *left* of the pole, $d > 0$ (Figure 4).

Now consider an arbitrary point $P = (r, \theta)$ in the plane, and let Q be the point at the foot of the perpendicular from P to the directrix D. Switching momentarily to Cartesian coordinates, so that $P = (x, y)$, $Q = (-d, y)$, $x = r \cos \theta$, and $y = r \sin \theta$ (Figure 5), we see that

$$|\overline{PQ}| = |d + x| = |d + r \cos \theta|$$

and $|\overline{PF}| = \sqrt{x^2 + y^2} = |r|$. By definition, P belongs to the conic if and only if

$$\frac{|\overline{PF}|}{|\overline{PQ}|} = e$$

that is,

$$\frac{|r|}{|d + r \cos \theta|} = e.$$

Figure 4

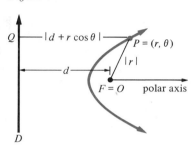

Therefore, an equation of the conic in polar form is

$$\left| \frac{r}{d + r \cos \theta} \right| = e$$

or

$$\frac{\pm r}{d + r \cos \theta} = e.$$

If $P = (r_1, \theta_1)$ satisfies the equation

$$\frac{-r}{d + r \cos \theta} = e,$$

then $P = (-r_1, \theta_1 + \pi)$ also satisfies the equation

$$\frac{r}{d + r \cos \theta} = e.$$

Hence, we lose no points on the conic by writing its equation as

$$\frac{r}{d + r \cos \theta} = e.$$

Solving for r, we obtain a polar equation for the graph in Figure 4:

$$r = \frac{ed}{1 - e \cos \theta}.$$

Figure 5

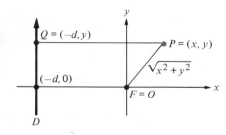

Similar derivations lead to the standard equations in the polar coordinate system for conics with vertical or horizontal directrices and a focus at the pole, as shown in the following table.

Table 1 **Standard Polar Equations for Conics with a Focus at the Pole and Eccentricity e**

Polar Equation	Direction and Location of Directrix
(i) $r = \dfrac{ed}{1 - e \cos \theta}$	Vertical, d units left of the pole
(ii) $r = \dfrac{ed}{1 + e \cos \theta}$	Vertical, d units right of the pole
(iii) $r = \dfrac{ed}{1 + e \sin \theta}$	Horizontal, d units above the pole
(iv) $r = \dfrac{ed}{1 - e \sin \theta}$	Horizontal, d units below the pole

In each of the cases (i) to (iv), the major axis of the ellipse, the axis of the parabola, or the transverse axis of the hyperbola is perpendicular to the directrix.

In Examples 3 and 4, identify and then graph the given conic.

Example 3 $r = \dfrac{12}{3 - 2 \cos \theta}$

Solution   Dividing numerator and denominator of the fraction by 3 gives

$$r = \frac{4}{1 - \frac{2}{3} \cos \theta} = \frac{\frac{2}{3}(6)}{1 - \frac{2}{3} \cos \theta}.$$

By Table 1, this is a polar equation of a conic with focus at the pole, a vertical directrix $d = 6$ units to the left of the pole, and eccentricity $e = \frac{2}{3}$. Since $e < 1$, the conic is an ellipse. By part (i) of Theorem 1,

$$a = \frac{ed}{1 - e^2} = \frac{4}{1 - \frac{4}{9}} = \frac{36}{5} \qquad \text{and} \qquad c = ae = \frac{36}{5} \cdot \frac{2}{3} = \frac{24}{5}.$$

Furthermore, since the curve is an ellipse,

$$b = \sqrt{a^2 - c^2} = \sqrt{\left(\frac{36}{5}\right)^2 - \left(\frac{24}{5}\right)^2} = \frac{\sqrt{36^2 - 24^2}}{5} = \frac{\sqrt{720}}{5} = \frac{12\sqrt{5}}{5}.$$

The pole O is the first focus. Since the directrix is vertical, the major axis of the ellipse is horizontal; hence, the center and the second focus lie along the polar axis. The center C is $c = \frac{24}{5}$ units to the right of O and the second focus is $c = \frac{24}{5}$ units to the right of C. Vertices V_1 and V_2 are $a = \frac{36}{5}$ units to the left and right of C and vertices V_3 and V_4 are $b = 12\sqrt{5}/5$ units below and above C (Figure 6). ■

Figure 6

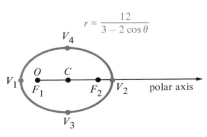

$r = \dfrac{12}{3 - 2 \cos \theta}$

polar axis

© **Example 4** $r = \dfrac{2\sqrt{3}}{1 + \sqrt{3}\,\sin\theta}$

Solution   The polar equation has the form

$$r = \frac{ed}{1 + e\,\sin\theta}$$

with $e = \sqrt{3}$ and $d = 2$; hence, by Table 1, its graph is a conic with eccentricity $\sqrt{3}$, focus F_1 at the origin, and directrix parallel to the polar axis and $d = 2$ units above it. Since $e > 1$, the conic is a hyperbola, and, since the directrix is horizontal, the transverse axis is vertical. By part (iii) of Theorem 1,

$$a = \frac{ed}{e^2 - 1} = \frac{\sqrt{3}(2)}{3 - 1} = \sqrt{3}$$

and $c = ae = \sqrt{3}\sqrt{3} = 3.$

Furthermore, since the curve is a hyperbola,

$$b = \sqrt{c^2 - a^2} = \sqrt{9 - 3} = \sqrt{6}.$$

The center C is $c = 3$ units above $F_1 = 0$ and the vertics V_1 and V_2 are $a = \sqrt{3}$ units below and above C, respectively (Figure 7). ∎

Figure 7

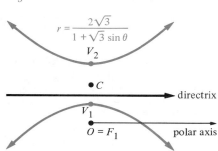

Parametric Equations

It is often useful to describe the location of a point (x, y) on a curve by writing x and y as functions of a third variable, say t, called a **parameter.** For instance, the equation

$$x^2 + y^2 = 1$$

of a circle can be rewritten as

$$\begin{cases} x = \cos t \\ y = \sin t, \end{cases}$$

where t is the parameter and $0 \le t \le 2\pi$. As t varies from 0 to 2π, the point $(x, y) = (\cos t, \sin t)$ traces out the circle $x^2 + y^2 = 1$.

In general, two equations

$$\begin{cases} x = f(t) \\ y = g(t), \end{cases}$$

used to express the coordinates of a point (x, y) on a curve as functions of a parameter t, are called **parametric equations** for the curve. Often, the parameter t is thought of as *time*, so that these equations describe how the x and y coordinates of a moving point vary in time as the point traces out the curve.

Example 5   Describe and sketch the curve with parametric equations

$$\begin{cases} x = 3 \cos t \\ y = 4 \sin t \end{cases} \quad \text{for } 0 \le t \le 2\pi.$$

Solution   We eliminate t by writing

$$\begin{cases} \dfrac{x}{3} = \cos t \\ \dfrac{y}{4} = \sin t, \end{cases}$$

from which it follows that

Figure 8

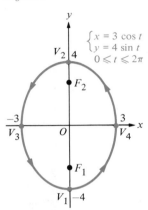

$$\left(\frac{x}{3}\right)^2 + \left(\frac{y}{4}\right)^2 = \cos^2 t + \sin^2 t = 1$$

or

$$\frac{x^2}{9} + \frac{y^2}{16} = 1.$$

Thus, the curve is an ellipse with center $(0,0)$, major axis on the y axis, and minor axis on the x axis (Figure 8). The condition $0 \le t \le 2\pi$ suggests that a point (x, y), tracing out the ellipse as t increases from 0 to 2π, starts at the point $(3 \cos 0, 4 \sin 0) = (3, 0)$, passes through the points $\left(3 \cos \dfrac{\pi}{2}, 4 \sin \dfrac{\pi}{2}\right) = (0, 4)$, $(3 \cos \pi, 4 \sin \pi) = (-3, 0)$, and $\left(3 \cos \dfrac{3\pi}{2}, 4 \sin \dfrac{3\pi}{2}\right) = (0, -4)$, and ends at the point $(3 \cos 2\pi, 4 \sin 2\pi) = (3, 0)$, thus tracing out the ellipse in a counterclockwise direction around the origin.   ▪

Problem Set 10.5

In Problems 1 to 6, find a Cartesian equation of the conic whose focus F is at the origin and whose eccentricity and directrix are given. Sketch the graph.

1. Eccentricity $e = \frac{2}{3}$, directrix $x = -\frac{5}{2}$.

2. Eccentricity $e = \frac{1}{2}$, directrix $x = -\frac{4}{5}$.

3. Eccentricity $e = 1$, directrix $x = -4$.

4. Eccentricity $e = 1$, directrix $x = -\frac{1}{3}$.

5. Eccentricity $e = 2$, directrix $x = -3$.

6. Eccentricity $e = \sqrt{5}$, directrix $x = -1$.

In Problems 7 to 16, find the eccentricity of each conic.

7. $\dfrac{x^2}{9} - \dfrac{y^2}{16} = 1$

8. $\dfrac{y^2}{4} - \dfrac{x^2}{2} = 1$

9. $2y^2 + 9x^2 = 18$

10. $\dfrac{(x + 1)^2}{9} + \dfrac{(y + 3)^2}{1} = 1$

11. $100x^2 + 36y^2 = 3600$

12. $(x - 3)^2 - 4(y + 2)^2 = 4$

13. $3x^2 - 5y^2 = 15$

14. $5x^2 - 4xy + 8y^2 = 144$

15. $y^2 - 3x = 0$

16. $16x^2 - 24xy + 9y^2 - 2x = 1$

C In Problems 17 to 26, identify each conic; find the eccentricity, the directrix (directrices), the center (if it exists), the focus (foci), and the vertex (vertices); and sketch the graph.

17. $r = \dfrac{16}{5 - 3 \cos \theta}$

18. $r = \dfrac{16}{5 + 3 \cos \theta}$

19. $r = \dfrac{6}{1 - \cos \theta}$

20. $r = \dfrac{24}{5 - 7 \cos \theta}$

21. $r = \dfrac{10}{1 - \sin \theta}$

22. $r = \dfrac{6}{1 - 2 \sin \theta}$

23. $r = \dfrac{1}{1 + 2 \sin \theta}$

24. $r = \dfrac{6}{10 + 5 \sin \theta}$

25. $r = \dfrac{4}{1 + \cos \theta}$

26. $r = \csc^2 \theta - \csc \theta \cot \theta$

In Problems 27 to 36, eliminate the parameter and write a Cartesian equation for the curve defined by the given parametric equations. Also, sketch the curve.

27. $\begin{cases} x = 3 \cos t \\ y = 3 \sin t \end{cases}$ for $0 \le t \le 2\pi$

28. $\begin{cases} x = 2t + 1 \\ y = 4t^2 - 1 \end{cases}$ for $-1 \le t \le 1$

29. $\begin{cases} x = t \\ y = 1 - t^2 \end{cases}$ for $-1 \le t \le 1$

30. $\begin{cases} x = -t \\ y = 1 - t^2 \end{cases}$ for $-1 \le t \le 1$

31. $\begin{cases} x = \cos t \\ y = 2 \sin t \end{cases}$ for $0 \le t \le 2\pi$

32. $\begin{cases} x = 2 + \cos t \\ y = 3 + \sin t \end{cases}$ for $0 \le t \le 2\pi$

33. $\begin{cases} x = 1 + \sin t \\ y = -1 + 2 \cos t \end{cases}$ for $0 \le t \le 2\pi$

34. $\begin{cases} x = \sin t \\ y = -\cos^2 t \end{cases}$ for $0 \le t \le \pi$

35. $\begin{cases} x = \sin t \\ y = \csc t \end{cases}$ for $0 < t < \pi$

36. $\begin{cases} x = e^t \\ y = e^{-2t} \end{cases}$ for $-\infty < t < \infty$

37. Using the equations in part (i) of Theorem 1, show that the eccentricity e of an ellipse is given by the ratio of the distance between the two foci to the length of the major axis.

38. By symmetry, it is clear that an ellipse has *two* directrices, one corresponding to each of its two foci (Figure 9). Using the equations in part (i) of Theorem 1, show that the distance between these two directrices is $2a^2/c$ units.

Figure 9

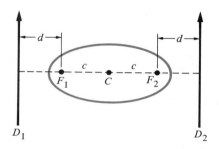

39. Using the equations in part (i) of Theorem 1, show that the eccentricity e of an ellipse is given by $e = \sqrt{1 - (b/a)^2}$.

40. By symmetry, it is clear that a hyperbola has *two* directrices, one corresponding to each of its two foci (Figure 10). Using the equations in part (iii) of Theorem 1, show that the distance between these two directrices is $2a^2/c$ units.

Figure 10

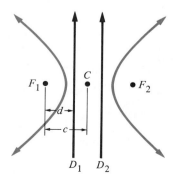

C **41.** Except for minor perturbations, the orbit of the earth is an ellipse with the sun at one focus. The least and greatest distances (at *perihelion* and *aphelion*) between the earth and the sun have a ratio of approximately $\frac{29}{30}$. Find the approximate eccentricity of the earth's orbit.

42. Let the focus F of a conic with eccentricity $e > 0$ be the origin O of a Cartesian coordinate system, and suppose that the directrix is $x = -d$, where $d > 0$. Using Definition 1 directly, prove that $(1 - e^2)x^2 - (2e^2 d)x + y^2 = e^2 d^2$ is a Cartesian equation of the conic. Using this result, prove Theorem 1.

REVIEW PROBLEM SET, CHAPTER 10

In Problems 1 to 4, determine whether the graph of the equation is a circle or an ellipse. If it is a circle, find the radius. If it is an ellipse, find the vertices. Sketch the graph.

1. $9x^2 + 9y^2 = 1$

2. $x^2 + \dfrac{y^2}{4} = 1$

3. $\dfrac{x^2}{16} + \dfrac{y^2}{9} = 1$

4. $3x^2 + 2y^2 = 1$

In Problems 5 to 8, find the equation in standard form for the ellipse that satisfies the indicated conditions.

5. Center at $(0, 0)$; horizontal major axis of length 16; minor axis of length 8.

6. Vertices $(0, 0)$, $(6, 0)$, $(3, -5)$, and $(3, 5)$.

7. Vertices $(-3, 1)$, $(5, 1)$, $(1, -4)$, and $(1, 6)$.

8. Center $(0, 0)$; major axis horizontal; ellipse contains the points $(4, 3)$ and $(6, 2)$.

In Problems 9 to 14, find the center and the vertices of each ellipse and sketch the graph.

9. $\dfrac{x^2}{8} + \dfrac{y^2}{12} = 1$

10. $144x^2 + 169y^2 = 24{,}336$

11. $9x^2 + 25y^2 + 18x - 50y = 191$

12. $3x^2 + 4y^2 - 28x - 16y + 48 = 0$

13. $9x^2 + 4y^2 + 72x - 48y + 144 = 0$

14. $9x^2 + 4y^2 + 36x - 24y = 252$

In Problems 15 to 20, find the vertex of each parabola, determine whether its axis is horizontal or vertical, and sketch the graph.

15. $y^2 = 4x$

16. $-x^2 = 5y$

17. $x^2 = -4(y - 1)$

18. $x^2 = 8(y + 1)$

19. $x + 8 - y^2 + 2y = 0$

20. $x + 4 + 2y = y^2$

In Problems 21 and 22, find the equation in standard form for the parabola that satisfies the indicated conditions.

21. Vertex at $(0, -6)$; horizontal axis; parabola contains the point $(-9, -3)$.

22. Vertical axis; parabola contains the points $(9, -1)$, $(3, -4)$, and $(-9, 8)$.

In Problems 23 to 28, find the center, the asymptotes, and the vertices of each hyperbola, and sketch the graph.

23. $x^2 - 9y^2 = 72$

24. $y^2 - 9x^2 = 54$

25. $x^2 - 4y^2 + 4x + 24y - 48 = 0$

26. $16x^2 - 9y^2 = 96x$

27. $4y^2 - x^2 - 24y + 2x + 34 = 0$

28. $11y^2 - 66y - 25x^2 + 100x = 276$

In Problems 29 and 30, find the equation in standard form for the hyperbola that satisfies the indicated conditions.

29. Asymptotes $y = -2x$ and $y = 2x$; hyperbola contains the point $(1, 1)$.

30. Vertices $(2, -8)$ and $(2, 2)$; asymptotes $25x - 9y = 77$ and $25x + 9y = 23$.

In Problems 31 to 36, rewrite each equation in standard form and identify the conic section.

31. $16x^2 + y^2 - 32x + 4y - 44 = 0$

32. $x^2 + y^2 - 4x + 2y = 4$

33. $4x^2 + 4y^2 + 4x - 4y + 1 = 0$

34. $4x^2 + 9y^2 + 16x = 18y + 11$

35. $4y^2 - x^2 - 8y + 2x + 7 = 0$

36. $9x^2 + 25y^2 + 18x = 50y + 191$

37. A point $P = (x, y)$ moves so that the product of the slopes of the line segments $\overline{PQ}$ and $\overline{PR}$ is -6, where $Q = (3, -2)$ and $R = (-2, 1)$. Find an equation of the curve traced out by P, identify the curve, and sketch it.

38. The cable of a suspension bridge assumes the shape of a parabola if the weight of the suspended roadbed (together with that of the cable) is uniformly distributed horizontally. Suppose that the towers of such a bridge are 240 meters apart and 60 meters high and that the lowest point of the cable is 20 meters above the roadway. Find the vertical distance from roadway to the cable at intervals of 20 meters.

In Problems 39 and 40, assume that a new $\bar{x}\bar{y}$ coordinate system is obtained by translating the old xy coordinate system so that the origin $\bar{O}$ of the new system has old coordinates $(3, -2)$.

39. Find the new $\bar{x}\bar{y}$ coordinates of the points whose old xy coordinates are:

(a) $(0, 0)$ (b) $(5, -7)$ (c) $(4, 4)$

(d) $(3, -2)$ (e) $(0, 1)$ (f) $(-1, 0)$.

40. Find the old xy coordinates of the points whose new $\bar{x}\bar{y}$ coordinates are:

(a) $(0, 0)$ (b) $(-3, 2)$ (c) $(3, 10)$

(d) $(-2, 3)$ (e) $(5, 0)$ (f) $(0, -7)$.

In Problems 41 to 44, find a translation of axes $x = \bar{x} + h$ and $y = \bar{y} + k$ that will reduce each equation to an equation involving no first-degree terms in $\bar{x}$ or $\bar{y}$. Identify the graph.

41. $4x^2 + 4y^2 + 8x - 4y + 1 = 0$

42. $x^2 - 4y^2 + 6x + 24y = 31$

43. $9x^2 - 4y^2 - 54x + 45 = 0$

44. $2x^2 + 8y^2 - 8x - 16y + 9 = 0$

In Problems 45 to 50, the old xy axis has been rotated counterclockwise about the origin through the angle ϕ to form a new $\bar{x}\bar{y}$ coordinate system. The point P has coordinates (x, y) in the old system and $(\bar{x}, \bar{y})$ in the new system. Supply the missing information.

45. $(x, y) = (-4, 6)$, $\phi = 30°$, $(\bar{x}, \bar{y}) = ?$

46. $(x, y) = (-1, 4)$, $\phi = -\pi/2$, $(\bar{x}, \bar{y}) = ?$

47. $(\bar{x}, \bar{y}) = (2, -4)$, $\phi = \pi/3$, $(x, y) = ?$

48. $(\bar{x}, \bar{y}) = (-2, -6)$, $\phi = 240°$, $(x, y) = ?$

49. $(x, y) = (4, -4)$, $\phi = 300°$, $(\bar{x}, \bar{y}) = ?$

50. $(x, y) = (-3, -4)$, $(\bar{x}, \bar{y}) = (-4, 3)$, $0 < \phi < \pi$, $\phi = ?$

In Problems 51 to 62, (a) find an angle ϕ through which to rotate the coordinate system so as to remove the mixed term from the equation, (b) find x and y in terms of $\bar{x}$ and $\bar{y}$, (c) find the new equation in terms of $\bar{x}$ and $\bar{y}$, and (d) sketch the graph.

51. $4x^2 + 4xy + 4y^2 = 24$

52. $3x^2 + 4\sqrt{3}xy - y^2 = 15$

53. $\sqrt{3}x^2 + xy = 11$ **54.** $y^2 + xy + 2x^2 = 7$

55. $13x^2 + 6\sqrt{3}xy + 7y^2 = 16$

56. $x^2 + 6xy + y^2 + 8 = 0$

57. $x^2 - 3xy + 5y^2 = 16$

58. $7x^2 + 2\sqrt{3}xy + 5y^2 - 1 = 0$

59. $41x^2 - 24xy + 34y^2 - 25 = 0$

60. $9x^2 + 24xy + 16y^2 + 80x - 60y = 0$

61. $x^2 + 4xy + 4y^2 = 9$

62. $x^2 + 6xy + 9y^2 - 30x + 10y = 0$

63. For a main cable in a suspension bridge, the dimensions H and L shown in Figure 1 are called the *sag* and the *span*, respectively. If the weight of the suspended roadbed (together with that of the cables) is uniformly distributed horizontally, the main cables form parabolas. Taking the origin of an xy

Figure 1

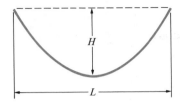

coordinate system at the lowest point of a parabolic main cable with sag H and span L, find an equation of the cable.

64. Two points are 800 meters apart. At one of these points the report of a cannon is heard 1 second later than at the other. Show that the cannon is somewhere on a certain hyperbola, and write an equation for the hyperbola after making a suitable choice of axes. (Consider the velocity of sound to be 332 meters per second.)

In Problems 65 to 70, identify the type of conic, give its eccentricity, and sketch the graph.

65. $r = \dfrac{17}{1 - \cos \theta}$

66. $r = \dfrac{15}{3 + 5 \sin \theta}$

67. $r = \dfrac{10}{5 - 2 \sin \theta}$

68. $r = \dfrac{3}{2} \csc^2 \dfrac{\theta}{2}$

69. $r = \dfrac{1}{\frac{1}{2} + \sin \theta}$

70. $r = \dfrac{1}{1 + \cos (\theta + \pi/4)}$

[c] **71.** The Yale Bowl, completed in 1914, is a football stadium modeled after the Colosseum in Rome. The concrete stands surround a central region in the shape of an ellipse with a major axis 148 meters

long and a minor axis 84 meters long. Find the eccentricity of this ellipse.

Yale Bowl

[c] **72.** The Ellipse in front of the White House in Washington, D.C., has a semimajor axis of 229 meters and a semiminor axis of 195 meters. What is its eccentricity?

In Problems 73 to 76, eliminate the parameter and write a Cartesian equation for the curve defined by the given parametric equations. Also sketch the curve.

73. $\begin{cases} x = 1 + \cos t \\ y = 1 - \sin t \end{cases}$ for $0 \le t \le 2\pi$

74. $\begin{cases} x = \cos t \\ y = 2 \sin t \end{cases}$ for $0 \le t \le 2\pi$

75. $\begin{cases} x = \cos t \\ y = 1 - \sin^2 t \end{cases}$ for $0 \le t \le \pi$

76. $\begin{cases} x = 2^t \\ y = 4^{-2t} \end{cases}$ for $-\infty < t < \infty$

Chapter 11

Additional Topics in Algebra

In this chapter, we give a brief introduction to several topics in algebra that extend, supplement, and round out the ideas presented in earlier chapters. These topics include mathematical induction, the binomial theorem, sequences, series, permutations and combinations, and probability. Here we can only give you the "flavor" of these topics—we urge you to consult more advanced textbooks for a more detailed and systematic presentation.

11.1 MATHEMATICAL INDUCTION

The principle of *mathematical induction* is often illustrated by a row of oblong wooden blocks, called dominoes, set on end as in Figure 1. If we are guaranteed two conditions,

(i) that the first domino is toppled over, and

(ii) that if any domino topples over, it will hit the next one and topple it over,

then we can be certain that *all* the dominoes will topple over.

Figure 1

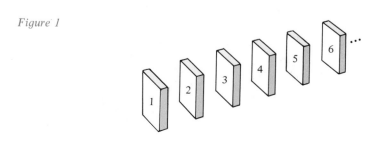

The principle of mathematical induction concerns a sequence of statements

$$S_1, S_2, S_3, S_4, S_5, S_6, \ldots.$$

Let's think of this sequence as corresponding to a row of dominoes; the first statement S_1 corresponds to the first domino, the second statement S_2 corresponds to the second domino, and so on. Each statement can be either true or false. If a statement proves to be true, let's think of the corresponding domino toppling over. The two conditions given above for the row of dominoes can then be interpreted as follows:

(i) The first statement S_1 is true.

(ii) If any statement S_k is true, then the next statement S_{k+1} is also true.

If these two conditions hold, then, by analogy with the toppling over of *all* the dominoes, we conclude that *all* statements $S_1, S_2, S_3, \ldots$ are true. That this argument is valid is the **principle of mathematical induction.**

Although the principle of mathematical induction can be established formally, the required argument is beyond the scope of this book. We ask you to accept this principle on the intuitive basis of the domino analogy. Thus, we have the following.

Procedure for Making a Proof by Mathematical Induction

Begin by clearly identifying the sequence $S_1, S_2, S_3, \ldots$ of statements to be proved. This is usually done by specifying the meaning of S_n, where n denotes an arbitrary positive integer, $n = 1, 2, 3, \ldots$. Then carry out the following two steps:

> **Step 1.** Show that S_1 is true.
>
> **Step 2.** Let k denote an arbitrary positive integer. Assume that S_k is true and show, on the basis of this assumption, that S_{k+1} is also true.

If both steps can be carried out, conclude that S_n is true for all positive integer values of n.

The assumption in step 2 that S_k is true is called the **induction hypothesis.** When you make the induction hypothesis, you're not saying that S_k is *in fact* true—you're just *supposing* that it is true to see if the truth of S_{k+1} follows from this supposition. It's as if you are checking to make sure that if the kth domino *were* to topple over, it *would* knock over the next domino.

In Examples 1 and 2, use mathematical induction to prove each result.

Example 1   For every positive integer n, $1 + 2 + 3 + \cdots + n = \dfrac{n(n+1)}{2}$.

Solution   Let S_n be the statement

$$S_n: \quad 1 + 2 + 3 + \cdots + n = \frac{n(n + 1)}{2}.$$

In other words, S_n asserts that the sum of the first n positive integers is $n(n + 1)/2$. For instance,

$$S_2 \text{ says that} \quad 1 + 2 = \frac{2(2 + 1)}{2} = 3$$

and

$$S_{20} \text{ says that} \quad 1 + 2 + 3 + \cdots + 20 = \frac{20(20 + 1)}{2} = 210.$$

Step 1. The statement

$$S_1: \quad 1 = \frac{1(1 + 1)}{2}$$

is clearly true.

Step 2. Here we must deal with the statements

$$S_k: \quad 1 + 2 + 3 + \cdots + k = \frac{k(k + 1)}{2}$$

and $$S_{k+1}: \quad 1 + 2 + 3 + \cdots + (k + 1) = \frac{(k + 1)[(k + 1) + 1]}{2}$$

obtained by replacing n in S_n by k and then by $k + 1$. The statement S_k is our induction hypothesis, and we assume, for the sake of argument, that it is true. Our goal is to prove that, on the basis of this assumption, S_{k+1} is true. To this end, we add $k + 1$ to both sides of the equation expressing S_k to obtain

$$1 + 2 + 3 + \cdots + k + (k + 1) = \frac{k(k + 1)}{2} + (k + 1)$$

$$= \frac{k(k + 1) + 2(k + 1)}{2}$$

$$= \frac{(k + 1)(k + 2)}{2}$$

$$= \frac{(k + 1)[(k + 1) + 1]}{2},$$

which is the equation expressing S_{k+1}. This completes the proof by mathematical induction.   ∎

Example 2   If a principal of P dollars is invested at a compound interest rate R per conversion period, the final value F dollars of the investment at the end of n conversion periods is given by $F = P(1 + R)^n$. [For *compound interest*, the interest is periodically calculated and added to the principal. The time interval between successive conversions of interest into principal is called the *conversion period*.]

Solution   Let S_n be the statement

$$S_n: \quad F = P(1 + R)^n \text{ at the end of } n \text{ conversion periods.}$$

Step 1. The statement

$$S_1: \quad F = P(1 + R) \text{ at the end of the first conversion period}$$

is true, because the interest on P dollars for one conversion period is PR dollars, so the final value of the investment at the end of the first conversion period is

$$P + PR = P(1 + R) \text{ dollars.}$$

Step 2. Assume that the statement

$$S_k: \quad F = P(1 + R)^k \text{ at the end of } k \text{ conversion periods}$$

is true. It follows that, over the $(k + 1)$st conversion period, interest at rate R is paid on $P(1 + R)^k$ dollars. Adding this interest, $P(1 + R)^k R$ dollars, to the value of the investment at the beginning of the $(k + 1)$st conversion period, we find that the value of the investment at the end of the $(k + 1)$st conversion period is given by

$$F = P(1 + R)^k + P(1 + R)^k R$$
$$= P(1 + R)^k (1 + R)$$
$$= P(1 + R)^{k+1} \text{ dollars.}$$

Hence, S_{k+1} is true, and the proof by mathematical induction is complete.   ∎

Problem Set 11.1

In each problem set, problems with colored numbers constitute a good representation of the main ideas of the section.

In Problems 1 to 14, use mathematical induction to prove that the assertion is true for all positive integers n.

1. $1 + 3 + 5 + \cdots + (2n - 1) = n^2$; in other words, the sum of the first n odd positive integers is n^2.

2. $2 + 4 + 6 + \cdots + 2n = n(n + 1)$; in other words, the sum of the first n even positive integers is $n(n + 1)$.

3. $1^2 + 2^2 + 3^2 + \cdots + n^2 = \frac{1}{6}n(n + 1)(2n + 1)$; that

is, the sum of the first n perfect squares is $\frac{1}{6}n(n + 1)(2n + 1)$.

4. $1 + 5 + 9 + \cdots + (4n - 3) = n(2n - 1)$

5. $1^2 + 3^2 + 5^2 + \cdots + (2n - 1)^2 = \frac{1}{3}n(2n - 1)(2n + 1)$

6. $1^3 + 2^3 + 3^3 + \cdots + n^3 = \left[\dfrac{n(n + 1)}{2}\right]^2$

7. $1 \cdot 2 + 2 \cdot 3 + 3 \cdot 4 + \cdots + n(n + 1) = \frac{1}{3}n(n + 1)(n + 2)$

8. $\dfrac{1}{1 \cdot 2} + \dfrac{1}{2 \cdot 3} + \dfrac{1}{3 \cdot 4} + \cdots + \dfrac{1}{n(n + 1)} = \dfrac{n}{n + 1}$

9. $(ab)^n = a^n b^n$

10. If $h \geq 0$, then $1 + nh \leq (1 + h)^n$.

11. $2 + 2^2 + 2^3 + \cdots + 2^n = 2(2^n - 1)$

12. $r + r^2 + r^3 + \cdots + r^n = \dfrac{r(r^n - 1)}{r - 1}$

13. $\cos(n\pi) = (-1)^n$

14. De Moivre's theorem:

$$[r(\cos\theta + i\sin\theta)]^n = r^n(\cos n\theta + i\sin n\theta)$$

C 15. Using a calculator, verify the identities in odd Problems 1 to 7 for $n = 15$ by directly calculating both sides.

16. The domino theory was a tenet of U.S. foreign policy subscribed to by the administrations of Presidents Eisenhower, Kennedy, Johnson, and Nixon. Explain the connection, if any, between the domino theory and mathematical induction.

11.2 THE BINOMIAL THEOREM

The principle of mathematical induction can be used to establish a general formula for the expansion of $(a + b)^n$, where the exponent n is an arbitrary positive integer. The expression $a + b$ is a binomial, so the formula is called the **binomial theorem.**

Of course, the expansion of $(a + b)^n$ for small values of n can be obtained by direct calculation. For instance,

$$(a + b)^1 = a + b$$
$$(a + b)^2 = a^2 + 2ab + b^2$$
$$(a + b)^3 = a^3 + 3a^2b + 3ab^2 + b^3$$
$$(a + b)^4 = a^4 + 4a^3b + 6a^2b^2 + 4ab^3 + b^4$$
$$(a + b)^5 = a^5 + 5a^4b + 10a^3b^2 + 10a^2b^3 + 5ab^4 + b^5,$$

and so forth.

Notice that we have written the expansions in descending powers of a and ascending powers of b. A certain pattern is already apparent. Indeed, in the expansion of $(a + b)^n$:

1. There are $n + 1$ terms, beginning with a^n and ending with b^n.

2. As we move from each term to the next, powers of a decrease by 1 and powers of b increase by 1. Therefore, in each term, the exponents of a and b add up to n.

3. The successive terms can be written in the form

$$(\text{numerical coefficient})a^{n-j}b^j$$

for $j = 0, 1, 2, \ldots, n$.

That statements 1 to 3 are true for every positive integer n is part of the binomial theorem. The remaining part concerns the values of the numerical coefficients, which are called the **binomial coefficients.** In order to obtain a useful formula for the binomial coefficients, we begin by introducing *factorial notation*.

Definition 1   **Factorial Notation**

> If n is a positive integer, we define $n!$, read n **factorial,** by
>
> $$n! = n(n-1)(n-2)\cdots 3 \cdot 2 \cdot 1.$$
>
> We also define
>
> $$0! = 1.$$

In words, *the factorial of a positive integer n is the product of n and all smaller positive integers.* For instance,

$$7! = 7 \cdot 6 \cdot 5 \cdot 4 \cdot 3 \cdot 2 \cdot 1 = 5040.$$

Table 1

$0! = 1$
$1! = 1$
$2! = 2$
$3! = 6$
$4! = 24$
$5! = 120$
$6! = 720$
$7! = 5040$

The special definition $0! = 1$ extends the usefulness of the factorial notation. As a direct consequence of Definition 1, we have the important **recursion formula** for factorials

$$(n + 1)! = (n + 1)n!,$$

which holds for every integer $n \geq 0$ (Problem 35). Using the recursion formula, you can make a table of values of $n!$ (Table 1). The table can be continued indefinitely; for instance, to obtain the next entry, use the recursion formula as follows:

$$8! = 8(7!) = 8(5040) = 40,320.$$

The factorials increase very rapidly. Indeed, 10! is over 3 **million,** and 70! is beyond the range of most calculators.

Using factorial notation, we now introduce special symbols for what, as we shall soon see, are the binomial coefficients.

Definition 2   **The Binomial Coefficients**

> If n and j are integers with $n \geq j \geq 0$, the symbol $\dbinom{n}{j}$ is defined as follows:
>
> $$\binom{n}{j} = \frac{n!}{(n-j)!\,j!}.$$

Example 1   Evaluate

(a) $\dbinom{5}{2}$ **(b)** $\dbinom{7}{6}$ **(c)** $\dbinom{n}{0}$ **(d)** $\dbinom{n}{n}$,

if n is a nonnegative integer.

Solution   Using Definition 2, we have:

(a) $\dbinom{5}{2} = \dfrac{5!}{(5-2)!2!} = \dfrac{5!}{3!2!} = \dfrac{5 \cdot 4 \cdot 3 \cdot 2 \cdot 1}{(3 \cdot 2 \cdot 1)(2 \cdot 1)} = \dfrac{5 \cdot 4}{2} = 10$

(b) $\dbinom{7}{6} = \dfrac{7!}{(7-6)!6!} = \dfrac{7!}{1!6!} = \dfrac{7!}{6!} = \dfrac{7(6!)}{6!} = 7$

(c) $\dbinom{n}{0} = \dfrac{n!}{(n-0)!0!} = \dfrac{n!}{n!0!} = \dfrac{1}{0!} = \dfrac{1}{1} = 1$

(d) $\dbinom{n}{n} = \dfrac{n!}{(n-n)!n!} = \dfrac{1}{0!} = \dfrac{1}{1} = 1.$

Example 2   If k and j are integers with $k \geq j > 0$, show that

$$\binom{k}{j-1} + \binom{k}{j} = \binom{k+1}{j}.$$

Solution   Using Definition 2 and the recursion formula for factorials, we have

$$\binom{k}{j-1} + \binom{k}{j} = \frac{k!}{[k-(j-1)]!(j-1)!} + \frac{k!}{(k-j)!j!}$$

$$= \frac{jk!}{(k-j+1)!j(j-1)!} + \frac{(k-j+1)k!}{(k-j+1)(k-j)!j!}$$

$$= \frac{jk!}{(k-j+1)!j!} + \frac{(k+1)k! - jk!}{(k-j+1)!j!}$$

$$= \frac{(k+1)k!}{(k-j+1)!j!} = \frac{(k+1)!}{[(k+1)-j]!j!}$$

$$= \binom{k+1}{j}.$$

We can now state and prove the binomial theorem.

Theorem 1   **The Binomial Theorem**

> If n is a positive integer and if a and b are any two numbers, then
>
> $$(a+b)^n = \binom{n}{0}a^n + \binom{n}{1}a^{n-1}b + \cdots + \binom{n}{j}a^{n-j}b^j + \cdots + \binom{n}{n}b^n.$$

Proof   The proof is by mathematical induction. Thus, for each positive integer n, let S_n be the statement

$$S_n: \quad (a+b)^n = \binom{n}{0}a^n + \binom{n}{1}a^{n-1}b + \cdots + \binom{n}{j}a^{n-j}b^j + \cdots + \binom{n}{n}b^n.$$

Step 1. The statement

$$S_1: \quad (a+b)^1 = \binom{1}{0}a^1 + \binom{1}{1}b^1$$

is true because $\dbinom{1}{0} = 1$ and $\dbinom{1}{1} = 1.$

Step 2. Assume the induction hypothesis

$$S_k: \quad (a+b)^k = \binom{k}{0}a^k + \binom{k}{1}a^{k-1}b + \cdots + \binom{k}{j}a^{k-j}b^j + \cdots + \binom{k}{k}b^k,$$

where k is a positive integer. Notice that the term immediately preceding $\binom{k}{j}a^{k-j}b^j$ in this equation is $\binom{k}{j-1}a^{k-(j-1)}b^{j-1}$; that is, $\binom{k}{j-1}a^{k-j+1}b^{j-1}$. We use this fact in the following computation:

$(a+b)^{k+1}$

$$= (a+b)(a+b)^k = a(a+b)^k + b(a+b)^k$$

$$= a\left[\binom{k}{0}a^k + \binom{k}{1}a^{k-1}b + \cdots + \binom{k}{j}a^{k-j}b^j + \cdots + \binom{k}{k}b^k\right]$$

$$+ b\left[\binom{k}{0}a^k + \binom{k}{1}a^{k-1}b + \cdots + \binom{k}{j-1}a^{k-j+1}b^{j-1} + \cdots + \binom{k}{k}b^k\right]$$

$$= \binom{k}{0}a^{k+1} + \binom{k}{1}a^k b + \cdots + \binom{k}{j}a^{k-j+1}b^j + \cdots + \binom{k}{k}ab^k$$

$$+ \binom{k}{0}a^k b + \cdots + \binom{k}{j-1}a^{k-j+1}b^j + \cdots + \binom{k}{k-1}ab^k + \binom{k}{k}b^{k+1}$$

$$= \binom{k}{0}a^{k+1} + \left[\binom{k}{1}+\binom{k}{0}\right]a^k b + \cdots + \left[\binom{k}{j}+\binom{k}{j-1}\right]a^{k+1-j}b^j$$

$$+ \cdots + \left[\binom{k}{k}+\binom{k}{k-1}\right]ab^k + \binom{k}{k}b^{k+1}.$$

This expression of $(a+b)^{k+1}$ will fulfill the requirements of statement S_{k+1}, provided that the following three conditions hold:

(i) $\binom{k}{0} = \binom{k+1}{0}$, **(ii)** $\binom{k}{j}+\binom{k}{j-1} = \binom{k+1}{j}$, **(iii)** $\binom{k}{k} = \binom{k+1}{k+1}$.

That (i) and (iii) hold is left as an exercise (Problem 36). We just proved (ii) in Example 2 above. Hence S_{k+1} is established, and the proof by induction is complete. ∎

By arranging the binomial coefficients in a triangle as shown in Figure 1a, we obtain the **Pascal triangle,** named in honor of its discoverer, the French mathematician Blaise Pascal (1623–1662). The numerical values of the entries in the first six horizontal rows of the Pascal triangle are shown in Figure 1b. Notice that:

> Each number in the Pascal triangle (other than those on the border, which are all equal to 1) can be obtained by adding the two numbers diagonally above it.

This is called the **additive property** of the Pascal triangle, and it is a consequence of the identity established in Example 2.

Figure 1

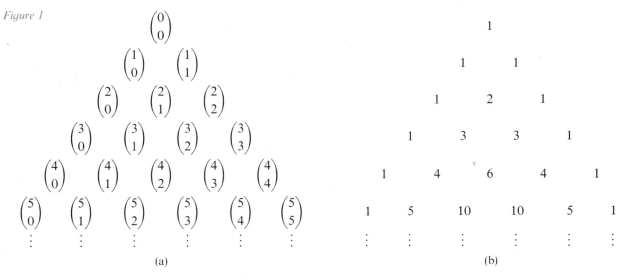

(a)   (b)

Example 3   Find the numerical entries in the seventh horizontal row of the Pascal triangle.

Solution   The seventh horizontal row is obtained from the sixth horizontal row in Figure 1b by using the additive property.

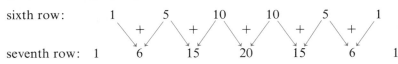

According to the binomial theorem:

> The binomial coefficients in the expansion of $(a + b)^n$ appear in row number $n + 1$ of the Pascal triangle.

For instance, using the result of Example 3, we have

$$(a + b)^6 = 1a^6 + 6a^5b + 15a^4b^2 + 20a^3b^3 + 15a^2b^4 + 6ab^5 + 1b^6.$$

In Examples 4 and 5, use the binomial theorem and the Pascal triangle to expand each of the given expressions.

Example 4 $(2x + y)^4$

Solution   In the binomial theorem (Theorem 1), let $a = 2x$, $b = y$, and $n = 4$. The appropriate binomial coefficients $(1, 4, 6, 4, 1)$ are found in the *fifth* horizontal row of the Pascal triangle (Figure 1b), since $n + 1 = 4 + 1 = 5$. Therefore,

$$(2x + y)^4 = 1(2x)^4 + 4(2x)^3y + 6(2x)^2y^2 + 4(2x)y^3 + 1y^4$$
$$= 16x^4 + 32x^3y + 24x^2y^2 + 8xy^3 + y^4.$$

Example 5 $(x - y)^5$

Solution   In the binomial theorem, let $a = x$, $b = -y$, and $n = 5$. The appropriate binomial coefficients $(1, 5, 10, 10, 5, 1)$ are found in the *sixth* horizontal row of the Pascal triangle. Thus,

$$(x - y)^5 = 1x^5 + 5x^4(-y) + 10x^3(-y)^2 + 10x^2(-y)^3 + 5x(-y)^4 + 1(-y)^5$$
$$= x^5 - 5x^4 y + 10x^3 y^2 - 10x^2 y^3 + 5xy^4 - y^5.$$

■

According to the binomial theorem (Theorem 1), the $(j + 1)$st term in the expansion of $(a + b)^n$ is given by the formula

$$(j + 1)\text{st term} = \binom{n}{j} a^{n-j} b^j \quad \text{for} \quad 0 \le j \le n.$$

You can use this formula to find a specified term in a binomial expansion.

Example 6   Find the tenth term in the expansion of $(x - \frac{1}{2}y)^{12}$.

Solution   The tenth term is obtained by putting $j = 9$, $a = x$, $b = -\frac{1}{2}y$, and $n = 12$ in the formula above. Thus,

$$10\text{th term} = \binom{12}{9} a^{12-9} b^9$$

$$= \binom{12}{9} x^3 (-\tfrac{1}{2}y)^9$$

$$= -\binom{12}{9} \frac{x^3 y^9}{2^9}.$$

Now,

$$\binom{12}{9} = \frac{12!}{(12 - 9)!9!} = \frac{12!}{3!9!} = \frac{12 \cdot 11 \cdot 10(9!)}{3!9!} = \frac{12 \cdot 11 \cdot 10}{6} = 220,$$

and

$$2^9 = 512,$$

so we have

$$10\text{th term} = -220 \frac{x^3 y^9}{512} = -\frac{55}{128} x^3 y^9.$$

■

Example 7   Find the term involving y^4 in the expansion of $(x + y)^8$.

Solution   With $a = x$, $b = y$, and $n = 8$, we have

$$(j + 1)\text{st term} = \binom{n}{j} a^{n-j} b^j = \binom{8}{j} x^{8-j} y^j.$$

Hence, the term involving y^4, obtained by putting $j = 4$, is the

$$\text{5th term} = \binom{8}{4} x^{8-4} y^4 = \binom{8}{4} x^4 y^4.$$

Now,

$$\binom{8}{4} = \frac{8!}{(8-4)!4!} = \frac{8!}{4!4!} = \frac{8 \cdot 7 \cdot 6 \cdot 5(4!)}{4!4!} = \frac{8 \cdot 7 \cdot 6 \cdot 5}{4 \cdot 3 \cdot 2 \cdot 1} = 2 \cdot 7 \cdot 5 = 70.$$

Therefore, the term involving y^4 is the

$$\text{5th term} = 70 x^4 y^4.$$

Problem Set 11.2

In Problems 1 to 10, find the numerical value of each expression.

1. $9!$

2. $10!$

3. $\dfrac{7!}{5!3!}$

4. $\dfrac{8!6!}{5!3!}$

5. $\dbinom{7}{3}$

6. $\dbinom{12}{10}$

7. $\dbinom{5}{5}$

8. $\dbinom{12}{2}$

9. $\dbinom{6}{3}$

10. $\dbinom{50}{49}$

11. Using the additive property of the Pascal triangle and the fact that the numerical entries in the seventh horizontal row are $(1, 6, 15, 20, 15, 6, 1)$, find the numerical entries in the eighth row.

12. Construct a Pascal triangle showing the numerical entries in the first ten horizontal rows.

In Problems 13 to 24, use the binomial theorem and Pascal triangle to expand each expression.

13. $(a + 3x)^4$

14. $(3a - 2b)^5$

15. $(x - y)^5$

16. $(x + y^2)^5$

17. $(c + 2)^6$

18. $(r - 3s)^5$

19. $(1 - c^3)^7$

20. $(b^2 + 10)^4$

21. $(\sqrt{x} - \sqrt{y})^6$

22. $\left(\dfrac{1}{x} + \dfrac{1}{y}\right)^6$

23. $\left(\dfrac{x}{2} - 2y\right)^6$

24. $(1 - 1)^6$

In Problems 25 to 32, find and simplify the specified term in the binomial expansion of the indicated expression.

25. The third term of $(s - t)^5$.

26. The fourth term of $(a^2 - b)^6$.

27. The fifth term of $\left(2x^2 + \dfrac{y^3}{4}\right)^{10}$.

28. The sixth term of $(\pi - \sqrt{2})^9$.

29. The term involving y^4 in $(x + y)^5$.

30. The term involving b^6 in $(a - 3b^3)^4$.

31. The term involving c^6 in $(c + 2d)^{10}$.

32. The term involving x^3 in $(\sqrt{x} + \sqrt{y})^8$.

33. Use mathematical induction to show that $2^{n+3} < (n + 3)!$ for all positive integers n.

C **34.** Using the recursion formula and a calculator, extend Table 1 to show values of $n!$ for $n = 0, 1, 2, \ldots, 15$.

35. Verify the recursion formula for factorials: $(n + 1)! = (n + 1)n!$ for every nonnegative integer n.

36. If k is a nonnegative integer, show that $\binom{k}{0} = \binom{k+1}{0}$ and that $\binom{k}{k} = \binom{k+1}{k+1}$.

37. Show that the Pascal triangle is symmetric about a vertical line through its apex by showing that for integers n and j with $n \geq j \geq 0$, $\binom{n}{j} = \binom{n}{n-j}$.

38. If n and j are integers with $n \geq j > 0$, show that

$$\binom{n}{j} = \frac{\overbrace{n(n-1) \cdots (n-j+1)}^{j \text{ factors}}}{\underbrace{j(j-1) \cdots 1}_{j \text{ factors}}}.$$

39. If n is a positive integer, show that

$$(1+x)^n = 1 + nx + \binom{n}{2}x^2 + \cdots + \binom{n}{j}x^j$$
$$+ \cdots + x^n.$$

40. Show that the sum of all the entries in row number $n + 1$ of the Pascal triangle is 2^n; that is,

$$\binom{n}{0} + \binom{n}{1} + \binom{n}{2} + \cdots + \binom{n}{n} = 2^n.$$

[*Hint:* Use the binomial theorem to expand $(1 + 1)^n$.]

41. If, in a term of the binomial expansion of $(a + b)^n$, we multiply the coefficient by the exponent of a and divide by the number of the term, show that we obtain the coefficient of the next term.

© **42. Stirling's approximation,**

$$n! \approx \sqrt{2n\pi}\left(\frac{n}{e}\right)^n\left(1 + \frac{1}{12n - 1}\right),$$

is often used to approximate $n!$ for large values of n.

(a) Use Stirling's approximation to estimate 15!, and compare the result with the true value.

(b) Use Stirling's approximation to estimate 50!.

11.3 SEQUENCES

A **sequence** is a function

$$n \longmapsto a_n$$

whose domain is the set of positive integers. Following tradition, we write the function values as a_n rather than as $a(n)$, and call them the **terms** of the sequence. Thus, a_1 is the **first term,** a_2 is the **second term,** and, in general, a_n is the **nth term** of the sequence $n \longmapsto a_n$. The symbolism

$$a_1, a_2, a_3, \ldots, a_n, \ldots$$

is often used to denote the sequence $n \longmapsto a_n$; it suggests a never-ending list in which the terms appear in order, with the nth term in the nth position for each positive integer n. In this notation, a set of three dots is read "and so on." The more compact notation $\{a_n\}$, in which the general term is enclosed in braces, is also used to denote the sequence.

To specify a particular sequence, we give a rule by which the nth term a_n is determined. This is often done by means of a formula.

Example 1   Find the first five terms of each sequence:

(a) $a_n = 2n - 1$ **(b)** $b_n = \frac{(-1)^n}{n(n+1)}$

Solution   **(a)** The first five terms of the sequence are found by substituting the positive integers 1, 2, 3, 4, and 5 for n in the formula for the general term. Thus,

$$a_1 = 2(1) - 1 = 1, \quad a_2 = 2(2) - 1 = 3; \quad a_3 = 2(3) - 1 = 5,$$

$$a_4 = 2(4) - 1 = 7, \quad a_5 = 2(5) - 1 = 9;$$

so the first five terms of the sequence $\{a_n\}$ are

$$1, 3, 5, 7, 9, \ldots .$$

(b) We have

$$b_1 = \frac{(-1)^1}{1(1 + 1)} = -\frac{1}{2}, \quad b_2 = \frac{(-1)^2}{2(2 + 1)} = \frac{1}{6}, \quad b_3 = \frac{(-1)^3}{3(3 + 1)} = -\frac{1}{12},$$

$$b_4 = \frac{(-1)^4}{4(4 + 1)} = \frac{1}{20}, \quad b_5 = \frac{(-1)^5}{5(5 + 1)} = -\frac{1}{30};$$

so the first five terms of the sequence $\{b_n\}$ are

$$-\tfrac{1}{2}, \tfrac{1}{6}, -\tfrac{1}{12}, \tfrac{1}{20}, -\tfrac{1}{30}, \ldots .$$

A formula that relates the general term a_n of a sequence to one or more of the terms that come before it is called a **recursion formula.** A sequence that is specified by giving the first term (or the first few terms) together with a recursion formula is said to be **defined recursively.**

Example 2   Find the first five terms of the sequence $\{a_n\}$ defined recursively by $a_1 = 1$ and $a_n = na_{n-1}$.

Solution   We are given $a_1 = 1$. By the recursion formula $a_n = na_{n-1}$, we have

$$a_2 = 2a_{2-1} = 2a_1 = 2(1) = 2.$$

Now, using the fact that $a_2 = 2$, we find that

$$a_3 = 3a_{3-1} = 3a_2 = 3(2) = 6.$$

Continuing in this way, we have

$$a_4 = 4a_{4-1} = 4a_3 = 4(6) = 24, \text{ and}$$

$$a_5 = 5a_{5-1} = 5a_4 = 5(24) = 120.$$

Thus, the first five terms of the sequence $\{a_n\}$ are

$$1, 2, 6, 24, 120, \ldots .$$

The pattern here is clear: $a_n = n!$. This can be proved by mathematical induction (Problem 30).

Arithmetic and **geometric** sequences (also called *progressions*) satisfy the following simple recursion formulas:

---

arithmetic sequence: $a_n = d + a_{n-1},$ where d is a fixed number called the **common difference**

geometric sequence: $a_n = ra_{n-1},$ where r is a fixed number known as the **common ratio**

In an arithmetic sequence, each term (after the first) is obtained by adding the fixed number d to the term just before it. Therefore, d is the common difference between successive terms; that is, $d = a_n - a_{n-1}$ for $n = 2, 3, 4, \ldots$. For instance, the sequence

$$5, 8, 11, \ldots, 3n + 2, \ldots$$

with general term $a_n = 3n + 2$, is an arithmetic sequence, and the common difference is

$$d = a_n - a_{n-1} = (3n + 2) - [3(n - 1) + 2]$$
$$= 3n + 2 - 3n + 3 - 2 = 3.$$

If $\{a_n\}$ is an arithmetic sequence with common difference d, then by the recursion formula $a_n = d + a_{n-1}$, we have

$$a_2 = a_1 + d$$

$$a_3 = a_2 + d = (a_1 + d) + d = a_1 + 2d$$

$$a_4 = a_3 + d = (a_1 + 2d) + d = a_1 + 3d,$$

and so on.

Evidently, for every positive integer n,

$$a_n = a_1 + (n - 1)d.$$

This formula for the nth term of an arithmetic sequence can be proved by mathematical induction (Problem 43).

Example 3   Find the seventeenth term of the arithmetic sequence $2, 5, 8, 11, \ldots$.

Solution   Here, the common difference of successive terms is $d = 3$ and the first term is $a_1 = 2$. Using the formula $a_n = a_1 + (n - 1)d$ with $n = 17$, we obtain

$$a_{17} = 2 + (17 - 1)(3) = 50. \qquad \blacksquare$$

In a geometric sequence, each term (after the first) is obtained by multiplying the term just before it by r. Therefore, if the terms are nonzero, r *is the common ratio of successive terms;* that is, $r = a_n/a_{n-1}$ for $n = 2, 3, 4, \ldots$.

For instance, the sequence

$$6, 12, 24, \ldots, 3(2^n), \ldots$$

with general term $a_n = 3(2^n)$ is a geometric sequence and the common ratio is

$$r = \frac{3(2^n)}{3(2^{n-1})} = 2^{n-(n-1)} = 2.$$

If $\{a_n\}$ is a geometric sequence with common ratio r, then by the recursion formula $a_n = ra_{n-1}$, we have

$$a_2 = ra_1$$
$$a_3 = ra_2 = r(ra_1) = r^2 a_1$$
$$a_4 = ra_3 = r(r^2 a_1) = r^3 a_1,$$

and so on. Evidently, for every positive integer n,

$$\boxed{a_n = r^{n-1} a_1.}$$

This formula for the nth term of a geometric sequence can be proved by mathematical induction (Problem 44).

Example 4   Find the sixth term of the geometric sequence $7, 21, 63, \ldots$.

Solution   Here, the common ratio of successive terms is $r = 3$ and the first term is $a_1 = 7$. Using the formula $a_n = r^{n-1} a_1$ with $n = 6$, we obtain

$$a_6 = 3^{6-1}(7) = 3^5(7) = 1701. \qquad \blacksquare$$

Problem Set 11.3

In Problems 1 to 12, find the first five terms of the sequence with the specified general term.

1. $a_n = \dfrac{1}{n+1}$

2. $b_n = \dfrac{(-1)^n}{n^2}$

3. $c_n = (n+1)^2$

4. $a_n = \dfrac{1}{n(n+1)}$

5. $b_n = (-1)^n n$

6. $c_n = \dfrac{1 + (-1)^n}{1 + 3n}$

7. $a_n = \dfrac{2n-1}{2n+1}$

8. $b_n = \left(-\dfrac{3}{2}\right)^{n-1}$

9. $c_n = \left(\dfrac{1}{3}\right)^n$

10. $a_n = n \cos n\pi$

11. $b_n = \dfrac{n^2 - 1}{n^2 + 1}$

12. $a_n = \dfrac{1}{n^2} \sin \dfrac{n\pi}{2}$

In Problems 13 to 18, find the first five terms of the recursively defined sequence $\{a_n\}$.

13. $a_1 = 2$ and $a_n = -3a_{n-1}$

14. $a_1 = 5$ and $a_n = 7 - 2a_{n-1}$

C 15. $a_1 = 2$ and $a_n = (a_{n-1})^{n-1}$

16. $a_1 = 1$, $a_2 = 2$, and $a_n = a_{n-1}a_{n-2}$ for $n \geq 3$

17. $a_1 = 1$ and $a_n = \dfrac{1}{n} a_{n-1}$

18. $a_1 = 0$, $a_2 = 1$, and $a_n = a_{n-1} - a_{n-2}$ for $n \geq 3$

In Problems 19 to 29, use the formula in the second box on page 593, to find the indicated term in each arithmetic sequence.

19. The sixth term of 3, 9, 15, . . .

20. The eighth term of 12, 24, 36, . . .

21. The fifteenth term of 2, 4, 6, . . .

22. The tenth term of 40, 50, 60, . . .

23. The twentieth term of 5, 10, 15, . . .

24. The twenty-fifth term of -20, -16, -12, . . .

25. The tenth term of -0.8, 0, 0.8, . . .

26. The fifteenth term of $5\sqrt{2}$, $7\sqrt{2}$, $9\sqrt{2}$, . . .

27. The eighth term of $m + r$, $m + 2r$, $m + 3r$, . . .

28. The tenth term of $2 + b$, $2 + 4b$, $2 + 7b$, . . .

29. The thirteenth term of $\sqrt{5}$, 0, $-\sqrt{5}$, . . .

30. If the sequence $\{a_n\}$ has the first term $a_1 = 1$ and satisfies the recursion formula $a_n = na_{n-1}$, use mathematical induction to prove that $a_n = n!$ for all positive integers n.

In Problems 31 to 42, use the boxed formula on page 594 to find the indicated term in each geometric sequence.

31. The eighth term of 2, 1, $\frac{1}{2}$, . . .

32. The fourth term of 8, 4, 2, . . .

33. The sixth term of $\frac{2}{3}$, 2, 6, . . .

34. The tenth term of $\frac{1}{4}$, $\frac{1}{8}$, $\frac{1}{16}$, . . .

35. The sixteenth term of 2, 4, 8, . . .

36. The ninth term of 3, 9, 27, . . .

37. The fifth term of -90, -9, -0.9, . . .

38. The seventh term of $\sqrt{2}/2$, -1, $\sqrt{2}$, . . .

39. The fifteenth term of $2x$, $2x^2$, $2x^3$, . . .

40. The tenth term of c^{-4}, $-c^{-2}$, 1, . . .

41. The sixth term of 0.3, 0.03, 0.003, . . .

42. The fifth term of 0.12, 0.012, 0.0012, . . .

43. Use mathematical induction to prove that for every positive integer n, the nth term of an arithmetic sequence $\{a_n\}$ with first term a_1 and common difference d is given by $a_n = a_1 + (n - 1)d$.

44. Use mathematical induction to prove that for every positive integer n, the nth term of a geometric sequence $\{a_n\}$ with first term a_1 and common ratio r is given by $a_n = r^{n-1} a_1$.

45. Suppose that the population of a certain city increases at the rate of 5% per year and that the present population is 300,000.

    (a) If a_n denotes the population of the city n years from now, show that $\{a_n\}$ is a geometric sequence.

    © (b) Find the population of the city in 5 years.

46. The **Fibonacci sequence** $\{a_n\}$, which was introduced by Leonardo Fibonacci in 1202, is defined recursively as follows:

    $$a_1 = 1, \ a_2 = 1, \text{ and } a_n = a_{n-1} + a_{n-2} \text{ for } n \geq 3.$$

    Find the first ten terms of the Fibonacci sequence.

11.4 SERIES

In many applications of mathematics, we have to find the sum of the terms of a sequence. Such a sum is called a **series.** Although the sum of the first n terms of a sequence $\{a_n\}$ can be written as

$$a_1 + a_2 + a_3 + \cdots + a_n,$$

a more compact notation is sometimes required. The capital Greek letter Σ (sigma), which corresponds to the letter S, is used for this purpose, and we write the sum in **sigma notation** as

$$\sum_{k=1}^{n} a_k = a_1 + a_2 + a_3 + \cdots + a_n.$$

Here, Σ indicates a sum and k is called the **index of summation.** That the summation begins with $k = 1$ and ends with $k = n$ is indicated by writing $k = 1$ below Σ and n above it. For some purposes, it is useful to allow sequences that begin with a "zeroth term" a_0, and to write

$$\sum_{k=0}^{n} a_k = a_0 + a_1 + a_2 + \cdots + a_n.$$

In any case, it is understood that the summation index runs through all *integer* values starting with the value shown below Σ and ending with the value shown above it.

In Examples 1 and 2, evaluate each sum.

Example 1 $\displaystyle\sum_{k=1}^{4} (2k + 1)$

Solution $\displaystyle\sum_{k=1}^{4} (2k + 1) = [2(1) + 1] + [2(2) + 1] + [2(3) + 1] + [2(4) + 1]$

$$= 3 + 5 + 7 + 9 = 24$$ ■

Example 2 $\displaystyle\sum_{k=0}^{3} \frac{2^k}{4k - 5}$

Solution $\displaystyle\sum_{k=0}^{3} \frac{2^k}{4k - 5} = \frac{2^0}{4(0) - 5} + \frac{2^1}{4(1) - 5} + \frac{2^2}{4(2) - 5} + \frac{2^3}{4(3) - 5}$

$$= \frac{1}{-5} + \frac{2}{-1} + \frac{4}{3} + \frac{8}{7}$$

$$= -\frac{21}{105} - \frac{210}{105} + \frac{140}{105} + \frac{120}{105}$$

$$= \frac{29}{105}$$ ■

If C is a constant, the notation $\displaystyle\sum_{k=1}^{n} C$ is understood to mean the sum of the first n terms of the sequence $C, C, C, \ldots, C, \ldots$. Thus,

$$\sum_{k=1}^{n} C = \overbrace{C + C + C + \cdots + C}^{n \text{ terms}} = nC.$$

Example 3   Evaluate $\displaystyle\sum_{k=1}^{7} 5$

Solution $\displaystyle\sum_{k=1}^{7} 5 = 7(5) = 35.$ ∎

In calculations involving sums in sigma notation, the following properties may be used.

Properties of Summation

If $\{a_n\}$ and $\{b_n\}$ are sequences and C is a constant, then:

(i) $\displaystyle\sum_{k=1}^{n} (a_k + b_k) = \sum_{k=1}^{n} a_k + \sum_{k=1}^{n} b_k$

(ii) $\displaystyle\sum_{k=1}^{n} (a_k - b_k) = \sum_{k=1}^{n} a_k - \sum_{k=1}^{n} b_k$

(iii) $\displaystyle\sum_{k=1}^{n} Ca_k = C \sum_{k=1}^{n} a_k$

(iv) $\displaystyle\sum_{k=1}^{n} C = nC$

Properties (i), (ii), and (iii) can be verified by expanding both sides of the expression or by using mathematical induction. We have already discussed Property (iv). Of course, Properties (i), (ii), and (iii) continue to hold if all summations begin with $k = 0$. In Property (iv), if we begin with $k = 0$, we obtain

$$\sum_{k=0}^{n} C = (n + 1)C$$

(Problem 42).

In addition to Properties (i) through (iv), we have

(v) $\displaystyle\sum_{k=1}^{n} k = \frac{n(n + 1)}{2}$

(vi) $\displaystyle\sum_{k=1}^{n} r^{k-1} = \frac{1 - r^n}{1 - r}$ if $r \neq 0, \quad r \neq 1.$

Property (v), which is a formula for the sum of the first n positive integers, was verified by mathematical induction in Example 1 of Section 11.1. Property (vi) can also be confirmed by mathematical induction, but the following argument is more interesting: Let $r \neq 0$, $r \neq 1$, and let $x = \displaystyle\sum_{k=1}^{n} r^{k-1}$. Then

$$x = 1 + r + r^2 + \cdots + r^{n-2} + r^{n-1},$$

so $$rx = r + r^2 + r^3 + \cdots + r^{n-1} + r^n.$$

If we subtract the second equation from the first, we find that

$$x - rx = 1 - r^n \qquad \text{or} \qquad (1 - r)x = 1 - r^n.$$

Since $r \neq 1$, the last equation can be rewritten as $x = \dfrac{1 - r^n}{1 - r}$, which is Property (vi).

Using Properties (i) to (vi), we can derive formulas for the sum of the first n terms of an arithmetic or geometric sequence. We begin with the arithmetic case.

Theorem 1   **Sum of the First n Terms of an Arithmetic Sequence**

> Let $\{a_n\}$ be an arithmetic sequence with first term a_1 and common difference d. Then, for every positive integer n,
>
> $$\sum_{k=1}^{n} a_k = na_1 + \frac{n}{2}(n - 1)d = \frac{n}{2}(a_1 + a_n).$$

Proof   We have already obtained the formula $a_k = a_1 + (k - 1)d$ for the kth term of $\{a_n\}$ (page 593). Thus,

$$\sum_{k=1}^{n} a_k = \sum_{k=1}^{n} [a_1 + (k - 1)d] = \sum_{k=1}^{n} a_1 + \sum_{k=1}^{n} (k - 1)d \quad \text{[by Properties (i)]}$$

$$= na_1 + d \sum_{k=1}^{n} (k - 1) \quad \text{[by Properties (iv) and (iii)]}$$

$$= na_1 + d[0 + 1 + 2 + \cdots + (n - 1)]$$

$$= na_1 + d\left[\frac{(n - 1)n}{2}\right] \quad \text{[by Property (v)]}$$

$$= na_1 + \frac{n}{2}(n - 1)d,$$

which proves the first part of the formula. To obtain the remaining part, we use the fact that $a_n = a_1 + (n - 1)d$. Thus,

$$na_1 + \frac{n}{2}(n - 1)d = \left(\frac{n}{2}a_1 + \frac{n}{2}a_1\right) + \frac{n}{2}(n - 1)d$$

$$= \frac{n}{2}[a_1 + a_1 + (n - 1)d]$$

$$= \frac{n}{2}(a_1 + a_n). \qquad \blacksquare$$

Example 4   Find the sum of the first six terms of the arithmetic sequence 4, 8, 12, 16, 20, 24, ...

Solution   We use the formula

$$\sum_{k=1}^{n} a_k = \frac{n}{2}(a_1 + a_n)$$

of Theorem 1. Here, $n = 6$, $a_1 = 4$, and $a_6 = 24$, so

$$\sum_{k=1}^{6} a_k = \frac{6}{2}(4 + 24) = 84.$$

■

Example 5   Find the sum of the first twelve terms of the arithmetic sequence $5, 1, -3, -7, \ldots$.

Solution   Here $d = -4$, $a_1 = 5$, and $n = 12$. Using the formula

$$\sum_{k=1}^{n} a_k = na_1 + \frac{n}{2}(n - 1)d$$

of Theorem 1, we have

$$\sum_{k=1}^{12} a_k = 12(5) + \frac{12}{2}(12 - 1)(-4) = 60 - 264 = -204.$$

■

Example 6   A paperboy receives \$30 from a newspaper route for the first month. During each succeeding month he earns \$2 more than he did the month before. If this pattern continues, how much will he earn over a period of 2 years?

Solution   The paperboy's monthly income forms an arithmetic sequence with $a_1 = 30$ dollars and $d = 2$ dollars. His total income over a 2-year period is given by the formula

$$\sum_{k=1}^{n} a_k = na_1 + \frac{n}{2}(n - 1)d$$

with $n = 24$. We have

$$\sum_{k=1}^{24} a_k = 24(30) + \frac{24}{2}(24 - 1)(2) = \$1272.$$

■

Now, we turn our attention to the geometric case.

Theorem 2   **Sum of the First n Terms of a Geometric Sequence**

Let $\{a_n\}$ be a geometric sequence with first term a_1 and common ratio $r \neq 0, r \neq 1$. Then, for every positive integer n,

$$\sum_{k=1}^{n} a_k = \sum_{k=1}^{n} a_1 r^{k-1} = a_1 \frac{1 - r^n}{1 - r}.$$

Proof   We have already obtained the formula $a_k = r^{k-1}a_1 = a_1 r^{k-1}$ for the kth term of a geometric sequence with common ratio r (page 594). Therefore,

$$\sum_{k=1}^{n} a_k = \sum_{k=1}^{n} a_1 r^{k-1} = a_1 \sum_{k=1}^{n} r^{k-1} \qquad \text{[by Property (iii)]}$$

$$= a_1 \frac{1-r^n}{1-r} \qquad \text{[by Property (vi)]},$$

and the proof is complete. ∎

Example 7   Find the sum of the first eight terms of the geometric sequence $3, -6, 12, -24, \ldots$.

Solution   We use the formula of Theorem 2, with $a_1 = 3$, $r = -2$, and $n = 8$. Thus,

$$\sum_{k=1}^{n} a_k = a_1 \frac{1-r^n}{1-r} = (3)\frac{1-(-2)^8}{1-(-2)} = (3)\frac{1-256}{3} = -255. \blacksquare$$

Ⓒ **Example 8**   For doing a certain job, you are offered 1¢ the first day, 3¢ the second day, 9¢ the third day, and so forth, so that your daily earnings form a geometric sequence with first term $a_1 = 1$ cent and common ratio $r = 3$. How much will you earn in 14 days of work?

Solution   By Theorem 2,

$$\sum_{k=1}^{n} a_1 r^{k-1} = \sum_{k=1}^{14} 3^{k-1} = \frac{1-3^{14}}{1-3}$$

$$= \frac{1-3^{14}}{-2} = \frac{3^{14}-1}{2}.$$

Using a calculator, we find that the total earnings amount to 2,391,484 cents or $23,914.84. (Not bad for two weeks' work!) ∎

Infinite Geometric Series

An indicated sum of all the terms of a sequence $\{a_n\}$, such as

$$a_1 + a_2 + a_3 + \cdots + a_n + \cdots$$

is called an **infinite series** or simply a **series**. Using the sigma notation, we can write such a series more compactly as

$$\sum_{k=1}^{\infty} a_k.$$

Although we cannot literally add an infinite number of terms, it is sometimes useful to assign a numerical value as the "sum" of an infinite series. This is accomplished by using the idea of a "limit," which is studied in calculus.

Although the general idea of the "sum" of an infinite series is beyond the scope of this textbook, you can easily get some feeling for this concept by considering an **infinite geometric series** of the form

$$\sum_{k=1}^{\infty} a_1 r^{k-1} = a_1 + a_1 r + a_1 r^2 + \cdots + a_1 r^{n-1} + \cdots$$

for the case in which $0 < |r| < 1$. By Theorem 2, the sum of the *first n terms* of this series is

$$\sum_{k=1}^{n} a_1 r^{k-1} = a_1 + a_1 r + \cdots + a_1 r^{n-1} = a_1 \frac{1 - r^n}{1 - r}.$$

Because $0 < |r| < 1$, it follows that r^n gets smaller and smaller as n gets larger and larger. (Try it on a calculator, say for $r = 0.75$ with $n = 10$, $n = 50$, $n = 100$, and so forth.) Therefore, as n gets larger and larger, r^n gets closer and closer to 0 and

$$a_1 \frac{1 - r^n}{1 - r} \qquad \text{comes closer and closer to} \qquad a_1 \frac{1 - 0}{1 - r} = \frac{a_1}{1 - r}.$$

In other words, as you add more and more terms of the geometric series, the sum comes closer and closer to $a_1/(1 - r)$. This suggests the following definition.

Definition 1   **Sum of an Infinite Geometric Series**

> If $|r| < 1$, we define
> $$\sum_{k=1}^{\infty} a_1 r^{k-1} = a_1 + a_1 r + a_1 r^2 + \cdots + a_1 r^{n-1} + \cdots = \frac{a_1}{1 - r}.$$

In Examples 9 and 10, find the sum of each infinite geometric series.

Example 9 $\dfrac{1}{2} + \dfrac{1}{4} + \dfrac{1}{8} + \cdots$

Solution   Here $a_1 = \frac{1}{2}$ and $r = \frac{1}{2}$, so by Definition 1,

$$\frac{1}{2} + \frac{1}{4} + \frac{1}{8} + \cdots = \frac{a_1}{1 - r} = \frac{\frac{1}{2}}{1 - \frac{1}{2}} = \frac{\frac{1}{2}}{\frac{1}{2}} = 1. \qquad \blacksquare$$

Example 10 $\displaystyle\sum_{k=1}^{\infty} 4\left(-\frac{1}{3}\right)^{k-1}$

Solution   Here $a_1 = 4$ and $r = -\frac{1}{3}$, so by Definition 1,

$$\sum_{k=1}^{\infty} 4\left(-\frac{1}{3}\right)^{k-1} = \frac{a_1}{1 - r}$$

$$= \frac{4}{1 - (-\frac{1}{3})} = \frac{4}{1 + \frac{1}{3}} = \frac{4}{\frac{4}{3}} = 3. \qquad \blacksquare$$

When we write an infinite decimal

$$x = 0.d_1 d_2 d_3 d_4 \cdots d_n \cdots$$

we are actually forming the sum of an infinite series

$$x = \frac{d_1}{10} + \frac{d_2}{100} + \frac{d_3}{1000} + \frac{d_4}{10,000} + \cdots + \frac{d_n}{10^n} + \cdots.$$

Thus, you can use the formula in Definition 1 to convert a repeating decimal to a quotient of integers.

Example 11   Rewrite the repeating decimal $0.\overline{31}$ as a quotient of integers.

Solution $$0.\overline{31} = 0.31313131\cdots = 0.31 + 0.0031 + 0.000031 + 0.00000031 + \cdots,$$

which is an infinite geometric series with first term $a_1 = 0.31$ and common ratio $r = 0.01$. Hence, by Definition 1,

$$0.\overline{31} = \frac{a_1}{1 - r} = \frac{0.31}{1 - 0.01} = \frac{0.31}{0.99} = \frac{31}{99}.$$

(Note that the same answer is obtained by using the method in Example 6 of Section 2.1.)   ∎

Problem Set 11.4

In Problems 1 to 6, write out and evaluate each sum.

1. $\displaystyle\sum_{k=1}^{4} (3k - 2)$

2. $\displaystyle\sum_{k=1}^{5} [2 + (-1)^k]$

3. $\displaystyle\sum_{k=1}^{5} \left(\frac{1}{4}k + 3\right)$

4. $\displaystyle\sum_{k=2}^{6} (-1)^k 3^k$

5. $\displaystyle\sum_{k=0}^{5} \frac{2^k}{3k + 1}$

6. $\displaystyle\sum_{j=0}^{4} \frac{5^j}{2j - 1}$

In Problems 7 to 12, use Properties (i) to (vi) of summation to evaluate each expression.

7. $\displaystyle\sum_{k=1}^{40} 3k$

8. $\displaystyle\sum_{k=1}^{30} (3 - 4k)$

9. $\displaystyle\sum_{k=1}^{24} (2^k + 1)$

10. $\displaystyle\sum_{k=0}^{19} 2^{-k}$

11. $\displaystyle\sum_{k=1}^{100} \frac{1}{10^k}$

12. $\displaystyle\sum_{j=0}^{100} (5^{j+1} - 5^j)$

In Problems 13 to 24, find the sum of the first n terms of each arithmetic sequence for the given value of n.

13. $1, 2, 3, \ldots$ for $n = 10$

14. $5, 9, 13, \ldots$ for $n = 13$

15. $7, 10, 13, \ldots$ for $n = 5$

16. $14, 19, 24, \ldots$ for $n = 21$

17. $36, 48, 60, \ldots$ for $n = 8$

18. $27, 33, 39, \ldots$ for $n = 30$

19. $2, -2, -6, \ldots$ for $n = 15$

20. $0.6, 0.4, 0.2, \ldots$ for $n = 16$

21. $-3t, -5t, -7t, \ldots$ for $n = 7$

22. $-2y, 0, 2y, \ldots$ for $n = 10$

23. $x, 0, -x, \ldots$ for $n = 10$

24. $5k - 4, 3k - 3, 2k - 1, \ldots$ for $n = 50$

25. A company had sales of \$200,000 during its first year of operation, and sales increased by \$30,000 per year during each successive year. Find the total sales of the company during its first 8 years.

26. A freely falling object dropped from rest falls $\frac{1}{2}g(2k - 1)$ units of distance during the kth second of its fall, where g is the acceleration of gravity. Find a formula for the total distance through which the body falls in n seconds.

27. A display of canned soup in a supermarket is to have the form of a pyramid with 20 cans in the bottom row, 19 cans in the next row, 18 cans in the next row, and so on, with a single can at the top. How many cans of soup will be required for the display?

28. If $\{a_n\}$ is an arithmetic sequence and if p and q are positive integers, show that

$$\sum_{k=p}^{p+q} a_k = \frac{q+1}{2}(a_p + a_{p+q}).$$

In Problems 29 to 38, find the sum of the first n terms of each geometric sequence for the given value of n.

29. $4, 40, 400, \ldots$ for $n = 5$

30. $-3, 15, -75, \ldots$ for $n = 6$

31. $-\frac{1}{3}, -\frac{1}{9}, -\frac{1}{27}, \ldots$ for $n = 10$

32. $-\frac{1}{16}, -\frac{1}{8}, -\frac{1}{4}, \ldots$ for $n = 8$

33. $5^{10}, 5^8, 5^6, \ldots$ for $n = 9$

34. $1, -1, 1, \ldots$ for $n = 10$

35. $100, 100, 100, \ldots$ for $n = 25$

36. $0.3, 0.03, 0.003, \ldots$ for $n = 6$

37. $c^4, c^6, c^8, \ldots$ for $n = 7$

38. $1, \frac{1}{10}, \frac{1}{100}, \ldots$ for $n = 15$

C **39.** Suppose that the distance traveled by a point on a pendulum in any swing is 5% less than in the previous swing. If the point travels 6 centimeters on the first swing, find the total distance traveled by the point in 7 swings.

40. If $\{a_n\}$ is a geometric sequence with common ratio r, show that $\sum_{k=1}^{n} a_k = \frac{a_1 - a_n r}{1 - r}$.

41. A ball is dropped from a height of 256 centimeters, and on each rebound it rises to $\frac{1}{2}$ the height from which it last fell. Find a formula for the total distance traveled by the ball when it hits the floor for the nth time.

42. Justify the formula $\sum_{k=0}^{n} C = (n + 1)C$. [*Hint:* Use $\sum_{k=0}^{n} a_k$ where $a_k = C$ for all values of k.]

In Problems 43 to 50, find the sum of each infinite geometric series.

43. $\displaystyle\sum_{k=1}^{\infty} \left(\frac{2}{3}\right)^{k-1}$

44. $\displaystyle\sum_{k=1}^{\infty} \left(-\frac{1}{3}\right)^{k-1}$

45. $\displaystyle\sum_{k=1}^{\infty} \left(-\frac{2}{5}\right)^{k-1}$

46. $\displaystyle\sum_{k=1}^{\infty} \left(\frac{3}{7}\right)^{k-1}$

47. $\displaystyle\sum_{k=1}^{\infty} \left(-\frac{1}{6}\right)^{k-1}$

48. $\displaystyle\sum_{k=0}^{\infty} \left(-\frac{2}{9}\right)^{k}$

49. $\displaystyle\sum_{k=1}^{\infty} 5\left(-\frac{3}{5}\right)^{k-1}$

50. $\displaystyle\sum_{k=0}^{\infty} ar^k$ for $|r| < 1$

In Problems 51 to 56, use the formula for the sum of an infinite geometric series to rewrite each infinite repeating decimal as a quotient of integers.

51. $0.\overline{4}$ **52.** $4.\overline{53}$ **53.** $7.\overline{27}$ **54.** $0.2\overline{79}$

55. $0.02\overline{34}$ **56.** $-1.9\overline{81}$

57. True or false: $0.\overline{9} < 1$. Explain your answer.

58. In the discussion leading up to Definition 1, it was assumed that $0 < |r| < 1$; however, in Definition 1, we only require that $|r| < 1$. Explain.

11.5 PERMUTATIONS AND COMBINATIONS

It is often necessary to calculate the number of different ways in which something can be done or can happen. For instance, in order to find the odds of being dealt a full house in a game of five-card poker, it is necessary to calculate the total number of different possible poker hands. In this section we discuss some of the methods used in making such calculations.

We begin by stating the additive and multiplicative principles of counting, which are the basic tools for the calculations made in this section.

The Additive Principle of Counting

Let A and B be two events that cannot occur simultaneously. Then, if A can occur in a ways and B can occur in b ways, it follows that the number of ways in which either A or B can occur is $a + b$.

The additive principle of counting can be extended to three or more events in an obvious way.

Example 1   Suppose you have six signal flags—two red, two green, and two blue flags. By displaying a single flag, you can send three different signals. By displaying two flags, one above the other, you can send nine different signals. How many different signals can you send using one or two flags?

Solution   Let A be the event that a one-flag signal is sent, so that A can occur in $a = 3$ ways. Let B be the event that a two-flag signal is sent, so that B can occur in $b = 9$ ways. Since a one-flag signal and a two-flag signal can't be sent at the same time, the additive principle of counting applies, and $a + b = 3 + 9 = 12$ different signals can be sent using either one or two flags. ∎

The Multiplicative Principle of Counting

If a first event A can occur in a ways and, independently, a second event B can occur in b ways, then the number of ways in which both A and B can occur is ab.

The multiplicative principle of counting can be extended to three or more events in an obvious way.

Example 2   A restaurant offers a meal consisting of one of three entrees and one of four desserts at a special price. How many different meals can be ordered at this price?

Solution   By the multiplicative principle of counting, $3 \cdot 4 = 12$ different meals can be ordered. ∎

The idea of a *permutation*, introduced in the following definition, is useful in many counting problems.

Definition 1   **Permutations**

> A **permutation** is an *ordered* arrangement of distinct objects in a row. Let S be a set of n distinct elements. An ordered arrangement of r of these elements in a row is called a *permutation of size r chosen from the n elements of S*. The symbol $_nP_r$ denotes the number of different permutations of size r chosen from a set of n elements.*

Example 3   Let $S = \{a, b, c, d\}$. List all permutations of size 2 chosen from the 4 elements of S, and count these permutations to find $_4P_2$.

Solution   We list the permutations according to which element is in the first position:

a first	b first	c first	d first
ab	ba	ca	da
ac	bc	cb	db
ad	bd	cd	dc

Thus, $_4P_2 = 12$.   ∎

Theorem 1   **A Formula for $_nP_r$**

> If n and r are positive integers with $n \geq r$, then
>
> $$_nP_r = \overbrace{n(n - 1)(n - 2) \cdots (n - r + 1)}^{r \text{ factors}}$$
>
> $$= \frac{n!}{(n - r)!}.$$

Proof   In forming a permutation of r elements chosen from a set of n elements, we have n choices for the element to be placed in the first position in the row. Once this choice is made, there remain $n - 1$ elements to be placed in the remaining $r - 1$ positions. Therefore, we have $n - 1$ choices for the element to be placed in the second position. By the multiplicative principle of counting, there are $n(n - 1)$ ways to choose distinct elements to place in the first two positions. Similarly, there are $n(n - 1)(n - 2)$ ways to choose distinct elements to place in the first three positions.

* The symbol $_nP_r$ is read as "*the number of permutations of n things taken r at a time.*"

Continuing in this way, we see that all r positions in the row can be filled with distinct elements in $n(n-1)(n-2)\cdots[n-(r-1)]$ ways. Therefore,

$$_nP_r = \overbrace{n(n-1)(n-2)\cdots(n-r+1)}^{r \text{ factors}}$$

$$= \frac{n(n-1)(n-2)\cdots(n-r+1)(n-r)!}{(n-r)!}$$

$$= \frac{n!}{(n-r)!}.$$

Example 4   Evaluate

(a) $_7P_1$ (b) $_5P_3$

(c) $_4P_4$ (d) $_9P_4$.

Solution   Using the formula $_nP_r = \overbrace{n(n-1)(n-2)\cdots(n-r+1)}^{r \text{ factors}}$, we have:

(a) $_7P_1 = 7$ (b) $_5P_3 = 5 \cdot 4 \cdot 3 = 60$

(c) $_4P_4 = 4 \cdot 3 \cdot 2 \cdot 1 = 24$ (d) $_9P_4 = 9 \cdot 8 \cdot 7 \cdot 6 = 3024$.

Example 5   A portfolio manager knows 13 stocks that meet her investment criteria. She will choose three of these and rank them first, second, and third in order of preference. In how many ways can she do this?

Solution $_{13}P_3 = 13 \cdot 12 \cdot 11 = 1716$ ways.

Suppose that we have n distinct objects that are to be arranged in a definite order. By Theorem 1, this can be done in

$$_nP_n = n(n-1)(n-2)\cdots 1 = n!$$

different ways.

ⓒ **Example 6**   In how many ways can 12 different books be arranged on a shelf?

Solution   Using a calculator, we find that there are

$$12! = 479,001,600$$

different arrangements.

The value of a poker hand doesn't depend on the order in which the cards are arranged in the hand, but only on which cards are present. Thus, the idea of a permutation, which involves *order*, isn't the right tool for counting poker hands. Such situations, in which order isn't important, suggest the following definition.

Definition 2   **Combinations**

> Let S be a set consisting of n distinct elements. A subset consisting of r of these elements is called a **combination of size r chosen from the n elements in S.** No importance is attached to the order of the r chosen elements. The symbol $_nC_r$ denotes the number of different combinations of size r chosen from a set of n elements.*

Example 7   Let $S = \{a, b, c, d\}$. List all combinations of size 2 chosen from the 4 elements of S, and count these combinations to find $_4C_2$.

Solution   The subsets of S that consist of two elements are

$$\{a, b\}, \quad \{a, c\}, \quad \{a, d\}, \quad \{b, c\}, \quad \{b, d\}, \quad \text{and} \quad \{c, d\}.$$

[Because order isn't important, the combination $\{a, b\}$ is the same as the combination $\{b, a\}$, the combination $\{a, c\}$ is the same as the combination $\{c, a\}$, and so on.] Thus, $_4C_2 = 6$. ∎

There is a simple relationship between permutations and combinations. To obtain this relationship, we consider a set S consisting of n distinct elements. To specify a permutation of size r chosen from these n elements, we can first select the r elements that will appear in the permutation, and then we can give the order in which the selected elements are to be arranged. The first step, the selection of a combination of r elements from S, can be done in $_nC_r$ ways; the second step, the ordering of these r elements, can be accomplished in $r!$ ways. Therefore, by the multiplicative principle of counting,

$$_nP_r = (_nC_r)\,r!.$$

Using this relationship, we can prove the following theorem.

Theorem 2   **A Formula for $_nC_r$**

> If n and r are integers with $n \geq r \geq 0$, then
>
> $$_nC_r = \frac{\overbrace{n(n-1)(n-2)\cdots(n-r+1)}^{r \text{ factors}}}{\underbrace{r(r-1)(r-2)\cdots 1}_{r \text{ factors}}} = \frac{n!}{(n-r)!\,r!}.$$

Proof   By the relationship established above and Theorem 1, we have

$$_nC_r = \frac{_nP_r}{r!} = \frac{n(n-1)(n-2)\cdots(n-r+1)}{r(r-1)(r-2)\cdots 1}.$$

* The symbol $_nC_r$ is read as "*the number of combinations of n things taken r at a time*," or as "*n choose r*."

Since $_nP_r = \dfrac{n!}{(n-r)!}$, we can also write

$$_nC_r = \frac{_nP_r}{r!} = \frac{n!}{(n-r)!r!}.$$

Notice that $_nC_r$, the number of combinations of size r chosen from a set of n elements, is the same as the binomial coefficient $\binom{n}{r}$ (Definition 2, page 585).

Example 8 Evaluate

(a) $_7C_3$ (b) $_7C_7$ (c) $_7C_1$

Solution Using the formula

$$_nC_r = \frac{n(n-1)(n-2)\cdots(n-r+1)}{r(r-1)(r-2)\cdots 1}$$

in Theorem 2, we have

(a) $_7C_3 = \dfrac{7\cdot 6\cdot 5}{3\cdot 2\cdot 1} = 35$ (b) $_7C_7 = \dfrac{7!}{7!} = 1$ (c) $_7C_1 = \dfrac{7}{1} = 7$

©Example 9 Find the number of different committees of four students each that can be formed from a group of 36 students.

Solution Here the set S consists of the 36 students, so $n = 36$. A committee of 4 of these students is a subset of S consisting of $r = 4$ elements. The order of students within a committee is not important, so each committee is a combination of size $r = 4$ chosen from the $n = 36$ elements of S. The number of such committees is

$$_{36}C_4 = \frac{36\cdot 35\cdot 34\cdot 33}{4\cdot 3\cdot 2\cdot 1} = \frac{\overset{3}{36}\cdot 35\cdot 34\cdot 33}{4\cdot 3\cdot 2} = \frac{3\cdot 35\cdot \overset{17}{34}\cdot 33}{2}$$

$$= 3\cdot 35\cdot 17\cdot 33 = 58{,}905.$$

©Example 10 Find the number of different hands in five-card poker.

Solution There are $n = 52$ cards in the deck, and the order of the cards in a hand isn't important. Thus, the number of different poker hands is given by

$$_{52}C_5 = \frac{52\cdot 51\cdot \overset{10}{50}\cdot 49\cdot 48}{5\cdot 4\cdot 3\cdot 2\cdot 1} = \frac{\overset{13}{52}\cdot 51\cdot 10\cdot 49\cdot 48}{4\cdot 3\cdot 2} = \frac{13\cdot 51\cdot 10\cdot 49\cdot \overset{8}{48}}{3\cdot 2}$$

$$= 13\cdot 51\cdot 10\cdot 49\cdot 8 = 2{,}598{,}960.$$

Suppose that we have r red balls, w white balls, and b black balls, and that balls within a color group are indistinguishable from one another. Altogether, we have $n = r + w + b$ balls. In how many *distinguishable ways* can we arrange these n balls in an ordered row; that is, how many **distinguishable permutations** are there of the n colored balls?

To specify a distinguishable arrangement of the balls in an ordered row, we first decide which of the n positions in the row will be occupied by the red balls. This amounts to specifying a subset of r positions among the n available positions, and it can be done in $_nC_r$ ways. Having made this decision, we place the red balls in the specified positions. There remain $n - r$ positions to be filled with the remaining white and black balls. We choose w of these positions for the white balls. This choice can be made in $_{n-r}C_w$ ways. The remaining $n - r - w$ positions will have to be filled with black balls—here we have no choice. Therefore, by the multiplicative principle of counting, we can specify $(_nC_r)(_{n-r}C_w)$ distinguishable ordered arrangements or permutations of the colored balls.

Using the formula in Theorem 2 and the fact that $n = r + w + b$, so that $n - r - w = b$, we have

$$(_nC_r)(_{n-r}C_w) = \frac{n!}{(n-r)!r!} \cdot \frac{(n-r)!}{[(n-r)-w]!w!}$$
$$= \frac{n!(n-r)!}{(n-r)!r!b!w!} = \frac{n!}{r!w!b!}.$$

This formula for the number of distinguishable permutations of the colored balls can be generalized as follows.

Theorem 3   **Distinguishable Permutations**

> Consider a set of n objects of which n_1 are alike of one kind, n_2 are alike of another kind, . . . , and n_k are alike of a last kind. Then, the number of distinguishable permutations of the objects is
>
> $$\frac{n!}{n_1!n_2! \cdots n_k!}.$$

Example 11   How many distinguishable permutations are there of the letters in the word CINCINNATI?

Solution   There are $n = 10$ letters altogether. Of these, there are $n_1 = 2$ C's, $n_2 = 3$ I's, $n_3 = 3$ N's, $n_4 = 1$ A, and $n_5 = 1$ T. By Theorem 3, the number of distinguishable permutations of these letters is given by

$$\frac{n!}{n_1!n_2!n_3!n_4!n_5!} = \frac{10!}{2!3!3!1!1!} = 50,400.$$

Problem Set 11.5

1. There are six ways of rolling a "7" with two dice, and there are two ways of rolling an "11" with two dice. In how many ways can you roll "7 or 11"?

2. There are three different roads connecting Newberry with New Mattoon, and four different roads connecting New Mattoon with Gainesburg. In how many different ways can a person drive from Newberry to Gainesburg, passing through New Mattoon on the way?

3. How many different two-letter "words" can be formed using the 26 letters of the alphabet if repeated letters are allowed? [The "words" need not be in a dictionary.]

4. How many different three-letter "words" can be formed using the 26 letters of the alphabet if repeated letters are allowed but the middle letter must be be a, e, i, o, or u?

5. How many positive three-digit integers less than 500 can be formed using only the digits 1, 3, 5, and 7 if repetitions are allowed?

6. A combination lock for a bicycle has three levers, each of which can be set in nine positions. A thief attempting to steal the bicycle begins trying all possible arrangements of the levers. If the thief tries two arrangements every second, what is the maximum time required to open the lock?

7. List and count all permutations of size 2 chosen from the elements of $S = \{a, b, c, d, e\}$, and thus determine $_5P_2$.

8. List and count all permutations of size 3 chosen from the elements of the set $S = \{a, b, c, d\}$, and thus determine $_4P_3$.

In Problems 9 to 16, find the value of the expression by using Theorem 1.

9. $_4P_3$ 10. $_8P_6$ C 11. $_9P_9$ 12. $_{12}P_1$

13. $_5P_2$ C 14. $_{13}P_{13}$ 15. $_{11}P_3$ 16. $_8P_2$

C 17. In how many ways can five of ten books be chosen and arranged next to each other on a shelf?

18. In how many ways can a left end, a right end, and a center for a football team be picked from among Dean, Dolores, Carlos, Carmine, Gus, and Olga?

C 19. In how many ways can seven people line up at a ticket window?

20. How many signals can be sent by using four distinguishable flags, one above the other, on a flag pole, if the flags can be used one, two, three, or four at a time?

C 21. In how many ways can a baseball team of nine players be arranged in batting order if tradition is followed and the pitcher must bat last, but there are no other restrictions?

22. A poll consists of five questions. In how many different orders can these questions be asked?

23. If ten runners are entered in a race for which first, second, and third prizes will be awarded, in how many different ways can the prizes be distributed? (Assume that there are no ties.)

24. How many integers that do not contain repeated digits are there between 1 and 1000, inclusive?

25. List and count all combinations of size 3 chosen from the elements of the set $S = \{a, b, c, d\}$, and thus determine $_4C_3$.

26. List and count all combinations of size 2 chosen from the elements of $S = \{a, b, c, d, e\}$, and thus determine $_5C_2$.

In Problems 27 to 35, find the value of the expression by using Theorem 2.

27. $_4C_4$ 28. $_{48}C_3$ 29. $_5C_2$ C 30. $_{52}C_{13}$

31. $_6C_3$ 32. $_{10}C_{10}$ 33. $_{10}C_9$ 34. $_{10}C_1$

35. $_{10}C_0$

36. Show that $_nC_0 = 1$ holds for every integer $n \geq 0$ and give an interpretation of this fact.

37. A total of how many games will be played by eight teams in a league if each team plays the other teams just once?

38. A dealer has 15 different models of television sets and wishes to display three models in the store window. Disregarding the arrangement of the three sets in the window, in how many ways can this be done?

39. An election is to be held to choose four delegates from among ten nominees to represent a district at

a political convention. In how many ways can the delegation be formed?

40. A department store plans to fill ten positions with four men and six women. In how many ways can these positions be filled if there are nine men and eleven women applicants?

Ⓒ In Problems 41 to 46, determine the number of distinguishable permutations of the letters in each word.

41. OHIO

42. MASSACHUSETTS

43. REARRANGEMENT

44. MISSISSIPPI

45. TENNESSEE

46. COMMITTEE

47. In how many distinguishable ways can eight people be arranged in a police lineup if two of them are identical twins and three are identical triplets?

48. There are three copies of a chemistry book, two copies of a biology book, and four copies of a sociology book to be placed on a shelf. In how many distinguishable ways can this be done?

11.6 PROBABILITY

The mathematical theory of probability has applications in nearly every area of human activity. These applications range from medical statistics to actuarial science, from statistical physics to law, and from decision making in the business world to gambling in Las Vegas. Here we can only touch the subject in the most superficial way and we can give only the simplest examples, many of which involve playing cards or dice.

Consider an experiment that has a finite number of **equally likely outcomes.** For instance, if you roll a balanced die (a small cube with different numbers of spots on its faces), the possible outcomes are 1, 2, 3, 4, 5, or 6, and they are equally likely. An **event** is something that may or may not occur, depending on which outcome is obtained. For instance, if you roll a die, the event "rolling an odd number" occurs if the outcome is 1, 3, or 5, and fails to occur if the outcome is 2, 4, or 6. The outcomes for which an event occurs are said to be **favorable** to the event. For instance, the outcomes 1, 3, and 5 are favorable to the event "rolling an odd number."

Definition 1   **The Probability of an Event**

> Let A be an event associated with an experiment with N equally likely outcomes. If $n(A)$ is the number of outcomes favorable to A, the **probability** of A, in symbols $P(A)$, is defined by
>
> $$P(A) = \frac{n(A)}{N}.$$

The number $P(A)$ in Definition 1 is the fraction of all possible outcomes that are favorable to A. Hence, $P(A)$ can be regarded as a numerical measure of the likelihood, on a scale from 0 to 1, that A will occur if the experiment is performed.

Example 1   Find the probability of rolling an odd number with a balanced die.

Solution   In Definition 1, let A denote the event "rolling an odd number." When a balanced die is rolled, there are $N = 6$ equally likely outcomes; $n(A) = 3$ of these outcomes are favorable to the event A, so

$$P(A) = \frac{n(A)}{N} = \frac{3}{6} = \frac{1}{2}.$$

We can also express $P(A)$ in decimal form as $P(A) = 0.5$, or as $P(A) = 50\%$.   ∎

Example 2   Find the probability of drawing a face card (jack, queen, or king) from a well-shuffled deck of 52 cards.

Solution   Here there are $N = 52$ possible outcomes of the experiment of drawing one card from the deck. Since the deck is well-shuffled, these outcomes are all equally likely. Of the 52 cards, there are 4 jacks, 4 queens, and 4 kings; hence, the number of face cards is $4 + 4 + 4 = 12$. Thus,

$$P(\text{"drawing a face card"}) = \frac{12}{52} = \frac{3}{13}.$$ ∎

Example 3   From a group of nine people, five male and four female, a committee of three is to be selected by chance. What is the probability that no females are on the committee?

Solution   Here the "experiment" is the selection of a committee and the outcome is the selected committee. There are

$$N = {}_9C_3 = 84$$

ways of selecting a committee of 3 from 9 people. Since the committee is to be selected by chance, all 84 of these outcomes are equally likely. The event in question, "no females on the committee," occurs if and only if the committee consists only of males. There are

$$n = {}_5C_3 = 10$$

ways of selecting a committee consisting entirely of males from among the 5 available. It follows that

$$P(\text{"no females on committee"}) = \frac{n}{N} = \frac{10}{84} \approx 11.9\%.$$ ∎

Example 4   What is the probability of being dealt a full house (three cards of one denomination—say, three 6's or three kings—and two cards of another denomination—say, two aces or two 9's) from a well-shuffled deck in 5-card poker?

Solution   Here an outcome is a 5-card hand. There are $N = {}_{52}C_5$ such hands, all of which are equally likely. Of these, we must calculate the number n of full houses. This is best done by asking ourselves in how many ways we could form a 5-card hand with 3 cards of one denomination and 2 cards of another. The common denomination of the 3 cards can be chosen in 13 ways, and the 3 cards themselves can be chosen from among the 4 cards of this denomination in ${}_4C_3$ ways. The common denomination of the other 2 cards can then be chosen in any of the remaining 12 ways, and the 2 cards themselves can be chosen from among the 4 cards of this denomination in $_4C_2$ ways. By the multiplicative principle of counting, we can therefore form full houses in

$$n = (13)(_4C_3)(12)(_4C_2)$$

different ways. It follows that

$$P(\text{"full house"}) = \frac{n}{N}$$

$$= \frac{(13)(_4C_3)(12)(_4C_2)}{_{52}C_5}$$

$$= \frac{(13)\left(\dfrac{4 \cdot 3 \cdot 2}{3 \cdot 2 \cdot 1}\right)(12)\left(\dfrac{4 \cdot 3}{2 \cdot 1}\right)}{\dfrac{52 \cdot 51 \cdot 50 \cdot 49 \cdot 48}{5 \cdot 4 \cdot 3 \cdot 2 \cdot 1}}$$

$$= \frac{13 \cdot 4 \cdot 12 \cdot 2 \cdot 3}{52 \cdot 51 \cdot 10 \cdot 49 \cdot 2}$$

$$= \frac{6}{4165}.$$

Example 5   What is the probability of rolling a "7" with two dice?

Solution   Call one of the dice the *first die* and the other the *second die*. An outcome of a toss of the two dice can be denoted by an ordered pair (x, y), where x is the number on the first die and y the number on the second. Here x can have any of 6 possible values and y can have any of 6 possible values, so, by the multiplicative principle of counting, there are $N = 6 \cdot 6 = 36$ possible outcomes. Of these, the outcomes favorable to the event "rolling a 7" are

$$(1, 6), \quad (2, 5), \quad (3, 4), \quad (4, 3), \quad (5, 2), \quad \text{and} \quad (6, 1).$$

Thus there are $n = 6$ outcomes favorable to this event. It follows that

$$P(\text{"rolling a 7"}) = \frac{n}{N} = \frac{6}{36} = \frac{1}{6}.$$

Problem Set 11.6

In Problems 1 to 14, find the probability of the event.

1. Obtaining "heads" when tossing a coin.

2. Drawing a king from a deck of 52 well-shuffled cards.

3. Drawing a green ball, blindfolded, from a hat containing six red balls and eight green balls, all of the same size.

4. Failing to draw one of the kings from a deck of 52 well-shuffled cards.

5. Rolling a "5" with a single die.

6. Rolling less than a "3" with a single die.

©**7.** Being dealt four of a kind (four cards of the same denomination) in a game of five-card poker.

©**8.** Being dealt a flush (all cards of the same suit) in a game of five-card poker.

©**9.** Being dealt a royal flush (ten, jack, queen, king, and ace, all of the same suit) in a game of five-card poker.

©**10.** Being dealt a straight (five cards whose denominations form a sequence such as 7, 8, 9, 10, jack) in a game of five-card poker. (The ace can count either as a 1 or as the denomination just above the king.)

11. Rolling a "6" with a pair of dice.

12. Rolling snake eyes ("2") with a pair of dice.

13. Rolling at least one "6" in two rolls of a single die.

14. Two people, chosen at random, having birthdays on the same day of the year. (Disregard leap year complications.)

15. A committee of five people is to be chosen at random from among ten men and three women. What is the probability that there will be three men and two women on the committee?

16. A committee of seven people is to be chosen at random from among five skilled workers, three unskilled workers, and four supervisory personnel. What is the probability that all four supervisory personnel and no unskilled workers are on the committee?

17. Two coins are tossed and you are told that at least one coin fell "heads." What is the probability that the other coin fell "heads"? [Be careful—consider the possible outcomes in the face of the given information and how many of these are favorable to the event in question.]

18. Four Eastern states have daily lotteries in which numbers from 0 to 999 are drawn. What is the probability that two (or more) of these states draw the same lucky number on the same day? [*Hint:* To count the number of outcomes favorable to the event, begin by counting the number of outcomes *unfavorable* to the event.]

19. Cards are drawn one at a time from a well-shuffled deck. They are not replaced after being drawn. What is the probability that the last ace in the deck is drawn before the last king in the deck?

20. Two cards are drawn from a well-shuffled deck. The first card is *not* replaced before the second card is drawn. If the second card is a face card, what is the probability that the first card was a face card?

REVIEW PROBLEM SET, CHAPTER 11

In Problems 1 to 4, use mathematical induction to prove that the assertion is true for all positive integers n.

1. $n^2 + 3n$ is an even integer

2. 3 is an exact integer divisor of $n^3 + 6n^2 + 11n$

3. $1 \cdot 2 \cdot 3 + 2 \cdot 3 \cdot 4 + 3 \cdot 4 \cdot 5 + \cdots + n(n+1)(n+2)$
$$= \frac{1}{4} n(n+1)(n+2)(n+3)$$

4. $2 \cdot 4 + 4 \cdot 6 + 6 \cdot 8 + \cdots + 2n(2n+2)$
$$= \frac{n}{3}(2n+2)(2n+4)$$

In Problems 5 to 10, use the binomial theorem and the Pascal triangle to expand each expression.

5. $(2+x)^5$ **6.** $(3x-4y)^4$

©**7.** $(3x^2 - 2y^2)^7$ **8.** $(2a+3b)^6$

©**9.** $\left(\frac{3}{2}x - 1\right)^7$ **10.** $\left(2x^2 + \frac{3}{y}\right)^3$

In Problems 11 to 14, find and simplify the specified term in the binomial expansion of the expression.

11. The fourth term of $(2x + y)^9$

12. The fifth term of $(x - 2y)^7$

13. The term containing x^5 in $(3x + y)^{10}$

14. The term containing x^{10} in $(2x^2 - 3)^{11}$

In Problems 15 and 16, find the first five terms of the sequence with the given general term.

15. $a_n = \dfrac{(-1)^n}{n + 1}$ **16.** $b_n = \dfrac{(-1)^{n+1}}{n^2 + 4n + 1}$

In Problems 17 and 18, find the first six terms of the recursively defined sequence $\{a_n\}$.

17. $a_1 = 0$, $a_2 = \dfrac{1}{2}$, and $a_n = \dfrac{1}{n} a_{n-2}$

18. $a_1 = 1$, $a_2 = 0$, and $(n + 1)(n + 2)a_{n+2} = -na_n$

In Problems 19 to 22, find the indicated term in each arithmetic sequence.

19. The twentieth term of $3, 5, 7, 9, \ldots$

20. The fiftieth term of $17, 14, 11, 8, \ldots$

21. The thirty-fourth term of $-5, -9, -13, -17, \ldots$

22. The hundredth term of $1, \frac{7}{2}, 6, \frac{17}{2}, \ldots$

In Problems 23 to 26, find the indicated term in each geometric sequence.

23. The tenth term of $2, 4, 8, 16, \ldots$

24. The ninth term of $12, 6, 3, \frac{3}{2}, \ldots$

25. The eighth term of $3, -6, 12, -24, \ldots$

26. The seventh term of $9, -3, 1, -\frac{1}{3}, \ldots$

In Problems 27 to 34, evaluate each sum.

27. $\displaystyle\sum_{k=1}^{3} (2k + 7)$ **28.** $\displaystyle\sum_{k=1}^{50} [3 + (-1)^k]$

29. $\displaystyle\sum_{k=1}^{20} \left(\dfrac{k}{3} + 2\right)$ **30.** $\displaystyle\sum_{k=1}^{20} (1 - 3k)$

31. $\displaystyle\sum_{k=0}^{3} \dfrac{7^k}{1 + 2^k}$ **32.** $\displaystyle\sum_{k=0}^{15} \dfrac{3}{2^k}$

33. $\displaystyle\sum_{k=0}^{2} \dfrac{(-5)^k}{1 + 3^k}$ **34.** $\displaystyle\sum_{k=0}^{25} (4^{k+1} - 4^k)$

In Problems 35 to 38, find the sum of the first n terms of each arithmetic sequence for the given value of n.

35. $2, 6, 10, 14, \ldots$ for $n = 15$

36. $12, 13.5, 15, 16.5, \ldots$ for $n = 20$

37. $3, \frac{8}{3}, \frac{7}{3}, 2, \ldots$ for $n = 11$

38. $3x - 2, -x + 1, \ldots$ for $n = 10$

© In Problems 39 to 42, find the sum of the first n terms of each geometric sequence for the given value of n.

39. $48, 96, 192, \ldots$ for $n = 8$

40. $-81, -27, -9, \ldots$ for $n = 12$

41. $\frac{3}{4}, 3, 12, \ldots$ for $n = 10$

42. $0.2, 0.002, 0.00002, \ldots$ for $n = 5$

43. A mathematics club raffles a calculator by selling 100 sealed tickets numbered in order $1, 2, 3, \ldots$. The tickets are drawn at random by purchasers who pay the number of cents indicated by the number on the ticket. How much money does the club receive if all tickets are sold?

44. A clock strikes on the hour. How many times does the clock strike between 12:00 noon on one day and 12:00 noon on the next day, inclusive?

45. Gus saved \$200 the first year he was employed. Each year thereafter, he saved \$50 more than the year before. How much did he save at the end of 8 years?

46. A car costs \$7670 and depreciates in value by 31% during the first year, by 26% during the second year, by 21% during the third year, and so on. What is the value of the car at the end of 5 years?

© **47.** The rungs of a ladder decrease uniformly (that is, linearly) in length from 32 inches to 18 inches. If there are 25 rungs in the ladder, find the total length of the wood in all of these rungs.

© **48.** In a lottery, the first ticket drawn will pay the ticket holder \$1, and each succeeding ticket will pay twice as much as the preceding one. If 15 tickets are drawn, what is the total amount paid in prize money?

49. A pyramid of cannonballs in an armaments museum stands on a square base having n

cannonballs on a side. How many cannonballs are there in the pile?

50. Work Problem 49 if the base has the form of an equilateral triangle.

In Problems 51 to 54, find the sum of each infinite geometric series.

51. $\displaystyle\sum_{k=1}^{\infty} 3\left(\frac{2}{3}\right)^{k-1}$

52. $\displaystyle\sum_{k=1}^{\infty} 3\left(-\frac{2}{3}\right)^{k-1}$

53. $\displaystyle\sum_{k=1}^{\infty} \left(\sqrt{\frac{5}{7}}\right)^{k-1}$

54. $\displaystyle\sum_{k=0}^{\infty} (-1)^k \left(\frac{3}{4}\right)^k$

55. Rewrite each infinite repeating decimal as a quotient of integers:

(a) $0.\overline{83}$ (b) $0.\overline{91}$ (c) $4.65\overline{223}$ (d) $3.1\overline{9}$

© **56.** Suppose that on each separate swing, a pendulum describes an arc whose length is 0.98 of the length of the preceding arc. If the length of the first arc is 24 centimeters, what total distance is covered by the pendulum before it comes to rest?

In Problems 57 to 66, find the value of the given expression.

57. $_6P_4$ **58.** $_7P_4$ **59.** $_{11}P_2$ **60.** $_8P_3$

61. $_{16}C_2$ © **62.** $_{16}C_7$ **63.** $_9C_3$ **64.** $_7C_5$

65. $_8C_8$ **66.** $_{19}C_0$

67. A woman has a choice of four airlines between Chicago and New York, and three airlines between New York and London. In how many ways can she fly from Chicago to London, if she stops in New York?

© **68.** (a) How many different four-letter "words" can be formed using the twenty-six letters of the alphabet, if repeated letters are allowed?

(b) In how many ways can the letters of the word WORTH be arranged?

69. How many different four-course meals can be ordered in a restaurant if there is a choice of three soups, four entrees, two salads, and five desserts?

© **70.** In how many ways can seven of twelve books be chosen and arranged next to each other on a shelf?

71. How many different signals can be sent by arranging up to five distinguishable flags, one above the other, on a flag pole with the condition that the flags can be used one, two, three, four, or five at a time?

72. How many straight lines can be drawn through pairs of ten points, if no three points are in a straight line?

73. Twelve people meet and shake hands. If everyone shakes hands with everyone else, how many handshakes are exchanged?

© **74.** A bridge hand consists of 13 cards dealt from a deck of 52. How many different bridge hands are there?

© **75.** A corporation owns 20 motels. If there are 25 people eligible to be managers of these motels, in how many ways can the motel managers be appointed?

76. In how many distinguishable ways can the letters of the word MINIMUM be arranged?

77. In how many distinguishable ways can three identical racquet balls, four identical golf balls, and five identical tennis balls be placed in a row?

78. If three dimes are tossed in the air, find the probability that all three fall "heads."

79. You meet a married couple and learn that they have two children. During the course of the conversation, it becomes clear that at least one of their children is a girl. What is the probability that the other child is a girl?

80. What is the probability of rolling an "11" with two dice?

81. Four married couples draw lots to decide who will be partners in a card game. If partners must be of opposite sexes, what is the probability that no woman is paired with her own husband?

82. A bag contains four white balls, six black balls, three red balls, and eight green balls. If two balls are selected blindly, what is the probability that they are of the same color?

83. Professor Grumbles has four suits of clothes, each consisting of a vest, trousers, and a jacket. If he dresses at random, what is the probability that his clothes match?

Tables of Logarithms and Exponentials

In this appendix, we present Table IA of natural logarithms, Table IB of common logarithms, Table IC of exponential functions, and examples illustrating their use.

Table IA gives values of ln x, rounded off to four decimal places, corresponding to values of x between 1 and 9.99 in steps of 0.01.

Example 1   Use Table IA to find the value of ln 3.47 rounded off to four decimal places.

Solution   We begin by locating the number 3.4 in the vertical column on the left side of the table. The numbers in the horizontal row to the right of 3.4 are the natural logarithms of the numbers from 3.40 to 3.49 in steps of 0.01. The entry in this horizontal row below the heading 0.07 is the natural logarithm of 3.47. Thus,

$$\ln 3.47 = 1.2442.$$ ■

For values of x lying between two numbers whose natural logarithms are shown in Table IA, it is possible to find the approximate value of ln x (again rounded off to four decimal places) by using **linear interpolation**. The basic idea of linear interpolation is that small changes in the independent variable cause approximately proportional changes in the values of a function.

Example 2   Use linear interpolation and Table IA to find the approximate value of ln 2.724.

Solution   We arrange the work as follows:

$$0.01 \left[0.004 \begin{bmatrix} \begin{array}{cc} x & \ln x \\ 2.72 & 1.0006 \\ 2.724 & \ln 2.724 \end{array} \end{bmatrix} d \\ \begin{array}{cc} 2.73 & 1.0043 \end{array} \right] 0.0037.$$

We have used Table IA to determine that $\ln 2.72 = 1.0006$ and $\ln 2.73 = 1.0043$. The notation

$$0.004\begin{bmatrix}2.72 \\ 2.724\end{bmatrix}$$

indicates that the difference between 2.72 and 2.724 (bottom number minus top number) is

$$2.724 - 2.72 = 0.004.$$

Other such differences are indicated similarly. For linear interpolation, we assume that corresponding differences are (approximately) proportional, so that

$$\frac{0.004}{0.01} = \frac{d}{0.0037} \qquad \text{or} \qquad d = 0.0037\left(\frac{0.004}{0.01}\right) = 0.00148.$$

Thus, rounded off to four decimal places, we have

$$\ln 2.724 = 1.0006 + d$$
$$= 1.0006 + 0.00148 = 1.0021. \qquad ■$$

Table IB gives values of $\log x$, rounded off to four decimal places, corresponding to values of x between 1 and 9.99 in steps of 0.01.

Example 3   Using Table IB, find the value of $\log 7.36$ (rounded off to four decimal places).

Solution $\log 7.36 = 0.8669$ ■

Example 4   Use linear interpolation and Table IB to find the approximate value of $\log 5.068$.

Solution   We have

$$0.01\left[0.008\begin{bmatrix} & x & \log x \\ 5.06 & 0.7042 \\ 5.068 & \log 5.068 \end{bmatrix}d \atop 5.07 \quad 0.7050\right]0.0008.$$

Thus, $\dfrac{0.008}{0.01} = \dfrac{d}{0.0008}$ or $d = 0.0008\left(\dfrac{0.008}{0.01}\right) = 0.00064;$

hence, rounded off to four decimal places,

$$\log 5.068 = 0.7042 + d$$
$$= 0.7042 + 0.00064$$
$$= 0.7048. \qquad ■$$

Now, suppose that x is a positive number less than 1 or greater than 9.99. To find $\log x$, begin by writing x in scientific notation

$$x = p \times 10^n,$$

where $1 \leq p < 10$ and n is an integer. (See Section 1.9, page 58.) Thus,

$$\log x = \log p + \log 10^n = \log p + n.$$

The quantity $\log p$ is called the **mantissa** of $\log x$, whereas the integer n is called the **characteristic** of $\log x$. Thus,

$$\log x = \text{mantissa} + \text{characteristic}.$$

You can find the mantissa by using Table IB.

In Examples 5 and 6, use Table IB to find the approximate value of each logarithm.

Example 5 $\log 371.4$

Solution   In scientific notation,

$$371.4 = 3.714 \times 10^2.$$

Here, the mantissa

$$\log 3.714 = 0.5698$$

is obtained by using linear interpolation and Table IB; the characteristic is $n = 2$, and so, rounded off to four decimal places,

$$\log 371.4 = 0.5698 + 2 = 2.5698.$$ ■

Example 6 $\log 0.05422$

Solution   In scientific notation

$$0.05422 = 5.422 \times 10^{-2}.$$

Here, the mantissa

$$\log 5.422 = 0.7342$$

is obtained by using linear interpolation and Table IB; the characteristic is $n = -2$, and so, rounded off to four decimal places,

$$\log 0.05422 = 0.7342 + (-2)$$
$$= -1.2658.$$

Although a scientific calculator gives $\log 0.05422 = -1.2658$ (rounded off to four decimal places), it is customary to leave the answer in the form

$$\log 0.05422 = 0.7342 + (-2)$$

when using a table of logarithms.   ■

By reading Table IB "backwards," you can find 10^x (sometimes called the **antilogarithm** of x).

In Examples 7 to 10, use Table IB to find each antilogarithm.

Example 7 $10^{0.9159}$

Solution   In the body of Table IB, we find 0.9159 and see (by reading the table "backwards") that it is the common logarithm of 8.24; hence,

$$\log 8.24 = 0.9159$$

or

$$10^{0.9159} = 8.24.$$ ■

Example 8 $10^{0.03}$

Solution   The number $0.03 = 0.0300$ does not appear in the body of Table IB. The numbers closest to it that do appear are 0.0294 and 0.0334. Thus, we must use linear interpolation. We have

$$
0.0040\left[0.0006\left[\begin{array}{cc} x & 10^x \\ 0.0294 & 1.07 \\ 0.0300 & 10^{0.03} \end{array}\right]d \atop 0.0334 \quad 1.08\right]0.01.
$$

Thus, $\dfrac{0.0006}{0.0040} = \dfrac{d}{0.01}$ or $d = 0.01\left(\dfrac{0.0006}{0.0040}\right) = 0.0015,$

so that, rounded off to four significant digits,

$$10^{0.03} = 1.07 + d = 1.07 + 0.0015 = 1.072.$$ ■

Example 9 $10^{4.3312}$

Solution $$10^{4.3312} = 10^{4+0.3312} = 10^4 \cdot 10^{0.3312}.$$

Using Table IB "backwards" with linear interpolation, we find that, rounded off to four significant digits,

$$10^{0.3312} = 2.144.$$

Therefore, $10^{4.3312} = 10^4 \cdot 10^{0.3312} = 10{,}000(2.144) = 21{,}440.$ ■

Example 10 $10^{-5.7074}$

Solution   We begin by writing -5.7074 in the form

$$-5.7074 = \text{mantissa} + \text{characteristic},$$

where the characteristic is negative and the mantissa is between 0 and 1. Evidently, the characteristic is -6; hence,

$$\text{mantissa} = -5.7074 - (-6) = 0.2926.$$

Using Table IB "backwards" with linear interpolation, we find that, rounded off to four significant digits,

$$10^{0.2926} = 1.961.$$

Therefore, rounded off to four significant digits,

$$10^{-5.7074} = 10^{0.2926 + (-6)} = 10^{0.2926} \cdot 10^{-6} = 1.961(10^{-6})$$
$$= 0.000,001,961.$$

■

By reading Table IA "backwards," you can find values of e^x; however, for convenience, we present a separate table of such values (Table IC).

Example 11 Use linear interpolation and Table IC to find the approximate value of $e^{1.96}$.

Solution We have

$$
0.1\left[0.06\left[\begin{array}{cc} x & e^x \\ 1.9 & 6.6859 \\ 1.96 & e^{1.96} \end{array}\right]d\right]0.7032
$$
$$
\begin{array}{cc} 2.0 & 7.3891 \end{array}
$$

Thus, $\dfrac{0.06}{0.1} = \dfrac{d}{0.7032}$ or $d = 0.7032\left(\dfrac{0.06}{0.1}\right) = 0.4219,$

so $e^{1.96} = 6.6859 + 0.4219 = 7.1078.$

Because Table IC is just a "short table" of values of e^x, linear interpolation may not be very accurate here. (A calculator gives $e^{1.96} = 7.0993$, rounded off to five significant digits.)

■

Table IA **Natural Logarithms**

x	0.00	0.01	0.02	0.03	0.04	0.05	0.06	0.07	0.08	0.09
1.0	0.0000	0.0100	0.0198	0.0296	0.0392	0.0488	0.0583	0.0677	0.0770	0.0862
1.1	0.0953	0.1044	0.1133	0.1222	0.1310	0.1398	0.1484	0.1570	0.1655	0.1740
1.2	0.1823	0.1906	0.1989	0.2070	0.2151	0.2231	0.2311	0.2390	0.2469	0.2546
1.3	0.2624	0.2700	0.2776	0.2852	0.2927	0.3001	0.3075	0.3148	0.3221	0.3293
1.4	0.3365	0.3436	0.3507	0.3577	0.3646	0.3716	0.3784	0.3853	0.3920	0.3988
1.5	0.4055	0.4121	0.4187	0.4253	0.4318	0.4383	0.4447	0.4511	0.4574	0.4637
1.6	0.4700	0.4762	0.4824	0.4886	0.4947	0.5008	0.5068	0.5128	0.5188	0.5247
1.7	0.5306	0.5365	0.5423	0.5481	0.5539	0.5596	0.5653	0.5710	0.5766	0.5822
1.8	0.5878	0.5933	0.5988	0.6043	0.6098	0.6152	0.6206	0.6259	0.6313	0.6366
1.9	0.6419	0.6471	0.6523	0.6575	0.6627	0.6678	0.6729	0.6780	0.6831	0.6881
2.0	0.6931	0.6981	0.7031	0.7080	0.7130	0.7178	0.7227	0.7275	0.7324	0.7372
2.1	0.7419	0.7467	0.7514	0.7561	0.7608	0.7655	0.7701	0.7747	0.7793	0.7839
2.2	0.7885	0.7930	0.7975	0.8020	0.8065	0.8109	0.8154	0.8198	0.8242	0.8286
2.3	0.8329	0.8372	0.8416	0.8459	0.8502	0.8544	0.8587	0.8629	0.8671	0.8713
2.4	0.8755	0.8796	0.8838	0.8879	0.8920	0.8961	0.9002	0.9042	0.9083	0.9123
2.5	0.9163	0.9203	0.9243	0.9282	0.9322	0.9361	0.9400	0.9439	0.9478	0.9517
2.6	0.9555	0.9594	0.9632	0.9670	0.9708	0.9746	0.9783	0.9821	0.9858	0.9895
2.7	0.9933	0.9969	1.0006	1.0043	1.0080	1.0116	1.0152	1.0188	1.0225	1.0260
2.8	1.0296	1.0332	1.0367	1.0403	1.0438	1.0473	1.0508	1.0543	1.0578	1.0613
2.9	1.0647	1.0682	1.0716	1.0750	1.0784	1.0818	1.0852	1.0886	1.0919	1.0953
3.0	1.0986	1.1019	1.1053	1.1086	1.1119	1.1151	1.1184	1.1217	1.1249	1.1282
3.1	1.1314	1.1346	1.1378	1.1410	1.1442	1.1474	1.1506	1.1537	1.1569	1.1600
3.2	1.1632	1.1663	1.1694	1.1725	1.1756	1.1787	1.1817	1.1848	1.1878	1.1909
3.3	1.1939	1.1970	1.2000	1.2030	1.2060	1.2090	1.2119	1.2149	1.2179	1.2208
3.4	1.2238	1.2267	1.2296	1.2326	1.2355	1.2384	1.2413	1.2442	1.2470	1.2499
3.5	1.2528	1.2556	1.2585	1.2613	1.2641	1.2669	1.2698	1.2726	1.2754	1.2782
3.6	1.2809	1.2837	1.2865	1.2892	1.2920	1.2947	1.2975	1.3002	1.3029	1.3056
3.7	1.3083	1.3110	1.3137	1.3164	1.3191	1.3218	1.3244	1.3271	1.3297	1.3324
3.8	1.3350	1.3376	1.3403	1.3429	1.3455	1.3481	1.3507	1.3533	1.3558	1.3584
3.9	1.3610	1.3635	1.3661	1.3686	1.3712	1.3737	1.3762	1.3788	1.3813	1.3838
4.0	1.3863	1.3888	1.3913	1.3938	1.3962	1.3987	1.4012	1.4036	1.4061	1.4085
4.1	1.4110	1.4134	1.4159	1.4183	1.4207	1.4231	1.4255	1.4279	1.4303	1.4327
4.2	1.4351	1.4375	1.4398	1.4422	1.4446	1.4469	1.4493	1.4516	1.4540	1.4563
4.3	1.4586	1.4609	1.4633	1.4656	1.4679	1.4702	1.4725	1.4748	1.4770	1.4793
4.4	1.4816	1.4839	1.4861	1.4884	1.4907	1.4929	1.4952	1.4974	1.4996	1.5019
4.5	1.5041	1.5063	1.5085	1.5107	1.5129	1.5151	1.5173	1.5195	1.5217	1.5239
4.6	1.5261	1.5282	1.5304	1.5326	1.5347	1.5369	1.5390	1.5412	1.5433	1.5454
4.7	1.5476	1.5497	1.5518	1.5539	1.5560	1.5581	1.5602	1.5623	1.5644	1.5665
4.8	1.5686	1.5707	1.5728	1.5748	1.5769	1.5790	1.5810	1.5831	1.5851	1.5872
4.9	1.5892	1.5913	1.5933	1.5953	1.5974	1.5994	1.6014	1.6034	1.6054	1.6074
5.0	1.6094	1.6114	1.6134	1.6154	1.6174	1.6194	1.6214	1.6233	1.6253	1.6273
5.1	1.6292	1.6312	1.6332	1.6351	1.6371	1.6390	1.6409	1.6429	1.6448	1.6467
5.2	1.6487	1.6506	1.6525	1.6544	1.6563	1.6582	1.6601	1.6620	1.6639	1.6658
5.3	1.6677	1.6696	1.6715	1.6734	1.6752	1.6771	1.6790	1.6808	1.6827	1.6845
5.4	1.6864	1.6882	1.6901	1.6919	1.6938	1.6956	1.6974	1.6993	1.7011	1.7029
5.5	1.7047	1.7066	1.7084	1.7102	1.7120	1.7138	1.7156	1.7174	1.7192	1.7210
5.6	1.7228	1.7246	1.7263	1.7281	1.7299	1.7317	1.7334	1.7352	1.7370	1.7387
5.7	1.7405	1.7422	1.7440	1.7457	1.7475	1.7492	1.7509	1.7527	1.7544	1.7561
5.8	1.7579	1.7596	1.7613	1.7630	1.7647	1.7664	1.7682	1.7699	1.7716	1.7733
5.9	1.7750	1.7766	1.7783	1.7800	1.7817	1.7834	1.7851	1.7867	1.7884	1.7901

x	0.00	0.01	0.02	0.03	0.04	0.05	0.06	0.07	0.08	0.09
6.0	1.7918	1.7934	1.7951	1.7967	1.7984	1.8001	1.8017	1.8034	1.8050	1.8066
6.1	1.8083	1.8099	1.8116	1.8132	1.8148	1.8165	1.8181	1.8197	1.8213	1.8229
6.2	1.8245	1.8262	1.8278	1.8294	1.8310	1.8326	1.8342	1.8358	1.8374	1.8390
6.3	1.8406	1.8421	1.8437	1.8453	1.8469	1.8485	1.8500	1.8516	1.8532	1.8547
6.4	1.8563	1.8579	1.8594	1.8610	1.8625	1.8641	1.8656	1.8672	1.8687	1.8703
6.5	1.8718	1.8733	1.8749	1.8764	1.8779	1.8795	1.8810	1.8825	1.8840	1.8856
6.6	1.8871	1.8886	1.8901	1.8916	1.8931	1.8946	1.8961	1.8976	1.8991	1.9006
6.7	1.9021	1.9036	1.9051	1.9066	1.9081	1.9095	1.9110	1.9125	1.9140	1.9155
6.8	1.9169	1.9184	1.9199	1.9213	1.9228	1.9242	1.9257	1.9272	1.9286	1.9301
6.9	1.9315	1.9330	1.9344	1.9359	1.9373	1.9387	1.9402	1.9416	1.9430	1.9445
7.0	1.9459	1.9473	1.9488	1.9502	1.9516	1.9530	1.9544	1.9559	1.9573	1.9587
7.1	1.9601	1.9615	1.9629	1.9643	1.9657	1.9671	1.9685	1.9699	1.9713	1.9727
7.2	1.9741	1.9755	1.9769	1.9782	1.9796	1.9810	1.9824	1.9838	1.9851	1.9865
7.3	1.9879	1.9892	1.9906	1.9920	1.9933	1.9947	1.9961	1.9974	1.9988	2.0001
7.4	2.0015	2.0028	2.0042	2.0055	2.0069	2.0082	2.0096	2.0109	2.0122	2.0136
7.5	2.0149	2.0162	2.0176	2.0189	2.0202	2.0215	2.0229	2.0242	2.0255	2.0268
7.6	2.0282	2.0295	2.0308	2.0321	2.0334	2.0347	2.0360	2.0373	2.0386	2.0399
7.7	2.0412	2.0425	2.0438	2.0451	2.0464	2.0477	2.0490	2.0503	2.0516	2.0528
7.8	2.0541	2.0554	2.0567	2.0580	2.0592	2.0605	2.0618	2.0631	2.0643	2.0665
7.9	2.0669	2.0681	2.0694	2.0707	2.0719	2.0732	2.0744	2.0757	2.0769	2.0782
8.0	2.0794	2.0807	2.0819	2.0832	2.0844	2.0857	2.0869	2.0882	2.0894	2.0906
8.1	2.0919	2.0931	2.0943	2.0956	2.0968	2.0980	2.0992	2.1005	2.1017	2.1029
8.2	2.1041	2.1054	2.1066	2.1078	2.1090	2.1102	2.1114	2.1126	2.1138	2.1150
8.3	2.1163	2.1175	2.1187	2.1199	2.1211	2.1223	2.1235	2.1247	2.1258	2.1270
8.4	2.1282	2.1294	2.1306	2.1318	2.1330	2.1342	2.1353	2.1365	2.1377	2.1389
8.5	2.1401	2.1412	2.1424	2.1436	2.1448	2.1459	2.1471	2.1483	2.1494	2.1506
8.6	2.1518	2.1529	2.1541	2.1552	2.1564	2.1576	2.1587	2.1599	2.1610	2.1622
8.7	2.1633	2.1645	2.1656	2.1668	2.1679	2.1691	2.1702	2.1713	2.1725	2.1736
8.8	2.1748	2.1759	2.1770	2.1782	2.1793	2.1804	2.1815	2.1827	2.1838	2.1849
8.9	2.1861	2.1872	2.1883	2.1894	2.1905	2.1917	2.1928	2.1939	2.1950	2.1961
9.0	2.1972	2.1983	2.1994	2.2006	2.2017	2.2028	2.2039	2.2050	2.2061	2.2072
9.1	2.2083	2.2094	2.2105	2.2116	2.2127	2.2138	2.2148	2.2159	2.2170	2.2181
9.2	2.2192	2.2203	2.2214	2.2225	2.2235	2.2246	2.2257	2.2268	2.2279	2.2289
9.3	2.2300	2.2311	2.2322	2.2332	2.2343	2.2354	2.2364	2.2375	2.2386	2.2396
9.4	2.2407	2.2418	2.2428	2.2439	2.2450	2.2460	2.2471	2.2481	2.2492	2.2502
9.5	2.2513	2.2523	2.2534	2.2544	2.2555	2.2565	2.2576	2.2586	2.2597	2.2607
9.6	2.2618	2.2628	2.2638	2.2649	2.2659	2.2670	2.2680	2.2690	2.2701	2.2711
9.7	2.2721	2.2732	2.2742	2.2752	2.2762	2.2773	2.2783	2.2793	2.2803	2.2814
9.8	2.2824	2.2834	2.2844	2.2854	2.2865	2.2875	2.2885	2.2895	2.2905	2.2915
9.9	2.2925	2.2935	2.2946	2.2956	2.2966	2.2976	2.2986	2.2996	2.3006	2.3016

Table IB **Common Logarithms**

x	0.00	0.01	0.02	0.03	0.04	0.05	0.06	0.07	0.08	0.09
1.0	0.0000	0.0043	0.0086	0.0128	0.0170	0.0212	0.0253	0.0294	0.0334	0.0374
1.1	0.0414	0.0453	0.0492	0.0531	0.0569	0.0607	0.0645	0.0682	0.0719	0.0755
1.2	0.0792	0.0828	0.0864	0.0899	0.0934	0.0969	0.1004	0.1038	0.1072	0.1106
1.3	0.1139	0.1173	0.1206	0.1239	0.1271	0.1303	0.1335	0.1367	0.1399	0.1430
1.4	0.1461	0.1492	0.1523	0.1553	0.1584	0.1614	0.1644	0.1673	0.1703	0.1732
1.5	0.1761	0.1790	0.1818	0.1847	0.1875	0.1903	0.1931	0.1959	0.1987	0.2014
1.6	0.2041	0.2068	0.2095	0.2122	0.2148	0.2175	0.2201	0.2227	0.2253	0.2279
1.7	0.2304	0.2330	0.2355	0.2380	0.2405	0.2430	0.2455	0.2480	0.2504	0.2529
1.8	0.2553	0.2577	0.2601	0.2625	0.2648	0.2672	0.2695	0.2718	0.2742	0.2765
1.9	0.2788	0.2810	0.2833	0.2856	0.2878	0.2900	0.2923	0.2945	0.2967	0.2989
2.0	0.3010	0.3032	0.3054	0.3075	0.3096	0.3118	0.3139	0.3160	0.3181	0.3201
2.1	0.3222	0.3243	0.3263	0.3284	0.3304	0.3324	0.3345	0.3365	0.3385	0.3404
2.2	0.3424	0.3444	0.3464	0.3483	0.3502	0.3522	0.3541	0.3560	0.3579	0.3598
2.3	0.3617	0.3636	0.3655	0.3674	0.3692	0.3711	0.3729	0.3747	0.3766	0.3784
2.4	0.3802	0.3820	0.3838	0.3856	0.3874	0.3892	0.3909	0.3927	0.3945	0.3962
2.5	0.3979	0.3997	0.4014	0.4031	0.4048	0.4065	0.4082	0.4099	0.4116	0.4133
2.6	0.4150	0.4166	0.4183	0.4200	0.4216	0.4232	0.4249	0.4265	0.4281	0.4298
2.7	0.4314	0.4330	0.4346	0.4362	0.4378	0.4393	0.4409	0.4425	0.4440	0.4456
2.8	0.4472	0.4487	0.4502	0.4518	0.4533	0.4548	0.4564	0.4579	0.4594	0.4609
2.9	0.4624	0.4639	0.4654	0.4669	0.4683	0.4698	0.4713	0.4728	0.4742	0.4757
3.0	0.4771	0.4786	0.4800	0.4814	0.4829	0.4843	0.4857	0.4871	0.4886	0.4900
3.1	0.4914	0.4928	0.4942	0.4955	0.4969	0.4983	0.4997	0.5011	0.5024	0.5038
3.2	0.5051	0.5065	0.5079	0.5092	0.5105	0.5119	0.5132	0.5145	0.5159	0.5172
3.3	0.5185	0.5198	0.5211	0.5224	0.5237	0.5250	0.5263	0.5276	0.5289	0.5302
3.4	0.5315	0.5328	0.5340	0.5353	0.5366	0.5378	0.5391	0.5403	0.5416	0.5428
3.5	0.5441	0.5453	0.5465	0.5478	0.5490	0.5502	0.5514	0.5527	0.5539	0.5551
3.6	0.5563	0.5575	0.5587	0.5599	0.5611	0.5623	0.5635	0.5647	0.5658	0.5670
3.7	0.5682	0.5694	0.5705	0.5717	0.5729	0.5740	0.5752	0.5763	0.5775	0.5786
3.8	0.5798	0.5809	0.5821	0.5832	0.5843	0.5855	0.5866	0.5877	0.5888	0.5899
3.9	0.5911	0.5922	0.5933	0.5944	0.5955	0.5966	0.5977	0.5988	0.5999	0.6010
4.0	0.6021	0.6031	0.6042	0.6053	0.6064	0.6075	0.6085	0.6096	0.6107	0.6117
4.1	0.6128	0.6138	0.6149	0.6160	0.6170	0.6180	0.6191	0.6201	0.6212	0.6222
4.2	0.6232	0.6243	0.6253	0.6263	0.6274	0.6284	0.6294	0.6304	0.6314	0.6325
4.3	0.6335	0.6345	0.6355	0.6365	0.6375	0.6385	0.6395	0.6405	0.6415	0.6425
4.4	0.6435	0.6444	0.6454	0.6464	0.6474	0.6484	0.6493	0.6503	0.6513	0.6522
4.5	0.6532	0.6542	0.6551	0.6561	0.6571	0.6580	0.6590	0.6599	0.6609	0.6618
4.6	0.6628	0.6637	0.6646	0.6656	0.6665	0.6675	0.6684	0.6693	0.6702	0.6712
4.7	0.6721	0.6730	0.6739	0.6749	0.6758	0.6767	0.6776	0.6785	0.6794	0.6803
4.8	0.6812	0.6821	0.6830	0.6839	0.6848	0.6857	0.6866	0.6875	0.6884	0.6893
4.9	0.6902	0.6911	0.6920	0.6928	0.6937	0.6946	0.6955	0.6964	0.6972	0.6981
5.0	0.6990	0.6998	0.7007	0.7016	0.7024	0.7033	0.7042	0.7050	0.7059	0.7067
5.1	0.7076	0.7084	0.7093	0.7101	0.7110	0.7118	0.7126	0.7135	0.7143	0.7152
5.2	0.7160	0.7168	0.7177	0.7185	0.7193	0.7202	0.7210	0.7218	0.7226	0.7235
5.3	0.7243	0.7251	0.7259	0.7267	0.7275	0.7284	0.7292	0.7300	0.7308	0.7316
5.4	0.7324	0.7332	0.7340	0.7348	0.7356	0.7364	0.7372	0.7380	0.7388	0.7396
5.5	0.7404	0.7412	0.7419	0.7427	0.7435	0.7443	0.7451	0.7459	0.7466	0.7474
5.6	0.7482	0.7490	0.7497	0.7505	0.7513	0.7520	0.7528	0.7536	0.7543	0.7551
5.7	0.7559	0.7566	0.7574	0.7582	0.7589	0.7597	0.7604	0.7612	0.7619	0.7627
5.8	0.7634	0.7642	0.7649	0.7657	0.7664	0.7672	0.7679	0.7686	0.7694	0.7701
5.9	0.7709	0.7716	0.7723	0.7731	0.7738	0.7745	0.7752	0.7760	0.7767	0.7774

x	0.00	0.01	0.02	0.03	0.04	0.05	0.06	0.07	0.08	0.09
6.0	0.7782	0.7789	0.7796	0.7803	0.7810	0.7818	0.7825	0.7832	0.7839	0.7846
6.1	0.7853	0.7860	0.7868	0.7875	0.7882	0.7889	0.7896	0.7903	0.7910	0.7917
6.2	0.7924	0.7931	0.7938	0.7945	0.7952	0.7959	0.7966	0.7973	0.7980	0.7987
6.3	0.7993	0.8000	0.8007	0.8014	0.8021	0.8028	0.8035	0.8041	0.8048	0.8055
6.4	0.8062	0.8069	0.8075	0.8082	0.8089	0.8096	0.8102	0.8109	0.8116	0.8122
6.5	0.8129	0.8136	0.8142	0.8149	0.8156	0.8162	0.8169	0.8176	0.8182	0.8189
6.6	0.8195	0.8202	0.8209	0.8215	0.8222	0.8228	0.8235	0.8241	0.8248	0.8254
6.7	0.8261	0.8267	0.8274	0.8280	0.8287	0.8293	0.8299	0.8306	0.8312	0.8319
6.8	0.8325	0.8331	0.8338	0.8344	0.8351	0.8357	0.8363	0.8370	0.8376	0.8382
6.9	0.8388	0.8395	0.8401	0.8407	0.8414	0.8420	0.8426	0.8432	0.8439	0.8445
7.0	0.8451	0.8457	0.8463	0.8470	0.8476	0.8482	0.8488	0.8494	0.8500	0.8506
7.1	0.8513	0.8519	0.8525	0.8531	0.8537	0.8543	0.8549	0.8555	0.8561	0.8567
7.2	0.8573	0.8579	0.8585	0.8591	0.8597	0.8603	0.8609	0.8615	0.8621	0.8627
7.3	0.8633	0.8639	0.8645	0.8651	0.8657	0.8663	0.8669	0.8675	0.8681	0.8686
7.4	0.8692	0.8698	0.8704	0.8710	0.8716	0.8722	0.8727	0.8733	0.8739	0.8745
7.5	0.8751	0.8756	0.8762	0.8768	0.8774	0.8779	0.8785	0.8791	0.8797	0.8802
7.6	0.8808	0.8814	0.8820	0.8825	0.8831	0.8837	0.8842	0.8848	0.8854	0.8859
7.7	0.8865	0.8871	0.8876	0.8882	0.8887	0.8893	0.8899	0.8904	0.8910	0.8915
7.8	0.8921	0.8927	0.8932	0.8938	0.8943	0.8949	0.8954	0.8960	0.8965	0.8971
7.9	0.8976	0.8982	0.8987	0.8993	0.8998	0.9004	0.9009	0.9015	0.9020	0.9025
8.0	0.9031	0.9036	0.9042	0.9047	0.9053	0.9058	0.9063	0.9069	0.9074	0.9079
8.1	0.9085	0.9090	0.9096	0.9101	0.9106	0.9112	0.9117	0.9122	0.9128	0.9133
8.2	0.9138	0.9143	0.9149	0.9154	0.9159	0.9165	0.9170	0.9175	0.9180	0.9186
8.3	0.9191	0.9196	0.9201	0.9206	0.9212	0.9217	0.9222	0.9227	0.9232	0.9238
8.4	0.9243	0.9248	0.9253	0.9258	0.9263	0.9269	0.9274	0.9279	0.9284	0.9289
8.5	0.9294	0.9299	0.9304	0.9309	0.9315	0.9320	0.9325	0.9330	0.9335	0.9340
8.6	0.9345	0.9350	0.9355	0.9360	0.9365	0.9370	0.9375	0.9380	0.9385	0.9390
8.7	0.9395	0.9400	0.9405	0.9410	0.9415	0.9420	0.9425	0.9430	0.9435	0.9440
8.8	0.9445	0.9450	0.9455	0.9460	0.9465	0.9469	0.9474	0.9479	0.9484	0.9489
8.9	0.9494	0.9499	0.9504	0.9509	0.9513	0.9518	0.9523	0.9528	0.9533	0.9538
9.0	0.9542	0.9547	0.9552	0.9557	0.9562	0.9566	0.9571	0.9576	0.9581	0.9586
9.1	0.9590	0.9595	0.9600	0.9605	0.9609	0.9614	0.9619	0.9624	0.9628	0.9633
9.2	0.9638	0.9643	0.9647	0.9652	0.9657	0.9661	0.9666	0.9671	0.9675	0.9680
9.3	0.9685	0.9689	0.9694	0.9699	0.9703	0.9708	0.9713	0.9717	0.9722	0.9727
9.4	0.9731	0.9736	0.9741	0.9745	0.9750	0.9754	0.9759	0.9763	0.9768	0.9773
9.5	0.9777	0.9782	0.9786	0.9791	0.9795	0.9800	0.9805	0.9809	0.9814	0.9818
9.6	0.9823	0.9827	0.9832	0.9836	0.9841	0.9845	0.9850	0.9854	0.9859	0.9863
9.7	0.9868	0.9872	0.9877	0.9881	0.9886	0.9890	0.9894	0.9899	0.9903	0.9908
9.8	0.9912	0.9917	0.9921	0.9926	0.9930	0.9934	0.9939	0.9943	0.9948	0.9952
9.9	0.9956	0.9961	0.9965	0.9969	0.9974	0.9978	0.9983	0.9987	0.9991	0.9996

Table IC **Exponential Functions**

x	e^x	e^{-x}	x	e^x	e^{-x}
0.00	1.0000	1.0000	3.0	20.086	0.0498
0.05	1.0513	0.9512	3.1	22.198	0.0450
0.10	1.1052	0.9048	3.2	24.533	0.0408
0.15	1.1618	0.8607	3.3	27.113	0.0369
0.20	1.2214	0.8187	3.4	29.964	0.0334
0.25	1.2840	0.7788	3.5	33.115	0.0302
0.30	1.3499	0.7408	3.6	36.598	0.0273
0.35	1.4191	0.7047	3.7	40.447	0.0247
0.40	1.4918	0.6703	3.8	44.701	0.0224
0.45	1.5683	0.6376	3.9	49.402	0.0202
0.50	1.6487	0.6065	4.0	54.598	0.0183
0.55	1.7333	0.5769	4.1	60.340	0.0166
0.60	1.8221	0.5488	4.2	66.686	0.0150
0.65	1.9155	0.5220	4.3	73.700	0.0136
0.70	2.0138	0.4966	4.4	81.451	0.0123
0.75	2.1170	0.4724	4.5	90.017	0.0111
0.80	2.2255	0.4493	4.6	99.484	0.0101
0.85	2.3396	0.4274	4.7	109.95	0.0091
0.90	2.4596	0.4066	4.8	121.51	0.0082
0.95	2.5857	0.3867	4.9	134.29	0.0074
1.0	2.7183	0.3679	5.0	148.41	0.0067
1.1	3.0042	0.3329	5.1	164.02	0.0061
1.2	3.3201	0.3012	5.2	181.27	0.0055
1.3	3.6693	0.2725	5.3	200.34	0.0050
1.4	4.0552	0.2466	5.4	221.41	0.0045
1.5	4.4817	0.2231	5.5	244.69	0.0041
1.6	4.9530	0.2019	5.6	270.43	0.0037
1.7	5.4739	0.1827	5.7	298.87	0.0033
1.8	6.0496	0.1653	5.8	330.30	0.0030
1.9	6.6859	0.1496	5.9	365.04	0.0027
2.0	7.3891	0.1353	6.0	403.43	0.0025
2.1	8.1662	0.1225	6.5	665.14	0.0015
2.2	9.0250	0.1108	7.0	1096.6	0.0009
2.3	9.9742	0.1003	7.5	1808.0	0.0006
2.4	11.023	0.0907	8.0	2981.0	0.0003
2.5	12.182	0.0821	8.5	4914.8	0.0002
2.6	13.464	0.0743	9.0	8103.1	0.0001
2.7	14.880	0.0672	9.5	13,360	0.00007
2.8	16.445	0.0608	10.0	22,026	0.00004
2.9	18.174	0.0550			

Tables of Trigonometric Functions

In this appendix, we present Table IIA and Table IIB of trigonometric functions of angles measured in degrees and radians, and we give examples illustrating their use.

Table IIA gives values of sine, tangent, cotangent, and cosine, rounded off to four or five significant digits, corresponding to angles between $0°$ and $90°$ in steps of $0.1°$. Angles between $0°$ and $45°$ are found in the vertical columns on the left sides of the tables, and the captions at the tops of the tables apply to these angles. Angles between $45°$ and $90°$ are found in the vertical columns on the right sides of the tables, and the captions at the bottoms of the tables apply to these angles.

In Examples 1 to 3, use Table II A to find the approximate value of each trigonometric function.

Example 1 $\cos 29.3°$

Solution   Since $29.3° < 45°$, we begin by locating 29.3 in the vertical column on the *left* side of the table. As can be seen from the captions *above* the table, the numbers to the *right* of 29.3 are the sine, tangent, cotangent, and cosine of $29.3°$. Looking in the vertical column with the caption "cosine," we find that

$$\cos 29.3° = 0.8721.$$

Example 2 $\tan 128.8°$

Solution   The angle $128.8°$ lies in quadrant II, so its tangent is negative. The reference angle corresponding to $128.8°$ is

$$180° - 128.8° = 51.2°.$$

Thus, $\tan 128.8° = -\tan 51.2°.$

Since $51.2° > 45°$, we locate 51.2 in the vertical column on the *right* side of the table. As can be seen from the captions *below* the table, the numbers to the *left* of 51.2 are the sine, tangent, cotangent, and cosine of 51.2°. Thus,

$$\tan 51.2° = 1.2437, \quad \text{so that,} \quad \tan 128.8° = -1.2437.$$ ∎

Example 3   csc 72.3°

Solution   Since values of the cosecant do not appear in the table, we use the fact that

$$\csc 72.3° = \frac{1}{\sin 72.3°}.$$

From the table,

$$\sin 72.3° = 0.9527, \quad \text{so} \quad \csc 72.3° = \frac{1}{0.9527} = 1.050,$$

(rounded off to four significant digits). ∎

Linear interpolation can be used to find the values of trigonometric functions of angles lying between two angles in Table IIA. The basic idea of linear interpolation is that small changes in the angle cause approximately proportional changes in the values of the function.

Example 4   Use linear interpolation and Table IIA to find the approximate value of sin 19.74°.

Solution   We arrange the work as follows:

$$0.1°\left[0.04°\left[\begin{matrix} 19.7° \\ 19.74° \end{matrix} \quad \begin{matrix} 0.3371 \\ \sin 19.74° \end{matrix}\right]d\right]0.0016.$$

$$\begin{matrix} x & \sin x \\ 19.8° & 0.3387 \end{matrix}$$

Here, we have used Table IIA to determine that $\sin 19.7° = 0.3371$ and that $\sin 19.8° = 0.3387$. The notation

$$0.04°\left[\begin{matrix} 19.7° \\ 19.74° \end{matrix}\right.$$

indicates that the difference between 19.7° and 19.74° (bottom number minus top number) is

$$19.74° - 19.7° = 0.04°.$$

Other differences are indicated similarly. For linear interpolation, we assume that corresponding differences are (approximately) proportional, so that

$$\frac{0.04°}{0.1°} = \frac{d}{0.0016} \quad \text{or} \quad d = 0.0016\left(\frac{0.04}{0.1}\right) = 0.00064.$$

Thus, rounded off to four significant digits,

$$\begin{aligned} \sin 19.74° &= 0.3371 + d \\ &= 0.3371 + 0.00064 \\ &= 0.3377. \end{aligned}$$ ∎

By reading Table IIA "backwards," you can find values in degrees of the inverse trigonometric functions.

In Examples 5 and 6, use Table IIA to find the value of each inverse trigonometric function rounded off to the nearest 0.1°.

Example 5 $\cos^{-1} 0.3714$

Solution   Looking in the body of the table, we find 0.3714 in the vertical column with the *lower* caption "cos." By reading the table "backwards," we see that

$$\cos 68.2° = 0.3714 \qquad \text{or} \qquad \cos^{-1} 0.3714 = 68.2°.$$ ∎

Example 6 $\sin^{-1} 0.7320$

Solution   The number 0.7320 does not appear in the vertical sine column in Table IIA. The numbers closest to it are

$$0.7314 = \sin 47.0° \qquad \text{and} \qquad 0.7325 = \sin 47.1°.$$

Using linear interpolation, we have

$$0.0011\left[0.0006\left[\begin{matrix} 0.7314 \\ 0.7320 \\ 0.7325 \end{matrix}\quad\begin{matrix} 47.0° \\ \sin^{-1}0.7320 \\ 47.1° \end{matrix}\right]d\right]0.1°.$$

Thus, $$\frac{0.0006}{0.0011} = \frac{d}{0.1°} \qquad \text{or} \qquad d = 0.1°\left(\frac{0.0006}{0.0011}\right) = 0.05°$$

(rounded off to the nearest 0.01°), so that

$$\begin{aligned} \sin^{-1} 0.7320 &= 47.0° + d \\ &= 47.0° + 0.05° \\ &= 47.05°. \end{aligned}$$ ∎

Table IIB gives values of the six trigonometric functions, rounded off to four or five significant digits, corresponding to angles between 0 and $\pi/2$ radians in steps of 0.01 radian.

In Examples 7 and 8, use Table IIB to find the approximate value of each trigonometric function.

Example 7   csc 1.02

Solution   By Table IIB

$$\csc 1.02 = 1.174.$$

∎

Example 8   sin 3.538

Solution   In radians, $\pi < 3.538 < (3\pi/2)$, so 3.538 is a quadrant III angle and its sine is negative. The reference angle corresponding to 3.538 is

$$3.538 - \pi = 0.396$$

(rounded off to three decimal places), so

$$\sin 3.538 = -\sin 0.396.$$

Using Table IIB and linear interpolation, we have

$$0.01\left[0.006\left[\begin{matrix}x & \sin x \\ 0.39 & 0.3802 \\ 0.396 & \sin 0.396 \\ 0.40 & 0.3894\end{matrix}\right]d\right]0.0092.$$

Thus, $\dfrac{0.006}{0.01} = \dfrac{d}{0.0092}$ or $d = 0.0092\left(\dfrac{0.006}{0.01}\right) = 0.0055;$

so that, rounded off to four decimal places,

$$\begin{aligned}\sin 0.396 &= 0.3802 + d \\ &= 0.3802 + 0.0055 \\ &= 0.3857.\end{aligned}$$

Therefore, $\sin 3.538 = -0.3857.$

(Because of all the rounding off and the linear interpolation, this value isn't very accurate—a calculator gives

$$\sin 3.538 = -0.3861.$$

However, it's the best we can do with our table.)

∎

By reading Table IIB "backwards," you can find values in radians of the inverse trigonometric functions.

In Examples 9 and 10, use Table IIB to find the value of each inverse trigonometric function rounded off to the nearest 0.01 radian.

Example 9   arctan 2.427

Solution   Looking in the body of the table in the vertical column for tangents, we find 2.427 in the position corresponding to 1.18 radians. Hence,

$$\arctan 2.427 = 1.18.$$

Example 10 $\cos^{-1} 0.8932$

Solution   Since 0.8932 does not appear in the body of the table in the vertical column for cosines, we must use linear interpolation. Thus,

$$-0.0045\left[-0.0029\left[\begin{matrix} 0.8961 \\ 0.8932 \\ 0.8916 \end{matrix}\begin{matrix} 0.46 \\ \cos^{-1}0.8932 \\ 0.47 \end{matrix}\right]d\right]0.01.$$

It follows that

$$\frac{-0.0029}{-0.0045} = \frac{d}{0.01} \qquad \text{or} \qquad d = 0.01\left(\frac{-0.0029}{-0.0045}\right) = 0.006$$

(rounded off to three decimal places), and so

$$\cos^{-1} 0.8932 = 0.46 + d$$
$$= 0.46 + 0.006$$
$$= 0.466.$$

Table IIA **Trigonometric Functions—Degree Measure**

Deg.	sin	tan	cot	cos		Deg.	sin	tan	cot	cos	
0.0	0.00000	0.00000	∞	1.0000	90.0	5.0	0.08716	0.08749	11.430	0.9962	85.0
.1	.00175	.00175	573.0	1.0000	89.9	.1	.08889	.08925	11.205	.9960	84.9
.2	.00349	.00349	286.5	1.0000	.8	.2	.09063	.09101	10.988	.9959	.8
.3	.00524	.00524	191.0	1.0000	.7	.3	.09237	.09277	10.780	.9957	.7
.4	.00698	.00698	143.24	1.0000	.6	.4	.09411	.09453	10.579	.9956	.6
.5	.00873	.00873	114.59	1.0000	.5	.5	.09585	.09629	10.385	.9954	.5
.6	.01047	.01047	95.49	0.9999	.4	.6	.09758	.09805	10.199	.9952	.4
.7	.01222	.01222	81.85	.9999	.3	.7	.09932	.09981	10.019	.9951	.3
.8	.01396	.01396	71.62	.9999	.2	.8	.10106	.10158	9.845	.9949	.2
.9	.01571	.01571	63.66	.9999	89.1	.9	.10279	.10334	9.677	.9947	84.1
1.0	0.01745	0.01746	57.29	0.9998	89.0	6.0	0.10453	0.10510	9.514	0.9945	84.0
.1	.01920	.01920	52.08	.9998	88.9	.1	.10626	.10687	9.357	.9943	83.9
.2	.02094	.02095	47.74	.9998	.8	.2	.10800	.10863	9.205	.9942	.8
.3	.02269	.02269	44.07	.9997	.7	.3	.10973	.11040	9.058	.9940	.7
.4	.02443	.02444	40.92	.9997	.6	.4	.11147	.11217	8.915	.9938	.6
.5	.02618	.02619	38.19	.9997	.5	.5	.11320	.11394	8.777	.9936	.5
.6	.02792	.02793	35.80	.9996	.4	.6	.11494	.11570	8.643	.9934	.4
.7	.02967	.02968	33.69	.9996	.3	.7	.11667	.11747	8.513	.9932	.3
.8	.03141	.03143	31.82	.9995	.2	.8	.11840	.11924	8.386	.9930	.2
.9	.03316	.03317	30.14	.9995	88.1	.9	.12014	.12101	8.264	.9928	83.1
2.0	0.03490	0.03492	28.64	0.9994	88.0	7.0	0.12187	0.12278	8.144	0.9925	83.0
.1	.03664	.03667	27.27	.9993	87.9	.1	.12360	.12456	8.028	.9923	82.9
.2	.03839	.03842	26.03	.9993	.8	.2	.12533	.12633	7.916	.9921	.8
.3	.04013	.04016	24.90	.9992	.7	.3	.12706	.12810	7.806	.9919	.7
.4	.04188	.04191	23.86	.9991	.6	.4	.12880	.12988	7.700	.9917	.6
.5	.04362	.04366	22.90	.9990	.5	.5	.13053	.13165	7.596	.9914	.5
.6	.04536	.04541	22.02	.9990	.4	.6	.13226	.13343	7.495	.9912	.4
.7	.04711	.04716	21.20	.9989	.3	.7	.13399	.13521	7.396	.9910	.3
.8	.04885	.04891	20.45	.9988	.2	.8	.13572	.13698	7.300	.9907	.2
.9	.05059	.05066	19.74	.9987	87.1	.9	.13744	.13876	7.207	.9905	82.1
3.0	0.05234	0.05241	19.081	0.9986	87.0	8.0	0.13917	0.14054	7.115	0.9903	82.0
.1	.05408	.05416	18.464	.9985	86.9	.1	.14090	.14232	7.026	.9900	81.9
.2	.05582	.05591	17.886	.9984	.8	.2	.14263	.14410	6.940	.9898	.8
.3	.05756	.05766	17.343	.9983	.7	.3	.14436	.14588	6.855	.9895	.7
.4	.05931	.05941	16.832	.9982	.6	.4	.14608	.14767	6.772	.9893	.6
.5	.06105	.06116	16.350	.9981	.5	.5	.14781	.14945	6.691	.9890	.5
.6	.06279	.06291	15.895	.9980	.4	.6	.14954	.15124	6.612	.9888	.4
.7	.06453	.06467	15.464	.9979	.3	.7	.15126	.15302	6.535	.9885	.3
.8	.06627	.06642	15.056	.9978	.2	.8	.15299	.15481	6.460	.9882	.2
.9	.06802	.06817	14.669	.9977	86.1	.9	.15471	.15660	6.386	.9880	81.1
4.0	0.06976	0.06993	14.301	0.9976	86.0	9.0	0.15643	0.15838	6.314	0.9877	81.0
.1	.07150	.07168	13.951	.9974	85.9	.1	.15816	.16017	6.243	.9874	80.9
.2	.07324	.07344	13.617	.9973	.8	.2	.15988	.16196	6.174	.9871	.8
.3	.07498	.07519	13.300	.9972	.7	.3	.16160	.16376	6.107	.9869	.7
.4	.07672	.07695	12.996	.9971	.6	.4	.16333	.16555	6.041	.9866	.6
.5	.07846	.07870	12.706	.9969	.5	.5	.16505	.16734	5.976	.9863	.5
.6	.08020	.08046	12.429	.9968	.4	.6	.16677	.16914	5.912	.9860	.4
.7	.08194	.08221	12.163	.9966	.3	.7	.16849	.17093	5.850	.9857	.3
.8	.08368	.08397	11.909	.9965	.2	.8	.17021	.17273	5.789	.9854	.2
.9	.08542	.08573	11.664	.9963	85.1	.9	.17193	.17453	5.730	.9851	80.1
5.0	0.08716	0.08749	11.430	0.9962	85.0	10.0	0.1736	0.1763	5.671	0.9848	80.0
	cos	cot	tan	sin	Deg.		cos	cot	tan	sin	Deg.

Deg.	sin	tan	cot	cos		Deg.	sin	tan	cot	cos	
10.0	0.1736	0.1763	5.671	0.9848	80.0	15.0	0.2588	0.2679	3.732	0.9659	75.0
.1	.1754	.1781	5.614	.9845	79.9	.1	.2605	.2698	3.706	.9655	74.9
.2	.1771	.1799	5.558	.9842	.8	.2	.2622	.2717	3.681	.9650	.8
.3	.1788	.1817	5.503	.9839	.7	.3	.2639	.2736	3.655	.9646	.7
.4	.1805	.1835	5.449	.9836	.6	.4	.2656	.2754	3.630	.9641	.6
.5	.1822	.1853	5.396	.9833	.5	.5	.2672	.2773	3.606	.9636	.5
.6	.1840	.1871	5.343	.9829	.4	.6	.2689	.2792	3.582	.9632	.4
.7	.1857	.1890	5.292	.9826	.3	.7	.2706	.2811	3.558	.9627	.3
.8	.1874	.1908	5.242	.9823	.2	.8	.2723	.2830	3.534	.9622	.2
.9	.1891	.1926	5.193	.9820	79.1	.9	.2740	.2849	3.511	.9617	74.1
11.0	0.1908	0.1944	5.145	0.9816	79.0	16.0	0.2756	0.2867	3.487	0.9613	74.0
.1	.1925	.1962	5.079	.9813	78.9	.1	.2773	.2886	3.465	.9608	73.9
.2	.1942	.1980	5.050	.9810	.8	.2	.2790	.2905	3.442	.9603	.8
.3	.1959	.1998	5.005	.9806	.7	.3	.2807	.2924	3.420	.9598	.7
.4	.1977	.2016	4.959	.9803	.6	.4	.2823	.2943	3.398	.9593	.6
.5	.1994	.2035	4.915	.9799	.5	.5	.2840	.2962	3.376	.9588	.5
.6	.2011	.2053	4.872	.9796	.4	.6	.2857	.2981	3.354	.9583	.4
.7	.2028	.2071	4.829	.9792	.3	.7	.2874	.3000	3.333	.9578	.3
.8	.2045	.2089	4.787	.9789	.2	.8	.2890	.3019	3.312	.9573	.2
.9	.2062	.2107	4.745	.9785	78.1	.9	.2907	.3038	3.291	.9568	73.1
12.0	0.2079	0.2126	4.705	0.9781	78.0	17.0	0.2924	0.3057	3.271	0.9563	73.0
.1	.2096	.2144	4.665	.9778	77.9	.1	.2940	.3076	3.251	.9558	72.9
.2	.2113	.2162	4.625	.9774	.8	.2	.2957	.3096	3.230	.9553	.8
.3	.2130	.2180	4.586	.9770	.7	.3	.2974	.3115	3.211	.9548	.7
.4	.2147	.2199	4.548	.9767	.6	.4	.2990	.3134	3.191	.9542	.6
.5	.2164	.2217	4.511	.9763	.5	.5	.3007	.3153	3.172	.9537	.5
.6	.2181	.2235	4.474	.9759	.4	.6	.3024	.3172	3.152	.9532	.4
.7	.2198	.2254	4.437	.9755	.3	.7	.3040	.3191	3.133	.9527	.3
.8	.2215	.2272	4.402	.9751	.2	.8	.3057	.3211	3.115	.9521	.2
.9	.2233	.2290	4.366	.9748	77.1	.9	.3074	.3230	3.096	.9516	72.1
13.0	0.2250	0.2309	4.331	0.9744	77.0	18.0	0.3090	0.3249	3.078	0.9511	72.0
.1	.2267	.2327	4.297	.9740	76.9	.1	.3107	.3269	3.060	.9505	71.9
.2	.2284	.2345	4.264	.9736	.8	.2	.3123	.3288	3.042	.9500	.8
.3	.2300	.2364	4.230	.9732	.7	.3	.3140	.3307	3.024	.9494	.7
.4	.2317	.2382	4.198	.9728	.6	.4	.3156	.3327	3.006	.9489	.6
.5	.2334	.2401	4.165	.9724	.5	.5	.3173	.3346	2.989	.9483	.5
.6	.2351	.2419	4.134	.9720	.4	.6	.3190	.3365	2.971	.9478	.4
.7	.2368	.2438	4.102	.9715	.3	.7	.3206	.3385	2.954	.9472	.3
.8	.2385	.2456	4.071	.9711	.2	.8	.3223	.3404	2.937	.9466	.2
.9	.2402	.2475	4.041	.9707	76.1	.9	.3239	.3424	2.921	.9461	71.1
14.0	0.2419	0.2493	4.011	0.9703	76.0	19.0	0.3256	0.3443	2.904	0.9455	71.0
.1	.2436	.2512	3.981	.9699	75.9	.1	.3272	.3463	2.888	.9449	70.9
.2	.2453	.2530	3.952	.9694	.8	.2	.3289	.3482	2.872	.9444	.8
.3	.2470	.2549	3.923	.9690	.7	.3	.3305	.3502	2.856	.9438	.7
.4	.2487	.2568	3.895	.9686	.6	.4	.3322	.3522	2.840	.9432	.6
.5	.2504	.2586	3.867	.9681	.5	.5	.3338	.3541	2.824	.9426	.5
.6	.2521	.2605	3.839	.9677	.4	.6	.3355	.3561	2.808	.9421	.4
.7	.2538	.2623	3.812	.9673	.3	.7	.3371	.3581	2.793	.9415	.3
.8	.2554	.2642	3.785	.9668	.2	.8	.3387	.3600	2.778	.9409	.2
.9	.2571	.2661	3.758	.9664	75.1	.9	.3404	.3620	2.762	.9403	70.1
15.0	0.2588	0.2679	3.732	0.9659	75.0	20.0	0.3420	0.3640	2.747	0.9397	70.0
	cos	cot	tan	sin	Deg.		cos	cot	tan	sin	Deg.

Table IIA **Trigonometric Functions—Degree Measure**

Deg.	sin	tan	cot	cos		Deg.	sin	tan	cot	cos	
20.0	0.3420	0.3640	2.747	0.9397	70.0	25.0	0.4226	0.4663	2.145	0.9063	65.0
.1	.3437	.3659	2.733	.9391	69.9	.1	.4242	.4684	2.135	.9056	64.9
.2	.3453	.3679	2.718	.9385	.8	.2	.4258	.4706	2.125	.9048	.8
.3	.3469	.3699	2.703	.9379	.7	.3	.4274	.4727	2.116	.9041	.7
.4	.3486	.3719	2.689	.9373	.6	.4	.4289	.4748	2.106	.9033	.6
.5	.3502	.3739	2.675	.9367	.5	.5	.4305	.4770	2.097	.9026	.5
.6	.3518	.3759	2.660	.9361	.4	.6	.4321	.4791	2.087	.9018	.4
.7	.3535	.3779	2.646	.9354	.3	.7	.4337	.4813	2.078	.9011	.3
.8	.3551	.3799	2.633	.9348	.2	.8	.4352	.4834	2.069	.9003	.2
.9	.3567	.3819	2.619	.9342	69.1	.9	.4368	.4856	2.059	.8996	64.1
21.0	0.3584	0.3839	2.605	0.9336	69.0	26.0	0.4384	0.4887	2.050	0.8988	64.0
.1	.3600	.3859	2.592	.9330	68.9	.1	.4399	.4899	2.041	.8980	63.9
.2	.3616	.3879	2.578	.9323	.8	.2	.4415	.4921	2.032	.8973	.8
.3	.3633	.3899	2.565	.9317	.7	.3	.4431	.4942	2.023	.8965	.7
.4	.3649	.3919	2.552	.9311	.6	.4	.4446	.4964	2.014	.8957	.6
.5	.3665	.3939	2.539	.9304	.5	.5	.4462	.4986	2.006	.8949	.5
.6	.3681	.3959	2.526	.9298	.4	.6	.4478	.5008	1.997	.8942	.4
.7	.3697	.3979	2.513	.9291	.3	.7	.4493	.5029	1.988	.8934	.3
.8	.3714	.4000	2.500	.9285	.2	.8	.4509	.5051	1.980	.8926	.2
.9	.3730	.4020	2.488	.9278	68.1	.9	.4524	.5073	1.971	.8918	63.1
22.0	0.3746	0.4040	2.475	0.9272	68.0	27.0	0.4540	0.5095	1.963	0.8910	63.0
.1	.3762	.4061	2.463	.9265	67.9	.1	.4555	.5117	1.954	.8902	62.9
.2	.3778	.4081	2.450	.9259	.8	.2	.4571	.5139	1.946	.8894	.8
.3	.3795	.4101	2.438	.9252	.7	.3	.4586	.5161	1.937	.8886	.7
.4	.3811	.4122	2.426	.9245	.6	.4	.4602	.5184	1.929	.8878	.6
.5	.3827	.4142	2.414	.9239	.5	.5	.4617	.5206	1.921	.8870	.5
.6	.3843	.4163	2.402	.9232	.4	.6	.4633	.5228	1.913	.8862	.4
.7	.3859	.4183	2.391	.9225	.3	.7	.4648	.5250	1.905	.8854	.3
.8	.3875	.4204	2.379	.9219	.2	.8	.4664	.5272	1.897	.8846	.2
.9	.3891	.4224	2.367	.9212	67.1	.9	.4679	.5295	1.889	.8838	62.1
23.0	0.3907	0.4245	2.356	0.9205	67.0	28.0	0.4695	0.5317	1.881	0.8829	62.0
.1	.3923	.4265	2.344	.9198	66.9	.1	.4710	.5340	1.873	.8821	61.9
.2	.3939	.4286	2.333	.9191	.8	.2	.4726	.5362	1.865	.8813	.8
.3	.3955	.4307	2.322	.9184	.7	.3	.4741	.5384	1.857	.8805	.7
.4	.3971	.4327	2.311	.9178	.6	.4	.4756	.5407	1.849	.8796	.6
.5	.3987	.4348	2.300	.9171	.5	.5	.4772	.5430	1.842	.8788	.5
.6	.4003	.4369	2.289	.9164	.4	.6	.4787	.5452	1.834	.8780	.4
.7	.4019	.4390	2.278	.9157	.3	.7	.4802	.5475	1.827	.8771	.3
.8	.4035	.4411	2.267	.9150	.2	.8	.4818	.5498	1.819	.8763	.2
.9	.4051	.4431	2.257	.9143	66.1	.9	.4833	.5520	1.811	.8755	61.1
24.0	0.4067	0.4452	2.246	0.9135	66.0	29.0	0.4848	0.5543	1.804	0.8746	61.0
.1	.4083	.4473	2.236	.9128	65.9	.1	.4863	.5566	1.797	.8738	60.9
.2	.4099	.4494	2.225	.9121	.8	.2	.4879	.5589	1.789	.8729	.8
.3	.4115	.4515	2.215	.9114	.7	.3	.4894	.5612	1.782	.8721	.7
.4	.4131	.4536	2.204	.9107	.6	.4	.4909	.5635	1.775	.8712	.6
.5	.4147	.4557	2.194	.9100	.5	.5	.4924	.5658	1.767	.8704	.5
.6	.4163	.4578	2.184	.9092	.4	.6	.4939	.5681	1.760	.8695	.4
.7	.4179	.4599	2.174	.9085	.3	.7	.4955	.5704	1.753	.8686	.3
.8	.4195	.4621	2.164	.9078	.2	.8	.4970	.5727	1.746	.8678	.2
.9	.4210	.4642	2.154	.9070	65.1	.9	.4985	.5750	1.739	.8669	60.1
25.0	0.4226	0.4663	2.145	0.9063	65.0	30.0	0.5000	0.5774	1.732	0.8660	60.0
	cos	cot	tan	sin	Deg.		cos	cot	tan	sin	Deg.

Deg.	sin	tan	cot	cos		Deg.	sin	tan	cot	cos	
30.0	0.5000	0.5774	1.7321	0.8660	60.0	35.0	0.5736	0.7002	1.4281	0.8192	55.0
.1	.5015	.5797	1.7251	.8652	59.9	.1	.5750	.7028	1.4229	.8181	54.9
.2	.5030	.5820	1.7182	.8643	.8	.2	.5764	.7054	1.4176	.8171	.8
.3	.5045	.5844	1.7113	.8634	.7	.3	.5779	.7080	1.4124	.8161	.7
.4	.5060	.5867	1.7045	.8625	.6	.4	.5793	.7107	1.4071	.8151	.6
.5	.5075	.5890	1.6977	.8616	.5	.5	.5807	.7133	1.4019	.8141	.5
.6	.5090	.5914	1.6909	.8607	.4	.6	.5821	.7159	1.3968	.8131	.4
.7	.5105	.5938	1.6842	.8599	.3	.7	.5835	.7186	1.3916	.8121	.3
.8	.5120	.5961	1.6775	.8590	.2	.8	.5850	.7212	1.3865	.8111	.2
.9	.5135	.5985	1.6709	.8581	59.1	.9	.5864	.7239	1.3814	.8100	54.1
31.0	0.5150	0.6009	1.6643	0.8572	59.0	36.0	0.5878	0.7265	1.3764	0.8090	54.0
.1	.5165	.6032	1.6577	.8563	58.9	.1	.5892	.7292	1.3713	.8080	53.9
.2	.5180	.6056	1.6512	.8554	.8	.2	.5906	.7319	1.3663	.8070	.8
.3	.5195	.6080	1.6447	.8545	.7	.3	.5920	.7346	1.3613	.8059	.7
.4	.5210	.6104	1.6383	.8536	.6	.4	.5934	.7373	1.3564	.8049	.6
.5	.5225	.6128	1.6319	.8526	.5	.5	.5948	.7400	1.3514	.8039	.5
.6	.5240	.6152	1.6255	.8517	.4	.6	.5962	.7427	1.3465	.8028	.4
.7	.5255	.6176	1.6191	.8508	.3	.7	.5976	.7454	1.3416	.8018	.3
.8	.5270	.6200	1.6128	.8499	.2	.8	.5990	.7481	1.3367	.8007	.2
.9	.5284	.6224	1.6066	.8490	58.1	.9	.6004	.7508	1.3319	.7997	53.1
32.0	0.5299	0.6249	1.6003	0.8480	58.0	37.0	0.6018	0.7536	1.3270	0.7986	53.0
.1	.5314	.6273	1.5941	.8471	57.9	.1	.6032	.7563	1.3222	.7976	52.9
.2	.5329	.6297	1.5880	.8462	.8	.2	.6046	.7590	1.3175	.7965	.8
.3	.5344	.6322	1.5818	.8453	.7	.3	.6060	.7618	1.3127	.7955	.7
.4	.5358	.6346	1.5757	.8443	.6	.4	.6074	.7646	1.3079	.7944	.6
.5	.5373	.6371	1.5697	.8434	.5	.5	.6088	.7673	1.3032	.7934	.5
.6	.5388	.6395	1.5637	.8425	.4	.6	.6101	.7701	1.2985	.7923	.4
.7	.5402	.6420	1.5577	.8415	.3	.7	.6115	.7729	1.2938	.7912	.3
.8	.5417	.6445	1.5517	.8406	.2	.8	.6129	.7757	1.2892	.7902	.2
.9	.5432	.6469	1.5458	.8396	57.1	.9	.6143	.7785	1.2846	.7891	52.1
33.0	0.5446	0.6494	1.5399	0.8387	57.0	38.0	0.6157	0.7813	1.2799	0.7880	52.0
.1	.5461	.6519	1.5340	.8377	56.9	.1	.6170	.7841	1.2753	.7869	51.9
.2	.5476	.6544	1.5282	.8368	.8	.2	.6184	.7869	1.2708	.7859	.8
.3	.5490	.6569	1.5224	.8358	.7	.3	.6198	.7898	1.2662	.7848	.7
.4	.5505	.6594	1.5166	.8348	.6	.4	.6211	.7926	1.2617	.7837	.6
.5	.5519	.6619	1.5108	.8339	.5	.5	.6225	.7954	1.2572	.7826	.5
.6	.5534	.6644	1.5051	.8329	.4	.6	.6239	.7983	1.2527	.7815	.4
.7	.5548	.6669	1.4994	.8320	.3	.7	.6252	.8012	1.2482	.7804	.3
.8	.5563	.6694	1.4938	.8310	.2	.8	.6266	.8040	1.2437	.7793	.2
.9	.5577	.6720	1.4882	.8300	56.1	.9	.6280	.8069	1.2393	.7782	51.1
34.0	0.5592	0.6745	1.4826	0.8290	56.0	39.0	0.6293	0.8098	1.2349	0.7771	51.0
.1	.5606	.6771	1.4770	.8281	55.9	.1	.6307	.8127	1.2305	.7760	50.9
.2	.5621	.6796	1.4715	.8271	.8	.2	.6320	.8156	1.2261	.7749	.8
.3	.5635	.6822	1.4659	.8261	.7	.3	.6334	.8185	1.2218	.7738	.7
.4	.5650	.6847	1.4605	.8251	.6	.4	.6347	.8214	1.2174	.7727	.6
.5	.5664	.6873	1.4550	.8241	.5	.5	.6361	.8243	1.2131	.7716	.5
.6	.5678	.6899	1.4496	.8231	.4	.6	.6374	.8273	1.2088	.7705	.4
.7	.5693	.6924	1.4442	.8221	.3	.7	.6388	.8302	1.2045	.7694	.3
.8	.5707	.6950	1.4388	.8211	.2	.8	.6401	.8332	1.2002	.7683	.2
.9	.5721	.6976	1.4335	.8202	55.1	.9	.6414	.8361	1.1960	.7672	50.1
35.0	0.5736	0.7002	1.4281	0.8192	55.0	40.0	0.6428	0.8391	1.1918	0.7660	50.0
	cos	cot	tan	sin	Deg.		cos	cot	tan	sin	Deg.

Table IIA **Trigonometric Functions—Degree Measure**

Deg.	sin	tan	cot	cos		Deg.	sin	tan	cot	cos	
40.0	0.6428	0.8391	1.1918	0.7660	50.0	42.5	0.6756	0.9163	1.0913	0.7373	0.5
.1	.6441	.8421	1.1875	.7649	49.9	.6	.6769	.9195	1.0875	.7361	.4
.2	.6455	.8451	1.1833	.7638	.8	.7	.6782	.9228	1.0837	.7349	.3
.3	.6468	.8481	1.1792	.7627	.7	.8	.6794	.9260	1.0799	.7337	.2
.4	.6481	.8511	1.1750	.7615	.6	.9	.6807	.9293	1.0761	.7325	47.1
.5	.6494	.8541	1.1708	.7604	.5	43.0	0.6820	0.9325	1.0724	0.7314	47.0
.6	.6508	.8571	1.1667	.7593	.4	.1	.6833	.9358	1.0686	.7302	46.9
.7	.6521	.8601	1.1626	.7581	.3	.2	.6845	.9391	1.0649	.7290	.8
.8	.6534	.8632	1.1585	.7570	.2	.3	.6858	.9424	1.0612	.7278	.7
.9	.6547	.8662	1.1544	.7559	49.1	.4	.6871	.9457	1.0575	.7266	.6
41.0	0.6561	0.8693	1.1504	0.7547	49.0	.5	.6884	.9490	1.0538	.7254	.5
.1	.6574	.8724	1.1463	.7536	48.9	.6	.6896	.9523	1.0501	.7242	.4
.2	.6587	.8754	1.1423	.7524	.8	.7	.6909	.9556	1.0464	.7230	.3
.3	.6600	.8785	1.1383	.7513	.7	.8	.6921	.9590	1.0428	.7218	.2
.4	.6613	.8816	1.1343	.7501	.6	.9	.6934	.9623	1.0392	.7206	46.1
.5	.6626	.8847	1.1303	.7490	.5	44.0	0.6947	0.9657	1.0355	0.7193	46.0
.6	.6639	.8878	1.1263	.7478	.4	.1	.6959	.9691	1.0319	.7181	45.9
.7	.6652	.8910	1.1224	.7466	.3	.2	.6972	.9725	1.0283	.7169	.8
.8	.6665	.8941	1.1184	.7455	.2	.3	.6984	.9759	1.0247	.7157	.7
.9	.6678	.8972	1.1145	.7443	48.1	.4	.6997	.9793	1.0212	.7145	.6
42.0	0.6691	0.9004	1.1106	0.7431	48.0	.5	.7009	.9827	1.0176	.7133	.5
.1	.6704	.9036	1.1067	.7420	47.9	.6	.7022	.9861	1.0141	.7120	.4
.2	.6717	.9067	1.1028	.7408	.8	.7	.7034	.9896	1.0105	.7108	.3
.3	.6730	.9099	1.0990	.7396	.7	.8	.7046	.9930	1.0070	.7096	.2
.4	.6743	.9131	1.0951	.7385	.6	.9	.7059	.9965	1.0035	.7083	45.1
42.5	0.6756	0.9163	1.0913	0.7373	0.5	45.0	0.7071	1.0000	1.0000	0.7071	45.0
	cos	cot	tan	sin	Deg.		cos	cot	tan	sin	Deg.

Table IIB **Trigonometric Functions—Radian Measure**

x	$\sin x$	$\cos x$	$\tan x$	$\cot x$	$\sec x$	$\csc x$
0.00	0.0000	1.0000	0.0000	—	1.000	—
0.01	0.0100	1.0000	0.0100	99.997	1.000	100.00
0.02	0.0200	0.9998	0.0200	49.993	1.000	50.00
0.03	0.0300	0.9996	0.0300	33.323	1.000	33.34
0.04	0.0400	0.9992	0.0400	24.987	1.001	25.01
0.05	0.0500	0.9988	0.0500	19.983	1.001	20.01
0.06	0.0600	0.9982	0.0601	16.647	1.002	16.68
0.07	0.0699	0.9976	0.0701	14.262	1.002	14.30
0.08	0.0799	0.9968	0.0802	12.473	1.003	12.51
0.09	0.0899	0.9960	0.0902	11.081	1.004	11.13
0.10	0.0998	0.9950	0.1003	9.967	1.005	10.02
0.11	0.1098	0.9940	0.1104	9.054	1.006	9.109
0.12	0.1197	0.9928	0.1206	8.293	1.007	8.353
0.13	0.1296	0.9916	0.1307	7.649	1.009	7.714
0.14	0.1395	0.9902	0.1409	7.096	1.010	7.166
0.15	0.1494	0.9888	0.1511	6.617	1.011	6.692
0.16	0.1593	0.9872	0.1614	6.197	1.013	6.277
0.17	0.1692	0.9856	0.1717	5.826	1.015	5.911
0.18	0.1790	0.9838	0.1820	5.495	1.016	5.586
0.19	0.1889	0.9820	0.1923	5.200	1.018	5.295
0.20	0.1987	0.9801	0.2027	4.933	1.020	5.033
0.21	0.2085	0.9780	0.2131	4.692	1.022	4.797
0.22	0.2182	0.9759	0.2236	4.472	1.025	4.582
0.23	0.2280	0.9737	0.2341	4.271	1.027	4.386
0.24	0.2377	0.9713	0.2447	4.086	1.030	4.207
0.25	0.2474	0.9689	0.2553	3.916	1.032	4.042
0.26	0.2571	0.9664	0.2660	3.759	1.035	3.890
0.27	0.2667	0.9638	0.2768	3.613	1.038	3.749
0.28	0.2764	0.9611	0.2876	3.478	1.041	3.619
0.29	0.2860	0.9582	0.2984	3.351	1.044	3.497
0.30	0.2955	0.9553	0.3093	3.233	1.047	3.384
0.31	0.3051	0.9523	0.3203	3.122	1.050	3.278
0.32	0.3146	0.9492	0.3314	3.018	1.053	3.179
0.33	0.3240	0.9460	0.3425	2.920	1.057	3.086
0.34	0.3335	0.9428	0.3537	2.827	1.061	2.999
0.35	0.3429	0.9394	0.3650	2.740	1.065	2.916
0.36	0.3523	0.9359	0.3764	2.657	1.068	2.839
0.37	0.3616	0.9323	0.3879	2.578	1.073	2.765
0.38	0.3709	0.9287	0.3994	2.504	1.077	2.696
0.39	0.3802	0.9249	0.4111	2.433	1.081	2.630
0.40	0.3894	0.9211	0.4228	2.365	1.086	2.568
0.41	0.3986	0.9171	0.4346	2.301	1.090	2.509
0.42	0.4078	0.9131	0.4466	2.239	1.095	2.452
0.43	0.4169	0.9090	0.4586	2.180	1.100	2.399
0.44	0.4259	0.9048	0.4708	2.124	1.105	2.348

Table IIB **Trigonometric Functions—Radian Measure**

x	$\sin x$	$\cos x$	$\tan x$	$\cot x$	$\sec x$	$\csc x$
0.45	0.4350	0.9004	0.4831	2.070	1.111	2.299
0.46	0.4439	0.8961	0.4954	2.018	1.116	2.253
0.47	0.4529	0.8916	0.5080	1.969	1.122	2.208
0.48	0.4618	0.8870	0.5206	1.921	1.127	2.166
0.49	0.4706	0.8823	0.5334	1.875	1.133	2.125
0.50	0.4794	0.8776	0.5463	1.830	1.139	2.086
0.51	0.4882	0.8727	0.5594	1.788	1.146	2.048
0.52	0.4969	0.8678	0.5726	1.747	1.152	2.013
$\pi/6$	0.5000	0.8660	0.5774	1.732	1.155	2.000
0.53	0.5055	0.8628	0.5859	1.707	1.159	1.978
0.54	0.5141	0.8577	0.5994	1.668	1.166	1.945
0.55	0.5227	0.8525	0.6131	1.631	1.173	1.913
0.56	0.5312	0.8473	0.6269	1.595	1.180	1.883
0.57	0.5396	0.8419	0.6410	1.560	1.188	1.853
0.58	0.5480	0.8365	0.6552	1.526	1.196	1.825
0.59	0.5564	0.8309	0.6696	1.494	1.203	1.797
0.60	0.5646	0.8253	0.6841	1.462	1.212	1.771
0.61	0.5729	0.8196	0.6989	1.431	1.220	1.746
0.62	0.5810	0.8139	0.7139	1.401	1.229	1.721
0.63	0.5891	0.8080	0.7291	1.372	1.238	1.697
0.64	0.5972	0.8021	0.7445	1.343	1.247	1.674
0.65	0.6052	0.7961	0.7602	1.315	1.256	1.652
0.66	0.6131	0.7900	0.7761	1.288	1.266	1.631
0.67	0.6210	0.7838	0.7923	1.262	1.276	1.610
0.68	0.6288	0.7776	0.8087	1.237	1.286	1.590
0.69	0.6365	0.7712	0.8253	1.212	1.297	1.571
0.70	0.6442	0.7648	0.8423	1.187	1.307	1.552
0.71	0.6518	0.7584	0.8595	1.163	1.319	1.534
0.72	0.6594	0.7518	0.8771	1.140	1.330	1.517
0.73	0.6669	0.7452	0.8949	1.117	1.342	1.500
0.74	0.6743	0.7385	0.9131	1.095	1.354	1.483
0.75	0.6816	0.7317	0.9316	1.073	1.367	1.467
0.76	0.6889	0.7248	0.9505	1.052	1.380	1.452
0.77	0.6961	0.7179	0.9697	1.031	1.393	1.437
0.78	0.7033	0.7109	0.9893	1.011	1.407	1.422
$\pi/4$	0.7071	0.7071	1.000	1.000	1.414	1.414
0.79	0.7104	0.7038	1.009	0.9908	1.421	1.408
0.80	0.7174	0.6967	1.030	0.9712	1.435	1.394
0.81	0.7243	0.6895	1.050	0.9520	1.450	1.381
0.82	0.7311	0.6822	1.072	0.9331	1.466	1.368
0.83	0.7379	0.6749	1.093	0.9146	1.482	1.355
0.84	0.7446	0.6675	1.116	0.8964	1.498	1.343
0.85	0.7513	0.6600	1.138	0.8785	1.515	1.331
0.86	0.7578	0.6524	1.162	0.8609	1.533	1.320
0.87	0.7643	0.6448	1.185	0.8437	1.551	1.308
0.88	0.7707	0.6372	1.210	0.8267	1.569	1.297
0.89	0.7771	0.6294	1.235	0.8100	1.589	1.287

x	$\sin x$	$\cos x$	$\tan x$	$\cot x$	$\sec x$	$\csc x$
0.90	0.7833	0.6216	1.260	0.7936	1.609	1.277
0.91	0.7895	0.6137	1.286	0.7774	1.629	1.267
0.92	0.7956	0.6058	1.313	0.7615	1.651	1.257
0.93	0.8016	0.5978	1.341	0.7458	1.673	1.247
0.94	0.8076	0.5898	1.369	0.7303	1.696	1.238
0.95	0.8134	0.5817	1.398	0.7151	1.719	1.229
0.96	0.8192	0.5735	1.428	0.7001	1.744	1.221
0.97	0.8249	0.5653	1.459	0.6853	1.769	1.212
0.98	0.8305	0.5570	1.491	0.6707	1.795	1.204
0.99	0.8360	0.5487	1.524	0.6563	1.823	1.196
1.00	0.8415	0.5403	1.557	0.6421	1.851	1.188
1.01	0.8468	0.5319	1.592	0.6281	1.880	1.181
1.02	0.8521	0.5234	1.628	0.6142	1.911	1.174
1.03	0.8573	0.5148	1.665	0.6005	1.942	1.166
1.04	0.8624	0.5062	1.704	0.5870	1.975	1.160
$\pi/3$	0.8660	0.5000	1.732	0.5774	2.000	1.155
1.05	0.8674	0.4976	1.743	0.5736	2.010	1.153
1.06	0.8724	0.4889	1.784	0.5604	2.046	1.146
1.07	0.8772	0.4801	1.827	0.5473	2.083	1.140
1.08	0.8820	0.4713	1.871	0.5344	2.122	1.134
1.09	0.8866	0.4625	1.917	0.5216	2.162	1.128
1.10	0.8912	0.4536	1.965	0.5090	2.205	1.122
1.11	0.8957	0.4447	2.014	0.4964	2.249	1.116
1.12	0.9001	0.4357	2.066	0.4840	2.295	1.111
1.13	0.9044	0.4267	2.120	0.4718	2.344	1.106
1.14	0.9086	0.4176	2.176	0.4596	2.395	1.101
1.15	0.9128	0.4085	2.234	0.4475	2.448	1.096
1.16	0.9168	0.3993	2.296	0.4356	2.504	1.091
1.17	0.9208	0.3902	2.360	0.4237	2.563	1.086
1.18	0.9246	0.3809	2.427	0.4120	2.625	1.082
1.19	0.9284	0.3717	2.498	0.4003	2.691	1.077
1.20	0.9320	0.3624	2.572	0.3888	2.760	1.073
1.21	0.9356	0.3530	2.650	0.3773	2.833	1.069
1.22	0.9391	0.3436	2.733	0.3659	2.910	1.065
1.23	0.9425	0.3342	2.820	0.3546	2.992	1.061
1.24	0.9458	0.3248	2.912	0.3434	3.079	1.057
1.25	0.9490	0.3153	3.010	0.3323	3.171	1.054
1.26	0.9521	0.3058	3.113	0.3212	3.270	1.050
1.27	0.9551	0.2963	3.224	0.3102	3.375	1.047
1.28	0.9580	0.2867	3.341	0.2993	3.488	1.044
1.29	0.9608	0.2771	3.467	0.2884	3.609	1.041
1.30	0.9636	0.2675	3.602	0.2776	3.738	1.038
1.31	0.9662	0.2579	3.747	0.2669	3.878	1.035
1.32	0.9687	0.2482	3.903	0.2562	4.029	1.032
1.33	0.9711	0.2385	4.072	0.2456	4.193	1.030
1.34	0.9735	0.2288	4.256	0.2350	4.372	1.027

Table IIB   **Trigonometric Functions—Radian Measure**

x	$\sin x$	$\cos x$	$\tan x$	$\cot x$	$\sec x$	$\csc x$
1.35	0.9757	0.2190	4.455	0.2245	4.566	1.025
1.36	0.9779	0.2092	4.673	0.2140	4.779	1.023
1.37	0.9799	0.1994	4.913	0.2035	5.014	1.021
1.38	0.9819	0.1896	5.177	0.1931	5.273	1.018
1.39	0.9837	0.1798	5.471	0.1828	5.561	1.017
1.40	0.9854	0.1700	5.798	0.1725	5.883	1.015
1.41	0.9871	0.1601	6.165	0.1622	6.246	1.013
1.42	0.9887	0.1502	6.581	0.1519	6.657	1.011
1.43	0.9901	0.1403	7.055	0.1417	7.126	1.010
1.44	0.9915	0.1304	7.602	0.1315	7.667	1.009
1.45	0.9927	0.1205	8.238	0.1214	8.299	1.007
1.46	0.9939	0.1106	8.989	0.1113	9.044	1.006
1.47	0.9949	0.1006	9.887	0.1011	9.938	1.005
1.48	0.9959	0.0907	10.983	0.0910	11.029	1.004
1.49	0.9967	0.0807	12.350	0.0810	12.390	1.003
1.50	0.9975	0.0707	14.101	0.0709	14.137	1.003
1.51	0.9982	0.0608	16.428	0.0609	16.458	1.002
1.52	0.9987	0.0508	19.670	0.0508	19.695	1.001
1.53	0.9992	0.0408	24.498	0.0408	24.519	1.001
1.54	0.9995	0.0308	32.461	0.0308	32.476	1.000
1.55	0.9998	0.0208	48.078	0.0208	48.089	1.000
1.56	0.9999	0.0108	92.620	0.0108	92.626	1.000
1.57	1.0000	0.0008	1255.8	0.0008	1255.8	1.000
$\pi/2$	1.0000	0.0000	—	0.0000	—	1.000

Review of Plane Geometry

This appendix is intended as a concise review of the plane geometry that you will need in the textbook. To facilitate your review, the material is presented in condensed "outline" form with little or no attempt to justify the basic results.

The line segment with endpoints A and B is denoted by $\overline{AB}$ and its length by $|\overline{AB}|$. The line L that is perpendicular to $\overline{AB}$ and intersects $\overline{AB}$ at its midpoint M is called the **perpendicular bisector** of $\overline{AB}$ (Figure 1). If P is any point on the perpendicular bisector L of $\overline{AB}$, then P is *equidistant* from the two endpoints A and B, that is, $|\overline{PA}| = |\overline{PB}|$.

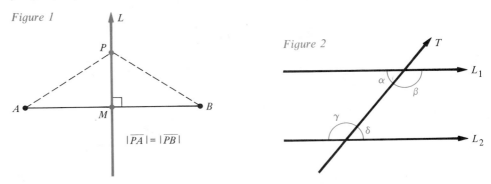

Figure 1

$|\overline{PA}| = |\overline{PB}|$

Figure 2

Two lines L_1 and L_2 that do not intersect are said to be **parallel**, and a third line T that intersects both L_1 and L_2 is called a **transversal** (Figure 2). The four angles α, β, γ, and δ, formed by T, L_1, and L_2, that lie between L_1 and L_2 are called **interior angles** (Figure 2). Interior angles, such as α and δ or β and γ, that lie on opposite sides of the transversal T are called **alternate** interior angles. In plane geometry, it is shown that

> Alternate interior angles are equal.

For instance, in Figure 2, $\alpha = \delta$ and $\beta = \gamma$.

Example 1   In Figure 3, L_1 and L_2 are parallel lines and the measures of angles α and δ are $(5x - 46)°$ and $(3x + 2)°$, respectively. Find x.

Figure 3

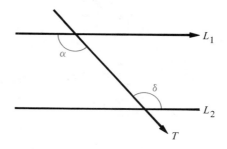

Solution   Because α and δ are alternate interior angles, they are equal. Therefore,

$$5x - 46 = 3x + 2$$
$$2x = 48$$
$$x = 24.$$

If the three vertex angles in a first triangle $A_1B_1C_1$ are equal to the corresponding vertex angles in a second triangle $A_2B_2C_2$, we say that the triangles are **similar** (Figure 4).

Corresponding side lengths of two similar triangles are proportional.

For instance, in Figure 4,

$$\frac{a_1}{a_2} = \frac{b_1}{b_2} = \frac{c_1}{c_2}.$$

Figure 4

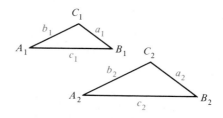

It follows that

$$\frac{a_1}{b_1} = \frac{a_2}{b_2} \quad \text{and} \quad \frac{a_1}{c_1} = \frac{a_2}{c_2} \quad \text{and} \quad \frac{b_1}{c_1} = \frac{b_2}{c_2}.$$

The sum of the three vertex angles of a triangle is always $180°$ (π radians).

It follows that two vertex angles of a triangle determine the third one; hence, if two vertex angles of one triangle are equal to two vertex angles of another triangle, then the triangles are similar.

If one triangle can be obtained from another by vertical shifting, horizontal shifting, rotation, or reflection across a line, the triangles are said to be **congruent**.

> If the lengths of the three sides of a first triangle are equal to the lengths of corresponding sides of a second triangle, then the two triangles are congruent.

If two triangles are congruent, the three vertex angles in either triangle are equal to the corresponding vertex angles in the other triangle.

> If two sides and the angle included between them in a first triangle are equal to two sides and the angle included between them in a second triangle, then the two triangles are congruent.

A triangle in which one of the vertex angles is a right angle (that is, a 90° angle) is called a **right triangle**. In a right triangle, the side opposite the right angle is called the **hypotenuse** (Figure 5). The word "hypotenuse" is also used to refer to the *length* of the side opposite the right angle. The **Pythagorean theorem** states that the square of the length of the hypotenuse is equal to the sum of the squares of the lengths of the other two sides. Thus, in Figure 5, by the Pythagorean theorem,

$$c^2 = a^2 + b^2.$$

The *converse of the Pythagorean theorem also holds:* If the three sides a, b, and c of a triangle satisfy $a^2 + b^2 = c^2$, then the angle opposite side c is a right angle.

Figure 5

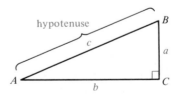

Example 2 In right triangle ACB (Figure 5), suppose $a = 7$ and $c = 25$. Find b.

Solution By the Pythagorean theorem,

$$b^2 = c^2 - a^2 = 25^2 - 7^2 = 625 - 49 = 576,$$

so $$b = \sqrt{576} = 24.$$

Example 3 The three sides of a triangle have lengths $a = 5$, $b = 12$, and $c = 13$. Is the triangle a right triangle?

Solution   Here,
$$a^2 + b^2 = 5^2 + 12^2 = 25 + 144 = 169 = 13^2 = c^2.$$

Hence, by the converse of the Pythagorean theorem, we have a right triangle with hypotenuse c. ∎

In a right triangle ACB (Figure 5), as in any triangle, the sum of the three vertex angles is $180°$. Because the angle at vertex C accounts for 90 of these 180 degrees, the sum of the remaining two angles must be 90 degrees. Thus,

vertex angle at A + vertex angle at $B = 90°$.

Therefore, either of these vertex angles determines the other. It follows that:

> Two right triangles are similar if a vertex angle (other than the right angle) in one triangle is equal to a vertex angle in the other triangle.

A perpendicular dropped from a vertex of a triangle to the opposite side (extended if necessary) is called an **altitude** of the triangle (Figure 6), its length h is called the **height** of the triangle as measured from that vertex, and the side of the triangle opposite to the vertex is the corresponding **base.** If b is the length of the base, then the **area** of the triangle is given by

$$\text{area} = \tfrac{1}{2}hb.$$

Figure 6

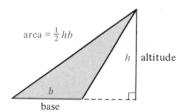

area $= \frac{1}{2} hb$

h | altitude

b

base

The ratio of the **circumference** c to the **diameter** d of a circle is the constant π (the Greek letter pi) (Figure 7). Because $\pi = c/d$, it follows that

$$c = \pi d \qquad \text{or} \qquad c = 2\pi r.$$

Figure 7

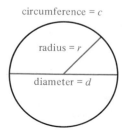

circumference $= c$

radius $= r$

diameter $= d$

where $r = d/2$ is the **radius** of the circle. The **area** of the circle is given by the familiar formula

$$\text{area} = \pi r^2.$$

Additional formulas for areas of various plane figures and volumes of certain solids are given inside the front cover of this book.

Computer Program to Locate Zeros by the Bisection Method

The following computer program, written in BASIC by Robert J. Weaver, uses the bisection method (Section 4.5 page 238) to locate zeros. For demonstration purposes, the function $F(X) = X^3 + X - 1$ is entered in line 100 of the program (see Example 4 in Section 4.5 page 239). To use the program for other functions, edit line 100 and replace $X^3 + X - 1$ by the function of your choice (written in BASIC). The program is especially useful for working the exercises in Problem Set 4.5 page 240.

```
100 DEF FN F(X) = X^3 + X - 1
110 PRINT "========== ENTER ENDPOINTS =========="
120 PRINT
130 PRINT "WHAT IS YOUR INITIAL LEFT ENDPOINT";
140 INPUT A
150 PRINT "WHAT IS YOUR INITIAL RIGHT ENDPOINT";
160 INPUT B
170 PRINT
180 IF A < B THEN GOTO 230
190 PRINT "THE LEFT ENDPOINT MUST BE LESS THAN"
200 PRINT "THE RIGHT ENDPOINT.  TRY AGAIN
210 PRINT
220 GOTO 110
230 PRINT "F(";A;") = "; FN F(A)
240 PRINT "F(";B;") = "; FN F(B)
250 PRINT
260 REM - VARIABLE 'S' DETERMINES A CHANGE OF SIGN
270 LET S = FN F(A)*FN F(B)
280 IF S < 0 THEN GOTO 410
290 REM - ENDPOINTS NOT APPROPRIATE
300 PRINT "================ NOTICE ================"
310 PRINT
320 PRINT "F DOES NOT HAVE DIFFERENT ALGEBRAIC"
330 PRINT "SIGN AT THE ENDPOINTS YOU HAVE CHOSEN."
340 PRINT
350 PRINT "DO YOU WISH TO TRY AGAIN?"
360 PRINT "'Y' OR 'N'";
```

```
370 INPUT A$
380 IF A$ = "N" THEN GOTO 920
390 PRINT
400 GOTO 110
410 PRINT "HOW MANY TIMES DO YOU WISH TO BISECT"
420 PRINT "THE INTERVAL FROM ";A;" TO ";B;
430 INPUT N
440 IF N = INT(N) AND N > 0 THEN GOTO 490
450 PRINT
460 PRINT "ENTER A POSITIVE INTEGER, PLEASE!"
470 PRINT
480 GOTO 410
490 PRINT
500 PRINT "DO YOU WISH TO PAUSE AFTER EACH"
510 PRINT "BISECTION? 'Y' OR 'N'";
520 INPUT P$
530 PRINT
540 PRINT "=============== BEGIN ==============="
550 PRINT
560 LET L = A
570 LET R = B
580 FOR I = 1 TO N
590 LET C = (L + R)/2
600 LET F1 = FN F(L)
610 LET F2 = FN F(C)
620 LET S = F1 * F2
630 IF S = 0 THEN GOTO 870
640 IF S > 0 THEN GOTO 680
650 REM - F CHANGES SIGN ON THE LEFT-HAND INTERVAL
660 LET R = C
670 GOTO 700
680 REM - F CHANGES SIGN ON THE RIGHT HAND INTERVAL
690 LET L = C
700 PRINT "AFTER BISECTION NUMBER ";I
710 PRINT "F(";L;") = "; FN F(L)
720 PRINT "F(";R;") = "; FN F(R)
730 IF P$ <> "Y" THEN GOTO 770
740 PRINT "CONTINUE - 'Y' OR 'N'";
750 INPUT Q$
760 IF Q$ = "N" THEN GOTO 790
770 PRINT "----------------------------------
780 NEXT I
790 REM - CALCULATE THE MIDPOINT OF THE LAST INTERVAL
800 LET C = (L + R)/2
810 PRINT
820 PRINT "THERE IS A ZERO IN THE INTERVAL FROM"
830 PRINT L;" TO ";R;"."
840 PRINT "THE MIDPOINT OF THIS INTERVAL IS "
850 PRINT C
860 GOTO 920
870 REM - COME HERE WHEN THERE IS A ZERO FOUND
880 REM - DIRECTLY FROM CALCULATION AT A MIDPOINT
890 PRINT " THERE IS A ZERO AT X = ";C
900 PRINT "THIS WAS FOUND AT A MIDPOINT AFTER"
910 PRINT I;" BISECTION(S)."
920 END
```

Answers to Selected Problems

CHAPTER 1

Problem Set 1.1, page 6

1. 5^8 **3.** $3^4 \cdot 4^3$ **5.** $x^4 y^3$ **7.** $(-x)^2(-y)^3$
9. $(2a + 1)^3$ **11.** $z = 2(x + y)$ **13.** $A = \pi r^2$
15. $x = 0.05n$ **17.** $A = 6x^2$ **19.** $A = \frac{1}{2}bh$
21. $A = prt$ **23.** \$6077.53
25. Commutative for $+$ **27.** Commutative for $\times$
29. Identity for $\times$ **31.** Associative for $+$
33. Identity for $+$ **35.** Distributive
37. Inverse for $+$ **39.** Associative for $\times$
41. Negation (i) **43.** Negation (ii)
45. Cancellation for $+$ **47.** Zero factor (ii)
49. Negative of a fraction property
51. $(3 + 5)^2 = 8^2 = 64$ **53.** No
55. By the reflexive property, $ca = ca$. Since $a = b$, substitution yields $ca = cb$.

Problem Set 1.2, page 14

1. $\{4,6,8,10\}$ **3.** $\{2,4\}$ **5.** False **7.** False
9. True **11.** True **13.** True **15.** False
17.

19. (a)

; (b)

(c)

; (d)

(e)

; (f)

(g)

; (h)

21.

23.

25. 0.8 **27.** -0.375 **29.** -0.26
31. -0.085 **33.** -0.625 **35.** $-1.\overline{6}$
37. $2.\overline{142857}$ **39.** $\frac{41}{100}$ **41.** $-\frac{4}{125}$ **43.** $-\frac{581}{1000}$
45. $-\frac{913}{1000}$ **47.** $\frac{10451}{10000}$ **49.** 0.11 **51.** 0.0103
53. 4.32 **55.** 0.000006 **57.** 50% **59.** 2.4%
61. $66.\overline{6}\%$ **63.** $28.\overline{571428}\%$
65. 15% **67.** 20% **69.** $16.\overline{6}\%$

Problem Set 1.3, page 21

1. 18 **3.** 400
5. Monomial; polynomial in x;
degree 2; coefficient -4
7. Trinomial; multinomial in x and y
9. Multinomial; polynomial in x and z; degree 3;
coefficients 3, $-\frac{6}{11}$, -1, -1, 2
11. Multinomial; polynomial in x; degree 5; coefficients
$\sqrt{2}$, π, $-\sqrt{\pi}$, $\frac{13}{13}$, -5 **13.** Multinomial; polynomial in
x, y, z, w; degree 3; coefficients 1, 1, 1, 1
15. $2x + 12$ **17.** $18x^2 + x + 5$ **19.** $4z^2 + 10z + 3$
21. $3n^3 - 5n^2 - n + 7$ **23.** $2x^2y + 4xy^2 + 11xy - 3$
25. $-7t^2 + 14t - 9$
27. $2x^2 - 5xy + 6y^2 - 2x + y - 3$ **29.** $6x^7$
31. 2^{10n} **33.** $(3x + y)^6$ **35.** t^{48} **37.** u^{5n^2}
39. $324v^6$ **41.** $675a^7b^8$ **43.** $864(x + 3y)^{22}$
45. $15x^4y - 20x^4y^2 + 20x^3yz$

47. $-12x^2 + 7xy + 10y^2$
49. $x^3 + 8y^3$ **51.** $625c^4 - d^4$
53. $3x^4 + 20x^3y - 9x^2y^2 + 8xy^3 - 15x^2y - 5xy^2 + 14xy - 2x^3 + 2x^2 - y^4 + y^3 - 2y^2$
55. $4p^4q^2 - 11p^3q^3 - 7p^3q^2 + 6p^2q^4 - p^2q^3 - 15p^2q^2$
57. $16 + 24x + 9x^2$
59. $16t^4 + 8t^2 + 1 - 8st^2 - 2s + s^2$
61. $9r^2 - 4s^2$ **63.** $4a^2 - 12ab + 9b^2 - 16c^2$
65. $8 + t^3$ **67.** $8x^6 + 36x^4y + 54x^2y^2 + 27y^3$
69. $4x^2 + y^2 + z^2 - 4xy + 4xz - 2yz$
71. $t^6 - 4t^5 + 14t^4 - 16t^3 + 17t^2 + 20t + 4$
77. $a^4 + 4a^3b + 6a^2b^2 + 4ab^3 + b^4$
79. $x^2 + y^2 = (a^2 - b^2)^2 + (2ab)^2 = a^4 - 2a^2b^2 + b^4 + 4a^2b^2 = a^4 + 2a^2b^2 + b^4 = (a^2 + b^2)^2 = z^2$

Problem Set 1.4, page 27

1. $5x^2y(2x - y)$ **3.** $a^2b(a + 2 + b)$
5. $2r(x - y)(r + h)$ **7.** $(a - b)(t + r)$
9. $(x + 6)(x - 6)$ **11.** $(pq + 8)(pq - 8)$
13. $(n + 6m)(n - 6m)$ **15.** $(x + y + 7z)(x + y - 7z)$
17. $(9y^2 + z^2)(3y + z)(3y - z)$
19. $8(x + 5)(x^2 - 5x + 25)$
21. $(a - 3b - c)(a^2 + 3ab + ac + 9b^2 + 6bc + c^2)$
23. $(4x^2 + p - q)(16x^4 - 4x^2p + 4x^2q + p^2 - 2pq + q^2)$
25. $(x - 5)(x - 3)$ **27.** $(y - 5)(y + 2)$
29. $(t - 9)(t + 2)$ **31.** $(x - 5y)(x + 4y)$
33. $(4x + 7)(2x - 1)$ **35.** $(2v - 5)(v + 4)$
37. $(2r - 3s)^2$ **39.** $(3x^2 + 2)(2x^2 + 3)$
41. $(9a - 6b + 7)(3a - 2b - 2)$ **43.** Prime
45. $(4r - 1)(3r - 2)$ **47.** Prime
49. $(3 + y)(x - 2)$ **51.** $(x + d)(a - 1)(a + 1)$
53. $(2x + 3 + 3y)(2x + 3 - 3y)$
55. $(3v - 1)(2u + 3)(2u - 3)$
57. $(x - 3y)(x + 3y + 1)$
59. $(x^2 + y^2 - xy)(x^2 + y^2 + xy)$
61. $(t - 2)^2(t + 2)^2$
63. $(2s + 1)(a - b)(a^2 + ab + b^2)$
65. $(x^4 + 16y^4)(x^2 + 4y^2)(x - 2y)(x + 2y)$
67. $y^3z(5x + y)(2x - 3y)$
69. (a) $3(3x^2 + 7)(5 - 3x)^2(-21x^2 + 20x - 21)$; (b) $4t(5t^2 + 1)(3t^4 + 2)^3(15t^4 + 12t^2 + 10)$

Problem Set 1.5, page 35

1. $\dfrac{by}{3}$ **3.** $\dfrac{c}{2}$ **5.** $\dfrac{-1}{5 + t}$ **7.** $\dfrac{1}{r + 6}$ **9.** $\dfrac{3y + 1}{y + 5}$

11. 1   **13.** $\dfrac{3c + 1}{2c - 1}$ **15.** $\dfrac{3r + 3t + 2}{2r + 2t - 3}$
17. $\dfrac{c + d - 3x}{c + d + x}$ **19.** $c + 2$ **21.** $\dfrac{1}{4u}$
23. $\dfrac{3(x^2 + 5)(x + 1)}{(x^2 + 6x + 15)(x - 1)}$ **25.** $1/(a + 1)$ **27.** $4/t$
29. $(x - y)^2$ **31.** $\dfrac{2u - 1}{u + 2}$ **33.** $\dfrac{2x + 3}{x + 3}$
35. $\dfrac{t + 5}{t - 6}$ **37.** $(3a - 1)/b$ **39.** $x - 1$
41. $\dfrac{3x - 8}{(x + 2)(x - 5)}$ **43.** $\dfrac{36u^2 - 40u - 13}{4(2u - 1)(2u + 1)}$
45. $\dfrac{2y^2 + 2y - 1}{(y - 2)(y + 1)}$ **47.** $\dfrac{x + 15}{(x - 3)(x + 3)^2}$ **49.** $\dfrac{4}{t - 2}$
51. $\dfrac{-u(3u^3 - u^2 + 8u + 2)}{(u^2 + 3)(u - 1)^2(u + 2)}$ **53.** $\dfrac{5x^2 - 4x + 1}{(2x - 1)^2x^3}$
55. $\dfrac{1}{cd(c + d)}$ **57.** $ab(a + b)$ **59.** $\dfrac{pq(p + q)}{p^2 + pq + q^2}$
61. -1 **63.** $\dfrac{x^2 - xy + 2y^2}{(x - 3y)(x + 2y)}$ **65.** $\dfrac{-2x - h}{x^2(x + h)^2}$

71. 4 isn't a factor of the numerator
73. c isn't a factor of the denominator
75. Middle terms $(a/b) + (b/a)$ missing

Problem Set 1.6, page 43

1. 11   **3.** $\frac{3}{4}$ **5.** Irrational   **7.** $\frac{2}{3}$ **9.** $\frac{1}{2}$
11. $\frac{3}{4}$ **13.** $3x\sqrt{3x}$ **15.** $5x^2y^4\sqrt{3y}$
17. $2ab\sqrt[3]{3ab^2}$ **19.** $(x - 1)^2$ **21.** $y/2x^4$
23. $\sqrt{15y}/6y$ **25.** $(-3x\sqrt[3]{x})/y^7$ **27.** -66
29. $5t$ **31.** $x^2y^4\sqrt[4]{y^3}$
33. $ab^2\sqrt[3]{b}$ **35.** 1   **37.** $(3/2y)\sqrt[3]{xz}$ **39.** $\frac{x}{2}\sqrt[3]{x^2y}$
41. a^2 **43.** 98   **45.** $13\sqrt[3]{3}$ **47.** $19t^3\sqrt{5t}$
49. $\frac{1}{6}\sqrt[4]{6}$ **51.** $xy\sqrt[3]{xy^2}$ **53.** $6\sqrt{2} - 6\sqrt{3}$
55. $-1 - \sqrt{3}$ **57.** $x + 3 + 2\sqrt{3x}$ **59.** -13
61. 4   **63.** $4 + 2\sqrt{6}$
65. $2a$ **67.** $8\sqrt{11}/11$ **69.** $\sqrt{2}$ **71.** $\sqrt{10}/3$
73. $\dfrac{\sqrt{2(x + 2y)}}{3(x + 2y)}$ **75.** $4\sqrt[3]{25}$ **77.** $\dfrac{5\sqrt{3} - 3}{22}$
79. $(7 + 2\sqrt{10})/3$ **81.** $\dfrac{12p + 3q + 13\sqrt{pq}}{16p - 9q}$
83. $\dfrac{2a + 5 + 2\sqrt{a^2 + 5a}}{5}$ **85.** $\dfrac{5(4 + 2\sqrt[3]{x} + \sqrt[3]{x^2})}{8 - x}$

87. $\dfrac{a - 2\sqrt[3]{a^2b} + 2\sqrt[3]{ab^2} - b}{a + b}$

89. $\dfrac{1}{\sqrt{x + h} + \sqrt{x}}$

91. $\dfrac{2x + h}{\sqrt{(x + h)^2 + 1} + \sqrt{x^2 + 1}}$

93. Use common denominator $\sqrt{x}\,\sqrt{x + a}$

Problem Set 1.7, page 50

1. 27　**3.** $\frac{1}{64}$　**5.** y^4/x^2z　**7.** -1　**9.** $-28c^2$
11. m^3　**13.** $y^2/2x^5$　**15.** $9t/(9 + t)$

17. $(a^2 + b^2)/ab$　**19.** $1/x^n$　**21.** $1/x$　**23.** $\dfrac{p^2 - 3}{p^2 - 1}$

25. $\dfrac{(t - 2)^2}{t + 2}$　**27.** $\left(\dfrac{a}{b}\right)^{3n}$　**29.** $(a + 8b)^{3+n}$

31. $\left(\dfrac{r + s}{c + d}\right)^4$　**33.** $-1/2q$　**35.** 27　**37.** -32

39. 8　**41.** Undefined　**43.** 4　**45.** 30
47. a^2　**49.** $y^{7/3}$　**51.** $16p^{12}$　**53.** $u^3/8$
55. $1/x^{3/2}$　**57.** $x^3/8$　**59.** $1/x^{11}$　**61.** $2r^3s^2/3$

63. y^4/x^2　**65.** $2p + q$　**67.** $\dfrac{3x + 2y}{4r + 3t}$

69. $\dfrac{1}{(m^2 - n^2)^{1/3}}$　**71.** $\dfrac{(x + y)^2}{\sqrt{x}}$　**73.** $\dfrac{-(p + 3)}{(p - 1)^3}$

75. $\dfrac{2(8t + 1)}{(4t - 1)^2(2t + 1)^2}$

77. $\dfrac{3x - 1}{x^{2/3}(x - 1)^{1/3}}$

87. $2(3x + x^{-1})(6x - 1)^4(63x - 3 + 9x^{-1} + x^{-2})$
89. $2(7y + 3)^{-3}(2y - 1)^3(14y + 19)$
91. Exponent not reduced to lowest terms
93. Can't apply rule when numbers are negative
95. Can't apply rule when number is negative

Problem Set 1.8, page 56

1. $\sqrt{2}i$　**3.** $3\sqrt{3}i$　**5.** $3 + 4i$　**7.** $0 + 30\sqrt{2}i$
9. $9 + i$　**11.** $10 - i$　**13.** $19 - 4i$　**15.** $-2 - 2i$
17. $-5 + 2i$　**19.** $-11 - 4i$　**21.** $-1 - 14i$
23. $-3 + 11i$　**25.** $-3 + 54i$　**27.** $-7 + 9i$
29. $15 - 23i$　**31.** $64 - 40i$　**33.** -20　**35.** $\frac{13}{36}$
37. i　**39.** i　**41.** $12 + 16i$　**43.** $-\frac{1}{2} + \frac{1}{2}\sqrt{3}i$
45. $\frac{2}{13} - \frac{3}{13}i$　**47.** $\frac{21}{53} - \frac{6}{53}i$　**49.** $\frac{1}{5} - \frac{1}{10}i$
51. $\frac{11}{34} + \frac{41}{34}i$　**53.** $\frac{23}{74} + \frac{27}{74}i$　**55.** $-\frac{13}{10} - \frac{19}{10}i$
57. $\frac{63}{290} - \frac{201}{290}i$　**59.** $\frac{6}{145} - \frac{43}{145}i$
61. (a) $2 - i$; (b) 4; (c) $2i$; (d) 5
63. (a) i; (b) 0; (c) $-2i$; (d) 1

65. (a) $-i$; (b) 0; (c) $2i$; (d) 1　**67.** (a) $-12 - 5i$;
(b) -24; (c) $10i$; (d) 169
69. (a) $-\frac{1}{5} + \frac{11}{10}i$; (b) $-\frac{2}{5}$; (c) $-\frac{11}{5}i$; (d) $\frac{5}{4}$
71. (a) 5; (b) 10; (c) 0; (d) 25
73. (c) Let $z = a + bi$; $w = c + di$. Then
$zw = (ac - bd) + (bc + ad)i$, $\overline{zw} =$
$(ac - bd) - (bc + ad)i$, and $\bar{z}\bar{w} = (a - bi)(c - di) =$
$(ac - bd) - (bc + ad)i$.
75. By 73(c), $\overline{zz^{-1}} = \bar{z}\,\overline{z^{-1}}$. But $\overline{zz^{-1}} = \bar{1} = 1$, so
$\bar{z}\,\overline{z^{-1}} = 1$, and it follows that $(\bar{z})^{-1} = \overline{z^{-1}}$.

Problem Set 1.9, page 62

1. 1.55×10^4　**3.** 5.8761×10^7　**5.** 1.86×10^{11}
7. 9.01×10^{-7}　**9.** 33,300　**11.** 0.00004102
13. 10,010,000　**15.** 6.2×10^{-2} second
17. 9.29×10^7 miles, 1.92×10^{13} miles
19. 2.51×10^{-10} second　**21.** 4.8×10^8
23. 7×10^{31}　**25.** 6.56×10^{-4}　**27.** 2　**29.** 5
31. 5300　**33.** 0.015　**35.** 110,000
37. 2.14×10^{-13}　**39.** 2.3102×10^2
41. 4.867×10^{-4}　**43.** 1.51　**45.** 1.06×10^{27}
47. 3.47×10^5
49. (a) $\$1.926 \times 10^{12}$; (b) 1.3×10^4 dollars per person
51. If p is rounded off to k decimal places, then p
consists of a single digit followed by k decimal places;
that is, x is rounded off to $1 + k$ significant digits.

Review Problem Set, Chapter 1, page 64

1. 5^8x^3　**3.** $(-4)^5y^6$　**5.** $w = 3xy/z$
7. $s = \frac{1}{2}(a + b + c)$　**9.** $P = 1000(3^n)$
11. 315 joules　**13.** Commutative for $\times$
15. Associative for $+$　**17.** Commutative for $\times$
19. Identity for $\times$　**21.** Inverse for $\times$
23. Property of negation (i)　**25.** Zero factor (i)
27. $\frac{1}{2} + \frac{1}{3} = \frac{3}{6} + \frac{2}{6} = \frac{5}{6}$　**29.** $\{1,2\}$

31. (a), (b), (c), (d), (e), (f), (g), (h)

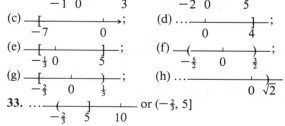

33. (or $(-\frac{2}{3}, 5]$)

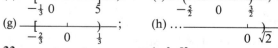

35. (a) 0.22; (b) 22%   **37.** (a) -0.085; (b) -8.5%
39. (a) 3.25; (b) 325%   **41.** (a) 0.495; (b) $\frac{99}{200}$
43. (a) 0.0043; (b) $\frac{43}{10000}$ **45.** (a) 1.4; (b) $\frac{7}{5}$
47. 6.25%   **49.** Monomial; polynomial; degree 2;
coefficient -2 **51.** Monomial; rational expression
53. Trinomial; multinomial; polynomial; degree 2;
coefficients 3, -5, -1
55. Binomial; multinomial; radical expression
57. Binomial; multinomial; polynomial; degree 3;
coefficients $\sqrt{2}$, $-\sqrt[5]{7}$
59. Binomial; multinomial; equivalent to the rational
expression $(\sqrt{\pi}y + x)/y$
61. $3x^3 - 3x^2 + 2x + 6$ **63.** $3a^3 - a^2b + 6ab^2 + 3b^3$
65. $20y^5$ **67.** $-x^6y^4$
69. p^8 **71.** $-x^{14}y^7$ **73.** $a^{n+1}b^{2n}$
75. $3x^5 + 6x^3 - 9x^2$ **77.** $2t^2 - 5t - 12$
79. $2x^2y^4 + 7xy^2 + 3$ **81.** $2p^3 - 5p^2 - 10p + 3$
83. $2x^3 - 3x^2 - 5x + 6$ **85.** $9x^2 + 30xy + 25y^2$
87. $4x^4 - 20x^2yz + 25y^2z^2$
89. $4x^2 + y^2 + 9z^2 - 4xy + 12xz - 6yz$
91. $9t^{2n} - 121$ **93.** $27x^9 - 54x^7y + 36x^5y^2 - 8x^3y^3$
95. $3xy^2(3x - 4y^2)$ **97.** $(a + b)(a + b - c^2)c^2$
99. $(6c - d^2)(6c + d^2)$ **101.** $(x - y - z)(x - y + z)$
103. $(x^n - y)(x^n + y)$ **105.** $x^2(5 - 7y)(5 + 7y)$
107. $(2p + 3q)(4p^2 - 6pq + 9q^2)$
109. $4b(3a^2 + 4b^2)$ **111.** $(x - 4)(x + 6)$
113. $(a^3 + 4)^2$ **115.** $(3x - 2y)(2x + 3y)$
117. $(4uv + 1)(uv - 2)$ **119.** $(5 - 2x)(4 + 3x)$
121. $(p + 3q - 2)(p + 3q + 2)$
123. $(a^2 + 2ab + b^2 + 1 - 3a - 3b)(a^2 + 2ab + b^2 +$
$$1 + 3a + 3b)$$

125. $\dfrac{x + 3}{5}$ **127.** $\dfrac{t + 3}{t + 2}$ **129.** $\dfrac{c - 3}{c - 2}$

131. $\dfrac{5(x - 3)}{3x}$ **133.** $\dfrac{2(x + 1)}{x + 2}$ **135.** $2y$

137. $\dfrac{2(t - 5)}{t + 1}$ **139.** $2/x$ **141.** $1/t$

143. $1/(c + 3)$ **145.** $1/(a - 3)$
147. $-1/(a + 5)$ **149.** 13   **151.** Irrational
153. Not real   **155.** $2x^2$ **157.** $(a + b)^3/3a$
159. p^7 **161.** $a^2\sqrt[3]{5a}$
163. $12\sqrt{2a}$ **165.** $13\sqrt[3]{2p}$ **167.** $5\sqrt{6}$
169. $2a - 3$ **171.** $2a + b - 2\sqrt{a^2 + ab}$

173. $(3\sqrt{2x})/x$ **175.** $\dfrac{a + \sqrt{ab}}{a - b}$

177. $\dfrac{\sqrt{a - 1} - a + 1}{2 - a}$ **179.** $5(\sqrt[3]{4} + \sqrt[3]{2} + 1)$

181. $\dfrac{9x - y}{5(3\sqrt{x} - \sqrt{y})}$ **183.** 1   **185.** $(x + y)/x^2y^2$
187. $(x^2 - 1)/x^4$ **189.** $3/a^6$ **191.** x^{13}/y^6
193. $1/25p^{10}$ **195.** $\frac{4}{9}$ **197.** $\frac{1}{512}$ **199.** y
201. a^3 **203.** $y^{3/2}$ **205.** $2^{14}b^{13}c^{2/3}$
207. $x^2y^{-12}(x + y)^{-4}$
209. $(y + 2)^{-2/3}(y + 1)^{-1/3}(3y + 5)$ **211.** $10 + 5i$
213. $1 - 9i$ **215.** $13 - 8i$ **217.** $47 - 45i$
219. $-24 - 2i$ **221.** $\frac{16}{13} + \frac{11}{13}i$ **223.** $\frac{48}{221} - \frac{46}{221}i$
225. $-i$ **227.** $\frac{1}{27}i$ **229.** $12 - 5i$
231. 5.712×10^{10} **233.** 7.14×10^{-7}
235. 17,320,000   **237.** 0.0000000312
239. 5×10^{-6} gram   **241.** 3.584×10^8 kilometers
243. 14.78976   **245.** 3   **247.** 2   **249.** 17,000
251. 7.23×10^5 **253.** 3.6×10^4 **255.** 2.65×10^{33}
257. 1.8×10^{-5}
259. Yes. Consider $a + b + c$, where $a = 1$, $b = 2.4$,
$c = 3.1$, and round off to the nearest unit.

CHAPTER 2

Problem Set 2.1, page 75

1. -2 **3.** $\frac{4}{3}$ **5.** $\frac{9}{2}$ **7.** -1.538 **9.** 42.57
11. 9   **13.** 3   **15.** 3   **17.** $-\frac{19}{2}$ **19.** 3
21. 4   **23.** 12   **25.** 7   **27.** $-\frac{5}{6}$ **29.** 1
31. (a) $\frac{7}{33}$; (b) $\frac{563}{165}$; (c) $\frac{13}{330}$; (d) $-\frac{9917}{9900}$; (e) $\frac{7}{900}$ **33.** 5

35. 4   **37.** $\frac{1}{3}$ **39.** $-\frac{5}{2}$ **41.** $\dfrac{5a + c}{b - 10}$, $b \neq 10$

43. $\dfrac{4(cd - b)}{4a - 1}$, $a \neq \dfrac{1}{4}$ **45.** $(a + md)/2$

47. $\dfrac{mn - 1}{bc + d}$, $d \neq -bc$, $x \neq 0$ **49.** $-b$, $b \neq 0$

51. $V/\pi r^2$ **53.** $(A - P)/Pr$ **55.** $pq/(p + q)$
57. $P_2V_2T_1/P_1V_1$ **59.** Identity
61. Conditional
63. Conditional
65. No, -6 is a root of the second equation, but not
the first.
67. Yes, they have exactly the same (real) roots. (If
complex roots are allowed, the answer is no.)
69. No, -1 is a solution of the first equation, but not
the second.

71. Both sides reduce to $\dfrac{ab + 8c}{a + 2}$ **73.** -18.1

Problem Set 2.2, page 84

1. 8, 20 **3.** $8500
5. $42,000 at 8.5%, $33,000 at 9.2%
7. $4760 solar, $1750 insulation **9.** 57 hours
11. 200,000 gallons of 9%, 100,000 gallons of 12%
13. 30 milliliters **15.** $5n, 12d, 6q, 9h$
17. 22 km/hr **19.** 2520 feet **21.** 27 hours
23. $4\frac{2}{7}$ hours **25.** 150 meters by 300 meters
27. $-40°C = -40°F$ **29.** $F = -7.\overline{27}°$
31. 3390 kilometers (to three significant digits)

Problem Set 2.3, page 94

1. 0, 7 **3.** ± 4 **5.** $-3, 1$ **7.** $-3, 7$
9. $-\frac{3}{2}, 5$ **11.** $-\frac{1}{3}, \frac{3}{2}$ **13.** $\frac{1}{3}, \frac{4}{3}$ **15.** $-\frac{27}{8}, 1$
17. $\frac{4}{3}, \frac{4}{3}$ **19.** $-5, 21$ **21.** $-\frac{5}{2}, 1$ **23.** $-\frac{5}{2}, \frac{7}{2}$
25. $x^2 + 6x + 9 = (x + 3)^2$
27. $x^2 - 5x + \frac{25}{4} = (x - \frac{5}{2})^2$
29. $x^2 + \frac{3}{4}x + \frac{9}{64} = (x + \frac{3}{8})^2$ **31.** $-2 \pm \sqrt{19}$
33. $3 \pm \sqrt{5}$ **35.** $-1 \pm (\sqrt{21}/3)$ **37.** $-\frac{3}{2}, 2$
39. $(-5 \pm \sqrt{13})/6$ **41.** $(1 \pm \sqrt{17})/4$
43. $(3 \pm \sqrt{57})/8$ **45.** $\frac{1}{3}, \frac{1}{3}$ **47.** $(5 \pm \sqrt{13})/3$
49. $-1.85, 0.38$ **51.** $1.42, 18.25$ **53.** $\pm \sqrt{5}$
55. $(1 \pm \sqrt{7})/4$ **57.** $-3, -3$ **59.** $-2 \pm \sqrt{3}$
61. $\frac{1}{3}, \frac{1}{3}$ **63.** $-4, 2$ **65.** 10 **67.** 9 rows
69. 40 feet and 30 feet **71.** 1.1 seconds
73. $\dfrac{-R \pm \sqrt{R^2 - (4L/C)}}{2L}$, $L \neq 0$, $R^2 \geq 4L/C$, $C \neq 0$
75. Real, unequal **77.** Rational, double root
79. Complex conjugates **81.** $(-1 \pm i\sqrt{71})/12$
83. $(2 \pm i\sqrt{6})/5$ **85.** $(-1 \pm i\sqrt{3})/2$
87. $(1 \pm i\sqrt{3})/4$

Problem Set 2.4, page 102

1. ± 3 **3.** $\pm \sqrt[6]{2}$ **5.** $\pm \frac{1}{2}$ **7.** 3 **9.** $-\sqrt[7]{4}$
11. -2 **13.** 8 **15.** -27 **17.** $3^{2/5}$
19. $\frac{33}{64}$ and $\frac{31}{64}$ **21.** $\frac{8}{3}$
23. $-1, 8, (7 + \sqrt{17})/2, (7 - \sqrt{17})/2$ **25.** 20
27. $-\frac{9}{4}$ **29.** 7 **31.** -2 **33.** 3 **35.** 8
37. 1 **39.** 27 **41.** $-1, -\frac{3}{4}$ **43.** 2 **45.** 2

47. $\pm \sqrt{3}, \pm 2$ **49.** $-2, 1$ **51.** $1, 3^6$ **53.** 1
55. $\pm \frac{1}{4}, \pm 3$ **57.** $-\frac{7}{2}, 6$ **59.** $-1 \pm \sqrt{13}$
61. $-19, 61$ **63.** $-\frac{3}{2}, 3$ **65.** $-1, -1, 1$
67. ± 1 **69.** 1 **71.** ± 1
73. $r = \sqrt{\dfrac{\sqrt{\pi^2 h^4 + 4A^2} - \pi h^2}{2\pi}}$

75. 5 cm by 12 cm **77.** $l = \dfrac{d^2 + d\sqrt{d^2 + a^2}}{a}$, $a \neq 0$

Problem Set 2.5, page 108

1. (a) Addition; (b) transitive; (c) multiplication;
(d) trichotomy
3. (a) $x + 4 > 1$; (b) $x - 4 > -7$; (c) $5x > -15$;
(d) $-5x < 15$
5. $(-\infty, 5)$, ...
7. $(-\infty, 5]$, ...
9. $(-\infty, 2)$, ...
11. $[4, \infty)$,
13. $[-\frac{22}{3}, \infty)$,
15. $(-\infty, \frac{24}{7})$, ...
17. $[-13, \infty)$,
19. $\mathbb{R}$, ...
21. $[2, \infty)$,
23. $[1, 5]$,
25. $[1, 2)$,
27. $[-3, 2)$,
29. $[-2, 1]$,
31. $(-4, 4]$,
33. $[\frac{10}{9}, 10]$,
35. $[\frac{1}{3}, 1]$,

37. $[53, 93)$ **39.** $5400 < R < 6600$ **41.** $(\frac{10}{3}, \frac{35}{6})$
43. $[13, 20]$ **47.** $-2 < 1$, but $(-2)^2 \not< 1^2$

Problem Set 2.6, page 115

1. 2,

$$|\leftarrow 2 \rightarrow|$$

with marks at 3 and 5

3. 2,

$$|\leftarrow 2 \rightarrow|$$

with marks at -5 and -3

5. 1,

$$|\leftarrow 1 \rightarrow|$$

with marks at $\frac{3}{2}$ and $\frac{5}{2}$

7. 6.043,

$$|\leftarrow 6.043 \rightarrow|$$

with marks at -3.311, 0, 2.732

9. 2.141592654,

$$|\leftarrow \pi - 1 \rightarrow|$$

with marks at 1 and π

11. 8

13. 32   **15.** 8   **17.** 0   **19.** 0   **21.** ± 4
23. ± 3 **25.** 0   **27.** $-4, 5$ **29.** 2, 6
31. $p \geq 0$ **33.** No solution   **35.** $\frac{4}{3}, \frac{5}{3}$
37. $-\frac{1}{5}, \frac{1}{5}$ **39.** $-4, -\frac{2}{3}$ **41.** $\frac{7}{4}, 3$ **43.** 1, 3
45. $(-2, 2)$, with marks at -2, 0, 2
47. $[-2, 2]$, with marks at -2, 0, 2
49. $(1, 5)$, with marks at 1, 5
51. $[-5, 5]$, with marks at -5, 0, 5
53. $(-\infty, \frac{3}{4}] \cup [\frac{9}{4}, \infty)$, ... with marks at 0, $\frac{3}{4}$, $\frac{9}{4}$
55. $(-\infty, -\frac{3}{2}) \cup (3, \infty)$, ... with marks at $-\frac{3}{2}$, 0, 3
57. $[\frac{6}{5}, 2]$, with marks at $\frac{6}{5}$, 2
59. $(-\infty, -\frac{13}{8}] \cup [-\frac{11}{8}, \infty)$, ... with marks at $-\frac{13}{8}$, $-\frac{11}{8}$
61. $[\frac{4}{7}, \frac{8}{7}]$, with marks at $\frac{4}{7}$, $\frac{8}{7}$
63. $(-\infty, 1] \cup [\frac{5}{2}, \infty)$, ... with marks at 0, 1, $\frac{5}{2}$
65. $(-\infty, 2) \cup (\frac{7}{2}, \infty)$, ... with marks at 2, $\frac{7}{2}$
67. $\mathbb{R}$, ... with mark at 0 **69.** $|x| < 1$
71. $|x| < 3$
73. If $x < 0$, then $0 < -x = |x|$ and $-|x| = -(-x) = x < 0$. If $x \geq 0$, then $-|x| = -x \leq 0 \leq x = |x|$.

75. $|x/y| = \sqrt{(x/y)^2} = \sqrt{x^2/y^2} = \sqrt{x^2}/\sqrt{y^2} = |x|/|y|$
83. If $x \geq y$, then $\frac{1}{2}(x + y + |x - y|) = \frac{1}{2}(x + y + x - y) = \frac{1}{2}(2x) = x$. If $x < y$, then $\frac{1}{2}(x + y + |x - y|) = \frac{1}{2}[x + y - (x - y)] = \frac{1}{2}(2y) = y$.
87. $(5.95, 7.65)$

Problem Set 2.7, page 123

1. $x^2 - 6x + 8 < 0$ **3.** $x^2 - 10x + 25 \geq 0$
5. $(2, 4)$ **7.** $(-\infty, -6) \cup (-4, \infty)$ **9.** $\{7\}$
11. $(-\infty, -5] \cup [-\frac{7}{3}, \infty)$ **13.** $[\frac{1}{3}, \frac{1}{2}]$ **15.** $\mathbb{R}$
17. $(-\infty, -\frac{4}{3}) \cup (2, \infty)$ **19.** $[-1, \frac{1}{2}]$ **21.** $(-\frac{3}{4}, \frac{1}{2})$
23. $\mathbb{R}$ **25.** $\left[\dfrac{1 - \sqrt{61}}{10}, 0\right] \cup \left[\dfrac{1 + \sqrt{61}}{10}, \infty\right)$

27. $(-1, 0) \cup (1, \infty)$ **29.** $\mathbb{R}$ **31.** $\dfrac{2x - 1}{x + 3} > 0$

33. $\dfrac{(x - 3)(x + 2)}{(x + 3)(x + 2)} \leq 0$

35. All real numbers, except -2 **37.** $(-3, \frac{3}{2}]$
39. $(-\infty, 0) \cup (\frac{3}{2}, \infty)$ **41.** $(-\infty, 2)$
43. $(-\infty, -2) \cup (-1, 1) \cup (2, \infty)$
45. $(-\infty, -3) \cup (-1, 2) \cup (3, \infty)$
47. $(-\infty, -3) \cup [-2, -1) \cup [1, 2) \cup [3, \infty)$
49. $(10, 20)$ **51.** $\$1.25$ **53.** 5
55. $(-\infty, 0.203] \cup [0.684, \infty)$
57. $(-\infty, -4.379) \cup (0.141, 0.794)$

Review Problem Set, Chapter 2, page 124

1. 5   **3.** 5   **5.** 15   **7.** -2 **9.** 32   **11.** 6
13. $-\frac{35}{2}$ **15.** -2 **17.** $(a + b)^2$
19. No solution unless $m = 0$, in which case, x is any nonzero number
21. 0.4650 **23.** -7.234×10^{-3} **25.** 4 liters
27. $\$6500$ at 7%, $\$3500$ at 8%   **29.** $\$750$
31. 30 meters   **33.** $x^2 + x + \frac{1}{4}$ **35.** $x^2 - 9x + \frac{81}{4}$
37. $-1, \frac{7}{2}$ **39.** $-\frac{1}{2}, \frac{3}{3}$ **41.** $(7 \pm \sqrt{89})/10$
43. $-2.867, 0.135$ **45.** $(3 \pm i\sqrt{39})/8$
47. $(1 \pm i\sqrt{15})/4$ **49.** $(-3 \pm \sqrt{13})/4$
51. Rational, unequal   **53.** Complex conjugates
55. 450 knots   **57.** 16 inches by 2 inches   **59.** 39
61. $(15 - \sqrt{33})/2$ **63.** $-\frac{10}{3}$ **65.** 3   **67.** -3
69. $\pm 1, \pm 2$ **71.** $\pm \sqrt{6}/3, \pm \frac{2}{3}\sqrt{3}$ **73.** 25
75. $-8, 1, 3 \pm \sqrt{17}$ **77.** $-\frac{1}{3}, \frac{1}{2}$ **79.** -1
81. -19 **83.** $1, \pm\sqrt{2}$ **85.** $-1, -1$

87. $-1, \pm 2^{-5/2}$ **89.** 2   **91.** $\frac{4}{3}$ **93.** $L^3 F/4 d^3 wY$
95. When the number r in the register doesn't change,
$r = (r/2) + (x/2r)$ or $2r^2 = r^2 + x$, so $r^2 = x$.
97. (a) $0 \le 5 - 5x < 7$; (b) $-\frac{4}{5} < x - \frac{2}{5} \le \frac{3}{5}$
99. $[5, \infty)$, ——|——→
       $\quad\quad\quad\quad 5$
101. $(3, \infty)$, ——(——→
       $\quad\quad\quad\quad 3$
103. $(\frac{24}{13}, \infty)$, ——(——→
       $\quad\quad\quad\quad \frac{24}{13}$
105. $[73, 98]$
107. 7, |← 7 →|
       ——+——+——+——
       $-3\quad 0\quad\quad 4$
109. 0.407, |← d →|
       ——+————+——
       $-\pi\quad\quad -2.735$
111. $\frac{19}{6}$, |← $\frac{19}{6}$ →|
       ——+——+——+——
       $-\frac{2}{3}\quad 0\quad\quad \frac{5}{2}$ **113.** True
115. True   **117.** True   **119.** $-\frac{3}{2}, \frac{9}{2}$ **121.** $\frac{1}{4}$
123. $(-\infty, 1) \cup (3, \infty)$, …—)——(—→
       $\quad\quad\quad\quad\quad 1\quad\quad 3$
125. $(-\infty, -\frac{1}{2}) \cup [2, \infty)$, …—)——[—→
       $\quad\quad\quad\quad\quad -\frac{1}{2}\quad\quad 2$
127. $[\frac{1}{3}, 1]$, —[——]—
       $\quad\quad\quad \frac{1}{3}\quad\quad 1$
129. $(\frac{3}{2}, \frac{11}{6})$, ——(——)——
       $\quad\quad\quad\quad \frac{3}{2}\quad\quad \frac{11}{6}$
131. $(-\frac{2}{3}, \frac{5}{2})$, ——(——)——
       $\quad\quad\quad\quad -\frac{2}{3}\quad\quad \frac{5}{2}$
133. $[-\frac{3}{2}, \frac{2}{3}]$, —[——]—
       $\quad\quad\quad -\frac{3}{2}\quad\quad \frac{2}{3}$
135. $(-5, \frac{1}{3})$, ——(——)——
       $\quad\quad\quad -5\quad\quad \frac{1}{3}$
137. $(-\infty, -5) \cup (-3, 0)$, …—)——(——)—→
       $\quad\quad\quad -5\quad -3\quad\quad 0$
139. $(-\infty, \frac{1}{3}) \cup (\frac{5}{2}, \infty)$, …—)——(—→
       $\quad\quad\quad\quad \frac{1}{3}\quad\quad \frac{5}{2}$
141. $(\frac{7}{5}, \frac{13}{7})$, ——(——)——
       $\quad\quad\quad \frac{7}{5}\quad\quad \frac{13}{7}$
143. $(-\infty, -3) \cup (-2, -1) \cup (0, 2)$,
       …—)——(——)——(——)—
       $\quad -3\ -2\ -1\ 0\quad\quad 2$
145. $(-\infty, -1) \cup (-1, 1] \cup (2, \infty)$,
       …—✕——|——(—→
       $\quad -1\quad\quad 1\quad\quad 2$
147. $[85, 105]$ **149.** $[55, 75]$
151. $[\$12, \$12.11]$ **153.** $A = 1, B = 0.02$

CHAPTER 3

Problem Set 3.1, page 135

1.

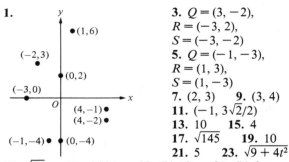

3. $Q = (3, -2)$,
$R = (-3, 2)$,
$S = (-3, -2)$
5. $Q = (-1, -3)$,
$R = (1, 3)$,
$S = (1, -3)$
7. $(2, 3)$ **9.** $(3, 4)$
11. $(-1, 3\sqrt{2}/2)$
13. 10   **15.** 4
17. $\sqrt{145}$ **19.** 10
21. 5   **23.** $\sqrt{9 + 4t^2}$
25. $\sqrt{33}$ **27.** 6.224   **29.** (b) 12   **31.** (b) 6
33. $|\overline{AB}| = |\overline{BC}| = |\overline{CD}| = |\overline{AD}| = \sqrt{29}$, so $ABCD$ is a
rhombus (4 sides equal). $|\overline{BD}|^2 = 58 = 29 + 29 =$
$|\overline{AD}|^2 + |\overline{AB}|^2$, so A is a right angle.
35. Yes   **37.** $x = -17$ or $x = 7$
39. $t = 3$ or $t = -2$ **41.** Yes   **43.** No
45. $x^2 + y^2 = 16$ **47.** $(x + 1)^2 + (y - 3)^2 = 4$
49. $(x - 1)^2 + (y - 3)^2 = 9$
51.

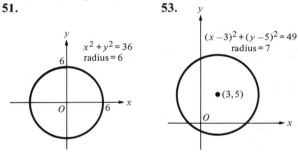

$x^2 + y^2 = 36$
radius $= 6$

53.

$(x - 3)^2 + (y - 5)^2 = 49$
radius $= 7$
$(3, 5)$

55.

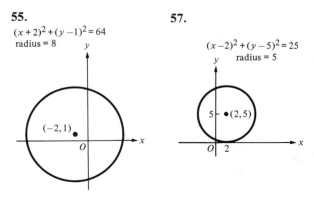

$(x + 2)^2 + (y - 1)^2 = 64$
radius $= 8$
$(-2, 1)$

57.

$(x - 2)^2 + (y - 5)^2 = 25$
radius $= 5$
$(2, 5)$

59.

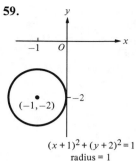

$(x+1)^2 + (y+2)^2 = 1$
radius = 1

61. 13 nautical miles

13.

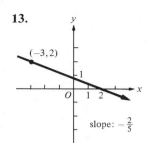

slope: $-\frac{2}{5}$

15.

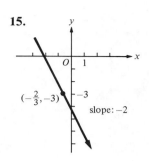

slope: -2

Problem Set 3.2, page 141

1.

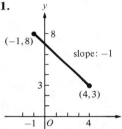

slope: -1

3.

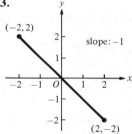

slope: -1

17.

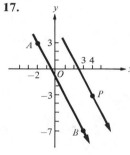

19. $-\frac{1}{5}$

5.

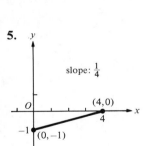

slope: $\frac{1}{4}$

7.

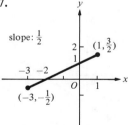

slope: $\frac{1}{2}$

21.

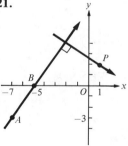

23.

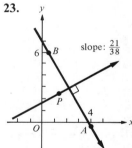

slope: $\frac{21}{38}$

9.

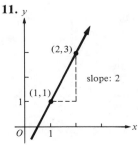

slope: $-\frac{42}{9}$

11.

slope: 2

27. $\dfrac{6-(-2)}{4-(-4)} = 1,\ \dfrac{-8-(-2)}{2-(-4)} = -1$

29. (a) $d = 1$; (b) $k = -\frac{10}{3}$

Problem Set 3.3, page 145

1. $y - 2 = \frac{3}{4}(x - 3)$
3. $y - 1 = -\frac{1}{7}(x - 4)$ or $y - 2 = -\frac{1}{7}(x + 3)$
5. $y - 5 = 1(x + 3)$

7. $y = \frac{3}{2}x - 3$, $m = \frac{3}{2}$, $b = -3$

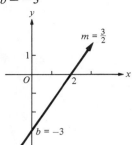

9. $y = 3x + 1$, $m = 3$, $b = 1$

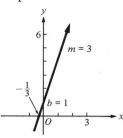

31. $y = 0.25x + 3.45$. In 1992, $y = \$5.45$ per share.

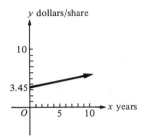

33. $y = -0.75x + 7$. Free of pollution in 9 years and 4 months—in 1993.

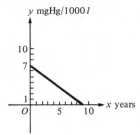

11. $y = -\frac{1}{3}x + 3$, $m = -\frac{1}{3}$, $b = 3$

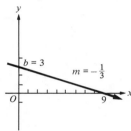

13. $y = 2x + 3$, $m = 2$, $b = 3$

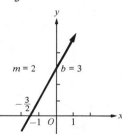

Problem Set 3.4, page 153

1. -5 **3.** 2   **5.** 3   **7.** -4 **9.** $-\frac{5}{11}$
11. 4.106   **13.** $-\frac{7}{18}$ **15.** 2   **17.** 3
19. $2a + 3$ **21.** $b^2 + 3b - 4$ **23.** $|8 + 5c|$
25. $|2 - (5/a)|$ **27.** $-2x + 1$ **29.** $\sqrt{3x^4 + 5}$
31. $x + 1$ **33.** x **35.** $(a - b)(a + b - 3)$
37. $\sqrt{3x + 5} - \sqrt{5}$ **39.** 2   **41.** $x \neq 0$ **43.** $\mathbb{R}$
45. $x \geq 0$ **47.** $x \neq 0$ **49.** $[-3, 3]$ **51.** $\mathbb{R}$
53. $(-\infty, -2) \cup (-2, 2) \cup (2, \infty)$ **55.** 4
57. $2x + h$ **59.** $-1/\sqrt{x}\,\sqrt{x + h}\,(\sqrt{x} + \sqrt{x + h})$
61.

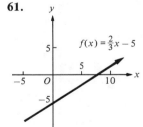

63.

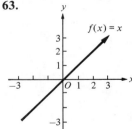

15. (a) $y - 2 = 4(x + 5)$; (b) $y = 4x + 22$; (c) $4x - y + 22 = 0$
17. (a) $y - 5 = -3(x - 0)$; (b) $y = -3x + 5$; (c) $-3x - y + 5 = 0$
19. (a) $y - 5 = -\frac{5}{3}(x - 0)$; (b) $y = -\frac{5}{3}x + 5$; (c) $5x + 3y - 15 = 0$
21. (a) $y + 4 = \frac{2}{5}(x - 4)$; (b) $y = \frac{2}{5}x - \frac{28}{5}$; (c) $2x - 5y - 28 = 0$
23. (a) $y - \frac{2}{3} = \frac{3}{5}(x + 3)$; (b) $y = \frac{3}{5}x + \frac{37}{15}$; (c) $9x - 15y + 37 = 0$
25. (a) $y - \frac{5}{7} = -\frac{7}{3}(x - \frac{2}{3})$; (b) $y = -\frac{7}{3}x + \frac{143}{63}$; (c) $147x + 63y - 143 = 0$
27. $x + 2y - 9 = 0$
29. $y = 22N + 0.2x$

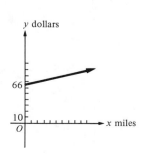

65.

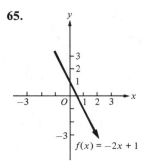

67. (a) and (c)
69. $f(0) = 32$ or $0 \xrightarrow{f} 32$, $f(15) = 59$ or $15 \xrightarrow{f} 59$, $f(-10) = 14$ or $-10 \xrightarrow{f} 14$, $f(55) = 131$ or $55 \xrightarrow{f} 131$

71. $T = -\frac{1}{250}h + 65$;
when $h = 30,000$ ft, $T = -55°F$

73. $A = x(12 - x)$, $x > 0$

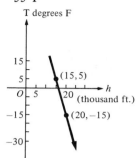

Problem Set 3.5, page 162

1.

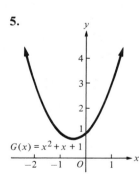

3.

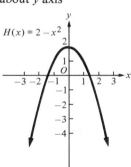

5.

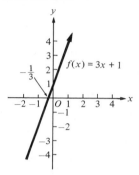

7.

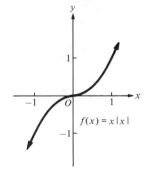

9. (a) Domain $\mathbb{R}$, range $(-\infty, 2]$, increasing on $(-\infty, -2]$ and $[0, 2]$, decreasing on $[-2, 0]$ and $[2, \infty)$, even; (b) Domain $[-5, 5]$, range $[-3, 3]$, increasing on $[-1, 1]$ and $[3, 5]$, decreasing on $[-5, -1]$ and $[1, 3]$, neither; (c) Domain $\left[-\frac{3\pi}{2}, \frac{3\pi}{2}\right]$, range $[-1, 1]$, increasing on $\left[-\frac{3\pi}{2}, -\frac{\pi}{2}\right]$ and $\left[\frac{\pi}{2}, \frac{3\pi}{2}\right]$, decreasing on

$\left[-\frac{\pi}{2}, \frac{\pi}{2}\right]$, odd; (d) Domain $\mathbb{R}$, range $[-2, 1]$, constant on $(-\infty, 0]$ and $[\pi, \infty)$, increasing on $[0, \pi]$, neither
11. Even, symmetric about y axis
13. Odd, symmetric about origin   **15.** Even, symmetric about y axis   **17.** Neither   **19.** Odd, symmetric about origin
21. Domain $\mathbb{R}$, range $\mathbb{R}$, increasing on $\mathbb{R}$, neither even nor odd
23. Domain $\mathbb{R}$, range $\{5\}$, constant on $\mathbb{R}$, even, symmetric about y axis

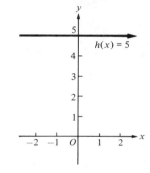

25. Domain $\mathbb{R}$, range $(-\infty, 2]$, increasing on $(-\infty, 0]$, decreasing on $[0, \infty)$, even, symmetric about y axis

27. Domain $\mathbb{R}$, range $\mathbb{R}$, increasing on $\mathbb{R}$, odd, symmetric about origin

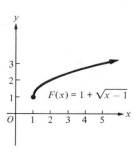

29. Domain $[1, \infty)$, range $[1, \infty)$, increasing on $[1, \infty)$, neither even nor odd

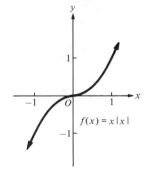

31. Domain $(-\infty, 0)$ and $(0, \infty)$, range $\{-1, 1\}$, constant on $(-\infty, 0)$ and on $(0, \infty)$, odd, symmetric about origin

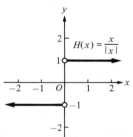

33. Domain $\mathbb{R}$, range $\mathbb{R}$, increasing on $\mathbb{R}$, odd, symmetric about origin

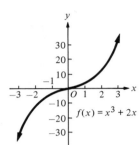

43.

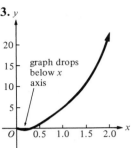

45.

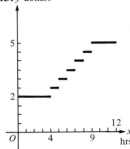

Problem Set 3.6, page 171

1.

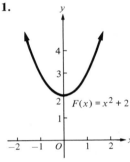

3.

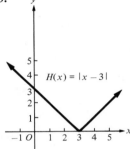

35. Domain $(-\infty, 0)$ and $(0, \infty)$, increasing on $(-\infty, 0)$, decreasing on $(0, \infty)$, even, symmetric about y axis

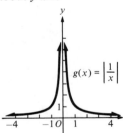

37.

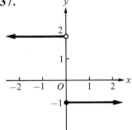

5.

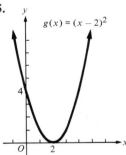

7.

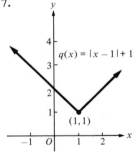

39.

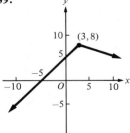

41.

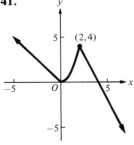

9.

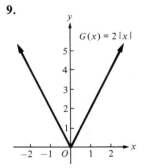

11.

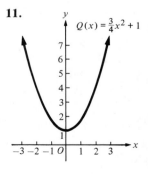

13.

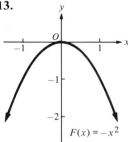

$F(x) = -x^2$

15.

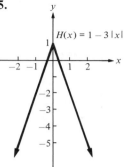

$H(x) = 1 - 3|x|$

25. *T* is obtained by shifting *t* by 2 units to the right, multiplying resulting ordinates by 2, then shifting up 4 units.

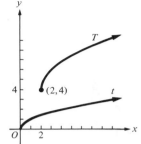

17.

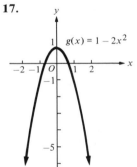

$g(x) = 1 - 2x^2$

19. *F* is obtained by shifting *f* up 2 units.

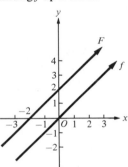

27.

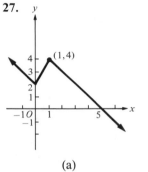

(a)

(b)

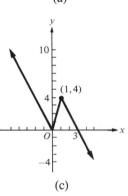

(c)

(d)

21. *P* is obtained by reflecting *p* about the *x* axis, then shifting up 1 unit.

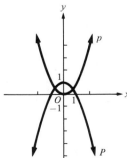

23. *R* is obtained by multiplying each ordinate of *r* by 2, then shifting up 1 unit.

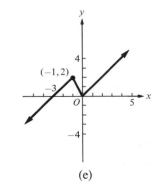

(e)

29.

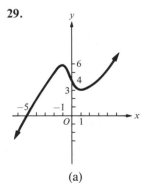

(a)

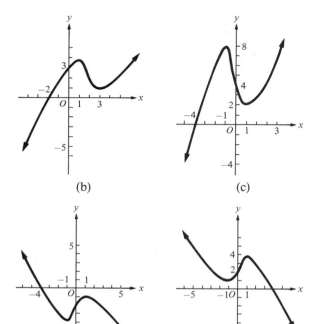

(b)   (c)

(d)   (e)

31. (x, y) belongs to graph of F if and only if $y = f(-x)$; that is, $(-x, y)$ belongs to graph of f

33.

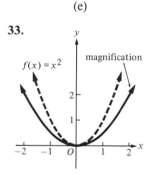

Problem Set 3.7, page 176

1. (a) $7x - 3$; (b) $3x + 7$; (c) $10x^2 - 21x - 10$; (d) $\dfrac{5x + 2}{2x - 5}$

3. (a) $x^2 + 4$; (b) $x^2 - 4$; (c) $4x^2$; (d) $x^2/4$

5. (a) -1; (b) 5; (c) -6; (d) $-\frac{2}{3}$

7. (a) $x^2 + 2x - 4$; (b) $-x^2 + 2x - 6$; (c) $2x^3 - 5x^2 + 2x - 5$; (d) $\dfrac{2x - 5}{x^2 + 1}$

9. (a) $\dfrac{8x + 1}{2x - 1}$; (b) 1 if $x \neq \frac{1}{2}$; (c) $\dfrac{15x^2 + 5x}{4x^2 - 4x + 1}$; (d) $\dfrac{5x}{3x + 1}$

11.

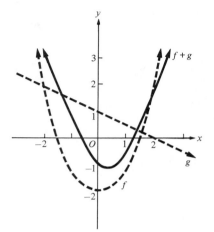

13.

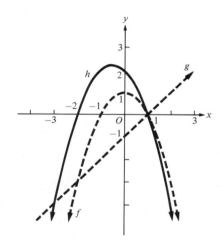

15. 17 **17.** 6.9929 **19.** -3 **21.** 2

23. (a) $3x + 3$; (b) $3x + 1$; (c) $9x$

25. (a) x; (b) $|x|$; (c) x^4

27. (a) $-x^2 - 1$; (b) $x^2 - 2x + 3$; (c) x

29. (a) $1 - 5|2x + 3|$; (b) $|5 - 10x|$; (c) $25x - 4$

31. (a) $\dfrac{1}{4x - 9}$; (b) $\dfrac{11 - 6x}{2x - 3}$; (c) $\dfrac{2x - 3}{11 - 6x}$

33. $h \circ g$ **35.** $h \circ f$ **37.** $f \circ h$ **39.** $h \circ h$

41. $f(x) = x^{-7}$, $g(x) = 2x^2 - 5x + 1$

43. $f(x) = x^5$, $g(x) = \dfrac{1 + x^2}{1 - x^2}$

45. $f(x) = \sqrt{x}$, $g(x) = \dfrac{x + 1}{x - 1}$

47. $f(x) = \dfrac{|x|}{x}$, $g(x) = x + 1$

49. (a) $R(x) = 10x$;
(b) $P(x) = 10x - 50,000 - 10,000\sqrt[3]{x+1}$;
(c) $\$56,550$

51. $\dfrac{\sqrt{3}p^2}{36}$ = area of equilateral triangle with perimeter p

53. (a) x; (b) x; (c) $\frac{1}{2}x^2 - 3x + \frac{15}{2}$; (d) $\frac{1}{2}x^2 - 3x + \frac{15}{2}$;
(e) $2x^2 + x + 3$
55. For example: $f(x) = 2x$, $g(x) = x^2 - 1$
57. $[(f \circ g) \circ h](x) = (f \circ g)(h(x)) = f(g(h(x))) = f((g \circ h)(x)) = [f \circ (g \circ h)](x)$
59. $\frac{3}{2}$ **61.** -1

Problem Set 3.8, page 183

1. $f[g(x)] = f[(x+3)/2] = 2[(x+3)/2] - 3 = x$ and
$g[f(x)] = g(2x - 3) = [(2x - 3) + 3]/2 = x$
3. For $x \neq 0$, $f[g(x)] = f(1/x) = 1/(1/x) = x$ and,
likewise, $g[f(x)] = x$
5. $f[g(x)] = f(x^3 - 8) = \sqrt[3]{(x^3 - 8) + 8} = x$
and $g[f(x)] = g(\sqrt[3]{x+8}) = (\sqrt[3]{x+8})^3 - 8 = x$
7. (a) Invertible; (b) Not invertible; (c) Not invertible

9.

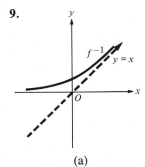

(a)

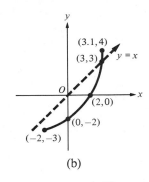

(b)

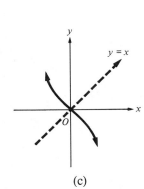

(c)

11. $f^{-1}(x) = \dfrac{x + 13}{7}$

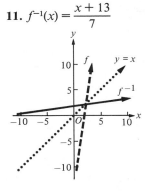

13. $f^{-1}(x) = -2x + 6$

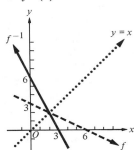

15. $f^{-1}(x) = -\sqrt{-x}$

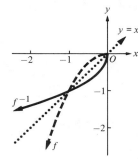

17. $f^{-1}(x) = (x - 1)^2$, $x \geq 1$

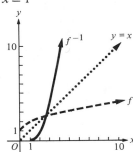

19. $f^{-1}(x) = 1 + \sqrt{x - 1}$

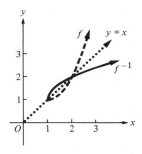

21. (a) $f^{-1}(x) = \dfrac{x + 5}{2}$

23. (a) $f^{-1}(x) = \sqrt[3]{x} - 2$

25. (a) $f^{-1}(x) = \dfrac{\sqrt[3]{1 - x}}{2}$

27. (a) $f^{-1}(x) = \dfrac{-1}{1 + x}$

29. (a) $f^{-1}(x) = \dfrac{-x - 7}{x - 3}$

31. Not invertible, fails horizontal line test

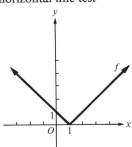

33. $f^{-1}(x) = \dfrac{b - dx}{cx - a}$

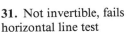

35. If (b, a) belongs to graph of g, $a = g(b)$, so $f(a) = f[g(b)] = b$; hence, (a, b) belongs to graph of f.
37. A horizontal line $y = b$ intersects the graph of f at (x_1, b) and $x_2, b)$ if and only if $f(x_1) = b = f(x_2)$.
39. $C = f^{-1}(t)$
41. The effect is to interchange the x and y axes.
43. $f^{-1}(x) = \dfrac{-B + \sqrt{B^2 - 4AC + 4Ax}}{2A}$

Problem Set 3.9, page 193

1. $\frac{3}{4}$ **3.** $\frac{9}{5}$ **5.** $\frac{3}{200}$ **7.** $-\frac{11}{32}$ **9.** $\frac{67}{3}$ **11.** $\frac{21}{2}$
13. $\frac{40}{21}$ **15.** $\frac{121}{2}$
17. From $a/b = c/d$ we have $ad = bc$. Dividing by ab, we obtain $d/b = c/a$ or $c/a = d/b$.
23. $y = kt$ **25.** $V = kr^3$ **27.** $P = kv^3$ **29.** $E = k(AT/d)$
31 $F = k(Q_1Q_2/d^2)$.   **33** $n = k\sqrt{F}/(l\sqrt{d})$
35. 6   **37.** $\frac{3}{2}$ **39.** $\frac{200}{9}$ **41.** 8   **43.** $\frac{28}{3}$
45. 54   **47.** $P = MS/m$ **49.** 40
51. $F = 32 + \frac{9}{5}C$ **53.** $P = \frac{2}{15}R - 4000$
55. $A = \frac{9}{4}N^{2/3}$ **57.** $\frac{15}{8}$ days
59. 2.36×10^{-3} newton   **61.** $n = (kT + K)/l$
63. $(\sqrt{5} - 1)/2 \approx 0.618$

Review Problem Set, Chapter 3, page 196

1. 5   **3.** 13   **5.** 8   **7.** 30.35
9. A circle of radius 5 with center $(2, -3)$
11. $m = -7$, $y - 2 = -7(x - 2)$
13. $m = \frac{3}{2}$, $y - 2 = \frac{3}{2}(x - 1)$ **15.** $y - 2 = -\frac{3}{2}(x - 5)$
17. (a) $y + 5 = \frac{3}{2}(x - 7)$; (b) $y + 5 = -\frac{2}{3}(x - 7)$
19. $y = \frac{4}{3}x + \frac{2}{3}$, $m = \frac{4}{3}$, $b = \frac{2}{3}$
21. (a) $y - 1 = 3(x + 7)$; (b) $y = 3x + 22$;
(c) $3x - y + 22 = 0$
23. (a) $y + 2 = \frac{7}{3}(x - 1)$; (b) $y = \frac{7}{3}x - \frac{13}{3}$;
(c) $7x - 3y - 13 = 0$
25. $y - b = -(a/b)(x - a)$
27. (a) $y - b = 2a(x - a)$;   (c)
(b) $y - 8 = 4(x - 2)$;

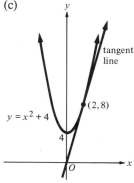
tangent line
(2, 8)
$y = x^2 + 4$
4
O

29. 23   **31.** 0
33. $3x^2 - 12$ **35.** $75x^2 - 180x + 104$
37. 12   **39.** $-1/[(x + k)x]$
41. $(-\infty, 1) \cup (1, \infty)$
43. $[-1, \infty)$ **45.** $6x + 3h - 2$
47. (a) Graph of a function; (b) Not the graph of a function   **49.** $f(74) = 99.26$, $f(75) = 99.63$, $f(76) = 100$
53. (a) Domain $\mathbb{R}$, range $[-1, 1]$, constant on $(-\infty, -2]$, decreasing on $[-2, 0]$, increasing on $[0, 2]$, constant on $[2, \infty)$, even, symmetric about y axis; (b) Domain $\mathbb{R}$, range $[-3, \infty)$, decreasing on $(-\infty, -3]$, increasing on $[-3, -\frac{2}{3}]$, decreasing on $[-\frac{2}{3}, 2]$, constant on $[2, 4]$, increasing on $[4, \infty)$, neither even nor odd
55. Odd, symmetric about origin   **57.** Neither
59. Even, symmetric about y axis   **61.** Domain $\mathbb{R}$, range $\mathbb{R}$, increasing on $\mathbb{R}$, neither even nor odd
63. Domain $[2, \infty)$, range $[0, \infty)$, increasing on $[2, \infty)$, neither even nor odd
65. Domain $\mathbb{R}$, range $\mathbb{R}$, increasing on $\mathbb{R}$, odd, symmetric about origin
67. (a) Shift up 1 unit; (b) Shift down 2 units; (c) Shift 1 unit to the right; (d) Shift 2 units to the left; (e) Reflect across x axis
69. (a) Shift f up 3 units; (b) Shift g down 5 units; (c) Reflect h about x axis, then shift up 1 unit; (d) Shift k one unit to the right; (e) Shift q 2 units to the left, then take half of each ordinate; (f) Shift r 1 unit to the left, take half of each ordinate, reflect about x axis, and shift up 1 unit
71. (a) Increasing; (b) $G(p) = ps$ or $G(p) = pg(p)$; (c) At that price, producers won't supply.
73. (a) $4x - 2$; (b) $-2x + 6$; (c) $3x^2 + 2x - 8$;
(d) $\dfrac{x + 2}{3x - 4}$; (e) $3x - 2$

75. (a) $\dfrac{2x}{x^2 - 1}$; (b) $\dfrac{2}{x^2 - 1}$; (c) $\dfrac{1}{x^2 - 1}$; (d) $\dfrac{x + 1}{x - 1}$;
(e) $\dfrac{x + 1}{-x}$

77. (a) $x^4 + \sqrt{x + 1}$; (b) $x^4 - \sqrt{x + 1}$; (c) $x^4\sqrt{x + 1}$;
(d) $\dfrac{x^4}{\sqrt{x + 1}}$; (e) $(x + 1)^2$

79. (a) $|x| - x$; (b) $|x| + x$; (c) $-x|x|$; (d) $\dfrac{|x|}{-x}$; (e) $|x|$
81. (a) $x^{2/3} + 1 + \sqrt{x}$; (b) $x^{2/3} + 1 - \sqrt{x}$; (c) $x^{7/6} + x^{1/2}$;
(d) $\dfrac{x^{2/3} + 1}{\sqrt{x}}$; (e) $x^{1/3} + 1$

83. 0.8693   **85.** -11.8623 **87.** $g \circ h$

89. $h \circ g$ **91.** $f \circ h$
93. $f(x) = x^{-3}$, $g(x) = 4x^3 - 2x + 5$

95. $f(x) = \dfrac{2x^2 + x}{\sqrt{x}}$, $g(x) = x^2 + 1$ **97.** (a) $2x$; (b) 2;

(c) $2x$; (d) 2
99. (a) $f(x) = \sqrt{8100 + x^2}$, $g(t) = 50t$; (b) $y = f[g(t)]$, so
$y = (f \circ g)(t)$ (c) $(f \circ g)(t) = 10\sqrt{81 + 25t^2}$
101. (a) $P = -\frac{84}{125}p^2 + 102p - 2750$; (b) \$875.20

103. $f^{-1}(x) = \dfrac{x + 1}{3}$ **105.** $f^{-1}(x) = 5x - 25$

107. $f^{-1}(x) = \frac{1}{4}(x + 1)^2$ **111.** $f^{-1}(x) = \dfrac{1}{1 - x}$

113. The identity function

CHAPTER 4

Problem Set 4.1, page 211

1.

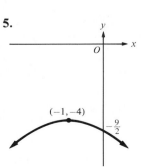

3.

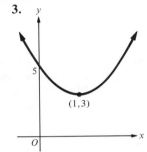

5.

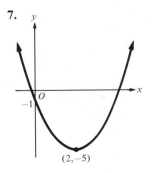

7.

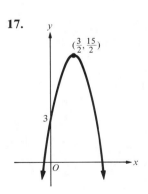

9.

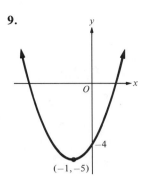

11.

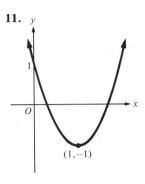

13.

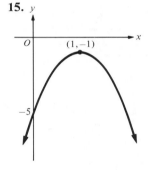

15.

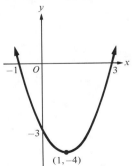

17.

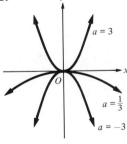

19. Vertex $(1, -4)$, y intercept -3, x intercepts -1, 3, opens upward, domain $\mathbb{R}$, range $[-4, \infty)$

21. Vertex $(-2, -4)$, y intercept 0, x intercepts 0, -4, opens upward, domain $\mathbb{R}$, range $[-4, \infty)$

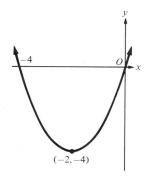

23. Vertex $(\frac{1}{4}, -\frac{119}{8})$, y intercept -15, x intercepts—none, opens downward, domain $\mathbb{R}$, range $(-\infty, -\frac{119}{8}]$

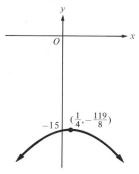

33. $(-\infty, -1] \cup [3, \infty)$

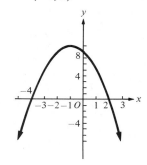

35. $(-4, 2)$

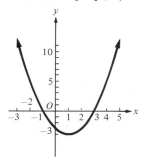

25. Vertex $(-\frac{1}{12}, -\frac{49}{24})$, y intercept -2, x intercepts $-\frac{2}{3}, \frac{1}{2}$, opens upward, domain $\mathbb{R}$, range $[-\frac{49}{24}, \infty)$

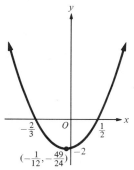

27. Vertex $(2, 27)$, y intercept 15, x intercepts -1, 5, opens downward, domain $\mathbb{R}$, range $(-\infty, 27]$

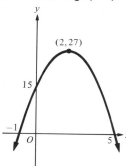

37. $(-\infty, 0] \cup [8, \infty)$

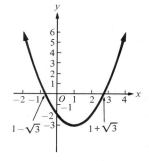

39. $[1 - \sqrt{3}, 1 + \sqrt{3}]$

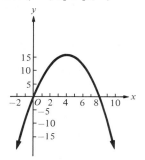

41. 50, 50 **43.** 150 meters by 150 meters
45. Maximum height $= 144$ feet; returns when $t = 6$ seconds **47.** $2AB^3/27$
49. 96 kilometers/hour **51.** $8/month

29. Vertex $(-1, \frac{3}{2})$, y intercept 2, x intercepts — none, opens upward, domain $\mathbb{R}$, range $[\frac{3}{2}, \infty)$

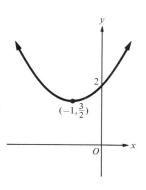

31. Vertex $(5, 7)$, y intercept 57, x intercepts — none, opens upward, domain $[0, 10]$, range $[7, 57]$

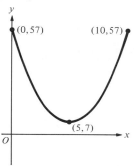

Problem Set 4.2, page 217

1. Polynomial **3.** Not **5.** Not
7.

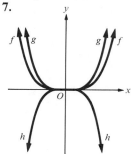

9.

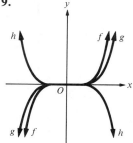

11. y intercept 1, x intercept—none

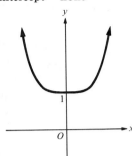

13. y intercept -1, x intercept $\sqrt[5]{\frac{1}{3}}$

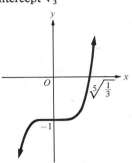

33. (a) $-1, 0, 1$; (c) $(-1, 0) \cup (1, \infty)$ (b)

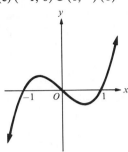

35. (a) $-2, 0, 1$; (c) $(-\infty, -2] \cup [0, 1]$ (b)

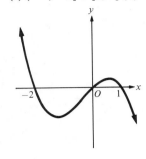

15. y intercept 2, x intercept -1

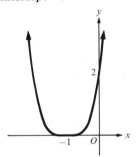

17. y intercept -1, x intercept—none

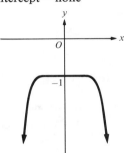

37. (a) $-1, 1, 3$; (c) $(-\infty, -1) \cup (1, 3)$ (b)

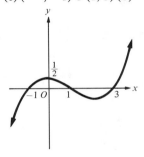

39. (a) $-4, -2, 1$; (c) $(-\infty, -4] \cup [-2, 1]$ (b)

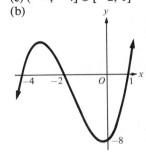

41. V is maximum for $x \approx 1.7$ inches

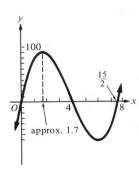

19. y intercept $-\frac{7}{8}$, x intercept $\frac{1}{2}$

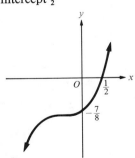

21. y intercept 2593, x intercept—none

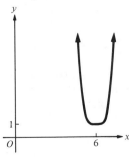

23. 4   **25.** $1, \frac{3}{2}, \frac{3}{2}$
27. $-\frac{1}{2}, \frac{1}{3}, \frac{1}{2}$ **29.** $-2, -\frac{5}{3}, \frac{5}{3}, 7$
31. $-1, 0, 1$

Problem Set 4.3, page 225

1. $Q: x + 8$, $R: 30$ **3.** $Q: x^2 + x + 1$, $R: 0$
5. $Q: 3x^2 + 6x - 5$, $R: -1$
7. $Q: 2x^3 - 3x^2 + 3x - 4$, $R: -12$
9. $Q: x + 3$, $R: -x - 13$
11. $Q: x^2 - x + 1$, $R: -x^2 - 1$
13. $Q: (x/2) + (1/2)$, $R: 3/2$

15. Q: $(x/3) + (1/9)$, R: $-(5/9)x + (8/9)$
17. Q: $2t + 3$, R: -10
19. $5x^2 + 23x + 91 + [366/(x - 4)]$

21. $5 + \dfrac{4x^2 - 28x - 16}{x^3 - 2x^2 - 8x}$ **23.** Q: $3x^2 + 4x + 7$, R: 18

25. Q: $x^4 - 2x^3 - x^2 + 2x - 3$, R: -10
27. Q: $3x^5 - 3x^4 + x^3 - x^2 + 2x - 2$, R: 2
29. Q: $-16x^2 - 20x - 8$, R: 3
31. Q: $x^2 + 2.1x + 3.31$, R: 4.641
33. Q: $5x^2 + 4x - 2$, R: -16
35. Q: $-2x^2 + 2x + 10$, R: -20 **37.** 3
39. 15   **41.** q is the zero polynomial, $r = f$

Problem Set 4.4, page 232

1. 0   **3.** -456 **5.** 4   **7.** 51   **9.** $\frac{181}{27}$
11. Yes   **13.** No   **15.** No   **17.** No   **19.** 2
21. 3   **23.** 3   **25.** (a) 2; (b) 1   **27.** (a) 3; (b) 2
29. (a) 2; (b) 3   **31.** Positive: 1, negative: 2 or 0
33. Positive: 3 or 1, negative: 0
35. Positive: 3 or 1, negative: 1
37. Positive: 1, negative: 2 or 0
39. Positive: 2 or 0, negative: 2 or 0
41. Positive: 2 or 0, negative: 1   **43.** $-1, 4$
45. $-3, 4$ **47.** $-3, 4$ **49.** $-1, 3$ **51.** $-2, 2$
53. $-3, 5$ **55.** Zeros of $f(-x)$ are negatives of zeros
of $f(x)$ **57.** Coefficients of odd powers change
signs   **59.** $(x - p)(x - q) = x^2 + (-p - q)x + pq$

Problem Set 4.5, page 240

1. 1, 3, 5   **3.** 3   **5.** $-\frac{1}{2}$ **7.** 2, 2   **9.** $-\frac{1}{2}, -\frac{1}{2}$
11. 3   **13.** $-\frac{3}{4}, \frac{1}{3}, 1$
15. $f(x) = (x - 6)(x + 1)(x - 1)$

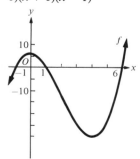

17. $Q(x) = (x - 2)(x - 1)(2x - 1)(x + 2)$

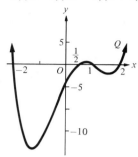

19. $f(1) < 0, f(2) > 0$ **21.** $f(1.5) < 0, f(1.6) > 0$
23. $h(-2) < 0, h(-1) > 0$ **25.** Zero in $[1.5, 1.75]$
27. Zero in $[-1.5, -1.375]$
29. Zeros in $[-2, -1.75]$, $[0.25, 0.50]$, and $[1.5, 1.75]$
31. -0.68 **33.** 1.29   **35.** 2.42
37. Width 5 inches, length 10 inches, height 7 inches
39. Sides: 3 kilometers, 4 kilometers, hypotenuse:
5 kilometers

Problem Set 4.6, page 249

1. $(-\infty, 0) \cup (0, \infty)$ **3.** $\mathbb{R}$ **5.** $\mathbb{R}$ **7.** $\mathbb{R}$
9. $(-\infty, -5) \cup (-5, -3) \cup (-3, \infty)$
11. Horizontal
asymptote x axis, vertical
asymptote y axis

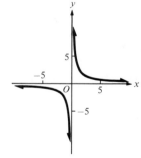

13. Horizontal
asymptote x axis, vertical
asymptote y axis

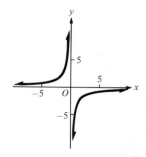

15. Horizontal asymptote x axis, vertical asymptote y axis

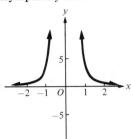

17. Horizontal asymptote x axis, vertical asymptote y axis

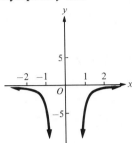

27. Horizontal asymptote $y = 6$, vertical asymptote $x = 2$

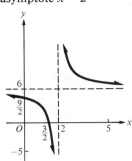

29.

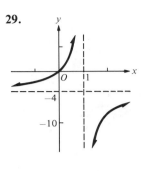

19. Horizontal asymptote $y = 1$, vertical asymptote y axis

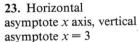
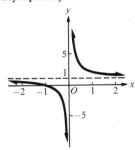

21. Horizontal asymptote $y = -4$, vertical asymptote y axis

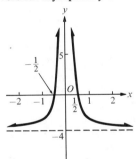

31.

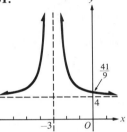

33.

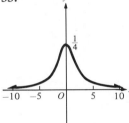

35.

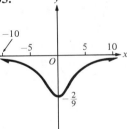

37.

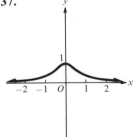

23. Horizontal asymptote x axis, vertical asymptote $x = 3$

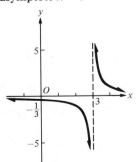

25. Horizontal asymptote x axis, vertical asymptote $x = -1$

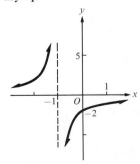

39.

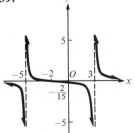

41.

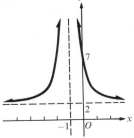

43.

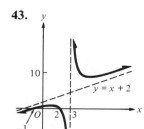

$y = x + 2$

45.

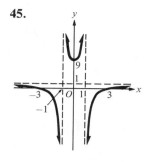

47.

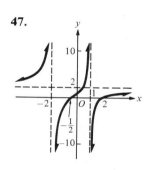

49.

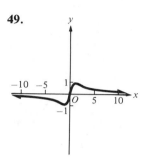

51.

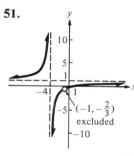

$(-1, -\frac{2}{3})$
excluded

53.

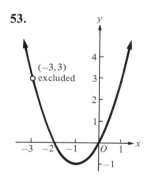

$(-3, 3)$
excluded

55.

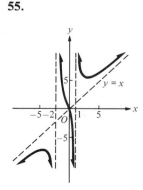

$y = x$

57. (a)

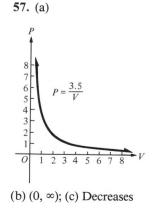

$P = \dfrac{3.5}{V}$

(b) $(0, \infty)$; (c) Decreases

59. (a)

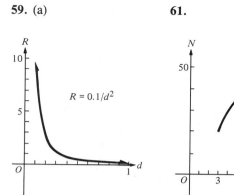

$R = 0.1/d^2$

(b) $(0, \infty)$; (c) Decreases

63. A rational function can't have more than *one* horizontal asymptote.

Problem Set 4.7, page 256

1. $z^3 - (6 + i)z^2 + (9 + 6i)z - 9i$

3. $z^3 + (-1 + 2i)z^2 - (1 + 2i)z + 1$

5. $z^4 + 4z^3 + 5z^2 + 4z + 4$

7. $z^4 - 3z^3 + 3z^2 - 3z + 2$

9. $z^4 + 2z^3 + 2z^2 + 2z + 1$

11. $z^6 - 2z^5 + 10z^4 - z^2 + 2z - 10$

13. $z^5 - 3z^4 + z^3 - 3z^2$

15. $z^5 - \frac{7}{3}z^4 + \frac{11}{3}z^3 - 3z^2 + \frac{8}{3}z - \frac{2}{3}$

17. $\pm\left(\dfrac{\sqrt{2}}{2} + \dfrac{\sqrt{2}}{2}i\right)$ **19.** $\pm\left(\dfrac{3\sqrt{2}}{2} - \dfrac{\sqrt{2}}{2}i\right)$

21. $\pm\left(\dfrac{\sqrt{2}}{2} - \dfrac{\sqrt{2}}{2}i\right)$

23. $\pm\left[\sqrt{(2 + \sqrt{3})/2} + i\sqrt{(2 - \sqrt{3})/2}\right]$ **25.** $1 \pm i$

27. $\dfrac{-1 \pm \sqrt{5}}{2}i$ **29.** $\dfrac{-i \pm \sqrt{3}}{2}$ **31.** $2, 1 + i$

33. (a) $1, -1, i, -i$; (b) $(z - 1)(z + 1)(z - i)(z + i)$;
(c) Each zero has multiplicity 1

35. (a) $-2, -2, 2i, -2i$; (b) $(z + 2)^2(z + 2i)(z - 2i)$;
(c) -2 has multiplicity 2; others 1

37. (a) $-\frac{1}{2}, -\frac{1}{2}, -\frac{1}{2}, -1, 3$; (b) $(2z + 1)^3(z + 1)(z - 3)$;
(c) $-\frac{1}{2}$ has multiplicity 3; others 1

39. (a) $1, i, -i$; (b) $(z - 1)(z + i)(z - i)$; (c) Each zero
has multiplicity 1

41. (a) $3, 1 + i, 1 - i$; (b) $(z - 3)(z - 1 - i)(z - 1 + i)$;
(c) Each zero has multiplicity 1

43. (a) $2, 2, (-1 + i\sqrt{3})/2, (-1 - i\sqrt{3})/2$,
(b) $(z - 2)^2(2z + 1 - \sqrt{3}i)(2z + 1 - \sqrt{3}i)$;
(c) 2 is a zero of multiplicity 2; others 1
45. (a) $-1, \frac{2}{3}, 1 + i, 1 - i$;
(b) $(z + 1)(3z - 2)(z - 1 - i)(z - 1 + i)$; (c) Each zero
has multiplicity 1
47. $az^2 + bz + c = a(z - z_1)(z - z_2) =$
$az^2 + a(-z_1 - z_2)z + az_1z_2$
49. Complex zeros come in *pairs*

51. 1 and $-\dfrac{1}{2} - \dfrac{\sqrt{3}}{2} i$

Review Problem Set, Chapter 4, page 257

1. Domain $\mathbb{R}$, range $[0, \infty)$, x intercept 0, y intercept 0,
vertex $(0, 0)$, opens upward
3. Domain $\mathbb{R}$, range $(-\infty, 0]$, x intercept 0, y intercept
0, vertex $(0, 0)$, opens downward
5. Domain $\mathbb{R}$, range $[1, \infty)$, y intercept 13, vertex $(2, 1)$,
opens upward
7. Domain $\mathbb{R}$, range $[-\frac{1}{4}, \infty)$, x intercepts 1 and 2, y
intercept 2, vertex $(\frac{3}{2}, -\frac{1}{4})$, opens upward
9. Domain $\mathbb{R}$, range $(-\infty, \frac{529}{24}]$, x intercepts $-\frac{2}{3}$ and $\frac{4}{3}$, y
intercept 20, vertex $(-\frac{7}{12}, \frac{529}{24})$, opens downward
11. Domain $\mathbb{R}$, range $(-\infty, 0]$, x intercept 5, y intercept
-25, vertex $(5, 0)$, opens downward
13. $(-\infty, 1] \cup [2, \infty)$ **15.** $(-\infty, -\frac{5}{2}] \cup [\frac{4}{3}, \infty)$
17. $\frac{21}{2}$ and $\frac{21}{2}$ **19.** $15°F$ **21.** 30 thousand
23. Polynomial   **25.** Not a polynomial
31. x intercept 0, y intercept 0
33. x intercept 1, y intercept $-\frac{21}{4}$
35. (a) $-3, 0, 3$; (c) $(-\infty, -3] \cup [0, 3]$
37. (a) $-5, -2, 0$; (c) $(-5, -2) \cup (0, \infty)$
39. (a) $-\frac{6}{5}, -1, \frac{3}{2}$; (c) $[-\frac{6}{5}, -1] \cup [\frac{3}{2}, \infty)$
41. (a) $-3, 2$; (c) $(-\infty, -3)$ **43.** $Q: x + 2, R: -4$
45. $Q: x^2, R: 8x^2 - 32$
47. $Q: 3x^4 - x^2 + 2, R: -1$

49. $4x^3 - 3x^2 + 3x - 5 + \dfrac{8}{x + 1}$

51. $x + 5 + \dfrac{3x - 10}{x^2 + x + 2}$

53. $Q: 3x^3 - 4x^2 + 9x - 17, R: 36$
55. $Q: x^2 - 4x + 11, R: -27$
57. $Q: 2x^2 + 7x + 28, R: 119$
59. $Q: x^4 - x^3 + x^2 - x + 6, R: -19$
61. 1637.75   **63.** -27 **65.** -137 **67.** -7
69. Yes   **71.** No   **73.** 2   **75.** 3
77. 1, 2; positive 1, negative 2 or 0; $-3, 1$

79. 2, 0; positive 2 or 0, negative 0; 0, 2
81. $-1, 2, 7$
83. $-3, 3, 2, 2$
85. $-1, -\frac{1}{2}, 1, \frac{3}{2}$
87. $(x + 2)(x + 1)(x - 1)$
89. $(x + 1)(x - 3)(x + 3)(x - 5)$
91. 0.71
93. 2.28
95. 8 centimeters
97. $f(x) = q(x)(x - a)(x - b) + Ax + B$, substitute
$x = a$ and $x = b$ and solve for A and B
99. Domain $(-\infty, 0) \cup (0, \infty)$, asymptotes $x = 0, y = 0$,
no intercepts
101. Domain $(-\infty, 0) \cup (0, \infty)$, asymptotes $x = 0$,
$y = 1$, no y intercept, x intercepts $\pm\sqrt{3}$
103. Domain $(-\infty, 1) \cup (1, \infty)$, asymptotes $x = 1$,
$y = 2$, y intercept 0, x intercept 0
105. Domain $(-\infty, -2) \cup (-2, 0) \cup (0, \infty)$, asymptotes
$x = 0, y = 1$, no intercepts
107. Domain $(-\infty, 0) \cup (0, 3) \cup (3, \infty)$, asymptotes
$x = 0, x = 3, y = 1$, no intercepts
109. Domain $(-\infty, -3) \cup (-3, 3) \cup (3, \infty)$, asymptotes
$x = -3, x = 3, y = 0$, y intercept $\frac{1}{9}$, x intercept 1
111. Domain $(-\infty, 2) \cup (2, \infty)$, asymptotes $x = 2, y = x$,
y intercept $\frac{3}{2}$, x intercepts $-1, 3$

113. (a)   (b) Domain $[0, \infty)$,
                                   range $[0, 1)$; (c) Increases

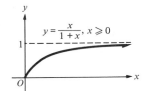

115. (a) $-2, 1 \pm \sqrt{6}$;
(b) $(z + 2)(z - 1 + \sqrt{6})(z - 1 - \sqrt{6})$; (c) Each zero has
multiplicity 1
117. (a) $-3, 1, \dfrac{-3 \pm i\sqrt{11}}{2}$;

(b) $(z - 1)(z + 3)\left(z + \dfrac{3}{2} - \dfrac{i\sqrt{11}}{2}\right)\left(z + \dfrac{3}{2} + \dfrac{i\sqrt{11}}{2}\right)$;

(c) Each zero has multiplicity 1
119. (a) $-2, 2, 3 \pm \sqrt{2}$;
(b) $(z - 2)(z + 2)(z - 3 + \sqrt{2})(z - 3 - \sqrt{2})$;
(c) Each zero has multiplicity 1
121. $z^4 - z^3 - 2z^2 + 6z - 4$
123. $z^5 - 6z^4 + 13z^3 - 14z^2 + 12z - 8$
125. $z^4 - 8z^3 + 26z^2 - 40z + 33$

CHAPTER 5

Problem Set 5.1, page 269

1.

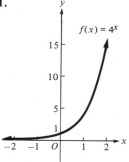

3.

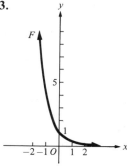

15. Domain ℝ, range $(0, \infty)$, increasing, asymptote $y = 0$

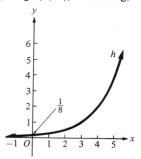

5. (a) 2.665144142;
(b) 0.3752142273;

(c) 8.824977830;
(d) 0.0119935487

17. Domain ℝ, range $(3, \infty)$, increasing, asymptote $y = 3$

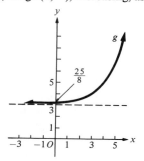

7. Domain ℝ, range $(1, \infty)$, increasing, asymptote $y = 1$

9. Domain ℝ, range $(-1, \infty)$, increasing, asymptote $y = -1$

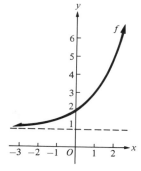

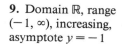

19. Domain ℝ, range $(2, \infty)$, decreasing, asymptote $y = 2$

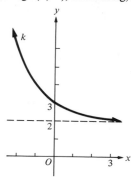

11. Domain ℝ, range $(0, \infty)$, increasing, asymptote $y = 0$

13. Domain ℝ, range $(-3, \infty)$, decreasing, asymptote $y = -3$

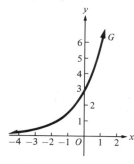

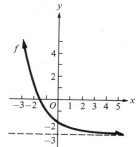
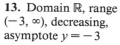
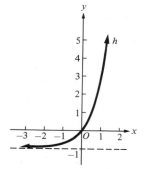

21. $a^x a^y = 38.80960174$, $a^{x+y} = 38.80960175$
23. $(a^x)^y = 0.01937829357$, $a^{xy} = 0.01937829357$
25. $(ab)^x = 42.67011803$, $a^x b^x = 42.67011804$
27. 4   **29.** (a) \$2409.85; (b) 7%
31. (a) \$4363.49; (b) 12%
33. (a) \$53,471.36; (b) 14.37%
35. (a) \$1166.40; (b) \$1169.86; (c) \$1171.66;
(d) \$1172.89; (e) \$1173.37; (f) \$1173.49; (g) \$1173.50
37. \$549.19   **39.** \$96.09

Problem Set 5.2, page 275

1. (a) 0.3678794412; (b) 0.1353352832;
(c) 20.08553692; (d) 23.14069264; (e) 9.356469012

3.

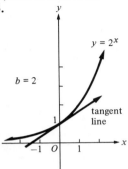

$y = 2^x$
$b = 2$
tangent line

$y = 3^x$
$b = 3$
tangent line

5.

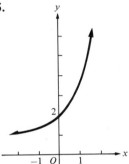

7.

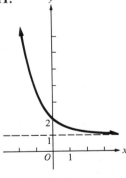

9.

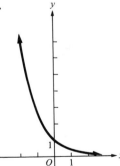

11.

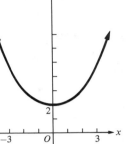

13. $e^x e^y = 23.24905230$, $e^{x+y} = 23.24905230$
15. $(e^x)^y = 0.3166675732$, $e^{xy} = 0.3166675733$
17. 0, 2   **19.** 0   **21.** 1
23. (a) Discrepancy in third decimal place;
(b) Discrepancy in third decimal place; (c) Discrepancy in first decimal place; (d) Discrepancy in first decimal place

25.

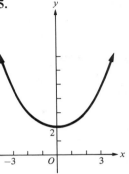

27. (a) $1645.31;
(b) $1648.72

29. (a)

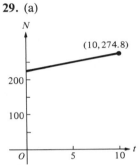

$(10, 274.8)$

(b) 292 bears

31. (a)

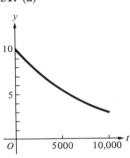

(b) 2.976 grams

33. (a) 2,000,000;
(b) 3,644,238

35. (a) 2000;
(b) 8,894,133; (d) 4 hours
(c)

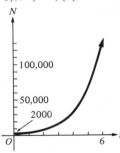

37.

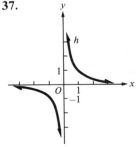

h

39.

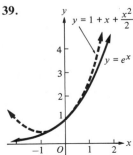

$y = 1 + x + \dfrac{x^2}{2}$
$y = e^x$

Problem Set 5.3, page 284

1. 3   **3.** $\frac{1}{2}$ **5.** 1   **7.** 0   **9.** $-2, 1$ **11.** 1
13. (a) 2; (b) 3; (c) 4; (d) 5; (e) -2; (f) -2; (g) 5
15. (a) $2^5 = 32$; (b) $16^{1/4} = 2$; (c) $9^{-1/2} = \frac{1}{3}$; (d) $e^1 = e$;
(e) $(\sqrt{3})^4 = 9$; (f) $10^n = 10^n$; (g) $x^5 = x^5$
17. (a) $\log_8 1 = 0$; (b) $\log_{10} 0.0001 = -4$;
(c) $\log_4 256 = 4$; (d) $\log_{27} \frac{1}{3} = -\frac{1}{3}$; (e) $\log_8 4 = \frac{2}{3}$;
(f) $\log_a y = c$
19. 2   **21.** 5   **23.** 7   **25.** $\frac{1}{8}$ **27.** $\frac{1}{8}$
29. -7 **31.** $-\frac{2}{3}$ **33.** $\frac{1}{4}$ **35.** $\frac{9}{2}$ **37.** $\frac{85}{3}$
39. $-2, -1$ **41.** $-\frac{5}{3}, 1$ **43.** 2
45. (a) 2.31; (b) 2.70; (c) -0.95; (d) 1.87; (e) 0.74;
(f) 0.95; (g) 1.63
47. $\log_3 7 + \log_3 t$ **49.** $\frac{1}{2} \log_5 p$
51. $\log_2 u + \log_2 v - \log_2 w$
53. $\log_b x + \log_b(x + 1)$
55. $2 \log_{10} x + \log_{10}(x + 1)$
57. $3 \log_3 x + 2 \log_3 y - \log_3 z$
59. $\frac{1}{2}[\log_e x + \log_e(x + 3)]$ **61.** $\log_3 x^9$

63. $\log_5 \frac{2}{3}$ **65.** $\log_e(x + 1)$ **67.** $\log_3 \dfrac{x + 9}{x + 5}$

69. 2   **71.** $\frac{1}{5}$ **73.** 2   **75.** 67   **77.** $\log_e 5 / \log_e 10$
79. $\log_a b = \log_b b / \log_b a = 1 / \log_b a$

Problem Set 5.4, page 291

1.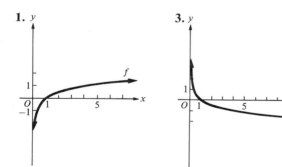

3.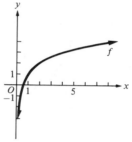

5.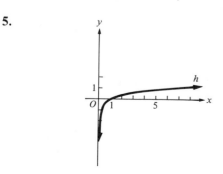

7. (a) 0.8043439185; (b) 3.090046322;
(c) -1.453333975; (d) 11.48387245; (e) -8.182963774
9. (a) $\log xy = 2.214956167$,
$\log x + \log y = 2.214956167$;
(b) $\log xy = 0.6476648705$,
$\log x + \log y = 0.6476648704$
11. (a) 8.325063694; (b) 0.9947321582;
(c) -3.208826489; (d) 20.41142767; (e) -15.56881884
13. (a) $\ln e^\pi = \ln 23.14069264 = 3.141592654$;
(b) $e^{\ln \pi} = e^{1.144729886} = 3.141592654$
15. (a) 4.643856190; (b) 0.6309297537;
(c) 0.4808983469; (d) 1.405954306; (e) -7.551524229
17. (a) $(2, \infty)$; (b) $(0, \infty)$; (c) $(-\infty, 4)$;
(d) $(-\infty, 0) \cup (0, \infty)$; (e) $(0, 1) \cup (1, \infty)$; (f) $\mathbb{R}$; (g) $\mathbb{R}$
19. Domain $(0, \infty)$, range
$\mathbb{R}$, x intercept 0.5,
increasing, asymptote
$x = 0$

21. Domain $(0, \infty)$, range
$\mathbb{R}$, x intercept 1,
decreasing, asymptote
$x = 0$

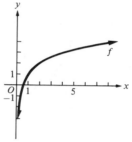
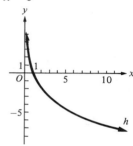

23. Domain $(0, \infty)$, range
$\mathbb{R}$, x intercept 1,
increasing, asymptote
$x = 0$

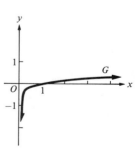

25. Domain $(1, \infty)$, range
$\mathbb{R}$, x intercept 2,
increasing, asymptote
$x = 1$

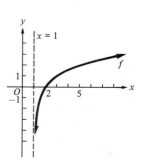

27. Domain $(0, \infty)$, range $\mathbb{R}$, x intercept 1, increasing, asymptote $x = 0$

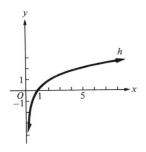

29. Domain $(0, \infty)$, range $\mathbb{R}$, x intercept 1, decreasing, asymptote $x = 0$

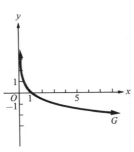

49. As base increases, graph rises less rapidly to the right of $(0, 1)$; all graphs contain $(0, 1)$

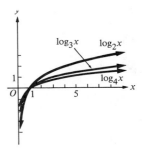

51.

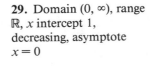

31. Domain $(1, \infty)$, range $\mathbb{R}$, x intercept $1 + e^{-2}$, increasing, asymptote $x = 1$

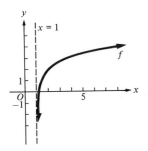

33. Domain $(0, \infty)$, range $\mathbb{R}$, x intercept 1, decreasing, asymptote $x = 0$

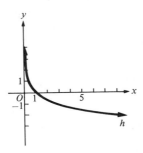

53. 2   **55.** Tangent line is $y = x - 1$
57. $e^{x \ln y} = (e^{\ln y})^x = y^x$
59. True only for $x > 0$

Problem Set 5.5, page 299

1. 0.7924812504   **3.** 3.886474732
5. -1.413390105 **7.** 4   **9.** $\frac{1}{27}$
11. 0.4785236727   **13.** -3.209511289
15. -2.328201576 **17.** -0.3420914978

19. $y = \dfrac{1 + \ln x}{2}$ **21.** $y = \ln(x + \sqrt{x^2 + 1})$

23. 11.559 years $\approx$ 11 years and 29 weeks
25. (a) 7.8; (b) 4.2; (c) 6.4   **27.** 8863 meters
29. 5545 meters   **31.** 113 decibels   **33.** 10^{12}
35. $10^{8/5} \approx 39.8$
37. (a) $y = y_0 e^{-0.00041857t}$; (b) 0.0418%; (c) 0.9917 gram
39. 20.35 years   **41.** (a) $k = 1/(2 \ln 4) \approx 0.361$;
(b) 0.354 or 35.4%

35. Domain $(-2, \infty)$, range $\mathbb{R}$, x intercept $e^2 - 2$, increasing, asymptote $x = -2$

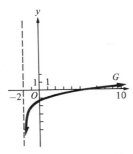

37. $e^0 = 1$ **39.** $e^1 = e$
41. $\log_e xy = \log_e x + \log_e y$
43. $\log_e x/y = \log_e x - \log_e y$
45. $f^{-1}(x) = (\ln x) - 2$

47. ln is the inverse of $f(x) = e^x$. Because the domain of f is $\mathbb{R}$, the range of ln is also $\mathbb{R}$.

Problem Set 5.6, page 306

1. (a) $N = 10^7 e^{kt}$; $k = \ln 1.03$; (b) 18,061,112;
(c) 23.45 years
3. 9.97 hours

5. (a) $N = 10e^{(t \ln 2)/20}$; (b) 80 *E. coli*   7. 41.42%
9. (a) $699.86 \approx 700$; (b) 421; (c) 2.88 years
11. Notice that the inflection point is not dramatic enough to show.
13. Graph rises more slowly than graph in Problem 11.

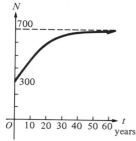

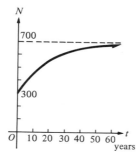

15. Graph rises more slowly than graph in Problem 11, more rapidly than graph in Problem 13.

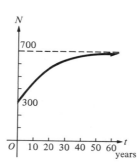

Review Problem Set, Chapter 5, page 307

1. (a) 27; (b) 1; (c) 5; (d) $\frac{1}{4}$; (e) $\sqrt{3}$; (f) $\sqrt{5}/5$
3. Domain $\mathbb{R}$, range $(0, \infty)$, asymptote $y = 0$, increasing
5. Domain $\mathbb{R}$, range $(0, \infty)$, asymptote $y = 0$, decreasing
7. Domain $\mathbb{R}$, range $(3, \infty)$, asymptote $y = 3$, decreasing
9. Domain $\mathbb{R}$, range $(-\infty, 1)$, asymptote $y = 1$, increasing
11. $6734.28   13. $839.62
15. (a) 27.56714844; (b) 0.0432139183
17. Domain $\mathbb{R}$, range $(3, \infty)$, asymptote $y = 3$, increasing
19. Domain $\mathbb{R}$, range $(0, \infty)$, asymptote $y = 0$, increasing
21. Domain $\mathbb{R}$, range $(2, \infty)$, asymptote $y = 2$, increasing
23. $11,082.57
25. (a) 66,079.39737; (b) 1505.971060;
(c) $5\sqrt{35}/7$; (d) 98.17%
27. $5^x = \frac{1}{125}, -3$ 29. $3^1 = 2 + x, 1$
31. $3^{5x} = 81, \frac{4}{5}$ 33. $x^2 = \frac{1}{49}, \frac{1}{7}$ 35. $3^{-2x} = \frac{1}{81}, 2$
37. $7^{|x|-1} = 49, \pm 3$
39. (a) $\log_3 729 = 6$; (b) $\log_2 \frac{1}{1024} = -10$;
(c) $\log_{64} 256 = \frac{4}{3}$; (d) $\log_x w = a$; (e) $\log y = x$;
(f) $\ln y = x$

41. (a) 4; (b) $\frac{1}{4}$; (c) $\frac{1}{2}$; (d) 0; (e) -3; (f) 33; (g) -1.4
43. (a) 0.4342944818; (b) 4.605170186;
(c) 3.505557397; (d) $1.443225879 \times 10^{-5}$;
(e) $5.923313615 \times 10^{-2}$; (f) 1.667389292;
(g) 5.574941522; (h) $1.340164240 \times 10^{18}$;
(i) 6.581412462
45. $(1/n)\log_b p - n \log_b R$
47. $\frac{1}{2} \log(4 - x) - \frac{1}{2} \log(4 + x)$ 49. $\log_b x^a y^{1/c}$
51. $\ln \dfrac{(x-2)(2x-1)}{(x-4)(x-5)}$ 53. 0   55. -2
57. (a) 3.459431619; (b) 11.07309365;
(c) 0.3156023436; (d) 0.4306765582
59. $\log \sqrt{x} = 0.7606002382$, $\frac{1}{2} \log x = 0.7606002380$
61. $\ln x = 4.344855520$, $\log x/\log e = 4.344855521$
63. $\log y^x = -28.59502498$, $x \log y = -28.59502498$
65. $\log \pi/\log e = 1.145$, $\log \pi - \log e = 0.063$
67. $1/\log(\frac{2}{3}) = -5.679$, $\log \frac{2}{3} = 0.176$
69. $\log_2 3 = 1.585$, $\log 8 = 0.903$
71. (a) $(\frac{3}{4}, \infty)$; (b) $(-\infty, -1) \cup (-1, \infty)$;
(c) $(-\infty, 2 - 2\sqrt{2}) \cup (2 + 2\sqrt{2}, \infty)$; (d) $\mathbb{R}$
73. Domain $(1, \infty)$, range $\mathbb{R}$, asymptote $x = 1$, x intercept $1 + (1/e)$, increasing
75. Domain $(-\infty, 4)$, range $\mathbb{R}$, asymptote $x = 4$, x intercept 3, y intercept log 2, decreasing
77. Domain $(-\infty, 2) \cup (2, \infty)$, range $\mathbb{R}$, asymptote $x = 2$; x intercepts 1 and 3, y intercept ln 2, decreasing on $(-\infty, 2)$, increasing on $(2, \infty)$
79. $x - 1$ 81. $x - 5$ 83. 1.301029996
85. -3.561615892 87. 0.6538311574
89. (a) 78; (b) 35.20; (c) 25.19 months
91. 123 decibels   93. Approximately 0.9752 gram
95. (a) In 12 years, you'll have $99,603.51; (b) In 12 years, you'll have $167,772.16. Choose plan b.
97. (a) $205.07; (b) $1382.37
99. (a) $514.31; (b) $7147.14
101. (a) 4255; (b) 198 days   103. (b) 2209 years
105. (a) 75,064; (b) 57,624;
(c) 8.07 years after introduction

CHAPTER 6

Problem Set 6.1, page 319

1. 60°   3. $-792°$ 5. $\frac{1}{8}$ counterclockwise
7. $\frac{1}{3}$ counterclockwise   9. $\frac{11}{12}$ clockwise
11. $\frac{1}{30}$ clockwise   13. $0.0\overline{3}°$ 15. $0.0008\overline{3}°$
17. $100.50\overline{5}°$ 19. $62°15'$ 21. $-0°7'30''$

23. 21°9′36″   **25.** 3.3 meters   **27.** $\frac{4}{5}$ radians
29. 10π/3 inches
31. (a)π/6; (b) π/4; (c) π/2; (d) 2π/3; (e) −5π/6;
(f) 26π/9; (g) $\frac{2}{3}$π; (h) 3π/8; (i) −11π/6; (j) 5π/2;
(k) 7π/60; (l) −2π
33. (a) 90°; (b) 60°; (c) 45°; (d) 30°; (e) 120°; (f)−180°;
(g) 108°; (h) −450°; (i) 405°; (j) −67.5°; (k) 1260°;
(l) −($\frac{90}{7}$)°
35. (a) −3π/4, −135°; (b) 25π/3, 1500°;
(c) −2π/3, −120°
37. 10π/3 meters   **39.** 11,057°
41. (a) 4π/3 radians per hour; (b) 10,240π ≈ 32,170
kilometers per hour
43. 760,320/13 ≈ 58,486 radians per hour
45. 146π/9 ≈ 50.96 centimeters per second
47. 1,550,000π/73 ≈ 66,700 miles per hour
49. (a) 21π/4 square centimeters; (b) 117π/2 square
inches

Problem Set 6.2, page 328

1. sin θ = $\frac{3}{5}$, cos θ = $\frac{4}{5}$, tan θ = $\frac{3}{4}$, cot θ = $\frac{4}{3}$, sec θ = $\frac{5}{4}$,
csc θ = $\frac{5}{3}$
3. sin θ = $\frac{3}{4}$, cos θ = $\sqrt{7}$/4, tan θ = 3$\sqrt{7}$/7, cot θ = $\sqrt{7}$/3,
sec θ = 4$\sqrt{7}$/7, csc θ = $\frac{4}{3}$
5. sin θ = $\frac{1}{2}$, cos θ = $\sqrt{3}$/2, tan θ = $\sqrt{3}$/3, cot θ = $\sqrt{3}$,
sec θ = 2$\sqrt{3}$/3, csc θ = 2
7. sin θ = $\sqrt{2}$/2, cos θ = $\sqrt{2}$/2, tan θ = 1, cot θ = 1,
sec θ = $\sqrt{2}$, csc θ = $\sqrt{2}$
9. 55°   **11.** 3π/10   **13.** π/12   **15.** 67.66°
17. cos 59°   **19.** cot(7π/18)   **21.** tan(π/10)
23. sec 12.97°
25. sin 48° ≈ 0.7431448255   csc 48° ≈ 1.345632730
cos 48° ≈ 0.6691306064   sec 48° ≈ 1.494476550
tan 48° ≈ 1.110612515   cot 48° ≈ 0.9004040442
27. sin 23°12′33″ ≈ 0.3940889557
cos 23°12′33″ ≈ 0.9190723013
tan 23°12′33″ ≈ 0.4287899387
csc 23°12′33″ ≈ 2.537498160
sec 23°12′33″ ≈ 1.088053680
cot 23°12′33″ ≈ 2.332144273
29. sin 16.19° ≈ 0.2788234989
cos 16.19° ≈ 0.9603423642
tan 16.19° ≈ 0.2903376018
csc 16.19° ≈ 3.586498283
sec 16.19° ≈ 1.041295310
cot 16.19° ≈ 3.444266240
31. sin(2π/7) ≈ 0.7818314825
cos(2π/7) ≈ 0.6234898018
tan(2π/7) ≈ 1.253960338

csc(2π/7) ≈ 1.279048008
sec(2π/7) ≈ 1.603875472
cot(2π/7) ≈ 0.7974733887
33. sin 0.7764 ≈ 0.7007155788
cos 0.7764 ≈ 0.7134407317
tan 0.7764 ≈ 0.9821636859
csc 0.7764 ≈ 1.427112555
sec 0.7764 ≈ 1.401658127
cot 0.7764 ≈ 1.018160226
35. cos 33° = sin 57° ≈ 0.8386705679
37. sin(2π/9) ≈ 0.6427876098,
cos(5π/18) ≈ 0.6427876096
39. Evaluate sin 30 on the calculator; if result is 0.50,
then 30 must be 30 degrees; and if result is not 0.50,
then calculator must be in radian mode.

43. cot θ = $\dfrac{\text{adj}}{\text{opp}}$ = $\dfrac{1}{\text{opp/adj}}$ = $\dfrac{1}{\tan θ}$

45. adj² + opp² = hyp², so $\left(\dfrac{\text{adj}}{\text{hyp}}\right)^2 + \left(\dfrac{\text{opp}}{\text{hyp}}\right)^2 = 1$; thus,
(cos θ)² + (sin θ)² = 1

Problem Set 6.3, page 336

1. Quadrant I   **3.** Quadrantal

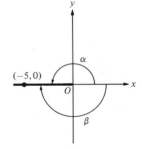

5. Quadrant IV   **7.** Quadrant II

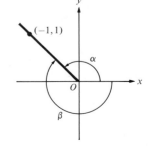

9. Quadrant III

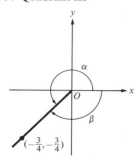

11. $420°, -300°, 780°$

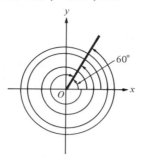

23. $\sin\theta = -\frac{4}{5}$,
$\cos\theta = -\frac{3}{5}$, $\tan\theta = \frac{4}{3}$,
$\cot\theta = \frac{3}{4}$, $\sec\theta = -\frac{5}{3}$,
$\csc\theta = -\frac{5}{4}$

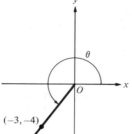

25. $\sin\theta = 3\sqrt{58}/58$,
$\cos\theta = 7\sqrt{58}/58$,
$\tan\theta = \frac{3}{7}$, $\cot\theta = \frac{7}{3}$,
$\sec\theta = \sqrt{58}/7$,
$\csc\theta = \sqrt{58}/3$

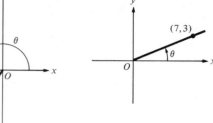

13. $7\pi/4, -9\pi/4, 15\pi/4$

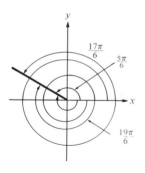

15. $-252°, 468°, 108°$

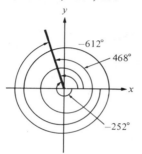

27. $\sin\theta = -\sqrt{10}/5$, $\cos\theta = -\sqrt{15}/5$, $\tan\theta = \sqrt{6}/3$,
$\cot\theta = \sqrt{6}/2$, $\sec\theta = -\sqrt{15}/3$, $\csc\theta = -\sqrt{10}/2$
29. No, since $\sin^2\theta + \cos^2\theta = 1$ implies that $\sin^2\theta \le 1$;
that is, $-1 \le \sin\theta \le 1$.
31. (a) I; (b) II; (c) II; (d) IV; (e) III; (f) III; (g) IV; (h) I
33. (a) Negative; (b) Negative; (c) Positive; (d) Negative;
(e) Positive; (f) Negative; (g) Negative

17. $17\pi/6, -7\pi/6,$
$-19\pi/6$

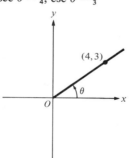

19. $\sin\theta = \frac{3}{5}$, $\cos\theta = \frac{4}{5}$,
$\tan\theta = \frac{3}{4}$, $\cot\theta = \frac{4}{3}$,
$\sec\theta = \frac{5}{4}$, $\csc\theta = \frac{5}{3}$

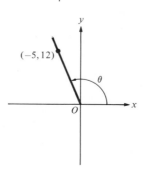

35. $\tan\theta = \dfrac{y}{x} = \dfrac{y/r}{x/r} = \dfrac{\sin\theta}{\cos\theta}$, $\cot\theta = \dfrac{1}{\tan\theta} = \dfrac{\cos\theta}{\sin\theta}$

37. (a) $\sec\theta = \dfrac{13}{12}$; (b) $\csc\theta = -\dfrac{13}{5}$; (c) $\tan\theta = -\dfrac{5}{12}$;

(d) $\cot\theta = -\dfrac{12}{5}$

21. $\sin\theta = \frac{12}{13}$,
$\cos\theta = -\frac{5}{13}$, $\tan\theta = -\frac{12}{5}$,
$\cot\theta = -\frac{5}{12}$, $\sec\theta = -\frac{13}{5}$,
$\csc\theta = \frac{13}{12}$

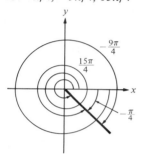

39. $\cos\theta = \frac{3}{5}$, $\tan\theta = \frac{4}{3}$, $\cot\theta = \frac{3}{4}$, $\sec\theta = \frac{5}{3}$, $\csc\theta = \frac{5}{4}$
41. $\cos\theta = -\sqrt{7}/4$, $\tan\theta = 3\sqrt{7}/7$, $\cot\theta = \sqrt{7}/3$,
$\sec\theta = -4\sqrt{7}/7$, $\csc\theta = -4/3$
43. $\sin\theta = -\sqrt{33}/7$, $\tan\theta = -\sqrt{33}/4$,
$\cot\theta = -4\sqrt{33}/33$, $\sec\theta = \frac{7}{4}$, $\csc\theta = -7\sqrt{33}/33$
45. $\sin\theta = \frac{2}{3}$, $\cos\theta = \sqrt{5}/3$, $\tan\theta = 2\sqrt{5}/5$,
$\cot\theta = \sqrt{5}/2$, $\sec\theta = 3\sqrt{5}/5$
47. $\sin\theta = \frac{4}{5}$, $\cos\theta = \frac{3}{5}$, $\cot\theta = \frac{3}{4}$, $\sec\theta = \frac{5}{3}$, $\csc\theta = \frac{5}{4}$
49. $\sin\theta = \frac{5}{13}$, $\cos\theta = -\frac{12}{13}$, $\tan\theta = -\frac{5}{12}$, $\sec\theta = -\frac{13}{12}$,
$\csc\theta = \frac{13}{5}$
51. $\sin\theta = 2\sqrt{2}/3$, $\cos\theta = -\frac{1}{3}$, $\tan\theta = -2\sqrt{2}$,
$\cot\theta = -\sqrt{2}/4$, $\csc\theta = 3\sqrt{2}/4$
53. $\cos\theta = \sqrt{1-\sin^2\theta} = 0.8829$,
$\tan\theta = 0.4695/0.8829 = 0.5318$, $\cot\theta = 1.8805$,
$\sec\theta = 1.1326$, $\csc\theta = 2.1299$
55. $\sin\theta = 1/\csc\theta = 0.3827$, $\cos\theta = -\sqrt{1-\sin^2\theta} = -0.9239$, $\tan\theta = -0.4142$, $\cot\theta = -2.4142$, $\sec\theta = -1.0824$
57. Yes, tan and cot

Problem Set 6.4, page 343

1. See Table 1   **3.** 0°, see Table 1
5. 180°, see Table 1
7. 30°, sin 30° = $\frac{1}{2}$, cos 30° = $\sqrt{3}/2$, tan 30° = $\sqrt{3}/3$,
cot 30° = $\sqrt{3}$, sec 30° = $2\sqrt{3}/3$, csc 30° = 2
9. 45°, sin 45° = $\sqrt{2}/2$, cos 45° = $\sqrt{2}/2$, tan 45° = 1,
cot 45° = 1, sec 45° = $\sqrt{2}$, csc 45° = $\sqrt{2}$
11. π, see Table 1
13. $\pi/3$, sin $\pi/3$ = $\sqrt{3}/2$, cos $\pi/3$ = $\frac{1}{2}$, tan $\pi/3$ = $\sqrt{3}$,
cot $\pi/3$ = $\sqrt{3}/3$, sec $\pi/3$ = 2, csc $\pi/3$ = $2\sqrt{3}/3$
17. θ_R = 30°, sin θ = $\frac{1}{2}$, cos θ = $-\sqrt{3}/2$, tan θ = $-\sqrt{3}/3$,
cot θ = $-\sqrt{3}$, sec θ = $-2\sqrt{3}/3$, csc θ = 2
19. θ_R = 60°, sin θ = $-\sqrt{3}/2$, cos θ = $-\frac{1}{2}$, tan θ = $\sqrt{3}$,
cot θ = $\sqrt{3}/3$, sec θ = -2, csc θ = $-2\sqrt{3}/3$
21. θ_R = 45°, sin θ = $-\sqrt{2}/2$, cos θ = $\sqrt{2}/2$, tan θ = -1,
cot θ = -1, sec θ = $\sqrt{2}$, csc θ = $-\sqrt{2}$
23. θ_R = 30°, sin θ = $-\frac{1}{2}$, cos θ = $-\sqrt{3}/2$, tan θ = $\sqrt{3}/3$,
cot θ = $\sqrt{3}$, sec θ = $-2\sqrt{3}/3$, csc θ = -2
25. θ_R = 60°, sin θ = $-\sqrt{3}/2$, cos θ = $\frac{1}{2}$, tan θ = $-\sqrt{3}$,
cot θ = $-\sqrt{3}/3$, sec θ = 2, csc θ = $-2\sqrt{3}/3$
27. θ_R = $\pi/4$, sin θ = $-\sqrt{2}/2$, cos θ = $\sqrt{2}/2$, tan θ = -1,
cot θ = -1, sec θ = $\sqrt{2}$, csc θ = $-\sqrt{2}$
29. θ_R = $\pi/3$, see Problem 19
31. θ_R = $\pi/4$, see Problem 21
33. θ_R = $\pi/3$, see Problem 25
35. θ_R = 60°, see Problem 25

37.

θ	sin θ	cos θ	tan θ
210°	$-\frac{1}{2}$	$-\sqrt{3}/2$	$\sqrt{3}/3$
225°	$-\sqrt{2}/2$	$-\sqrt{2}/2$	1
240°	$-\sqrt{3}/2$	$-\frac{1}{2}$	$\sqrt{3}$
300°	$-\sqrt{3}/2$	$\frac{1}{2}$	$-\sqrt{3}$
315°	$-\sqrt{2}/2$	$\sqrt{2}/2$	-1
330°	$-\frac{1}{2}$	$\sqrt{3}/2$	$-\sqrt{3}/3$

θ	cot θ	sec θ	csc θ
210°	$\sqrt{3}$	$-2\sqrt{3}/3$	-2
225°	1	$-\sqrt{2}$	$-\sqrt{2}$
240°	$\sqrt{3}/3$	-2	$-2\sqrt{3}/3$
300°	$-\sqrt{3}/3$	2	$-2\sqrt{3}/3$
315°	-1	$\sqrt{2}$	$-\sqrt{2}$
330°	$-\sqrt{3}$	$2\sqrt{3}/3$	-2

39. sin 46° ≈ 0.7193398003, cos 46° ≈ 0.6946583705,
tan 46° ≈ 1.035530314, cot 46° ≈ 0.9656887746,
sec 46° ≈ 1.439556540, csc 46° ≈ 1.390163591

41. sin 143° ≈ 0.6018150232,
cos 143° ≈ −0.7986355100,
tan 143° ≈ −0.7535540501, cot 143° ≈ −1.327044822,
sec 143° ≈ −1.252135658, csc 143° ≈ 1.661640141
43. sin(−61.37°) ≈ −0.8777322126,
cos(−61.37°) ≈ 0.4791515031,
tan(−61.37°) ≈ −1.831846936,
cot(−61.37°) ≈ −0.5458971382,
sec(−61.37°) ≈ 2.087022567,
csc(−61.37°) ≈ −1.139299647
45. sin 61°35′ ≈ 0.8795101821,
cos 61°35′ ≈ 0.4758800684, tan 61°35′ ≈ 1.848176128,
cot 61°35′ ≈ 0.5410739728, sec 61°35′ ≈ 2.101369791,
csc 61°35′ ≈ 1.136996501
47. sin(−97°9′8″) ≈ −0.9922188692,
cos(−97°9′8″) ≈ −0.1245058859,
tan(−97°9′8″) ≈ 7.969252714,
cot(−97°9′8″) ≈ 0.1254822799,
sec(−97°9′8″) ≈ −8.031748803,
csc(−97°9′8″) ≈ −1.007842152
49. sin(−2π/7) ≈ −0.7818314825,
cos(−2π/7) ≈ 0.6234898018,
tan(−2π/7) ≈ −1.253960338,
cot(−2π/7) ≈ −0.7974733887,
sec(−2π/7) ≈ 1.603875472,
csc(−2π/7) ≈ −1.279048008
51. sin(1.67) ≈ 0.9950833498,
cos(1.67) ≈ −0.0990410366,
tan(1.67) ≈ −10.04718230,
cot(1.67) ≈ −0.0995303927,
sec(1.67) ≈ −10.09682485, csc(1.67) ≈ 1.004940943
53. sin(−2.436) ≈ −0.6484850875,
cos(−2.436) ≈ −0.7612273585,
tan(−2.436) ≈ 0.8518940896,
cot(−2.436) ≈ 1.173854840,
sec(−2.436) ≈ −1.313667972,
csc(−2.436) ≈ −1.542055506
55. sin(−10.79) ≈ 0.9789439170,
cos(−10.79) ≈ −0.2041293889,
tan(−10.79) ≈ −4.795702973,
cot(−10.79) ≈ 1.021508978,
sec(−10.79) ≈ −4.898853641,
csc(−10.79) ≈ −0.2085200034
57. sin(9.673) ≈ −0.2456808808,
cos(9.673) ≈ −0.9693507646,
tan(9.673) ≈ 0.2534488957, cot(9.673) ≈ 3.945568582,
sec(9.673) ≈ −1.031618313,
csc(9.673) ≈ −4.070320803
59. tan 35° ≈ 0.7002075382,

$$\frac{\sin 35°}{\cos 35°} \approx \frac{0.5735764364}{0.8191520443} \approx 0.700207583$$

61. Check that $[\sin(3\pi/5)]^2 + [\cos(3\pi/5)]^2 = 1$.

63. $\dfrac{\sqrt{0}}{2} = 0, \dfrac{\sqrt{1}}{2} = \dfrac{1}{2}, \dfrac{\sqrt{4}}{2} = 1$

Problem Set 6.5, page 353

1.
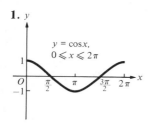
$y = \cos x,$
$0 \leqslant x \leqslant 2\pi$

3. $\sin(-4.203) \approx 0.8730426306,$
$-\sin(4.203) \approx 0.8730426306$
5. $\sin(2.771 + 3.141592654) =$
$\sin(5.912592654) \approx -0.3621679152,$
$-\sin(2.771) \approx -0.3621679156$
7. $-3\pi/2, \pi/2, 1$ **9.** $-3\pi/2, -\pi/2, \pi/2, 3\pi/2$
11. $-\pi/2, 3\pi/2, -1$
13. Consider the *number* to be a number of radians.
15. Amplitude 2, period 2π, phase shift 0, axis $y = 0$ **17.** Amplitude $\frac{1}{2}$, period 2π, phase shift 0, axis $y = 0$

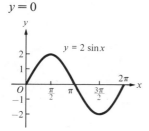
$y = 2 \sin x$

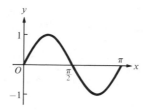

19. Amplitude $\frac{2}{3}$, period 2π, one cycle starts at $x = 0$ and ends with $x = 2\pi$, phase shift 0, axis $y = 1$ **21.** Amplitude 1, period π, phase shift 0, axis $y = 0$

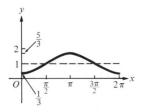

23. Amplitude 1, period $\pi/3$, phase shift 0, axis $y = 0$ **25.** Amplitude 1, period 4π, phase shift 0, axis $y = 1$

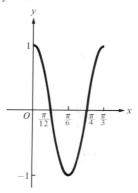

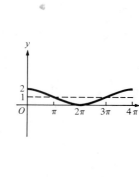

27. Amplitude π, period 2, one cycle starts at $x = 0$ and ends with $x = 2$, phase shift 0, axis $y = 2$ **29.** Amplitude 1, period 2π, one cycle starts at $x = \pi/6$ and ends with $x = 13\pi/6$, phase shift $\pi/6$, axis $y = 0$

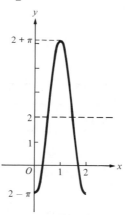

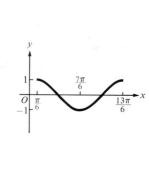

31. Amplitude 2, period 2π, one cycle starts at $x = \pi/3$ and ends with $x = 7\pi/3$, phase shift $\pi/3$, axis $y = 0$ **33.** Amplitude 1, period 2π, one cycle starts at $x = \pi/2$ and ends with $x = 5\pi/2$, phase shift $\pi/2$, axis $y = 1$

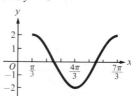

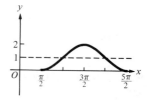

35. Amplitude 3, period 8π, one cycle starts at $x = 4\pi$ and ends with $x = 12\pi$, phase shift 4π, axis $y = 0$

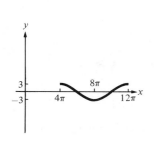

37. Amplitude 3, period $\pi/2$, one cycle starts at $x = \pi/4$ and ends with $x = 3\pi/4$, phase shift $\pi/4$, axis $y = 1$

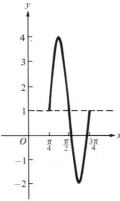

39. Amplitude $\frac{1}{2}$, period $2\pi/3$, one cycle starts at $-\pi/4$ and ends with $5\pi/12$, phase shift $-\pi/4$, axis $y = 0$

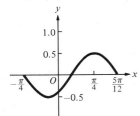

41. Amplitude $\frac{1}{2}$, period $8\pi/3$, one cycle starts at $-\pi/3$ and ends with $7\pi/3$, phase shift $-\pi/3$, axis $y = 1$

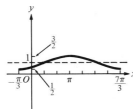

(b)

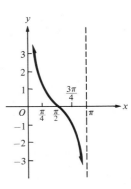

11. $-2\pi, -\pi, 0, \pi, 2\pi$ **13.** $-2\pi, 0, 2\pi$; 1
15. π **17.** $\sec(-x) = 1/\cos(-x) = 1/\cos x = \sec x$

19. Multiply the ordinates of points on the graph of $y = \tan x$ by 2.

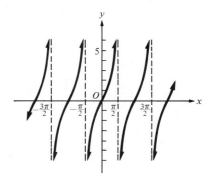

21. Multiply the ordinates of points on the graph of $y = \cot x$ by $\frac{1}{2}$, reflect the resulting graph about the x axis, and then shift this graph 1 unit upward.

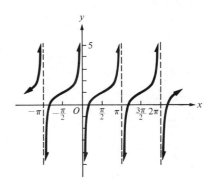

Problem Set 6.6, page 359

1. (a)

x	$\dfrac{\pi}{12}$	$\dfrac{\pi}{6}$	$\dfrac{\pi}{4}$	$\dfrac{\pi}{3}$	$\dfrac{5\pi}{12}$	$\dfrac{\pi}{2}$
$\cot x$	3.73	1.73	1	0.58	0.27	0

x	$\dfrac{7\pi}{12}$	$\dfrac{2\pi}{3}$	$\dfrac{3\pi}{4}$	$\dfrac{5\pi}{6}$	$\dfrac{11\pi}{12}$
$\cot x$	-0.27	-0.58	-1	-1.73	-3.73

23. Multiply the ordinates of points on the graph of $y = \sec x$ by $\frac{1}{2}$ and shift the resulting graph 1 unit upward.

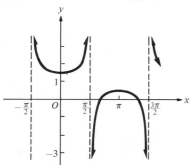

25. Shift the graph of $y = \tan x$ exactly $\pi/4$ units to the right.

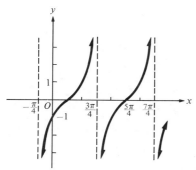

27. Multiply the ordinates of points on the graph of $y = \cot x$ by $\frac{2}{3}$ and shift the resulting graph $\pi/2$ units to the right.

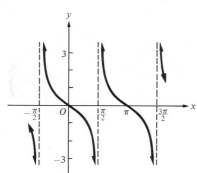

29. Shift the graph of $y = \sec x$ exactly $\pi/6$ units to the left.

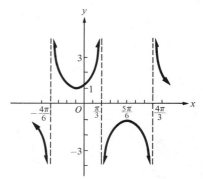

31. Multiply the ordinates of points on the graph of $y = \csc x$ by 2 and then shift the resulting graph $\pi/3$ units to the right.

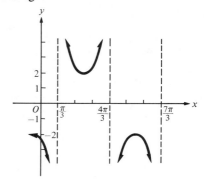

Problem Set 6.7, page 364

1. (a) $a = 2$; (b) $\omega = 1$;
(c) $T = 2\pi$; (d) $\phi = \pi/3$;
(e) $S = \pi/3$; (f) $k = 1$;
(g)

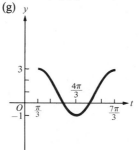

3. (a) $a = 4$; (b) $\omega = 1$;
(c) $T = 2\pi$; (d) $\phi = -\pi/4$;
(e) $S = -\pi/4$; (f) $k = 2$;
(g)

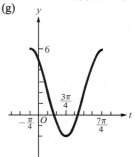

5. (a) $a = 3$; (b) $\omega = 3$;
(c) $T = 2\pi/3$;
(d) $\phi = -5\pi/2$;
(e) $S = -5\pi/6$; (f) $k = 0$;
(g)

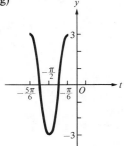

7. (a) $a = \frac{3}{4}$; (b) $\omega = 4$;
(c) $T = \pi/2$; (d) $\phi = -12$;
(e) $S = -3$; (f) $k = -\frac{3}{4}$;
(g)

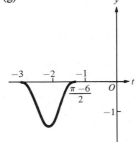

21. (a) Axis is $y = \dfrac{d+c}{2}$, $a = d - \dfrac{d+c}{2} = \dfrac{d-c}{2}$;

(b) $T = \beta - \alpha$, so $\omega = \dfrac{2\pi}{T} = \dfrac{2\pi}{\beta - \alpha}$; (c) $S = \alpha$,

$\phi = S\omega = \alpha\left(\dfrac{2\pi}{\beta - \alpha}\right) = \dfrac{2\pi\alpha}{\beta - \alpha}$;

(d) $k = d - a = d - \left(\dfrac{d-c}{2}\right) = \dfrac{d+c}{2}$

23. (a) See Problem 21(a); (b) See Problem 21(b);

(c) $S = \alpha + \dfrac{\beta - \alpha}{4} = \dfrac{3\alpha + \alpha\beta}{4}$, so $\phi = S\omega =$

$\left(\dfrac{3\alpha + \beta}{4}\right)\left(\dfrac{2\pi}{\beta - \alpha}\right) = \dfrac{(3\alpha + \beta)\pi}{2(\beta - \alpha)}$; (d) See Problem 21(d)

9. (a) $a = 110$;
(b) $\omega = 120\pi$; (c) $T = \frac{1}{60}$;
(d) $\phi = -3\pi/2$;
(e) $S = -\frac{1}{80}$; (f) $k = 0$;

(g)

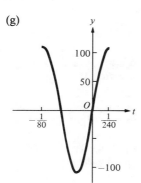

25.

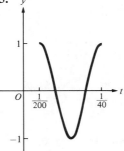

27.

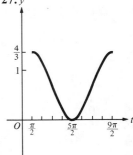

11. Problem 1: $v = 1/2\pi$, Problem 3: $v = 1/2\pi$, Problem 5: $v = 3/2\pi$, Problem 7: $v = 2/\pi$, Problem 9: $v = 60$

13. (a) $a = 2$; (b) $T = 2\pi$; (c) $S = \dfrac{2\pi}{3}$; (d) $\omega = 1$;

(e) $\phi = \dfrac{2\pi}{3}$; (f) $k = 3$; (g) $y = 2\cos\left(t - \dfrac{2\pi}{3}\right) + 3$

15. (a) $a = \frac{3}{4}$; (b) $T = 4$; (c) $S = \frac{3}{4}$; (d) $\omega = \dfrac{\pi}{2}$;

(e) $\phi = \dfrac{3\pi}{8}$; (f) $k = 0$; (g) $y = \frac{3}{4}\cos\left(\dfrac{\pi}{2}t - \dfrac{3\pi}{8}\right)$

17. (a) $a = 11$; (b) $T = 4$; (c) $S = 8$; (d) $\omega = \dfrac{\pi}{2}$;

(e) $\phi = 4\pi$; (f) $k = 11$; (g) $y = 11\cos\left(\dfrac{\pi}{2}t - 4\pi\right)$ or

$y = 11\cos\left(\dfrac{\pi}{2}t\right)$

19. Problem 13: $v = 1/2\pi$, Problem 15: $v = \frac{1}{4}$, Problem 17: $v = \frac{1}{4}$

29.

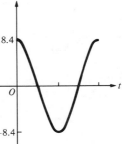

31. $T = \dfrac{\pi}{4}$ second, $v = \dfrac{4}{\pi}$

hertz, $\omega = 8$,
$y = 0.1\cos 8t$
33. Make a pendulum of length ≈ 9.7 inches

35.

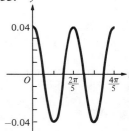

37. $E = 100 \cos(10\sqrt{2}t)$,
$T = (\sqrt{2}/10)\pi$ second,
$v = 5\sqrt{2}/\pi$ hertz
39.

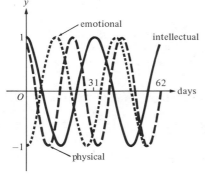

Review Problem Set, Chapter 6, page 368

1. $36°$ **3.** $-288°$ **5.** $\frac{1}{6}$ counterclockwise
7. 4 clockwise **9.** $\frac{31}{12}$ counterclockwise
11. $2.0500°$ **13.** 55.7597 **15.** $87°21'$
17. $-24°31'48''$ **19.** 2.85 meters **21.** 3 feet
23. $\pi/2$ radians **25.** $4\pi/9$ **27.** $-71\pi/36$
29. $-31\pi/18$ **31.** $72°$ **33.** $-\frac{315°}{2}$
35. $2295°$
37. (a) 0.0873; (b) 0.4844; (c) -0.2997; (d) 0.6158
39. 6.28 feet **41.** $45°$
43. $625\pi/12 \approx 163.6$ square centimeters
45. $0.08\pi \approx 0.25$ meter
47. $120\pi \approx 377$ feet per minute
49. 1207 miles above the earth
51. 2094 centimeters per minute
53. $\sin\theta = \sqrt{91}/10$, $\cos\theta = \frac{3}{10}$, $\tan\theta = \sqrt{91}/3$,
$\cot\theta = 3\sqrt{91}/91$, $\sec\theta = \frac{10}{3}$, $\csc\theta = 10\sqrt{91}/91$
55. (a) $\cos 40°$; (b) $\sin(5\pi/14)$; (c) $\csc 1°$; (d) $\tan(11\pi/28)$
57. (a) Quadrant I; (b) Quadrant II; (c) Quadrant III;
(d) Quadrantal; (e) Quadrant IV; (f) Quadrantal
59. (a) $345°$, $705°$, $-375°$; (b) $100°$, $-620°$, $-260°$;
(c) $530°$, $-190°$, $-550°$; (d) $-620°$, $-260°$, $100°$;
(e) $11\pi/3$, $-(\pi/3)$, $-(7\pi/3)$
61. (a) $\sin\theta = 5\sqrt{34}/34$, $\cos\theta = -3\sqrt{34}/34$,
$\tan\theta = -\frac{5}{3}$, $\cot\theta = -\frac{3}{5}$, $\sec\theta = -\sqrt{34}/3$, $\csc\theta = \sqrt{34}/5$;
(b) $\sin\theta = -3\sqrt{13}/13$, $\cos\theta = 2\sqrt{13}/13$, $\tan\theta = -\frac{3}{2}$,
$\cot\theta = -\frac{2}{3}$, $\sec\theta = \sqrt{13}/2$, $\csc\theta = -\sqrt{13}/3$; (c) $\sin\theta =$
$-\frac{4}{5}$, $\cos\theta = -\frac{3}{5}$, $\tan\theta = \frac{4}{3}$, $\cot\theta = \frac{3}{4}$, $\sec\theta = -\frac{5}{3}$,
$\csc\theta = -\frac{5}{4}$; (d) $\sin\theta = -\frac{1}{2}$, $\cos\theta = \sqrt{3}/2$,
$\tan\theta = -\sqrt{3}/3$, $\cot\theta = -\sqrt{3}$, $\sec\theta = 2\sqrt{3}/3$, $\csc\theta = -2$
63. (a) II; (b) III; (c) IV; (d) IV
65. $\tan\theta = -\frac{4}{3}$, $\cot\theta = -\frac{3}{4}$, $\sec\theta = \frac{5}{3}$, $\csc\theta = -\frac{5}{4}$

67. $\cos\theta = \frac{12}{13}$, $\tan\theta = -\frac{5}{12}$, $\cot\theta = -\frac{12}{5}$, $\sec\theta = \frac{13}{12}$,
$\csc\theta = -\frac{13}{5}$
69. $\sin\theta = \frac{12}{13}$, $\cos\theta = -\frac{5}{13}$, $\tan\theta = -\frac{12}{5}$, $\cot\theta = -\frac{5}{12}$,
$\sec\theta = -\frac{13}{5}$
71. $\cos\theta = -\frac{4}{5}$, $\tan\theta = -\frac{3}{4}$, $\cot\theta = -\frac{4}{3}$, $\sec\theta = -\frac{5}{4}$,
$\csc\theta = \frac{5}{3}$
73. $\tan\theta = 1.6003$, $\sec\theta = -1.8871$, $\cos\theta = -0.5299$,
$\sin\theta = -0.8480$, $\csc\theta = -1.1792$
75. (a) $\sin(-180°) = 0$, $\cos(-180°) = -1$,
$\tan(-180°) = 0$, $\sec(-180°) = -1$; (b) same as for part
(a); (c) $\sin 990° = -1$, $\cos 990° = 0$, $\cot 990° = 0$,
$\csc 990° = -1$; (d) $\sin(-360°) = 0$, $\cos(-360°) = 1$,
$\tan(-360°) = 0$, $\sec(-360°) = 1$; (e) same as for part
(c); (f) same as for part (a); (g) same as for part (a);
(h) same as for part (d)
77. (a) $\sin(-150°) = -\frac{1}{2}$, $\cos(-150°) = -\sqrt{3}/2$,
$\tan(-150°) = \sqrt{3}/3$, $\cot(-150°) = \sqrt{3}$,
$\sec(-150°) = -2\sqrt{3}/3$, $\csc(-150°) = -2$;
(b) $\sin(-315°) = \sqrt{2}/2$, $\cos(-315°) = \sqrt{2}/2$,
$\tan(-315°) = 1$, $\cot(-315°) = 1$, $\sec(-315°) = \sqrt{2}$,
$\csc(-315°) = \sqrt{2}$; (c) $\sin 780° = \sqrt{3}/2$, $\cos 780° = \frac{1}{2}$,
$\tan 780° = \sqrt{3}$, $\cot 780° = \sqrt{3}/3$, $\sec 780° = 2$,
$\csc 780° = 2\sqrt{3}/3$; (d) $\sin(13\pi/3) = \sqrt{3}/2$,
$\cos(13\pi/3) = 1/2$, $\tan(13\pi/3) = \sqrt{3}$, $\cot(13\pi/3) = \sqrt{3}/3$,
$\sec(13\pi/3) = 2$, $\csc(13\pi/3) = 2\sqrt{3}/3$;
(e) $\sin(-15\pi/4) = \sqrt{2}/2$, $\cos(-15\pi/4) = \sqrt{2}/2$,
$\tan(-15\pi/4) = 1$, $\cot(-15\pi/4) = 1$, $\sec(-15\pi/4) = \sqrt{2}$,
$\csc(-15\pi/4) = \sqrt{2}$
79. 0.4591664533 **81.** 0.5952436037
83. 1.940926426 **85.** 1.042572391
87. 1.701301619 **89.** -26.02388181
91. 0.9346780153 **93.** -0.9545616245
95. 0.4440158399
97. Multiply the ordinates of points on the graph of
$y = \sin x$ by $\frac{1}{3}$ and shift the resulting graph 1 unit upward.
99. Multiply the ordinates of points on the graph of
$y = \cos x$ by $\frac{1}{2}$, shift the resulting graph $\pi/2$ units to the
right, and then shift this resulting graph 2 units upward.
101. Shift the graph of $y = \tan x$ exactly $\pi/6$ unit to the
right and then reflect the resulting graph about the x axis.
103. Shift the graph of $y = \sec x$ exactly π units to the
right.
105. Multiply ordinates of points on the graph of
$y = \cot x$ by $\frac{2}{3}$ and shift the resulting graph $\pi/2$ unit to
the right.
107. Maximum value 1 is reached at $x = -2\pi$, at
$x = 0$, and at $x = 2\pi$; minimum value -1 is reached at
$x = -\pi$ and at $x = \pi$.
109. Sketch the graph of cosine function as usual, then
shift this graph $\pi/3$ unit to the right for graph of the sine
function.

111. (a) 2; (b) π; (c) 2; (d) $-\pi$; (e) -1; (f) 1; (g) $\frac{1}{2}$;
(h) One cycle starts at $x = -1$ and ends at $x = 1$, with trough at $x = 0$
113. (a) 3; (b) $\frac{1}{2}$; (c) 4π; (d) $\pi/2$; (e)π; (f) -3; (g) $1/4\pi$;
(h) One cycle starts at $x = \pi$ and ends at $x = 5\pi$, with trough at $x = 3\pi$
115. (a) 0.2; (b) 0.25; (c) 8π; (d) π; (e) 4π; (f) 0;
(g) $1/8\pi$; (h) One cycle starts at $x = 0$ and ends at $x = 8\pi$, with crest at $x = 4\pi$
117. $T \approx 4.6$ seconds, $v \approx 0.22$ hertz, $\omega \approx 1.4$, $y = 0.2 \cos(1.4t)$; one cycle starts at $t = 0$ and ends at $t = T$, and the next cycle starts at $t = T$ and ends at $t = 2T$; the troughs are at $x = T/2$ and $x = 3T/2$
119. $y = 0.03 \cos(880\pi t - \pi/2) + 10^5$
121. 10,626 days ≈ 29.1 years

CHAPTER 7

Problem Set 7.1, page 377

1. $-\sin \theta \cos \theta$ **3.** 0   **5.** $\dfrac{1 - \csc \alpha}{1 + \cot \beta}$ **7.** $\tan \theta$

9. $\csc v$ **11.** $\cot \beta$ **13.** -1 **15.** $\cot^2 u$
17. 1   **19.** 1   **21.** 1   **23.** $\sin x \cos x$
25. $\tan t$ **27.** $2 \csc t$ **29.** $\sin x$ **31.** -1

33. $\sin^2 \alpha$ **35.** $\cos \theta + \sin \theta$ **37.** $\dfrac{\sin^2 u}{\cos u}$

39. $\pm \dfrac{\sqrt{1 - \cos^2 \theta}}{\cos^3 \theta}$ **41.** $\dfrac{1}{\cos^2 x}$ **43.** $-\sin \alpha$

45. $a \sec \theta$ **47.** $8 \sin^3 t$
49.

Problem Set 7.2, page 381

1. $\sin \theta \sec \theta = \sin \theta(1/\cos \theta) = \tan \theta$
3. $\tan x \cos x = (\sin x/\cos x)\cos x = \sin x$

5. $\csc(-t)\tan(-t) = (-1/\sin t)(-\sin t/\cos t) = 1/\cos t = \sec t$

7. $\tan \alpha \sin \alpha + \cos \alpha = \dfrac{\sin \alpha}{\cos \alpha} \sin \alpha + \cos \alpha =$

$\dfrac{\sin^2 \alpha + \cos^2 \alpha}{\cos \alpha} = \dfrac{1}{\cos \alpha} = \sec \alpha$

53. Not true for $\theta = \pi/4$ **55.** Not true for $t = \pi/3$
57. Not true for $u = \pi/2$ **59.** Not true for $t = 2$

Problem Set 7.3, page 389

1. $(\sqrt{2}/4)(\sqrt{3} + 1)$ **3.** $(-\sqrt{2}/4)(\sqrt{3} - 1)$
5. $(-\sqrt{2}/4)(\sqrt{3} - 1)$ **7.** $(\sqrt{2}/4)(\sqrt{3} - 1)$
9. $2 - \sqrt{3}$ **11.** $-(2 + \sqrt{3})$ **13.** $\sin \theta$
15. $-\cos \alpha$ **17.** $-\sin s$ **19.** $-\tan t$
21. $-\sec \beta$ **23.** $\frac{1}{2}$ **25.** $\frac{1}{2}$ **27.** $\cos x$
29. $\cos \alpha$ **31.** $\sqrt{3}$ **33.** $\tan x$
35. (a) $\frac{3}{5}$; (b) $-\frac{5}{13}$; (c) $\frac{16}{65}$; (d) $-\frac{63}{65}$; (e) $-\frac{56}{65}$; (f) $-\frac{16}{63}$;
(g) Quadrant II   **37.** $\frac{3}{5}$
51. $\cos(\alpha - \beta) = 0.3090169938$, $\cos \alpha \cos \beta + \sin \alpha \sin \beta = 0.3090169947$
53. $\tan(\alpha + \beta) = 0.830215996$, $(\tan \alpha + \tan \beta)/(1 - \tan \alpha \tan \beta) = 0.830215995$
55. $\sin(s - t) = -0.428651799$, $\sin s \cos t - \cos s \sin t = -0.428651799$

Problem Set 7.4, page 396

1. $2 \sin 38° \cos 38°$ **3.** $\cos^2 72° - \sin^2 72°$

5. $2 \sin \dfrac{\pi}{9} \cos \dfrac{\pi}{9}$ **7.** $\sin 58°$ **9.** $\cos 5\theta$

11. $\cos \dfrac{2\pi}{17}$ **13.** (a) $\frac{24}{25}$; (b) $-\frac{7}{25}$; (c) $-\frac{24}{7}$

15. (a) $\frac{120}{169}$; (b) $-\frac{119}{169}$; (c) $-\frac{120}{119}$
17. (a) $-\frac{120}{169}$; (b) $-\frac{119}{169}$; (c) $\frac{120}{119}$ **19.** $\cos x$

21. $1 + 2 \sin 2t$ **23.** $\dfrac{\sin 2x}{2}$

25. $\dfrac{1 - \cos 2\theta}{2}$ or $\sin^2 \theta$

27. (a) $\sqrt{\dfrac{1 + \cos 30°}{2}}$; (b) $\dfrac{\sqrt{2 + \sqrt{3}}}{2}$

29. (a) $-\sqrt{\dfrac{1 + \cos(5\pi/4)}{2}}$; (b) $-\dfrac{\sqrt{2 - \sqrt{2}}}{2}$

31. (a) $\dfrac{1 - \cos 315°}{\sin 315°}$; (b) $1 - \sqrt{2}$

33. (a) $-\sqrt{\dfrac{1 - \cos 405°}{2}}$; (b) $-\dfrac{\sqrt{2 - \sqrt{2}}}{2}$

35. (a) $\dfrac{3}{5}$; (b) $\dfrac{4}{5}$; (c) $\dfrac{3}{4}$

37. (a) $\dfrac{3\sqrt{13}}{13}$; (b) $-\dfrac{2\sqrt{13}}{13}$; (c) $-\dfrac{3}{2}$

39. $\cos 125°$ **41.** $\tan 3x$ **43.** $\cos \dfrac{\pi}{5}$ **57.** $\dfrac{1 + z^2}{1 - z^2}$

59. $\dfrac{1 - z^2}{2z}$ **61.** $b \cos^2 ct = b\,\dfrac{1 + \cos 2ct}{2}$

Problem Set 7.5, page 405

1. $\frac{1}{2} \sin 145° + \frac{1}{2} \sin 65°$
3. $\cos(3\pi/4) + \cos(\pi/2)$ or $-\sqrt{2}/2$
5. $\frac{1}{2} \sin 8\theta - \frac{1}{2} \sin 2\theta$
7. $\frac{1}{2} \cos(7s - 5t) - \frac{1}{2} \cos(7s + 5t)$
9. $\frac{1}{2} \sin 4x - \frac{1}{2} \sin 2x$ **11.** $2 \sin 50° \cos 30°$
13. $2 \cos(\pi/4) \cos(\pi/8)$ **15.** $2 \sin 3\theta \cos \theta$

17. $2 \sin 3t \sin t$ **19.** $2 \cos \dfrac{2\alpha + \beta}{2} \sin \dfrac{\beta}{2}$

21. (a) $-\dfrac{2 \sin 45° \sin 30°}{2 \sin 45° \cos 30°} = -\dfrac{\sqrt{3}}{3}$;

(b) $\dfrac{2 \sin 45° \cos 35°}{2 \cos 45° \cos 35°} = 1$

23. $\sin 50° + \sin 10° = 2 \sin 30° \cos 20° = \cos 20° = $ $\sin 70°$

39.

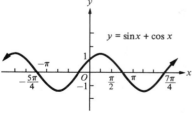

41.

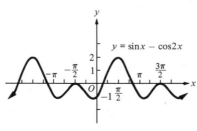

43.

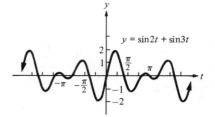

45.

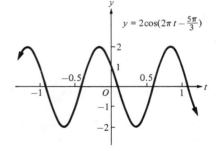

47.

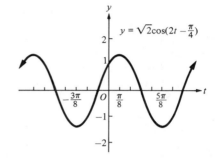

49.

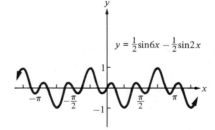

51.

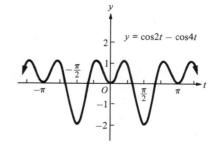

53.

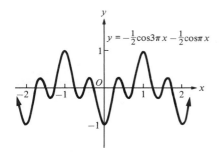

$$y = -\tfrac{1}{2}\cos 3\pi x - \tfrac{1}{2}\cos \pi x$$

55. 4   **57.** $2\sqrt{2}\pi$ **59.**

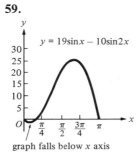

$$y = 19\sin x - 10\sin 2x$$

graph falls below x axis

61. No, maximum occurs a bit *to the right* of $2\pi/3$, at approximately $x = 2.104$

Problem Set 7.6, page 412

1. $\pi/2$ **3.** $-\pi/4$ **5.** 0   **7.** $\pi/4$ **9.** $\pi/3$
11. $-\pi/4$ **13.** 0.6999768749
15. 0.4391814802   **17.** 1.107148718
19. -1.270032196 **21.** -0.5829630403
23. 0.5931997761   **25.** $\tfrac{3}{4}$ **27.** $\pi/6$ **29.** $\tfrac{4}{5}$
31. $3\sqrt{10}/10$ **33.** $\tfrac{4}{3}$ **35.** $\tfrac{10}{7}$
45. (a) $-1 \le x \le 1$; (b) $0 \le x \le \pi$
47. (a) All real values of x; (b) $-\pi/2 < x < \pi/2$
49. (a)

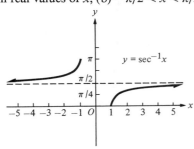

$$y = \sec^{-1} x$$

51. (a)

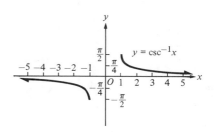

$$y = \csc^{-1} x$$

Problem Set 7.7, page 420

1. $90°, 270°$ **3.** $45°, 225°$ **5.** $\pi/6, 5\pi/6$
7. $2\pi/3, 5\pi/3$ **9.** $3\pi/2$
11. $k \cdot 180°, k = 0, \pm 1, \pm 2, \ldots$
13. $(\pi/4) + 2\pi k, (7\pi/4) + 2\pi k, k = 0, \pm 1, \pm 2, \ldots$
15. $(\pi/3) + k\pi, k = 0, \pm 1, \pm 2, \ldots$
17. $48.5904°, 131.4096°$ **19.** $0.8411, 5.4421$
21. $1.2900 + k\pi, k = 0, \pm 1, \pm 2, \ldots$
23. $15°, 75°, 195°, 255°$
25. $20°, 80°, 140°, 200°, 260°, 320°$
27. $2\pi/3, 4\pi/3$ **29.** $3\pi/2$ **31.** $210°, 330°$
33. $0, \pi/6, 5\pi/6, \pi$ **35.** $30°, 150°, 210°, 330°$
37. $\pi/4, 3\pi/4, 5\pi/4, 7\pi/4$ **39.** $0, \pi/3, \pi, 4\pi/3$
41. $0°, 45°, 180°, 225°$
43. $\pi/4, 3\pi/4, 5\pi/4, 7\pi/4$ **45.** $90°, 150°, 210°$
47. $\pi/3, \pi, 5\pi/3$ **49.** $\pi/6, \pi/2, 5\pi/6$
51. $0.4429, 1.5708, 2.6987$
53. $54.7356°, 90°, 270°, 305.2644°$
55. 1.288×10^{-3} second   **57.** $11.20°$
59. (a) $t = (2n + 1) \cdot 2\pi$ and $t = (4\pi/3) + (2n + 1) \cdot 2\pi$, $n = 0, \pm 1, \pm 2, \ldots$; (b) 30 or 45 wild pigs

Review Problem Set, Chapter 7, page 422

1. $-\tan \theta$ **3.** $\sin x$ **5.** $\tan^2 t$ **7.** $1 + \tan u$
9. $\tan \theta$ **11.** $125 \cos^3 \theta$ **27.** $\cos \theta$
29. $-\cos \alpha$ **31.** $\sin t$ **33.** $\sqrt{3}/2$ **35.** $\sin(x + y)$
37. (a) $\dfrac{\sqrt{2}(1 + \sqrt{3})}{4}$; (b) $\dfrac{\sqrt{2}(1 - \sqrt{3})}{4}$; (c) $-(2 + \sqrt{3})$
39. $-\tfrac{36}{85}$ **41.** $-\tfrac{21}{221}$ **43.** $-\tfrac{304}{425}$ **45.** $\tfrac{297}{304}$
47. $-\tfrac{36}{325}$ **57.** (a) $\tfrac{336}{625}$; (b) $-\tfrac{527}{625}$; (c) $\tfrac{4}{3}$; (d) $\tfrac{3}{4}$
59. $\cos 4x$ **61.** $\sin t$ **63.** 1   **65.** $\tfrac{1}{2}\tan 2\omega t$
67. 1   **85.** $\tfrac{1}{2}\sin 2x + \tfrac{1}{2}\sin x$
87. $\tfrac{1}{2}\cos 30° - \tfrac{1}{2}\cos 45°$ or $(\sqrt{3}/4) - (\sqrt{2}/4)$
89. $2 \sin 30° \cos 25°$ or $\cos 25°$
91. $2 \cos(5\beta/2) \sin(3\beta/2)$

105. $y = 2 \sin(7t/3) - 2 \sin(t/3)$
107. $y = 2\sqrt{2} \cos[(\pi/3)t - (7\pi/4)]$
109. $y = 2 \cos[2t - 3 + (\pi/6)]$ **111.** 30
113. $y = \cos 4\pi t + \cos 2\pi t$, $T = 1$ **115.** $-\pi/6$
117. $\pi/3$ **119.** 0.3843967745
121. 0.3836622700   **123.** 0.9588938924
125. -0.8480620790 **127.** 2.711892987
129. $\frac{3}{5}$ **131.** $-\frac{12}{13}$ **133.** $-5\pi/14$
141. 30°, 210°   **143.** $\pi/6$, $5\pi/6$, $7\pi/6$, $11\pi/6$
145. 0°, 180°, 225°, 315°   **147.** $7\pi/6$, $3\pi/2$, $11\pi/6$
149. 15°, 60°, 105°, 150°, 195°, 240°, 285°, 330°
151. 0°, 90°, 180°, 270°
153. 26.5651°, 153.4349°, 206.5651°, 333.4349°
155. 0.5236, 2.6180, 3.8713, 5.5535   **157.** 8.135

CHAPTER 8

Problem Set 8.1, page 432

1. $c = \dfrac{10\sqrt{3}}{3}$, $b = \dfrac{5\sqrt{3}}{3}$, $\beta = 30°$

3. $b = 10$, $\alpha = 45°$, $\beta = 45°$
5. $a = 1031$, $c = 1988$, $\beta = 58.77°$
7. $b = 5.826$, $\alpha = 50.76°$, $\beta = 39.24°$
9. $c = 4331$, $\alpha = 0.04°$, $\beta = 89.96°$
11. $a = 9299$, $c = 9647$, $\alpha = 74.55°$
13. $a = 1.387 \times 10^{-2}$ meter, $c = 1.498 \times 10^{-2}$ meter; $\beta = 22°15'$
15. $a = 70.79$ miles, $\alpha = 66.53°$, $\beta = 23.47°$
17. $c = 6.414 \times 10^{-6}$ meter, $\alpha = 50.26°$, $\beta = 39.74°$
19. 22 feet   **21.** 34 meters   **23.** 44 meters
25. 35.4°   **27.** 98.3 meters   **29.** 100 meters
31. 56°   **33.** 162 feet   **35.** 196 feet
37. 24.5 miles   **39.** S20.56°W

Problem Set 8.2, page 441

1. $\alpha = 85°$, $b = 26.94$, $c = 19.78$
3. $\alpha = 121°$, $a = 33.41$, $c = 13.97$
5. $\beta = 73.87°$, $a = 36.77$, $c = 41.67$
7. $\gamma = 23.55°$, $b = 52.91$, $c = 51.08$
9. $\beta = 25.1°$, $\gamma = 109.9°$, $c = 66.49$

11. $\beta = 67.21°$, $\gamma = 52.79°$, $c = 28.51$ or $\beta = 112.79°$, $\gamma = 7.21°$, $c = 4.49$
13. No triangle
15. $\beta = 52.02°$, $\gamma = 16.48°$, $c = 19.97$
17. $\gamma = 75.65°$, $a = 40.28$ feet, $c = 57.38$ feet
19. $\beta = 51.44°$, $\gamma = 69.23°$, $c = 119.58$ kilometers
21. $\beta = 89.53°$, $\gamma = 68.47°$, $c = 5190$ meters or $\beta = 90.47°$, $\gamma = 67.53°$, $c = 5156$ meters
23. No triangle
25. $|\overline{AC}| = 316.6$ meters, $|\overline{CB}| = 345.5$ meters
27. 91.43 meters   **29.** 72.86 meters
31. 0.9906 kilometer   **33.** 2.11 kilometers
35. 88 minutes, 16 seconds
41. $\gamma = 38°$, $c = 5.935$, $b = 9.219$
43. $\beta = 32°$, $b = 8.048$, $c = 13.98$

Problem Set 8.3, page 445

1. 8.888   **3.** 44.99   **5.** 3.336   **7.** 104.48°
9. 44.89°   **11.** 33.72°
13. $\alpha = 63.95°$, $\beta = 43.60°$, $c = 347.0$
15. $\beta = 60.75°$, $\gamma = 85.50°$, $c = 559.9$ or $\beta = 119.25°$, $\gamma = 27.00°$, $c = 255.9$
17. $\alpha = 26.75°$, $\beta = 39.75°$, $\gamma = 113.50°$
19. 69.14 meters   **21.** 815.9 meters
23. 744.5 kilometers   **25.** 57.49 miles
27. 17.32 cm²   **29.** 98.06 ft²   **31.** 74.22 km²
33. 85,733 m²   **41.** $a + b > c$, $a + c > b$, $b + c > a$

Problem Set 8.4, page 454

1. $4\sqrt{2}$; 135°   **3.** 5; 36.86989765°   **5.** 2; 300°
7. $\frac{5}{8}$; 53.13010235°   **9.** 4; 123.9878436°
11. 5; 53.13010235°   **13.** $\langle 5\sqrt{2}/2, 5\sqrt{2}/2 \rangle$
15. $\langle -4, 4 \rangle$ **17.** $\langle -\sqrt{3}, -1 \rangle$
19. $\langle 1.789851696, 6.798332951 \rangle$
21. (a) $\langle 2, 8 \rangle$; (b) $\langle 4, 0 \rangle$; (c) $\langle 5, 28 \rangle$; (d) $\langle 9, -4 \rangle$
23. (a) $\langle -\frac{1}{2}, \frac{13}{3} \rangle$; (b) $\langle -\frac{11}{2}, -\frac{1}{3} \rangle$; (c) $\langle 1, \frac{46}{3} \rangle$; (d) $\langle -\frac{27}{2}, -3 \rangle$
25. (a) $\langle 11, -21 \rangle$; (b) $\langle -7, 9 \rangle$; (c) $\langle 42, -78 \rangle$; (d) $\langle -23, 33 \rangle$
27. (a) $-\mathbf{i} + 12\mathbf{j}$; (b) $5\mathbf{i} - 2\mathbf{j}$; (c) $-6\mathbf{i} + 43\mathbf{j}$; (d) $13\mathbf{i} - 11\mathbf{j}$
29. (a) $10\mathbf{i} - 11\mathbf{j}$; (b) $2\mathbf{i} + \mathbf{j}$; (c) $34\mathbf{i} - 39\mathbf{j}$; (d) $8\mathbf{j}$
43. (a) 13; (b) 18; (c) -7; (d) 0; (e) 1

Problem Set 8.5, page 460

1–19.

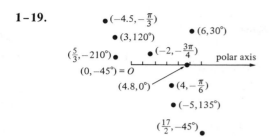

21. (a) $(-5, 225°)$; (b) $(5, -315°)$; (c) $(-5, -135°)$
23. $(2\sqrt{2}, 2\sqrt{2})$ **25.** $(-5\sqrt{2}/2, -5\sqrt{2}/2)$
27. $(\sqrt{2}/2, -\sqrt{2}/2)$ **29.** $(2\sqrt{3}, -2)$
31. $(-1.495, -2.326)$ **33.** $(14.17, -10.71)$
35. $(2, 45°)$ **37.** $(2, 60°)$ **39.** $(3, 180°)$
41. $(13, 112.62°)$ **43.** $(20.29, -50.12°)$
45. $(4, -\pi/6)$ **47.** $(2, -\pi/2)$ **49.** $(116.3, -2.239)$
51. Use the facts that $\cos(\theta + \pi) = -\cos\theta$ and $\sin(\theta + \pi) = -\sin\theta$

Problem Set 8.6, page 467

1. (a) 60.33 pounds; (b) 27.96°
3. (a) 29.10 newtons; (b) 79.11°
5. (a) 16.44 pounds; (b) 12.29°   **7.** (a) 816.5 kilometers/hour; (b) 2.97°; (c) N92.03°E
9. (a) 3.6 meters/second; (b) 56.3°
11. 109.83 newtons, N32.87°W
13.

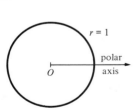

15.

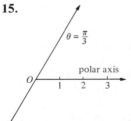

17.

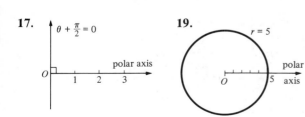

19.

21. $r = 6/(3\cos\theta - 2\sin\theta)$

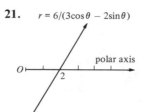

23.

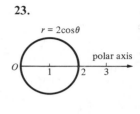

$r = 2\cos\theta$

25. $x^2 + y^2 = 1$

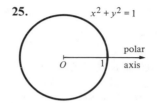

27.

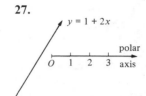

$y = 1 + 2x$

29. $y = x^2$

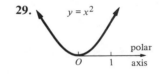

31. $r = 2(1 - \sin\theta)$

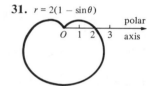

33.

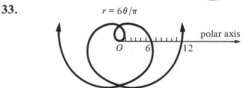

$r = 6\theta/\pi$

Problem Set 8.7, page 474

1. $2\sqrt{3} + 2i$ **3.** $-\frac{7}{2}\sqrt{2} + \frac{7}{2}\sqrt{2}i$
5. $5.909 + 1.042i$ **7.** $\sqrt{3} + i$
9. $\sqrt{2}[\cos(3\pi/4) + i\sin(3\pi/4)]$
11. $3[\cos(\pi/2) + i\sin(\pi/2)]$ or $3i$
13. $5.292(\cos 19.11° + i\sin 19.11°)$
15. $2[\cos(7\pi/6) + i\sin(7\pi/6)]$
17. $5(\cos\pi + i\sin\pi)$ or -5
19. (a) $8(\cos 110° + i\sin 110°)$;
(b) $2(\cos 30° + i\sin 30°)$
21. (a) $98[\cos(11\pi/4) + i\sin(11\pi/4)]$;
(b) $2[\cos(\pi/4) + i\sin(\pi/4)]$
23. (a) $18(\cos 135° + i\sin 135°)$;
(b) $2(\cos 45° + i\sin 45°)$
25. (a) $2\sqrt{2}[\cos(7\pi/12) + i\sin(7\pi/12)]$;
(b) $(\sqrt{2}/2)[\cos(23\pi/12) + i\sin(23\pi/12)]$
27. $16\sqrt{2} + 16\sqrt{2}i$ **29.** -243

31. $-256\sqrt{2} + 256\sqrt{2}i$ **33.** $-16\sqrt{3} - 16i$
35. $11{,}753 - 10{,}296i$ **37.** $(\sqrt{2}/2) - (\sqrt{2}/2)i$
39. $(3\sqrt{3}/2) - \frac{3}{2}i, -(3\sqrt{3}/2) + \frac{3}{2}i$
41. $(\sqrt[3]{4}/2)(1 + i), (\sqrt[3]{4}/4)[(-1 - \sqrt{3}) + (\sqrt{3} - 1)i],$
$(\sqrt[3]{4}/4)[(-1 + \sqrt{3}) + (-\sqrt{3} - 1)i]$
43. $[(\sqrt{6} + \sqrt{2})/2] - [(\sqrt{6} - \sqrt{2})/2]i, -[(\sqrt{6} + \sqrt{2})/2] +$
$[(\sqrt{6} - \sqrt{2})/2]i, [(\sqrt{6} - \sqrt{2})/2] + [(\sqrt{6} + \sqrt{2})/2]i,$
$-[(\sqrt{6} - \sqrt{2})/2] - [(\sqrt{6} + \sqrt{2})/2]i$
45. $(\sqrt{2}/2)(1 + i), (\sqrt{2}/2)(-1 + i), (\sqrt{2}/2)(1 - i),$
$(\sqrt{2}/2)(-1 - i)$
47. $\sqrt{2}(1 + i)$, etc.
51. (a) $1, -1$; (b) $1, -\frac{1}{2} + (\sqrt{3}/2)i, -\frac{1}{2} - (\sqrt{3}/2)i$; (c) $1,$
$-1, i, -i$
53. (a) Reflection of z about real axis; (b) Reflection of z through origin; (c) Reflection of z about imaginary axis; (d) Rotation of z by 90° clockwise about origin; (e) Distance between z and w

Review Problem Set, Chapter 8, page 476

1. $\beta = 58.75°$, $a = 103.8$ meters, $b = 171.0$ meters
3. $\alpha = 33.85°$, $a = 31.74$ microns, $c = 56.99$ microns
5. 25.49 meters   **7.** 658.2 meters
9. 264 meters/minute   **11.** 318.2 meters
13. 85.48 nautical miles; S13°10′W
15. 36.73 feet   **17.** 17.94 miles   **19.** 21.05°
21. $\alpha = 90°$, $a = 100\sqrt{3} \approx 173.2$ feet, $\gamma = 60°$ or
$\alpha = 30°$, $a = 50\sqrt{3} \approx 86.6$ feet, $\gamma = 120°$
23. $\alpha = 76.28°$, $a = 66.99$ centimeters, $\beta = 46.47°$
25. No triangle   **27.** 60°   **29.** 9.006 kilometers
31. 16.96 inches
33. $\alpha = 23.94°$, $c = 80.44$ kilometers, $\beta = 115.01°$
35. $\beta = 55.58°$, $b = 0.9869$ micron, $\alpha = 69.42°$ or
$\beta = 14.42°$, $b = 0.2979$ micron, $\alpha = 110.58°$
37. $\gamma = 105.40°$, $a = 1.788$ nautical miles, $c = 2.555$ nautical miles
39. 448.0 m²   **41.** 392,100 square miles
43. 2046 meters   **45.** 1:15 P.M.
47. 1275.3 meters   **49.** 83.51°
51. $\langle 10, -5 \rangle$ **53.** $\langle 5, 2 \rangle$
55. $\langle 13, 7 \rangle$ **57.** $\langle -13, 5 \rangle$
59. $4\sqrt{5}$ **61.** $2\sqrt{26}$
63. (a) $\langle 25\sqrt{3}, 25 \rangle$; (b) $\langle 33.94, 61.22 \rangle$;
(c) $\langle 16.07, -19.15 \rangle$; (d) $\langle 125\sqrt{3}, -125 \rangle$
65. (a) 2, 60°; (b) 19.21, 51.34°; (c) 50, 126.87°;
(d) 0.3125, 53.13°; (e) 4, 41.41°
67. (a) 30.96 pounds, 33.35°; (b) 221.2 newtons,
44.42°; (c) 7.598 tonnes, 76°; (d) 540.8 dynes, 303.69°

69. (a) 486.6 knots; (b) 7.99°; (c) N37.01°W
71. S29°26′E
73. (a) $(\sqrt{2}, \sqrt{2})$; (b) $(-\sqrt{2}, -\sqrt{2})$; (c) $(-0.1736, 0.9848)$;
(d) $(1.5, 0)$; (e) $(2\sqrt{3}, -2)$; (f) $(3\sqrt{2}, -3\sqrt{2})$
75. (a) $(1, \pi)$; (b) $(7\sqrt{2}, -3\pi/4)$; (c) $(13, 1.97)$;
(d) $(5, -0.93)$; (e) $(17, 2.06)$; (f) $(34, -2.65)$
77. $r \sin \theta = 3r \cos \theta + 1$
79. $r = 4 \sin \theta$ **81.** $r \sin^2 \theta = 6 \cos \theta$
83. $r^2(3 \cos^2 \theta + 1) = 4$
85. $(x - \frac{1}{2})^2 + (y - \frac{1}{2})^2 = \frac{1}{2}$ **87.** $x^2 - y^2 = 2$
89. Hyperbola, horizontal principal axis, $a = \sqrt{2}, b = \sqrt{2}$

91.

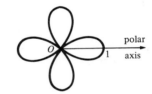

93.

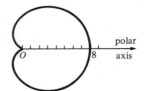

95. $\dfrac{3\sqrt{2}}{2} + \dfrac{3\sqrt{2}}{2} i$

97. $8 - 8\sqrt{3}i$ **99.** 5
101. 8   **103.** $\sqrt{13}/5$
105. $4\sqrt{2}[\cos(3\pi/4) + i \sin(3\pi/4)]$
107. $2[\cos(2\pi/3) + i \sin(2\pi/3)]$
109. (a) $24(\cos 30° + i \sin 30°)$;
(b) $\frac{3}{2}(\cos 14° + i \sin 14°)$
111. (a) $48(\cos 360° + i \sin 360°)$;
(b) $3(\cos 110° + i \sin 110°)$
113. (a) $50[\cos(7\pi/12) + i \sin(7\pi/12)]$;
(b) $2[\cos(5\pi/36) + i \sin(5\pi/36)]$
115. (a) $4\sqrt{3}[\cos(\pi/3) + i \sin(\pi/3)]$;
(b) $\sqrt{3}(\cos 0 + i \sin 0)$
117. $16 + 16\sqrt{3}i$
119. -64 **121.** $8i$
123. $-i, (\sqrt{3}/2) + \frac{1}{2}i, (-\sqrt{3}/2) + \frac{1}{2}i$
125. $[(\sqrt{6} + \sqrt{2})/4] - [(\sqrt{6} - \sqrt{2})/4]i, [(\sqrt{6} - \sqrt{2})/4] +$
$[(\sqrt{6} + \sqrt{2})/4]i, -[(\sqrt{6} + \sqrt{2})/4] + [(\sqrt{6} - \sqrt{2})/4]i,$
$-[(\sqrt{6} - \sqrt{2})/4] - [(\sqrt{6} + \sqrt{2})/4]i$
127. $\pm 1, \pm i$

CHAPTER 9

Problem Set 9.1, page 487

1. Consistent

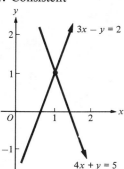

$3x - y = 2$

$4x + y = 5$

3. Dependent

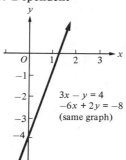

$3x - y = 4$
$-6x + 2y = -8$
(same graph)

5. Inconsistent

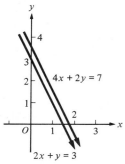

$4x + 2y = 7$

$2x + y = 3$

7. Inconsistent

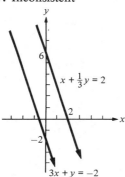

$x + \frac{1}{3}y = 2$

$3x + y = -2$

9. Consistent

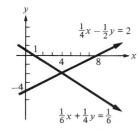

$\frac{1}{4}x - \frac{1}{2}y = 2$

$\frac{1}{6}x + \frac{1}{4}y = \frac{1}{6}$

11. $(1, 1)$, all points on the graph of $y = 3x - 4$, no solution
13. $(1, 3)$ **15.** $(4, -1)$ **17.** $\left(\frac{7}{11}, -\frac{10}{11}\right)$
19. $(1, -1, -2)$ **21.** $(3, 4)$ **23.** $(3, 4)$
25. $(1, 2, 3)$ **27.** $(8, 4, 0)$
29. (a) $x + y = 39, 280x + 315y = 11{,}375$;
(b) $x = 26, y = 13$

Problem Set 9.2, page 492

1. $0, 5, -3$ **3.** 2 **5.** $\begin{bmatrix} \frac{3}{4} & -\frac{2}{3} & \frac{1}{7} \\ -1 & 5 & 6 \end{bmatrix}$

7. $\begin{bmatrix} 40 & 22 & -1 & -17 \\ 0 & 1 & 1 & 0 \\ -13 & 17 & 2 & 5 \end{bmatrix}$

9. $x + 3y = 0, 2x - 4y = 1$
11. $2x + 5y + 3z = 1, -3x + 7y + \frac{1}{2}z = \frac{3}{4}, \frac{2}{3}y = -\frac{4}{5}$
13. $\left(\frac{24}{5}, -\frac{4}{5}\right)$ **15.** $(5, 0)$ **17.** $(2, 3)$
19. $(4, -2)$ **21.** No solution   **23.** $(3, 0)$
25. $(4, -2, 1)$ **27.** $(3, -1, -2)$ **29.** $\left(\frac{1}{4}, \frac{2}{3}, \frac{1}{6}\right)$
31. $\left(-\frac{11}{4}, 1, -\frac{1}{4}\right)$ **33.** $(0, 0, 0)$ **35.** No solution
37. $x = \frac{24}{11} - \frac{13}{11}t, y = -\frac{3}{11} + \frac{14}{11}t, z = t$

Problem Set 9.3, page 502

1. (a) $\begin{bmatrix} 8 & 1 \\ 8 & -1 \end{bmatrix}$; (b) $\begin{bmatrix} -4 & -7 \\ 2 & 3 \end{bmatrix}$;

(c) $\begin{bmatrix} -6 & 9 \\ -15 & -3 \end{bmatrix}$; (d) $\begin{bmatrix} 6 & 17 \\ -9 & -7 \end{bmatrix}$

3. (a) $\begin{bmatrix} 1 & 5 \\ 1 & 6 \\ 6 & -1 \\ -3 & 4 \end{bmatrix}$; (b) $\begin{bmatrix} 5 & -1 \\ -5 & 4 \\ -2 & 3 \\ -5 & 4 \end{bmatrix}$;

(c) $\begin{bmatrix} -9 & -6 \\ 6 & -15 \\ -6 & -3 \\ 12 & -12 \end{bmatrix}$; (d) $\begin{bmatrix} -13 & 0 \\ 12 & -13 \\ 2 & -7 \\ 14 & -12 \end{bmatrix}$

5. (a) $\begin{bmatrix} 4 & -6 & 2 & -1 \\ 0 & 4 & 0 & 6 \\ 4 & -1 & -2 & 4 \end{bmatrix}$;

(b) $\begin{bmatrix} 0 & 0 & 2 & -5 \\ -6 & 0 & 2 & -4 \\ 4 & 3 & -4 & 4 \end{bmatrix}$;

(c) $\begin{bmatrix} -6 & 9 & -6 & 9 \\ 9 & -6 & -3 & -3 \\ -12 & -3 & 9 & -12 \end{bmatrix}$;

(d) $\begin{bmatrix} -2 & 3 & -6 & 13 \\ 15 & -2 & -5 & 7 \\ -12 & -7 & 11 & -12 \end{bmatrix}$

7. (a) $\begin{bmatrix} 2 & -\frac{2}{3} & \sqrt{2} \\ \frac{17}{6} & \pi+1 & -2 \\ 4 & \frac{5}{2} & \frac{5}{3} \end{bmatrix}$; (b) $\begin{bmatrix} 0 & 1 & -\sqrt{2} \\ -\frac{1}{6} & \pi-1 & -2 \\ -2 & -\frac{5}{2} & \frac{5}{3} \end{bmatrix}$;

(c) $\begin{bmatrix} -3 & -\frac{1}{2} & 0 \\ -4 & -3\pi & 6 \\ -3 & 0 & -5 \end{bmatrix}$; (d) $\begin{bmatrix} -1 & -\frac{13}{6} & 2\sqrt{2} \\ -1 & 2-3\pi & 6 \\ 3 & 5 & -5 \end{bmatrix}$

21. $\begin{bmatrix} 14 & -3 \\ 6 & -2 \end{bmatrix}$ **23.** $\begin{bmatrix} 2 & 6 \\ 5 & 10 \end{bmatrix}$ **25.** $\begin{bmatrix} -2 & 9 \\ -3 & 1 \end{bmatrix}$

27. A **29.** $\begin{bmatrix} 2 & -5 & 2 \\ -2 & 0 & 6 \\ 7 & -14 & 1 \end{bmatrix}$

31. $\begin{bmatrix} -2 & 7 & 6 \\ 8 & -2 & -4 \\ 1 & -10 & -8 \end{bmatrix}$ **33.** $\begin{bmatrix} 5 & -10 & 17 \\ 10 & -14 & 30 \\ 8 & -20 & 30 \end{bmatrix}$

35. $\begin{bmatrix} 17 & 36 \\ 8 & 14 \end{bmatrix}$ **37.** $\begin{bmatrix} -6 & 4 \\ -23 & 16 \\ -20 & -2 \end{bmatrix}$

39. $\begin{bmatrix} 13 & -6 \\ 7 & -4 \end{bmatrix}$ **41.** Undefined **45.** $\begin{bmatrix} \frac{4}{5} & \frac{1}{5} \\ \frac{1}{5} & -\frac{1}{5} \end{bmatrix}$

47. $\begin{bmatrix} -\frac{5}{34} & \frac{1}{17} \\ \frac{1}{17} & \frac{3}{17} \end{bmatrix}$ **49.** $\begin{bmatrix} \frac{1}{4} & \frac{1}{12} \\ -\frac{3}{4} & \frac{1}{12} \end{bmatrix}$ **51.** No inverse

53. $\begin{bmatrix} \frac{3}{10} & -\frac{2}{5} & \frac{1}{2} \\ \frac{3}{10} & \frac{3}{5} & -\frac{1}{2} \\ -\frac{1}{10} & \frac{4}{5} & -\frac{1}{2} \end{bmatrix}$ **55.** $\begin{bmatrix} \frac{1}{3} & \frac{1}{3} & 0 \\ 0 & -\frac{3}{10} & \frac{1}{10} \\ -\frac{2}{3} & -\frac{4}{15} & \frac{1}{5} \end{bmatrix}$

57. $\begin{bmatrix} -\frac{3}{2} & \frac{9}{4} & -\frac{5}{2} \\ 1 & -1 & 1 \\ -\frac{1}{2} & \frac{3}{4} & -\frac{1}{2} \end{bmatrix}$ **59.** $(\frac{24}{5}, -\frac{4}{5})$

61. $(-\frac{9}{34}, \frac{12}{17})$ **63.** $(\frac{8}{3}, -\frac{10}{3})$ **65.** $(-1, 2, -3)$
67. $(-\frac{11}{4}, 1, -\frac{1}{4})$
69. (a) The entry in the ith row of TX is the number of units of commodity number i used in unit time in the production of all other commodities. (b) The entry in the ith row of $X - TX$ is the surplus number of units of commodity number i produced in unit time.

Problem Set 9.4, page 511

1. 5 **3.** 54 **5.** 54 **7.** 4 **9.** $(1, -1)$
11. $(2, 1)$ **13.** $(\frac{5}{4}, -\frac{1}{3})$ **15.** 28 **17.** 38
19. 9 **21.** $-\frac{53}{24}$ **23.** $(1, -1, 2)$
25. $(-\frac{10}{11}, \frac{18}{11}, \frac{38}{11})$ **27.** No solution **29.** -14
31. -80 **33.** 22 **35.** 25
37. -130.9347340 **39.** -12 **41.** 0
43. -1 **45.** 3 **47.** -3 **49.** -3
57. $(2 \pm \sqrt{58})/3$
59. Discriminant $= (a - c)^2 + b^2 \geq 0$

Problem Set 9.5, page 516

1. $\dfrac{3}{x-3} - \dfrac{3}{x-2}$ **3.** $\dfrac{\frac{1}{2}}{x+5} + \dfrac{\frac{1}{2}}{x-1}$

5. $\dfrac{\frac{3}{4}}{x} - \dfrac{\frac{9}{8}}{x-2} + \dfrac{\frac{11}{8}}{x+2}$ **7.** $\dfrac{-2}{x} + \dfrac{5}{x-1} - \dfrac{3}{x+1}$

9. $3 + \dfrac{\frac{1}{6}}{x} + \dfrac{\frac{1}{3}}{x+3} + \dfrac{\frac{1}{2}}{x-2}$

11. $\dfrac{-4}{x-3} + \dfrac{2}{x-1} + \dfrac{1}{(x-1)^2}$

13. $1 - \dfrac{1}{x} - \dfrac{1}{x^2} + \dfrac{1}{x-1}$

15. $\dfrac{4}{x-1} - \dfrac{1}{x+2} + \dfrac{3}{(x+2)^2}$

17. $\dfrac{\frac{5}{8}}{x+2} - \dfrac{\frac{13}{24}}{3x-2} + \dfrac{\frac{8}{3}}{(3x-2)^2}$

19. $\dfrac{\frac{9}{2}}{t+1} + \dfrac{(-\frac{9}{2})t + (\frac{11}{2})}{t^2+1}$ **21.** $x^2 + \dfrac{\frac{9}{2}}{x} - \dfrac{(\frac{9}{2})x}{x^2+9}$

23. $\dfrac{3}{t} + \dfrac{-3t+1}{t^2+1}$ **25.** $\dfrac{\frac{1}{4}}{u-1} + \dfrac{\frac{1}{4}}{u+1} - \dfrac{(\frac{1}{2})u}{u^2+1}$

27. $A = \frac{3}{2}, B = 0, C = -2, D = -\frac{3}{2}$
29. $A = 1, B = -1, C = -3, D = 3$
31. 225 adults, 700 children
33. $(50°, 40°, 90°)$
35. $(45, 20, 15)$ **37.** $p = 30, q = 90$
39. $x = \$8000, y = \$9000$
41. Coffee: \$2.80/pound, milk: \$0.60/quart, tuna: \$1.20/can

Problem Set 9.6, page 522

1. $(2, 2), (-5, \frac{25}{2})$ **3.** $(3, 3), (\frac{3}{4}, -\frac{3}{2})$

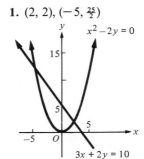

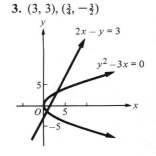

5. $(0, -2), (\frac{8}{5}, -\frac{6}{5})$

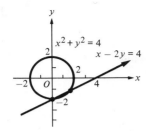

7. $\left(\dfrac{-1 + \sqrt{11}}{2}, \sqrt{11} - 2\right), \left(\dfrac{-1 - \sqrt{11}}{2}, -\sqrt{11} - 2\right)$

9. No real solution
11. $(2, 3), (2, -3), (-2, 3), (-2, -3)$
13. $(\sqrt{7}, \frac{1}{2}\sqrt{6}), (-\sqrt{7}, \frac{1}{2}\sqrt{6}), (\sqrt{7}, -\frac{1}{2}\sqrt{6}), (-\sqrt{7}, -\frac{1}{2}\sqrt{6})$
15. $(1, 2), (1, -2), (-1, 2), (-1, -2)$
17. $(3, 7), (-1, -1)$
19. $(2, 2), (-2, -2), (2\sqrt{2}, \sqrt{2}), (-2\sqrt{2}, -\sqrt{2})$
21. $(25, 4)$ **23.** $(16, 2)$
25. $(\frac{5}{2}, -\frac{3}{2}), (-\frac{3}{2}, \frac{5}{2}), (\frac{3}{2}, -\frac{5}{2}), (-\frac{5}{2}, \frac{3}{2})$
27. $(-1, 1 + \log_6 2)$ **29.** No solution
31. $(5, 1)$ **33.** 18 and 48   **35.** $b = 6, h = 10$
37. 4000   **39.** (a) $p = 7, q = 1$; (b) \$7000

Problem Set 9.7, page 530

1.

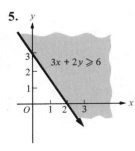

3.

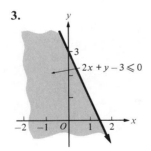

5.

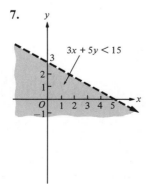

7.

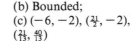

9.

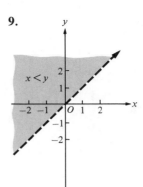

11. (a)

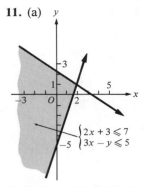

(b) Unbounded; (c) $(2, 1)$

13. (a)

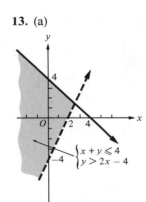

(b) Unbounded; (c) $(\frac{8}{3}, \frac{4}{3})$

15. (a)

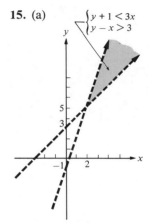

(b) Unbounded; (c) $(2, 5)$

17. (a)

(b) Bounded;
(c) $(-6, -2), (\frac{21}{2}, -2),$
$(\frac{21}{13}, \frac{40}{13})$

19. (a)

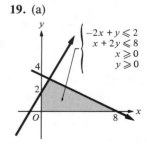

(b) Bounded; (c) $(0, 0),$
$(0, 2), (8, 0), (\frac{4}{5}, \frac{18}{5})$

21. (a)

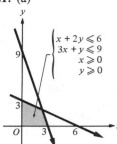

$$\begin{cases} x + 2y \leq 6 \\ 3x + y \leq 9 \\ x \geq 0 \\ y \geq 0 \end{cases}$$

(b) Bounded; (c) $(0, 0)$, $(3, 0)$, $(\frac{12}{5}, \frac{9}{5})$, $(0, 3)$

23.

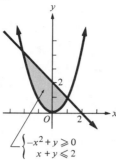

$$\begin{cases} -x^2 + y \geq 0 \\ x + y \leq 2 \end{cases}$$

25.

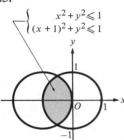

$$\begin{cases} x^2 + y^2 \leq 1 \\ (x + 1)^2 + y^2 \leq 1 \end{cases}$$

27.

$$\begin{cases} \sqrt{x} - y > 0 \\ x - 4y \leq 0 \end{cases}$$

$(16, 4)$

29.

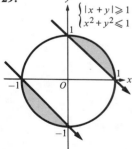

$$\begin{cases} |x + y| \geq 1 \\ x^2 + y^2 \leq 1 \end{cases}$$

31.

$$\begin{cases} 0 \leq x \\ x \leq 7 - 2y \\ 0 \leq y \\ 8y \leq 5x + 3 \end{cases}$$

$(\frac{25}{9}, \frac{19}{9})$

$$\begin{cases} 0 \leq x \leq 7 - 2y \\ 0 \leq 8y \leq 5x + 3 \end{cases}$$

33. max 5, min 2
35. max 25, min 6
37. max 32, min -28
39. max 11, min 5
41. max 30, min -10
43. 50 units of A, 40 units of B
45. 80 units of A, 0 units of B
47. Center I open 6 days, Center II open 2 days
49. $2b = 3a$

Review Problem Set, Chapter 9, page 532

1. Inconsistent   **3.** Consistent
5. Dependent   **7.** $(2, 1)$
9. $(\frac{11}{7}, \frac{8}{7})$ **11.** Inconsistent
13. $(1, 1)$ **15.** $(3, 4, -6)$ **17.** $(1, -1)$
19. $(5, 6)$ **21.** $(1, -3, 2)$ **23.** $(-1, 2, -2)$

25. $\begin{bmatrix} -16 & 4 \\ 8 & 5 \end{bmatrix}, \begin{bmatrix} -18 & 2 \\ 4 & 10 \end{bmatrix}$

27. $\begin{bmatrix} 6 & 16 \\ 26 & 0 \\ -16 & 2 \end{bmatrix}, \begin{bmatrix} -5 & -8 \\ -7 & 8 \\ 8 & -11 \end{bmatrix},$

29. $\begin{bmatrix} 0 & 3 \\ -1 & 20 \end{bmatrix}, \begin{bmatrix} 4 & 1 & 5 \\ 16 & 6 & 6 \\ 15 & 5 & 10 \end{bmatrix}$

31. $\begin{bmatrix} 7 & -7 \\ -5 & -4 \\ 9 & 10 \end{bmatrix}$ **33.** $\begin{bmatrix} \frac{7}{26} & \frac{2}{13} \\ -\frac{3}{26} & \frac{1}{13} \end{bmatrix}$

35. $\begin{bmatrix} \frac{2}{3} & 0 & -\frac{1}{3} \\ \frac{1}{3} & 0 & -\frac{2}{3} \\ -\frac{2}{3} & 1 & \frac{1}{3} \end{bmatrix}$ **37.** 6   **41.** $(3, 1)$

43. $(-1, 3, 4)$
45. $x = \frac{1}{13}(4a - b - 6c)$, $y = \frac{1}{13}(-4a + b - 7c)$, $z = \frac{1}{13}(5a + 2b + 12c)$
47. $\frac{11}{3}$ **49.** 133   **51.** 0   **53.** $2x^2 - 3x$
55. -20 **57.** 37   **59.** 2
61. 16   **63.** -2 **65.** $(\frac{26}{19}, \frac{29}{19})$
67. $(3, -3, 5)$ **69.** $(1, -1, 2)$

73. $\dfrac{1}{x - 1} + \dfrac{1}{x + 1}$ **75.** $\dfrac{3}{x} - \dfrac{1}{x + 3} + \dfrac{2}{x - 1}$

77. $\dfrac{4}{x} - \dfrac{3x}{x^2 + 1}$ **79.** $(\frac{14}{5}, \frac{6}{5})$

81. \$13,000 at 8.5%, \$27,000 at 11.2%
83. Byron: \$6, Jason: \$4, Adrian: \$2

85. $\left(\dfrac{\sqrt{13} - 2}{3}, 2\sqrt{\sqrt{13} - 2}\right), \left(\dfrac{\sqrt{13} - 2}{3}, -2\sqrt{\sqrt{13} - 2}\right)$

87. $(-1, 3)$, $(\frac{3}{2}, -2)$ **89.** $(1, 2)$, $(-1, -2)$
91. No real solution

93. $\left(\dfrac{-1 + \sqrt{11}}{2}, \sqrt{11} - 2\right), \left(\dfrac{-1 - \sqrt{11}}{2}, -\sqrt{11} - 2\right)$

95. 7 and 24   **97.** 100 meters by 75 meters
99. (b) Unbounded; (c) $(0, 5)$
101. (b) Unbounded; (c) $(-\frac{1}{4}, \frac{3}{4})$
103. (b) Bounded; (c) $(2, 0)$, $(-\frac{2}{3}, 0)$, $(\frac{2}{5}, \frac{16}{5})$
105. (b) Bounded; (c) $(0, 5)$, $(10, 5)$, $(10, -1)$, $(2, 3)$

107.

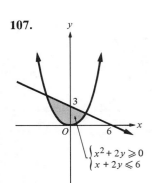

$$\begin{cases} x^2 + 2y \geqslant 0 \\ x + 2y \leqslant 6 \end{cases}$$

109.

no solution $\begin{cases} x^2 + y^2 \leqslant 1 \\ x^2 - y > 3 \end{cases}$

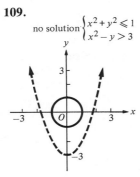

111. max 60, min 36   **113.** max 20, min 0
115. 400 units A, 800 units B

7. vertices: $(-\frac{2}{3}, 0)$, $(\frac{2}{3}, 0)$, $(0, -\frac{1}{3})$, $(0, \frac{1}{3})$;
foci: $\left(-\frac{\sqrt{3}}{3}, 0\right)$, $\left(\frac{\sqrt{3}}{3}, 0\right)$

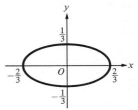

9. $\dfrac{x^2}{25} + \dfrac{y^2}{9} = 1$ **11.** $\dfrac{x^2}{25} + \dfrac{y^2}{169} = 1$

13. Center: $(1, -2)$;
vertices: $(-2, -2)$,
$(4, -2)$, $(1, -4)$, $(1, 0)$;
foci: $(1 - \sqrt{5}, -2)$,
$(1 + \sqrt{5}, 2)$

15. Center: $(-3, 0)$;
vertices: $(-6, 0)$, $(0, 0)$,
$(-3, -6)$, $(-3, 6)$; foci:
$(-3, -3\sqrt{3})$, $(-3, 3\sqrt{3})$

CHAPTER 10

Problem Set 10.1, page 545

1. vertices: $(-4, 0)$,
$(4, 0)$, $(0, -2)$, $(0, 2)$;
foci: $(-2\sqrt{3}, 0)$, $(2\sqrt{3}, 0)$

3. vertices: $(-2, 0)$,
$(2, 0)$, $(0, -4)$, $(0, 4)$;
foci: $(0, -2\sqrt{3})$, $(0, 2\sqrt{3})$

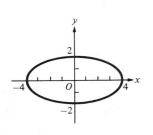

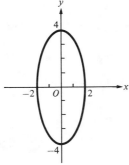

5. vertices: $(-4, 0)$,
$(4, 0)$, $(0, -1)$, $(0, 1)$;
foci: $(-\sqrt{15}, 0)$, $(\sqrt{15}, 0)$

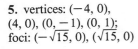

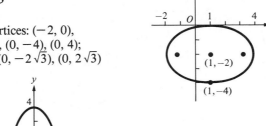

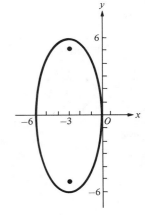

17. Center: $(-3, 0)$;
vertices: $(-3 - \sqrt{2}, 0)$,
$(-3 + \sqrt{2}, 0)$, $(-3, -1)$,
$(-3, 1)$; foci: $(-4, 0)$,
$(-2, 0)$

19. Center: $(-5, 3)$;
vertices: $(-5 - \sqrt{10}, 3)$,
$(-5 + \sqrt{10}, 3)$, $(-5, 1)$,
$(-5, 5)$;
foci: $(-5 - \sqrt{6}, 3)$,
$(-5 + \sqrt{6}, 3)$

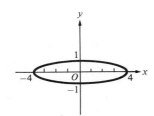

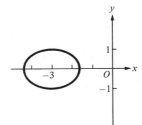

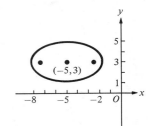

21. (a) If q is the length of the focal chord, then $(c, q/2)$ belongs to the ellipse, $c = \sqrt{a^2 - b^2}$. Substitute into the equation of the ellipse and solve for q. (b) $\frac{9}{2}$ units
23. Length $20 + 3\sqrt{39}$ meters, $2c = 3\sqrt{39}$ meters

25. Curve is the ellipse $\dfrac{x^2}{12} + \dfrac{(y-8)^2}{16} = 1$

27. $8\sqrt{901} \approx 240$ meters   **29.** 3,107,000 miles

11. $V = (4, -7)$,
$F = (4, -4)$, $D: y = -10$,
F.C. $= 12$

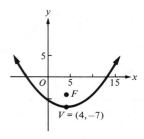

Problem Set 10.2, page 551

1. $V = (0, 0)$, $F = (1, 0)$,
$D: x = -1$, F.C. $= 4$

3. $V = (0, 0)$, $F = (0, \frac{1}{4})$,
$D: y = -\frac{1}{4}$, F.C. $= 1$

13. $V = (-3, 4)$,
$F = (-\frac{3}{2}, 4)$, $D: x = -\frac{9}{2}$,
F.C. $= 6$

15. $V = (3, -1)$,
$F = (3, 1)$, $D: y = -3$,
F.C. $= 8$

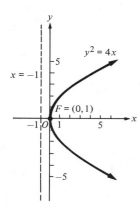

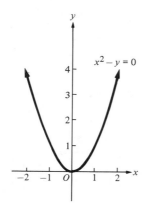

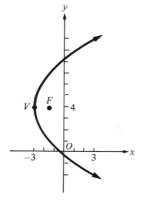

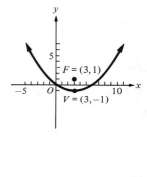

5. $V = (0, 0)$,
$F = (0, -\frac{9}{4})$, $D: y = \frac{9}{4}$,
F.C. $= 9$

9. $V = (-3, 2)$,
$F = (-1, 2)$, $D: x = -5$,
F.C. $= 8$

17. $x - 5 = -\frac{1}{4}(y-2)^2$ **19.** $x + 6 = \frac{1}{32}(y+5)^2$
21. $x + \frac{1}{2} = \frac{1}{8}(y+1)^2$
23. 58, 28, 10, 4, 10, 28, and 58 meters

25. For $y = \dfrac{1}{4p}x^2$, the focus is $(0, p)$. Let $y = p$ and
solve for x to get two endpoints $(-2p, p)$ and $(2p, p)$ of
the focal chord; therefore, length $= 4p$.

27. Equation of the parabola is $y = \dfrac{1}{4p}x^2$, and $\left(\dfrac{a}{2}, b\right)$
is a point on the parabola. Substitute $x = \dfrac{a}{2}$, $y = b$ into
the equation and solve for p to get $p = \dfrac{a^2}{16b}$.

29. (a) Any two parabolas are similar in the sense that
one is a magnification of the other. (b) Two ellipses are
similar only if they have the same ratio of semimajor to
semiminor axes.

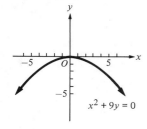

7. $y = \frac{1}{12}x^2$

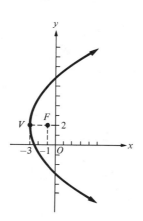

Problem Set 10.3, page 558

1. Vertices: (3, 0), (−3, 0); foci: ($\sqrt{13}$, 0), (−$\sqrt{13}$, 0); asymptotes: $y = \pm\frac{2}{3}x$

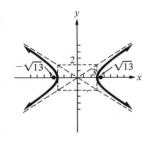

3. Vertices: (0, 4), (0, −4); foci: (0, −2$\sqrt{5}$), (0, 2$\sqrt{5}$); asymptotes: $y = \pm 2x$

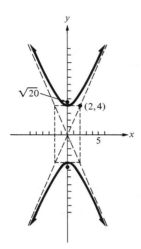

9. $\dfrac{x^2}{16} - \dfrac{y^2}{20} = 1$

11. $\dfrac{x^2}{16} - \dfrac{y^2}{25} = 1$

13. Center: (1, −2); vertices: (4, −2), (−2, −2); foci: (1 + $\sqrt{13}$, −2), (1 − $\sqrt{13}$, −2); asymptotes: $3y - 2x + 8 = 0$, $3y + 2x + 4 = 0$

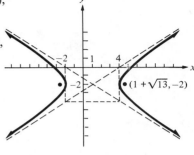

15. Center: (−2, −1); vertices: (−2, 3), (−2, −5); foci: (−2, −1 − $\sqrt{41}$), (−2, −1 + $\sqrt{41}$); asymptotes: $5y - 4x - 3 = 0$, $5y + 4x + 13 = 0$

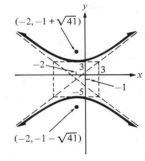

5. Vertices: (4, 0), (−4, 0); foci: (2$\sqrt{5}$, 0), (−2$\sqrt{5}$, 0); asymptotes: $y = \pm\frac{1}{2}x$

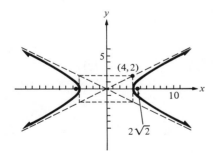

17. Center: (2, −1); vertices: (4, −1), (0, −1); foci: (2 − $\sqrt{5}$, −1), (2 + $\sqrt{5}$, −1); asymptotes: $2y - x + 4 = 0$, $2y + x = 0$

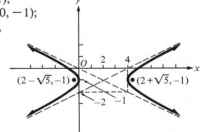

7. Vertices: (0, −$\sqrt{10}$), (0, $\sqrt{10}$); foci: (0, −$\sqrt{46}$), (0, $\sqrt{46}$); asymptotes:
$y = \pm\dfrac{\sqrt{10}}{6}x$

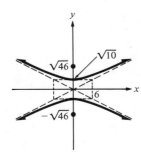

19. Center: (−4, −2); vertices: (−4, 1), (−4, −5); foci: (−4, −2 − $\sqrt{34}$), (−4, −2 + $\sqrt{34}$); asymptotes: $5y - 3x - 2 = 0$, $5y + 3x + 22 = 0$

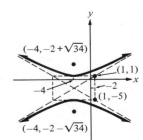

21. (a) $\dfrac{(x-4)^2}{1} - \dfrac{(y+1)^2}{8} = 1$;

(b) $\dfrac{(x+2)^2}{4} - \dfrac{(y-3)^2}{\left(\frac{9}{4}\right)} = 1$; (c) $\dfrac{(y-3)^2}{25} - \dfrac{(x-2)^2}{11} = 1$

23. $x^2 - y^2 = a^2$

25. The hyperbola begins to look more and more like the pair of intersecting lines

27. Points on right-hand branch of hyperbola

$\dfrac{b^2x^2}{h^2s^2} - \dfrac{b^2y^2}{h^2(b^2-s^2)} = 1$ (except at vertex, where bullet

will pass)

29. Hyperbola

Problem Set 10.4, page 568

1. (a) $(1, -2)$; (b) $(-1, -1)$; (c) $(4, -5)$; (d) $(-2, -4)$;
(e) $(6, 3)$; (f) $(7, -2)$

3. $\bar{x} = x + 2$, $\bar{y} = y - 1$, circle

5. $\bar{x} = x + 1$, $\bar{y} = y - 1$, ellipse

7. $\bar{x} = x - 2$, $\bar{y} = y - 3$, ellipse

9. $\bar{x} = x - 1$, $\bar{y} = y$, hyperbola

11. $\bar{x} = x + 1$, $\bar{y} = y - 5$, parabola

13. $\bar{x} = x + 3$, $\bar{y} = y - 1$, hyperbola

15. $\bar{x} = x - 2$, $\bar{y} = y + 1$, hyperbola

17. $(-7, -4)$　**19.** $\left(-\dfrac{3}{2} + \dfrac{3\sqrt{3}}{2}, -\dfrac{3\sqrt{3}}{2} - \dfrac{3}{2}\right)$

21. $(-2\sqrt{3} + 1, -2 - \sqrt{3})$　**23.** $(4, 0)$

25. (a) $(\bar{x} + \sqrt{3}\bar{y})^2 = 6(\sqrt{3}\bar{x} - \bar{y})$;

(b) $\left(\dfrac{3\sqrt{3}}{2} + \dfrac{1}{2}\right)x + \left(\dfrac{3}{2} - \dfrac{\sqrt{3}}{2}\right)y = 0$; (c) $x^2 + y^2 = 1$;

(d) $\left(\dfrac{5\sqrt{3}}{2} - \dfrac{1}{2}\right)\bar{x} - \left(\dfrac{5}{2} + \dfrac{\sqrt{3}}{2}\right)\bar{y} = 4$; (e) $\bar{x}^2 + \bar{y}^2 = 1$;

(f) $(\sqrt{3}\bar{x} - \bar{y})^2 = 100$

27. (a) $\phi = \dfrac{1}{2}\cot^{-1}\dfrac{3}{4} \approx 26.6°$;　(d)

(b) $x = \dfrac{\sqrt{5}}{5}(2\bar{x} - \bar{y})$,

$y = \dfrac{\sqrt{5}}{5}(\bar{x} + 2\bar{y})$;

(c) $\dfrac{\bar{x}^2}{6} - \dfrac{\bar{y}^2}{4} = 1$;

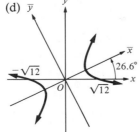

29. (a) $\phi = 45°$;　(d)

(b) $x = \dfrac{\sqrt{2}}{2}(\bar{x} - \bar{y})$,

$y = \dfrac{\sqrt{2}}{2}(\bar{x} + \bar{y})$;

(c) $\bar{x} = \pm\dfrac{\sqrt{2}}{2}$;

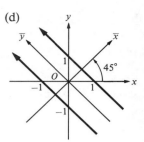

31. (a) $\phi = \dfrac{1}{2}\cot^{-1}\dfrac{7}{24} \approx$
$36.9°$; (b) $x = \dfrac{4}{5}\bar{x} - \dfrac{3}{5}\bar{y}$,
$y = \dfrac{3}{5}\bar{x} + \dfrac{4}{5}\bar{y}$; (c) $\bar{y} = \pm\dfrac{12}{5}$;　(d)

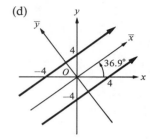

33. (a) $\phi \approx 18.4°$;　(d)

(b) $x = \dfrac{3\sqrt{10}}{10}\bar{x} - \dfrac{\sqrt{10}}{10}\bar{y}$,

$y = \dfrac{\sqrt{10}}{10}\bar{x} + \dfrac{3\sqrt{10}}{10}\bar{y}$;

(c) $\dfrac{\bar{x}^2}{9} + \dfrac{\bar{y}^2}{3} = 1$;

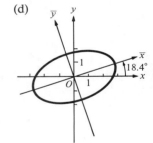

35. (a) $\phi \approx 18.4°$;　(d)

(b) $x = \dfrac{3\sqrt{10}}{10}\bar{x} - \dfrac{\sqrt{10}}{10}\bar{y}$,

$y = \dfrac{\sqrt{10}}{10}\bar{x} + \dfrac{3\sqrt{10}}{10}\bar{y}$;

(c) $\dfrac{(\bar{x} + \frac{3}{20})^2}{\frac{321}{60}} -$

$\dfrac{(\bar{y} - \frac{1}{20})^2}{\frac{321}{140}} = 1$;

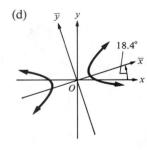

37. $\bar{x} = x\cos\phi + y\sin\phi$, $\bar{y} = -x\sin\phi + y\cos\phi$

39. (b) $x^2 + y^2 = -1$; (c) $(x - y)^2 = 0$;
(d) $4(x + 2)^2 = (y - 1)^2$; (e) $(x - y)^2 = 18$;
(f) $(x - 3)^2 + (y + 4)^2 = 0$

41. By multiplying the equation by -1 if necessary, we can assume $A, B > 0$. Complete the squares and rewrite the equation as $A(x-h)^2 + C(y-k)^2 = K$, where $h = -D/(2A)$, $k = -E/(2C)$, and $K = D^2/(4A) + E^2/(4C) - F$. If $K > 0$, the graph is an ellipse or a circle; if $K = 0$, the graph is the single point (h, k); if $K < 0$, the graph is the empty set.

43. Complete the squares as in Problem 41

45. Consider cases $A = 0$, $C \ne 0$; $A \ne 0$, $C = 0$; and $A = C = 0$ separately

Problem Set 10.5, page 575

1.
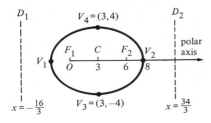
$$\frac{(x-2)^2}{9} + \frac{y^2}{5} = 1$$

3.
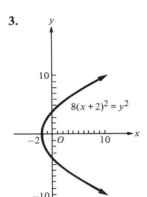
$8(x+2)^2 = y^2$

5.
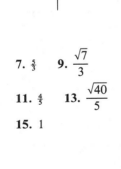
$$\frac{(x+4)^2}{4} - \frac{y^2}{12} = 1$$

7. $\frac{5}{3}$ **9.** $\dfrac{\sqrt{7}}{3}$

11. $\frac{4}{5}$ **13.** $\dfrac{\sqrt{40}}{5}$

15. 1

17. Ellipse; eccentricity $\frac{3}{5}$; directrices $x = -\frac{16}{3}$, $x = \frac{34}{3}$; center $(3, 0)$; foci $(0, 0)$, $(6, 0)$; vertices $(-2, 0)$, $(8, 0)$ $(3, -4)$, $(3, 4)$ (All in Cartesian coordinates)

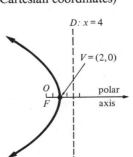

19. Parabola; eccentricity 1; directrix $x = -6$; focus $(0, 0)$; vertex $(-3, 0)$ (All in Cartesian coordinates)

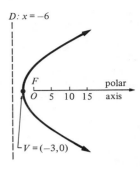

21. Parabola; eccentricity 1; directrix $y = -10$; focus $(0, 0)$; vertex $(0, -5)$ (All in Cartesian coordinates)

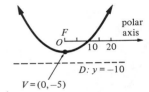

23. Hyperbola; eccentricity 2; directrices $y = \frac{1}{2}$, $y = \frac{5}{6}$; center $(0, \frac{2}{3})$; foci $(0, \frac{4}{3})$, $(0, 0)$; vertices $(0, 1)$, $(0, \frac{1}{3})$ (All in Cartesian coordinates)

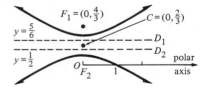

25. Parabola; eccentricity 1; directrix $x = 4$; focus $(0, 0)$; vertex $(2, 0)$ (All in Cartesian coordinates)

27. $x^2 + y^2 = 9$

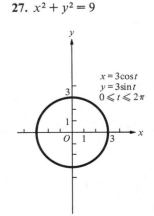

29. $y = 1 - x^2$, for
$-1 \le x \le 1$

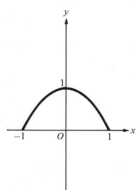

31. $x^2 + \dfrac{y^2}{4} = 1$

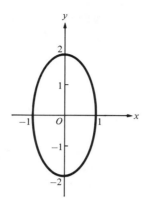

33. $(x-1)^2 +$
$\dfrac{(y+1)^2}{4} = 1$

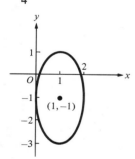

35. $y = \dfrac{1}{x}$, for $0 < x < 1$

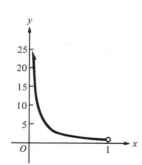

37. $\dfrac{2c}{2a} = \dfrac{c}{a} = e$

39. $e = \dfrac{c}{a} = \dfrac{\sqrt{a^2 - b^2}}{a} = \sqrt{\dfrac{a^2 - b^2}{a^2}} = \sqrt{1 - \left(\dfrac{b}{a}\right)^2}$

41. $e \approx 0.017$

Review Problem Set, Chapter 10, page 577

1. Circle, radius $\frac{1}{3}$ **3.** Ellipse; $a = 1$, $b = 2$

5. $\dfrac{x^2}{64} + \dfrac{y^2}{16} = 1$ **7.** $\dfrac{(x-1)^2}{16} + \dfrac{(y-1)^2}{25} = 1$

9. Center $(0, 0)$; vertices $(\pm 2\sqrt{2}, 0)$, $(0, \pm 2\sqrt{3})$

11. Center $(-1, 1)$; vertices $(-6, 1)$, $(4, 1)$, $(-1, -2)$, $(-1, 4)$

13. Center $(-4, 6)$; vertices $(-8, 6)$, $(0, 6)$, $(-4, 0)$, $(-4, 12)$

15. Vertex $(0, 0)$, horizontal axis, opens to right

17. Vertex $(0, 1)$, vertical axis, opens downward

19. Vertex $(-9, 1)$, horizontal axis, opens to right

21. $x = -(y + 6)^2$

23. Center $(0, 0)$, asymptotes $y = \pm\frac{1}{3}x$, vertices $(\pm 6\sqrt{2}, 0)$

25. Center $(-2, 3)$; asymptotes $y = 3 \pm \frac{1}{2}(x + 2)$; vertices $(-6, 3)$, $(2, 3)$

27. Center $(1, 3)$; asymptotes $y = 3 \pm \frac{1}{2}(x - 1)$; vertices $(1, \frac{5}{2})$, $(1, \frac{7}{2})$

29. $\dfrac{x^2}{\frac{3}{4}} - \dfrac{y^2}{4} = 1$

31. $\dfrac{(x-1)^2}{4} + \dfrac{(y+2)^2}{64} = 1$, ellipse

33. $(x + \frac{1}{2})^2 + (y - \frac{1}{2})^2 = \frac{1}{4}$, circle

35. $\dfrac{(x-1)^2}{4} - \dfrac{(y-1)^2}{1} = 1$, hyperbola

37. $\dfrac{(x - \frac{1}{2})^2}{\frac{53}{8}} + \dfrac{(y + \frac{1}{2})^2}{\frac{159}{4}} = 1$, ellipse

39. (a) $(-3, 2)$; (b) $2, -5)$; (c) $(1, 6)$; (d) $(0, 0)$; (e) $(-3, 3)$; (f) $(-4, 2)$

41. $x = \bar{x} - 1$, $y = \bar{y} + \frac{1}{2}$; $\bar{x}^2 + \bar{y}^2 = 1$; circle; center $(0, 0)$; radius 1

43. $x = \bar{x} + 3$, $y = \bar{y}$; $\dfrac{\bar{x}^2}{4} - \dfrac{\bar{y}^2}{9} = 1$; hyperbola; center $(0, 0)$

45. $(3 - 2\sqrt{3}, 2 + 3\sqrt{3})$ **47.** $(1 + 2\sqrt{3}, \sqrt{3} - 2)$

49. $(2 + 2\sqrt{3}, 2\sqrt{3} - 2)$

51. (a) $\phi = \dfrac{\pi}{4}$; (b) $x = \dfrac{\sqrt{2}}{2}(\bar{x} - \bar{y})$, $y = \dfrac{\sqrt{2}}{2}(\bar{x} + \bar{y})$; (c) $\dfrac{\bar{x}^2}{4} + \dfrac{\bar{y}^2}{12} = 1$; (d) Ellipse

53. (a) $\phi = 15°$; (b) $x = \bar{x}\cos 15° - \bar{y}\sin 15°$, $y = \bar{x}\sin 15° + \bar{y}\cos 15°$; (c) $\dfrac{\bar{x}^2}{22(2 - \sqrt{3})} - \dfrac{\bar{y}^2}{22(2 + \sqrt{3})} = 1$; (d) Hyperbola

55. (a) $\phi = 30°$; (b) $x = \frac{1}{2}(\sqrt{3}\bar{x} - \bar{y})$, $y = \frac{1}{2}(x + \sqrt{3}\bar{y})$; (c) $\dfrac{\bar{x}^2}{1} + \dfrac{\bar{y}^2}{4} = 1$; (d) Ellipse

57. (a) $\phi = \sin^{-1} \dfrac{\sqrt{10}}{10} \approx 18.4°$; (b) $x = \dfrac{\sqrt{10}}{10}(3\bar{x} - \bar{y})$,

$y = \dfrac{\sqrt{10}}{10}(\bar{x} + 3\bar{y})$; (c) $\dfrac{\bar{x}^2}{32} + \dfrac{\bar{y}^2}{\frac{32}{11}} = 1$; (d) Ellipse

59. (a) $\phi = \sin^{-1} \frac{4}{5} \approx 53.13°$; (b) $x = \frac{3}{5}\bar{x} - \frac{4}{5}\bar{y}$,

$y = \frac{4}{5}\bar{x} + \frac{3}{5}\bar{y}$; (c) $\dfrac{\bar{x}^2}{1} + \dfrac{\bar{y}^2}{\frac{1}{2}} = 1$; (d) Ellipse

61. (a) $\phi = \sin^{-1} \dfrac{2\sqrt{5}}{5} \approx 63.43°$;

(b) $x = \dfrac{\sqrt{5}}{5}(\bar{x} - 2\bar{y})$, $y = \dfrac{\sqrt{5}}{5}(2\bar{x} + \bar{y})$;

(c) $\bar{x} = \pm \dfrac{3\sqrt{5}}{5}$; (d) Pair of parallel lines

63. $y = \dfrac{4H}{L^2} x^2$ **65.** Parabola, $e = 1$

67. Ellipse, $e = \frac{2}{3}$ **69.** Hyperbola, $e = 2$

71. $\dfrac{4\sqrt{58}}{37} \approx 0.8233$

73. $(x - 1)^2 + (y - 1)^2 = 1$; circle of radius 1, center $(1, 1)$

75. Portion of the parabola $y = x^2$, for $-1 \le x \le 1$

CHAPTER 11

Problem Set 11.2, page 590

1. 362,880   **3.** 7   **5.** 35
7. 1   **9.** 20
11. $(1, 7, 21, 35, 35, 21, 7, 1)$
13. $a^4 + 12a^3x + 54a^2x^2 + 108ax^3 + 81x^4$
15. $x^5 - 5x^4y + 10x^3y^2 - 10x^2y^3 + 5xy^4 - y^5$
17. $c^6 + 12c^5 + 60c^4 + 160c^3 + 240c^2 + 192c + 64$
19. $1 - 7c^3 + 21c^6 - 35c^9 + 35c^{12} - 21c^{15} + 7c^{18} - c^{21}$
21. $x^3 - 6x^2\sqrt{xy} + 15x^2y - 20xy\sqrt{xy} + 15xy^2 - 6y^2\sqrt{xy} + y^3$
23. $\frac{1}{64}x^6 - \frac{3}{8}x^5y + \frac{15}{4}x^4y^2 - 20x^3y^3 + 60x^2y^4 - 96xy^5 + 64y^6$
25. $10s^3t^2$ **27.** $\frac{105}{2}x^{12}y^{12}$ **29.** $5xy^4$
31. $3360c^6d^4$

Problem Set 11.3, page 594

1. $\frac{1}{2}, \frac{1}{3}, \frac{1}{4}, \frac{1}{5}, \frac{1}{6}$ **3.** 4, 9, 16, 25, 36
5. $-1, 2, -3, 4, -5$ **7.** $\frac{1}{3}, \frac{3}{5}, \frac{5}{7}, \frac{7}{9}, \frac{9}{11}$
9. $\frac{1}{3}, \frac{1}{9}, \frac{1}{27}, \frac{1}{81}, \frac{1}{243}$ **11.** $0, \frac{3}{5}, \frac{4}{5}, \frac{15}{17}, \frac{12}{13}$
13. $2, -6, 18, -54, 162$
15. $2, 2, 4, 64, 16, 777, 216$ **17.** $1, \frac{1}{2}, \frac{1}{6}, \frac{1}{24}, \frac{1}{120}$
19. 33   **21.** 30   **23.** 100   **25.** 6.4
27. $m + 8r$ **29.** $-11\sqrt{5}$ **31.** $\frac{1}{64}$ **33.** 162
35. 2^{16} or 65,536   **37.** -0.009 **39.** $2x^{15}$
41. 0.000003   **45.** (a) $a_1 = 300,000$, $r = 1.05$;
(b) 382,884

Problem Set 11.4, page 602

1. $1 + 4 + 7 + 10 = 22$ **3.** $\frac{13}{4} + \frac{7}{4} + \frac{15}{4} + 4 + \frac{17}{4} = \frac{75}{4}$
5. $1 + \frac{1}{2} + \frac{4}{3} + \frac{4}{3} + \frac{16}{13} + 2 = \frac{5553}{910}$ **7.** 2460
9. $2^{25} + 22$ or 33,554,454   **11.** $\frac{1}{9}[1 - (1/10)^{100}]$
13. 55   **15.** 65   **17.** 624   **19.** -390
21. $-63t$ **23.** $-35x$ **25.** \$2,440,000
27. 210   **29.** 44,444   **31.** $-\frac{1}{2}[1 - (1/3)^{10}]$
33. $(5^{12}/24)[1 - (1/5)^{18}]$ **35.** 2500
37. $c^4(1 - c^{14})/(1 - c^2)$ **39.** 36.2 cm
41. $768 - 2^{10-n}$ **43.** 3   **45.** $\frac{5}{7}$ **47.** $\frac{6}{7}$
49. $\frac{25}{8}$ **51.** $\frac{4}{9}$ **53.** $\frac{80}{11}$ **55.** $\frac{58}{2475}$
57. False, $0.\overline{9} = 1$

Problem Set 11.5, page 610

1. 8   **3.** 676   **5.** 32
7. $ab, ac, ad, ae, ba, bc, bd, be, ca, cb, cd, ce, da, db, dc, de, ea, eb, ec, ed$; 20
9. 24   **11.** 9! or 362,880   **13.** 20   **15.** 990
17. $_{10}P_5 = 30,240$ **19.** $7! = 5040$ **21.** 40,320
23. $_{10}P_3 = 720$ **25.** $\{a, b, c\}, \{a, b, d\}, \{a, c, d\}$, $\{b, c, d\}$; 4   **27.** 1   **29.** 10   **31.** 20   **33.** 10
35. 1   **37.** $_8C_2 = 28$ **39.** $_{10}C_4 = 210$ **41.** 12
43. 43,243,200   **45.** 3780   **47.** 3360

Problem Set 11.6, page 613

1. $\frac{1}{2}$ **3.** $\frac{4}{7}$ **5.** $\frac{1}{6}$ **7.** $(13 \cdot 48)/_{52}C_5 = 1/4165$
9. $4/_{52}C_5 = 1/649,740$ **11.** $\frac{5}{36}$ **13.** $\frac{11}{36}$
15. $[(_{10}C_3)(_3C_2)]/_{13}C_5 = 40/143$ **17.** $\frac{1}{3}$ **19.** $\frac{1}{2}$

Review Problem Set, Chapter 11, page 614

5. $32 + 80x + 80x^2 + 40x^3 + 10x^4 + x^5$

7. $2187x^{14} - 10,206x^{12}y^2 + 20,412x^{10}y^4 - 22,680x^8y^6 + 15,120x^6y^8 - 6048x^4y^{10} + 1344x^2y^{12} - 128y^{14}$

9. $\frac{2187}{128}x^7 - \frac{5103}{64}x^6 + \frac{5103}{32}x^5 - \frac{2835}{16}x^4 + \frac{945}{8}x^3 - \frac{189}{4}x^2 + \frac{21}{2}x - 1$

11. $5376x^6y^3$ **13.** $61,236x^5y^5$

15. $-\frac{1}{2}, \frac{1}{3}, -\frac{1}{4}, \frac{1}{5}, -\frac{1}{6}$ **17.** $0, \frac{1}{2}, 0, \frac{1}{8}, 0, \frac{1}{48}$

19. 41   **21.** -137 **23.** $2^{10} = 1024$

25. -384 **27.** 33   **29.** 110   **31.** $\frac{4567}{90}$

33. $\frac{7}{4}$ **35.** 450   **37.** $\frac{44}{3}$ **39.** 12,240

41. $(4^{10} - 1)/4$ **43.** \$50.50   **45.** \$3000

47. 625 inches or 52 feet, 1 inch

49. $\frac{1}{6}n(n + 1)(2n + 1)$ **51.** 9   **53.** $(7 + \sqrt{35})/2$

55. (a) $\frac{83}{99}$; (b) $\frac{91}{99}$; (c) $\frac{232,379}{49,950}$; (d) $\frac{16}{5}$ **57.** 360

59. 110   **61.** 120   **63.** 84   **65.** 1   **67.** 12

69. 120   **71.** 325   **73.** 66

75. $_{25}P_{20} \approx 1.2926 \times 10^{23}$

77. 27,720   **79.** $\frac{1}{3}$

81. $\frac{3}{7}$ **83.** $\frac{1}{16}$

Indexes

Index of Applications

Index

Trigonometric Functions

Acute Angles

$$\sin \theta = \frac{\text{opp}}{\text{hyp}} \qquad \csc \theta = \frac{\text{hyp}}{\text{opp}}$$

$$\cos \theta = \frac{\text{adj}}{\text{hyp}} \qquad \sec \theta = \frac{\text{hyp}}{\text{adj}}$$

$$\tan \theta = \frac{\text{opp}}{\text{adj}} \qquad \cot \theta = \frac{\text{adj}}{\text{opp}}$$

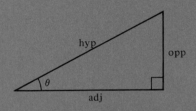

General Angles

$$\sin \theta = \frac{y}{r} \qquad \csc \theta = \frac{r}{y}$$

$$\cos \theta = \frac{x}{r} \qquad \sec \theta = \frac{r}{x}$$

$$\tan \theta = \frac{y}{x} \qquad \cot \theta = \frac{x}{y}$$

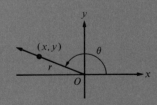

Trigonometric Identities

Fundamental Identities

1 $\csc \theta = \dfrac{1}{\sin \theta}$

2 $\sec \theta = \dfrac{1}{\cos \theta}$

3 $\cot \theta = \dfrac{1}{\tan \theta}$

4 $\tan \theta = \dfrac{\sin \theta}{\cos \theta}$

5 $\cot \theta = \dfrac{\cos \theta}{\sin \theta}$

6 $\cos^2 \theta + \sin^2 \theta = 1$

7 $1 + \tan^2 \theta = \sec^2 \theta$

8 $1 + \cot^2 \theta = \csc^2 \theta$

Even–Odd Identities

1 $\sin(-\theta) = -\sin \theta$

2 $\cos(-\theta) = \cos \theta$

3 $\tan(-\theta) = -\tan \theta$

4 $\cot(-\theta) = -\cot \theta$

5 $\sec(-\theta) = \sec \theta$

6 $\csc(-\theta) = -\csc \theta$

Addition Formulas

1 $\sin(\alpha + \beta) = \sin \alpha \cos \beta + \sin \beta \cos \alpha$

2 $\cos(\alpha + \beta) = \cos \alpha \cos\beta - \sin \alpha \sin \beta$

3 $\tan(\alpha + \beta) = \dfrac{\tan \alpha + \tan \beta}{1 - \tan \alpha \tan \beta}$

Subtraction Formulas

1 $\sin(\alpha - \beta) = \sin \alpha \cos \beta - \sin \beta \cos \alpha$

2 $\cos(\alpha - \beta) = \cos \alpha \cos \beta + \sin \alpha \sin \beta$

3 $\tan(\alpha - \beta) = \dfrac{\tan \alpha - \tan \beta}{1 + \tan \alpha \tan \beta}$